Advances in Experimental Medicine and Biology

Volume 1042

More information about this series at http://www.springer.com/series/5584

Hisao Masai • Marco Foiani
Editors

DNA Replication

From Old Principles to New Discoveries

Springer

Editors
Hisao Masai
Department of Genome Medicine
Tokyo Metropolitan Institute
of Medical Science
Tokyo, Japan

Marco Foiani
Istituto FIRC di Oncologia Molecolare
Milan, Italy

ISSN 0065-2598 ISSN 2214-8019 (electronic)
Advances in Experimental Medicine and Biology
ISBN 978-981-10-6954-3 ISBN 978-981-10-6955-0 (eBook)
https://doi.org/10.1007/978-981-10-6955-0

Library of Congress Control Number: 2017960795

Printed on acid-free paper

This Springer imprint is published by Springer Nature
The registered company is Springer Nature Singapore Pte Ltd.
The registered company address is: 152 Beach Road, #21-01/04 Gateway East, Singapore 189721, Singapore

Preface

The discovery of the double-helix structure in 1953 provided the basic concept of how genetic materials are duplicated. However, this also triggered the quest for understanding the whole picture of DNA replication. A semi-conservative replication was demonstrated in a historical experiment by Messelson and Stahl, and Arthur Kornberg discovered an enzyme responsible for synthesizing DNA. At that time, no one envisioned such complicated systems required to make copies of DNA.

The proposal of the replicon hypothesis by Jacob made a major impact in pointing the direction in which research of DNA replication in the following years would be led. In fact, the results of genetic studies in bacteria provided evidence for the presence of the factors (initiator and replicator) hypothesized in the model. This was striking and revealing, and the research of DNA replication in the following 50 years tried to recapitulate this finding in different organisms. The elegant single-molecule analyses of replicating DNA in mammalian cells done by Huberman and Riggs in 1966 were interpreted in the framework of the replicon model, becoming the basis for the "multiple replicon" hypothesis for the eukaryotic genomes.

Molecular genetic studies in *Escherichia coli*, in combination with development of recombinant DNA technology, clarified the structures of the replication origin (*oriC*), and "resolution and reconstitution" studies of the single-stranded DNA phage and eventually those of *oriC* replication elucidated the mechanisms of DNA chain elongation and initiation at the bacterial chromosome replicator.

A similar approach was taken for eukaryotes, and the studies using viruses as a model significantly contributed to the elucidation of eukaryotic DNA replication machinery. Genetic studies, in conjunction with newly developed methods for physical mapping of origins, in a unicellular eukaryote, yeast, led to identification of specific sequences that could serve as replicators. In contrast, the "initiator" remained elusive until Bell and Stillman discovered ORC (*O*rigin *R*ecognition *C*omplex), which appears to fulfill all the requirement to be qualified as the "initiator". All the data pointed to the adherence to the old "replicon" principle even in eukaryotes.

Fifty-four years after the replicon hypothesis, almost the entire process for eukaryotic DNA replication was reconstituted with purified proteins, and the detailed mechanisms are bound to be discovered in the ensuing years. Given the large extent of conservation of most of the core replication factors between yeast and higher eukaryotes, the basic mechanisms of origin activation and assembly of a replisome would be conserved through evolution.

Compared to the replication machinery, the sequences that define the "replicator" appear to have great divergence between species. A genome-wide approach with NGS (Next Generation Sequencing) generated an enormous amount of new information on the profiles of replication origins in higher eukaryotes, but consensus sequences, similar to those found in the yeast replication origins, have not yet been discovered. An additional complication is the absence of sequence-specific DNA binding activity in mammalian ORC. These results suggest the existence of a determinant other than nucleotide sequences that dictate the assembly of a replication complex.

This book compiles various timely topics in DNA replication. The volume starts with a historical description on studies of eukaryotic DNA replication by Professor Thomas Kelly. Professor Kelly has made major contributions in this field for the past over 40 years through his studies on viral DNA replication, human genome replication, and yeast replication. The history of Professor Kelly's research itself represents how this field evolved and materialized into the current understanding of eukaryotic DNA replication. We are very honored and excited to have this chapter at the beginning of the book. This chapter is followed by two chapters describing replication origins in higher eukaryotes. Dr. Mirit Aladjem describes various features that define metazoan replication origins; those include sequence bias, open chromatin, and histone modifications. Drs. Nozomi Sugimoto and Masatoshi Fujita focus on chromatin remodeling factors that determine origin activity though promotion of replication licensing.

Studies in *E. coli*, starting from those on phage replicons to those of *oriC* plasmid, have led the field of DNA replication. In spite of evolutional distance from eukaryotes, the expertise from the bacterial systems have been proven to play leading roles in elucidating the mechanisms of eukaryotic DNA replication. Dr. Tsutomu Katayama describes the most recent discovery on the detailed mechanism of *E. coli oriC* replication and various modes of its regulation. Studies in Archaea have filled unique roles in shedding new light on regulation of eukaryotic DNA replication. Dr. Stephen Bell has contributed a chapter describing initiation and its regulation in Archaea replicons.

In the following two chapters, replication machinery for DNA chain elongation is discussed. Drs. Joseph Stodola and Peter Burgers deal with the mechanism of lagging strand DNA replication in eukaryotes, which is a critical step for DNA chain elongation and is also crucial for stable maintenance of genome, while Drs. Eiji Ohashi and Toshiki Tsurimoto discuss the multiple clamp and its loaders, which are now known to play major roles in coordinating the process of DNA replication with various other chromosome maintenance systems, including checkpoint/DNA damage repair systems, epigenetic transmission, and chromosome partition.

Replication termination is a recent hot topic. In bacteria, replication termination occurs when two replication forks collide 180° from the origin, and their progression is arrested at the *ter* signals bound with tus protein. On the eukaryotic chromosomes, combined actions of topoisomerase and a ubiquitin ligase are required to dislodge the replisome at the termination site.

The recent technical revolution in cryoelectron microscopy has enabled structural analyses of huge protein complexes at a resolution close to that achieved by X-ray crystallography. Application of this new technology has revealed the complex structures and operation mechanism of MCM (*m*ini*c*hromosome *m*aintenance) and CMG helicase as well as more complex replisome assembly. Drs. Yuanliang Zhai and Bik-Kwoon Tye; Drs. Lin Bai, Zuanning Yuan, Jingchuan Sun, Roxana Georgescu, Michael O'Donnell, and Huilin Li describe their state-of-the-art analyses of yeast MCM2–7 double hexameter and the replisome complex, respectively.

Long-standing questions about DNA replication timing regulation are now being addressed in detail at a genome-wide level and detailed landscape of replication domains has been presented in various cell types. Drs. Peiyao Zhao, Juan Carlos Rivera-Mulia, and David Gilbert, a leading group in this area, discuss their current model on how replication domains are related to genome compartmentalization and chromatin architecture. The conserved Rif1 protein, originally identified as a telomere binding factor in yeast, was rediscovered as a critical regulator of replication timing in fission yeast. Rif1 also plays a major role in organizing the replication domains in mammalian cells. Dr. Sara Buonomo describes how chromatin architecture is regulated by Rif1 in mammalian cells to define the replication domains. Drs. Kenji Moriyama, Mong Sing Lai, and Hisao Masai describe functions of Rif1 in both fission yeast and mammalian cells, with particular emphasis on its ability to specifically recognize the G-quadruplex structure (G4), and also the potential roles of G4 in the regulation of chromosome functions. The presence of G4 is one of the most prevalent features associated with replication origins from higher eukaryotes. Dr. Marie-Noëlle Prioleau describes the genetic experiments that show the requirement for G4 in origin activity and discusses the potential roles of G4 in origin regulation.

Chromatin structures play central roles in almost all the metabolism of DNA. Drs. Constance Alabert, Zuzana Jasencakova, and Anja Groth discuss how chromatin is inherited during the course of DNA replication.

DNA replication is a part of cell cycle events. Therefore, how it is integrated in the global cell cycle regulation is an important issue, especially from the point of view that cell proliferation is regulated by the extracellular stimuli, and that it is this pathway that is often deregulated in cancer cells. Drs. Gavin Grant and Jeanette Cook discuss cell cycle regulation of the S phase, with particular emphasis on the events in G1 that are crucial for regulated execution of the S phase.

Modification of proteins with small polypeptides such as ubiquitin or SUMO permits rapid and reversible regulation of various biological reactions, and DNA replication is no exception. Drs. Tarek Abbas and Anindya Dutta describe how unperturbed DNA replication is regulated by protein ubiquitination and its relation to diseases. Drs. Sara Villa-Hernández, Avelino Bueno, and Rodrigo Bermejo describe protein ubiquitylation in cellular responses to perturbed DNA replication or DNA damages. SUMOylation also recently has been shown to play important roles during DNA replication, which is covered by Drs. Lei Wei and Xiaolan Zhao.

DNA replication and transcription take place on the same template DNA, and how they are coordinated has been an intriguing issue. It is usually assumed that collision of replication and transcription can cause genomic instability and needs to be avoided if possible. However, especially on the eukaryotic genome where there are many origins, this is inevitable. Drs. Yathish Achar and Marco Foiani describe how cells coordinate replication with transcription.

The processes of DNA replication and its regulation are directly linked to maintenance of genomic stability. Overexpression of Cyclin E has been known to induce aberrant DNA replication that leads to genome instability. Drs. Leonardo Teixeira and Steven Reed describe how cyclin E overexpression induces genome instability. Repetitive DNA sequences and sequences capable of forming unusual DNA structures are ubiquitous on the human genome, and increasing evidence points to their pathogenic nature for various diseases. Chromosomal fragile sites, often composed of unusual repetitive sequences, have long been known and implicated in genome instability. Drs. Wenyi Feng and Arijita Chakraborty provide a comprehensive historical account of chromosomal fragile sites and provide a detailed discussion of their disease association and then how their expression is related to DNA replication. Advaitha Madireddy and Jeannine Gerhardt discuss the consequence of replication through repetitive DNA elements and how they could lead to specific diseases.

With the astounding amount of new information on replication origins and protein associations on the chromatin and protein–protein/inter-chromatin networks and with the long-awaited reconstitution system of eukaryotic DNA replication in hand, we are at one of the most exciting moments in the field of DNA replication. While we will gain detailed mechanistic insight into the molecular basis of eukaryotic DNA replication, the basic principle of initiation of eukaryotic chromosomal replication remains elusive. It is becoming clear that initiation of replication in higher eukaryotes is quite distinct from that in bacteria in that it can be initiated almost anywhere, albeit with varied efficiency, and the initiation event may be quite stochastic, pointing to the possibility that the bacterial replicator–initiator concept may not apply to regulation in higher eukaryotes.

We hope that this book will help readers to get a taste of the newest trends in this exciting field and will trigger a new wave of research in search of a new principle of DNA replication.

Tokyo, Japan — Hisao Masai
Milan, Italy — Marco Foiani

Contents

About the Editors

Hisao Masai Hisao Masai graduated from the University of Tokyo in 1981 and received his Ph.D. from the same university in 1986 for his work on DNA replication in *E. coli*, which he conducted at the DNAX Research Institute (Palo Alto, California) under the supervision of Dr. Ken-ichi Arai. He did postdoctoral studies at the same institute and became an assistant professor (1990) and then associate professor (1995) of the Institute of Medical Science, the University of Tokyo. In 2000, he moved to the Tokyo Metropolitan Institute of Medical Science to be the head of the Department of Cell Biology. He is currently a vice director of the institute, leading a laboratory of genome dynamics. He continues to work on the mechanisms of DNA replication using *E. coli*, fission yeast, mammalian cells, and animal models. One of his current interests is to elucidate the biological significance of G (guanine)-quadruplex in various chromosome transactions including DNA replication. Dr. Masai served on the editorial board of *The Journal of Biological Chemistry* (2009–2012). He is currently an editor for *Biochemical and Biophysical Research Communications* and an associate editor for *Genes to Cells*. The awards he has received include the Tokyo Metropolitan Governor's Award (2006), A-IMBN (Asian International Molecular Biology Network) Arthur Kornberg Memorial Award (2012), the Prize for Science and Technology (Research Category), and the Commendation for Science and Technology from the Minister of Education, Culture, Sports, Science and Technology (2017).

Marco Foiani Marco Foiani obtained his Ph.D. in molecular and cell biology from the University of Milan in 1985. After conducting postdoctoral research at the University of Milan and at NIH-NICHD, USA (laboratory of Dr. Alan Hinnebusch), he became a faculty member at the University of Milan in 1990, and in 2000, he became the head of the Genome Integrity Laboratory at IFOM, Milan. He also became a full professor of molecular biology at the University of Milan in 2002. Since 2009 he has been the scientific director of IFOM, an international cancer center that hosts 20 groups and 300 scientists. At IFOM, Dr. Foiani is responsible for strategic research planning, the development of programs aimed at results reduction into practice, and the establishment of national and international cooperation programs and joint ventures. He is the founder of the European Nanomedicine Foundation (CEN) that aims to support multidisciplinary projects/teams in biomedicine. He is cofounder of the IFOM-IEO Campus, hosting IFOM, the European Institute of Oncology, the European School of Molecular Medicine, the Italian Institute of Technology, and Cogentech. Since 2012 he has directed a cancer genetics diagnostic laboratory within Cogentech. In 2004 he was elected an EMBO member and in 2010 a member of the Academia Europaea. He is currently a member of the editorial board of *Cell*.

Chapter 1
Historical Perspective of Eukaryotic DNA Replication

Thomas Kelly

Abstract The replication of the genome of a eukaryotic cell is a complex process requiring the ordered assembly of multiprotein replisomes at many chromosomal sites. The process is strictly controlled during the cell cycle to ensure the complete and faithful transmission of genetic information to progeny cells. Our current understanding of the mechanisms of eukaryotic DNA replication has evolved over a period of more than 30 years through the efforts of many investigators. The aim of this perspective is to provide a brief history of the major advances during this period.

Keywords DNA replication • Eukaryotes • Viral models • Origin of DNA replication • Prereplicative complex • Helicase • Initiator • Replisome

1.1 Introduction

The quest to understand how our genomes are duplicated began in earnest with the description of the double helix in 1953. The self-complementary structure of DNA immediately suggested how the information in the sequence of nucleotides could be copied during DNA replication and repaired after DNA damage. The first enzyme activities capable of synthesizing DNA were described within 5 years of the publication of the DNA structure (Bessman et al. 1956), but it would take many more years to achieve even a basic understanding of the complex machinery required to replicate genomes and to elucidate some of the mechanisms that control it. As one of the fundamental processes of life, DNA replication has been a central focus of molecular biology from the very beginning and remains so today.

Early studies of DNA replication focused on prokaryotic systems because of their relative simplicity. The work of many investigators established most of the basic principles that govern DNA replication in all organisms from prokaryotes to archaea and eukaryotes. In 1958 Meselson and Stahl demonstrated that the

T. Kelly (✉)
Sloan Kettering Institute, Memorial Sloan Kettering Cancer Center, New York, NY, USA
e-mail: tkelly@mskcc.org

H. Masai, M. Foiani (eds.), *DNA Replication*, Advances in Experimental Medicine and Biology 1042, https://doi.org/10.1007/978-981-10-6955-0_1

replication process in bacteria is semiconservative (Meselson and Stahl 1958). A few years later, Cairns showed that replication in *E. coli* begins at a single site in the circular chromosome and that DNA synthesis occurs on two arms of a forked structure (Cairns 1963). Taken together, these observations implied the existence of a replication machine that is loaded on the genomic DNA at an "origin" and subsequently moves along the DNA, unwinding the two parental strands and synthesizing new complementary strands. One of the key features of replication machines was discovered in 1968 when Okazaki showed by pulse-labeling experiments that newly synthesized DNA in *E. coli* cells consists of short fragments (Okazaki et al. 1968). Further analysis in vitro suggested that DNA replication is semi-discontinuous at each growing point (Olivera and Bonhoeffer 1972; Herrmann et al. 1972). One strand (the leading strand) is synthesized continuously 5′–3′ in the same direction as the fork moves, while the other strand (the lagging strand) is synthesized discontinuously 5′–3′ in the direction opposite to fork movement, producing short DNA fragments that are subsequently joined together. In the 30 years following the publication of the DNA structure, many of the components of prokaryotic replication machines were identified and characterized, including origin recognition proteins, DNA polymerases, processivity factors, primases, single-stranded DNA-binding proteins, topoisomerases, etc. This saga is the main subject of the superb monographs on DNA replication by Kornberg and more recent reviews (e.g., Kornberg 1981; Kornberg and Baker 1992; Johnson and O'Donnell 2005; Lewis et al. 2016). It was (correctly) surmised that many of the components of the eukaryotic replication machinery would have similar functions to their prokaryotic counterparts.

The identification and characterization of essential prokaryotic replication factors were largely accomplished by classical biochemical approaches – developing assays specific for putative replication functions and then purifying the active proteins(s) to near homogeneity from crude cell extracts. Initial work focused on viruses of *E. coli*, such as T4, T7, and φX174, because their genomes are relatively small and readily obtained. Some of these viruses (e.g., T4, T7) encode most of the factors required for the replication of their genomes, while others (e.g., ϕX174) rely largely on *E. coli* replication proteins. The isolation of many bacterial and phage mutants with conditional defects in DNA replication accelerated the identification and purification of replication proteins by in vitro complementation assays. Other proteins were purified by straightforward fractionation and reconstitution approaches. By 1980 most of the proteins required for propagation of a bacterial replication fork had been identified, and their functions characterized through studies of viruses and plasmids (Kornberg 1981). Understanding of the mechanisms involved in initiation of bacterial DNA replication during the subsequent decade required the development and analysis of cell-free systems capable of replicating plasmids containing *oriC*, the *E. coli* origin of DNA replication (Kornberg and Baker 1992). These studies established that initiation of bacterial DNA replication is largely controlled at the level of occupancy of the *oriC* origin by the DnaA initiator protein, consistent with the replicon model of Jacob et al. (1964).

It was clear early on that eukaryotic DNA replication was likely to be much more complex than prokaryotic DNA replication because eukaryotic genomes can be orders of magnitude larger than their prokaryotic counterparts. It follows that to complete chromosome duplication in a timely fashion, DNA replication must start at many sites along the chromosomal DNA. Using fiber autoradiographic methods similar to those of Cairns, two graduate students at Caltech, Joel Huberman and Arthur Riggs, demonstrated in 1968 that a mammalian cell utilizes tens of thousands of origins and that replication is bidirectional from each origin (Huberman and Riggs 1968). It was also known that eukaryotic chromosomes are duplicated precisely once each cell cycle, so the requirement for multiple origins raised a number of regulatory issues that were not apparent in bacterial systems with single origins per chromosome. When *E. coli* cells are growing rapidly in rich media, they initiate DNA replication at *oriC* prior to the complete duplication of the chromosome. As a result, DNA synthesis is continuous throughout the cell cycle, and segments of the bacterial chromosome proximal to *oriC* are present in more than two copies (Skarstad and Katayama 2013). This scenario does not happen in eukaryotes where each segment of DNA is duplicated once and only once during a defined period of each cell cycle. These considerations strongly implied that the logic of replication control in eukaryotes would prove to be substantially different from that of prokaryotes and that the biochemical mechanism of initiation of eukaryotic DNA replication would be correspondingly complex. Because of this complexity, the pathway of initiation of eukaryotic DNA replication was not worked out in detail and recapitulated in cell-free systems until comparatively recently.

As described in the succeeding chapters of this volume, the complexity of eukaryotic DNA replication extends to the organization of the machinery at the replication fork. A number of the core components of the eukaryotic (and archaeal) replisomes are unrelated, or only distantly related, to their bacterial counterparts, suggesting that the prokaryotic and eukaryotic/archaeal replication machineries may have evolved largely independently (Edgell and Doolittle 1997; Leipe et al. 1999; Makarova and Koonin 2013). The eukaryotic replisome contains multiple DNA polymerases and a replicative helicase with a considerably more complex subunit composition than that of *E. coli.* In addition, several factors associated with the replisome have no clear prokaryotic counterparts. The reasons for this extra complexity are not yet completely clear, but may be related to the need for greater regulation of DNA chain elongation in large genomes. For example, the packaging of the nuclear DNA into nucleosomes and higher-order structures likely creates unique problems for the elongation machinery. In addition, the replication machinery must deal with multiple sources of endogenous and exogenous DNA damage and potentially other barriers to DNA synthesis, such as large transcriptional units, DNA-binding proteins, etc. While these obstacles are not unique to eukaryotes, the dimension of the problem may be magnified by the large number of replicons involved in the replication of eukaryotic genomes.

In this perspective, I describe some of the major milestones in the study of eukaryotic DNA replication over the past 35 years. There were numerous contributors to this story, and, generally speaking, each of the major advances described here was built on discoveries made in many different laboratories. Given the span of time covered and the great progress that has been made, this perspective can't be comprehensive and of necessity must be somewhat idiosyncratic. While many primary references are provided, they represent a tiny fraction of the important publications of the last three or four decades, so I have also cited review articles that contain more comprehensive reference lists in particular areas of investigation. In writing this historical review, I was struck by the number of different lines of investigation that have converged to give us our current picture of how genomes are duplicated in eukaryotic cells. The field has grown enormously over the years and is now relatively mature, but the excitement remains.

1.2 The Beginning: Viral Models for Eukaryotic DNA Replication

The study of eukaryotic DNA replication was hampered for many years by the primitive state of genetic approaches in most eukaryotes and the lack of simple systems for biochemical analysis. The impasse was overcome by the development and characterization of cell-free systems capable of replicating the genomes of animal viruses. This advance, inspired by the success of previous studies of bacterial virus DNA replication, opened a viable pathway for identifying and purifying cellular replication proteins and characterizing their mechanisms of action. The result was a rapid acceleration of the study of eukaryotic DNA replication.

The replication of the human adenoviruses (Ad) was the first to be established in a completely soluble cell-free system (Challberg and Kelly 1979). The genome of these viruses consists of a linear duplex DNA molecule of about 35 kb with two features that are significant for viral DNA replication: the ends of the genome have identical sequences, and the 5′ terminus of each DNA strand is covalently attached to a virus-encoded protein of 55 kDa (the terminal protein or TP). Extracts of adenovirus-infected human cell nuclei are capable of carrying out the complete replication of exogenously added viral DNA molecules by a mechanism that closely resembles viral DNA synthesis in vivo. Analysis of replication in vitro led to the identification of the minimal essential requirements for the reaction and defined the basic mechanisms involved in initiation and chain elongation (Challberg and Kelly 1982; Stillman 1983; Nagata et al. 1983; Sussenbach and van der Vliet 1984; de Jong et al. 2003). Three viral proteins are required for adenovirus DNA replication: an 80 kDa precursor to the adenovirus 55 kDa terminal protein (pTP), a DNA polymerase (Ad Pol), and a single-stranded DNA-binding protein (Ad DBP). DNA replication is initiated by a novel protein-priming mechanism in which the first nucleotide in the adenovirus genome is covalently linked to a serine residue in the pTP (Rekosh et al. 1977; Challberg et al. 1980, 1982; Desiderio and Kelly 1981; Enomoto et al. 1981; Tamanoi and Stillman 1982; King and van der Vliet 1994). The biochemical mechanism of the initiation reaction is quite interesting although not directly germane to cellular DNA replication. One novel feature of the reaction is the requirement for two cellular transcription factors, nuclear factor I (NF-1/CTF) and Oct-1 (NF-III/Otf-1/Oct-1), for efficient initiation (Nagata et al. 1982; Tamanoi and Stillman 1983; Pruijn et al. 1986; Rosenfeld and Kelly 1986; O'Neill et al. 1988). These two factors bind to sequences at the ends of the viral genome and act to stabilize the binding of a complex of the pTP and Ad Pol and to facilitate the initial unwinding of the DNA. Together these cellular factors increase the efficiency of initiation by more than 100-fold. After initiation, daughter strand synthesis proceeds in the 5′–3′ direction by a strand displacement mechanism mediated by the adenovirus DNA polymerase and the Ad DBP (Challberg and Kelly 1982; Stillman 1983; Sussenbach and van der Vliet 1984). No separate helicase is required for duplex unwinding during chain elongation. The energy provided by hydrolysis of the nucleotide precursors and by the cooperative binding of the Ad DBP to single-stranded DNA is sufficient to drive strand displacement, which is further facilitated by a cellular topoisomerase. The products of this first round of DNA replication are a daughter duplex and a displaced parental single strand. Annealing of the self-complementary ends of the displaced single strand generates a duplex segment identical to the ends of the original viral genome. Following a second initiation event by the same protein-priming mechanism, complementary strand synthesis proceeds from one end of the genome to the other, generating the second daughter duplex.

Many features of adenovirus DNA replication differ from those of cellular DNA replication. The protein-priming mechanism represents an efficient solution to the end replication problem of linear DNA molecules but is not utilized for the ends of eukaryotic chromosomes. The adenovirus replisome is a remarkably efficient machine, requiring only a DNA polymerase and a single-stranded DNA binding protein, plus topoisomerase activity. At each growing point, only one of the two strands is synthesized, so there is no need for the complexities inherent in discontinuous DNA synthesis. On the other hand, the mechanism is unsuitable for cellular DNA replication because it exposes long regions of single-stranded DNA which is more sensitive to many DNA-damaging agents. Thus, the study of adenovirus DNA replication did not provide much insight into the normal mechanism of cellular DNA replication. However, the extremely rapid progress in defining the adenovirus replication mechanism after the initial development of the in vitro system clearly demonstrated the potential of the fractionation-reconstitution approach for studying DNA replication in eukaryotic cells and provided strong motivation to develop more informative models.

1.3 The SV40 DNA Replication System

A major turning point in the study of eukaryotic DNA replication was the discovery that extracts of primate cells infected with the polyomavirus SV40 could support the complete replication of added viral genomes (Li and Kelly 1984). Previous studies had indicated that SV40 offered many advantages as a model system. The viral genome is only about 5 kb and exists in infected cells as a minichromosome with a nucleoprotein structure like that of cellular chromatin. DNA replication initiates at a single origin and proceeds bidirectionally, similar to a single cellular replicon (Danna and Nathans 1972). Importantly, most of the proteins required for SV40 DNA replication are provided by the host cell, so biochemical dissection of the cell-free system provided a powerful approach for identifying cellular replication proteins and characterizing their mechanisms of action (Fig. 1.1a).

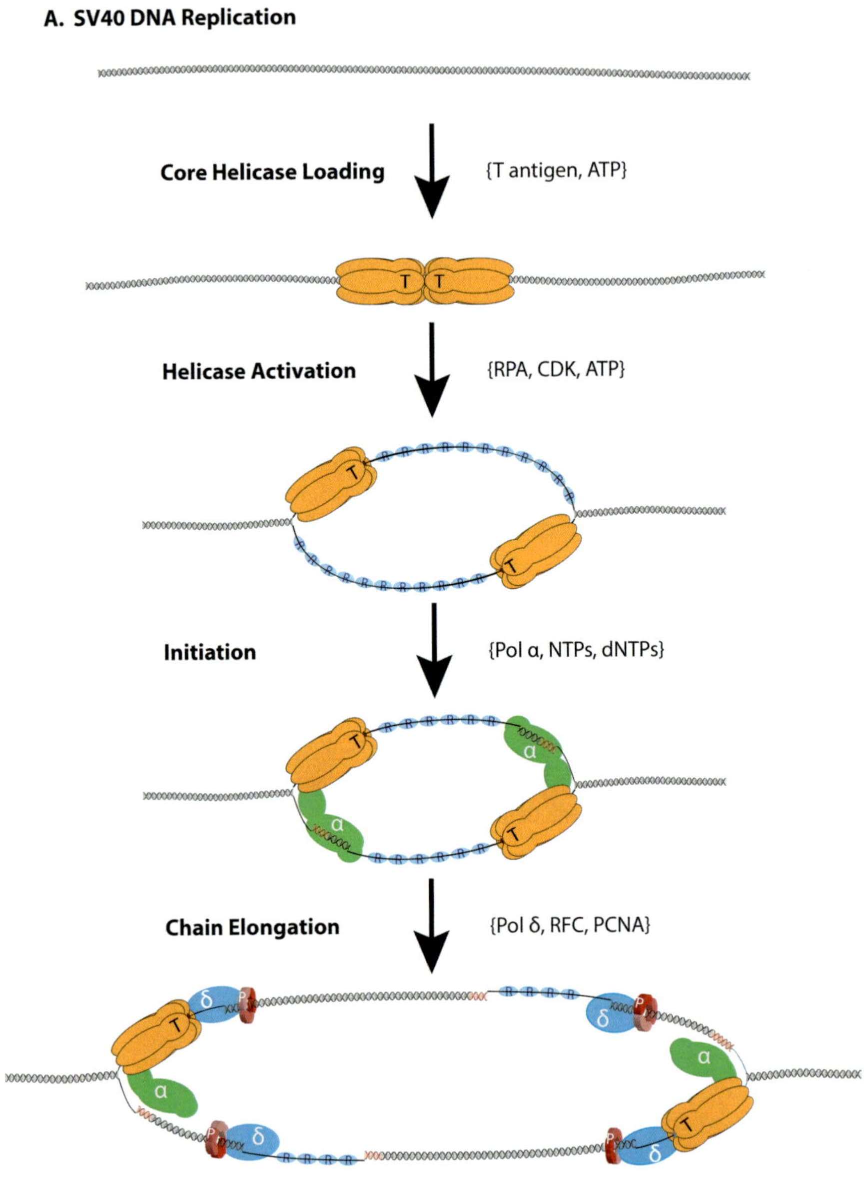

Fig. 1.1 SV40 and cellular DNA replication. A diagrammatic representation of the major steps in the viral and cellular replication pathways and the protein requirements for each step. See the text for details. (The steps involved in maturation of Okazaki fragments, relaxation of supercoiling, replication termination, and decatenation are left out for clarity.) (**a**) SV40 DNA replication. T, T antigen; α, DNA polymerase alpha-primase; R, RPA; δ, DNA polymerase delta; P, PCNA. (**b**) Cellular DNA replication. M, MCM2-7; 45, Cdc45; G, GINS, ε, DNA polymerase epsilon; α, DNA polymerase alpha-primase; R, RPA; δ, DNA polymerase delta; P, PCNA

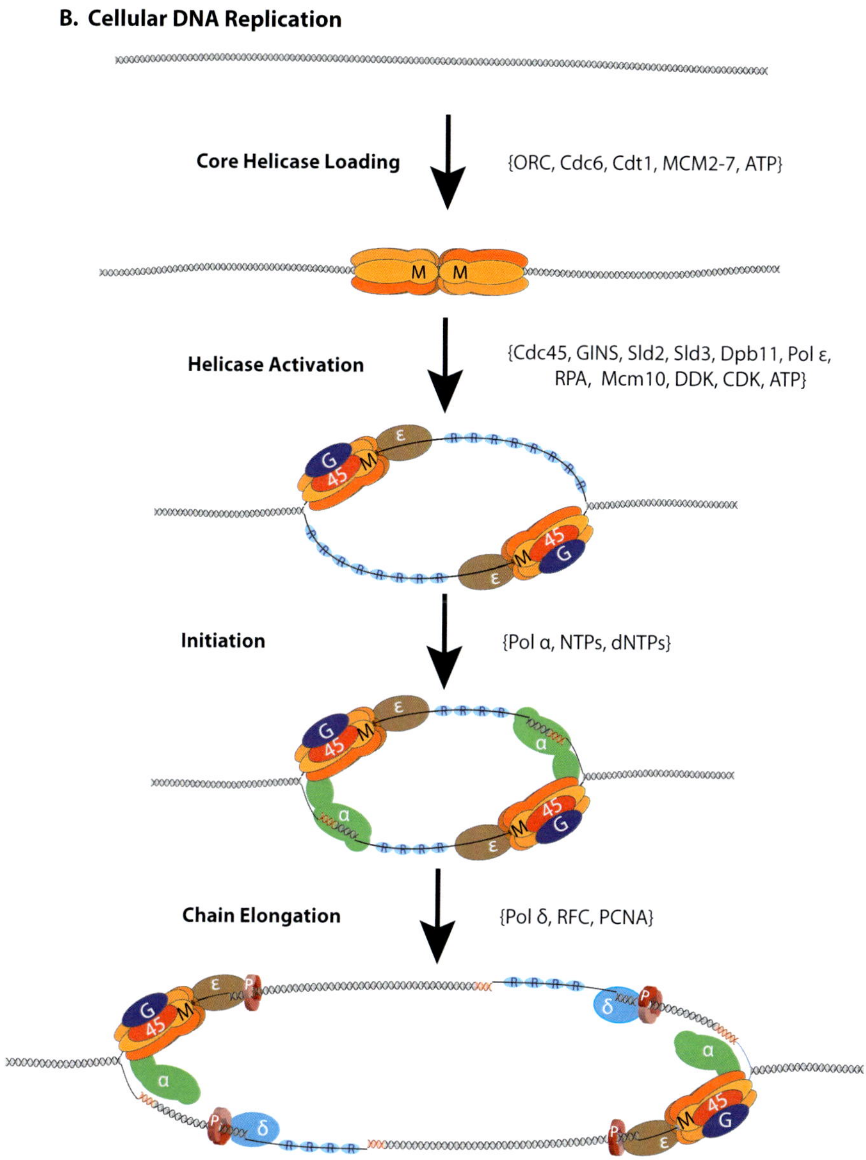

Fig. 1.1 (continued)

The only SV40-encoded protein required for DNA replication in vivo is the viral T antigen, which binds to the origin of DNA replication and serves both as the initiator protein and as the replicative helicase (Tegtmeyer 1972; Tjian 1978; Delucia et al. 1983). In the initial description of the SV40 cell-free DNA replication system, it was demonstrated that extracts from uninfected primate cells, supplemented with purified T antigen, were sufficient for replication of DNA molecules containing the

SV40 origin of DNA replication (Li and Kelly 1984). Detailed studies demonstrated that T antigen monomers assemble into a double hexamer around the origin DNA in a reaction dependent upon ATP (Dean et al. 1992; Valle et al. 2000). The hexamers interact head-to-head via the N-terminal origin binding domains of T antigen and alter the structure of the origin DNA (Borowiec and Hurwitz 1988). In the presence of ATP and a single-stranded DNA-binding protein, each T antigen hexamer functions as a helicase to unwind the template DNA (Stahl et al. 1986; Dean et al. 1987; Wold et al. 1987). Helicase activity is dependent upon a C-terminal AAA+ module, which binds and hydrolyzes ATP to drive translocation of the hexamer in the 3′–5′ direction along the leading strand template at each replication fork (Fig. 1.1a). Structural studies of the T antigen helicase revealed a double-ring with one tier containing the AAA+ motor domains (Li et al. 2003a). The central channel contains basic residues that can interact with the DNA. The precise mechanism of T antigen helicase translocation on DNA is not yet understood, although structural studies have inspired some interesting models. In addition to serving as the replicative helicase, T antigen interacts with other replication proteins to organize the replisome and coordinate its activities (see more below).

Biochemical analysis of the cell-free system derived from human cells demonstrated that a number of cellular proteins were required for SV40 DNA replication in vitro. All of these proteins proved to be involved in cellular DNA replication and were subjects of extensive biochemical and structural studies in the ensuing years (Reviewed in Kelly 1988; Challberg and Kelly 1989; Stillman 1989; Hurwitz et al. 1990; Brush et al. 1995; Waga and Stillman 1998).

The first cellular protein identified by fractionation of the SV40 system was RPA, the eukaryotic single-stranded DNA-binding protein (Wobbe et al. 1987; Wold and Kelly 1988; Fairman and Stillman 1988; Wold 1997; Chen and Wold 2014). The three nonidentical subunits of RPA contain multiple OB folds that bind single-stranded DNA. RPA was subsequently found to be subject to multiple post-translational modifications and to interact with many cellular proteins involved in a wide range of transactions involving DNA (Chen and Wold 2014). The contrast of the multi-subunit structure of RPA with the simpler bacterial single-stranded DNA-binding protein (SSB) was an early hint that the eukaryotic replisome would prove to be much more complex than that of prokaryotes.

An unexpected discovery that emerged from analysis of the SV40 system was that viral DNA synthesis in vitro is dependent on more than one cellular DNA polymerase. The eukaryotic DNA polymerase α had been discovered many years prior to the development of the SV40 system and was thought to be the major, perhaps the only, replicative DNA polymerase in eukaryotic cells on the basis of many indirect lines of evidence (Campbell 1986; Lehman and Kaguni 1989). After initial difficulties with biochemical characterization, the enzyme was eventually shown to contain four subunits, one of which harbors the polymerase activity. The two smallest subunits comprise a primase enzyme (Tseng and Ahlem 1982; Kaguni et al. 1983; Plevani et al. 1985). The primase catalyzes de novo synthesis of RNA primers on single-stranded DNA templates that can be further extended into nascent DNA chains by the DNA polymerase activity of the enzyme. By antibody depletion and fractionation/reconstitution experiments, it was demonstrated that DNA polymerase α is absolutely required for SV40 DNA replication in vitro. In the presence of T antigen and RPA, which are sufficient to drive extensive DNA unwinding, DNA polymerase α can initiate DNA synthesis on DNA molecules containing the SV40 origin and catalyze DNA synthesis on both the leading and lagging strand templates (Li and Kelly 1984; Murakami et al. 1986, 1992; Wold et al. 1988; Fig. 1.1a).

DNA polymerase δ, a second eukaryotic DNA polymerase, was identified in the 1970s but had been largely ignored (Byrnes et al. 1976). The activity of DNA polymerase δ was initially distinguished from that of DNA polymerase α because it contained an associated 3′–5′ proofreading exonuclease activity, which was lacking in DNA polymerase α. The first clue that this polymerase might play a role in SV40 DNA replication was the discovery that PCNA, a 37 kDa protein essential for SV40 DNA replication in vitro, was identical to a previously identified factor that increased the processivity of DNA polymerase δ (Tan et al. 1986; Prelich et al. 1987). The requirement for DNA polymerase δ was subsequently confirmed by direct reconstitution of the SV40 replication reaction with purified proteins (Lee et al. 1989; Weinberg and Kelly 1989; Tsurimoto et al. 1990). As described in more detail below, DNA polymerase δ catalyzes the bulk of DNA synthesis on both the leading and lagging strands of SV40 (Fig. 1.1a).

The PCNA processivity factor required for efficient DNA synthesis by DNA polymerase δ was reminiscent of the sliding clamps previously described in the prokaryotic T4 and *E. coli* systems (Tan et al. 1986; Prelich et al. 1987; Tsurimoto and Stillman 1990). It was ultimately shown by elegant biochemical and structural studies that PCNA, the *E. coli* β-clamp, the T4 gp45 sliding clamp, and archaeal sliding clamps are ring-shaped proteins with pseudo sixfold symmetry that accommodate duplex DNA in a topological linkage (Jeruzalmi et al. 2002). The general structure of the rings and the process by which they are loaded onto DNA have been highly conserved in evolution. In addition to its role in mediating processive DNA synthesis, PCNA plays major roles in other processes, such as Okazaki fragment maturation, DNA repair, recombination, chromatin assembly, cell cycle control, etc. (Moldovan et al. 2007).

The eukaryotic clamp loader RF-C, identified initially as a fraction required for SV40 DNA replication, consists of five subunits, each of which contains an AAA+ domain (Tsurimoto and Stillman 1989; Lee et al. 1991; Cai et al. 1996; Bowman et al. 2005). RF-C, like prokaryotic and archaeal clamp loaders, functions as a machine that couples the energy of ATP hydrolysis to open PCNA and load it at a primer terminus (Bowman et al. 2005; Yao et al. 2006). After dissociation of the RF-C loader, DNA polymerase δ associates with the loaded PCNA to form a highly processive complex. Much has been learned about the structural basis of the specificity of RF-C for primer termini and about the biochemical mechanism of clamp loading (Yao and O'Donnell 2012). On the lagging strand, RF-C and other clamp loaders can also function to unload PCNA rings from double-stranded DNA after completion of Okazaki fragment synthesis and ligation so that they can be recycled.

Studies of SV40 DNA replication in vitro also allowed the detailed analysis of the roles of DNA topoisomerases (Yang et al. 1987). It was shown that topoisomerase activity is required during DNA chain elongation to remove supercoils generated by DNA unwinding and that this function can be mediated by either type I or type II topoisomerase. Human topoisomerase I binds to T antigen and may play the predominant role in relieving topological stress at replication forks (Simmons et al. 1996). Interestingly, yeast topoisomerase I co-purifies with the cellular helicase, suggesting that it may also travel with the replication fork (Gambus et al. 2006). Topoisomerase activity is additionally required for the decatenation and segregation of completed daughter DNA molecules synthesized in vitro, but this activity can only be provided by type II topoisomerase. These observations are entirely consistent with studies of SV40 DNA replication in vivo (Sundin and Varshavsky 1980, 1981).

Analysis of the interactions among the proteins required for SV40 DNA replication provided insights into the functional organization of the replication fork, many of which are relevant to understanding the cellular replication fork (Reviewed in Kelly 1988; Challberg and Kelly 1989; Stillman 1989; Hurwitz et al. 1990; Waga and Stillman 1998; Fig. 1.1a). Movement of the replication fork, driven by the T antigen helicase motor, generates single-stranded DNA bound by RPA. During this process, the complex of DNA polymerase δ with PCNA advances, synthesizing the leading strand, while a region of the lagging strand template accumulates in single-stranded form prior to synthesis of a primer by DNA polymerase α. Interestingly, it was observed that DNA polymerase α, in the absence of other factors, is completely unable to initiate primer synthesis on DNA templates that are coated with bound RPA (Collins and Kelly 1991; Melendy and Stillman 1993). This observation led to the discovery that T antigen promotes primer synthesis via specific interactions with DNA polymerase α and RPA that presumably destabilize bound RPA and allow access of primase to the template (Collins and Kelly 1991; Dornreiter et al. 1992; Collins et al. 1993; Melendy and Stillman 1993; Zhou et al. 2012). Thus, the T antigen helicase-DNA polymerase α complex constitutes a mobile primosome that is active in primer synthesis and likely limits priming to the vicinity of replication forks. Presumably, functionally similar interactions involving the cellular helicase or associated factors are required for priming by DNA polymerase α at cellular replication forks.

Another important phenomenon discovered in the SV40 system is switching among DNA polymerases (Lee et al. 1989; Weinberg and Kelly 1989; Tsurimoto et al. 1990; Waga et al. 1994; Waga and Stillman 1994, 1998; Fig. 1.1a). Polymerase switching is unique to eukaryotic DNA replication and occurs in both viral and cellular systems. DNA synthesis is initiated on both the leading and lagging strand templates by the primase activity of DNA polymerase α. After synthesis of a short RNA primer, DNA polymerase α extends it into an "initiator DNA" chain. In the case of SV40, DNA polymerase α is subsequently replaced by the complex of PCNA and DNA polymerase δ, which carries out the bulk of DNA synthesis. The mechanism of this polymerase switch involves a competition between DNA polymerase α and RF-C for the primer terminus (Tsurimoto et al. 1990; Tsurimoto and Stillman 1991a, b; Waga et al. 1994). Since DNA polymerase α is not highly processive, RF-C will gain access to the primer terminus at some point during synthesis of the initiator DNA. When this happens, RF-C loads PCNA and then dissociates, allowing the binding of DNA polymerase δ to PCNA. The PCNA-DNA polymerase δ complex is stable, creating a highly processive polymerase engine capable of synthesizing long nascent DNA strands. On the leading strand, the polymerase switch is only required once, while on the lagging strand, one switching event occurs for each Okazaki fragment. DNA polymerase δ efficiently completes the synthesis of Okazaki fragments, after which they are joined together by DNA ligase in a maturation process that involves removal of the RNA primer and much of the initiator DNA (Waga and Stillman 1994; Balakrishnan and Bambara 2013). DNA polymerase switching plays an important role in maintaining the integrity of the genome. DNA polymerase δ has a proofreading exonuclease, but DNA polymerase α does not, so the polymerase switch, together with the removal of the primer and much of the initiator DNA during completion of Okazaki fragment synthesis, helps ensure the fidelity of DNA replication. Subsequent genetic and biochemical studies in yeast revealed the existence of a third eukaryotic DNA polymerase, DNA polymerase ε (Pol ε), that functions in cellular DNA replication but apparently not in SV40 DNA replication (Budd et al. 1989; Morrison et al. 1990; Araki et al. 1992; Zlotkin et al. 1996; Pospiech et al. 1999). Multiple recent studies indicate that DNA polymerase ε plays the major role in the synthesis of the leading strand during cellular DNA replication, while DNA polymerase δ synthesizes most of the DNA on the lagging strand (Burgers et al. 2016). It follows that a switch from DNA polymerase α to DNA polymerase ε occurs at the terminus of the leading strand shortly after synthesis of the first primer. The mechanism of this switch is not yet known. However, yeast DNA polymerase ε is quite processive, even in the absence of RF-C and PCNA, so it may simply replace the less processive DNA polymerase α after synthesis of a short segment of initiator DNA. In the yeasts *S. cerevisiae* and *S. pombe*, the catalytic activity of DNA polymerase ε is not essential, although DNA replication is somewhat abnormal in its absence (Dua et al. 1999; Kesti et al. 1999; Feng and D'Urso 2001). In this circumstance DNA polymerase δ is presumably responsible for most DNA synthesis on both the leading and lagging strand templates, as is the case with SV40.

The SV40 system also focused attention on the potential for regulation of DNA replication by protein phosphorylation. Viral DNA replication in vivo and in vitro was found to be completely dependent upon phosphorylation of T antigen on Thr-124 by cyclin-dependent kinase (McVey et al. 1989, 1993; Moarefi et al. 1993). The unphosphorylated protein is competent to bind to the origin and to induce structural distortions in origin DNA and is also competent to catalyze unidirectional unwinding in helicase assays with artificial DNA substrates containing single- and double-stranded DNA. However, the unphosphorylated protein is completely inactive in bidirectional helicase activity from the SV40 origin, indicating that phosphorylation controls a step subsequent to assembly of the T antigen double hexamer on the DNA, possibly an interaction between T antigen hexamers that is required for the initial unwinding of the origin. Interestingly, it was observed that phosphorylation of other residues in T antigen, notably Ser-120 and Ser-123, is inhibitory to initiation of DNA replication and, again, the modifications appear to regulate a step subsequent to assembly of the double hexamer (Virshup et al. 1989, 1992; Scheidtmann et al. 1991; Cegielska and Virshup 1993; Cegielska et al. 1994). Thus, T antigen helicase function is regulated both positively and negatively by phosphorylation. The biological role of this regulation is still unclear, but it may function in part to coordinate SV40 replication with the cell cycle, to enhance the efficiency of production of viral genomes. Consistent with this idea, human CDK can promote SV40 DNA replication in G1 extracts that are normally deficient in replication activity (D'Urso et al. 1990). These discoveries presaged the description of the (much more complex) control of the activation of the cellular helicase by protein phosphorylation (see below).

As the foregoing indicates, analysis of SV40 DNA replication in vitro by many investigators resulted in the identification of key cellular replication proteins and generated many insights into their mechanisms of action. Naturally, these advances kindled great interest in identifying the cellular counterparts of T antigen – the molecules that are required for origin recognition in cellular chromosomes and the molecules that comprise the cellular helicase. A related issue of great interest was how initiation of replication in cells is controlled to ensure that each segment of the genome is faithfully duplicated once each cell cycle. These are the main issues discussed in the remainder of this perspective. To attack these problems, attention shifted from SV40 to other model systems. The analysis of DNA replication in budding yeast proved particularly fruitful because of the availability of highly developed genetic approaches and the existence of defined origins of replication. Fission yeast provided many novel insights as well and, in many cases, complemented the work in budding yeast. Many advances also came from studying replication in extracts of eggs of *Xenopus laevis*. The *Xenopus* system lacks genetics and is not particularly amenable to large-scale fractionation/reconstitution approaches. However, it can be used effectively to test specific requirements for cellular replication by antibody depletion-reconstitution experiments. It is also especially useful for analyzing the regulation of cellular DNA replication.

Because of methodological advances, there was also a gradual shift away from classical approaches for identifying cellular replication proteins by fractionation of crude extracts toward more efficient methods. The products of putative replication genes identified in genetic screens or homology searches were expressed as tagged variants that could be quickly purified and biochemically characterized. But reconstitution with purified proteins remained the gold standard for defining the minimal constellation of proteins required for eukaryotic DNA replication.

1.4 Cellular Origins and Initiators

The first cellular origins of DNA replication to be characterized were those of the budding yeast *S. cerevisiae* (Hsiao and Carbon 1979; Stinchcomb et al. 1979; Newlon 1988; Fangman and Brewer 1991; Campbell and Newlon 1991). In early genetic studies, it was discovered that certain chromosomal DNA segments, called autonomously replicating sequences or ARSs, could increase the transformation efficiency and replication stability of yeast plasmids. It was subsequently shown by direct physical assays that ARS elements represent the start sites for initiation of bidirectional replication in budding yeast (Brewer and Fangman 1987; Huberman et al. 1987). Detailed analysis revealed that ARS elements share a common sequence of 11 bp called the ARS consensus sequence, as well as adjacent, less highly conserved sequence elements (Broach et al. 1983; Celniker et al. 1984; Van Houten and Newlon 1990; Marahrens and Stillman 1992; Breier et al. 2004). The strong sequence specificity and other properties of *S. cerevisiae* origins of DNA replication were similar to the familiar prokaryotic origins of DNA replication and the origins of eukaryotic viruses like SV40. Importantly, this sequence specificity would prove crucial for identifying the cellular proteins that assemble at origins of replication prior to the onset of DNA synthesis. Central among these is the origin recognition complex (ORC), which was detected as a protein that binds specifically to ARS elements and is essential for initiation of DNA replication (Bell and Stillman 1992; Bell et al. 1993). The discovery of ORC was a major advance that opened the way to study the early events in cellular DNA replication.

ORC was found to have six subunits, five of which are related to the AAA+ protein family of ATPases (Bell et al. 1995; Bell and Stillman 1992; Bell 1995; Bell and Dutta 2002; Speck et al. 2005; Clarey et al. 2006). High-affinity binding of ORC to ARS elements is dependent upon ATP binding to the largest subunit, Orc1. In the years following the identification of ORC, a general picture of the first steps in cellular replication emerged from studies in yeast cells and the *Xenopus* egg extract system. These steps result in the loading of the core replicative helicase onto the origin DNA. An early finding that opened a path toward defining the helicase loading reaction was the identification of a protein assembly at yeast origins of DNA replication that is referred to as the prereplicative complex or pre-RC (Diffley and Cocker 1992; Diffley et al. 1994). The pre-RC was originally characterized as a distinct pattern of nuclease protection in genomic footprinting experiments that was observed at yeast origins during the G1 phase of the cell cycle. The requirements for pre-RC formation and the protein composition of the complex were not immediately apparent but were rapidly determined by a variety of experimental approaches (Cocker et al. 1996; Rowles et al. 1996; Coleman et al. 1996; Romanowski et al. 1996b; Donovan et al. 1997; Tanaka et al. 1997; Aparicio et al. 1997; Nishitani et al. 2000; Maiorano et al. 2000; Labib et al. 2001; Devault et al. 2002; Tanaka and Diffley 2002). These experiments demonstrated that in the initial step of pre-RC formation, ORC recruits the Cdc6 protein to the origin. Cdc6 is a AAA+ protein in the same clade as the Orc1-5 subunits. The complex of ORC and Cdc6 then recruits the Cdt1 protein and the MCM2-7 complex. Cdt1, a factor first identified in fission yeast, binds to MCM2-7 and facilitates its interaction with ORC-Cdc6 bound at the origin. The MCM2-7 complex is the core of the eukaryotic replicative helicase. It is loaded onto the origin DNA by ORC-Cdc6-Cdt1 in a reaction requiring ATP hydrolysis (Fig. 1.1b). The nature of the association of MCM2-7 with DNA was not clear from these early experiments but was eventually determined by elegant biochemical studies with purified components (Evrin et al. 2009; Remus et al. 2009; Kawasaki et al. 2006). Two hexameric rings of MCM2-7 are loaded in a head-to-head configuration with duplex DNA passing through a central channel. In the loading reaction, ORC and Cdc6 function as a molecular machine to close a gate between the Mcm2 and Mcm5 subunits, leaving the MCM2-7 rings topologically linked to the DNA. The current view is that the two hexamers are loaded in separate steps which are dependent on distinct Cdc6 and Cdt1 molecules (Bell and Kaguni 2013; Sun et al. 2014; Ticau et al. 2015). Once the MCM double hexamer is loaded on the DNA, subsequent events in initiation of DNA replication do not require Cdc6, Cdt1, or ORC, and in vitro studies suggest that these factors may dissociate from the DNA (Rowles et al. 1999; Hua and Newport 1998; Gros et al. 2014; On et al. 2014; Ticau et al. 2015). Thus, loaded double hexamers of MCM2-7 mark the potential sites of initiation of DNA replication in chromosomal DNA at the end of the G1 phase of the cell cycle. The transformation of MCM2-7 into an active helicase and the assembly of the other replisome components begin at the onset of S phase and require the CDK and DDK protein kinases (more about this below).

It was quickly established that the early events in DNA replication just described are largely conserved in all eukaryotes. ORC molecules were identified in eukaryotic species other than *S. cerevisiae*, including humans and other metazoa, and the general requirements for pre-RC assembly proved to be universal (Bell and Dutta 2002). However, studies in other systems did not reveal the strong sequence specificity of budding yeast ARS elements. It gradually became apparent that DNA replication origins in most eukaryotes conform to a paradigm that is somewhat different from that of *S. cerevisiae*. A well-characterized example is the fission yeast *S. pombe*. It was possible to isolate segments from *S. pombe* chromosomal DNA that, like ARS elements in budding yeast, confer stable plasmid replication in fission yeast cells (Dubey et al. 1994, 1996; Clyne and Kelly 1995; Dai et al. 2005). However, the properties of the active segments are different from those of budding yeast: they are much larger than budding yeast ARS elements, and they exhibit little conservation of nucleotide sequence. The average AT content of the sequences active as origins is greater than that of the fission yeast genome (Segurado et al. 2003; Dai et al. 2005), an observation that was explained, at least in part, by the discovery that *S. pombe* ORC binds to chromosomal DNA via multiple copies of a DNA-binding motif called the AT hook (Chuang and Kelly 1999; Kong and DePamphilis 2001). But fission yeast origins do not share a common consensus sequence. In fact, the best predictors of the ability of a segment of the fission yeast genome to function as a plasmid origin are AT content and length (Segurado et al. 2003; Dai et al. 2005). At least half the intergenic regions in the fission yeast genome can exhibit detectable origin activity in plasmid assays. It was suggested that the properties of fission yeast origins are best explained by a stochastic model in which ORC can bind and drive initiation of DNA replication at many potential sites in the genome with little intrinsic sequence specificity, and during each cell cycle these potential sites are chosen largely at random with perhaps some preference for AT-rich intergenic DNA (Dai et al. 2005). Subsequent DNA combing experiments demonstrated that the distribution of distances between start sites in a fission yeast genome is exponential, which is completely consistent with a stochastic model for the spatial distribution of potential origins (Patel et al. 2006; Kaykov and Nurse 2015). The firing of potential origins appears to be stochastic in time as well as space, although the rates of firing per unit time may vary from origin to origin (Heichinger et al. 2006; Kaykov and Nurse 2015).

The nature of origins of DNA replication in metazoan organisms is still not well understood, but the available evidence suggests that there are many potential origin sites and that such sites lack strong sequence determinants (Leonard and Mechali 2013). Consistent with this impression is the observation that purified metazoan ORC molecules, such as those of *Drosophila* or human cells, do not exhibit a strong preference for any particular DNA sequence (Vashee et al. 2003; Remus et al. 2004). Human ORC can drive replication of essentially any DNA molecule in a replication system derived from *Xenopus* eggs (Vashee et al. 2003).

It is important to recognize that even in the case of *S. cerevisiae,* the highly specific interactions between ORC and the ARS consensus sequences are not essential for the fundamental mechanism by which ORC promotes initiation of DNA replication. Specific binding by ORC as observed in budding yeast simply affects the distribution of ORC molecules over the genome. In cell-free systems with *S. cerevisiae* proteins, ORC can efficiently initiate DNA replication on DNA molecules lacking ARS elements (Gros et al. 2014; On et al. 2014). Moreover, deleting chromosomal ARS elements from an *S. cerevisiae* chromosome has surprisingly little effect on the efficiency of DNA replication (Dershowitz et al. 2007). Thus, it seems likely that initiation of DNA replication in most, if not all, eukaryotes may occur on essentially any DNA molecule with some efficiency. *S. cerevisiae* may have acquired the capacity to initiate DNA replication at more specific sites because its chromosomes are relatively small. If replication initiation were purely random, there would be some probability that a small chromosome would not be duplicated in a significant fraction of cell cycles.

Why have eukaryotes largely departed from the prokaryotic paradigm of highly specific origin sequences enshrined in the replicon model? One possibility is that the distribution of chromosomal sites that are available for ORC binding and initiation may vary considerably in eukaryotic cells in different physiological states (e.g., different states of development, different states of differentiation, different tissue environments, etc.) because the composition and organization of chromatin and the pattern of gene expression are different in such states. Thus, the ability to initiate DNA replication in almost any accessible region may be advantageous in the face of the increased complexity of eukaryotic biology.

Because of the many potential origins of replication, the dynamics of DNA replication in an individual eukaryotic cell is extremely complex and is different from one cell cycle to the next. The factors that determine the sites where replication is initiated and the timing of the replication of each genomic sequence are not yet understood in detail in any organism. The overall pattern of DNA replication is established by events that occur at two stages of initiation: the loading of the core helicase in G1 and the activation of the helicase at the onset of S phase. The distribution of loaded MCMs obviously depends upon the relative affinity of ORC for particular DNA sequences, which may be affected to some degree by other components of the pre-RC (e.g., Cdc6) or by interactions with other cellular proteins (e.g., Beall et al. 2002; Speck and Stillman 2007). The distribution of loaded MCMs is also greatly influenced by competition with the multitude of other chromatin factors that bind to chromosomal DNA. This competition affects the accessibility of chromosomal loci to the helicase loading factors and probably explains why many studies have correlated initiation of DNA replication with so-called open chromatin domains (Berbenetz et al. 2010; Eaton et al. 2011; Leonard and Mechali 2013). Once S phase begins, the loading of MCMs is precluded (see below), so their number and their approximate locations are fixed (although there is some recent evidence that they may have limited mobility) (Gros et al. 2015). Therefore, subsequent events are largely determined by the pattern of activation of loaded MCMs by the CDK and DDK protein kinases to generate functional replisomes. This pattern is also extremely complex. Not all loaded MCMs are activated in a given cell cycle – many are functionally inactivated when they are replicated passively by a replisome originating at another site (Friedman et al. 1997; Santocanale et al. 1999; Vujcic et al. 1999). Thus, the determination of which loaded MCMs are activated and which are not is highly dependent upon their order of activation, and this is likely to be largely stochastic in a given cell cycle. There is no convincing evidence that the ordering of helicase activation in the cell cycle is deterministic, but, as discussed later in this perspective, the relative rates of activation of loaded MCMs at different sites can differ, so that on average activation at some sites occurs earlier than at others. In the end, the timing of replication of a particular sequence in the genome during a given cell cycle is a function of (1) its distance from the flanking initiation sites, (2) the times of initiation at those sites, and (3) the rates at which the forks move after initiation. These factors, particularly the first and second, are subject to variation from one cell cycle to the next. This is an active area of research, and it is expected that further insights will come from increasing the spatial and temporal resolution of single molecule and ensemble approaches to studying replication timing and from improving the methods of analysis in silico.

1.5 CMG the Cellular Helicase

Following the discovery of ORC, the cellular initiator protein, another 15 years were required to definitively identify the active form of the cellular helicase. This advance was driven by two separate lines of investigation that eventually came together. Genetic studies in budding yeast defined a group of genes required for the stable maintenance of plasmids as autonomously replicating units in cells (Moir et al. 1982; Maine et al. 1984). Although isolated by several different assays, it was eventually appreciated that the corresponding proteins form a complex, which was named the MCM2-7 complex (for minichromosome maintenance 2-7). The six subunits of the MCM2-7 complex were shown to be related members of the AAA+ superfamily of ATPases (Chong et al. 1996), and a number of studies suggested that they are required for initiation of DNA replication (Yan et al. 1991; Hennessy et al.

1990; Chong et al. 1995; Madine et al. 1995; Romanowski et al. 1996a). The first clue regarding their potential function was the discovery that a sub-complex of MCM4-6-7 purified from human cells possessed ATP-dependent helicase activity in vitro (Ishimi 1997). This insight was followed by the demonstration that the single archaeal MCM protein forms multimeric complexes with robust helicase activity (Kelman et al. 1999; Chong et al. 2000). Analysis of the distribution of MCM proteins in yeast chromosomes during S phase strongly suggested that the MCM complex travels with replication forks, and it was subsequently shown that depletion of MCM subunits interrupts replication fork progression (Aparicio et al. 1997; Labib et al. 2000). Experiments in *Xenopus* egg extracts confirmed that inhibition of MCM activity prevents efficient DNA unwinding (Pacek and Walter 2004). In the course of these studies, it was observed that Cdc45, another protein thought to be involved in initiation of DNA replication, interacts with the MCM complex and is also required for both DNA unwinding and fork progression (Hopwood and Dalton 1996; Aparicio et al. 1997; Pacek and Walter 2004). While these studies, taken together, established the likelihood that the MCM complex was a required component of the cellular helicase, it did not seem to be sufficient for helicase activity, since purified yeast MCM2-7, containing all six subunits, did not have detectable helicase activity under the usual assay conditions.

A second line of investigation that proved critical for the identification of the cellular helicase was a product of the discovery that DNA polymerase ε is required for DNA replication in yeast (Morrison et al. 1990; Araki et al. 1992). A network of proteins that exhibited genetic and physical interactions with each other and with DNA polymerase ε was discovered and characterized. The key players in this network were Dpb11, Sld2, Sld3, and a complex of four proteins called the GINS complex (Araki et al. 1995; Kamimura et al. 1998, 2001; Takayama et al. 2003). It was observed that GINS colocalizes with Cdc45 and MCM2-7 at sites of DNA unwinding during DNA replication and is required to maintain the association of Cdc45 with MCM2-7 during DNA chain elongation (Aparicio et al. 1997; Gambus et al. 2006; Pacek et al. 2006). The definitive biochemical definition of the cellular helicase was a result of experiments aimed at purifying the Cdc45 protein from extracts of *Drosophila* embryos (Moyer et al. 2006; Ilves et al. 2010). The purification yielded a complex of 11 polypeptides that included MCM2-7 and GINS, as well as Cdc45, and is now referred to as CMG. Strikingly, the purified CMG complex possessed robust 3′–5′ helicase activity on partially duplex DNA molecules. Reconstitution of the complex with recombinant proteins revealed that the presence of GINS and Cdc45 dramatically changed the properties of the MCM2-7 complex, increasing its affinity for DNA and stimulating its ATPase activity by two orders of magnitude (Ilves et al. 2010; Costa et al. 2011, 2014). Since GINS and Cdc45 do not have ATPase active sites, the observed stimulation is due to allosteric remodeling of the core MCM2-7 helicase engine. More recent studies have examined the structure of the CMG in the presence of single-stranded DNA and non-hydrolysable ATP (Costa et al. 2014). In this complex the AAA+ motor domains of the MCM2-7 hexamer form a cracked-ring, right-handed spiral with the crack at the MCM2-5 interface as predicted from biochemical studies (Bochman and Schwacha 2008). GINS and Cdc45 bridge the gap creating a topologically closed toroid. The CMG prefers to bind to single- stranded DNA as expected from previous work demonstrating that a single strand of DNA (the leading strand template) is enclosed by the central channel of the replicative helicase at the replication fork (Fu et al. 2011). The structure and biochemical properties of the CMG are consistent with a steric exclusion model in which unwinding occurs as a result of exclusion of the lagging strand template from the central channel as the helicase engine translocates in the 3′–5′ direction on the leading strand template (Kaplan et al. 2003; Fu et al. 2011; Fig. 1.1b). The molecular details of DNA unwinding by the CMG helicase are not yet clear. Much of the thinking about this issue has come from detailed structural studies of the homohexameric helicases of archaea, SV40 and papillomavirus (Gai et al. 2004; Singleton et al. 2007; Enemark and Joshua-Tor 2008; Hauk and Berger 2016). In these cases, it has been suggested that cooperative structural changes in the AAA+ motors that are mediated by cycles of sequential ATP binding and hydrolysis around the ring drive the motion of DNA-binding loops in the central channel to propel the helicase along the DNA. The same general mechanism may hold for the heterohexameric CMG as well, but there are some data that are not completely consistent with it. For example, only two of the six ATP-binding sites in CMG are absolutely essential for helicase activity (Ilves et al. 2010). In recent structural studies, two alternative states of the CMG have been observed, one compact and the other extended (Yuan et al. 2016; Abid Ali et al. 2016). This observation has led to the proposal that the helicase moves by a ratcheting or inchworm-like motion. The precise mechanism of this critical feature of replisome function awaits further analysis.

1.6 Activation of the Cellular Helicase and Building the Replisome

The realization that the core replicative helicase, MCM2-7, was loaded onto DNA in a form completely inactive in DNA unwinding raised the pivotal question of how this structure is converted into an active CMG helicase within a functional replisome during S phase. The transformation of the double-hexamer encircling duplex DNA into an active CMG helicase capable of translocating on single-stranded DNA and the recruitment of additional components to form the complete replisome are still not understood completely, but great progress has been made (Fig. 1.1b). The process is quite complex, involving multiple protein-protein interactions regulated by protein phosphorylation. All of the essential factors are now known, but many aspects of the mechanisms involved in the process remain mysterious (Tanaka and Araki 2013; Yeeles et al. 2015, 2017; Devbhandari et al. 2017; Kurat et al. 2017).

CMG assembly and activation are promoted by two protein kinases, cyclin-dependent kinase (CDK) and Dbf4-dependent kinase (DDK). Each kinase has a regulatory subunit – a cyclin in the case of CDK and Dbf4 in the case of DDK – whose abundance increases at the onset of S phase. The kinases and their regulatory subunits were discovered by multiple lines of investigation, and their positive role in DNA replication has long been known (Siddiqui et al. 2013). The identification of the key substrates of these enzymes and the elucidation of their roles in CMG assembly and activation were products of a number of independent studies that have only recently converged.

In addition to the protein kinases, the efficient assembly and activation of the replicative helicase are dependent upon several proteins that are not present in the final active helicase. These proteins, Sld2, Sld3, Sld7, and Dpb11 were identified, along with GINS, as factors that interact directly or indirectly with DNA polymerase ε, which becomes associated with the CMG during its assembly (Araki et al. 1995; Kamimura et al. 1998, 2001; Takayama et al. 2003; Tanaka and Araki 2013). Many interactions have been identified among the proteins involved in CMG assembly and probably more will be found (Tanaka and Araki 2013; Deegan et al. 2016). Detailed understanding of the overall pathway will require extensive biochemical and structural studies. At this point in time, the evidence suggests that assembly of the CMG proceeds in (at least) two steps driven by DDK and CDK. DDK phosphorylation of the MCM2-7 double hexamer drives its association with Sld3, Sld7, and Cdc45 (Hardy et al. 1997; Kamimura et al. 2001; Randell et al. 2010; Sheu and Stillman 2010; Heller et al. 2011). In this reaction Sld3 binds to phosphorylated Mcm4 or Mcm6 and then recruits Cdc45 (Deegan et al. 2016). Importantly, recruitment of Sld3 and Cdc45 to MCM2-7 is not dependent upon CDK activity, but the stability of the complex is increased following CDK activation. Sld7 is not essential for this reaction but may increase its efficiency. In budding yeast, it was observed that association of Sld3, Cdc45, and MCM2-7 can occur to a limited extent in cells blocked in G1 phase with alpha factor presumably because of the presence of a low level of active DDK (Aparicio et al. 1999; Kanemaki and Labib 2006; Tanaka et al. 2011; Heller et al. 2011). Subsequent analysis of a cell-free yeast replication system demonstrated clearly that purified DDK can induce the association of the three proteins in extracts derived from alpha factor blocked cells (Heller et al. 2011).

The final assembly of the CMG helicase is driven by CDK phosphorylation of Sld2 and Sld3, which allows the formation of a complex of the two proteins with Dpb11 (Masumoto et al. 2002; Tak et al. 2006; Zegerman and Diffley 2007; Tanaka et al. 2007). Dpb11 contains four BRCT domains. The amino-terminal pair binds phospho-Sld3 and the carboxyl-terminal binds phospho-Sld2. Phosphorylated Sld2 is present in a complex containing Dpb11, GINS, and DNA polymerase ε, which is referred to as the pre-loading complex (Muramatsu et al. 2010). The protein-protein interactions required to form the pre-loading complex are not known in detail, but there is good evidence that the Dpb2 subunit of DNA polymerase ε plays an important structural role in the assembly of the CMG helicase and in maintaining association of the helicase with the leading strand DNA polymerase at the fork (Muramatsu et al. 2010; Sengupta et al. 2013). The binding of Sld2 and Sld3 to Dpb11 thus brings together all of the components required to form the CMG. The initiation of DNA unwinding at an origin requires one additional protein, MCM10, whose role at the molecular level is not yet clear (Kanke et al. 2012; van Deursen et al. 2012; Watase et al. 2012; Yeeles et al. 2015). In vivo and in vitro studies indicate that the stable association of MCM10 with origins occurs after MCM2-7 loading, and its accumulation is dependent upon DDK and CDK. DNA unwinding also requires the single-stranded DNA-binding protein RPA and, for circular templates, a DNA topoisomerase. How the interactions among this multitude of proteins promote the assembly of the active CMG helicase, remodel the association of the helicase with the two DNA strands, and initiate DNA unwinding are fascinating issues that will be a major focus of future work.

Following the assembly of an active helicase capable of unwinding DNA, other factors are recruited to create a complete replisome at each replication fork (Fig. 1.1b). Exploration of replisome organization and function are at an early stage, but some general features are becoming evident from increasingly sophisticated reconstitution experiments and improved structural studies by cryo-EM. A major milestone was the reconstitution of initiation of *S. cerevisiae* DNA replication with a minimal set of purified proteins (Yeeles et al. 2015). More recently the replication of both naked DNA and chromatin templates was reconstituted in vitro (Kurat et al. 2017; Yeeles et al. 2017; Devbhandari et al. 2017). These advances represent the culmination of years of efforts by many laboratories to define the requirements for eukaryotic DNA replication. Importantly, the work has confirmed the minimal protein requirements for the several stages of DNA replication predicted by previous work (Fig. 1.1b): (1) the loading of the MCM2-7 complex requires ORC, Cdc6, and Cdt1; (2) the formation and activation of the CMG helicase additionally requires Cdc45, GINS, Sld3/7, Sld2, Dpb11, DNA polymerase ε, DDK, CDK, and MCM10; (3) the initiation of DNA synthesis additionally requires RPA and DNA polymerase α; (4) the efficient elongation of leading and lagging strands additionally requires RFC, PCNA, topoisomerase I or II, and DNA polymerase δ; and (5) the maturation of Okazaki fragments additionally requires Fen1 and DNA ligase. All together eukaryotic DNA replication on naked DNA templates requires a minimum more than 40 distinct polypeptide chains.

The reconstitution studies have opened the way to a deeper understanding of the fundamental mechanisms of replication and have already provided several interesting mechanistic details (Devbhandari et al. 2017; Yeeles et al. 2017). For example, it was demonstrated directly that rapid leading strand DNA synthesis at rates comparable to those observed in vivo requires DNA polymerase ε together with the processivity factor PCNA. DNA polymerase δ can support leading strand synthesis when DNA polymerase ε is defective in catalysis, but it functions at a significantly slower rate. On the other hand, DNA polymerase δ in complex with PCNA is essential for the complete synthesis and efficient joining of nascent Okazaki fragments on the lagging strand. Only DNA polymerase δ can carry out the strand displacement synthesis required for primer removal by the Fen1 endonuclease during Okazaki fragment maturation (Devbhandari et al. 2017). Reconstitution studies have also demonstrated that maximal rates of DNA replication require the nonessential protein factors Mrc1 and Csm3/Tof1, which had previously been shown to be associated with active replisomes. The functional role of these factors is not yet clear, but it has been suggested that they accelerate the rate of DNA unwinding by the CMG helicase (Yeeles et al. 2017).

The replication of chromatin templates has also been explored in the reconstituted yeast replication systems (Kurat et al. 2017; Devbhandari et al. 2017). Interestingly, chromatin appears to enforce the origin specificity of *S. cerevisiae* DNA replication. As mentioned earlier, the replication of naked DNA templates in vitro does not exhibit the strong origin dependence characteristic of replication in vivo. However, assembly of the template into chromatin significantly increases the dependence of replication on a canonical yeast origin (Kurat et al. 2017; Devbhandari et al. 2017). The evidence suggests that this effect is due to suppression of ORC binding to non-specific sites (Kurat et al. 2017). Binding of ORC to the origin is less affected by nucleosomes because of its high affinity for specific origin sequences and because origin sequences tend to exclude nucleosomes even in the absence of ORC. The latter phenomenon is reminiscent of the observation that initiation of DNA replication in mammalian cells is correlated with open chromatin domains. With the minimal set of replication proteins listed above, assembly of the CMG helicase readily occurs on chromatin templates, but replisome progression is blocked. A recent study demonstrated that FACT (facilitates chromatin transcription), a dimeric histone chaperone, is sufficient to overcome the block and allow replication to proceed (Kurat et al. 2017). The precise role of FACT is unclear, but it has been suggested that it may function by displacing nucleosomes ahead of the fork or by facilitating the transfer of nucleosomes to the nascent DNA behind the fork. The addition of FACT alone to the minimal reconstituted system, though sufficient for replication, does not restore the full rate of DNA replication observed with naked DNA templates. Several additional factors, including the chaperone Nhp6; the chromatin remodelers, INO80 and ISW1A; and the histone acetyltransferases, pNuA4 and pSAGA, further stimulate the rate of fork movement (Kurat et al. 2017). These seminal studies indicate that detailed analysis of the factors and mechanisms that allow the replication machinery to progress through chromatin (and other protein-DNA complexes) is now feasible, and these issues will clearly be the focus of much future work.

Like the replisomes of prokaryotes and SV40, protein-protein interactions between the replicative helicase and other replisome components are a major organizational principle of the eukaryotic replisome. As described above, the leading strand DNA polymerase, Pol ε, associates with the components of the CMG during helicase assembly. Studies of replisomes isolated from yeast or reconstituted in vitro indicate that DNA polymerase epsilon is bound directly to the CMG (Langston et al. 2014; Georgescu et al. 2015a, b; O'Donnell and Li 2016). An interaction between Psf1 in GINS with the Dpb2 subunit of Pol ε is required to maintain association of the polymerase with the replisome, and crosslinking studies suggest additional contacts with MCM proteins and Cdc45 (Sengupta et al. 2013; Sun et al. 2015). DNA polymerase α-primase may also be tethered to the core CMG. In this case the tether is provided, at least in part, by the protein Ctf4, which was originally identified in a screen for factors affecting the fidelity of chromosome segregation (Zhu et al. 2007; Gambus et al. 2009; Tanaka et al. 2009; Simon et al. 2014; Samora et al. 2016; Sutani and Shirahige 2016; Villa et al. 2016). Recent structural studies have shown that Ctf4 forms a trimer capable of binding to similar motifs present in the Sld5 subunit of GINS, the catalytic subunit of DNA Pol α, and other proteins (Simon et al. 2014). Linking of Pol α to the CMG likely increases the efficiency of priming of Okazaki fragments on the lagging strand of replication forks. Interactions of the lagging strand DNA polymerase δ and the RF-C clamp loader with other components of the replisome have not yet been observed but may well be important for replisome function.

1.7 Replication Control

As noted at the outset of this review, a distinguishing feature of eukaryotic DNA replication is an elaborate and finely tuned set of control mechanisms that ensure the accurate and timely completion of DNA replication and coordinate replication with other events that may occur during a cell cycle. The problem of control is complicated by the existence of thousands of potential sites of initiation in a genome, none of which can fire more than once. Moreover, the replication apparatus must deal with various kinds of damage and other threats to the fidelity of the replication process. While there has been considerable progress, there remains much to be learned about the control of DNA replication, particularly in metazoan organisms.

1.7.1 The Two-State Model for Initiation

Cell fusion experiments performed by Rao and Johnson published in 1970 revealed some of the features of replication control (Rao and Johnson 1970). Their work established that factors present in S phase cells can trigger the initiation of DNA replication in G1 phase nuclei but not G2 phase nuclei. Thus, there appeared to be (at least) two chromosomal states: one that is competent for initiation of DNA replication upon exposure to the S phase activator(s) and one that is not. On the basis of these experiments and subsequent studies in the *Xenopus* egg extract system, it was recognized in a general way that if some factor(s) required for establishment of the competent chromosomal state in G1 phase (a "licensing" factor(s)) were to be inactivated as a consequence of initiation of DNA synthesis and only restored in the following G1 phase, replication of the genome would be limited to a single round in each cell cycle (e.g., Blow and Laskey 1988; Blow 1993). However, the molecular processes responsible for controlling DNA replication remained obscure until the 1990s when the factors that establish the permissive chromosomal state in G1 and block it afterward were largely identified by the convergence of several lines of investigation in budding and fission yeast and the *Xenopus* egg extract system.

An important observation that shaped thinking about replication control was that yeast origins exist in distinct chromatin states before and after the onset of DNA replication (Diffley et al. 1994). In the post-replicative state, the genomic footprint produced by DNase digestion closely resembled that produced by ORC in vitro, while in the prereplicative state the genomic footprint was larger, suggesting the presence of an additional protein(s). The significance of this observation was two-fold: first it suggested that the prereplicative state might be the physical correlate of the state of initiation competence observed by Rao and Johnson, and second, it focused attention on the question of how the formation of the prereplicative state is controlled. It gradually became clear that the establishment of this state corresponds to loading the core MCM2-7 helicase on DNA and that this event is restricted to the G1 phase of the cell cycle by a number of regulatory mechanisms. Since activation of the loaded core helicase is restricted to S phase when further helicase loading is blocked, the outcome is a single round of genome duplication.

Experiments over the last two decades have shown that inhibition of the helicase loading reaction can be coupled to the onset of DNA replication in a number of different ways. Early studies led to the appreciation that CDK, in addition to its positive role in triggering initiation of DNA replication, also functions to inhibit loading of new MCM2-7 complexes on the DNA. Work in fission yeast demonstrated that several kinds of genetic manipulations that reduced CDK activity could induce re-replication in a single cell cycle (Broek et al. 1991; Hayles et al. 1994; Moreno and Nurse 1994; Jallepalli and Kelly 1996). For example, when mutants with a temperature-sensitive allele of *cdc2*$^+$, synchronized in G2 phase, were shifted to high temperature to inactivate CDK activity and then returned to a permissive temperature, they underwent a second round of DNA replication prior to mitosis. Similar results were obtained by manipulating the level of an inhibitor of CDK activity. In perhaps the most striking example of this phenomenon, deletion of Cdc13, the major mitotic B cyclin in fission yeast, led to multiple rounds of DNA replication in the absence of mitosis (Hayles et al. 1994). This work, taken together, demonstrated quite clearly that CDK plays a negative role in controlling initiation of DNA replication. Studies in *S. cerevisiae* confirmed this conclusion and, importantly, showed that CDK acts by suppressing the assembly of prereplicative complexes (Dahmann et al. 1995). Thus, by 1996 the basic logic of control of DNA replication was understood (Reviewed in Diffley 1996). ORC, Cdc6, and Cdt1 generate prereplicative complexes from the end of mitosis through G1 when CDK activity is low. An increase in CDK activity activates prereplicative complexes at the onset of S phase (Li et al. 2003b; Takeda et al. 2005; Sugimoto et al. 2004), and the very same kinase activity acts to prevent further assembly of prereplicative complexes. Very soon after these seminal studies in yeast, it was demonstrated in the *Xenopus* egg extract system that similar control mechanisms are operative in vertebrates (Hua et al. 1997). The identification of the targets of negative regulation by CDK became the next key issue, and it was found that all of the players required for loading the core helicase are potential CDK targets in budding yeast (Tanaka et al. 1997; Drury et al. 1997; Nguyen et al. 2000, 2001; Li et al. 2003b; Arias and Walter 2007; Yeeles et al. 2015; Labib et al. 1999). In S phase CDK phosphorylation of Cdc6 marks it for ubiquitin-mediated proteolysis, and the MCM2-7 complex is exported from the nuclease by a CDK-dependent mechanism. In addition, CDK phosphorylation of yeast ORC subunits inhibits loading of the core helicase by a mechanism that is not yet clear. In fission yeast CDK phosphorylation of Cdc18 (homologue of Cdc6) drives its destruction by ubiquitin-mediated proteolysis, but this mechanism is redundant with a CDK-independent mechanism that targets Cdt1 (Nishitani and Nurse 1995; Jallepalli et al. 1997; Nishitani et al. 2000; Gopalakrishnan et al. 2001; Guarino et al. 2011). In human cells Cdt1 is targeted for proteolysis by a CDK-dependent pathway (Liu et al. 2004).

Studies of the control of DNA replication in the *Xenopus* system uncovered additional mechanisms that are independent of CDK (See Arias and Walter 2007 for review). In metazoans, loading of MCM2-7 is inhibited in S phase by a protein factor called geminin that binds to Cdt1 and sequesters it in an inactive form (McGarry and Kirschner 1998; Wohlschlegel et al. 2000; Tada et al. 2001; Lee et al. 2004; Cook et al. 2004; Lutzmann et al. 2006). Cdt1 is also targeted for replication-dependent proteolysis by an interesting mechanism that is dependent upon specific interaction of Cdt1 with the replication processivity factor, PCNA (Arias and Walter 2005, 2006; Senga et al. 2006). PCNA-dependent proteolysis of Cdt1 is also observed in fission yeast (Ralph et al. 2006; Guarino et al. 2011). Thus, it has become apparent that negative control of replication can involve several partially redundant mechanisms and these may differ to some extent in different eukaryotes. So far, it appears that metazoans inhibit core helicase loading in S phase largely by degrading or blocking the activity of Cdt1.

1.7.2 Replication Timing

The development of methods to physically map sites of initiation in the budding yeast genome led to the discovery that initiation events occur throughout S phase and that the average time of activation of a particular yeast origin depends in part on its chromosomal context (Fangman and Brewer 1991). Why is the activation of loaded MCM2-7s distributed across S phase rather than being concentrated at the beginning of S phase? One possibility is to ensure the completion of DNA replication without sacrificing efficiency. Since loading of additional core helicases is prevented during S phase, DNA replication will be incomplete if forks converging from adjacent initiation sites should stall or collapse. The probability of such an event can be reduced by increasing the number of loaded MCM2-7 hexamers across the genome. However, activating all such hexamers at the same time would be wasteful of resources, since many of them, perhaps the majority, will not be needed. By spreading the activation of loaded hexamers over time, many will be inactivated by passive replication, but those that lie in regions that are, for whatever reason, late to replicate can still be activated to complete replication. Since the loci of stalled or collapsed forks cannot be known a priori, there is no obvious penalty for an activation mechanism that is completely stochastic in time. Thus, it is possible that the average number of potential origins (loaded MCM2-7 hexamers) in cells is determined by a trade-off between reducing the chance of leaving a segment of DNA unreplicated and using more resources than necessary to load and activate helicases.

Recent evidence suggests that the activation of loaded MCM2-7 hexamers in S phase is largely stochastic in time (see, e.g., Bechhoefer and Rhind 2012). It follows that the average time of initiation at a particular site reflects the relative rate constant for activation of loaded MCM complexes at that site. Other things being equal, an increase or decrease in the probability of activation per unit time would be expected to shift the average time of replication of the neighboring DNA to an earlier or later time, respectively. Variation in initiation timing appears to hold for eukaryotes other than budding yeast, but analysis is more difficult in organisms with less defined origins of DNA replication. The mechanisms responsible for modulating the probability of activation of loaded MCMs per unit time are under active investigation. Emerging evidence suggests that regulation of the activity of DDK in the vicinity of potential origins may be one major mechanism for altering initiation timing. For example, it has been reported that the pericentromeric regions of the *S. pombe* genome initiate replication early in S phase on average, and this phenomenon has been explained by recruitment of DDK by the pericentromeric protein HP1 (Kim and Huberman 2001; Hayashi et al. 2009). On the other hand, subtelomeric sequences in fission yeast (and other eukaryotes) initiate DNA synthesis later than the bulk of the genome, and recent work indicates that the telomere-binding protein Rif1 is required for late replication (Hayano et al. 2012). It has been suggested that Rif1 recruits a cellular phosphatase that inhibits activation of loaded MCMs to replisomes by the DDK protein kinase (Dave et al. 2014; Hiraga et al. 2014; Mattarocci et al. 2014). Other mechanisms that affect replication timing of particular regions of the genome have been described, and it is likely that their numbers will grow.

A very important example of the control of the efficiency of activation of loaded MCMs is the so-called intra-S phase checkpoint. This signal transduction mechanism recognizes when replication forks are slowed, which can occur, for example, if the levels of nucleotide precursors are suboptimal. In this situation a checkpoint-dependent protein kinase is activated and phosphorylates Sld3 and Dbf4, two proteins required for the transition of loaded MCMs to active replisomes (Santocanale and Diffley 1998; Shirahige et al. 1998; Lopez-Mosqueda et al. 2010; Zegerman and Diffley 2010). These phosphorylation events inhibit further activation of loaded MCMs and thus, preferentially, affect origins that normally fire later in S phase. The resulting reduction in the number of active replicons dynamically lessens the cellular requirement for precursors and other replication factors. These examples all suggest that DDK and the protein-protein interactions that it facilitates during helicase activation may be central targets of the mechanisms controlling the timing of initiation within the genome. The biological meaning of timing control seems fairly obvious in some cases – e.g., checkpoint control of initiation in response to perturbations of DNA replication – but remains obscure in other cases, e.g., late replication of telomere proximal DNA and early replication of pericentromeric DNA.

1.8 Termination of DNA Synthesis and Disassembly of the Replisome

Termination of DNA synthesis occurs when two convergent forks meet. Early studies of the termination of SV40 DNA replication focused on the roles of DNA topoisomerases as forks converge (Sundin and Varshavsky 1980, 1981). Topoisomerases are required to prevent the buildup of positive writhe in the DNA as the parental strands are unwound. Positive writhe increases the energetic cost of unwinding and, if not removed, can eventually slow or stop fork progression. If the replication fork is free to rotate in space, positive writhe can be manifested as supercoils in the unreplicated DNA in front of the fork or as interlinks (pre-catenanes) of the two daughter duplexes behind the fork. The former can be removed by the action of either a type I or type II topoisomerase, while the latter can only be removed by a type II topoisomerase. As two forks converge, however, the target size for topoisomerase action in front of the fork is progressively reduced. Studies of SV40 DNA replication revealed that replication forks slow as they approach one another and suggested that completion of DNA synthesis at converging forks is highly dependent upon type II topoisomerase, presumably functioning behind the replication fork (Levine et al. 1970; Tapper and DePamphilis 1978; Sundin and Varshavsky 1981; Ishimi et al. 1992). Subsequent work in yeast and the *Xenopus* egg extract system has not revealed a general requirement for type II topoisomerase for termination of cellular DNA replication, although the enzyme may play a special role at barriers to fork progression (DiNardo et al. 1984; Lucas et al. 2001; Baxter and Diffley 2008; Fachinetti et al. 2010). Like SV40, both type I and type II topoisomerases function to remove positive writhe during DNA chain elongation, but DNA replication can be completed in a timely fashion in the absence of type II topoisomerase. Type I topoisomerase appears to be sufficient for complete duplication of the genome. Following completion of replication, type II topoisomerase is uniquely required for decatenation of newly synthesized daughter duplexes to allow chromosome segregation during mitosis (DiNardo et al. 1984).

Another event that occurs when replication terminates at converging forks is the disassembly of the CMG helicase, the core of the replisome. As discussed earlier in this review, the helicase is loaded onto DNA and activated in a complex and highly regulated series of reactions. The CMG is quite stable and presumably remains associated with the replication fork until the termination of DNA synthesis. Early studies revealed that MCM proteins are progressively lost from chromatin during S phase, and more recent work indicates that disassembly of the CMG following the convergence of replication forks is an active process that is also highly regulated (Maric et al. 2014). The CMG helicase is ubiquitylated on the MCM7 subunit after the completion of DNA synthesis. The modification requires a ubiquitin ligase containing the F-box protein, Dia2, which is essential for CMG disassembly. Disassembly is also dependent upon the Cdc48 segregase, a AAA+ family member that associates with ubiquitylated CMG. The action of Cdc48 results in the dissociation of GINS and Cdc45 from the CMG helicase, after which the MCM2-7 hexamer dissociates from the DNA. It is clearly important that the disassembly reaction only occurs after completion of DNA synthesis and not during active DNA chain elongation. The signal responsible for activating the disassembly machinery has not yet been identified. One interesting speculation is that ubiquitylation of CMG is triggered when completion of DNA synthesis leaves the CMG encircling double-stranded DNA instead of single-stranded DNA (Dewar et al. 2015).

1.9 The Future

After more than 35 years of research, the basic features of DNA replication in eukaryotic cells are reasonably well understood. The minimal set of proteins is known, many of their interactions and functions have been defined, and some aspects of regulation of DNA replication are understood at least in outline. Many important and interesting issues remain, some of which have implications for human disease. It is expected that the field will continue to build on the legacy of success that it has enjoyed over the last three to four decades. The following are some areas that are likely to see important advances in the next decade.

The temporal program of replication in individual eukaryotic cells remains of considerable interest, and the central issues are how potential origins are distributed and how they are activated in space and time. There are now excellent tools that can be used to measure replication of individual segments of the genome at different points in the cell cycle under different physiological conditions. These methods will improve in resolution and will likely be coupled to increasingly sophisticated in silico modeling approaches to gain a deeper appreciation of the processes that affect the pattern of replication across the genome. This issue is of particular importance for understanding the mechanisms that underlie the correlation between replication timing and the frequency of mutations in the human genome (Stamatoyannopoulos et al. 2009; Lang and Murray 2011; Woo and Li 2012).

Much remains to be learned about the details of helicase activation and replisome assembly and function. Structural studies by X-ray diffraction and cryo-EM coupled with mechanistic biochemical approaches, including single-molecule experiments, will play an increasingly important role. The path to deeper understanding has been opened by the recent reconstitution of core replication reactions with purified yeast proteins. It will be important in the coming years to extend these approaches to higher eukaryotes. Experiments in the *Xenopus* egg extract system have revealed a number of differences in the regulation of DNA replication between vertebrates and yeast, and it is likely that more differences will be uncovered.

Another question of vital interest is how the replication machinery coexists with other nuclear processes, including transcription, chromatin assembly, and various repair processes. A closely related issue is how replication deals with proteins and RNAs that compete with replication proteins for access to the DNA substrate. These factors can clearly affect both the initiation of DNA replication and the elongation of replication forks and can now be studied in vitro with purified proteins. As illustrated by recent studies of Csm3/Tof1 and Mrc1, the recent reconstitution of yeast DNA replication will make it possible to define the functional roles of proteins that normally travel with the replisome but are not absolutely essential for DNA replication (Yeeles et al. 2017). Such proteins may function to allow forks to deal with barriers to chain elongation or other factors that may perturb ongoing DNA replication. It is expected that progress in understanding the role of these additional factors will now be rapid.

Finally, a question that is of great significance for disease and aging is how the replication machinery deals with DNA damage and how replication is integrated with the cellular DNA damage response. It has become apparent that most somatic mutations are the result of encounters of the replication apparatus with damage of various kinds in the template. With the tools in hand, it should be possible to gain a much deeper understanding of the results of such encounters. There is no doubt that such understanding would contribute to building better approaches to prevention and therapy of cancer and perhaps provide insights into the origins of somatic mutations that contribute to aging.

References

Abid Ali F, Renault L, Gannon J, Gahlon HL, Kotecha A, Zhou JC, Rueda D, Costa A (2016) Cryo-EM structures of the eukaryotic replicative helicase bound to a translocation substrate. Nat Commun 7:10708

Aparicio OM, Weinstein DM, Bell SP (1997) Components and dynamics of DNA replication complexes in S. cerevisiae: redistribution of MCM proteins and Cdc45p during S phase. Cell 91(1):59–69

Aparicio OM, Stout AM, Bell SP (1999) Differential assembly of cdc45p and DNA polymerases at early and late origins of DNA replication [in process citation]. Proc Natl Acad Sci U S A 96(16):9130–9135

Araki H, Ropp PA, Johnson AL, Johnston LH, Morrison A, Sugino A (1992) DNA polymerase II, the probable homolog of mammalian DNA polymerase epsilon, replicates chromosomal DNA in the yeast Saccharomyces cerevisiae. EMBO J 11:733

Araki H, Leem SH, Phongdara A, Sugino A (1995) Dpb11, which interacts with DNA polymerase II (epsilon) in Saccharomyces cerevisiae, has a dual role in S-phase progression and at a cell cycle checkpoint. Proc Natl Acad Sci U S A 92(25):11791–11795

Arias EE, Walter JC (2005) Replication-dependent destruction of Cdt1 limits DNA replication to a single round per cell cycle in Xenopus egg extracts. Genes Dev 19(1):114–126

Arias EE, Walter JC (2006) PCNA functions as a molecular platform to trigger Cdt1 destruction and prevent re-replication. Nat Cell Biol 8(1):84–90

Arias EE, Walter JC (2007) Strength in numbers: preventing rereplication via multiple mechanisms in eukaryotic cells. Genes Dev 21(5):497–518

Balakrishnan L, Bambara RA (2013) Okazaki fragment metabolism. Cold Spring Harb Perspect Biol 5(2), a010173

Baxter J, Diffley JF (2008) Topoisomerase II inactivation prevents the completion of DNA replication in budding yeast. Mol Cell 30(6):790–802

Beall EL, Manak JR, Zhou S, Bell M, Lipsick JS, Botchan MR (2002) Role for a Drosophila Myb-containing protein complex in site-specific DNA replication. Nature 420(6917):833–837

Bechhoefer J, Rhind N (2012) Replication timing and its emergence from stochastic processes. Trends Genet 28(8):374–381

Bell SP (1995) Eukaryotic replicators and associated protein complexes. Curr Opin Genet Dev 5(2):162–167
Bell SP, Dutta A (2002) DNA replication in eukaryotic cells. Annu Rev Biochem 71:333–374
Bell SP, Kaguni JM (2013) Helicase loading at chromosomal origins of replication. Cold Spring Harb Perspect Biol 5(6), a010124
Bell SP, Stillman B (1992) ATP-dependent recognition of eukaryotic origins of DNA replication by a multiprotein complex. Nature 357(6374):128–134
Bell SP, Kobayashi R, Stillman B (1993) Yeast origin recognition complex functions in transcription silencing and DNA replication. Science 262(5141):1844–1849
Bell SP, Mitchell J, Leber J, Kobayashi R, Stillman B (1995) The multidomain structure of Orc1p reveals similarity to regulators of DNA replication and transcriptional silencing. Cell 83(4):563–568
Berbenetz NM, Nislow C, Brown GW (2010) Diversity of eukaryotic DNA replication origins revealed by genome-wide analysis of chromatin structure. PLoS Genet 6(9), e1001092
Bessman MJ, Kornberg A, Lehman IR, Simms ES (1956) Enzymic synthesis of deoxyribonucleic acid. Biochim Biophys Acta 21(1):197–198
Blow JJ (1993) Preventing re-replication of DNA in a single cell cycle: evidence for a replication licensing factor. J Cell Biol 122(5):993–1002
Blow JJ, Laskey RA (1988) A role for the nuclear envelope in controlling DNA replication within the cell cycle. Nature 332:546–548
Bochman ML, Schwacha A (2008) The Mcm2-7 complex has in vitro helicase activity. Mol Cell 31(2):287–293
Borowiec JA, Hurwitz J (1988) Localized melting and structural changes in the SV40 origin of replication induced by T-antigen. EMBO J 7(10):3149–3158
Bowman GD, Goedken ER, Kazmirski SL, O'Donnell M, Kuriyan J (2005) DNA polymerase clamp loaders and DNA recognition. FEBS Lett 579(4):863–867
Breier AM, Chatterji S, Cozzarelli NR (2004) Prediction of Saccharomyces cerevisiae replication origins. Genome Biol 5(4):R22
Brewer BJ, Fangman WL (1987) The localization of replication origins on ARS plasmids in S. cerevisiae. Cell 51(3):463–471
Broach JR, Li YY, Feldman J, Jayaram M, Abraham J, Nasmyth KA, Hicks JB (1983) Localization and sequence analysis of yeast origins of DNA replication. Cold Spring Harb Symp Quant Biol 47(Pt 2):1165–1173
Broek D, Bartlett R, Crawford K, Nurse P (1991) Involvement of p34cdc2 in establishing the dependency of S phase on mitosis. Nature 349(6308):388–393
Brush GS, Kelly TJ, Stillman B (1995) Identification of eukaryotic DNA replication proteins using simian virus 40 in vitro replication system. Methods Enzymol 262:522–548
Budd ME, Sitney KC, Campbell JL (1989) Purification of DNA polymerase II, a distinct DNA polymerase, from Saccharomyces cerevisiae. J Biol Chem 264(11):6557–6565
Burgers PM, Gordenin D, Kunkel TA (2016) Who is leading the replication fork, Pol epsilon or Pol delta? Mol Cell 61(4):492–493
Byrnes JJ, Downey KM, Black VL, So AG (1976) A new mammalian DNA polymerase with 3′ to 5′ exonuclease activity: DNA polymerase delta. Biochemistry 15(13):2817–2823
Cai J, Uhlmann F, Gibbs E, Flores-Rozas H, Lee CG, Phillips B, Finkelstein J, Yao N, O'Donnell M, Hurwitz J (1996) Reconstitution of human replication factor C from its five subunits in baculovirus-infected insect cells. Proc Natl Acad Sci U S A 93(23):12896–12901
Cairns J (1963) The bacterial chromosome and its manner of replication as seen by autoradiography. J Mol Biol 6:208–213
Campbell JL (1986) Eukaryotic DNA replication. Annu Rev Biochem 55:733–771
Campbell JL, Newlon CS (1991) Chromosomal DNA replication. In: Broach JR, Pringle JR, Jones EW (eds) The molecular and cellular biology of the yeast saccharomyces, vol 1. Cold Spring Harbor Laboratory Press, Cold Spring Harbor, pp 41–146
Cegielska A, Virshup DM (1993) Control of simian virus 40 DNA replication by the HeLa cell nuclear kinase casein kinase I. Mol Cell Biol 13(2):1202–1211

Cegielska A, Moarefi I, Fanning E, Virshup DM (1994) T-antigen kinase inhibits simian virus 40 DNA replication by phosphorylation of intact T antigen on serines 120 and 123. J Virol 68(1):269–275
Celniker SE, Sweder K, Srienc F, Bailey JE, Campbell JL (1984) Deletion mutations affecting autonomously replicating sequence ARS1 of Saccharomyces cerevisiae. Mol Cell Biol 4(11):2455–2466
Challberg MD, Kelly TJ (1979) Adenovirus DNA replication in vitro. Proc Natl Acad Sci U S A 76(2):655–659
Challberg MD, Kelly TJ (1982) Eukaryotic DNA replication: viral and plasmid model systems. Annu Rev Biochem 51:901–934
Challberg MD, Kelly TJ (1989) Animal virus DNA replication. Annu Rev Biochem 58(671):671–717
Challberg MD, Desiderio SV, Kelly TJ (1980) Adenovirus DNA replication in vitro: characterization of a protein covalently linked to nascent DNA strands. Proc Natl Acad Sci U S A 77(9):5105–5109
Challberg MD, Ostrove JM, Kelly TJ Jr (1982) Initiation of adenovirus DNA replication: detection of covalent complexes between nucleotide and the 80-kilodalton terminal protein. J Virol 41(1):265–270
Chen R, Wold MS (2014) Replication protein a: single-stranded DNA's first responder: dynamic DNA-interactions allow replication protein A to direct single-strand DNA intermediates into different pathways for synthesis or repair. BioEssays 36(12):1156–1161
Chong JP, Mahbubani HM, Khoo CY, Blow JJ (1995) Purification of an MCM-containing complex as a component of the DNA replication licensing system [see comments]. Nature 375(6530):418–421
Chong JP, Thommes P, Blow JJ (1996) The role of MCM/P1 proteins in the licensing of DNA replication. Trends Biochem Sci 21(3):102–106
Chong JP, Hayashi MK, Simon MN, Xu RM, Stillman B (2000) A double-hexamer archaeal minichromosome maintenance protein is an ATP-dependent DNA helicase. Proc Natl Acad Sci U S A 97(4):1530–1535
Chuang R, Kelly TJ (1999) The fission yeast homologue of Orc4p binds to replication origin DNA via multiple AT-hooks. Proc Natl Acad Sci U S A 96:2656–2661
Clarey MG, Erzberger JP, Grob P, Leschziner AE, Berger JM, Nogales E, Botchan M (2006) Nucleotide-dependent conformational changes in the DnaA-like core of the origin recognition complex. Nat Struct Mol Biol 13(8):684–690
Clyne RK, Kelly TJ (1995) Genetic analysis of an ARS element from the fission yeast Schizosaccharomyces pombe. EMBO J 14(24):6348–6357
Cocker JH, Piatti S, Santocanale C, Nasmyth K, Diffley JF (1996) An essential role for the Cdc6 protein in forming the pre-replicative complexes of budding yeast. Nature 379(6561):180–182
Coleman TR, Carpenter PB, Dunphy WG (1996) The Xenopus Cdc6 protein is essential for the initiation of a single round of DNA replication in cell-free extracts. Cell 87(1):53–63
Collins KL, Kelly TJ (1991) Effects of T antigen and replication protein A on the initiation of DNA synthesis by DNA polymerase alpha-primase. Mol Cell Biol 11(4):2108–2115
Collins KL, Russo AA, Tseng BY, Kelly TJ (1993) The role of the 70 kDa subunit of human DNA polymerase alpha in DNA replication. EMBO J 12(12):4555–4566
Cook JG, Chasse DA, Nevins JR (2004) The regulated association of Cdt1 with minichromosome maintenance proteins and Cdc6 in mammalian cells. J Biol Chem 279(10):9625–9633
Costa A, Ilves I, Tamberg N, Petojevic T, Nogales E, Botchan MR, Berger JM (2011) The structural basis for MCM2-7 helicase activation by GINS and Cdc45. Nat Struct Mol Biol 18(4):471–477
Costa A, Renault L, Swuec P, Petojevic T, Pesavento JJ, Ilves I, MacLellan-Gibson K, Fleck RA, Botchan MR, Berger JM (2014) DNA binding polarity, dimerization, and ATPase ring remodeling in the CMG helicase of the eukaryotic replisome. Elife 3, e03273
D'Urso G, Marraccino RL, Marshak DR, Roberts JM (1990) Cell cycle control of DNA replication by a homologue from human cells of the p34cdc2 protein kinase. Science 250(4982):786–791
Dahmann C, Diffley JF, Nasmyth KA (1995) S-phase-promoting cyclin-dependent kinases prevent re-replication by inhibiting the transition of replication origins to a pre-replicative state. Curr Biol 5(11):1257–1269

Dai J, Chuang RY, Kelly TJ (2005) DNA replication origins in the Schizosaccharomyces pombe genome. Proc Natl Acad Sci U S A 102(2):337–342

Danna KJ, Nathans D (1972) Bidirectional replication of Simian Virus 40 DNA. Proc Natl Acad Sci U S A 69(11):3097–3100

Dave A, Cooley C, Garg M, Bianchi A (2014) Protein phosphatase 1 recruitment by Rif1 regulates DNA replication origin firing by counteracting DDK activity. Cell Rep 7(1):53–61

Dean FB, Bullock P, Murakami Y, Wobbe CR, Weissbach L, Hurwitz J (1987) Simian virus 40 (SV40) DNA replication: SV40 large T antigen unwinds DNA containing the SV40 origin of replication. Proc Natl Acad Sci U S A 84:16–20

Dean FB, Borowiec JA, Eki T, Hurwitz J (1992) The simian virus 40 T antigen double hexamer assembles around the DNA at the replication origin. J Biol Chem 267(20):14129–14137

Deegan TD, Yeeles JT, Diffley JF (2016) Phosphopeptide binding by Sld3 links Dbf4-dependent kinase to MCM replicative helicase activation. EMBO J 35(9):961–973

Delucia A, Lewton B, Tjian R, Tegtmeyer P (1983) Topography of simian virus 40 A protein-DNA complexes: arrangement of pentanucleotide interaction sites at the origin of replication. J Virol 46:143–150

Dershowitz A, Snyder M, Sbia M, Skurnick JH, Ong LY, Newlon CS (2007) Linear derivatives of Saccharomyces cerevisiae chromosome III can be maintained in the absence of autonomously replicating sequence elements. Mol Cell Biol 27(13):4652–4663

Desiderio SV, Kelly TJ (1981) Structure of the linkage between adenovirus DNA and the 55,000 molecular weight terminal protein. J Mol Biol 145(2):319–337

van Deursen F, Sengupta S, De Piccoli G, Sanchez-Diaz A, Labib K (2012) Mcm10 associates with the loaded DNA helicase at replication origins and defines a novel step in its activation. EMBO J 31(9):2195–2206

Devault A, Vallen EA, Yuan T, Green S, Bensimon A, Schwob E (2002) Identification of Tah11/Sid2 as the ortholog of the replication licensing factor Cdt1 in Saccharomyces cerevisiae. Curr Biol 12(8):689–694

Devbhandari S, Jiang J, Kumar C, Whitehouse I, Remus D (2017) Chromatin constrains the initiation and elongation of DNA replication. Mol Cell 65(1):131–141

Dewar JM, Budzowska M, Walter JC (2015) The mechanism of DNA replication termination in vertebrates. Nature 525(7569):345–350

Diffley JF (1996) Once and only once upon a time: specifying and regulating origins of DNA replication in eukaryotic cells. Genes Dev 10(22):2819–2830

Diffley JF, Cocker JH (1992) Protein-DNA interactions at a yeast replication origin. Nature 357(6374):169–172

Diffley JF, Cocker JH, Dowell SJ, Rowley A (1994) Two steps in the assembly of complexes at yeast replication origins in vivo. Cell 78(2):303–316

DiNardo S, Voelkel K, Sternglanz R (1984) DNA topoisomerase II mutant of Saccharomyces cerevisiae: topoisomerase II is required for segregation of daughter molecules at the termination of DNA replication. Proc Natl Acad Sci U S A 81(9):2616–2620

Donovan S, Harwood J, Drury LS, Diffley JF (1997) Cdc6p-dependent loading of Mcm proteins onto pre-replicative chromatin in budding yeast. Proc Natl Acad Sci U S A 94(11):5611–5616

Dornreiter I, Erdile LF, Gilbert IU, von Winkler D, Kelly TJ, Fanning E (1992) Interaction of DNA polymerase alpha-primase with cellular replication protein A and SV40 T antigen. EMBO J 11(2):769–776

Drury LS, Perkins G, Diffley JF (1997) The Cdc4/34/53 pathway targets Cdc6p for proteolysis in budding yeast. EMBO J 16(19):5966–5976

Dua R, Levy DL, Campbell JL (1999) Analysis of the essential functions of the C-terminal protein/protein interaction domain of saccharomyces cerevisiae pol epsilon and its unexpected ability to support growth in the absence of the DNA polymerase domain [in process citation]. J Biol Chem 274(32):22283–22288

Dubey DD, Zhu J, Carlson DL, Sharma K, Huberman JA (1994) Three ARS elements contribute to the ura4 replication origin region in the fission yeast, Schizosaccharomyces pombe. EMBO J 13(15):3638–3647

Dubey DD, Kim SM, Todorov IT, Huberman JA (1996) Large, complex modular structure of a fission yeast DNA replication origin. Curr Biol 6(4):467–473
Eaton ML, Prinz JA, MacAlpine HK, Tretyakov G, Kharchenko PV, MacAlpine DM (2011) Chromatin signatures of the Drosophila replication program. Genome Res 21(2):164–174
Edgell DR, Doolittle FW (1997) Archaea and the origin(s) of DNA replication proteins. Cell 89:995–998
Enemark EJ, Joshua-Tor L (2008) On helicases and other motor proteins. Curr Opin Struct Biol 18(2):243–257
Enomoto T, Lichy JH, Ikeda JE, Hurwitz J (1981) Adenovirus DNA replication in vitro: purification of the terminal protein in a functional form. Proc Natl Acad Sci U S A 78(11):6779–6783
Evrin C, Clarke P, Zech J, Lurz R, Sun J, Uhle S, Li H, Stillman B, Speck C (2009) A double-hexameric MCM2-7 complex is loaded onto origin DNA during licensing of eukaryotic DNA replication. Proc Natl Acad Sci U S A 106(48):20240–20245
Fachinetti D, Bermejo R, Cocito A, Minardi S, Katou Y, Kanoh Y, Shirahige K, Azvolinsky A, Zakian VA, Foiani M (2010) Replication termination at eukaryotic chromosomes is mediated by Top2 and occurs at genomic loci containing pausing elements. Mol Cell 39(4):595–605
Fairman MP, Stillman B (1988) Cellular factors required for multiple stages of SV40 DNA replication in vitro. EMBO J 7(4):1211–1218
Fangman WL, Brewer BJ (1991) Activation of replication origins within yeast chromosomes. Annu Rev Cell Biol 7(375):375–402
Feng W, D'Urso G (2001) Schizosaccharomyces pombe cells lacking the amino-terminal catalytic domains of DNA polymerase epsilon are viable but require the DNA damage checkpoint control. Mol Cell Biol 21(14):4495–4504
Friedman KL, Brewer BJ, Fangman WL (1997) Replication profile of Saccharomyces cerevisiae chromosome VI. Genes Cells 2(11):667–678
Fu YV, Yardimci H, Long DT, Ho TV, Guainazzi A, Bermudez VP, Hurwitz J, van Oijen A, Scharer OD, Walter JC (2011) Selective bypass of a lagging strand roadblock by the eukaryotic replicative DNA helicase. Cell 146(6):931–941
Gai D, Zhao R, Li D, Finkielstein CV, Chen XS (2004) Mechanisms of conformational change for a replicative hexameric helicase of SV40 large tumor antigen. Cell 119(1):47–60
Gambus A, Jones RC, Sanchez-Diaz A, Kanemaki M, van Deursen F, Edmondson RD, Labib K (2006) GINS maintains association of Cdc45 with MCM in replisome progression complexes at eukaryotic DNA replication forks. Nat Cell Biol 8(4):358–366
Gambus A, van Deursen F, Polychronopoulos D, Foltman M, Jones RC, Edmondson RD, Calzada A, Labib K (2009) A key role for Ctf4 in coupling the MCM2-7 helicase to DNA polymerase alpha within the eukaryotic replisome. EMBO J 28(19):2992–3004
Georgescu R, Langston L, O'Donnell M (2015a) A proposal: evolution of PCNA's role as a marker of newly replicated DNA. DNA Repair (Amst) 29:4–15
Georgescu RE, Schauer GD, Yao NY, Langston LD, Yurieva O, Zhang D, Finkelstein J, O'Donnell ME (2015b) Reconstitution of a eukaryotic replisome reveals suppression mechanisms that define leading/lagging strand operation. Elife 4, e04988
Gopalakrishnan V, Simancek P, Houchens C, Snaith HA, Frattini MG, Sazer S, Kelly TJ (2001) Redundant control of rereplication in fission yeast. Proc Natl Acad Sci U S A 98(23):13114–13119
Gros J, Devbhandari S, Remus D (2014) Origin plasticity during budding yeast DNA replication in vitro. EMBO J 33(6):621–636
Gros J, Kumar C, Lynch G, Yadav T, Whitehouse I, Remus D (2015) Post-licensing specification of eukaryotic replication origins by facilitated Mcm2-7 sliding along DNA. Mol Cell 60(5):797–807
Guarino E, Shepherd ME, Salguero I, Hua H, Deegan RS, Kearsey SE (2011) Cdt1 proteolysis is promoted by dual PIP degrons and is modulated by PCNA ubiquitylation. Nucleic Acids Res 39(14):5978–5990

Hardy CF, Dryga O, Seematter S, Pahl PM, Sclafani RA (1997) mcm5/cdc46-bob1 bypasses the requirement for the S phase activator Cdc7p. Proc Natl Acad Sci U S A 94(7):3151–3155
Hauk G, Berger JM (2016) The role of ATP-dependent machines in regulating genome topology. Curr Opin Struct Biol 36:85–96
Hayano M, Kanoh Y, Matsumoto S, Renard-Guillet C, Shirahige K, Masai H (2012) Rif1 is a global regulator of timing of replication origin firing in fission yeast. Genes Dev 26(2):137–150
Hayashi MT, Takahashi TS, Nakagawa T, Nakayama J, Masukata H (2009) The heterochromatin protein Swi6/HP1 activates replication origins at the pericentromeric region and silent mating-type locus. Nat Cell Biol 11(3):357–362
Hayles J, Fisher D, Woollard A, Nurse P (1994) Temporal order of S phase and mitosis in fission yeast is determined by the state of the p34cdc2-mitotic B cyclin complex. Cell 78(5):813–822
Heichinger C, Penkett CJ, Bahler J, Nurse P (2006) Genome-wide characterization of fission yeast DNA replication origins. EMBO J 25(21):5171–5179
Heller RC, Kang S, Lam WM, Chen S, Chan CS, Bell SP (2011) Eukaryotic origin-dependent DNA replication in vitro reveals sequential action of DDK and S-CDK kinases. Cell 146(1):80–91
Hennessy KM, Clark CD, Botstein D (1990) Subcellular localization of yeast CDC46 varies with the cell cycle. Genes Dev 4(12B):2252–2263
Herrmann R, Huf J, Bonhoeffer F (1972) Cross hybridization and rate of chain elongation of the two classes of DNA intermediates. Nat New Biol 240(103):235–237
Hiraga S, Alvino GM, Chang F, Lian HY, Sridhar A, Kubota T, Brewer BJ, Weinreich M, Raghuraman MK, Donaldson AD (2014) Rif1 controls DNA replication by directing Protein Phosphatase 1 to reverse Cdc7-mediated phosphorylation of the MCM complex. Genes Dev 28(4):372–383
Hopwood B, Dalton S (1996) Cdc45p assembles into a complex with Cdc46p/Mcm5p, is required for minichromosome maintenance, and is essential for chromosomal DNA replication. Proc Natl Acad Sci U S A 93(22):12309–12314
Hsiao CL, Carbon J (1979) High-frequency transformation of yeast by plasmids containing the cloned yeast ARG4 gene. Proc Natl Acad Sci U S A 76(8):3829–3833
Hua XH, Newport J (1998) Identification of a preinitiation step in DNA replication that is independent of origin recognition complex and cdc6, but dependent on cdk2. J Cell Biol 140:271–281
Hua XH, Yan H, Newport J (1997) A role for Cdk2 kinase in negatively regulating DNA replication during S phase of the cell cycle. J Cell Biol 137(1):183–192
Huberman JA, Riggs AD (1968) On the mechanism of DNA replication in mammalian chromosomes. J Mol Biol 32(2):327–341
Huberman JA, Spotila LD, Nawotka KA, el Assouli S, Davis LR (1987) The in vivo replication origin of the yeast 2 microns plasmid. Cell 51(3):473–481
Hurwitz J, Dean FB, Kwong AD, Lee SH (1990) The in vitro replication of DNA containing the SV40 origin. J Biol Chem 265(30):18043–18046
Ilves I, Petojevic T, Pesavento JJ, Botchan MR (2010) Activation of the MCM2-7 helicase by association with Cdc45 and GINS proteins. Mol Cell 37(2):247–258
Ishimi Y (1997) A DNA helicase activity is associated with an MCM4, -6, and -7 protein complex [published erratum appears in J Biol Chem 1998 Sep 4;273(36):23616]. J Biol Chem 272(39):24508–24513
Ishimi Y, Sugasawa K, Hanaoka F, Eki T, Hurwitz J (1992) Topoisomerase II plays an essential role as a swivelase in the late stage of SV40 chromosome replication in vitro. J Biol Chem 267(1):462–466
Jacob F, Brenner S, Cuzin F (1964) On the regulation of DNA replication in bacteria. Cold Spring Harb Symp Quant Biol 28:329–348
Jallepalli PV, Kelly TJ (1996) Rum1 and Cdc18 link inhibition of cyclin-dependent kinase to the initiation of DNA replication in Schizosaccharomyces pombe. Genes Dev 10(5):541–552
Jallepalli PV, Brown GW, Muzi-Falconi M, Tien D, Kelly TJ (1997) Regulation of the replication initiator protein p65cdc18 by CDK phosphorylation. Genes Dev 11(21):2767–2779
Jeruzalmi D, O'Donnell M, Kuriyan J (2002) Clamp loaders and sliding clamps. Curr Opin Struct Biol 12(2):217–224

Johnson A, O'Donnell M (2005) Cellular DNA replicases: components and dynamics at the replication fork. Annu Rev Biochem 74:283–315

de Jong RN, van der Vliet PC, Brenkman AB (2003) Adenovirus DNA replication: protein priming, jumping back and the role of the DNA binding protein DBP. Curr Top Microbiol Immunol 272:187–211

Kaguni LS, Rossignol JM, Conaway RC, Lehman IR (1983) Isolation of an intact DNA polymerase-primase from embryos of Drosophila melanogaster. Proc Natl Acad Sci U S A 80(8):2221–2225

Kamimura Y, Masumoto H, Sugino A, Araki H (1998) Sld2, which interacts with Dpb11 in Saccharomyces cerevisiae, is required for chromosomal DNA replication. Mol Cell Biol 18(10):6102–6109

Kamimura Y, Tak YS, Sugino A, Araki H (2001) Sld3, which interacts with Cdc45 (Sld4), functions for chromosomal DNA replication in Saccharomyces cerevisiae. EMBO J 20(8):2097–2107

Kanemaki M, Labib K (2006) Distinct roles for Sld3 and GINS during establishment and progression of eukaryotic DNA replication forks. EMBO J 25(8):1753–1763

Kanke M, Kodama Y, Takahashi TS, Nakagawa T, Masukata H (2012) Mcm10 plays an essential role in origin DNA unwinding after loading of the CMG components. EMBO J 31(9):2182–2194

Kaplan DL, Davey MJ, O'Donnell M (2003) Mcm4,6,7 uses a "pump in ring" mechanism to unwind DNA by steric exclusion and actively translocate along a duplex. J Biol Chem 278(49):49171–49182

Kawasaki Y, Kim HD, Kojima A, Seki T, Sugino A (2006) Reconstitution of Saccharomyces cerevisiae prereplicative complex assembly in vitro. Genes Cells 11(7):745–756

Kaykov A, Nurse P (2015) The spatial and temporal organization of origin firing during the S-phase of fission yeast. Genome Res 25(3):391–401

Kelly TJ (1988) SV40 DNA replication. J Biol Chem 263:17889–17892

Kelman Z, Lee JK, Hurwitz J (1999) The single minichromosome maintenance protein of Methanobacterium thermoautotrophicum DeltaH contains DNA helicase activity. Proc Natl Acad Sci U S A 96(26):14783–14788

Kesti T, Flick K, Keranen S, Syvaoja JE, Wittenberg C (1999) DNA polymerase epsilon catalytic domains are dispensable for DNA replication, DNA repair, and cell viability. Mol Cell 3(5):679–685

Kim SM, Huberman JA (2001) Regulation of replication timing in fission yeast. EMBO J 20(21):6115–6126

King AJ, van der Vliet PC (1994) A precursor terminal protein-trinucleotide intermediate during initiation of adenovirus DNA replication: regeneration of molecular ends in vitro by a jumping back mechanism. EMBO J 13(23):5786–5792

Kong D, DePamphilis ML (2001) Site-specific DNA binding of the Schizosaccharomyces pombe origin recognition complex is determined by the Orc4 subunit. Mol Cell Biol 21(23):8095–8103

Kornberg A (1981) DNA replication, 1st edn. W.H. Freeman and Co, New York

Kornberg A, Baker TA (1992) DNA replication, 2nd edn. Freeman and Co., New York

Kurat CF, Yeeles JT, Patel H, Early A, Diffley JF (2017) Chromatin controls DNA replication origin selection, lagging-strand synthesis, and replication fork rates. Mol Cell 65(1):117–130

Labib K, Diffley JF, Kearsey SE (1999) G1-phase and B-type cyclins exclude the DNA-replication factor Mcm4 from the nucleus. Nat Cell Biol 1(7):415–422

Labib K, Tercero JA, Diffley JF (2000) Uninterrupted MCM2-7 function required for DNA replication fork progression. Science 288(5471):1643–1647

Labib K, Kearsey SE, Diffley JF (2001) MCM2-7 proteins are essential components of prereplicative complexes that accumulate cooperatively in the nucleus during G1-phase and are required to establish, but not maintain, the S-phase checkpoint. Mol Biol Cell 12(11):3658–3667

Lang GI, Murray AW (2011) Mutation rates across budding yeast chromosome VI are correlated with replication timing. Genome Biol Evol 3:799–811

Langston LD, Zhang D, Yurieva O, Georgescu RE, Finkelstein J, Yao NY, Indiani C, O'Donnell ME (2014) CMG helicase and DNA polymerase epsilon form a functional 15-subunit holoenzyme for eukaryotic leading-strand DNA replication. Proc Natl Acad Sci U S A 111(43):15390–15395

Lee S-H, Eki T, Hurwitz J (1989) Synthesis of DNA containing the simian virus 40 origin of replication by the combined action of DNA polymerases alpha and delta. Proc Natl Acad Sci U S A 86(19):7361–7365

Lee SH, Kwong AD, Pan ZQ, Hurwitz J (1991) Studies on the activator 1 protein complex, an accessory factor for proliferating cell nuclear antigen-dependent DNA polymerase delta. J Biol Chem 266(1):594–602

Lee C, Hong B, Choi JM, Kim Y, Watanabe S, Ishimi Y, Enomoto T, Tada S, Kim Y, Cho Y (2004) Structural basis for inhibition of the replication licensing factor Cdt1 by geminin. Nature 430(7002):913–917

Lehman IR, Kaguni LS (1989) DNA polymerase alpha. J Biol Chem 264:4265–4268

Leipe DD, Aravind L, Koonin EV (1999) Did DNA replication evolve twice independently? Nucleic Acids Res 27(17):3389–3401

Leonard AC, Mechali M (2013) DNA replication origins. Cold Spring Harb Perspect Biol 5(10):a010116

Levine AJ, Kang HS, Billheimer FE (1970) DNA replication in SV40 infected cells. I. Analysis of replicating SV40 DNA. J Mol Biol 50(2):549–568

Lewis JS, Jergic S, Dixon NE (2016) The E. coli DNA replication fork. Enzyme 39:31–88

Li JJ, Kelly TJ (1984) Simian virus 40 DNA replication in vitro. Proc Natl Acad Sci U S A 81:6973–6977

Li D, Zhao R, Lilyestrom W, Gai D, Zhang R, DeCaprio JA, Fanning E, Jochimiak A, Szakonyi G, Chen XS (2003a) Structure of the replicative helicase of the oncoprotein SV40 large tumour antigen. Nature 423(6939):512–518

Li X, Zhao Q, Liao R, Sun P, Wu X (2003b) The SCF(Skp2) ubiquitin ligase complex interacts with the human replication licensing factor Cdt1 and regulates Cdt1 degradation. J Biol Chem 278(33):30854–30858

Liu E, Li X, Yan F, Zhao Q, Wu X (2004) Cyclin-dependent kinases phosphorylate human Cdt1 and induce its degradation. J Biol Chem 279(17):17283–17288

Lopez-Mosqueda J, Maas NL, Jonsson ZO, Defazio-Eli LG, Wohlschlegel J, Toczyski DP (2010) Damage-induced phosphorylation of Sld3 is important to block late origin firing. Nature 467(7314):479–483

Lucas I, Germe T, Chevrier-Miller M, Hyrien O (2001) Topoisomerase II can unlink replicating DNA by precatenane removal. EMBO J 20(22):6509–6519

Lutzmann M, Maiorano D, Mechali M (2006) A Cdt1-geminin complex licenses chromatin for DNA replication and prevents rereplication during S phase in Xenopus. EMBO J 25(24):5764–5774

Madine MA, Khoo CY, Mills AD, Laskey RA (1995) MCM3 complex required for cell cycle regulation of DNA replication in vertebrate cells [see comments]. Nature 375(6530):421–424

Maine GT, Sinha P, Tye B-K (1984) Mutants of *S. cerevisiae* defective in the maintenance of mini-chromosomes. Genetics 106:365–385

Maiorano D, Moreau J, Mechali M (2000) XCDT1 is required for the assembly of pre-replicative complexes in Xenopus laevis [see comments]. Nature 404(6778):622–625

Makarova KS, Koonin EV (2013) Archaeology of eukaryotic DNA replication. Cold Spring Harb Perspect Biol 5(11):a012963

Marahrens Y, Stillman B (1992) A yeast chromosomal origin of DNA replication defined by multiple functional elements. Science 255(5046):817–823

Maric M, Maculins T, De Piccoli G, Labib K (2014) Cdc48 and a ubiquitin ligase drive disassembly of the CMG helicase at the end of DNA replication. Science 346(6208):1253596

Masumoto H, Muramatsu S, Kamimura Y, Araki H (2002) S-Cdk-dependent phosphorylation of Sld2 essential for chromosomal DNA replication in budding yeast. Nature 415(6872):651–655

Mattarocci S, Shyian M, Lemmens L, Damay P, Altintas DM, Shi T, Bartholomew CR, Thoma NH, Hardy CF, Shore D (2014) Rif1 controls DNA replication timing in yeast through the PP1 phosphatase Glc7. Cell Rep 7(1):62–69

McGarry TJ, Kirschner MW (1998) Geminin, an inhibitor of DNA replication, is degraded during mitosis. Cell 93(6):1043–1053

McVey D, Brizuela L, Mohr I, Marshak DR, Gluzman Y, Beach D (1989) Phosphorylation of large tumour antigen by cdc2 stimulates SV40 DNA replication. Nature 341:503–507
McVey D, Ray S, Gluzman Y, Berger L, Wildeman AG, Marshak DR, Tegtmeyer P (1993) cdc2 phosphorylation of threonine 124 activates the origin-unwinding functions of simian virus 40 T antigen. J Virol 67(9):5206–5215
Melendy T, Stillman B (1993) An interaction between replication protein A and SV40 T antigen appears essential for primosome assembly during SV40 DNA replication. J Biol Chem 268(5):3389–3395
Meselson M, Stahl F (1958) The replication of DNA in Escherichia coli. Proc Natl Acad Sci U S A 44(7):671–682
Moarefi IF, Small D, Gilbert I, Hopfner M, Randall SK, Schneider C, Russo AA, Ramsperger U, Arthur AK, Stahl H, Kelly TJ, Fanning E (1993) Mutation of the cyclin-dependent kinase phosphorylation site in simian virus 40 (SV40) large T antigen specifically blocks SV40 origin DNA unwinding. J Virol 67(8):4992–5002
Moir D, Stewart SE, Osmond BC, Botstein D (1982) Cold-sensitive cell-division-cycle mutants of yeast: isolation, properties, and pseudoreversion studies. Genetics 100(4):547–563
Moldovan GL, Pfander B, Jentsch S (2007) PCNA, the maestro of the replication fork. Cell 129(4):665–679
Moreno S, Nurse P (1994) Regulation of progression through the G1 phase of the cell cycle by the rum1+ gene. Nature 367(6460):236–242
Morrison A, Araki H, Clark AB, Hamatake RK, Sugino A (1990) A third essential DNA polymerase in S. cerevisiae. Cell 62(6):1143–1151
Moyer SE, Lewis PW, Botchan MR (2006) Isolation of the Cdc45/Mcm2-7/GINS (CMG) complex, a candidate for the eukaryotic DNA replication fork helicase. Proc Natl Acad Sci U S A 103(27):10236–10241
Murakami Y, Wobbe CR, Weissbach L, Dean FB, Hurwitz J (1986) Role of DNA polymerase a and DNA primase in simian virus 40 DNA replication in vitro. Proc Natl Acad Sci U S A 83:2869–2873
Murakami Y, Eki T, Hurwitz J (1992) Studies on the initiation of simian virus 40 replication in vitro: RNA primer synthesis and its elongation. Proc Natl Acad Sci U S A 89(3):952–956
Muramatsu S, Hirai K, Tak YS, Kamimura Y, Araki H (2010) CDK-dependent complex formation between replication proteins Dpb11, Sld2, Pol (epsilon), and GINS in budding yeast. Genes Dev 24(6):602–612
Nagata K, Guggenheimer RA, Enomoto T, Lichy JH, Hurwitz J (1982) Adenovirus DNA replication in vitro: identification of a host factor that stimulates synthesis of the preterminal protein-dCMP complex. Proc Natl Acad Sci U S A 79(21):6438–6442
Nagata K, Guggenheimer RA, Hurwitz J (1983) Adenovirus DNA replication in vitro: synthesis of full-length DNA with purified proteins. Proc Natl Acad Sci U S A 80(14):4266–4270
Newlon CS (1988) Yeast chromosome replication and segregation. Microbiol Rev 52(4):568–601
Nguyen VQ, Co C, Irie K, Li JJ (2000) Clb/Cdc28 kinases promote nuclear export of the replication initiator proteins Mcm2-7. Curr Biol 10(4):195–205
Nguyen VQ, Co C, Li JJ (2001) Cyclin-dependent kinases prevent DNA re-replication through multiple mechanisms. Nature 411(6841):1068–1073
Nishitani H, Nurse P (1995) p65cdc18 plays a major role controlling the initiation of DNA replication in fission yeast. Cell 83(3):397–405
Nishitani H, Lygerou Z, Nishimoto T, Nurse P (2000) The Cdt1 protein is required to license DNA for replication in fission yeast [see comments]. Nature 404(6778):625–628
O'Donnell M, Li H (2016) The eukaryotic replisome goes under the microscope. Curr Biol 26(6):R247–R256
O'Neill EA, Fletcher C, Burrow CR, Heintz N, Roeder RG, Kelly TJ (1988) Transcription factor OTF-1 is functionally identical to the DNA replication factor NF-III. Science 241:1210–1213
Okazaki R, Okazaki T, Sakabe K, Sugimoto K, Sugino A (1968) Mechanism of DNA chain growth. I. Possible discontinuity and unusual secondary structure of newly synthesized chains. Proc Natl Acad Sci U S A 59(2):598–605

Olivera BM, Bonhoeffer F (1972) Discontinuous DNA replication in vitro. I. Two distinct size classes of intermediates. Nat New Biol 240(103):233–235

On KF, Beuron F, Frith D, Snijders AP, Morris EP, Diffley JF (2014) Prereplicative complexes assembled in vitro support origin-dependent and independent DNA replication. EMBO J 33(6):605–620

Pacek M, Walter JC (2004) A requirement for MCM7 and Cdc45 in chromosome unwinding during eukaryotic DNA replication. EMBO J 23(18):3667–3676

Pacek M, Tutter AV, Kubota Y, Takisawa H, Walter JC (2006) Localization of MCM2-7, Cdc45, and GINS to the site of DNA unwinding during eukaryotic DNA replication. Mol Cell 21(4):581–587

Patel PK, Arcangioli B, Baker SP, Bensimon A, Rhind N (2006) DNA replication origins fire stochastically in fission yeast. Mol Biol Cell 17(1):308–316

Plevani P, Foiani M, Valsasnini P, Badaracco G, Cheriathundam E, Chang LM (1985) Polypeptide structure of DNA primase from a yeast DNA polymerase-primase complex. J Biol Chem 260(11):7102–7107

Pospiech H, Kursula I, Abdel-Aziz W, Malkas L, Uitto L, Kastelli M, Vihinen-Ranta M, Eskelinen S, Syvaoja JE (1999) A neutralizing antibody against human DNA polymerase epsilon inhibits cellular but not SV40 DNA replication. Nucleic Acids Res 27(19):3799–3804

Prelich G, Tan CK, Kostura M, Mathews MB, So AG, Downey KM, Stillman B (1987) Functional identity of proliferating cell nuclear antigen and a DNA polymerase-delta auxiliary protein. Nature 326(6112):517–520

Pruijn GJ, van Driel W, van der Vliet PC (1986) Nuclear factor III, a novel sequence-specific DNA-binding protein from HeLa cells stimulating adenovirus DNA replication. Nature 322(6080):656–659

Ralph E, Boye E, Kearsey SE (2006) DNA damage induces Cdt1 proteolysis in fission yeast through a pathway dependent on Cdt2 and Ddb1. EMBO Rep 7(11):1134–1139

Randell JC, Fan A, Chan C, Francis LI, Heller RC, Galani K, Bell SP (2010) Mec1 is one of multiple kinases that prime the Mcm2-7 helicase for phosphorylation by Cdc7. Mol Cell 40(3):353–363

Rao PN, Johnson RT (1970) Mammalian cell fusion: studies on the regulation of DNA synthesis and mitosis. Nature 225(228):159–164

Rekosh DM, Russell WC, Bellet AJ, Robinson AJ (1977) Identification of a protein linked to the ends of adenovirus DNA. Cell 11(2):283–295

Remus D, Beall EL, Botchan MR (2004) DNA topology, not DNA sequence, is a critical determinant for Drosophila ORC-DNA binding. EMBO J 23(4):897–907

Remus D, Beuron F, Tolun G, Griffith JD, Morris EP, Diffley JF (2009) Concerted loading of Mcm2-7 double hexamers around DNA during DNA replication origin licensing. Cell 139(4):719–730

Romanowski P, Madine MA, Laskey RA (1996a) XMCM7, a novel member of the Xenopus MCM family, interacts with XMCM3 and colocalizes with it throughout replication [see comments]. Proc Natl Acad Sci U S A 93(19):10189–10194

Romanowski P, Madine MA, Rowles A, Blow JJ, Laskey RA (1996b) The Xenopus origin recognition complex is essential for DNA replication and MCM binding to chromatin. Curr Biol 6(11):1416–1425

Rosenfeld PJ, Kelly TJ (1986) Purification of nuclear factor I by DNA recognition site affinity chromatography. J Biol Chem 261(3):1398–1408

Rowles A, Chong JP, Brown L, Howell M, Evan GI, Blow JJ (1996) Interaction between the origin recognition complex and the replication licensing system in Xenopus. Cell 87(2):287–296

Rowles A, Tada S, Blow JJ (1999) Changes in association of the Xenopus origin recognition complex with chromatin on licensing of replication origins. J Cell Sci 112(Pt 12):2011–2018

Samora CP, Saksouk J, Goswami P, Wade BO, Singleton MR, Bates PA, Lengronne A, Costa A, Uhlmann F (2016) Ctf4 links DNA replication with sister chromatid cohesion establishment by recruiting the Chl1 helicase to the replisome. Mol Cell 63(3):371–384

Santocanale C, Diffley JF (1998) A Mec1- and Rad53-dependent checkpoint controls late-firing origins of DNA replication. Nature 395(6702):615–618

Santocanale C, Sharma K, Diffley JF (1999) Activation of dormant origins of DNA replication in budding yeast. Genes Dev 13(18):2360–2364

Scheidtmann KH, Virshup DM, Kelly TJ (1991) Protein phosphatase 2A dephosphorylates SV40 large T antigen specifically at residues involved in regulation of the DNA binding activity. J Virol 65:2098–2101

Segurado M, de Luis A, Antequera F (2003) Genome-wide distribution of DNA replication origins at A+T-rich islands in Schizosaccharomyces pombe. EMBO Rep 4(11):1048–1053

Senga T, Sivaprasad U, Zhu W, Park JH, Arias EE, Walter JC, Dutta A (2006) PCNA is a cofactor for Cdt1 degradation by CUL4/DDB1-mediated N-terminal ubiquitination. J Biol Chem 281(10):6246–6252

Sengupta S, van Deursen F, de Piccoli G, Labib K (2013) Dpb2 integrates the leading-strand DNA polymerase into the eukaryotic replisome. Curr Biol 23(7):543–552

Sheu YJ, Stillman B (2010) The Dbf4-Cdc7 kinase promotes S phase by alleviating an inhibitory activity in Mcm4. Nature 463(7277):113–117

Shirahige K, Hori Y, Shiraishi K, Yamashita M, Takahashi K, Obuse C, Tsurimoto T, Yoshikawa H (1998) Regulation of DNA-replication origins during cell-cycle progression. Nature 395(6702):618–621

Siddiqui K, On KF, Diffley JF (2013) Regulating DNA replication in eukarya. Cold Spring Harb Perspect Biol 5(9), a012922

Simmons DT, Melendy T, Usher D, Stillman B (1996) Simian virus 40 large T antigen binds to topoisomerase I. Virology 222(2):365–374

Simon AC, Zhou JC, Perera RL, van Deursen F, Evrin C, Ivanova ME, Kilkenny ML, Renault L, Kjaer S, Matak-Vinkovic D, Labib K, Costa A, Pellegrini L (2014) A Ctf4 trimer couples the CMG helicase to DNA polymerase alpha in the eukaryotic replisome. Nature 510(7504):293–297

Singleton MR, Dillingham MS, Wigley DB (2007) Structure and mechanism of helicases and nucleic acid translocases. Annu Rev Biochem 76:23–50

Skarstad K, Katayama T (2013) Regulating DNA replication in bacteria. Cold Spring Harb Perspect Biol 5(4), a012922

Speck C, Stillman B (2007) Cdc6 ATPase activity regulates ORC x Cdc6 stability and the selection of specific DNA sequences as origins of DNA replication. J Biol Chem 282(16):11705–11714

Speck C, Chen Z, Li H, Stillman B (2005) ATPase-dependent cooperative binding of ORC and Cdc6 to origin DNA. Nat Struct Mol Biol 12(11):965–971

Stahl H, Droge P, Knippers R (1986) DNA helicase activity of SV40 large tumor antigen. EMBO J 5:1939–1944

Stamatoyannopoulos JA, Adzhubei I, Thurman RE, Kryukov GV, Mirkin SM, Sunyaev SR (2009) Human mutation rate associated with DNA replication timing. Nat Genet 41(4):393–395

Stillman BW (1983) The replication of adenovirus DNA with purified proteins. Cell 35(1):7–9

Stillman B (1989) Initiation of eukaryotic DNA replication in vitro. Annu Rev Cell Biol 5(197):197–245

Stinchcomb DT, Struhl K, Davis RW (1979) Isolation and characterisation of a yeast chromosomal replicator. Nature 282(5734):39–43

Sugimoto N, Tatsumi Y, Tsurumi T, Matsukage A, Kiyono T, Nishitani H, Fujita M (2004) Cdt1 phosphorylation by cyclin A-dependent kinases negatively regulates its function without affecting geminin binding. J Biol Chem 279(19):19691–19697

Sun J, Fernandez-Cid A, Riera A, Tognetti S, Yuan Z, Stillman B, Speck C, Li H (2014) Structural and mechanistic insights into Mcm2-7 double-hexamer assembly and function. Genes Dev 28(20):2291–2303

Sun J, Shi Y, Georgescu RE, Yuan Z, Chait BT, Li H, O'Donnell ME (2015) The architecture of a eukaryotic replisome. Nat Struct Mol Biol 22(12):976–982

Sundin O, Varshavsky A (1980) Terminal stages of SV40 DNA replication proceed via multiply intertwined catenated dimers. Cell 21(1):103–114

Sundin O, Varshavsky A (1981) Arrest of segregation leads to accumulation of highly intertwined catenated dimers: dissection of the final stages of SV40 DNA replication. Cell 25(3):659–669
Sussenbach JS, van der Vliet PC (1984) The mechanism of adenovirus DNA replication and the characterization of replication proteins. Curr Top Microbiol Immunol 109:53–73
Sutani T, Shirahige K (2016) Attaching accessory devices to the replisome. Mol Cell 63(3):347–348. https://doi.org/10.1016/j.molcel.2016.07.017
Tada S, Li A, Maiorano D, Mechali M, Blow JJ (2001) Repression of origin assembly in metaphase depends on inhibition of RLF- B/Cdt1 by geminin. Nat Cell Biol 3(2):107–113
Tak YS, Tanaka Y, Endo S, Kamimura Y, Araki H (2006) A CDK-catalysed regulatory phosphorylation for formation of the DNA replication complex Sld2-Dpb11. EMBO J 25(9):1987–1996
Takayama Y, Kamimura Y, Okawa M, Muramatsu S, Sugino A, Araki H (2003) GINS, a novel multiprotein complex required for chromosomal DNA replication in budding yeast. Genes Dev 17(9):1153–1165
Takeda DY, Parvin JD, Dutta A (2005) Degradation of Cdt1 during S phase is Skp2-independent and is required for efficient progression of mammalian cells through S phase. J Biol Chem 280(24):23416–23423
Tamanoi F, Stillman BW (1982) Function of adenovirus terminal protein in the initiation of DNA replication. Proc Natl Acad Sci U S A 79(7):2221–2225
Tamanoi F, Stillman BW (1983) Initiation of adenovirus DNA replication in vitro requires a specific DNA sequence. Proc Natl Acad Sci U S A 80(21):6446–6450
Tan CK, Castillo C, So AG, Downey KM (1986) An auxiliary protein for DNA polymerase-delta from fetal calf thymus. J Biol Chem 261(26):12310–12316
Tanaka S, Araki H (2013) Helicase activation and establishment of replication forks at chromosomal origins of replication. Cold Spring Harb Perspect Biol 5(12), a010371
Tanaka S, Diffley JF (2002) Interdependent nuclear accumulation of budding yeast Cdt1 and Mcm2-7 during G1 phase. Nat Cell Biol 4(3):198–207
Tanaka T, Knapp D, Nasmyth K (1997) Loading of an Mcm protein onto DNA replication origins is regulated by Cdc6p and CDKs. Cell 90(4):649–660
Tanaka S, Umemori T, Hirai K, Muramatsu S, Kamimura Y, Araki H (2007) CDK-dependent phosphorylation of Sld2 and Sld3 initiates DNA replication in budding yeast. Nature 445(7125):328–332
Tanaka H, Katou Y, Yagura M, Saitoh K, Itoh T, Araki H, Bando M, Shirahige K (2009) Ctf4 coordinates the progression of helicase and DNA polymerase alpha. Genes Cells 14(7):807–820
Tanaka S, Nakato R, Katou Y, Shirahige K, Araki H (2011) Origin association of Sld3, Sld7, and Cdc45 proteins is a key step for determination of origin-firing timing. Curr Biol 21(24):2055–2063
Tapper DP, DePamphilis ML (1978) Discontinuous DNA replication: accumulation of Simian virus 40 DNA at specific stages in its replication. J Mol Biol 120(3):401–422
Tegtmeyer P (1972) Simian virus 40 deoxyribonucleic acid synthesis: the viral replicon. J Virol 10(4):591–598
Ticau S, Friedman LJ, Ivica NA, Gelles J, Bell SP (2015) Single-molecule studies of origin licensing reveal mechanisms ensuring bidirectional helicase loading. Cell 161(3):513–525
Tjian R (1978) The binding site on SV40 DNA for a T antigen-related protein. Cell 13:165–179
Tseng BY, Ahlem CN (1982) DNA primase activity from human lymphocytes. Synthesis of oligoribonucleotides that prime DNA synthesis. J Biol Chem 257(13):7280–7283
Tsurimoto T, Stillman B (1989) Purification of a cellular replication factor, RF-C, that is required for coordinated synthesis of leading and lagging strands during simian virus 40 DNA replication in vitro. Mol Cell Biol 9(2):609–619
Tsurimoto T, Stillman B (1990) Functions of replication factor C and proliferating-cell nuclear antigen: functional similarity of DNA polymerase accessory proteins from human cells and bacteriophage T4. Proc Natl Acad Sci U S A 87:1023–1027
Tsurimoto T, Stillman B (1991a) Replication factors required for SV40 DNA replication in vitro. I. DNA structure-specific recognition of a primer-template junction by eukaryotic DNA polymerases and their accessory proteins. J Biol Chem 266(3):1950–1960

Tsurimoto T, Stillman B (1991b) Replication factors required for SV40 DNA replication in vitro. II. Switching of DNA polymerase alpha and delta during initiation of leading and lagging strand synthesis. J Biol Chem 266(3):1961–1968
Tsurimoto T, Melendy T, Stillman B (1990) Sequential initiation of lagging and leading strand synthesis by two different polymerase complexes at the SV40 DNA replication origin. Nature 346(6284):534–539
Valle M, Gruss C, Halmer L, Carazo JM, Donate LE (2000) Large T-antigen double hexamers imaged at the simian virus 40 origin of replication. Mol Cell Biol 20(1):34–41
Van Houten J, Newlon CS (1990) Mutational analysis of the consensus sequence of a replication origin from yeast chromosome III. Mol Cell Biol 10(8):3917–3925
Vashee S, Cvetic C, Lu W, Simancek P, Kelly TJ, Walter JC (2003) Sequence-independent DNA binding and replication initiation by the human origin recognition complex. Genes Dev 17(15):1894–1908
Villa F, Simon AC, Ortiz Bazan MA, Kilkenny ML, Wirthensohn D, Wightman M, Matak-Vinkovic D, Pellegrini L, Labib K (2016) Ctf4 is a hub in the eukaryotic replisome that links multiple CIP-box proteins to the CMG helicase. Mol Cell 63(3):385–396
Virshup DM, Kauffman MG, Kelly TJ (1989) Activation of SV40 DNA replication in vitro by cellular protein phosphatase 2A. EMBO J 8(12):3891–3898
Virshup DM, Russo AA, Kelly TJ (1992) Mechanism of activation of simian virus 40 DNA replication by protein phosphatase 2A. Mol Cell Biol 12(11):4883–4895
Vujcic M, Miller CA, Kowalski D (1999) Activation of silent replication origins at autonomously replicating sequence elements near the HML locus in budding yeast. Mol Cell Biol 19(9):6098–6109
Waga S, Stillman B (1994) Anatomy of a DNA replication fork revealed by reconstitution of SV40 DNA replication in vitro. Nature 369(6477):207–212
Waga S, Stillman B (1998) The DNA replication fork in eukaryotic cells. Annu Rev Biochem 67:721–751
Waga S, Bauer G, Stillman B (1994) Reconstitution of complete SV40 DNA replication with purified replication factors. J Biol Chem 269(14):10923–10934
Watase G, Takisawa H, Kanemaki MT (2012) Mcm10 plays a role in functioning of the eukaryotic replicative DNA helicase, Cdc45-Mcm-GINS. Curr Biol 22(4):343–349
Weinberg DH, Kelly TJ (1989) Requirement for two DNA polymerases in the replication of simian virus 40 DNA in vitro. Proc Natl Acad Sci U S A 86:9742–9746
Wobbe CR, Weissbach L, Borowiec JA, Dean FB, Murakami Y, Bullock P, Hurwitz J (1987) Replication of SV40 origin-containing DNA with purified proteins. Proc Natl Acad Sci U S A 84:1834–1838
Wohlschlegel JA, Dwyer BT, Dhar SK, Cvetic C, Walter JC, Dutta A (2000) Inhibition of eukaryotic DNA replication by geminin binding to cdt1. Science 290(5500):2309–2312
Wold MS (1997) Replication protein A: a heterotrimeric, single-stranded DNA-binding protein required for eukaryotic DNA metabolism. Annu Rev Biochem 66:61–92
Wold MS, Kelly TJ (1988) Purification and characterization of replication protein A, a cellular protein required for in vitro replication of simian virus 40 DNA. Proc Natl Acad U S A 85:2523–2527
Wold MS, Li JJ, Kelly TJ (1987) Initiation of simian virus 40 DNA replication in vitro: large-tumor-antigen- and origin-dependent unwinding of the template. Proc Natl Acad Sci U S A 84:3643–3647
Wold MS, Li JJ, Weinberg DH, Virshup DM, Sherley JM, Verheyen E, Kelly T (1988) Cellular proteins required for SV40 DNA replication in vitro. In: Kelly T, Stillman B (eds) Eukaryotic DNA replication. Cancer Cells. Cold Spring Harbor Laboratory, Cold Spring Harbor, pp 133–142
Woo YH, Li WH (2012) DNA replication timing and selection shape the landscape of nucleotide variation in cancer genomes. Nat Commun 3:1004
Yan H, Gibson S, Tye BK (1991) Mcm2 and Mcm3, two proteins important for ARS activity, are related in structure and function. Genes Dev 5(6):944–957

Yang L, Wold MS, Li JJ, Kelly TJ, Liu LF (1987) Roles of DNA topoisomerases in simian virus 40 DNA replication in vitro. Proc Natl Acad Sci U S A 84(4):950–954
Yao NY, O'Donnell M (2012) The RFC clamp loader: structure and function. Subcell Biochem 62:259–279
Yao NY, Johnson A, Bowman GD, Kuriyan J, O'Donnell M (2006) Mechanism of proliferating cell nuclear antigen clamp opening by replication factor C. J Biol Chem 281(25):17528–17539
Yeeles JT, Deegan TD, Janska A, Early A, Diffley JF (2015) Regulated eukaryotic DNA replication origin firing with purified proteins. Nature 519(7544):431–435
Yeeles JT, Janska A, Early A, Diffley JF (2017) How the eukaryotic replisome achieves rapid and efficient DNA replication. Mol Cell 65(1):105–116
Yuan Z, Bai L, Sun J, Georgescu R, Liu J, O'Donnell ME, Li H (2016) Structure of the eukaryotic replicative CMG helicase suggests a pumpjack motion for translocation. Nat Struct Mol Biol 23(3):217–224
Zegerman P, Diffley JF (2007) Phosphorylation of Sld2 and Sld3 by cyclin-dependent kinases promotes DNA replication in budding yeast. Nature 445(7125):281–285
Zegerman P, Diffley JF (2010) Checkpoint-dependent inhibition of DNA replication initiation by Sld3 and Dbf4 phosphorylation. Nature 467(7314):474–478
Zhou B, Arnett DR, Yu X, Brewster A, Sowd GA, Xie CL, Vila S, Gai D, Fanning E, Chen XS (2012) Structural basis for the interaction of a hexameric replicative helicase with the regulatory subunit of human DNA polymerase alpha-primase. J Biol Chem 287(32):26854–26866
Zhu W, Ukomadu C, Jha S, Senga T, Dhar SK, Wohlschlegel JA, Nutt LK, Kornbluth S, Dutta A (2007) Mcm10 and And-1/CTF4 recruit DNA polymerase alpha to chromatin for initiation of DNA replication. Genes Dev 21(18):2288–2299
Zlotkin T, Kaufmann G, Jiang Y, Lee MY, Uitto L, Syvaoja J, Dornreiter I, Fanning E, Nethanel T (1996) DNA polymerase epsilon may be dispensable for SV40- but not cellular-DNA replication. EMBO J 15(9):2298–2305

Chapter 2
Regulation of Replication Origins

Anna B. Marks, Haiqing Fu, and Mirit I. Aladjem

Abstract In eukaryotes, genome duplication starts concomitantly at many replication initiation sites termed replication origins. The replication initiation program is spatially and temporally coordinated to ensure accurate, efficient DNA synthesis that duplicates the entire genome while maintaining other chromatin-dependent functions. Unlike in prokaryotes, not all potential replication origins in eukaryotes are needed for complete genome duplication during each cell cycle. Instead, eukaryotic cells vary the use of initiation sites so that only a fraction of potential replication origins initiate replication each cell cycle. Flexibility in origin choice allows each eukaryotic cell type to utilize different initiation sites, corresponding to unique nuclear DNA packaging patterns. These patterns coordinate replication with gene expression and chromatin condensation. Budding yeast replication origins share a consensus sequence that marks potential initiation sites. Metazoan origins, on the other hand, lack a consensus sequence. Rather, they are associated with a collection of structural features, chromatin packaging features, histone modifications, transcription, and DNA-DNA/DNA-protein interactions. These features confer cell type-specific replication and expression and play an essential role in maintaining genomic stability.

Keywords DNA replication • Cell cycle regulation • Replication origin licensing • Chromatin organization • Histone modification • Replication timing

2.1 Initiation of DNA Replication

Origins of replication are defined as chromosomal sites where double-stranded DNA unwinds to form single-stranded DNA templates for genome duplication. Genetically, the *cis*-acting sequences that determine the locations of replication initiation events are termed replicators (Jacob et al. 1963). Replicators can confer the ability to start replication when transferred from their original locations to ectopic sites. Replicators

A.B. Marks • H. Fu • M.I. Aladjem (✉)
Developmental Therapeutics Branch, Center for Cancer Research, NCI, NIH, Bethesda, MD, USA
e-mail: aladjemm@mail.nih.gov

H. Masai, M. Foiani (eds.), *DNA Replication*, Advances in Experimental Medicine and Biology 1042, https://doi.org/10.1007/978-981-10-6955-0_2

interact with trans-acting factors, termed initiators, to facilitate DNA replication. In eukaryotes, initiators are highly conserved, as all eukaryotes share a group of essential DNA-binding protein complexes that form pre-replication complexes. Pre-replication complexes assemble on chromatin in a process termed "replication licensing," and the components of pre-replication complexes are orthologous in all eukaryotes (Aladjem 2007; DePamphilis 1999; Fragkos et al. 2015; Masai et al. 2010; Remus and Diffley 2009). Conversely, the chromatin features associated with eukaryotic replicators vary and are often cell type and/or developmental specific (Aladjem 2007; Besnard et al. 2012; Cayrou et al. 2015; Smith and Aladjem 2014; Smith et al. 2016).

The conserved proteins that form pre-replication complexes (Fig. 2.1) include a DNA-binding origin recognition complex (ORC) that serves as a platform to recruit

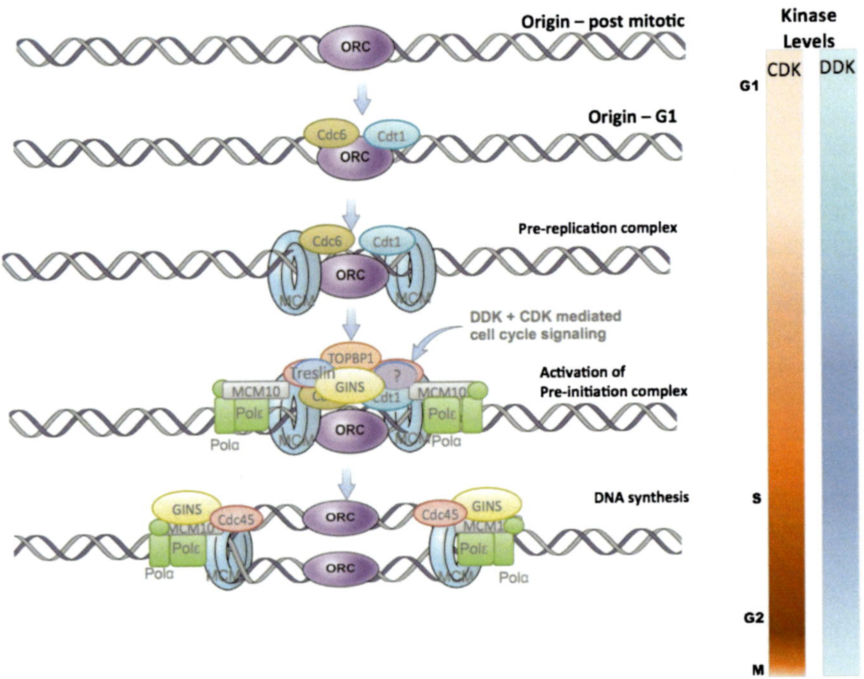

Fig. 2.1 Pre-replication complex proteins bind in a stepwise manner throughout G1. Recruited during the M to G1 transition, the origin recognition complex (ORC) is a platform to recruit a conserved group of helicases, polymerases, and accessory proteins that catalyze the initiation of DNA replication. ORC binds to chromatin as cells emerge from mitosis. Licensing factors Cdc6 and Cdt1 bind to ORC, followed by the inactive form of the replicative helicase MCM2-7. Additional proteins are required to activate the MCM helicase and initiate DNA replication. Specifically, the inactive MCM2-7 helicase then interacts with CDC45, MCM10, and GINS (Sld5, Psf1, Psf2, Psf3) to form the complete helicase (CMG) complex. Cyclin-dependent kinase (CDK)- and Dbf4-dependent kinase (DDK)-mediated phosphorylation activates proteins and allows Cdc45 interacts with Treslin (Sld3 in yeast). Sld2/RecQL4 and DPB11/TopBP1 are then recruited to the complex. Chromatin-associated DNA polymerases (Pol-α and pol-δ), replication protein A (RPA), CMG, and Dpb11/TopBP1 then initiate DNA replication

a conserved group of helicases, polymerases, and accessory proteins that catalyze the initiation of DNA replication. Assembly of the conserved pre-replication and pre-initiation complexes at potential replication origins occurs stepwise during the G1 phase of each the cell cycle (Aladjem 2007; DePamphilis 1999; Fragkos et al. 2015; Remus and Diffley 2009). First, ORC binds to chromatin as cells emerge from mitosis. Then, licensing factors CDC6 and CDT1 bind to ORC, which allows for the binding of the inactive form of the replicative helicase MCM2-7. The resulting pre-replication complex recruits additional proteins required to activate the MCM helicase and initiate DNA replication (Boos et al. 2013; Sansam et al. 2015; Sheu et al. 2016; Tanaka and Araki 2013). The chromatin-bound but inactive MCM2-7 helicase then interacts with additional components, CDC45, MCM10, and GINS (Sld5, Psf1, Psf2, Psf3), to form the complete helicase (CMG) complex. Prompted by cyclin-dependent kinase (CDK)- and Dbf4-dependent kinase (DDK)-mediated phosphorylation, CDC45 interacts with Treslin (Sld3 in yeast), which recruits Sld2/RecQL4 and DPB11/TopBP1 (Bruck et al. 2015; Depamphilis et al. 2012; Masai et al. 2010; Remus and Diffley 2009; Zegerman and Diffley 2007). Chromatin-associated DNA polymerases (Pol-α and pol-δ), replication protein A (RPA), CMG, and Dpb11/TopBP1 then initiate DNA replication (Abid and Costa 2016; Kanemaki and Labib 2006; Takayama et al. 2003).

The assembly of pre-replication complexes and the subsequent initiation are tightly coupled with cell cycle progression by the phosphorylation activities of two kinases, CDK and DDK. Helicase recruitment, in an inactive form, can only occur at low kinase levels (Bell 2002; Remus and Diffley 2009), whereas the activities of DDK and CDK are required at subsequent steps to activate the helicase and initiate replication (Boos et al. 2013; Remus and Diffley 2009; Sansam et al. 2015; Sheu et al. 2016; Tanaka and Araki 2013). The need for low kinase levels in the early stages of pre-replication complex assembly implies that such complexes cannot be assembled once DNA replication has started, insuring orderly cell cycle progression as well as preventing re-replication of cellular DNA, a hallmark of genomic instability (Abbas et al. 2013; Hanlon and Li 2015; Remus and Diffley 2009; Richardson and Li 2014).

Although the events that lead to initiation of DNA replication occur at all potential replication origins, replication initiation occurs with a remarkably consistent order in most cells (Besnard et al. 2012; Cayrou et al. 2011; Martin et al. 2011) to create a coordinated replication timing program (Koren et al. 2014; Mukhopadhyay et al. 2014; Rhind and Gilbert 2013b). The binding patterns of pre-replication complexes do not provide clues to the principles of origin choice as ORC does not exhibit sequence-specific DNA binding (Miotto et al. 2016) and replication origins in most eukaryotes do not share a clear common consensus (Aladjem 2007; Bartholdy et al. 2015; Leonard and Mechali 2013; Masai et al. 2010). Interactions between replication origins and components of pre-replication complexes, therefore, are essential for initiation but cannot intuitively explain the consistent replication patterns observed in most mitotic cell cycles.

2.2 Genetic Features and Local Determinants of Replication Origins

Replication origins in viruses and in some prokaryotes and eukaryotes exhibit distinct sequence features that facilitate interactions with unique initiator proteins. In DNA tumor viruses, replication origins colocalize with replicator sequences that bind specialized initiators (e.g., SV40 T-antigen and BPV E1 protein) to catalyze DNA unwinding and recruit the host replication machinery (Fanning and Zhao 2009). In budding yeast, replication origins contain an AT-rich, 11 bp consensus ORC binding sequence. High frequency of initiation from yeast replication origins also requires accessory sequences that affect chromatin structure by dictating the efficiency of ORC binding as well as directing initiation in unique chromosomal environments (Hoggard et al. 2013; Marahrens and Stillman 1992; Palacios DeBeer et al. 2003).

In metazoa, replication origin sequences exhibit high heterogeneity (Aladjem 2007; Bartholdy et al. 2015; DePamphilis 1999; Fragkos et al. 2015) and do not share a clear consensus, consistent with the observation that not all potential replication origins initiate replication in all cells each cell cycle. There are two sources of replication origin heterogeneity. First, a large fraction of replication origins exhibit consistent initiation in particular cell types and not in other cells (Besnard et al. 2012; Cayrou et al. 2015; Martin et al. 2011; Smith et al. 2016). For example, less than 50% of all origins identified in a survey of eight cell lines initiated replication in all cells within that cohort, and a large fraction (about 15%) only initiated replication in a single cell line (Smith et al. 2016). Second, within populations of cells of the same type, most replication origins initiate replication stochastically in a fraction of cell cycles. In most somatic metazoan cells, only 10–20% of all potential origins actually initiate replication each cell cycle, suggesting that most origins exhibit flexible initiation patterns (Cayrou et al. 2015). When such flexible origins remain "dormant" and do not initiate replication, they replicate passively from adjacent replication forks (Fig. 2.2). Notably, however, the presence of excess replication origins plays a role in genome preservation: a marked reduction in the frequency of licensed but dormant origins, achieved either by targeted genetic deletions of origins or by mutating replication licensing factors, increases genomic instability (Abbas et al. 2013; Besnard et al. 2012; Blow et al. 2011; Fragkos et al. 2015; Kawabata et al. 2011; Marks et al. 2016). These observations suggest that flexible or consistent dormant origins, which rarely or never initiate replication during normal cell cycle progression, might serve as backup origins when replication forks stall.

The exact sequence features that mark replication origins and determine the frequency of initiation at each origin remain to be elucidated. Budding yeast potential replication origins share a consensus sequence that is recognized by the ORC complex. Yeast origins can all initiate replication on plasmids (Marahrens and Stillman 1992; Masai et al. 2010), but chromatin context plays a role in determining the activation rates of particular chromosomal origins (Hoggard et al. 2013; Knott et al. 2012). Metazoan replication origins do not exhibit a prominent single consensus sequence (Aladjem 2007; Fragkos et al. 2015; Mechali et al. 2013; Urban et al.

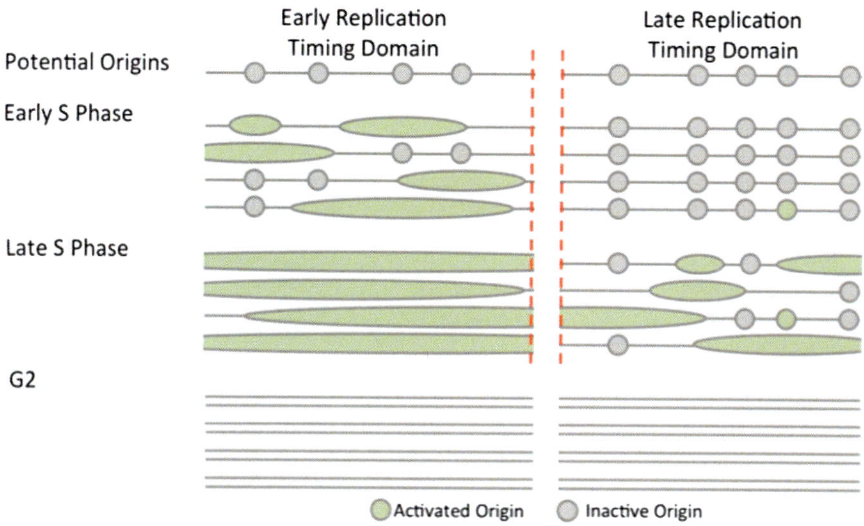

Fig. 2.2 While the profiles of activated replication origins are similar within cell line, specific replication origin chosen by cells within that population varies in location and replication time. Most origins exhibit flexible initiation patterns, where the specific origins activated differ between cells. When such flexible origins remain "dormant" and do not initiate replication, they replicate passively from adjacent replication forks

2015) but share several sequence features (Fig. 2.3a) including regions that exhibit strand asymmetry, CpG islands, G-quadruplexes, transcription start sites, origin G-rich repeated elements (OGREs), and regions of DNase hypersensitivity (Besnard et al. 2012; Cayrou et al. 2015; Foulk et al. 2015; Martin et al. 2011; Mukhopadhyay et al. 2014; Rao et al. 2014). Of those, genetic association studies on phased genome revealed strong high association with strand asymmetry (Bartholdy et al. 2015).

The primary DNA sequence at replication origins can determine the ability to initiate replication at ectopic sites (Aladjem et al. 1998; Liu et al. 2003), possibly via affecting chromatin modifications (Chen et al. 2013; Conner and Aladjem 2012; Fu et al. 2006; Liu et al. 2003). For example, the replication origin at the human *HBB* locus can create an open chromatin conformation at ectopic sites (Fu et al. 2006). A group of replication origins colocalizes with ubiquitous chromatin opening elements (UCOES), which maintain open chromatin structure by recruiting transcription factors (Conner and Aladjem 2012; Majocchi et al. 2014) and protecting transcriptional activity despite local repressive epigenetic features (Flickinger 2015). Hence, recruiting chromatin modifiers could allow replication origins to alter the local environment and create a context permissive for both transcriptional activity and replication initiation (Aladjem 2007; Hassan-Zadeh et al. 2012; Huang et al. 2011).

In agreement, local histone modifications correlate with, and can determine, replication origin locations (Feng et al. 2006; Leonard and Mechali 2013; Mechali et al. 2013; Rhind and Gilbert 2013b; Smith and Aladjem 2014; Smith et al. 2016; Vogelauer et al. 2002). Comparisons of initiation sites with histone features and

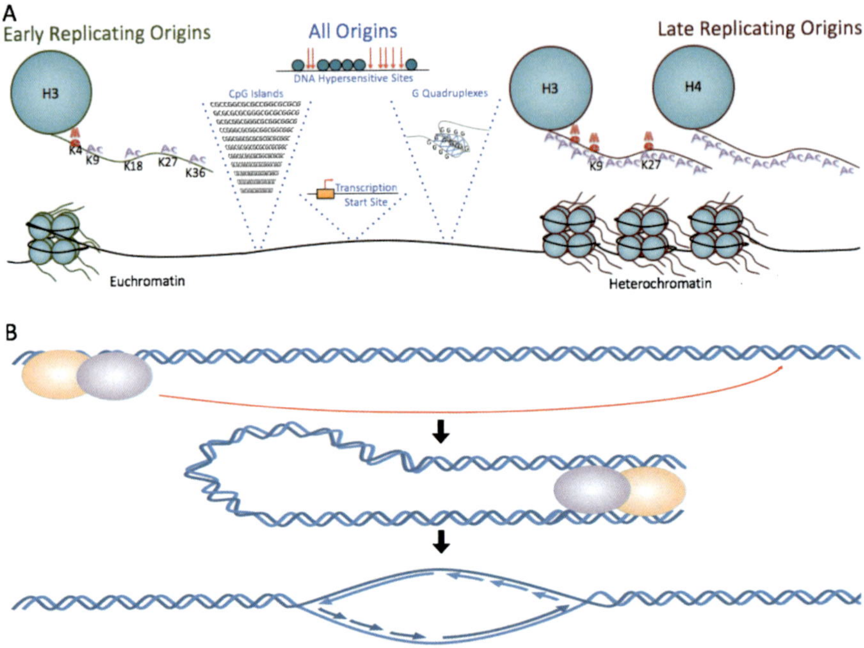

Fig. 2.3 (**a**) Metazoan replication origins share several sequence features. Origins generally associate with regions that exhibit strand asymmetry, CpG islands, G-quadruplexes, transcription start sites, origin G-rich-repeated elements (OGREs), and regions of DNase hypersensitivity. In agreement, local histone modifications correlate with and can determine replication origin locations and timing. Early replicating regions associate with H3K4me1/2/3, H3K9ac, H3K18ac, H3K36me3, and H3K27ac. These histone modifications also associate with open chromatin and are enriched in moderately active transcription start sites. Late replicating regions tend to associate with H3 and H4 hypoacetylation, H3K9 and H3K27 methylation, and are found in heterochromatic regions. (**b**) Distal DNA sequences affect origin activity and transcriptional through long-distance interactions. Such interactions can be mediated via protein interaction with enhancers and locus control regions, by chromatin remodeling factors and transcriptional activators that bind enhancers and locus control regions, and by long noncoding RNAs

replication timing domains identify certain histone modifications as strong indicators of origin utilization (Fig. 2.3a). For example, early replicating regions associate with H3K4me1/2/3, H3K9ac, H3K18ac, H3K36me3, and H3K27ac (Smith et al. 2016). Origins that associate with those chromatin features also localize in open chromatin and are enriched in moderately active transcription start sites (Besnard et al. 2012; Martin et al. 2011). Replication origins that initiate replication early during S-phase tend to associate with open chromatin features and are often activated in many cell types. In contrast, late replicating regions tend to associate with H3 and H4 hypoacetylation and H3K9 and H3K27 methylation and often initiate replication in a cell type-specific manner (Cayrou et al. 2015; Mechali et al. 2013; Smith et al. 2016). The tendency to initiate replication in most cells or in a particular cell type does not depend on cancer status, as common and cell type-specific initiation activity correlates with cellular lineages rather than with cancer or noncancer (Smith et al. 2016).

Transcriptional activity of local genes and cellular differentiation state can alter the replication timing program. For example, *Xenopus* early embryos do not exhibit a strong preference for initiation sites, correlated with the absence of transcription (Mechali and Kearsey 1984). In those embryos, induced transcription either through development or by tethering specific transcription factors resulted in increased localized initiation (Fragkos et al. 2015; Mechali et al. 2013). In somatic dividing cells, replication origins often associate with transcription start sites at active genes (Valenzuela et al. 2011) and, in particular, with transcription start sites at moderately active transcribed regions (Martin et al. 2011; Sequeira-Mendes et al. 2009). However, this association is diminished in highly transcribed regions, suggesting that transcription and replication regulate each other to avoid disruptions and polymerase collision events (Martin et al. 2011; Sequeira-Mendes et al. 2009). In agreement, transcriptional activity coordinates with the replication timing program (Rivera-Mulia et al. 2015), and replication delays often accompany gene silencing.

2.3 Influences of Global Chromatin Organization on Replication Initiation

On a larger scale, replication timing domains, each containing multiple replication origins that replicate concomitantly (Bartholdy et al. 2015; Mukhopadhyay et al. 2014), exhibit high concordance with large-scale chromatin organization units termed topologically associated domains that encompass several hundred kilobases to megabases (Hiratani et al. 2010; Lieberman-Aiden et al. 2009; Mattarocci et al. 2014; Moindrot et al. 2012; Rao et al. 2014; Rhind and Gilbert 2013b; Yaffe et al. 2010). This association suggests that the time of activation of replication origins reflects a fundamental structural property of the nucleus (Hiratani et al. 2010; Moindrot et al. 2012; Pope et al. 2014; Rhind and Gilbert 2013b; Yaffe et al. 2010). In agreement, replication origins are known to associate nuclear structural features such as matrix attachment sites (MARS), scaffold attachment sites (SARs), and stabilizing anti-repressor elements (STARs) (Mechali et al. 2013; Smith and Aladjem 2014) as well as with lamins and cohesins (Cayrou et al. 2015; Smith and Aladjem 2014).

High-resolution whole-genome analyses reveal that replication timing domains often reflect chromatin modifications (Dileep et al. 2015; Pope et al. 2014). Early replicating regions often associate with transcriptionally active topological domains, whereas late replicating origins often associate with heterochromatin. The effects of the primary sequence on replication origin activity and replication timing were assessed using analyses on phased genomes, which permit identification of paternal vs. maternal origins to characterize the effects of specific sequence variations on origin activity (Bartholdy et al. 2015; Mukhopadhyay et al. 2014), and by identifying inherited alleles that affect replication timing following the sequencing of 161 individual proliferating cell samples (Koren et al. 2014). These analyses have demonstrated that *cis*-acting genetic elements determine, at least in part, the locations of megabase-scale replication timing domains. In cancer cells, the replication of entire chromosomes could be delayed in a sequence-specific manner, and interactions with long noncoding RNAs could alter the timing of replication for entire chromosomes (Donley et al. 2015).

Proteins that catalyze distinct histone modifications can facilitate or modulate initiation in groups of replication origins. HBO1, a histone acetyltransferase that modifies H4K5 and H4K12, binds near origins of replication by associating with ORC1 and Cdt1 (Iizuka et al. 2006; Miotto and Struhl 2010). The chromatin decondensation promoted by HBO1 is enhanced near H3K4me3 and reduced near H3K20me1/2/3 (Huang et al. 2006; Kim et al. 2006; Saksouk et al. 2009). While histone acetyltransferase HBO1 is associated with early replicating origins, histone methyltransferase PR-Set7 and heterochromatin-associated proteins ORC-associated protein (ORCA) and HP1 are associated with late replicating origins. ORCA/LRWD1 stabilizes ORC on origins and promotes late replication by perpetuating histone compaction near the repressive methylation of H3K9, H4K20, and H4K27 (Chakraborty et al. 2011; Giri et al. 2015). HP1 is also known to stabilize ORC on origins via interaction with ORC2 and ORC3 and to bind to methylated H3K9 to establish late replicating domains (Chakraborty et al. 2011; Schwaiger et al. 2010). H4K20me1 serves as a binding domain for other histone modifiers, like Suv4, promoting further chromatin compaction (Tardat et al. 2010). Cells depleted in further methylation of H4K20 are also shown have reduced ORCA and ORC1 binding (Sherstyuk et al. 2014).

In addition to modifying histones near origins of replication, some *trans*-acting factors, like Rif1, Taz1, and FKH1/2, facilitate the recruitment of replication factors to origins. Taz1 and Rif1 help promote replication initiation in heterochromatic, telomeric regions (Cornacchia et al. 2012; Hayano et al. 2012; Tazumi et al. 2012; Yamazaki et al. 2012). Both Taz1, which prevents early-S replication activation, and Rif1, which recruits protein phosphatase 1 (PP1) and modulates the chromatin binding of pre-initiation complex components (Dave et al. 2014; Foti et al. 2016; Hiraga et al. 2014; Kanoh et al. 2015), are associated with late replication. Taz1 and Rif1 might delay replication by interfering with DDK phosphorylation of Mcm2-7 and associating with nuclear architectures that anchor heterochromatin (Foti et al. 2016; Tazumi et al. 2012). Inhibition of MCM phosphorylation subsequently interferes with CDC45 and Sld3 loading (Francis et al. 2009; Tazumi et al. 2012). The replication and DNA repair features regulated by Rif1 are conserved across eukaryotes (Mattarocci et al. 2014). Rif1 organizes replication timing domains by associating with G-quadruplexes to suppress replication (Foti et al. 2016; Kanoh et al. 2015; Mattarocci et al. 2014).

Unlike Taz1 and Rif1, FKH1/2 can promote early replication by recruiting replication factors to early replicating DNA (Knott et al. 2012). In addition, FKH1/2 overexpression advances the replication time of late replicating origins (Knott et al. 2012). These proteins help either activate or repress replication by facilitating interchromosomal interactions (Musialek and Rybaczek 2015). FKH1/2 advance the time of initiation by acting during the late G1 phase of the cell cycle (Peace et al. 2016), indicating that replication timing can be reset subsequently to origin licensing.

Distal DNA sequences affect transcriptional activity and origin activity through long-distance interactions (Aladjem et al. 1998; Gerhardt et al. 2014; Norio et al. 2005). Such interactions (Fig. 2.3b) can be mediated via protein interaction with enhancers and locus control regions (Huang et al. 2011) or by chromatin remodeling factors and transcriptional activators that bind enhancers and locus control regions (Aladjem 2007; Fragkos et al. 2015). RepID, a protein that interacts with a group of replication origins, is associated with an origin-activating chromatin loop between the origin and the locus control region at the human *HBB* locus (Zhang et al. 2016). Long-distance interactions that modulate replication timing can also be mediated by long noncoding RNAs (lncRNAs) such as *Xist* and *HOTAIR*, which guide histone and chromatin remodeling proteins to specific DNA sequences and facilitate chromatin interactions (Fragkos et al. 2015; Nagano and Fraser 2011). lncRNAs can stabilize ORC to origins in viruses (Fragkos et al. 2015; Nagano and Fraser 2011) and can affect the timing of replication for entire chromosomes in cancer cells (Donley et al. 2015).

2.4 Role of Replication Origins

The apparent excess of potential replication origins and the absence of sequence-specific initiation during early embryogenesis both suggest that a consistent replication initiation program is not a mechanistic requirement for genome duplication. Replication also proceeds stochastically with no apparent replication timing domains within the human inactive X chromosome (Koren et al. 2014), again suggesting that a consistent replication timing program is not required merely to insure genome duplication. The replication program could be established to coordinate replication with other chromatin transactions, primarily transcription. Consistent replication initiation sites could facilitate genome integrity by coordinating replication with transcription and chromatin assembly on the shared chromatin template. The consistent replication timing programs establish regions that replicate late during S-phase and might serve to establish and maintain specific nuclear compartments, such as heterochromatin. Late replication of heterochromatin can be required to preserve the structural integrity of the nucleus by preventing rapid, massive chromatin decondensation and re-condensation that could and lead to DNA damage (Bustin and Misteli 2016).

Although the severe effects of changes in replication timing support a critical role for replication timing regulation in maintaining genomic stability, recent mathematical models suggest that the relative efficiencies of initiation at replication origins are sufficient to determine the organization of replication timing domains. A mathematical model can predict replication timing with high accuracy in human cells without assuming any "replication timing factor," using two variables: the known distribution of replication origins as correlates of DNase hypersensitive sites and the assumption that replication initiation is restricted by the availability of a single rate-limiting activator (Gindin et al. 2014). A second model (Lob et al. 2016) was also able to predict the general progression of DNA replication and in addition predicted the three-dimensional spatial organization of replication events based on higher chromatin organization, assuming spontaneous stochastic initiation within euchromatin and facultative heterochromatin. Again, this model did not assume a replication timing factor, and replication timing could be deduced without such a factor assuming concomitant inhibition of replication initiation at distances below the size of chromatin loops and a domino-like effect by which replication at a particular origin would induce initiation from adjacent origins. A third model was able to predict replication timing in yeast with high accuracy relying on the density of the MCM replicative helicase, assuming a high level of MCMs at early origins (Das et al. 2015). Together, all models suggest that the spatial distribution of replication origins determines the temporal organization of replication.

The ability to modify the spatial and temporal initiation profile also allows a cell to accommodate its specific transcription program. A large fraction of the human genome exhibits changes in the order of replication during nuclear reorganization associated with differentiation and development (Pope et al. 2014; Rhind and Gilbert 2013a; Rivera-Mulia and Gilbert 2016). Consistent with the need to activate initiation at distinct times to accommodate changes in transcription, cell type-specific replication origins are often located in regions that exhibit differentiation-specific and tissue-specific gene expression (Gerhardt et al. 2014; Norio et al. 2005). Origins that are activated in a narrow set of distinct cell types tend to initiate replication late in S-phase in regions that contain few and sparse origins, whereas origins that are commonly activated in many cell type initiate replication throughout S-phase (Smith et al. 2016). Since chromatin and histone modifications influence transcription and replication patterns, varying recruitment of modifiers and other proteins by transcription factors can markedly influence the replication program and vice versa (Bar-Ziv et al. 2016). Massive alterations in replication initiation patterns can be programmed, associated with activation of a differentiation cascade leading to changes in gene expression patterns (Gerhardt et al. 2014; Norio et al. 2005). Conversely, since transcription can hinder initiation of DNA replication on the common chromatin template (Martin et al. 2011), regions that exhibit massive differentiation-induced transcription might contain fewer and sparser replication origins due to the paucity of genetic elements that can support initiation. The precise timing of origin activation within replication timing domains is determined anew after each mitotic cell division (Wu and Gilbert 1996), facilitating dynamic and flexible changes in replication order (Rhind and Gilbert 2013a).

The excess of replication origins might also play a regulatory role to facilitate genomic stability by allowing for timely, accurate replication under stress. Activation of stress responses in actively proliferating cells, including changes in the rate of replication fork progression, can signal for changes in the utilization of certain origins. Conversely, alterations in the frequency of initiation can affect the rate of DNA synthesis. In yeast, replication can proceed upon depletion of the most or all replication origins in specific chromosomes, but those cells exhibit elevated chromosome loss rates (Dershowitz et al. 2007). In addition, a lower number of potential origins can increase the abundance of DNase hypersensitive regions, chromosome fragility, and chromosomal rearrangements (Huang and Koshland 2003; Lengronne and Schwob 2002).

Mammalian cells often exhibit increased frequency of replication initiation events (activation of "dormant origins") in response to events that slow the progression of replication forks, including changes in nucleotide pool levels (Anglana et al. 2003), exposure to histone deacetylase inhibitors (Conti et al. 2010) diminished homologous recombination (Daboussi et al. 2008) and dysfunctional DNA modifying enzymes such as topoisomerase I (Tuduri et al. 2009) and Mus81 endonuclease (Fu et al. 2015). Since the enhanced frequency of initiation in those cases associates with a mild decrease in replication fork progression rates, it is unclear whether the overall increase in initiation frequency indicates a global compensatory mechanism linking replication fork rates and origin activity or reflects local changes in a group of loci (e.g., fragile sites) that are particularly prone to potentially genotoxic lesions under conditions of slow replication fork progression.

2.5 Conclusions and Future Questions

Cells duplicate their genomes starting from many origins and proceeding along a well-established program that sequentially replicate the entire genome. Consistent replication origins are evident in most cells, although they are not essential for complete genome duplication. Origin activation dynamics might therefore primarily play a role in establishing local and global chromatin structure and facilitate the cellular response to adverse events that perturb the replication process.

Although the ability to initiate DNA replication can partially be conferred by the primary DNA sequence, most replication origins exhibit flexible initiation, as their activation in a fraction of cells is affected by the chromatin environment and by interactions with distal DNA elements. This flexible initiation program facilitates coordination between replication and transcription and preserves genome stability by maintaining a group of "reserve" potential replication origins that can be activated if replication at adjacent replicons stalls.

Understanding the molecular interactions at replication origins is critical for establishing a complete picture of how cells coordinate chromatin transactions, including transcription, chromatin decondensation and compaction, and DNA synthesis. Despite rapid progress in mapping the locations and timing of replication initiation events, we have yet to identify the corresponding molecular interactions that dictate initiation of DNA replication at particular sites. Future progress in addressing these issues will be achieved by identifying the exact combination of DNA-binding proteins or chromatin modifiers that activate DNA replication. The evident flexibility of the replication program also necessitates studies that characterize cell cycle signaling pathways that repress replication from origins that remain "dormant" during particular cell cycles and modulate replication initiation to coordinate with changes in the transcription program. Finally, future studies characterizing how cell cycle checkpoint pathways affect molecular interactions at replication origins could lead to a better understanding of cellular responses to potentially genotoxic stress.

References

Abbas T, Keaton MA, Dutta A (2013) Genomic instability in cancer. Cold Spring Harb Perspect Biol 5:a012914

Abid A, Costa A (2016) The MCM helicase motor of the eukaryotic replisome. J Mol Biol 428:10

Aladjem MI (2007) Replication in context: dynamic regulation of DNA replication patterns in metazoans. Nat Rev Genet 8:588–600

Aladjem MI, Rodewald LW, Kolman JL, Wahl GM (1998) Genetic dissection of a mammalian replicator in the human beta-globin locus. Science 281:1005–1009

Anglana M, Apiou F, Bensimon A, Debatisse M (2003) Dynamics of DNA replication in mammalian somatic cells: nucleotide pool modulates origin choice and interorigin spacing. Cell 114:385–394

Bartholdy B, Mukhopadhyay R, Lajugie J, Aladjem MI, Bouhassira EE (2015) Allele-specific analysis of DNA replication origins in mammalian cells. Nat Commun 6:7051

Bar-Ziv R, Voichek Y, Barkai N (2016) Chromatin dynamics during DNA replication. Genome Res 26:1245–1256. https://doi.org/10.1101/gr.201244.115

Bell SP (2002) The origin recognition complex: from simple origins to complex functions. Genes Dev 16:659–672

Besnard E, Babled A, Lapasset L, Milhavet O, Parrinello H, Dantec C, Marin JM, Lemaitre JM (2012) Unraveling cell type-specific and reprogrammable human replication origin signatures associated with G-quadruplex consensus motifs. Nat Struct Mol Biol 19:837–844

Blow JJ, Ge XQ, Jackson DA (2011) How dormant origins promote complete genome replication. Trends Biochem Sci 36:405–414

Boos D, Yekezare M, Diffley JF (2013) Identification of a heteromeric complex that promotes DNA replication origin firing in human cells. Science 340:981–984

Bruck I, Perez-Arnaiz P, Colbert MK, Kaplan DL (2015) Insights into the initiation of eukaryotic DNA replication. Nucleus 6:449–454

Bustin M, Misteli T (2016) Nongenetic functions of the genome. Science 352:aad6933

Cayrou C, Coulombe P, Vigneron A, Stanojcic S, Ganier O, Peiffer I, Rivals E, Puy A, Laurent-Chabalier S, Desprat R et al (2011) Genome-scale analysis of metazoan replication origins reveals their organization in specific but flexible sites defined by conserved features. Genome Res 21:1438–1449

Cayrou C, Ballester B, Peiffer I, Fenouil R, Coulombe P, Andrau JC, van Helden J, Mechali M (2015) The chromatin environment shapes DNA replication origin organization and defines origin classes. Genome Res 25:1873–1885

Chakraborty A, Shen Z, Prasanth SG (2011) "ORCanization" on heterochromatin: linking DNA replication initiation to chromatin organization. Epigenetics 6:665–670

Chen X, Liu G, Leffak M (2013) Activation of a human chromosomal replication origin by protein tethering. Nucleic Acids Res 41:6460–6474

Conner AL, Aladjem MI (2012) The chromatin backdrop of DNA replication: lessons from genetics and genome-scale analyses. Biochim Biophys Acta 1819:794–801

Conti C, Leo E, Eichler GS, Sordet O, Martin MM, Fan A, Aladjem MI, Pommier Y (2010) Inhibition of histone deacetylase in cancer cells slows down replication forks, activates dormant origins, and induces DNA damage. Cancer Res 70:4470–4480

Cornacchia D, Dileep V, Quivy JP, Foti R, Tili F, Santarella-Mellwig R, Antony C, Almouzni G, Gilbert DM, Buonomo SB (2012) Mouse Rif1 is a key regulator of the replication-timing programme in mammalian cells. EMBO J 31:3678–3690

Daboussi F, Courbet S, Benhamou S, Kannouche P, Zdzienicka MZ, Debatisse M, Lopez BS (2008) A homologous recombination defect affects replication-fork progression in mammalian cells. J Cell Sci 121:162–166

Das SP, Borrman T, Liu VW, Yang SC, Bechhoefer J, Rhind N (2015) Replication timing is regulated by the number of MCMs loaded at origins. Genome Res 25:1886–1892

Dave A, Cooley C, Garg M, Bianchi A (2014) Protein phosphatase 1 recruitment by Rif1 regulates DNA replication origin firing by counteracting DDK activity. Cell Rep 7:53–61

DePamphilis ML (1999) Replication origins in metazoan chromosomes: fact or fiction? BioEssays 21:5–16

Depamphilis ML, de Renty CM, Ullah Z, Lee CY (2012) "The Octet": eight protein kinases that control mammalian DNA replication. Front Physiol 3:368

Dershowitz A, Snyder M, Sbia M, Skurnick JH, Ong LY, Newlon CS (2007) Linear derivatives of Saccharomyces cerevisiae chromosome III can be maintained in the absence of autonomously replicating sequence elements. Mol Cell Biol 27:4652–4663

Dileep V, Ay F, Sima J, Vera DL, Noble WS, Gilbert DM (2015) Topologically associating domains and their long-range contacts are established during early G1 coincident with the establishment of the replication-timing program. Genome Res 25:1104–1113

Donley N, Smith L, Thayer MJ (2015) ASAR15, A cis-acting locus that controls chromosome-wide replication timing and stability of human chromosome 15. PLoS Genet 11:e1004923

Fanning E, Zhao K (2009) SV40 DNA replication: from the A gene to a nanomachine. Virology 384:352–359

Feng YQ, Desprat R, Fu H, Olivier E, Lin CM, Lobell A, Gowda SN, Aladjem MI, Bouhassira EE (2006) DNA methylation supports intrinsic epigenetic memory in mammalian cells. PLoS Genet 2:e65

Flickinger RA (2015) Possible role of H1 histone in replication timing. Develop Growth Differ 57:1–9

Foti R, Gnan S, Cornacchia D, Dileep V, Bulut-Karslioglu A, Diehl S, Buness A, Klein FA, Huber W, Johnstone E et al (2016) Nuclear architecture organized by Rif1 underpins the replication-timing program. Mol Cell 61:260–273

Foulk MS, Urban JM, Casella C, Gerbi SA (2015) Characterizing and controlling intrinsic biases of lambda exonuclease in nascent strand sequencing reveals phasing between nucleosomes and G-quadruplex motifs around a subset of human replication origins. Genome Res 25:725–735

Fragkos M, Ganier O, Coulombe P, Mechali M (2015) DNA replication origin activation in space and time. Nat Rev Mol Cell Biol 16:360–374

Francis LI, Randell JC, Takara TJ, Uchima L, Bell SP (2009) Incorporation into the prereplicative complex activates the Mcm2-7 helicase for Cdc7-Dbf4 phosphorylation. Genes Dev 23:643–654

Fu H, Wang L, Lin CM, Singhania S, Bouhassira EE, Aladjem MI (2006) Preventing gene silencing with human replicators. Nat Biotechnol 24:572–576

Fu H, Martin MM, Regairaz M, Huang L, You Y, Lin CM, Ryan M, Kim R, Shimura T, Pommier Y et al (2015) The DNA repair endonuclease Mus81 facilitates fast DNA replication in the absence of exogenous damage. Nat Commun 6:6746

Gerhardt J, Tomishima MJ, Zaninovic N, Colak D, Yan Z, Zhan Q, Rosenwaks Z, Jaffrey SR, Schildkraut CL (2014) The DNA replication program is altered at the FMR1 locus in fragile X embryonic stem cells. Mol Cell 53:19–31

Gindin Y, Valenzuela MS, Aladjem MI, Meltzer PS, Bilke S (2014) A chromatin structure-based model accurately predicts DNA replication timing in human cells. Mol Syst Biol 10:722

Giri S, Aggarwal V, Pontis J, Shen Z, Chakraborty A, Khan A, Mizzen C, Prasanth KV, Ait-Si-Ali S, Ha T et al (2015). The preRC protein ORCA organizes heterochromatin by assembling histone H3 lysine 9 methyltransferases on chromatin. Elife 4

Hanlon SL, Li JJ (2015) Re-replication of a centromere induces chromosomal instability and aneuploidy. PLoS Genet 11:e1005039
Hassan-Zadeh V, Chilaka S, Cadoret JC, Ma MK, Boggetto N, West AG, Prioleau MN (2012) USF binding sequences from the HS4 insulator element impose early replication timing on a vertebrate replicator. PLoS Biol 10:e1001277
Hayano M, Kanoh Y, Matsumoto S, Renard-Guillet C, Shirahige K, Masai H (2012) Rif1 is a global regulator of timing of replication origin firing in fission yeast. Genes Dev 26:137–150
Hiraga S, Alvino GM, Chang F, Lian HY, Sridhar A, Kubota T, Brewer BJ, Weinreich M, Raghuraman MK, Donaldson AD (2014) Rif1 controls DNA replication by directing protein phosphatase 1 to reverse Cdc7-mediated phosphorylation of the MCM complex. Genes Dev 28:372–383
Hiratani I, Ryba T, Itoh M, Rathjen J, Kulik M, Papp B, Fussner E, Bazett-Jones DP, Plath K, Dalton S et al (2010) Genome-wide dynamics of replication timing revealed by in vitro models of mouse embryogenesis. Genome Res 20:155–169
Hoggard T, Shor E, Muller CA, Nieduszynski CA, Fox CA (2013) A link between ORC-origin binding mechanisms and origin activation time revealed in budding yeast. PLoS Genet 9:e1003798
Huang D, Koshland D (2003) Chromosome integrity in Saccharomyces cerevisiae: the interplay of DNA replication initiation factors, elongation factors, and origins. Genes Dev 17:1741–1754
Huang Y, Fang J, Bedford MT, Zhang Y, Xu RM (2006) Recognition of histone H3 lysine-4 methylation by the double tudor domain of JMJD2A. Science 312:748–751
Huang L, Fu H, Lin CM, Conner AL, Zhang Y, Aladjem MI (2011) Prevention of transcriptional silencing by a replicator-binding complex consisting of SWI/SNF, MeCP1, and hnRNP C1/C2. Mol Cell Biol 31:3472–3484
Iizuka M, Matsui T, Takisawa H, Smith MM (2006) Regulation of replication licensing by acetyltransferase Hbo1. Mol Cell Biol 26:1098–1108
Jacob F, Brenner J, Cuzin F (1963) On the regulation of DNA replication in bacteria. Cold Spring Harbor Dymp Quant Biol 28:329
Kanemaki M, Labib K (2006) Distinct roles for Sld3 and GINS during establishment and progression of eukaryotic DNA replication forks. EMBO J 25:1753–1763
Kanoh Y, Matsumoto S, Fukatsu R, Kakusho N, Kono N, Renard-Guillet C, Masuda K, Iida K, Nagasawa K, Shirahige K et al (2015) Rif1 binds to G quadruplexes and suppresses replication over long distances. Nat Struct Mol Biol 22:889–897
Kawabata T, Luebben SW, Yamaguchi S, Ilves I, Matise I, Buske T, Botchan MR, Shima N (2011) Stalled fork rescue via dormant replication origins in unchallenged S phase promotes proper chromosome segregation and tumor suppression. Mol Cell 41:543–553
Kim J, Daniel J, Espejo A, Lake A, Krishna M, Xia L, Zhang Y, Bedford MT (2006) Tudor, MBT and chromo domains gauge the degree of lysine methylation. EMBO Rep 7:397–403
Knott SR, Peace JM, Ostrow AZ, Gan Y, Rex AE, Viggiani CJ, Tavare S, Aparicio OM (2012) Forkhead transcription factors establish origin timing and long-range clustering in S. cerevisiae. Cell 148:99–111
Koren A, Handsaker RE, Kamitaki N, Karlic R, Ghosh S, Polak P, Eggan K, McCarroll SA (2014) Genetic variation in human DNA replication timing. Cell 159:1015–1026
Lengronne A, Schwob E (2002) The yeast CDK inhibitor Sic1 prevents genomic instability by promoting replication origin licensing in late G(1). Mol Cell 9:1067–1078
Leonard AC, Mechali M (2013) DNA replication origins. Cold Spring Harb Perspect Biol 5:a010116
Lieberman-Aiden E, van Berkum NL, Williams L, Imakaev M, Ragoczy T, Telling A, Amit I, Lajoie BR, Sabo PJ, Dorschner MO et al (2009) Comprehensive mapping of long-range interactions reveals folding principles of the human genome. Science 326:289–293
Liu G, Malott M, Leffak M (2003) Multiple functional elements comprise a mammalian chromosomal replicator. Mol Cell Biol 23:1832–1842

Lob D, Lengert N, Chagin VO, Reinhart M, Casas-Delucchi CS, Cardoso MC, Drossel B (2016) 3D replicon distributions arise from stochastic initiation and domino-like DNA replication progression. Nat Commun 7:11207
Majocchi S, Aritonovska E, Mermod N (2014) Epigenetic regulatory elements associate with specific histone modifications to prevent silencing of telomeric genes. Nucleic Acids Res 42:193–204
Marahrens Y, Stillman B (1992) A yeast chromosomal origin of DNA replication defined by multiple functional elements. Science 255:817–823
Marks AB, Smith OK, Aladjem MI (2016) Replication origins: determinants or consequences of nuclear organization? Curr Opin Genet Dev 37:67–75
Martin MM, Ryan M, Kim R, Zakas AL, Fu H, Lin CM, Reinhold WC, Davis SR, Bilke S, Liu H et al (2011) Genome-wide depletion of replication initiation events in highly transcribed regions. Genome Res 21:1822–1832
Masai H, Matsumoto S, You Z, Yoshizawa-Sugata N, Oda M (2010) Eukaryotic chromosome DNA replication: where, when, and how? Annu Rev Biochem 79:89–130
Mattarocci S, Shyian M, Lemmens L, Damay P, Altintas DM, Shi T, Bartholomew CR, Thoma NH, Hardy CF, Shore D (2014) Rif1 controls DNA replication timing in yeast through the PP1 phosphatase Glc7. Cell Rep 7:62–69
Mechali M, Kearsey S (1984) Lack of specific sequence requirement for DNA replication in Xenopus eggs compared with high sequence specificity in yeast. Cell 38:55–64
Mechali M, Yoshida K, Coulombe P, Pasero P (2013) Genetic and epigenetic determinants of DNA replication origins, position and activation. Curr Opin Genet Dev 23:124–131
Miotto B, Struhl K (2010) HBO1 histone acetylase activity is essential for DNA replication licensing and inhibited by Geminin. Mol Cell 37:57–66
Miotto B, Ji Z, Struhl K (2016) Selectivity of ORC binding sites and the relation to replication timing, fragile sites, and deletions in cancers. Proc Natl Acad Sci U S A 113:E4810-9
Moindrot B, Audit B, Klous P, Baker A, Thermes C, de Laat W, Bouvet P, Mongelard F, Arneodo A (2012) 3D chromatin conformation correlates with replication timing and is conserved in resting cells. Nucleic Acids Res 40:9470–9481
Mukhopadhyay R, Lajugie J, Fourel N, Selzer A, Schizas M, Bartholdy B, Mar J, Lin CM, Martin MM, Ryan M et al (2014) Allele-specific genome-wide profiling in human primary erythroblasts reveal replication program organization. PLoS Genet 10:e1004319
Musialek MW, Rybaczek D (2015) Behavior of replication origins in Eukaryota – spatio-temporal dynamics of licensing and firing. Cell Cycle 14:2251–2264
Nagano T, Fraser P (2011) No-nonsense functions for long noncoding RNAs. Cell 145:178–181
Norio P, Kosiyatrakul S, Yang Q, Guan Z, Brown NM, Thomas S, Riblet R, Schildkraut CL (2005) Progressive activation of DNA replication initiation in large domains of the immunoglobulin heavy chain locus during B cell development. Mol Cell 20:575–587
Palacios DeBeer MA, Muller U, Fox CA (2003) Differential DNA affinity specifies roles for the origin recognition complex in budding yeast heterochromatin. Genes Dev 17:1817–1822
Peace JM, Villwock SK, Zeytounian JL, Gan Y, Aparicio OM (2016) Quantitative BrdU immunoprecipitation method demonstrates that Fkh1 and Fkh2 are rate-limiting activators of replication origins that reprogram replication timing in G1 phase. Genome Res 26:365–375
Pope BD, Ryba T, Dileep V, Yue F, Wu W, Denas O, Vera DL, Wang Y, Hansen RS, Canfield TK et al (2014) Topologically associating domains are stable units of replication-timing regulation. Nature 515:402–405
Rao SS, Huntley MH, Durand NC, Stamenova EK, Bochkov ID, Robinson JT, Sanborn AL, Machol I, Omer AD, Lander ES et al (2014) A 3D map of the human genome at kilobase resolution reveals principles of chromatin looping. Cell 159:1665–1680
Remus D, Diffley JF (2009) Eukaryotic DNA replication control: lock and load, then fire. Curr Opin Cell Biol 21:771–777
Rhind N, Gilbert DM (2013a) DNA replication timing. Cold Spring Harb Perspect Med 3:1–26
Rhind N, Gilbert DM (2013b) DNA replication timing. Cold Spring Harb Perspect Biol 5:a010132

Richardson CD, Li JJ (2014) Regulatory mechanisms that prevent re-initiation of DNA replication can be locally modulated at origins by nearby sequence elements. PLoS Genet 10:e1004358
Rivera-Mulia JC, Gilbert DM (2016) Replication timing and transcriptional control: beyond cause and effect-part III. Curr Opin Cell Biol 40:168–178
Rivera-Mulia JC, Buckley Q, Sasaki T, Zimmerman J, Didier RA, Nazor K, Loring JF, Lian Z, Weissman S, Robins AJ et al (2015) Dynamic changes in replication timing and gene expression during lineage specification of human pluripotent stem cells. Genome Res 25:1091–1103
Saksouk N, Avvakumov N, Champagne KS, Hung T, Doyon Y, Cayrou C, Paquet E, Ullah M, Landry AJ, Côté V et al (2009) HBO1 HAT complexes target chromatin throughout gene coding regions via multiple PHD finger interactions with histone H3 tail. Mol Cell 33:257–265
Sansam CG, Goins D, Siefert JC, Clowdus EA, Sansam CL (2015) Cyclin-dependent kinase regulates the length of S phase through TICRR/TRESLIN phosphorylation. Genes Dev 29:555–566
Schwaiger M, Kohler H, Oakeley EJ, Stadler MB, Schubeler D (2010) Heterochromatin protein 1 (HP1) modulates replication timing of the Drosophila genome. Genome Res 20:771–780
Sequeira-Mendes J, Diaz-Uriarte R, Apedaile A, Huntley D, Brockdorff N, Gomez M (2009) Transcription initiation activity sets replication origin efficiency in mammalian cells. PLoS Genet 5:e1000446
Sherstyuk VV, Shevchenko AI, Zakian SM (2014) Epigenetic landscape for initiation of DNA replication. Chromosoma 123:183–199
Sheu YJ, Kinney JB, Stillman B (2016) Concerted activities of Mcm4, Sld3, and Dbf4 in control of origin activation and DNA replication fork progression. Genome Res 26:315–330
Smith OK, Aladjem MI (2014) Chromatin structure and replication origins: determinants of chromosome replication and nuclear organization. J Mol Biol 426:3330–3341
Smith OK, Kim R, Fu H, Martin MM, Lin CM, Utani K, Zhang Y, Marks AB, Lalande M, Chamberlain S et al (2016) Distinct epigenetic features of differentiation-regulated replication origins. Epigenetics Chromatin 9:18
Takayama Y, Kamimura Y, Okawa M, Muramatsu S, Sugino A, Araki H (2003) GINS, a novel multiprotein complex required for chromosomal DNA replication in budding yeast. Genes Dev 17:1153–1165
Tanaka S, Araki H (2013) Helicase activation and establishment of replication forks at chromosomal origins of replication. Cold Spring Harb Perspect Biol 5:a010371
Tardat M, Brustel J, Kirsh O, Lefevbre C, Callanan M, Sardet C, Julien E (2010) The histone H4 Lys 20 methyltransferase PR-Set7 regulates replication origins in mammalian cells. Nat Cell Biol 12:1086–1093
Tazumi A, Fukuura M, Nakato R, Kishimoto A, Takenaka T, Ogawa S, Song JH, Takahashi TS, Nakagawa T, Shirahige K et al (2012) Telomere-binding protein Taz1 controls global replication timing through its localization near late replication origins in fission yeast. Genes Dev 26:2050–2062
Tuduri S, Crabbe L, Conti C, Tourriere H, Holtgreve-Grez H, Jauch A, Pantesco V, De Vos J, Thomas A, Theillet C et al (2009) Topoisomerase I suppresses genomic instability by preventing interference between replication and transcription. Nat Cell Biol 11:1315–1324
Urban JM, Foulk MS, Casella C, Gerbi SA (2015) The hunt for origins of DNA replication in multicellular eukaryotes. F1000prime Rep 7:30
Valenzuela MS, Chen Y, Davis S, Yang F, Walker RL, Bilke S, Lueders J, Martin MM, Aladjem MI, Massion PP et al (2011) Preferential localization of human origins of DNA replication at the 5′-ends of expressed genes and at evolutionarily conserved DNA sequences. Plos One 6:e17308
Vogelauer M, Rubbi L, Lucas I, Brewer BJ, Grunstein M (2002) Histone acetylation regulates the time of replication origin firing. Mol Cell 10:1223–1233
Wu JR, Gilbert DM (1996) A distinct G1 step required to specify the Chinese hamster DHFR replication origin. Science 271:1270–1272

Yaffe E, Farkash-Amar S, Polten A, Yakhini Z, Tanay A, Simon I (2010) Comparative analysis of DNA replication timing reveals conserved large-scale chromosomal architecture. PLoS Genet 6:e1001011

Yamazaki S, Ishii A, Kanoh Y, Oda M, Nishito Y, Masai H (2012) Rif1 regulates the replication timing domains on the human genome. EMBO J 31:3667–3677

Zegerman P, Diffley JF (2007) Phosphorylation of Sld2 and Sld3 by cyclin-dependent kinases promotes DNA replication in budding yeast. Nature 445:281–285

Zhang Y, Huang L, Fu H, Smith OK, Lin CM, Utani K, Rao M, Reinhold WC, Redon CE, Ryan M et al (2016) A replicator-specific binding protein essential for site-specific initiation of DNA replication in mammalian cells. Nat Commun 7:11748

Chapter 3
Molecular Mechanism for Chromatin Regulation During MCM Loading in Mammalian Cells

Nozomi Sugimoto and Masatoshi Fujita

Abstract DNA replication is a fundamental process required for the accurate and timely duplication of chromosomes. During late mitosis to G1 phase, the MCM2-7 complex is loaded onto chromatin in a manner dependent on ORC, CDC6, and Cdt1, and chromatin becomes licensed for replication. Although every eukaryotic organism shares common features in replication control, there are also some differences among species. For example, in higher eukaryotic cells including human cells, no strict sequence specificity has been observed for replication origins, unlike budding yeast or bacterial replication origins. Therefore, elements other than beyond DNA sequences are important for regulating replication. For example, the stability and precise positioning of nucleosomes affects replication control. However, little is known about how nucleosome structure is regulated when replication licensing occurs. During the last decade, histone acetylation enzyme HBO1, chromatin remodeler SNF2H, and histone chaperone GRWD1 have been identified as chromatin-handling factors involved in the promotion of replication licensing. In this review, we discuss how the rearrangement of nucleosome formation by these factors affects replication licensing.

Keywords Replication • Pre-RC • Cdt1 • MCM • HBO1 • SNF2H • GRWD1 • Histone • Nucleosome

3.1 Introduction

One of the fundamental events in the cell cycle is complete and precise duplication of the genome. Although major features of replication are well conserved among all organisms, there are also some differences. In all eukaryotes, DNA replication is thought to begin at origins of replication. In *S. cerevisiae*, autonomously replicating

N. Sugimoto (✉) • M. Fujita (✉)
Department of Cellular Biochemistry, Graduate School of Pharmaceutical Sciences, Kyushu University, Fukuoka, Japan
e-mail: sugimoto@phar.kyushu-u.ac.jp; mfujita@phar.kyushu-u.ac.jp

H. Masai, M. Foiani (eds.), *DNA Replication*, Advances in Experimental Medicine and Biology 1042, https://doi.org/10.1007/978-981-10-6955-0_3

sequences (ARSs) have been identified as origins (Stinchcomb et al. 1979; Rao et al. 1994; Wyrick et al. 2001). In *S. pombe*, although no consensus sequence has been identified, AT-rich regions serve as potential origins (Segurado et al. 2003; Yompakdee and Huberman 2004; Dai et al. 2005; Hayashi et al. 2007). In mammalian cells, DNA replication origins do not have such consensus sequences (Vashee et al. 2003; Schaarschmidt et al. 2004; Martin et al. 2011; Valenzuela et al. 2011; Besnard et al. 2012; Petryk et al. 2016). Although metazoan replication origins have been shown to possess some characteristic features, genome-wide studies have shown that most efficient origins in mammalian cells are strongly associated with human CpG island (CGI) promoters (Cadoret et al. 2008). The association with CGI has also been confirmed in mouse cells (Sequeira-Mendes et al. 2009; Cayrou et al. 2011). A conserved G-rich motif named OGRE (origin G-rich repeated elements) is present in most of mouse and *Drosophila* origins, and OGRE motifs can form a G-quadruplex (G4) (Cayrou et al. 2011, 2012). A genome-wide association between origins and G4 motifs has also been found in human cells (Besnard et al. 2012). In fact, some G4 motifs could promote replication initiation (Valton et al. 2014), and human ORC binds preferentially to G4 motifs on single-stranded DNA (Hoshina et al. 2013). Therefore, G4 structures could mediate ORC recruitment to initiation sites.

To initiate DNA replication, it is essential to form prereplicative complexes (pre-RCs). In late M to early G1 phase, pre-RCs are formed on chromatin (Fig. 3.1a). The pre-RC assembly, known as "licensing," involves the binding of ORC to DNA. In budding yeast, ORC interacts with ARSs in a sequence-specific manner (Bell and Stillman 1992). In fission yeast, Orc4, a subunit of the ORC complex, has unique sequence motifs called "AT hooks" and binds to replication origin DNA via these motifs (Chuang and Kelly 1999). In metazoan, including human cells, ORC binds to DNA without sequence specificity and tends to interact with open chromatin regions containing active chromatin marks such as histone H3 acetylation and H3K4 methylation (Vashee et al. 2003; Schaarschmidt et al. 2004; Dellino et al. 2013; Miotto et al. 2016).

ORC bound to origins, in conjunction with CDC6 and Cdt1, loads MCM double hexamers onto DNA. Once MCM complexes are loaded, the origin becomes "licensed" and is ready to be activated. As cells enter S phase, the licensing functions of ORC, CDC6, and Cdt1 are suppressed to inhibit MCM reloading (Fig. 3.1b). In addition, to activate MCM replicative helicase, two kinases, CDK and CDC7, phosphorylate fork components and facilitate CDC45 and GINS recruitment (Araki 2011; Heller et al. 2011; Yeeles et al. 2015; Bleichert et al. 2017).

Generally, eukaryotic DNA is packaged into nucleosomes. The nucleosome contains 146 bp of DNA wrapped around an octameric complex of H2A, H2B, H3, and H4 (Horn and Peterson 2002). The N-terminal tails of the histones protrude from the nucleosome. These histone tails are important for higher-order chromatin folding. Because nucleosomes prevent DNA-related processes, they must first be disassembled or moved transiently to allow DNA replication, DNA repair, and transcription machineries to access the DNA. At the replication fork during S phase, two histone chaperones, CAF1 and ASF1, aid histone eviction and redeposition (MacAlpine and Almouzni 2013). The FACT histone chaperone binds to several replisome components

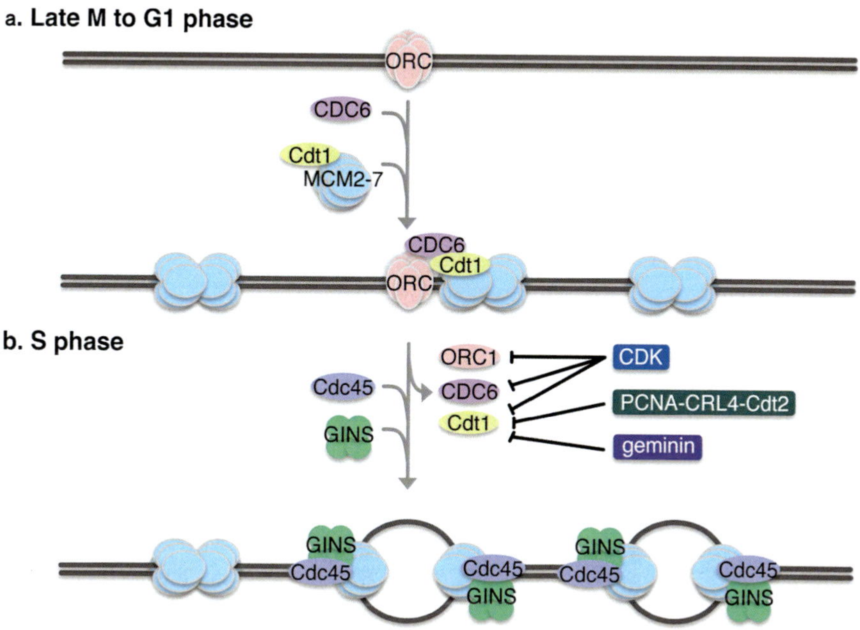

Fig. 3.1 A model for regulation of pre-RC formation and its firing in human cells. (**a**) During late M to G1 phase, the ORC, CDC6, Cdt1, and MCM2-7 proteins are sequentially assembled on replication origins (including dormant origins). The MCM complexes are loaded in excess onto chromatin and may be distributed to locations distant from the ORC binding sites. These ORC-distal MCMs may also function as origins (including dormant ones). (**b**) When cells enter S phase, CDK and CDC7 kinases are activated and facilitate loading of CDC45, GINS, and other proteins to activate the helicase activity of MCM. This triggers the unwinding of DNA and the subsequent loading of DNA polymerases and other factors onto chromatin to start DNA replication. To prevent relicensing (i.e., reloading of MCM) during S phase, it is important to downregulate the function of MCM loaders. A major pathway for this purpose is through regulation of Cdt1. Cdt1 activity is regulated by ubiquitin-mediated degradation and its specific inhibitor, geminin. Part of the mechanisms for inhibition of relicensing is regulation of CDC6 and ORC1. CDC6 is exported from the nucleus in a CDK phosphorylation-dependent manner. ORC1 is phosphorylated by CDK and is subsequently degraded by ubiquitin-mediated proteolysis

and travels with the replication fork (Wittmeyer and Formosa 1997; Gambus et al. 2006; Tan et al. 2006; VanDemark et al. 2006). Conditional knockout of SSRP1, a small subunit of FACT, reduces fork speed in chicken DT40 cells (Abe et al. 2011). In in vitro replication with purified yeast replication proteins and fully chromatinized templates, replisome progression requires the FACT (Kurat et al. 2017). In addition, several studies have shown that MCM2 interacts with histones H3-H4 and histone chaperone ASF1 and acts as a histone chaperone to assist disassembly and assembly of nucleosomes during fork progression (Ishimi et al. 2001; Groth et al. 2007; Jasencakova et al. 2010; Huang et al. 2015; Richet et al. 2015). However, little is known about nucleosomal regulation during pre-RC formation, especially in human cells. This review will focus on recent findings addressing this issue.

3.2 Factors That Promote Pre-RC Formation

Broadly, three families of functionally distinct protein complexes have been implicated in transient nucleosome eviction, disassembly, and/or movement: (1) ATP-dependent chromatin remodeling proteins, (2) histone acetyltransferases (HAT), and (3) histone chaperones (Aalfs and Kingston 2000; Hammond et al. 2017). "Chromatin openness" is important not only for transcription but also for replication licensing, origin firing, and other nuclear transactions. "Open chromatin regions" may represent transiently generated nucleosome-free or nucleosome-low DNA regions (Fig. 3.2). In general, when activating transcription, transcription factors promote chromatin openness by recruiting these chromatin-handling factors.

In yeast, ARS-binding factor 1 (ABF1), a transcription factor, is implicated in formation of nucleosome-free regions at its binding sites and activates transcription of many genes (Springer et al. 1997; Badis et al. 2008; Hartley and Madhani 2009). ABF1 binds to several ARSs and alters local chromatin structure (Venditti et al. 1994; Hu et al. 1999). In an in vitro system, ABF1-dependent nucleosomal arrangement of *ARS1* is required for efficient ORC loading (Lipford and Bell 2001). Furthermore, ABF1 binds to the promoters of many ribosomal genes and recruits Esa1, an essential histone acetyltransferase (Reid et al. 2000).

In higher eukaryotes, numerous studies have shown that transcription factors contribute to replication licensing. In *Drosophila*, Myb, a sequence-specific transcription factor, is required for site-specific replication at ACE3 and Ori-β (Beall et al. 2002). In human, mouse, and *Xenopus*, c-myc interacts with pre-RC components and regulates origin activity (Dominguez-Sola et al. 2007). Binding of *Drosophila* ORC to origin DNA is facilitated by E2F1-Rb (Bosco et al. 2001). These studies indicate that transcription factors bind to specific DNA sequences, thereby efficiently recruiting ORC via physical interactions. However, they could also promote MCM loading by recruitment of chromatin-handling factors. As another example, ORCA recruits ORC to chromatin at human heterochromatic regions (Shen et al. 2010).

3.3 Cdt1 Strongly Stimulates Pre-RC Formation and Its Activity Is Tightly Regulated by Multiple Mechanisms in Human Cells

Cdt1 is essential for loading of the MCM2-7 complex onto chromatin and physically interacts with several subunits of the MCM complex (Tanaka and Diffley 2002; Randell et al. 2006; You and Masai 2008). In human cells, Cdt1 strongly promotes MCM loading (Vaziri et al. 2003; Sugimoto et al. 2009), and accordingly, its activity is tightly regulated by multiple mechanisms (Fig. 3.1b). It was originally shown that geminin directly binds to Cdt1 and inhibits its activity (McGarry and Kirschner 1998; Wohlschlegel et al. 2000; Tada et al. 2001). Geminin is a substrate

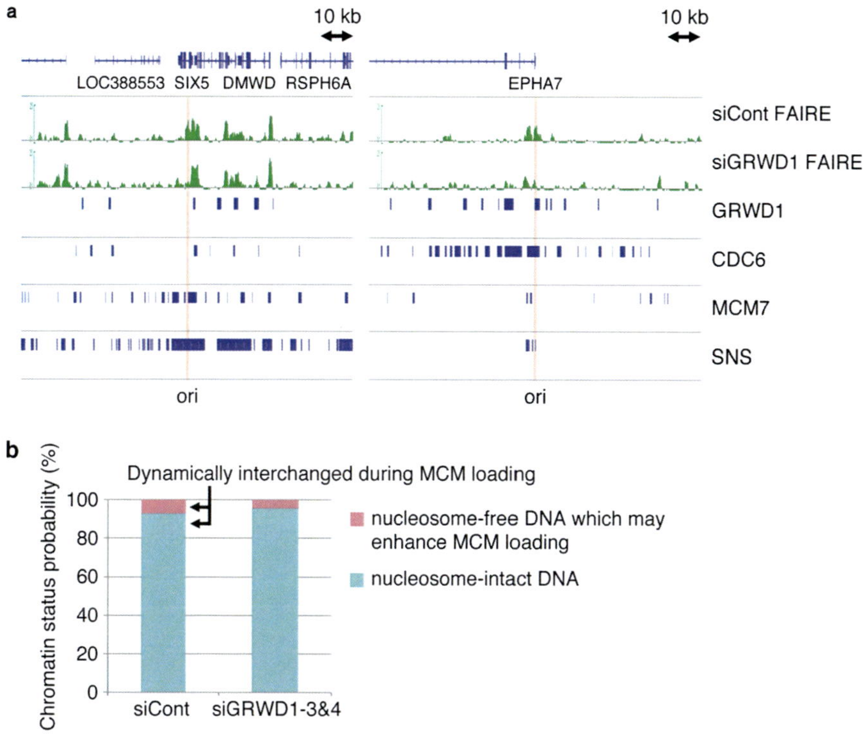

Fig. 3.2 "Open chromatin region" may have transient nucleosome-free or nucleosome-low DNA at a low frequency. (**a**) Open chromatin regions are associated with DNA replication origins. Here, chromatin openness is estimated by FAIRE (formaldehyde-assisted isolation of regulatory elements), a method of isolating genomic regions with no or depleted nucleosomes (Sugimoto et al. 2015). ChIP-seq data for CDC6 and MCM7 are also from Sugimoto et al. (2015). Short nascent strand (SNS) data are from Besnard et al. (2012). *SIX5* and *EPHA7* origins (Liu et al. 2012) are shown as examples of origins having GRWD1-dependent open chromatin structure. (**b**) In quantitative FAIRE-qPCR assays at several replication origins, the percentages of FAIRE signals are ~10% or less (Sugimoto et al. 2015). Therefore, theoretically, only a small fraction of DNA may be in a "nucleosome-free" form even in such "open chromatin" regions. Such nucleosome-free DNA may appear transiently and dynamically, which may generate spaces for efficient MCM loading. At a subset of replication origins, maintenance of "chromatin openness" is dependent on GRWD1 (Sugimoto et al. 2015). Given that most DNA is organized into nucleosomes even in such "open chromatin" regions, it would be difficult to directly detect changes in histone levels at replication origins

of the APC/C ubiquitin ligase and is destabilized during the G1 phase and accumulates during the S, G2, and M phases (McGarry and Kirschner 1998; Wohlschlegel et al. 2000; Tada et al. 2001; Vodermaier 2004). The protein level of Cdt1 is also strictly controlled during the cell cycle, being high in G1 phase, low in S phase, and high again at the M-to-G1 transition (Nishitani et al. 2001). In human cells, two E3 ubiquitin ligases, SCF^{Skp2} and CUL4-$DDB1^{Cdt2}$, redundantly mediate proteolysis of Cdt1 (Fujita 2006; Arias and Walter 2007; Coleman et al. 2015). Human Cdt1 interacts with the SCF^{Skp2} ubiquitin ligase when phosphorylated by cyclin A-dependent kinases and is then degraded (Li et al. 2003; Liu et al. 2004; Sugimoto et al. 2004; Takeda et al. 2005). By contrast, the binding of CUL4-DDB1^{Cdt2} to Cdt1 is dependent on prior interaction with chromatin-bound PCNA (Arias and Walter 2006; Jin et al. 2006; Nishitani et al. 2006; Ishii et al. 2010). These tight regulations of Cdt1 ensure that MCM is recruited only in G1 phase and that replication occurs only once during the cell cycle.

3.4 Cdt1 Interacts with Three Chromatin-Handling Factors to Promote Pre-RC Formation

How does Cdt1 promote MCM loading so strongly in human cells? One explanation is that Cdt1 might interact with other factors that enhance MCM recruitment. In human cells, Cdt1 interacts with histone acetyltransferase HBO1, ATP-dependent chromatin remodeler SNF2H, and novel histone chaperone GRWD1 (Miotto and Struhl 2008; Sugimoto et al. 2008, 2011, 2015; Miotto and Struhl 2010; Aizawa et al. 2016). As mentioned above, the regulation of local nucleosomal structure during replication licensing may be important. Thus, Cdt1 may recruit these chromatin-handling factors to facilitate MCM loading by altering chromatin accessibility. In support of this notion, LacI-Cdt1 tethered to LacO induces large-scale chromatin decondensation that may be required for MCM recruitment in G1 phase (Wong et al. 2010).

3.4.1 Histone Acetylation and HBO1 (Also Known as MYST2 or KAT7)

Highly conserved lysine residues present on all four core histones serve as the targets of acetylation. When histone tails are acetylated, inter-nucleosomal interactions are reduced, resulting in unfolding of the chromatin fiber (Tse et al. 1998; Annunziato and Hansen 2000; Shogren-Knaak et al. 2006; Robinson et al. 2008).

The acetylation of histone lysine residues has been implicated in origin activation during early development in *Xenopus* and at the chorion gene loci in *Drosophila* follicle cells (Aggarwal and Calvi 2004; Danis et al. 2004; Hartl et al. 2007). In *S. cerevisiae*, histone deacetylases Sir2 and Rpd3 control replication timing by regulating silencing at the rDNA array (Yoshida et al. 2014). In *Drosophila*, Rpd3 suppresses origin activity, whereas histone acetyltransferase Chameau promotes origin activation (Aggarwal and Calvi 2004). HBO1 is the mammalian homolog of Chameau and was originally identified through its physical interactions with human ORC1 (Iizuka and Stillman 1999). Subsequent work has shown that HBO1 interacts with Cdt1 and acetylates histone H4 tails at origin regions during G1, and this is required for efficient MCM loading (Miotto and Struhl 2008, 2010). In addition, HBO1 stimulates Cdt1-dependent re-replication (Miotto and Struhl 2008) and interacts with Kaposi's sarcoma-associated herpes virus terminal repeats and promotes replication of the viral genome (Stedman et al. 2004). It was recently reported that the HBO1-BRPF3 complex regulates origin activation through H3K14 acetylation and CDC45 loading (Feng et al. 2016). Therefore, these findings suggest that HBO1 has two distinct functions in promoting replication; during G1 phase, it promotes MCM2-7 loading, whereas during S phase, it promotes CDC45 loading. However, whether HBO1 is essential for replication remains to be clarified (see below).

3.4.2 SNF2H, an ATP-Dependent Chromatin Remodeler

Chromatin remodeling complexes that use energy derived from ATP hydrolysis alter chromatin structure by sliding, evicting, and/or modifying nucleosomes. This large group of complexes can be subdivided into four subfamilies that include the SWI/SNF-type complex, the ISWI-type complex, the INO80-type complex, and the CHD-type complex (Varga-Weisz 2001; Tsukiyama 2002; Corona and Tamkun 2004; Dirscherl and Krebs 2004; Eberharter and Becker 2004). In yeast, the function of the SWI/SNF remodeling complex is not limited to transcriptional regulation. For example, the SWI/SNF complex is required for efficient maintenance of a minichromosome containing ARS121, a yeast replication origin (Flanagan and Peterson 1999). In addition, ISW2 and INO80 facilitate replication fork progression in the presence of replication stress (Vincent et al. 2008; Au et al. 2011).

In human cells, SNF2H, an ISWI-type factor, has been implicated in regulation of DNA replication. The dyad symmetry (DS) region of EB viral origin of plasmid replication (*oriP*) is flanked by nucleosomes that undergo chromatin remodeling by SNF2H (Zhou et al. 2005). Furthermore, SNF2H depletion by siRNAs reduces MCM3 loading and replication at *oriP* (Zhou et al. 2005). It has been also reported that SNF2H is enriched at two genomic replication origins in G1 phase, and silencing of SNF2H suppresses MCM7 and MCM3 loading at the origins (Sugimoto et al. 2011). These studies provide evidence that chromatin remodeling may be required to move nucleosomes around the replication origin to unmask the pre-RC formation site. If chromatin remodeling complexes are required to promote pre-RC formation at the replication origins, a mechanism must exist to ensure they are recruited there. One mechanism may be through interaction with pre-RC components. In this regard, SNF2H and WSTF are identified as Cdt1-binding proteins and are recruited to origins through interaction with Cdt1 (Sugimoto et al. 2008, 2011). Another potential mechanism is the direct binding of chromatin remodeling complexes to origins, which could be mediated either through DNA binding or by recognition of replication licensing-coupled histone code(s) such as acetylation (see below). The involvement of other classes of ATP-dependent chromatin remodeling complexes in licensing remains to be clarified.

3.4.3 *Novel Histone Chaperone GRWD1*

Histone chaperones interact with histones and play crucial roles in mediating nucleosome assembly and disassembly. In *S. cerevisiae*, the histone chaperone FACT is ubiquitinated by E3 ubiquitin ligase Rtt101 and promotes MCM loading (Han et al. 2010). However, the mechanism of FACT recruitment to origins remains unclear. In human cells, Cdt1-binding protein GRWD1 is a novel histone chaperone that controls chromatin openness at pre-RC sites (Sugimoto et al. 2008, 2015; Aizawa et al. 2016) (Fig. 3.2). GRWD1 is recruited to replication origins in a CDC6- and Cdt1-dependent manner in G1 phase and promotes MCM loading (Sugimoto et al. 2015) (Fig. 3.2). Although GRWD1 is highly conserved among species (Gratenstein et al. 2005), it is not clear whether GRWD1 homologs in other species have a similar function. However, several findings link the budding yeast homolog of GRWD1, Rrb1, to replication licensing. Rrb1 is essential for growth, is involved in early ribosome assembly, and genetically interacts with Orc6 (Iouk et al. 2001; Schaper et al. 2001; Killian et al. 2004). In addition, Rrb1 interacts with Yph1, which functions cooperatively with ORC and MCM (Du and Stillman 2002). Taken together, these observations suggest that GRWD1 plays an important role in pre-RC formation at DNA replication origins.

GRWD1 has an acidic domain and promotes chromatin openness at replication origins and efficient MCM loading (Sugimoto et al. 2015). Recently, we used an in vitro reconstituted system to further show that GRWD1 can evict H2A-H2B dimers from nucleosomes (Aizawa et al. 2016). Interestingly, the acidic domain of GRWD1 is important for these activities (Aizawa et al. 2016), consistent with previous reports that acidic domains are found in numerous histone chaperones and play roles in histone chaperone activity (De Koning et al. 2007). In vitro studies have also shown that many acidic transcriptional activators can stimulate DNA replication by competing with the repressive effects of nucleosomes (Cheng and Kelly 1989; Cheng et al. 1992; Li and Botchan 1994). The acidic domain of GRWD1 is conserved among its homologs, including those in mouse, *Drosophila*, *C. elegans*, and *S. cerevisiae* (Gratenstein et al. 2005). Thus, this domain could be functionally important in other species.

3.5 Are HBO1, SNF2H, and GRWD1 Essential for Minimal Licensing to Support Normal Cell Growth?

Although H3K14 acetylation is reduced in HBO1 knockout mouse cells, both DNA replication and cell proliferation proceeded normally (Kueh et al. 2011). Also in flies, proliferation occurs with reduced levels of HBO1 (Grienenberger et al. 2002). Yeasts do not have obvious structural homologs of HBO1 (Avvakumov and Cote 2007; Lafon et al. 2007). Regarding SNF2H, DNA replication appears not to be affected by immunodepletion of SNF2H-WSTF complexes in *Xenopus* egg extract system (MacCallum et al. 2002). In addition, in human cells, the siRNA-mediated silencing of SNF2H, HBO1, or GRWD1 does not impede re-replication induced by overexpression of the degradation-resistant Cdt1 mutant Cy+D1m (Sugimoto et al. 2008, 2009, 2011; our unpublished data). How can these findings be reconciled with the observations suggesting that these chromatin-handling factors play important roles in promotion of MCM loading? It is possible that SNF2H-, HBO1-, and GRWD1-mediated chromatin regulation plays an important role in efficient MCM loading in human cells, but is not essential for minimal origin firing. Alternatively, SNF2H-, HBO1-, and GRWD1-mediated enhancement of licensing may be species or cell type specific. Since HBO1, SNF2H, and GRWD1 are overexpressed in cancer cell lines (Iizuka et al. 2009; Jin et al. 2015; Sugimoto et al. 2015), regulations of DNA replication by these factors could be important in cancer cell growth.

It is important to consider the fact that excess MCM loading is crucial for maintaining genome stability. Even if the number of MCM2-7 molecules loaded is reduced, normal replication rates are maintained (Edwards et al. 2002; Cortez et al. 2004; Tsao et al. 2004; Ge et al. 2007; Ibarra et al. 2008). However, depletion of MCM causes hypersensitivity to replicative stress and a defect in Rad17-dependent ATR-mediated checkpoint activation (Tsao et al. 2004; Woodward et al. 2006; Ge et al. 2007). A hypomorphic mutation in MCM4 termed *Chaos3* (chromosome aberrations occurring spontaneously 3) causes severe genomic instability, and Chaos3 females are highly prone to mammary adenocarcinoma (Shima et al. 2007). Although mutant mice expressing low levels of MCM2 grow normally, they develop T- and B-cell lymphomas (Pruitt et al. 2007). A reduction of MCM levels causes DNA damage involving ATR and ATM activation (Orr et al. 2010). These studies demonstrate that excess MCM loading is critical for toleration of replication stress and activation of the checkpoint. Therefore, the recruitment of HBO1, SNF2H, and GRWD1 by Cdt1 and subsequent enhancement of MCM loading may be important for the maintenance of genome stability. On the other hand, Das et al. suggest that the number of loaded MCMs at origins may regulate replication timing in budding yeast (Das et al. 2015).

3.6 Relationship Between HBO1, SNF2H, and GRWD1 in Nucleosome Regulation and Its Involvement in the Promotion of MCM Loading

We now favor a model in which HBO1, SNF2H, and GRWD1 function cooperatively in the promotion of pre-RC formation (Fig. 3.3). However, each factor may act individually on specific origins or under specific conditions. SNF2H-WSTF complexes bind to acetylated histones in chromatin (Hakimi et al. 2002). Therefore, it is possible that HBO1 is first recruited to replication origins via interaction with Cdt1 and acetylates histone H4, and then SNF2H is recruited to the origin through interaction with both Cdt1 and acetylated histone H4, where these factors together with GRWD1 alter chromatin plasticity (Fig. 3.3). In contrast to this model, SNF2H, MCM, and HDAC1/2 (histone deacetylase) co-localize at EB virus *OriP* (Zhou et al. 2005). Also, in an in vitro system, SNF2H preferentially interacts with

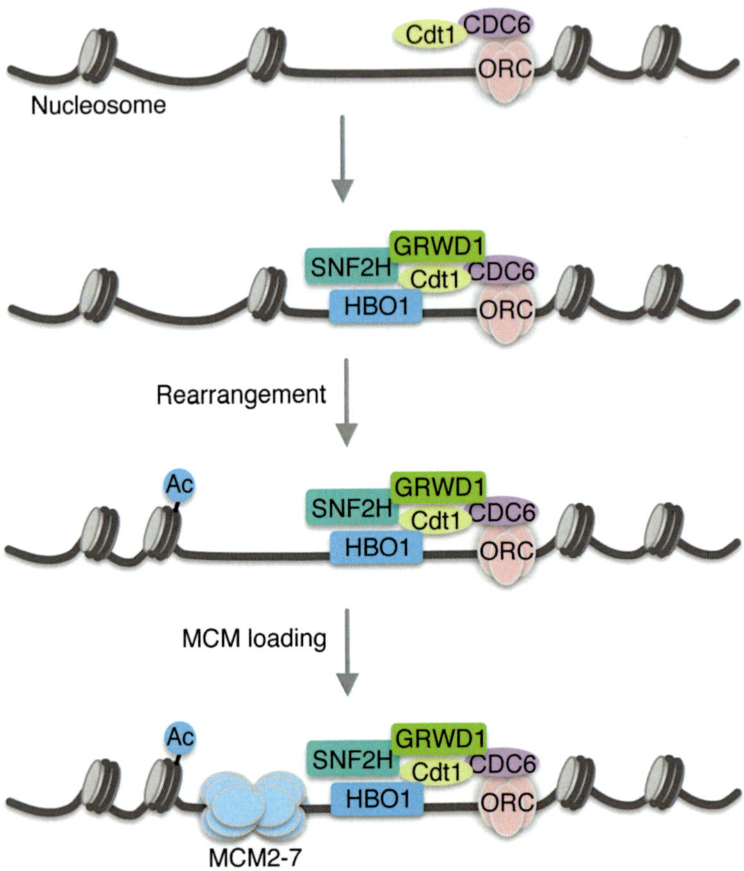

Fig. 3.3 A model for enhancement of MCM loading by the three chromatin-handling factors in the context of chromatin. Since nucleosome structure interferes with MCM loading, it should be rearranged during this process. Histone acetyltransferase HBO1, chromatin remodeler SNF2H, and histone chaperone GRWD1 are involved in this reaction. These factors are recruited to origins in a Cdt1-dependent manner. It is currently unknown whether they are recruited interdependently

unacetylated histone H4 tails (Alenghat et al. 2006). The reason(s) for this discrepancy is unknown. Nucleosome eviction activity of SWI/SNF is enhanced by transcription factors (Gutierrez et al. 2007). Thus, other cellular cofactors could also be required to effectively load MCM complexes.

There is a strong association between promotion of replication licensing and transcriptional activation. For example, HBO1 also acts as a transcriptional coactivator for hormone receptors and AP-1 proteins (Georgiakaki et al. 2006; Miotto and Struhl 2006; Miotto et al. 2006), and many studies have shown that SNF2H is involved in transcriptional regulation (Varga-Weisz 2001; Tsukiyama 2002; Corona and Tamkun 2004; Dirscherl and Krebs 2004; Eberharter and Becker 2004). Our recent results suggest that GRWD1 functions not only in replication licensing but also in transcription and other chromosome transactions (Sugimoto et al. 2015; our unpublished data). Furthermore, chromatin immunoprecipitation coupled with high-throughput sequencing (ChIP-seq) in mammalian cells shows that HBO1, SNF2H, and GRWD1 are highly enriched near the transcription start sites (Morris et al. 2014; Sugimoto et al. 2015; Feng et al. 2016). These results indicate that the Cdt1-interacting chromatin-handling proteins also act in transcriptional control. The dual roles of HBO1, SNF2H, and GRWD1 as coactivators for transcriptional regulation and for DNA licensing suggest the possibility that these factors might integrate internal and external stimuli to coordinate transcriptional responses with initiation of DNA replication.

3.7 Conclusion and Perspectives

Although increasing evidence suggests that HBO1, SNF2H, and GRWD1 promote MCM loading by regulating nucleosome structure in human cells, the detailed molecular mechanisms remain unclear, mainly because of the lack of useful cell-free in vitro reconstitution systems. With budding yeast proteins, pre-RC formation has been reconstituted on naked DNA with ARSs (Seki and Diffley 2000; Bowers et al. 2004; Kawasaki et al. 2006) or on the fully chromatinized templates (Kurat et al. 2017). In human cells, there is no sequence specificity for assembly of pre-RC, and numerous factors may provide either backup or fine-tuning mechanisms for the regulation of pre-RC assembly. Therefore, it may take some more time and effort if in vitro pre-RC reconstitution will be established using human proteins. Nevertheless, it should be examined whether the Cdt1-binding chromatin-handling factors have synergistic capabilities to efficiently remodel nucleosomes and/or evict histones in in vitro reconstitution assays. In the future, it will be very tempting to combine these factors with pre-RC reconstitution on nucleosome templates. Under these conditions, the histone-handling activity of MCM2 might also have a crucial role.

Finally, whereas it seems clear that "open chromatin" structure facilitates DNA replication, it remains unclear whether chromatin openness mainly promotes efficient MCM "loading" or efficient MCM "activation" by enhancing recruitment of CDC45 and GINS. Some suggest that efficient pre-RC formation leads to efficient DNA replication initiation (Das et al. 2015). However, it has also been suggested that "open chromatin" more strongly stimulates the latter process (Feng et al. 2016; our unpublished data). It will be interesting to address this important issue in human cells.

Acknowledgment We thank all members of our research group for the helpful discussion and critical reading of the manuscript. This work was supported in part by Grants to Sugimoto and Fujita from the Ministry of Education, Culture, Sports, Science and Technology of Japan (17080013, 21370084, 25291027, 26114713, 09J07233).

Conflicts of Interests The authors declare no conflicts of interest in relation to this article.

References

Aalfs JD, Kingston RE (2000) What does 'chromatin remodeling' mean? Trends Biochem Sci 25:548–555

Abe T, Sugimura K, Hosono Y, Takami Y, Akita M, Yoshimura A, Tada S, Nakayama T, Murofushi H, Okumura K, Takeda S, Horikoshi M, Seki M, Enomoto T (2011) The histone chaperone facilitates chromatin transcription (FACT) protein maintains normal replication fork rates. J Biol Chem 286:30504–30512

Aggarwal BD, Calvi BR (2004) Chromatin regulates origin activity in Drosophila follicle cells. Nature 430:372–376

Aizawa M, Sugimoto N, Watanabe S, Yoshida K, Fujita M (2016) Nucleosome assembly and disassembly activity of GRWD1, a novel Cdt1-binding protein that promotes pre-replication complex formation. Biochim Biophys Acta 1863:2739–2748

Alenghat T, Yu J, Lazar MA (2006) The N-CoR complex enables chromatin remodeler SNF2H to enhance repression by thyroid hormone receptor. EMBO J 25:3966–3974

Annunziato AT, Hansen JC (2000) Role of histone acetylation in the assembly and modulation of chromatin structures. Gene Expr 9:37–61

Araki H (2011) Initiation of chromosomal DNA replication in eukaryotic cells; contribution of yeast genetics to the elucidation. Genes Genet Syst 86:141–149

Arias EE, Walter JC (2006) PCNA functions as a molecular platform to trigger Cdt1 destruction and prevent re-replication. Nat Cell Biol 8:84–90

Arias EE, Walter JC (2007) Strength in numbers: preventing rereplication via multiple mechanisms in eukaryotic cells. Genes Dev 21:497–518

Au TJ, Rodriguez J, Vincent JA, Tsukiyama T (2011) ATP-dependent chromatin remodeling factors tune S phase checkpoint activity. Mol Cell Biol 31:4454–4463

Avvakumov N, Cote J (2007) The MYST family of histone acetyltransferases and their intimate links to cancer. Oncogene 26:5395–5407

Badis G, Chan ET, van Bakel H, Pena-Castillo L, Tillo D, Tsui K, Carlson CD, Gossett AJ, Hasinoff MJ, Warren CL, Gebbia M, Talukder S, Yang A, Mnaimneh S, Terterov D, Coburn D, Li Yeo A, Yeo ZX, Clarke ND, Lieb JD, Ansari AZ, Nislow C, Hughes TR (2008) A library of yeast transcription factor motifs reveals a widespread function for Rsc3 in targeting nucleosome exclusion at promoters. Mol Cell 32:878–887

Beall EL, Manak JR, Zhou S, Bell M, Lipsick JS, Botchan MR (2002) Role for a Drosophila Myb-containing protein complex in site-specific DNA replication. Nature 420:833–837

Bell SP, Stillman B (1992) ATP-dependent recognition of eukaryotic origins of DNA replication by a multiprotein complex. Nature 357:128–134

Besnard E, Babled A, Lapasset L, Milhavet O, Parrinello H, Dantec C, Marin JM, Lemaitre JM (2012) Unraveling cell type-specific and reprogrammable human replication origin signatures associated with G-quadruplex consensus motifs. Nat Struct Mol Biol 19:837–844

Bleichert F, Botchan MR, Berger JM (2017) Mechanisms for initiating cellular DNA replication. Science 355:eaah6317

Bosco G, Du W, Orr-Weaver TL (2001) DNA replication control through interaction of E2F-RB and the origin recognition complex. Nat Cell Biol 3:289–295

Bowers JL, Randell JC, Chen S, Bell SP (2004) ATP hydrolysis by ORC catalyzes reiterative Mcm2-7 assembly at a defined origin of replication. Mol Cell 16:967–978

Cadoret JC, Meisch F, Hassan-Zadeh V, Luyten I, Guillet C, Duret L, Quesneville H, Prioleau MN (2008) Genome-wide studies highlight indirect links between human replication origins and gene regulation. Proc Natl Acad Sci U S A 105:15837–15842

Cayrou C, Coulombe P, Vigneron A, Stanojcic S, Ganier O, Peiffer I, Rivals E, Puy A, Laurent-Chabalier S, Desprat R, Mechali M (2011) Genome-scale analysis of metazoan replication origins reveals their organization in specific but flexible sites defined by conserved features. GenomeRes 21:1438–1449

Cayrou C, Coulombe P, Puy A, Rialle S, Kaplan N, Segal E, Mechali M (2012) New insights into replication origin characteristics in metazoans. Cell Cycle 11:658–667
Cheng L, Kelly TJ (1989) Transcriptional activator nuclear factor I stimulates the replication of SV40 minichromosomes in vivo and in vitro. Cell 59:541–551
Cheng LZ, Workman JL, Kingston RE, Kelly TJ (1992) Regulation of DNA replication in vitro by the transcriptional activation domain of GAL4-VP16. Proc Natl Acad SciUSA 89:589–593
Chuang RY, Kelly TJ (1999) The fission yeast homologue of Orc4p binds to replication origin DNA via multiple AT-hooks. Proc Natl Acad Sci U S A 96:2656–2661
Coleman KE, Grant GD, Haggerty RA, Brantley K, Shibata E, Workman BD, Dutta A, Varma D, Purvis JE, Cook JG (2015) Sequential replication-coupled destruction at G1/S ensures genome stability. GenesDev 29:1734–1746
Corona DF, Tamkun JW (2004) Multiple roles for ISWI in transcription, chromosome organization and DNA replication. Biochim Biophys Acta 1677:113–119
Cortez D, Glick G, Elledge SJ (2004) Minichromosome maintenance proteins are direct targets of the ATM and ATR checkpoint kinases. Proc Natl Acad Sci USA 101:10078–10083
Dai J, Chuang RY, Kelly TJ (2005) DNA replication origins in the Schizosaccharomyces pombe genome. Proc Natl Acad Sci U S A 102:337–342
Danis E, Brodolin K, Menut S, Maiorano D, Girard-Reydet C, Mechali M (2004) Specification of a DNA replication origin by a transcription complex. Nat Cell Biol 6:721–730
Das SP, Borrman T, Liu VW, Yang SC, Bechhoefer J, Rhind N (2015) Replication timing is regulated by the number of MCMs loaded at origins. Genome Res 25:1886–1892
De Koning L, Corpet A, Haber JE, Almouzni G (2007) Histone chaperones: an escort network regulating histone traffic. Nat Struct Mol Biol 14:997–1007
Dellino GI, Cittaro D, Piccioni R, Luzi L, Banfi S, Segalla S, Cesaroni M, Mendoza-Maldonado R, Giacca M, Pelicci PG (2013) Genome-wide mapping of human DNA-replication origins: levels of transcription at ORC1 sites regulate origin selection and replication timing. Genome Res 23:1–11
Dirscherl SS, Krebs JE (2004) Functional diversity of ISWI complexes. Biochem Cell Biol 82:482–489
Dominguez-Sola D, Ying CY, Grandori C, Ruggiero L, Chen B, Li M, Galloway DA, Gu W, Gautier J, Dalla-Favera R (2007) Non-transcriptional control of DNA replication by c-Myc. Nature 448:445–451
Du YC, Stillman B (2002) Yph1p, an ORC-interacting protein: potential links between cell proliferation control, DNA replication, and ribosome biogenesis. Cell 109:835–848
Eberharter A, Becker PB (2004) ATP-dependent nucleosome remodelling: factors and functions. J Cell Sci 117:3707–3711
Edwards MC, Tutter AV, Cvetic C, Gilbert CH, Prokhorova TA, Walter JC (2002) MCM2-7 complexes bind chromatin in a distributed pattern surrounding the origin recognition complex in Xenopus egg extracts. J Biol Chem 277:33049–33057
Feng Y, Vlassis A, Roques C, Lalonde ME, Gonzalez-Aguilera C, Lambert JP, Lee SB, Zhao X, Alabert C, Johansen JV, Paquet E, Yang XJ, Gingras AC, Cote J, Groth A (2016) BRPF3-HBO1 regulates replication origin activation and histone H3K14 acetylation. EMBO J 35:176–192
Flanagan JF, Peterson CL (1999) A role for the yeast SWI/SNF complex in DNA replication. Nucleic Acids Res 27:2022–2028
Fujita M (2006) Cdt1 revisited: complex and tight regulation during the cell cycle and consequences of deregulation in mammalian cells. Cell Div 1:22
Gambus A, Jones RC, Sanchez-Diaz A, Kanemaki M, van Deursen F, Edmondson RD, Labib K (2006) GINS maintains association of Cdc45 with MCM in replisome progression complexes at eukaryotic DNA replication forks. Nat Cell Biol 8:358–366
Ge XQ, Jackson DA, Blow JJ (2007) Dormant origins licensed by excess Mcm2-7 are required for human cells to survive replicative stress. GenesDev 21:3331–3341

Georgiakaki M, Chabbert-Buffet N, Dasen B, Meduri G, Wenk S, Rajhi L, Amazit L, Chauchereau A, Burger CW, Blok LJ, Milgrom E, Lombes M, Guiochon-Mantel A, Loosfelt H (2006) Ligand-controlled interaction of histone acetyltransferase binding to ORC-1 (HBO1) with the N-terminal transactivating domain of progesterone receptor induces steroid receptor coactivator 1-dependent coactivation of transcription. Mol Endocrinol 20:2122–2140

Gratenstein K, Heggestad AD, Fortun J, Notterpek L, Pestov DG, Fletcher BS (2005) The WD-repeat protein GRWD1: potential roles in myeloid differentiation and ribosome biogenesis. Genomics 85:762–773

Grienenberger A, Miotto B, Sagnier T, Cavalli G, Schramke V, Geli V, Mariol MC, Berenger H, Graba Y, Pradel J (2002) The MYST domain acetyltransferase Chameau functions in epigenetic mechanisms of transcriptional repression. Curr Biol 12:762–766

Groth A, Corpet A, Cook AJ, Roche D, Bartek J, Lukas J, Almouzni G (2007) Regulation of replication fork progression through histone supply and demand. Science 318:1928–1931

Gutierrez JL, Chandy M, Carrozza MJ, Workman JL (2007) Activation domains drive nucleosome eviction by SWI/SNF. EMBO J 26:730–740

Hakimi MA, Bochar DA, Schmiesing JA, Dong Y, Barak OG, Speicher DW, Yokomori K, Shiekhattar R (2002) A chromatin remodelling complex that loads cohesin onto human chromosomes. Nature 418:994–998

Hammond CM, Strømme CB, Huang H, Patel DJ, Groth A (2017) Histone chaperone networks shaping chromatin function. Nat Rev Mol Cell Biol 18:141–158

Han J, Li Q, McCullough L, Kettelkamp C, Formosa T, Zhang Z (2010) Ubiquitylation of FACT by the cullin-E3 ligase Rtt101 connects FACT to DNA replication. Genes Dev 24:1485–1490

Hartl T, Boswell C, Orr-Weaver TL, Bosco G (2007) Developmentally regulated histone modifications in Drosophila follicle cells: initiation of gene amplification is associated with histone H3 and H4 hyperacetylation and H1 phosphorylation. Chromosoma 116:197–214

Hartley PD, Madhani HD (2009) Mechanisms that specify promoter nucleosome location and identity. Cell 137:445–458

Hayashi M, Katou Y, Itoh T, Tazumi A, Yamada Y, Takahashi T, Nakagawa T, Shirahige K, Masukata H (2007) Genome-wide localization of pre-RC sites and identification of replication origins in fission yeast. EMBO J 26:1327–1339

Heller RC, Kang S, Lam WM, Chen S, Chan CS, Bell SP (2011) Eukaryotic origin-dependent DNA replication in vitro reveals sequential action of DDK and S-CDK kinases. Cell 146:80–91

Horn PJ, Peterson CL (2002) Molecular biology. Chromatin higher order folding – wrapping up transcription. Science 297:1824–1827

Hoshina S, Yura K, Teranishi H, Kiyasu N, Tominaga A, Kadoma H, Nakatsuka A, Kunichika T, Obuse C, Waga S (2013) Human origin recognition complex binds preferentially to G-quadruplex-preferable RNA and single-stranded DNA. J Biol Chem 288:30161–30171

Hu YF, Hao ZL, Li R (1999) Chromatin remodeling and activation of chromosomal DNA replication by an acidic transcriptional activation domain from BRCA1. GenesDev 13:637–642

Huang H, Stromme CB, Saredi G, Hodl M, Strandsby A, Gonzalez-Aguilera C, Chen S, Groth A, Patel DJ (2015) A unique binding mode enables MCM2 to chaperone histones H3-H4 at replication forks. Nat Struct Mol Biol 22:618–626

Ibarra A, Schwob E, Mendez J (2008) Excess MCM proteins protect human cells from replicative stress by licensing backup origins of replication. Proc Natl Acad SciA 105:8956–8961

Iizuka M, Stillman B (1999) Histone acetyltransferase HBO1 interacts with the ORC1 subunit of the human initiator protein. J Biol Chem 274:23027–23034

Iizuka M, Takahashi Y, Mizzen CA, Cook RG, Fujita M, Allis CD, Frierson HF Jr, Fukusato T, Smith MM (2009) Histone acetyltransferase Hbo1: catalytic activity, cellular abundance, and links to primary cancers. Gene 436:108–114

Iouk TL, Aitchison JD, Maguire S, Wozniak RW (2001) Rrb1p, a yeast nuclear WD-repeat protein involved in the regulation of ribosome biosynthesis. Mol Cell Biol 21:1260–1271

Ishii T, Shiomi Y, Takami T, Murakami Y, Ohnishi N, Nishitani H (2010) Proliferating cell nuclear antigen-dependent rapid recruitment of Cdt1 and CRL4Cdt2 at DNA-damaged sites after UV irradiation in HeLa cells. J Biol Chem 285:41993–42000

Ishimi Y, Komamura-Kohno Y, Arai K, Masai H (2001) Biochemical activities associated with mouse Mcm2 protein. J Biol Chem 276:42744–42752
Jasencakova Z, Scharf AN, Ask K, Corpet A, Imhof A, Almouzni G, Groth A (2010) Replication stress interferes with histone recycling and predeposition marking of new histones. Mol Cell 37:736–743
Jin J, Arias EE, Chen J, Harper JW, Walter JC (2006) A family of diverse Cul4-Ddb1-interacting proteins includes Cdt2, which is required for S phase destruction of the replication factor Cdt1. Mol Cell 23:709–721
Jin Q, Mao X, Li B, Guan S, Yao F, Jin F (2015) Overexpression of SMARCA5 correlates with cell proliferation and migration in breast cancer. Tumour Biol 36:1895–1902
Kawasaki Y, Kim HD, Kojima A, Seki T, Sugino A (2006) Reconstitution of Saccharomyces cerevisiae prereplicative complex assembly in vitro. Genes Cells 11:745–756
Killian A, Le Meur N, Sesboue R, Bourguignon J, Bougeard G, Gautherot J, Bastard C, Frebourg T, Flaman JM (2004) Inactivation of the RRB1-Pescadillo pathway involved in ribosome biogenesis induces chromosomal instability. Oncogene 23:8597–8602
Kueh AJ, Dixon MP, Voss AK, Thomas T (2011) HBO1 is required for H3K14 acetylation and normal transcriptional activity during embryonic development. Mol Cell Biol 31:845–860
Kurat CF, Yeeles JT, Patel H, Early A, Diffley JF (2017) Chromatin controls DNA replication origin selection, lagging-strand synthesis, and replication fork rates. Mol Cell 65:117–130
Lafon A, Chang CS, Scott EM, Jacobson SJ, Pillus L (2007) MYST opportunities for growth control: yeast genes illuminate human cancer gene functions. Oncogene 26:5373–5384
Li R, Botchan MR (1994) Acidic transcription factors alleviate nucleosome-mediated repression of DNA replication of bovine papillomavirus type 1. Proc Natl Acad Sci U S A 91:7051–7055
Li X, Zhao Q, Liao R, Sun P, Wu X (2003) The SCF(Skp2) ubiquitin ligase complex interacts with the human replication licensing factor Cdt1 and regulates Cdt1 degradation. J Biol Chem 278:30854–30858
Lipford JR, Bell SP (2001) Nucleosomes positioned by ORC facilitate the initiation of DNA replication. Mol Cell 7:21–30
Liu E, Li X, Yan F, Zhao Q, Wu X (2004) Cyclin-dependent kinases phosphorylate human Cdt1 and induce its degradation. J Biol Chem 279:17283–17288
Liu G, Chen X, Gao Y, Lewis T, Barthelemy J, Leffak M (2012) Altered replication in human cells promotes DMPK (CTG) (n) · (CAG)(n) repeat instability. Mol Cell Biol 32:1618–1632
MacAlpine DM, Almouzni G (2013) Chromatin and DNA replication. Cold Spring Harb Perspect Biol 5:a010207
MacCallum DE, Losada A, Kobayashi R, Hirano T (2002) ISWI remodeling complexes in Xenopus egg extracts: identification as major chromosomal components that are regulated by INCENP-aurora B. Mol Biol Cell 13:25–39
Martin MM, Ryan M, Kim R, Zakas AL, Fu H, Lin CM, Reinhold WC, Davis SR, Bilke S, Liu H, Doroshow JH, Reimers MA, Valenzuela MS, Pommier Y, Meltzer PS, Aladjem MI (2011) Genome-wide depletion of replication initiation events in highly transcribed regions. Genome Res 21:1822–1832
McGarry TJ, Kirschner MW (1998) Geminin, an inhibitor of DNA replication, is degraded during mitosis. Cell 93:1043–1053
Miotto B, Struhl K (2006) Differential gene regulation by selective association of transcriptional coactivators and bZIP DNA-binding domains. Mol Cell Biol 26:5969–5982
Miotto B, Struhl K (2008) HBO1 histone acetylase is a coactivator of the replication licensing factor Cdt1. GenesDev 22:2633–2638
Miotto B, Struhl K (2010) HBO1 histone acetylase activity is essential for DNA replication licensing and inhibited by Geminin. Mol Cell 37:57–66
Miotto B, Sagnier T, Berenger H, Bohmann D, Pradel J, Graba Y (2006) Chameau HAT and DRpd3 HDAC function as antagonistic cofactors of JNK/AP-1-dependent transcription during Drosophila metamorphosis. Genes Dev 20:101–112
Miotto B, Ji Z, Struhl K (2016) Selectivity of ORC binding sites and the relation to replication timing, fragile sites, and deletions in cancers. Proc Natl Acad Sci U S A 113:E4810–E4819

Morris SA, Baek S, Sung MH, John S, Wiench M, Johnson TA, Schiltz RL, Hager GL (2014) Overlapping chromatin-remodeling systems collaborate genome wide at dynamic chromatin transitions. Nat Struct Mol Biol 21:73–81
Nishitani H, Taraviras S, Lygerou Z, Nishimoto T (2001) The human licensing factor for DNA replication Cdt1 accumulates in G1 and is destabilized after initiation of S-phase. J Biol Chem 276:44905–44911
Nishitani H, Sugimoto N, Roukos V, Nakanishi Y, Saijo M, Obuse C, Tsurimoto T, Nakayama KI, Nakayama K, Fujita M, Lygerou Z, Nishimoto T (2006) Two E3 ubiquitin ligases, SCF-Skp2 and DDB1-Cul4, target human Cdt1 for proteolysis. EMBO J 25:1126–1136
Orr SJ, Gaymes T, Ladon D, Chronis C, Czepulkowski B, Wang R, Mufti GJ, Marcotte EM, Thomas NS (2010) Reducing MCM levels in human primary T cells during the G(0)-->G(1) transition causes genomic instability during the first cell cycle. Oncogene 29:3803–3814
Petryk N, Kahli M, d'Aubenton-Carafa Y, Jaszczyszyn Y, Shen Y, Silvain M, Thermes C, Chen CL, Hyrien O (2016) Replication landscape of the human genome. NatCommun 7:10208
Pruitt SC, Bailey KJ, Freeland A (2007) Reduced Mcm2 expression results in severe stem/progenitor cell deficiency and cancer. Stem Cells 25:3121–3132
Randell JC, Bowers JL, Rodriguez HK, Bell SP (2006) Sequential ATP hydrolysis by Cdc6 and ORC directs loading of the Mcm2-7 helicase. Mol Cell 21:29–39
Rao H, Marahrens Y, Stillman B (1994) Functional conservation of multiple elements in yeast chromosomal replicators. Mol Cell Biol 14:7643–7651
Reid JL, Iyer VR, Brown PO, Struhl K (2000) Coordinate regulation of yeast ribosomal protein genes is associated with targeted recruitment of Esa1 histone acetylase. Mol Cell 6:1297–1307
Richet N, Liu D, Legrand P, Velours C, Corpet A, Gaubert A, Bakail M, Moal-Raisin G, Guerois R, Compper C, Besle A, Guichard B, Almouzni G, Ochsenbein F (2015) Structural insight into how the human helicase subunit MCM2 may act as a histone chaperone together with ASF1 at the replication fork. Nucleic Acids Res 43:1905–1917
Robinson PJ, An W, Routh A, Martino F, Chapman L, Roeder RG, Rhodes D (2008) 30 nm chromatin fibre decompaction requires both H4-K16 acetylation and linker histone eviction. J Mol Biol 381:816–825
Schaarschmidt D, Baltin J, Stehle IM, Lipps HJ, Knippers R (2004) An episomal mammalian replicon: sequence-independent binding of the origin recognition complex. EMBO J 23:191–201
Schaper S, Fromont-Racine M, Linder P, de la Cruz J, Namane A, Yaniv M (2001) A yeast homolog of chromatin assembly factor 1 is involved in early ribosome assembly. Curr Biol 11:1885–1890
Segurado M, de Luis A, Antequera F (2003) Genome-wide distribution of DNA replication origins at A+T-rich islands in Schizosaccharomyces pombe. EMBO Rep 4:1048–1053
Seki T, Diffley JF (2000) Stepwise assembly of initiation proteins at budding yeast replication origins in vitro. Proc Natl Acad Sci U S A 97:14115–14120
Sequeira-Mendes J, Diaz-Uriarte R, Apedaile A, Huntley D, Brockdorff N, Gomez M (2009) Transcription initiation activity sets replication origin efficiency in mammalian cells. PLoS Genet 5:e1000446
Shen Z, Sathyan KM, Geng Y, Zheng R, Chakraborty A, Freeman B, Wang F, Prasanth KV, Prasanth SG (2010) A WD-repeat protein stabilizes ORC binding to chromatin. Mol Cell 40:99–111
Shima N, Alcaraz A, Liachko I, Buske TR, Andrews CA, Munroe RJ, Hartford SA, Tye BK, Schimenti JC (2007) A viable allele of Mcm4 causes chromosome instability and mammary adenocarcinomas in mice. Nat Genet 39:93–98
Shogren-Knaak M, Ishii H, Sun JM, Pazin MJ, Davie JR, Peterson CL (2006) Histone H4-K16 acetylation controls chromatin structure and protein interactions. Science 311:844–847
Springer C, Krappmann S, Kunzler M, Zmasek C, Braus GH (1997) Regulation of the yeast HIS7 gene by the global transcription factor Abf1p. Mol Gen Genet 256:136–146
Stedman W, Deng Z, Lu F, Lieberman PM (2004) ORC, MCM, and histone hyperacetylation at the Kaposi's sarcoma-associated herpesvirus latent replication origin. J Virol 78:12566–12575
Stinchcomb DT, Struhl K, Davis RW (1979) Isolation and characterisation of a yeast chromosomal replicator. Nature 282:39–43

Sugimoto N, Tatsumi Y, Tsurumi T, Matsukage A, Kiyono T, Nishitani H, Fujita M (2004) Cdt1 phosphorylation by cyclin A-dependent kinases negatively regulates its function without affecting geminin binding. J Biol Chem 279:19691–19697

Sugimoto N, Kitabayashi I, Osano S, Tatsumi Y, Yugawa T, Narisawa-Saito M, Matsukage A, Kiyono T, Fujita M (2008) Identification of novel human Cdt1-binding proteins by a proteomics approach: proteolytic regulation by APC/CCdh1. Mol Biol Cell 19:1007–1021

Sugimoto N, Yoshida K, Tatsumi Y, Yugawa T, Narisawa-Saito M, Waga S, Kiyono T, Fujita M (2009) Redundant and differential regulation of multiple licensing factors ensures prevention of re-replication in normal human cells. J Cell Sci 122:1184–1191

Sugimoto N, Yugawa T, Iizuka M, Kiyono T, Fujita M (2011) Chromatin remodeler sucrose nonfermenting 2 homolog (SNF2H) is recruited onto DNA replication origins through interaction with Cdc10 protein-dependent transcript 1 (Cdt1) and promotes pre-replication complex formation. J Biol Chem 286:39200–39210

Sugimoto N, Maehara K, Yoshida K, Yasukouchi S, Osano S, Watanabe S, Aizawa M, Yugawa T, Kiyono T, Kurumizaka H, Ohkawa Y, Fujita M (2015) Cdt1-binding protein GRWD1 is a novel histone-binding protein that facilitates MCM loading through its influence on chromatin architecture. Nucleic Acids Res 43:5898–5911

Tada S, Li A, Maiorano D, Mechali M, Blow JJ (2001) Repression of origin assembly in metaphase depends on inhibition of RLF-B/Cdt1 by geminin. Nat Cell Biol 3:107–113

Takeda DY, Parvin JD, Dutta A (2005) Degradation of Cdt1 during S phase is Skp2-independent and is required for efficient progression of mammalian cells through S phase. J Biol Chem 280:23416–23423

Tan BC, Chien CT, Hirose S, Lee SC (2006) Functional cooperation between FACT and MCM helicase facilitates initiation of chromatin DNA replication. EMBO J 25:3975–3985

Tanaka S, Diffley JF (2002) Interdependent nuclear accumulation of budding yeast Cdt1 and Mcm2-7 during G1 phase. Nat Cell Biol 4:198–207

Tsao CC, Geisen C, Abraham RT (2004) Interaction between human MCM7 and Rad17 proteins is required for replication checkpoint signaling. EMBO J 23:4660–4669

Tse C, Fletcher TM, Hansen JC (1998) Enhanced transcription factor access to arrays of histone H3/H4 tetramer. DNA complexes in vitro: implications for replication and transcription. Proc Natl Acad Sci U S A 95:12169–12173

Tsukiyama T (2002) The in vivo functions of ATP-dependent chromatin-remodelling factors. Nat Rev Mol CellBiol 3:422–429

Valenzuela MS, Chen Y, Davis S, Yang F, Walker RL, Bilke S, Lueders J, Martin MM, Aladjem MI, Massion PP, Meltzer PS (2011) Preferential localization of human origins of DNA replication at the 5′-ends of expressed genes and at evolutionarily conserved DNA sequences. PLoS One 6:e17308

Valton AL, Hassan-Zadeh V, Lema I, Boggetto N, Alberti P, Saintome C, Riou JF, Prioleau MN (2014) G4 motifs affect origin positioning and efficiency in two vertebrate replicators. EMBOJ 33:732–746

VanDemark AP, Blanksma M, Ferris E, Heroux A, Hill CP, Formosa T (2006) The structure of the yFACT Pob3-M domain, its interaction with the DNA replication factor RPA, and a potential role in nucleosome deposition. Mol Cell 22:363–374

Varga-Weisz P (2001) ATP-dependent chromatin remodeling factors: nucleosome shufflers with many missions. Oncogene 20:3076–3085

Vashee S, Cvetic C, Lu W, Simancek P, Kelly TJ, Walter JC (2003) Sequence-independent DNA binding and replication initiation by the human origin recognition complex. Genes Dev 17:1894–1908

Vaziri C, Saxena S, Jeon Y, Lee C, Murata K, Machida Y, Wagle N, Hwang DS, Dutta A (2003) A p53-dependent checkpoint pathway prevents rereplication. Mol Cell 11:997–1008

Venditti P, Costanzo G, Negri R, Camilloni G (1994) ABFI contributes to the chromatin organization of Saccharomyces cerevisiae ARS1 B-domain. Biochim Biophys Acta 1219:677–689

Vincent JA, Kwong TJ, Tsukiyama T (2008) ATP-dependent chromatin remodeling shapes the DNA replication landscape. Nat Struct Mol Biol 15:477–484

Vodermaier HC (2004) APC/C and SCF: controlling each other and the cell cycle. Curr Biol 14:R787–R796

Wittmeyer J, Formosa T (1997) The Saccharomyces cerevisiae DNA polymerase alpha catalytic subunit interacts with Cdc68/Spt16 and with Pob3, a protein similar to an HMG1-like protein. Mol Cell Biol 17:4178–4190

Wohlschlegel JA, Dwyer BT, Dhar SK, Cvetic C, Walter JC, Dutta A (2000) Inhibition of eukaryotic DNA replication by geminin binding to Cdt1. Science 290:2309–2312

Wong PG, Glozak MA, Cao TV, Vaziri C, Seto E, Alexandrow M (2010) Chromatin unfolding by Cdt1 regulates MCM loading via opposing functions of HBO1 and HDAC11-geminin. Cell Cycle 9:4351–4363

Woodward AM, Gohler T, Luciani MG, Oehlmann M, Ge X, Gartner A, Jackson DA, Blow JJ (2006) Excess Mcm2-7 license dormant origins of replication that can be used under conditions of replicative stress. J Cell Biol 173:673–683

Wyrick JJ, Aparicio JG, Chen T, Barnett JD, Jennings EG, Young RA, Bell SP, Aparicio OM (2001) Genome-wide distribution of ORC and MCM proteins in S. cerevisiae: high-resolution mapping of replication origins. Science 294:2357–2360

Yeeles JT, Deegan TD, Janska A, Early A, Diffley JF (2015) Regulated eukaryotic DNA replication origin firing with purified proteins. Nature 519:431–435

Yompakdee C, Huberman JA (2004) Enforcement of late replication origin firing by clusters of short G-rich DNA sequences. J Biol Chem 279:42337–42344

Yoshida K, Bacal J, Desmarais D, Padioleau I, Tsaponina O, Chabes A, Pantesco V, Dubois E, Parrinello H, Skrzypczak M, Ginalski K, Lengronne A, Pasero P (2014) The histone deacetylases sir2 and rpd3 act on ribosomal DNA to control the replication program in budding yeast. MolCell 54:691–697

You Z, Masai H (2008) Cdt1 forms a complex with the minichromosome maintenance protein (MCM) and activates its helicase activity. J Biol Chem 283:24469–24477

Zhou J, Chau CM, Deng Z, Shiekhattar R, Spindler MP, Schepers A, Lieberman PM (2005) Cell cycle regulation of chromatin at an origin of DNA replication. EMBO J 24:1406–1417

Chapter 4
Initiation of DNA Replication at the Chromosomal Origin of *E. coli*, *oriC*

Tsutomu Katayama

Abstract The *Escherichia coli* chromosomal origin consists of a duplex-unwinding region and a region bearing a DNA-bending protein, IHF-binding site, and clusters of binding sites for the initiator protein DnaA. ATP-DnaA molecules form highly organized oligomers in a process stimulated by DiaA, a DnaA-binding protein. The resultant ATP-DnaA complexes promote local unwinding of *oriC* with the aid of IHF, for which specific interaction of DnaA with the single-stranded DNA is crucial. DnaA complexes also interact with DnaB helicases bound to DnaC loaders, promoting loading of DnaB onto the unwound DNA strands for bidirectional replication. Initiation of replication is strictly regulated during the cell cycle by multiple regulatory systems for *oriC* and DnaA. The activity of *oriC* is regulated by its methylation state, whereas that of DnaA depends on the form of the bound nucleotide. ATP-DnaA can be yielded from initiation-inactive ADP-DnaA in a timely manner depending on specific chromosomal DNA elements termed *DARS* (DnaA-reactivating sequences). After initiation, DnaA-bound ATP is hydrolyzed by two systems, yielding ADP-DnaA. In this review, these and other mechanisms of initiation and its regulation in *E. coli* are described.

Keywords *oriC* • DnaA • IHF • DiaA • Hda • *DARS* • *datA* • Methylation • AAA+ • In vitro reconstitution

4.1 Introduction

Escherichia coli is a model bacterium for many aspects of molecular biology. Replication initiation of the *E. coli* chromosome occurs in a complex consisting of the initiator protein DnaA, the DnaA-binding protein DiaA, the DNA-bending protein IHF, and the chromosomal origin *oriC* (Costa et al. 2013; Kaguni 2011;

T. Katayama (✉)
Department of Molecular Biology, Graduate School of Pharmaceutical Science, Kyushu University, Fukuoka, Japan
e-mail: katayama@phar.kyushu-u.ac.jp

H. Masai, M. Foiani (eds.), *DNA Replication*, Advances in Experimental Medicine and Biology 1042, https://doi.org/10.1007/978-981-10-6955-0_4

Katayama et al. 2010; Leonard and Grimwade 2015; Wolański et al. 2015). DnaA tightly binds ATP or ADP, and ATP-DnaA rather than ADP-DnaA efficiently assembles specific oligomers on *oriC*, forming an initiation complex that is competent for localized DNA unwinding. DiaA is thought to be dissociated from DnaA to enable DnaB helicase-DnaC loader complexes to bind to the initiation complex (Katayama et al. 2010). For bidirectional replication, a pair of DnaB helicases is loaded onto the single-stranded DNA region via dynamic interactions with DnaC and DnaA. The loaded DnaB helicases form mobile complexes with DnaG primases, and the DnaB-DnaG complexes proceed on the single-stranded DNA in a 5–3′ direction, with duplex DNA unwinding and synthesis of primer RNAs for loading of DNA polymerase III holoenzymes (O'Donnell et al. 2013).

Replication initiation at *oriC* is highly regulated so that it occurs only once per *oriC* at a specific time in the cell cycle. This strict regulation is sustained by positive and negative regulatory systems for DnaA and *oriC* (Katayama et al. 2010; Wolański et al. 2015). In this review, we give an overall picture of the mechanisms of initiation as well as the multiple coordinated systems for regulating initiation in *E. coli*.

4.2 Basic Structure of *oriC*

The 245 bp *oriC* sequence has two functional domains, the duplex-unwinding element (DUE) and DnaA-oligomerization region (DOR) (Leonard and Grimwade 2015; Wolański et al. 2015) (Fig. 4.1a). The DUE contains three AT-rich repeats of a 13-mer sequence with the consensus GATCTnTTnTTTT. It should be noted that this consensus includes a GATC sequence and a T-stretch in one strand and an A-stretch in the other strand (see below). The DOR contains a specific binding site for the DNA-bending protein IHF (IHF-binding site; IBS) with the consensus (A/T) ATCAAnnnnTT(A/G). IHF binding effectively stimulates DUE unwinding in vitro (Hwang and Kornberg 1992; Ozaki and Katayama 2012). HU protein, a structural homolog of IHF, can substitute for IHF in this role in vitro (Hwang and Kornberg 1992). In vivo, a single mutation in IHF or HU causes moderate inhibition of initiation, whereas double mutations of IHF and HU are lethal (Kano and Imamoto 1990), consistent with the roles of these proteins identified in vitro.

The DOR is subdivided into at least two regions, the left-half and middle–right-half regions, which contain sets of DnaA boxes with opposite orientations (Rozgaja et al. 2011; Ozaki and Katayama 2012; Ozaki et al. 2012a; Noguchi et al. 2015; Shimizu et al. 2016) (Fig. 4.1a). DnaA box R1 and box R4, which reside at opposite ends of DOR, have high affinity for DnaA. DnaA box R2 has moderate affinity and comprises the middle region (Rozgaja et al. 2011; Shimizu et al. 2016). Other DnaA boxes are low-affinity sites and form clusters: τ1-I2 in the left half and C3-C1 in the right half (McGarry et al. 2004; Kawakami et al. 2005; Rozgaja et al. 2011; Shimizu et al. 2016). These low-affinity sites bear moderate similarities to the DnaA box consensus. DnaA binding to these low-affinity sites is supported by cooperative binding of ATP-DnaA

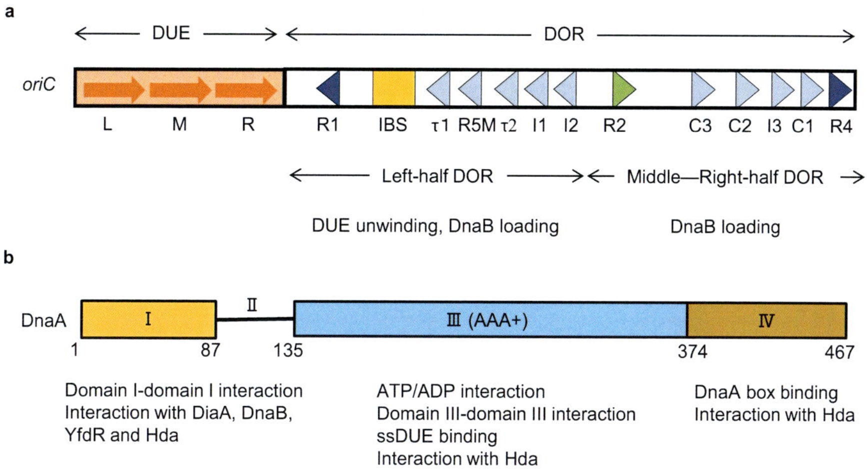

Fig. 4.1 Basic structures of *E. coli oriC* and DnaA. (**a**) *E. coli oriC* consists of the duplex-unwinding element (*DUE*) and DnaA-oligomerization region (*DOR*) domains. The DUE includes 13-mer repeats (L, M, and R). DnaA boxes (*triangles*) are shown with their directionality. Functions of the left-half and middle–right-half DOR are indicated (the middle region includes only the DnaA box R2). *IBS* IHF-binding sequence. (**b**) *E. coli* DnaA consists of four domains. Amino acid numbers are indicated for each domain. See text for details

(see below). ATP-DnaA binding to τ1 might be specific for linear form *oriC* and might not occur in supercoiled *oriC* (Kawakami et al. 2005; Rozgaja et al. 2011).

Functional differences exist between the left-half DOR and the middle–right-half DOR (Fig. 4.1a). When the left-half DOR binds IHF and ATP-DnaA in vitro, the flanking DUE region is unwound even without the middle–right-half DOR (Ozaki and Katayama 2012). Efficient DnaB loading requires both subregions of the DOR; the DUE together with the left-half DOR has only moderate activity for DnaB loading (Ozaki and Katayama 2012). Consistently, in vivo, cells bearing deletion of the DnaA box R4 in the chromosomal *oriC* initiate replication but with a moderate inhibition (Bates et al. 1987).

4.3 Basic Structure of DnaA

DnaA protein consists of 473 amino acid residues, which comprise four functional domains (Ozaki and Katayama 2009; Kaguni 2011) (Figs. 4.1b and 4.2a). The N-terminal domain I has at least two specific sites for protein binding. The first site contains Trp6 and supports DnaA domain I-domain I interaction with weak affinity (Felczak et al. 2005) (Table 4.1). The second site, containing Phe46, is used for binding to at least three proteins, DnaB helicase, DiaA protein, and YfdR protein (Abe et al. 2007; Keyamura et al. 2009; Noguchi and Katayama 2016) (Table 4.1, Fig. 4.2a, b). This residue forms a patch with Glu21, Trp25, and Trp50 on the

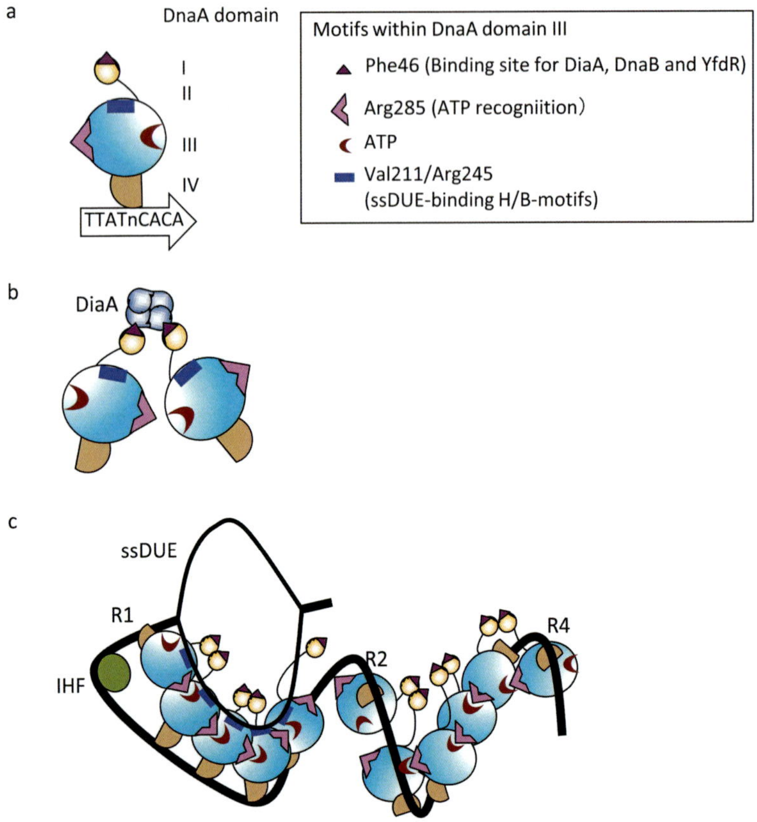

Fig. 4.2 Binding structures of DnaA. (**a**) DnaA-DnaA box binding. The arginine finger side of DnaA orients to the 5′ end of the DnaA box consensus. Important motifs are indicated. (**b**) DnaA-DiaA binding. Each protomer of a DiaA homotetramer can bind DnaA. For simplicity, only two DnaA molecules are shown. (**c**) A model for DUE unwinding by a DnaA-*oriC* complex. ATP-DnaA molecules and IHF bind to the DOR and can unwind the DUE. In the left-half DnaA sub-complex, DnaA boxes R1, R5M, τ2, I1, and I2 are occupied with DnaA, and in the ssDUE-recruitment model, DnaA bound to these sites bind the ssDUE. Even if ADP-DnaA binds to the R1 and R4 boxes, unwinding activity is sustained. See text for details

surface of domain I, which should interact with a specific site of DiaA (Keyamura et al. 2007, 2009). Affinity of DnaA domain I for DiaA is high, but affinity for DnaB and YfdR is weaker. DiaA is a stimulator of DnaA assembly on *oriC* (see below) (Fig. 4.2b). YfdR is encoded by a cryptic prophage (termed CPS-53) in the *E. coli* K12 chromosome and inhibits the DnaA-DnaB interaction. YfdR might be expressed under specific growth conditions to inhibit initiation at *ori*C (Noguchi and Katayama 2016). In addition, HU protein interacts with domain I, but its specific binding site has not been determined (Chodavarapu et al. 2008). Domain II is a flexible linker (Abe et al. 2007; Nozaki and Ogawa 2008).

Table 4.1 Important functional residues in DnaA

Residue	Domain	Function
Trp6	I	Domain I-domain I interaction
Asn44	I	Interaction with Hda
Phe46	I	Binding to DnaB, DiaA, and YfdR
Glu143	III	AAA+ N-linker, stable ATP/ADP binding
Glu173–Thr179	III	Walker A motif
Val211	III	ssDUE-binding H motif
Asp236–Asp237	III	Walker B motif
Arg227	III	AID1 motif, domain III-domain III interaction
Arg245	III	ssDUE-binding B motif
Asp269	III	AAA+ sensor I, stable ATP/ADP binding
Arg281	III	AAA+ box VII, domain III-domain III interaction
Arg285	III	AAA+ Arg finger, recognition of ATP, domain III-domain III interaction
Leu290	III	AID2 motif, domain III-domain III interaction
Arg334	III	AAA+ sensor II, ATP hydrolysis in RIDA, and DDAH
Lys366	III	Modulation of DnaA complex
Arg399–Lys443	IV	DNA-binding HTH motif
Leu422, Pro423	IV	Interaction with Hda

Domain III consists of the AAA+ domain that contains the Walker-type ATP(ADP)-binding sites and specific motifs for ATP binding/hydrolysis and for domain III-domain III interactions (Neuwald et al. 1999; Erzberger et al. 2002; Nishida et al. 2002; Iyer et al. 2004; Katayama et al. 2010) (Figs. 4.1b and 4.2a, Table 4.1). In addition to the typical Walker A motif (the P-loop), the AAA+ sensor 1 and N-linker motifs of DnaA support high-affinity ATP(ADP) binding (Kawakami et al. 2006; Ozaki et al. 2012b) (Table 4.1). ATP binding induces a conformational change in domains II–III (Saxena et al. 2015).

The sensor 2 Arg334 residue specifically supports ATP hydrolysis and is not required for the replication-initiation activity of DnaA (Nishida et al. 2002) (Table 4.1). The arginine finger motif Arg285 is the key residue in ATP activation of DnaA for initiation at *oriC* (Fig. 4.2a, Table 4.1); in common with typical AAA+ family proteins, DnaA domain III forms homo-oligomers by head-to-tail interaction, and at the interface, Arg285 of one protomer interacts with ATP bound to the flanking protomer (Neuwald et al. 1999; Iyer et al. 2004; Kawakami et al. 2005; Erzberger et al. 2006; Noguchi et al. 2015). DnaA is monomeric in solution, and these specific domain III-domain III interactions occur when multiple ATP-DnaA molecules are bound to *oriC*, which results in ATP-DnaA-specific, initiation-competent complexes (Felczak and Kaguni 2004; Kawakami et al. 2005; Noguchi et al. 2015; Shimizu et al. 2016) (Fig. 4.2c). In addition, AID (ATP-DnaA-specific interactive locus for DUE unwinding) motifs corresponding to Arg227 and Leu290 in domain III support specific DnaA-DnaA interaction for construction of DUE unwinding-competent DnaA oligomers (Ozaki et al. 2012a) (Table 4.1).

The single-stranded DUE (ssDUE)-binding H (hydrophobic)/B (basic) motifs of domain III correspond to Val211 and Arg245 residues and have a crucial role in DUE unwinding (Ozaki et al. 2008) (Fig. 4.2a, c, Table 4.1). These residues would be exposed on the surface of the central pore when DnaA domain III oligomers form a helical configuration as in typical AAA+ proteins (Ozaki et al. 2008; Ozaki and Katayama 2009; Duderstadt et al. 2011) (see below). The C-terminus of domain III is an amphipathic α-helix, in which Lys366 is important for modulation of *oriC*-DnaA complexes (Garner and Crooke 1996; Saxena et al. 2011) (Table 4.1). Also, a short loop connecting to the N-terminus of this α-helix allows domain IV to rotate at a certain extent (Erzberger et al. 2002; Shimizu et al. 2016).

Domain IV contains the sequence-specific DNA-binding helix-turn-helix motif (Erzberger et al. 2002; Obita et al. 2002; Yoshida et al. 2003; Fujikawa et al. 2003) (Figs. 4.1b and 4.2a, Table 4.1). The DnaA box consensus sequence is the nonamer TTATNCACA. As this sequence is asymmetric, the orientation of bound DnaA is determined by the directionality of the sequence; the arginine finger side of DnaA is oriented toward the 5′ end of the nonamer, whereas the ATP/ADP-binding site is oriented toward the 3′ end (Noguchi et al. 2015) (Fig. 4.2a).

4.4 Basic Structure of DiaA

The 196 amino acid residue protein DiaA forms homotetramers with structural symmetry (Keyamura et al. 2007). Each protomer contains a DnaA-binding site that includes Leu190. This site binds tightly to the DnaA domain I Phe46 region, and, theoretically, four DnaA molecules can simultaneously bind to a single DiaA molecule (Fig. 4.2b). Experimentally, binding of at least three DnaA molecules has been demonstrated (Keyamura et al. 2009). As such, DiaA acts as a bridge between DnaA molecules, exerting the so-called linkage effect, which drastically enhances cooperative binding of ATP-DnaA molecules (Stauffer and Chazin 2004; Katayama 2008). Thus, DiaA enhances specific assembly of ATP-DnaA molecules on *oriC* and DUE unwinding (Ishida et al. 2004; Keyamura et al. 2007). Even in the presence of DiaA, the requirement for ATP-DnaA (not ADP-DnaA) in construction of active *oriC*-DnaA complexes is preserved (Keyamura et al. 2007, 2009).

As DiaA binds tightly to the DnaA domain I Phe46 site, and this site is also the primary, but weak, binding site of the DnaB helicase, DiaA binding to DnaA results in inhibition of DnaA-DnaB binding and DnaB loading onto unwound DUE strands (Keyamura et al. 2009). The question of how this inhibition is resolved has to be explored in the future.

DiaA deletion mutations, as well as a DiaA L190A mutation, cause moderate inhibition of the initiation of chromosome replication, consistent with the role for this protein in the stimulation of initiation that was revealed in vitro (Ishida et al. 2004; Keyamura et al. 2007). Oversupply of DiaA also causes moderate inhibition of initiation (Ishida et al. 2004; Flåtten et al. 2015), which might be caused by inhibition of DnaA-DnaB interaction.

4.5 Structure and Dynamics of the Initiation Complex

4.5.1 Assembly of DnaA on oriC

Assembly of DnaA can be subdivided into several stages, although further analyses are required to fully understand this process. The cellular level of ATP-DnaA fluctuates during the cell cycle, with a peak at the time of replication initiation (Kurokawa et al. 1999). For a certain period before replication initiation, ADP-DnaA predominates, and only high-affinity DnaA boxes R1 and R4 and moderate-affinity DnaA box R2 bind DnaA, according to the results of in vivo footprinting experiments (Samitt et al. 1989; Miller et al. 2009). As the DiaA-DnaA interaction is stable, DnaA molecules bound at the R1, R4, and R2 boxes might be accompanied by DiaA (Fig. 4.2b), which would enhance the next step of ATP-DnaA assembly on the clusters of low-affinity DnaA boxes. As the ATP/ADP-binding sites of DnaA protomers that are bound to the R1 and R4 boxes orient toward the outer edges of *oriC* (Fig. 4.2c), whether ADP-DnaA or ATP-DnaA binds at these sites is not important for the assembly of ATP-DnaA to the low-affinity regions (Noguchi et al. 2015).

When the level of ATP-DnaA increases and becomes predominant, ATP-DnaA cooperatively binds to the low-affinity DOR sites, with head-to-tail domain III-domain III interactions (Fig. 4.2c). This process depends on the ATP-arginine finger intermolecular interaction and is enhanced by DiaA, as described above. In addition, IHF binding also enhances ATP-DnaA binding to the I1–3 sites (Grimwade et al. 2000; McGarry et al. 2004), which may be an indirect consequence of IHF-dependent DNA bending and the resultant DnaA-DnaA interactions. Results from studies of structural biology suggest that domain III homo-oligomers have a helical configuration, which is a common structural feature in the AAA+ family proteins (Erzberger et al. 2006; Duderstadt et al. 2011) (Fig. 4.2c).

Finally, three DnaA subcomplexes are constructed on *oriC* (Rozgaja et al. 2011; Ozaki and Katayama 2012; Ozaki et al. 2012a; Noguchi et al. 2015; Shimizu et al. 2016). The subcomplex that binds to the left-half *oriC* sustains the full activity in DUE unwinding and basal activity for DnaB loading, whereas the subcomplexes that bind to the R2 site and the region from the R4 site to the C3 site in the middle–right-half *oriC* is not required for DUE unwinding but enhances the DnaB loading activity (Fig. 4.2c).

4.5.2 DUE Unwinding

DUE unwinding is enhanced by IHF binding to *oriC* (Hwang and Kornberg 1992). The possibility that modulation of superhelicity by IHF is the main cause of this stimulation has been considered. However, it has now been demonstrated that even in the absence of superhelicity, IHF binding drastically stimulates DUE unwinding in the presence of ATP-DnaA (Ozaki and Katayama 2012). This observation supports the idea that local structural change, in the form of sharp DNA bending between DUE and DnaA box R1, is crucial for stimulation of DUE unwinding (see below) (Fig. 4.2c).

DUE unwinding requires binding of DnaA to the ssDUE. ATP-DnaA complexes (but not ADP-DnaA complexes) assembled on the DOR bind specifically to the ssDUE T-rich strand, but not the A-rich strand (Ozaki et al. 2008). In the absence of DOR binding, DnaA does not bind ssDUE efficiently. The DnaA Val211 and Arg245 residues (H/B motifs) located in domain III are required for DUE unwinding as well as binding to the ssDUE T-rich strand (Ozaki et al. 2008). As the corresponding residues are conserved in DnaA orthologs, they are named ssDUE-binding H (hydrophobic) motif (Val211 in *E. coli*) and B (basic) motif (Arg245 in *E. coli*) (Table 4.1). These motifs are exposed on the inner surface of the central pore of the DnaA domain III helical oligomer in a model structure (Fig. 4.2c). A study of the crystal structure of a hyperthermophilic bacterial DnaA ortholog domains III–IV supports the specific structure and role of these residues (Duderstadt et al. 2011).

Overall structure models of DnaA complexes binding the ssDUE have been proposed. In the ssDUE-recruitment model (Fig. 4.2c), the DnaA subcomplex constructed on the left-half DOR binds the ssDUE T-rich strand via DNA bending by IHF (Ozaki et al. 2008; Ozaki and Katayama 2009; Noguchi et al. 2015). This model explains the importance of IHF binding, ATP-DnaA complex formation on the DOR and structural features of *oriC*, such as strict conservation of the spacing between the DUE and the DnaA box R1 (Ozaki and Katayama 2009). Another model postulates formation of a continuous DnaA fiber from DnaA box R1 to the DUE region (Duderstadt et al. 2011). This model also supports the importance of H/B motifs but is less able to explain the roles of IHF and specific DnaA complex formation on the DOR in the unwinding mechanism. In addition, when ADP-DnaA is bound to box R1 (the origin of the postulated DnaA fiber that expands to the DUE in the continuous-fiber model), activity in DUE unwinding is sustained similarly to the case of ATP-DnaA binding to R1 (Noguchi et al. 2015). This fact is consistent with the ssDUE-recruitment model, but not with the continuous DnaA fiber model, as construction of a DnaA fiber requires the intermolecular ATP-arginine finger interaction, especially on DNA without high-affinity DnaA-binding sites. In addition, structures deduced from molecular dynamics simulation of *oriC*-IHF-DnaA complexes are consistent with the ssDUE-recruitment model (Shimizu et al. 2016).

4.5.3 Interaction with Helicase

DnaA has specific binding sites for DnaB helicase; the primary site includes Phe46 (Abe et al. 2007; Keyamura et al. 2009) (Fig. 4.2a, Table 4.1). The secondary, weaker site is suggested to reside at a region spanning the domain II-C-terminus to the domain III-N-terminus but has not been determined at the amino acid residue level (Marszalek et al. 1996; Seitz et al. 2000). Because affinity between DnaA monomers and DnaB is low (Sutton et al. 1998), formation of DnaA oligomers is required for functional interaction with DnaB helicase. DnaB helicase is a homo-hexamer, so oligomerization of DnaA would provide multiple binding sites for a single DnaB hexamer, stabilizing DnaA-DnaB binding by the linkage effect (Abe et al. 2007; Keyamura et al. 2009). A pair of DnaB helicases is proposed to bind to the two DnaA oligomers assembled on the DOR in opposite directions, enabling bidirectional loading on the single-stranded DNA region (Ozaki and Katayama 2009; Noguchi et al. 2015) (Fig. 4.2c). This hypothesis is consistent with structures deduced from molecular dynamics simulation of *oriC*-IHF-DnaA complexes (Shimizu et al. 2016). The mechanism of DnaB loading onto the single-stranded DNA has yet to be determined.

4.6 Regulation Systems for DnaA and *oriC* in *E. coli*

4.6.1 Overview

Regulation of initiation during the *E. coli* cell cycle is achieved by multiple systems targeting DnaA or *oriC* (Katayama et al. 2010; Saxena et al. 2013; Skarstad and Katayama 2013; Riber et al. 2016). These systems operate in different or overlapping periods of the replication cycle, ensuring robust regulation of initiation. For DnaA, multiple systems promote timely inactivation of DnaA by DnaA-ATP hydrolysis or reactivation of DnaA by exchange of bound ADP for ATP. Specific protein and DNA factors support these systems as key elements. The *dnaA* transcription is cell cycle dependent. For *oriC*, a specific protein binds to the nascent *oriC* copies, inhibiting reinitiation.

In addition, specific acetylation of DnaA occurs in the stationary phase, which might be important to regulate replication (Zhang et al. 2016). Interaction of DnaA with acidic phospholipids can stimulate exchange of bound nucleotide of DnaA in vitro, and changes in components of membrane phospholipids in cells influence initiation of replication (Sekimizu and Kornberg 1988; Yung and Kornberg 1988; Crooke et al. 1992; Aranovich et al. 2006; Fingland et al. 2012; Saxena et al. 2013). DnaA-phospholipid interaction also might participate in regulation of replication; a possibility that phospholipids affect initiation in vivo in an indirect unknown manner is not mutually exclusive.

4.6.2 The Replicative Clamp-Dependent, Negative Feedback for DnaA

This system, termed RIDA (regulatory inactivation of DnaA), depends on the DNA-loaded form of the replicative clamp (i.e., the β-subunit of the DNA polymerase III holoenzyme) as well as ADP-Hda protein (Katayama et al. 1998; Kurokawa et al.1999; Kato and Katayama 2001; Su'etsugu et al. 2008). As DNA-free clamps are inactive in RIDA, this system is activated upon replication initiation and loading of DNA polymerase III holoenzyme. During lagging-strand replication, Okazaki fragments are repeatedly synthesized, which leaves the used clamps on the nascent DNA regions. These DNA-bound clamps interact with ADP-Hda protein, and the resultant complex then interacts with ATP-DnaA, promoting ATP hydrolysis. RIDA operates throughout chromosomal replication.

Hda consists of a short N-terminal region bearing a clamp-binding motif and an AAA+ domain (Kato and Katayama 2001; Su'etsugu et al. 2005, 2008). Hda binds ADP, but not ATP, resulting in RIDA-active, monomeric Hda. The arginine finger motif of the Hda AAA+ domain is required for DnaA-ATP hydrolysis. A head-to-tail complex of DnaA domain III and the Hda AAA+ domain assembles. In addition, Hda interacts with DnaA domain I and domain IV with low affinity (Keyamura and Katayama 2011; Su'etsugu et al. 2013) (Fig. 4.1b, Table 4.1). These multiple weak interactions would support repeated interaction of a single ADP-Hda-clamp-DNA complex with ATP-DnaA molecules, catalytically promoting ATP hydrolysis. Whereas the cellular level of DnaA is about 2000 monomers per cell (Sekimizu et al. 1988), that of Hda is only about 100 monomers, and oversupply of Hda is very toxic to cells (Su'etsugu et al. 2005; Baxter and Sutton 2012).

RIDA is the predominant system for inactivating DnaA (Camara et al. 2005; Kasho and Katayama 2013). *Hda* mutations increase the cellular level of ATP-DnaA and cause over-initiation, which results in inhibition of cell division and cell growth (Kato and Katayama 2001; Fujimitsu et al. 2008; Charbon et al. 2014), although suppressor mutations frequently occur. It should also be noted that, in the presence of RIDA, an increase in the total cellular amount of DnaA causes only slight over-initiation (Atlung et al. 1987; Flåtten et al. 2015). By contrast, unlike expression of wild-type DnaA, expression of a RIDA-insensitive DnaA variant, which substantially constitutively takes on the ATP form, causes severe over-initiation, leading to inhibition of cell growth (Nishida et al. 2002; Keyamura and Katayama 2011).

*4.6.3 The Specific DNA Element (***datA***)-Dependent Timely Inactivation of DnaA*

The *datA* locus of the *E. coli* chromosome also has a role in repressing extra initiations (Kitagawa et al. 1998) (Fig. 4.3). This locus contains a DnaA box cluster, in which three boxes with the same direction (DnaA boxes 7, 2, and 3) are essential for function, and one box with the opposite direction (DnaA box 4) is stimulatory (Ogawa et al. 2002; Kasho and Katayama 2013; Kasho et al. 2017) (Fig. 4.3b). In addition, this locus contains a single IHF-binding site, which is essential for *datA* function (Nozaki et al. 2009b).

The IHF-*datA* complex promotes DnaA-ATP hydrolysis, yielding ADP-DnaA (Kasho and Katayama 2013). This function is termed DDAH (*datA*-dependent DnaA-ATP hydrolysis). The detailed mechanism of DDAH is under investigation, but specific DnaA domain III-domain III interaction is important for activating the

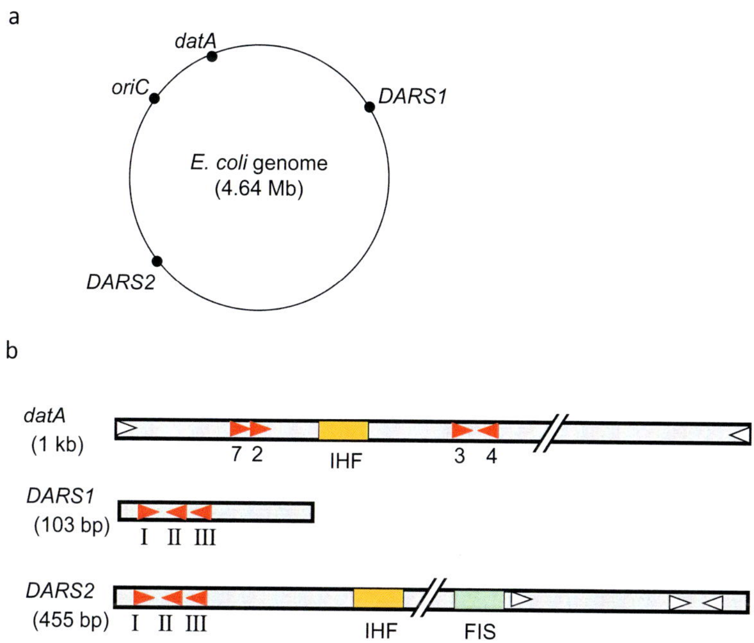

Fig. 4.3 Basic structures of *datA*, *DARS1*, and *DARS2*. (**a**) Chromosomal positions of *oriC*, *datA*, *DARS1*, and *DARS2*. (**b**) Overall structures of *datA*, *DARS1*, and *DARS2* are shown. Important DnaA boxes are shown by filled *triangles*. Those are DnaA boxes 2, 3, 4, and 7 of *datA*, and DnaA boxes I, II, and III of *DARS1* and *DARS2*. IHF and Fis binding sites are also indicated

ATPase activity of DnaA (Table 4.1). Supercoiled structure of the *datA* region stimulates assembly of DnaA and IHF, thereby enhancing DDAH (Kasho et al. 2017). It might be important to investigate if the chromosomal *datA* locus is regulated specifically in superhelicity.

Binding of IHF to *datA* is cell cycle specific; whereas binding is repressed before initiation, it increases after initiation, peaking 15 min after initiation (Kasho and Katayama 2013). DDAH is independent of RIDA and assists in repressing untimely initiations. *datA*-deleted cells can grow but experience a moderate level of untimely initiations (Kitagawa et al. 1998; Morigen et al. 2003, 2005; Nozaki et al. 2009b).

Fluorescent labeling experiments and genome conformation analysis suggest that in a cell, *oriC* is colocalized with *datA* (Nozaki et al. 2009a; Cagliero et al. 2013). The genomic locus of *datA* is near *oriC* (i.e., distance between the two is ~470 kb) (Fig. 4.3a), and both two are included in a compactly folded domain (i.e., Ori macrodomain) of the chromosome (Niki et al. 2000; Valens et al. 2004). Biological significance of the colocalization of *oriC* and *datA* remains unclear. However, relocation of *datA* to a site near *terC* (the replication termination site of the chromosome) inhibits function of *datA* in rapidly growing cells, which could be caused by delay of duplication of *datA* in a replication cycle or enlarged spacing itself between the two in a cell (Kitagawa et al. 1998; Frimodt-Møller et al. 2016).

4.6.4 *Sequestration:* oriC *DNA Methylation-Dependent Regulation*

This system depends on the Dam (DNA adenine methyltransferase) and SeqA proteins (Waldminghaus and Skarstad 2009). Dam methylates the N-6 position of adenine in the sequence GATC. As this sequence is palindromic, both A residues in the duplex DNA sequence are methylated, resulting in a fully methylated state. When this sequence is replicated, a hemi-methylated state (where the nascent strand is not yet methylated) exists before further action of Dam. The hemi-methylated GATC sites are binding targets of SeqA (Lu et al. 1994; Slater et al. 1995). *SeqA*-deleted cells can grow but experience a moderate level of untimely initiations.

SeqA contains GATC-binding and self-oligomerization domains (Fujikawa et al. 2004; Odsbu et al. 2005; Guarné et al. 2005; Chung et al. 2009). The *oriC* region includes a cluster of GATC sites (11 in total), and hemi-methylated *oriC* is bound by SeqA oligomers; this binding is maintained for about 10 min after initiation in cells with a doubling time of 30 min (Lu et al. 1994). Cooperative SeqA binding to the supercoiled DNA causes DNA topological changes (Torheim and Skarstad 1999; Kang et al. 2003), and within *oriC* it inhibits DnaA binding to the low-affinity boxes, thereby inhibiting initiation (Nievera et al. 2006). Subcellular localization of SeqA is dynamically changed during the replication cycle: observation of fluorescence-labeled SeqA in live cells suggests that SeqA molecules are assembled on *oriC* during the sequestration period, and thereafter those molecules are dissociated from *oriC* and assembled on the nascent DNA regions flanking the replication forks (Fossum-Raunehaug et al. 2014; Helgesen et al. 2015). Dissociation of SeqA from a hemi-methylated *oriC* region spontaneously occurs independently of Dam in vitro (Kang et al. 1999).

4.6.5 *Specific DNA Element (*DARS1 *and* DARS2*)-Dependent Timely Reactivation of DnaA*

The *E. coli* chromosome contains two sites which specifically interact with ADP-DnaA and reactivate it by exchanging ADP to ATP and producing ATP-DnaA (Fujimitsu et al. 2009). These DnaA-reactivating sequence sites are termed *DARS1* and *DARS2* (Fig. 4.3a). These sites both include a cluster of three DnaA boxes (Fig. 4.3b), which promote specific domain III-dependent DnaA-DnaA interactions, thereby enhancing dissociation of DnaA-bound nucleotide. These trios of DnaA boxes are connected to regulatory regions which structurally differ between *DARS1* and *DARS2*. The regulatory region of *DARS2* bears specific binding sites for IHF and Fis, which are required for activating *DARS2* function (Kasho et al. 2014) (Fig. 4.3b; also see below). However, the mechanisms by which the regulatory regions stimulate the DnaA-reactivating function are not yet known.

Both *DARS1* and *DARS2* are required for timely initiation during the cell cycle (Fujimitsu et al. 2009). In vivo, *DARS2* has a predominant role and *DARS1* has a supporting role. Activation of *DARS2* is regulated by timely binding of Fis and IHF; Fis binds to *DARS2* in log-phase cells, but not in stationary-phase cells, whereas IHF binds to *DARS2* in a pre-initiation stage of the cell cycle (Kasho et al. 2014). At the time of initiation, the ATP-DnaA level reaches its maximum, and IHF is dissociated from *DARS2*. Mechanisms involved in the timely binding/dissociation of IHF to/from *DARS2* should be explored in the future. The chromosomal location of *DARS2*, in a central region between *oriC* and the replication termination site (Fig. 4.3a), is important for regulating timely initiation in cells growing at 42 °C in LB medium (Inoue et al. 2016; Frimodt-Møller et al. 2016). In addition, genomic conformation analysis suggests that *oriC* is colocalized with *DARS2* in a cell (Cagliero et al. 2013). The biological significance of this colocalization remains to be investigated.

4.6.6 *Cell Cycle-Dependent Transcription of the* dnaA *Gene*

Transcription of the *dnaA* gene is regulated in a cell cycle-specific manner; it increases before initiation and decreases after initiation (Theisen et al. 1993). This fluctuation depends on Dam and SeqA (Bogan and Helmstetter 1997) and is important to sustain timely initiation (Riber and Løbner-Olesen 2005), probably because de novo-synthesized DnaA will bind ATP (which is much more abundant in the cell than ADP) to form ATP-DnaA. In addition, *dnaA* gene transcription is autoregulated (Speck et al. 1999).

oriC is flanked by two genes, *gidA* and *mioC*, which are transcribed in the same direction (*oriC* is downstream of *mioC*). The pattern of change of *gidA* transcription is similar to that of *dnaA*, whereas fluctuation of *mioC* transcription is opposite to that of *gidA* and *dnaA* (Theisen et al. 1993; Ogawa and Okazaki 1994). However, transcription of *gidA* and *mioC* is not required for regulation of initiation (Bates et al. 1987), although constitutive transcription of *mioC* moderately inhibits initiation (Su'etsugu et al. 2003).

4.7 Conservation in Eubacterial Species

This review focuses on mechanisms and regulation of replication in *E. coli*. However, many important studies on DNA replication initiation have been conducted in other bacteria, phages, and plasmids. Evidence from these studies indicates that the overall structure of *oriC* might be fundamentally similar in relation to the locations of the DUE and DOR, although the detailed structures vary in different species (Wolański et al. 2015). ssDUE-binding activity of DnaA has been reported in *Bacillus subtilis* (Richardson et al. 2016). In the hyperthermophilic eubacterium *Thermotoga maritima*, DnaA specifically binds to its cognate ssDUE when it forms homo-oligomers on its cognate DOR (Ozaki and Katayama 2012). In plasmid RK2, the origin includes iterons, which are repeats of the initiator protein TrfA-binding sequence and its flanking DUE region. TrfA forms a homo-oligomer on the iteron region, and the resultant complex binds the ssDUE-bottom (but not the DUE-top) strand, which is followed by loading of replisomes (Wegrzyn et al. 2014; Wawrzycka et al. 2015). This mechanism might be similar to the ssDUE-recruitment mechanism of *E. coli oriC*.

In *Helicobacter pylori*, HobA, the DiaA functional homolog, has a crucial role in assembly of the cognate DnaA at the replication origin (Zawilak-Pawlik et al. 2007; Natrajan et al. 2009). In *Caulobacter crescentus*, HdaA, a structural and functional homolog of *E. coli* Hda, is important for regulation of the cognate DnaA and replication initiation (Collier and Shapiro 2009; Fernandez-Fernandez et al. 2011). In addition, *C. crescentus* CtrA binds to the cognate *oriC* in a timely manner, inhibiting extra initiation. In *B. subtilis*, YabA, the functional counterpart of Hda, binds both the replicative clamp and DnaA, sequestrating DnaA from *oriC* and repressing untimely initiation (Noirot-Gros et al. 2006; Soufo et al. 2008). In addition, in *B. subtilis*, binding of DnaA domain I by SirA and domain III by Soj/ParA, is important for the repression of untimely initiation for sporulation (Jameson et al. 2014; Scholefield et al. 2012).

The *B. subtilis* and *Streptomyces coelicolor* chromosomes have specific DnaA box clusters that can repress untimely initiations (Smulczyk-Krawczyszyn et al. 2006; Okumura et al. 2012). Sequences corresponding to *DARS* are highly conserved in *Gammaproteobacteria* (Fujimitsu et al. 2009), but except for *E. coli*, these sequences have not yet been functionally analyzed.

Acknowledgments This work was supported by JSPS KAKENHI Grant Numbers 26291004, 16H00775, 17H03656, and 26650127. I apologize that many papers cannot be cited because of the limitation of space.

References

Abe Y, Jo T, Matsuda Y, Matsunaga C, Katayama T, Ueda T (2007) Structure and function of DnaA N-terminal domains: specific sites and mechanisms in inter-DnaA interaction and in DnaB helicase loading on *oriC*. J Biol Chem 282:17816–17827

Aranovich A, Gdalevsky GY, Cohen-Luria R, Fishov I, Parola AH (2006) Membrane-catalyzed nucleotide exchange on DnaA: effect of surface molecular crowding. J Biol Chem 281:12526–12534

Atlung T, Løbner-Olesen A, Hansen FG (1987) Overproduction of DnaA protein stimulates initiation of chromosome and minichromosome replication in *Escherichia coli*. Mol Gen Genet 206:51–59

Bates DB, Boye E, Asai T, Kogoma T (1987) The absence of effect of *gid* or *mioC* transcription on the initiation of chromosomal replication in *Escherichia coli*. Proc Natl Acad Sci USA 94:12497–12502

Baxter JC, Sutton MD (2012) Evidence for roles of the *Escherichia coli* Hda protein beyond regulatory inactivation of DnaA. Mol Microbiol 85:648–668

Bogan JA, Helmstetter CE (1997) DNA sequestration and transcription in the *oriC* region of *Escherichia coli*. Mol Microbiol 26:889–896

Cagliero C, Grand RS, Jones MB, Jin DJ, O'Sullivan JM (2013) Genome conformation capture reveals that the *Escherichia coli* chromosome is organized by replication and transcription. Nucleic Acids Res 41:6058–6071

Camara JE, Breier AM, Brendler T, Austin S, Cozzarelli NR, Crooke E (2005) Hda inactivation of DnaA is the predominant mechanism preventing hyperinitiation of *Escherichia coli* DNA replication. EMBO Rep 6:736–741

Charbon G, Bjørn L, Mendoza-Chamizo B, Frimodt-Møller J, Løbner-Olesen A (2014) Oxidative DNA damage is instrumental in hyperreplication stress-induced inviability of Escherichia coli. Nucleic Acids Res 42(21):13228–13241

Chodavarapu S, Felczak MM, Rouvière-Yaniv J, Kaguni JM (2008) *Escherichia coli* DnaA interacts with HU in initiation at the *E. coli* replication origin. Mol Microbiol 67:781–792

Chung YS1, Brendler T, Austin S, Guarné A (2009) Structural insights into the cooperative binding of SeqA to a tandem GATC repeat. Nucleic Acids Res 37:3143–3152

Collier J, Shapiro L (2009) Feedback control of DnaA-mediated replication initiation by replisome-associated HdaA protein in *Caulobacter*. J Bacteriol 191:5706–5716

Costa A, Hood IV, Berger JM (2013) Mechanisms for initiating cellular DNA replication. Annu Rev Biochem 82:25–54

Crooke E, Castuma CE, Kornberg A (1992) The chromosome origin of *Escherichia coli* stabilizes DnaA protein during rejuvenation by phospholipids. J Biol Chem 267:16779–16782

Duderstadt KE, Chuang K, Berger JM (2011) DNA stretching by bacterial initiators promotes replication origin opening. Nature 478:209–213

Erzberger JP, Pirruccello MM, Berger JM (2002) The structure of bacterial DnaA: implications for general mechanisms underlying DNA replication initiation. EMBO J 21:4763–4773

Erzberger JP, Mott ML, Berger JM (2006) Structural basis for ATP-dependent DnaA assembly and replication-origin remodeling. Nat Struct Mol Biol 13:676–683

Felczak MM, Kaguni JM (2004) The box VII motif of *Escherichia coli* DnaA protein is required for DnaA oligomerization at the *E. coli* replication origin. J Biol Chem 279:51156–51162

Felczak MM, Simmons LA, Kaguni JM (2005) An essential tryptophan of *Escherichia coli* DnaA protein functions in oligomerization at the *E. coli* replication origin. J Biol Chem 280:24627–24633

Fernandez-Fernandez C, Gonzalez D, Collier J (2011) Regulation of the activity of the dual-function DnaA protein in *Caulobacter crescentus*. PLoS One 6:e26028

Fingland N, Flåtten I, Downey CD, Fossum-Raunehaug S, Skarstad K, Crooke E (2012) Depletion of acidic phospholipids influences chromosomal replication in *Escherichia coli*. Microbiology 1:450–466

Flåtten I, Fossum-Raunehaug S, Taipale R, Martinsen S, Skarstad K (2015) The DnaA protein is not the limiting factor for initiation of replication in *Escherichia coli*. PLoS Genet 11:e1005276

Fossum-Raunehaug S, Helgesen E, Stokke C, Skarstad K (2014) *Escherichia coli* SeqA structures relocalize abruptly upon termination of origin sequestration during multifork DNA replication. PLoS One 9:e110575

Frimodt-Møller J, Charbon G, Krogfelt KA, Løbner-Olesen A (2016) DNA replication control is linked to genomic positioning of control regions in *Escherichia coli*. PLoS Genet 12:e1006286

Fujikawa N, Kurumizaka H, Nureki O, Terada T, Shirouzu M, Katayama T, Yokoyama S (2003) Structural basis of replication origin recognition by the DnaA protein. Nucleic Acids Res 31:2077–2086

Fujikawa N, Kurumizaka H, Nureki O, Tanaka Y, Yamazoe M, Hiraga S, Yokoyama S (2004) Structural and biochemical analyses of hemimethylated DNA binding by the SeqA protein. Nucleic Acids Res 32:82–92

Fujimitsu K, Su'etsugu M, Yamaguchi Y, Mazda K, Fu N, Kawakami H, Katayama T (2008) Modes of overinitiation, *dnaA* gene expression, and inhibition of cell division in a novel cold-sensitive *hda* mutant of *Escherichia coli*. J Bacteriol 190:5368–5381

Fujimitsu K, Senriuchi T, Katayama T (2009) Specific genomic sequences of *E. coli* promote replicational initiation by directly reactivating ADP-DnaA. Genes Dev 23:1221–1233

Garner J, Crooke E (1996) Membrane regulation of the chromosomal replication activity of *E. coli* DnaA requires a discrete site on the protein. EMBO J 15:3477–3485

Grimwade JE, Ryan VT, Leonard AC (2000) IHF redistributes bound initiator protein, DnaA, on supercoiled *oriC* of *Escherichia coli*. Mol Microbiol 35:835–844

Guarné A, Brendler T, Zhao Q, Ghirlando R, Austin S, Yang W (2005) Crystal structure of a SeqA-N filament: implications for DNA replication and chromosome organization. EMBO J 24:1502–1511

Helgesen E, Fossum-Raunehaug S, Sætre F, Schink KO, Skarstad K (2015) Dynamic *Escherichia coli* SeqA complexes organize the newly replicated DNA at a considerable distance from the replisome. Nucleic Acids Res 43:2730–2743

Hwang DS, Kornberg A (1992) Opening of the replication origin of *Escherichia coli* by DnaA protein with protein HU or IHF. J Biol Chem 267:23083–23086

Inoue Y, Tanaka H, Kasho K, Fujimitsu K, Oshima T, Katayama T (2016) Chromosomal location of the DnaA-reactivating sequence *DARS2* is important to regulate timely initiation of DNA replication in *Escherichia coli*. Genes Cells 21:1015–1023

Ishida T, Akimitsu N, Kashioka T, Hatano M, Kubota T, Ogata Y, Sekimizu K, Katayama T (2004) DiaA, a novel DnaA-binding protein, ensures the initiation timing of *E. coli* chromosome replication. J Biol Chem 279:45546–45555

Iyer LM, Leipe DD, Koonin EV, Aravind L (2004) Evolutionary history and higher order classification of AAA+ ATPases. J Struct Biol 146:11–31

Jameson KH, Rostami N, Fogg MJ, Turkenburg JP, Grahl A, Murray H, Wilkinson AJ (2014) Structure and interactions of the *Bacillus subtilis* sporulation inhibitor of DNA replication, SirA, with domain I of DnaA. Mol Microbiol 93:975–991

Kaguni JM (2011) Replication initiation at the *Escherichia coli* chromosomal origin. Curr Opin Chem Biol 15:606–613

Kang S, Lee H, Han JS, Hwang DS (1999) Interaction of SeqA and Dam methylase on the hemimethylated origin of *Escherichia coli* chromosomal DNA replication. J Biol Chem 274:11463–11468

Kang S, Han JS, Park JH, Skarstad K, Hwang DS (2003) SeqA protein stimulates the relaxing and decatenating activities of Topoisomerase IV. J Biol Chem 278:48779–48785

Kano Y, Imamoto F (1990) Requirement of integration host factor (IHF) for growth of *Escherichia coli* deficient in HU protein. Gene 89:133–137

Kasho K, Katayama T (2013) DnaA binding locus *datA* promotes ATP-DnaA hydrolysis to enable cell cycle-coordinated replication initiation. Proc Natl Acad Sci USA 110:936–941

Kasho K, Fujimitsu K, Matoba T, Oshima T, Katayama T (2014) Timely binding of IHF and Fis to *DARS2* regulates ATP-DnaA production and replication initiation. Nucleic Acids Res 42:13134–13149

Kasho K, Tanaka H, Sakai R, Katayama T (2017) Cooperative DnaA binding to the negatively supercoiled locus stimulates DnaA-ATP hydrolysis. J Biol Chem 292:1251–1266

Katayama T (2008) Roles for the AAA+ motifs of DnaA in the initiation of DNA replication. Biochem Soc Trans 36:78–82

Katayama T, Kubota T, Kurokawa K, Crooke E, Sekimizu K (1998) The initiator function of DnaA protein is negatively regulated by the sliding clamp of the *E. coli* chromosomal replicase. Cell 94:61–71

Katayama T, Ozaki S, Keyamura K, Fujimitsu K (2010) Regulation of the replication cycle: conserved and diverse regulatory systems for DnaA and *oriC*. Nat Rev Microbiol 8:163–170

Kato J, Katayama T (2001) Hda, a novel DnaA-related protein, regulates the replication cycle in *Escherichia coli*. EMBO J 20:4253–4262

Kawakami H, Keyamura K, Katayama T (2005) Formation of an ATP-DnaA-specific initiation complex requires DnaA Arginine 285, a conserved motif in the AAA+ protein family. J Biol Chem 280:27420–27430

Kawakami H, Ozaki S, Suzuki S, Nakamura K, Senriuchi T, Su'etsugu M, Fujimitsu K, Katayama T (2006) The exceptionally tight affinity of DnaA for ATP/ADP requires a unique aspartic acid residue in the AAA+ sensor 1 motif. Mol Microbiol 62:1310–1324

Keyamura K, Katayama T (2011) DnaA protein DNA-binding domain binds to Hda protein to promote inter-AAA+ domain interaction involved in regulatory inactivation of DnaA. J Biol Chem 286:29336–29346

Keyamura K, Fujikawa N, Ishida T, Ozaki S, Su'etsugu M, Fujimitsu K, Kagawa W, Yokoyama S, Kurumizaka H, Katayama T (2007) The interaction of DiaA and DnaA regulates the replication cycle in *E. coli* by directly promoting ATP-DnaA-specific initiation complexes. Genes Dev 21:2083–2099

Keyamura K, Abe Y, Higashi M, Ueda T, Katayama T (2009) DiaA dynamics are coupled with changes in initial origin complexes leading to helicase loading. J Biol Chem 284:25038–25050

Kitagawa R, Ozaki T, Moriya S, Ogawa T (1998) Negative control of replication initiation by a novel chromosomal locus exhibiting exceptional affinity for *Escherichia coli* DnaA protein. Genes Dev 12:3032–3043

Kurokawa K, Nishida S, Emoto A, Sekimizu K, Katayama T (1999) Replication cycle-coordinated change of the adenine nucleotide-bound forms of DnaA protein in *Escherichia coli*. EMBO J 18:6642–6652

Leonard AC, Grimwade JE (2015) The orisome: structure and function. Front Microbiol 6:545

Lu M, Campbell JL, Boye E, Kleckner N (1994) SeqA: a negative modulator of replication initiation in *E. coli*. Cell 77:413–426

Marszalek J, Zhang W, Hupp TR, Margulies C, Carr KM, Cherry S, Kaguni JM (1996) Domains of DnaA protein involved in interaction with DnaB protein, and in unwinding the *Escherichia coli* chromosomal origin. J Biol Chem 271:18535–18542

McGarry KC, Ryan VT, Grimwade JE, Leonard AC (2004) Two discriminatory binding sites in the *Escherichia coli* replication origin are required for DNA strand opening by initiator DnaA-ATP. Proc Natl Acad Sci USA 101:2811–2816

Miller DT, Grimwade JE, Betteridge T, Rozgaja T, Torgue JJ, Leonard AC (2009) Bacterial origin recognition complexes direct assembly of higher-order DnaA oligomeric structures. Proc Natl Acad Sci USA 106:18479–18484

Morigen, Løbner-Olesen A, Skarstad K (2003) Titration of the *Escherichia coli* DnaA protein to excess *datA* sites causes destabilization of replication forks, delayed replication initiation and delayed cell division. Mol Microbiol 50:349–362

Morigen, Molina F, Skarstad K (2005) Deletion of the *datA* site does not affect once-per-cell-cycle timing but induces rifampin-resistant replication. J Bacteriol 187:3913–3920

Natrajan G, Noirot-Gros MF, Zawilak-Pawlik A, Kapp U, Terradot L (2009) The structure of a DnaA/HobA complex from *Helicobacter pylori* provides insight into regulation of DNA replication in bacteria. Proc Natl Acad Sci U S A 106:21115–21120

Neuwald AF, Aravind L, Spouge JL, Koonin EV (1999) AAA+: a class of chaperone-like ATPases associated with the assembly, operation, and disassembly of protein complexes. Genome Res 9:27–43

Nievera C, Torgue JJ, Grimwade JE, Leonard AC (2006) SeqA blocking of DnaA-*oriC* interactions ensures staged assembly of the *E coli* pre-RC. Mol Cell 24:581–592

Niki H, Yamaichi Y, Hiraga S (2000) Dynamic organization of chromosomal DNA in *Escherichia coli*. Genes Dev 14:212–223

Nishida S, Fujimitsu K, Sekimizu K, Ohmura T, Ueda T, Katayama T (2002) A nucleotide switch in the *Escherichia coli* DnaA protein initiates chromosomal replication: evidence from a mutant DnaA protein defective in regulatory ATP hydrolysis in vitro and in vivo. J Biol Chem 277:14986–14995

Noguchi Y, Katayama T (2016) The *Escherichia coli* cryptic prophage protein YfdR binds to DnaA and initiation of chromosomal replication is inhibited by overexpression of the gene cluster *yfdQ-yfdR-yfdS-yfdT*. Front Microbiol 7:239

Noguchi Y, Sakiyama Y, Kawakami H, Katayama T (2015) The Arg fingers of key DnaA protomers are oriented inward within the replication origin *oriC* and stimulate DnaA subcomplexes in the initiation complex. J Biol Chem 290:20295–20312

Noirot-Gros MF, Velten M, Yoshimura M, McGovern S, Morimoto T, Ehrlich SD, Ogasawara N, Polard P, Noirot P (2006) Functional dissection of YabA, a negative regulator of DNA replication initiation in *Bacillus subtilis*. Proc Natl Acad Sci USA 103:2368–2373

Nozaki S, Ogawa T (2008) Determination of the minimum domain II size of *Escherichia coli* DnaA protein essential for cell viability. Microbiology 154:3379–3384

Nozaki S, Niki H, Ogawa T (2009a) Replication initiator DnaA of *Escherichia coli* changes its assembly form on the replication origin during the cell cycle. J Bacteriol 191:4807–4814

Nozaki S, Yamada Y, Ogawa T (2009b) Initiator titration complex formed at *datA* with the aid of IHF regulates replication timing in *Escherichia coli*. Genes Cells 14:329–341

O'Donnell M, Langston L, Stillman B (2013) Principles and concepts of DNA replication in bacteria, archaea, and eukarya. Cold Spring Harb Perspect Biol 5:a010108

Obita T, Iwura T, Su'etsugu M, Yoshida Y, Tanaka Y, Katayama T, Ueda T, Imoto T (2002) Determination of the secondary structure in solution of the *Escherichia coli* DnaA DNA-binding domain. Biochem Biophys Res Commun 299:42–48

Odsbu I, Klungsoyr HK, Fossum S, Skarstad K (2005) Specific N-terminal interactions of the *Escherichia coli* SeqA protein are required to form multimers that restrain negative supercoils and form foci. Genes Cells 10:1039–1049

Ogawa T, Okazaki T (1994) Cell cycle-dependent transcription from the *gid* and *mioC* promoters of *Escherichia coli*. J Bacteriol 176:1609–1615

Ogawa T, Yamada Y, Kuroda T, Kishi T, Moriya S (2002) The *datA* locus predominantly contributes to the initiator titration mechanism in the control of replication initiation in *Escherichia coli*. Mol Microbiol 44:1367–1375

Okumura H, Yoshimura M, Ueki M, Oshima T, Ogasawara N, Ishikawa S (2012) Regulation of chromosomal replication initiation by *oriC* proximal DnaA-box clusters in *Bacillus subtilis*. Nucleic Acids Res 40:220–234

Ozaki S, Katayama T (2009) DnaA structure, function, and dynamics in the initiation at the chromosomal origin. Plasmid 62:71–82

Ozaki S, Katayama T (2012) Highly organized DnaA-*oriC* complexes recruit the single-stranded DNA for replication initiation. Nucleic Acids Res 40:1648–1665

Ozaki S, Kawakami H, Nakamura K, Fujikawa N, Kagawa W, Park SY, Yokoyama S, Kurumizaka H, Katayama T (2008) A common mechanism for the ATP-DnaA-dependent formation of open complexes at the replication origin. J Biol Chem 283:8351–8362

Ozaki S, Noguchi Y, Hayashi Y, Miyazaki E, Katayama T (2012a) Differentiation of the DnaA-*oriC* subcomplex for DNA unwinding in a replication initiation complex. J Biol Chem 287:37458–37471

Ozaki S, Noguchi Y, Nishimura M, Katayama T (2012b) Stable nucleotide binding to DnaA requires a specific glutamic acid residue within the AAA+ box II motif. J Struct Biol 179:242–250

Riber L, Løbner-Olesen A (2005) Coordinated replication and sequestration of *oriC* and *dnaA* are required for maintaining controlled once-per-cell-cycle initiation in *Escherichia coli*. J Bacteriol 187:5605–5613

Riber L, Frimodt-Møller J, Charbon G, Løbner-Olesen A (2016) Multiple DNA binding proteins contribute to timing of chromosome replication in *E. coli*. Front Mol Biosci 3:29

Richardson TT, Harran O, Murray H (2016) The bacterial DnaA-trio replication origin element specifies single-stranded DNA initiator binding. Nature 534:412–416

Rozgaja TA, Grimwade JE, Iqbal M, Czerwonka C, Vora M, Leonard AC (2011) Two oppositely oriented arrays of low-affinity recognition sites in *oriC* guide progressive binding of DnaA during *Escherichia coli* pre-RC assembly. Mol Microbiol 82:475–488

Samitt CE, Hansen FG, Miller JF, Schaechter M (1989) In vivo studies of DnaA binding to the origin of replication of *Escherichia coli*. EMBO J 8:989–993

Saxena R, Rozgaja T, Grimwade J, Crooke E (2011) Remodeling of nucleoprotein complexes is independent of the nucleotide state of a mutant AAA+ protein. J Biol Chem 286:33770–33777

Saxena R, Fingland N, Patil D, Sharma AK, Crooke E (2013) Crosstalk between DnaA protein, the initiator of *Escherichia coli* chromosomal replication, and acidic phospholipids present in bacterial membranes. Int J Mol Sci 14:8517–8537

Saxena R, Vasudevan S, Patil D, Ashoura N, Grimwade JE, Crooke E (2015) Nucleotide-induced conformational changes in *Escherichia coli* DnaA protein are required for bacterial ORC to pre-RC conversion at the chromosomal origin. Int J Mol Sci 16:27897–27911

Scholefield G, Errington J, Murray H (2012) Soj/ParA stalls DNA replication by inhibiting helix formation of the initiator protein DnaA. EMBO J 31:1542–1555

Seitz H, Weigel C, Messer W (2000) The interaction domains of the DnaA and DnaB replication proteins of *Escherichia coli*. Mol Microbiol 37:1270–1279

Sekimizu K, Kornberg A (1988) Cardiolipin activation of dnaA protein, the initiation protein of replication in *Escherichia coli*. J Biol Chem 263:7131–7135

Sekimizu K, Yung BY, Kornberg A (1988) The dnaA protein of *Escherichia coli*. Abundance, improved purification, and membrane binding. J Biol Chem 263:7136–7140

Shimizu M, Noguchi Y, Sakiyama Y, Kawakami H, Katayama T, Takada S (2016) Near-atomic structural model for bacterial DNA replication initiation complex and its functional insights. Proc Natl Acad Sci U S A 113:E8021–E8030

Skarstad K, Katayama T (2013) Regulating DNA replication in bacteria. Cold Spring Harb Perspect Biol 5:a012922

Slater S, Wold S, Lu M, Boye E, Skarstad K, Kleckner N (1995) *E. coli* SeqA protein binds *oriC* in two different methyl-modulated reactions appropriate to its roles in DNA replication initiation and origin sequestration. Cell 82:927–936

Smulczyk-Krawczyszyn A, Jakimowicz D, Ruban-Osmialowska B, Zawilak-Pawlik A, Majka J, Chater K, Zakrzewska-Czerwinska J (2006) Cluster of DnaA boxes involved in regulation of *Streptomyces* chromosome replication: from in silico to in vivo studies. J Bacteriol 188:6184–6194

Soufo CD, Soufo HJ, Noirot-Gros MF, Steindorf A, Noirot P, Graumann PL (2008) Cell-cycle-dependent spatial sequestration of the DnaA replication initiator protein in *Bacillus subtilis*. Dev Cell 15:935–941

Speck C, Weigel C, Messer W (1999) ATP- and ADP-DnaA protein, a molecular switch in gene regulation. EMBO J 18:6169–6176

Stauffer ME, Chazin WJ (2004) Structural mechanisms of DNA replication, repair, and recombination. J Biol Chem 279:30915–30918

Su'etsugu M, Emoto A, Fujimitsu K, Keyamura K, Katayama T (2003) Transcriptional control for initiation of chromosomal replication in *Escherichia coli*: fluctuation of the level of origin transcription ensures timely initiation. Genes Cells 8:731–745

Su'etsugu M, Nakamura K, Keyamura K, Kudo Y, Katayama T (2008) Hda monomerization by ADP binding promotes replicase clamp mediated DnaA-ATP hydrolysis. J Biol Chem 283:36118–36131

Su'etsugu M, Shimuta TR, Ishida T, Kawakami H, Katayama T (2005) Protein associations in DnaA-ATP hydrolysis mediated by the Hda-replicase clamp complex. J Biol Chem 280:6528–6536

Su'etsugu M, Harada Y, Keyamura K, Matsunaga C, Kasho K, Abe Y, Ueda T, Katayama T (2013) The DnaA N-terminal domain interacts with Hda to facilitate replicase clamp-mediated inactivation of DnaA. Environ Microbiol 15:3183–3195

Sutton MD, Carr KM, Vicente M, Kaguni JM (1998) *Escherichia coli* DnaA protein: the N-terminal domain and loading of DnaB helicase at the *E. coli* chromosomal origin. J Biol Chem 273:34255–34262

Theisen PW, Grimwade JE, Leonard AC, Bogan JA, Helmstetter CE (1993) Correlation of gene transcription with the time of initiation of chromosome replication in *Escherichia coli*. Mol Microbiol 10:575–584

Torheim NK, Skarstad K (1999) *Escherichia coli* SeqA protein affects DNA topology and inhibits open complex formation at *oriC*. EMBO J 18:4882–4888

Valens M, Penaud S, Rossignol M, Cornet F, Boccard F (2004) Macrodomain organization of the *Escherichia coli* chromosome. EMBO J 23:4330–4341

Waldminghaus T, Skarstad K (2009) The *Escherichia coli* SeqA protein. Plasmid 61:141–150

Wawrzycka A, Gross M, Wasaznik A, Konieczny I (2015) Plasmid replication initiator interactions with origin 13-mers and polymerase subunits contribute to strand-specific replisome assembly. Proc Natl Acad Sci USA 112:E4188–E4296

Wegrzyn K, Fuentes-Perez ME, Bury K, Rajewska M, Moreno-Herrero F, Konieczny I (2014) Sequence-specific interactions of Rep proteins with ssDNA in the AT-rich region of the plasmid replication origin. Nucleic Acids Res 42:7807–7818

Wolański M, Donczew R, Zawilak-Pawlik A, Zakrzewska-Czerwińska J (2015) *oriC*-encoded instructions for the initiation of bacterial chromosome replication. Front Microbiol 5:735

Yoshida Y, Obita T, Kokusho Y, Ohmura T, Katayama T, Ueda T, Imoto T (2003) Identification of the region in *Escherichia coli* DnaA protein required for specific recognition of the DnaA box. Cell Mol Life Sci 60:1998–2008

Yung BY, Kornberg A (1988) Membrane attachment activates dnaA protein, the initiation protein of chromosome replication in *Escherichia coli*. Proc Natl Acad Sci U S A 85:7202–7205

Zawilak-Pawlik A, Kois A, Stingl K, Boneca IG, Skrobuk P, Piotr J, Lurz R, Zakrzewska-Czerwińska J, Labigne A (2007) HobA – a novel protein involved in initiation of chromosomal replication in *Helicobacter pylori*. Mol Microbiol 65:9799–9794

Zhang Q, Zhou A, Li S, Ni J, Tao J, Lu J, Wan B, Li S, Zhang J, Zhao S, Zhao GP, Shao F, Yao YF (2016) Reversible lysine acetylation is involved in DNA replication initiation by regulating activities of initiator DnaA in *Escherichia coli*. Sci Rep 6:30837

Chapter 5
Initiation of DNA Replication in the Archaea

Stephen D. Bell

Abstract Organisms within the archaeal domain of life possess a simplified version of the eukaryotic DNA replication machinery. While some archaea possess a bacterial-like mode of DNA replication with single origins of replication per chromosome, the majority of species characterized to date possess chromosomes with multiple replication origins. Genetic, structural, and biochemical studies have revealed the nature of archaeal origin specification. Recent work has begun to shed light on the mechanisms of replication initiation in these organisms.

Keywords DNA replication • Initiator protein • Helicase • Replication origin • Archaea • *Sulfolobus*

The archaea are a diverse range of microorganisms that share more recent evolutionary history with eukaryotes than do the bacteria (Woese and Fox 1977). The precise timing of the divergence of the archaeal and eukaryotic lineages is the subject of considerable debate, with some studies even suggesting that eukaryotes arose from within the archaeal domain of life (Williams et al. 2013; Rivera and Lake 2004; Forterre 2015). A number of phyla have been identified within the Archaea; again controversy exists regarding the precise nature of the taxonomic divisions between archaeal phyla. With increased sampling, particularly at the metagenomic level, some degree of consensus is being established. It is generally accepted that there is a broad divide between the phylum of the Euryarchaea and those of the Thaumarchaea, Aigarchaea, Crenarchaea, and Korarchaea. The latter four taxonomic groupings appear more closely related to one another and have been termed the "TACK superphylum" (Guy and Ettema 2011; Brochier-Armanet et al. 2008; Forterre 2015). At the morphological level, archaea are prokaryotes; most species have a single cell membrane and are devoid of any organellar structures.

S.D. Bell (✉)
Department of Molecular and Cellular Biochemistry, Department of Biology, Indiana University, Bloomington, IN, USA
e-mail: stedbell@indiana.edu

H. Masai, M. Foiani (eds.), *DNA Replication*, Advances in Experimental Medicine and Biology 1042, https://doi.org/10.1007/978-981-10-6955-0_5

Table 5.1 Taxonomic distribution of archaeal species described in the text

Phylum	Species	Origins in main chromosome	Copy number
Crenarchaea	*Sulfolobus islandicus*	3	1C-2C
	Sulfolobus solfataricus	3	1C-2C
	Aeropyrum pernix	2	1C-2C
	Pyrobaculum calidifontis	4	1C-2C
Euryarchaea	*Pyrococcus abyssi*	1	Polyploid
	Haloferax volcanii	3 (or 4)	Polyploid
	Haloferax mediterranei	3	Polyploid
	Methanothermobacter thermautotrophicum	one mapped thus far	Polyploid
	Thermococcus kodakarensis	ND	Polyploid

The ambiguity in origin number in *Haloferax volcanii* lies in the description of the integration of an extrachromosomal element into the 3-origin main chromosome in a lab-adapted strain (Hawkins et al. 2013). While one origin has been mapped in *M. thermautotrophicum* (Majernik and Chong 2008), no genome-wide repication profiling has been performed on this organism. It is possible, therefore, that additional origins exist in this species
ND not determined

Thus far, all archaea characterized have circular chromosomes; however, the chromosome copy number shows considerable variation across taxonomic divides. To a broad approximation, euryarchaea appear to be generally oligoploid or polyploid, while the members of TACK that have been studied have cell cycles that oscillate between one and two copies of their chromosome (Table 5.1) (Samson and Bell 2014; Breuert et al. 2006; Hildenbrand et al. 2011). Flow cytometry studies have revealed that the TACK superphylum organisms, such as members of the *Sulfolobus* genus of hyperthermophilic acidophiles, have cell cycles that contain defined gap phases separating DNA replication and cell division (Lundgren et al. 2008; Pelve et al. 2013). These observations have led to the adoption of the G1, S, G2, and M phase nomenclatures established in studies of the eukaryotic cell cycle to describe the analogous stages of archaeal cell cycle progression. It must be emphasized, however, that there is no evidence that archaeal chromosome segregation is in any way related to eukaryotic mitosis. Interestingly, in marked contrast to the orchestrated cell cycles of crenarchaea, the euryarchaea that have been studied appear to lack obvious gap phases, perhaps hinting that cell division can occur during ongoing rounds of replication of the multiple copies of the chromosome, in a manner somewhat reminiscent of fast-growing *E. coli* (Sherratt 2003).

5.1 The Replication Machinery of Archaea

With the availability of whole genome sequences of archaeal species in the 1990s, it became apparent that archaea possess clear orthologs of eukaryotic DNA replication-associated proteins (Edgell and Doolittle 1997). In general, and in keeping with the organizational simplicity of the organisms, the archaeal replication proteins are simplified versions of their eukaryotic counterparts (Barry and Bell 2006; Kelman and Kelman 2014). For example, the eukaryotic MCM(2-7) replicative helicase has six distinct subunits. However, all six subunits are related to each other in sequence, suggesting derivation from a common ancestor. Indeed, the majority of present-day archaea encode a single *mcm* gene the product of which homo-multimerizes to form a homohexameric assembly (Costa and Onesti 2009; Bochman and Schwacha 2009). Similarly, almost all archaea encode a protein that is related to both Cdc6 and Orc1 component of eukaryotic origin recognition complex, ORC (Bell 2012). Interestingly, early branching eukaryotes, such as trypanosomes, also encode an archaeal-like Orc1/Cdc6 protein, suggesting that the gene duplication and sequence diversification leading to "higher" eukaryotic Orc1 and Cdc6 occurred within the eukaryotic lineage (Samson and Bell 2016; Tiengwe et al. 2012). Importantly, the bacterial replication machinery, although, ultimately, performing the same function, is largely non-orthologous to the shared archaeal/eukaryotic replication apparatus. The key exceptions lie in the clamp/loader and sliding clamp that facilitate DNA polymerization, leading to the proposal that the elongation machinery is fundamentally conserved and thus ancestral, even though the rest of the replisome components are not conserved between bacteria and archaea/eukarya (Yao and O'Donnell 2016).

5.2 Archaeal Replication Initiation

The first archaeon in which the replication mode was experimentally determined, a euryarchaeon *Pyrococcus abyssi*, revealed a single origin of replication. The origin, *oriC*, is located in a gene environment that contained genes for several replication-associated proteins, including the candidate initiator protein *orc1/cdc6* gene (Myllykallio et al. 2000; Bell 2012). The *orc1/cdc6* nomenclature is cumbersome, and orthologs in archaeal genomes have been variously annotated as *orc1* or *cdc6* on an apparently random basis. In this chapter, for simplicity's sake, I will refer to these genes as *orc1*. Many archaea encode multiple Orc1 paralogs, and I will refer to these as Orc1-1, Orc1-2, etc.

Interestingly, the single-origin paradigm in *Pyrococcus* species actually appears to be atypical among the archaea, and it is now known that many archaea from both euryarchaea and TACK species have multiple replication origins per chromosome (Robinson and Bell 2007; Robinson et al. 2004; Robinson et al. 2007; Lundgren et al. 2004; Norais et al. 2007; Hawkins et al. 2013; Yang et al. 2015; Pelve et al. 2012). The highest number of origins reported is four per chromosome for lab strains of the euryarchaeon *Haloferax volcanii* and also the crenarchaeon *Pyrobaculum calidifontis* (Pelve et al. 2012; Hawkins et al. 2013). For most species, while origin number and location have been established, the extent to which each origin is used remains poorly resolved. The exception to this lies in *Sulfolobus* species where three origins have been mapped, and these have been experimentally determined to fire once per cell cycle (Duggin et al. 2008). Studies with synchronized cell populations have revealed that two of the origins, *oriC1* and *oriC3*, fire synchronously, thereby defining the start of S phase. Notably, *oriC2* fires a few minutes later. As will be discussed below, this temporal delay is likely linked to the expression of the initiator protein that defines this origin.

Many archaea encode multiple Orc1 paralogs. In the case of *Sulfolobus*, three such proteins, Orc1-1, Orc1-2, and Orc1-3, are encoded in the 2.2–3 megabase-pair genome. *Sulfolobus* also encodes a further candidate initiator protein, WhiP, that is a distant homolog of another eukaryotic replication initiation protein, the helicase co-loader, Cdt1 (Robinson and Bell 2007).

5.3 Origin Specification

Genetic studies in *Sulfolobus islandicus* have revealed a simple one-to-one relationship between the location of initiator protein genes (Fig. 5.1) and the origins that they specify (Samson et al. 2013). More specifically, Orc1-1 is encoded adjacent to, and specifies origin function at, *oriC1*; Orc1-3 is adjacent to *oriC2* and is required for function at that origin, and finally the gene for WhiP is beside *oriC3*, and the WhiP gene product is necessary for *oriC3* function. Furthermore, the initiator protein encoded adjacent to each origin is both necessary and sufficient for its cognate origin function. What then of Orc1-2? The *orc1-2* gene is not encoded immediately adjacent to any of the three origins, and deletion of *orc1-2* does not affect firing at any of the three origins. A range of biochemical and transcriptomic analyses have implicated Orc1-2 as a negative regulator of replication (Robinson et al. 2004; Maaty et al. 2009; Frols et al. 2007; Gotz et al. 2007). However, its role in this regard remains to be firmly established. Thus, the *Sulfolobus islandicus* chromosome is a mosaic of three distinct replicons, each origin having its own specific

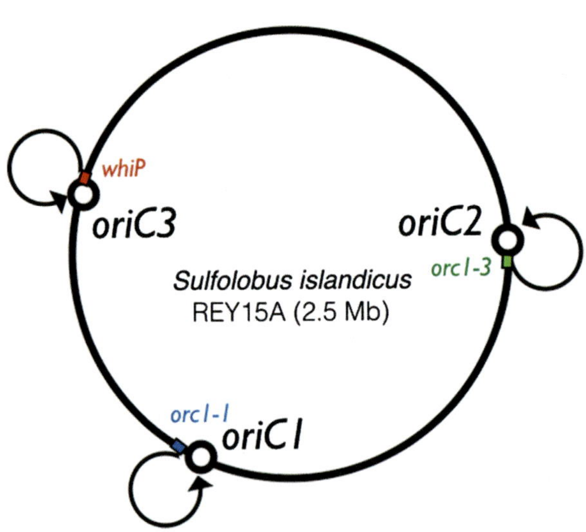

Fig. 5.1 Diagram of the organization of the 2.5 Mb chromosome of *Sulfolobus islandicus*. The relative positions of the three origins are indicated along with their cognate initiator proteins (Samson et al. 2013). Genetic dependence of the origin upon initiators is indicated by the circular arrows

initiator. Analyses of the phyletic distribution of the initiator proteins reveal that Orc1-1 is highly conserved across a broad range of archaeal species. For example, the single *orc1* gene encoded by *Pyrococcus* is most closely related to *Sulfolobus* Orc1-1. Indeed, it was demonstrated that *Sulfolobus solfataricus* Orc1-1 can bind specifically to conserved sequence elements, termed ORB (origin recognition box), in the *Pyrococcus oriC* in vitro (Robinson et al. 2004). ORB elements are conserved across the archaeal domain of life and possess a dyad symmetric element flanked uniquely on one side by a G-rich element. Interestingly, all characterized *oriC1* origins in archaea possess at least two ORB elements in inverted orientation and separated by an AT-rich candidate duplex unwinding element (see Samson and Bell 2016 for a review). The nature of Orc1-1 interaction with ORB elements is discussed below.

In contrast to the near universality of Orc1-1, Orc1-3 appears to be restricted to the Sulfolobales, and WhiP is found in both Sulfolobales and Desulfurococcales. This patchy distribution of the initiators suggests that the *oriC2* and *oriC3* origin/initiator cassettes are relatively recent acquisitions, and it has been proposed that they have been acquired by incorporation of extrachromosomal elements into an ancestral *oriC1*/Orc1-1 containing chromosome (McGeoch and Bell 2008; Robinson and Bell 2007; Samson and Bell 2014).

Direct evidence for functional incorporation of extrachromosomal origins has been documented in the halophilic euryarchaeon *Haloferax volcanii* where a lab strain differs from the parental strain by incorporation of a large plasmid, pHV4, into the host chromosome (Hawkins et al. 2013). Importantly, the origin on the plasmid remains functional in its new integrated location. The malleability of the replicon architecture of *H. volcanii* main chromosome is underscored by the remarkable observation that its replication can be maintained even in the apparent absence of active replication origins. More specifically, experiments to delete all four origins in the lab strain of *H. volcanii*'s main chromosome were successful, and, very strangely, the resultant "zero origin" strain actually outcompeted the wild type in coculture experiments. The zero origin strain was highly dependent on the RAD51/RecA ortholog, RadA, suggesting a recombination-based mechanism was able to drive genome duplication (Hawkins et al. 2013). How universal this remarkable observation is is not yet clear (Michel and Bernander 2014). When similar experiments were performed in the closely related *H. mediterranei*, the main chromosome of which normally has three active origins (Fig. 5.2), deletion of the three origins led to activation of a cryptic novel origin of replication (Yang et al. 2015). It is possible that the high ploidy, sexual promiscuity (as manifested by high levels of intraspecies and even interspecies genetic exchange mediated by this organism), and natural competence, i.e., ability to uptake DNA from the media, may be contributory to *H. volcanii*'s remarkable genomic plasticity (Zerulla et al. 2014; Zerulla and Soppa 2014; Naor et al. 2012).

Genetic studies in *Sulfolobus islandicus* (Sis) reveal that at least one replication origin is essential for viability and that each origin has a unique initiator protein. Intriguingly, this simple binary relationship of origin and initiator is not conserved across the *Sulfolobus* genus. Studies in *Sulfolobus solfataricus* (Sso) have revealed

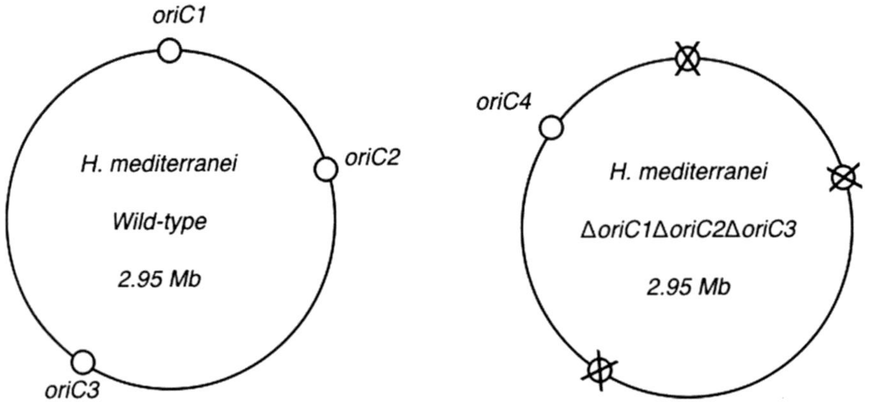

Fig. 5.2 Diagram of the organization of the 2.95 Mb main chromosome of *Haloferax mediterranei*. The locations of the three active origins in wild-type cells are shown in the left-hand panel. The right-hand panel indicated that, upon deletion *of oriC1*, *oriC2*, and *oriC3*, cell viability is maintained by activation of a novel cryptic origin, *oriC4*. For details see Yang et al. (2015)

that *oriC2* in that species is bound by both Orc1-1 and Orc1-3. While the genetic dependence of this origin on both initiators has not been tested, a range of chromatin immunoprecipitation and biochemical and structural studies have demonstrated that this origin is bound by both Orc1-1 and Orc1-3 (Robinson et al. 2004; Dueber et al. 2007; Dueber et al. 2011). The Orc1-1 and Orc1-3 binding sites at this origin are immediately adjacent, and the two proteins have a 360 $Å^2$ protein-protein interface (Dueber et al. 2007). A biochemical comparison of Orc1-1 and *oriC2* between *S. islandicus* and *S. solfataricus* revealed that both origin sequence and protein sequence have evolved to allow the binding of *S. solfataricus* Orc1-1 to *oriC2* in that species (Samson et al. 2013). This enhanced complexity of origin specification may give insight into the evolutionary transitions that drove the evolution of the multi-subunit present-day ORC complex found in eukaryotes.

5.4 Orc1 Protein Structure and Function

The structural studies of *Sulfolobus* Orc1-1 and Orc1-3 bound to *oriC2*, in conjunction with the work from the Wigley lab on *Aeropyrum pernix* Orc1-1 bound to its cognate *oriC1*, revealed some general principles of Orc1 protein/DNA interactions (Dueber et al. 2007; Gaudier et al. 2007). The archaeal Orc1 proteins are approximately 43 kDa in size and possess an N-terminal AAA+ domain and a C-terminal winged-helix (wH) DNA-binding domain (Fig. 5.3). While mutational studies had demonstrated the importance of the wH domain in DNA binding, the structural studies revealed that the AAA+ domain also contacted the DNA (Gaudier et al. 2007; Robinson et al. 2004; Dueber et al. 2007; Dueber et al. 2011). The contact between the AAA+ domain and DNA is mediated by a signature embellishment to

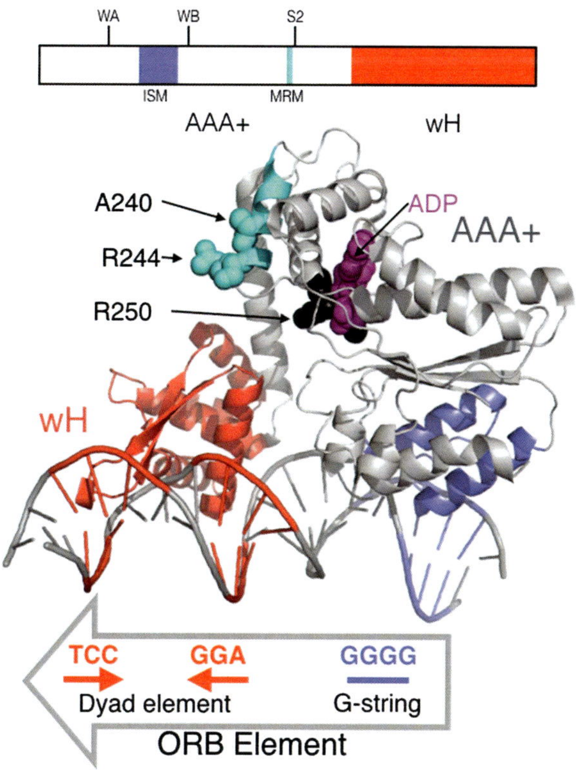

Fig. 5.3 Structure of the Orc1 proteins. The upper panel is a linear representation of the protein. The N-terminal two-thirds are a AAA+ domain, and the positions of the Walker A (WA), Walker B (WB), and Sensor 2 (S2) motifs are indicated. The ISM is the signature initiator-specific motif embellishment to the AAA+ fold found in the initiator clade of AAA+ proteins. MRM indicates the location of the MCM recruitment motif. The C-terminal third of the protein forms a winged-helix (wH) fold. The crystal structure shown below is of Orc1-1 bound to an ORB element (PDB Accession Number 2V1U). The ORB element is shown by a large gray arrow with internal dyad element and G-string element indicated. The orientation of the arrow is the same as that in Fig. 5.5. The wH domain in red interacts with the dyad symmetric element of the DNA. The ISM, in blue, mediates contacts a G-rich element, the so-called G-string. ADP is present in the active site of the AAA+ domain and is shown in magenta. The residues highlighted in cyan have been demonstrated to be essential for recruitment of MCM by Orc1-1 (residue numbering based on the *S. islandicus* Orc1-1 protein)

the classical AAA+ fold found in the initiator clade of AAA+ proteins, termed the initiator-specific motif (ISM). Thus, the *orc1* proteins make extended bipartite interactions with the origin DNA (Fig. 5.3). It had been demonstrated that Orc1-1 bound to conserved sequence elements, termed ORBs, at *oriC1* (Robinson et al. 2004). ORB elements contain a dyad symmetric element flanked on one side only by a string of G-C base pairs. The wH domain recognizes the dyad element, and the G-string interacts with the ISM (Fig. 5.3). Despite the presence of the dyad element, only a single Orc1-1 molecule binds per ORB element. The structural studies revealed that binding of Orc1-1 substantially distorts and underwinds the DNA to the extent that a second Orc1-1 molecule is unable to recognize the symmetry-related binding site (Gaudier et al. 2007). The preferred orientation of Orc1-1 on an ORB element is presumably defined by the unique ISM-G-string interaction.

Thus, at *oriC1*, Orc1-1 binds to ORB elements as a monomer. Another key feature in the structural studies was that the active site of the AAA+ ATPase domain was occupied by ADP. As no nucleotide was supplemented during purification and crystallization, this presumably reflects ATP bound during expression in *E. coli* and hydrolyzed during the expression and purification processes. Biochemical studies have confirmed that ADP is extremely stably bound to Orc1 proteins. Indeed, protein denaturation and extensive and subsequent re-folding are required to obtain nucleotide-free protein with which to perform ATPase studies (Grainge et al. 2006; Samson et al. 2013). Such studies have revealed that Orc1-1 undergoes a single-turnover ATP hydrolysis event leaving ADP stably bound in the active site. While bacterial DnaA is also active in its ATP-bound state, this activation is manifested in a fundamentally distinct manner from that of Orc1-1. ATP facilitates multimerization of DnaA, ultimately resulting in direct remodeling and melting of the origin DNA (see Bleichert et al. 2017 for a review). In contrast, Orc1-1 remains monomeric when ATP bound and undergoes a subtle conformational change (Samson et al. 2013) that facilitates interaction with MCM, as described below.

Studies using mutated versions of Orc1-1 in vivo and in vitro have revealed that stabilization of the ATP-bound form of the protein by substitution of the so-called Walker B glutamic acid residue by alanine results in a highly active form of the protein (Samson et al. 2013; Samson et al. 2016). In contrast, the ADP-bound form of Orc1-1 is inactive in MCM loading in vitro. On the biochemical level, ATP binding did not alter either the affinity or stoichiometry of Orc1-1 binding to DNA. Rather, ATP binding simply induced a modest conformational change in the protein, as detected by analytical ultracentrifugation and protease sensitivity assays. Despite these modest changes, the constitutively ATP-bound form of the protein was far more active in vitro than the ADP form (Samson et al. 2013). Thus, it appears that ATP binding and not hydrolysis is required for Orc1-1 function. Importantly, expression of the Walker B mutant form of the protein in vivo resulted in an overreplication phenotype, suggesting that ATP hydrolysis serves as an off switch. In this regard, it is significant that the *orc1-1* gene shows cell cycle-dependent regulation of its expression with transcript levels highest in cells about to enter G1 (Samson et al. 2013). Thus, the cell cycle dependence of *orc1-1* expression, coupled with the single-turnover ATP hydrolysis activity, indicates that Orc1-1 is acting as a molecular switch, permitting MCM recruitment in the Orc1-1•ATP state and inhibiting it in the Orc1-1•ADP state. Such a binary switch behavior is likely important for ensuring once-per-cell cycle regulation of origin activity (Fig. 5.4). The timing of expression of the initiator protein gene thus helps define a permissive window for initiator function. As mentioned above, *oriC2* fires a few minutes later in the cell cycle than does *oriC1*. This is reflected in the later peak of transcription of the Orc1-3 mRNA, relative to that for Orc1-1 (Samson et al. 2013). How the ADP-bound form of the initiator is removed from the origin at the end of the cell cycle is currently unknown. Possible

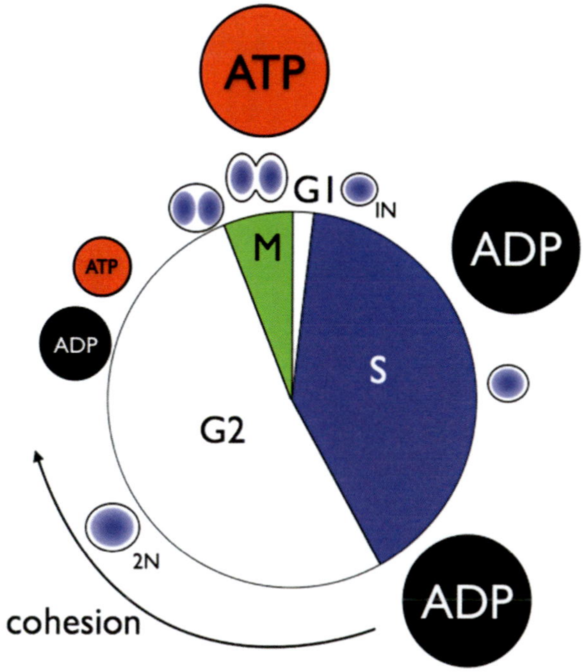

Fig. 5.4 Cartoon of the *Sulfolobus* cell cycle. The red and black spheres labeled ATP and ADP represent the nucleotide status of Orc1 proteins. A pulse of transcription of *orc1-1* at the time of cell division will produce Orc1-1 associated with ATP; subsequently ATP will be hydrolyzed to ADP in a single-turnover event. ADP will thus remain stably associated with Orc1-1 for the rest of the cell cycle

explanations include an ATP exchange factor or, perhaps more likely given the new wave of *orc1-1* transcription, targeted destruction of Orc1-1•ADP at cell division.

5.5 MCM Recruitment to Archaeal Replication Origins

As alluded to above, Orc1-1 is able to recruit MCM to *oriC1* in a defined reaction using recombinant proteins purified from *E. coli* (Samson et al. 2016). These experiments reveal that, in addition to Orc1-1 sharing sequence homology with Orc1 and Cdc6 of eukaryotes, Orc1-1 also shares Orc1 and Cdc6's respective functions of origin binding and helicase recruitment. Orc1-1•ATP was shown to contact MCM's C-terminal wH domain via a conserved motif in the lid domain of the AAA+ domain (the MRM – MCM recruitment motif; see Fig. 5.3). The basis of the ATP dependence of Orc1-1's functionality was ascribed to the Sensor 2 motif. This conserved arginine residue has the capacity to coordinate the gamma phosphate of ATP and in doing so modulate the relative positions of the two subdomains of the AAA+ module. Importantly, mutation of the Sensor 2 residue led to a protein that bound ATP but had substantially reduced ATPase activity. However, unlike the Walker B mutant that has similar ATPase-null behavior, the Sensor 2 mutant Orc1-1 was unable to recruit MCM to the origin in vitro and did not support origin firing in vivo. Thus, the

Sensor 2 residue may act to transduce the information of the nucleotide status of Orc1-1 to the conformation of the MCM recruitment site (Samson et al. 2016).

5.6 Active Loading or Passive Recruitment of MCM?

Classical views of the MCM helicase portray it as a ring-shaped hexamer (Costa and Onesti 2009). However, structural studies of both eukaryotic MCM2-7 and archaeal MCM have revealed a range of conformations. With regard to the archaeal MCMs, single and double hexamers and heptamers have been described, as have open-ring and even filamentous forms of the protein (Chen et al. 2005; Pape et al. 2003; Slaymaker et al. 2013; Samson and Bell 2016; Samson et al. 2016). There has been considerable debate about how the MCM ring might be opened to allow loading onto DNA (Yardimci and Walter 2014; Sakakibara et al. 2009). With regard to the archaeal protein, a notable electron microscopy study demonstrated that simply heating the MCM of *Methanothermobacter thermautotrophicum* to its normal physiological growth temperature resulted in greater than half of the particles adopting an open-ring conformation (Chen et al. 2005). Similarly, heat treatment of *Sulfolobus* MCM resulted in substantial elevation of recruitment of MCM by Orc1-1 to *oriC1* in vitro (Samson et al. 2016). Thus, based on Orc1-1's monomeric behavior, single-turnover ATP hydrolysis, activity when ATP bound, switch-off upon ATP hydrolysis, and the thermodynamically favored opening of MCM, we have proposed that Orc1-1 is acting as a conditional platform for MCM recruitment to replication origins. Importantly, *oriC1* possesses ORB elements aligned in inverted orientation flanking a ~90 bp AT-rich region. Replication initiation has been mapped at the boundary of this candidate duplex unwinding element, and so it is believed that two hexamers of MCM are loaded into this region by Orc1-1 bound to the flanking ORB elements (Fig. 5.5).

5.7 The Archaeal CMG Complex

The molecular basis of how initial DNA unwinding at replication origins is effected remains unknown at this time in both archaeal and eukaryotic systems. In eukaryotes, it is well established that the ultimate activation of the MCM helicase is tightly regulated and involves the facilitated recruitment of Cdc45 and GINS to form an active helicase assembly, termed CMG, that is capable of driving replication fork progression (Bell and Labib 2016).

Eukaryotic GINS is composed of four distinct subunits, Psf1, Psf2, Psf3, and Sld5 (Labib and Gambus 2007; MacNeill 2010). The subunits fall into two classes, related to each other by circular permutation. Psf2 and Psf3 have a domain order BA with a beta-strand domain followed by an alpha-helical domain. In Psf1 and Sld5, the order of the domains is switched to AB. The archaeal orthologs were initially

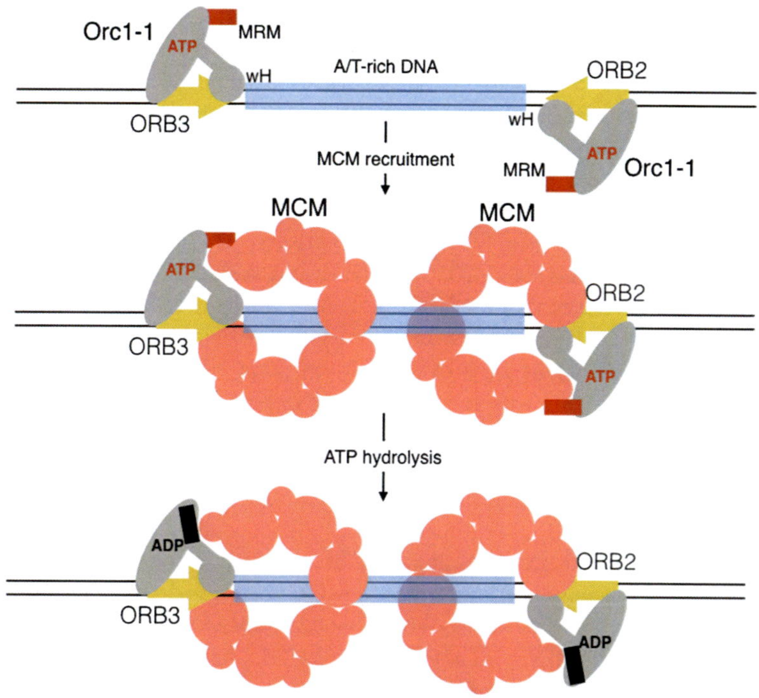

Fig. 5.5 Model of the ATP-dependent recruitment of MCM by Orc1 proteins. ATP-bound Orc1-1 associates with inverted ORB elements at *oriC1*. As illustrated in Fig. 5.3, Orc1-1 binds to ORB elements as a monomer with a defined polarity – the AAA+ module contacting a G-rich element and the wH domain binding a short inverted repeat. The region between the inverted ORB elements, colored in blue, is highly AT rich. The MRM is positioned such that it can interact with MCM, leading to MCM's recruitment to the origin with both hexamers encircling double-stranded DNA. Subsequent hydrolysis of ATP to ADP repositions the MRM (shown in black in the "off" state), preventing further rounds of MCM recruitment

identified by virtue of their ability to interact with the N-terminal domains of MCM in a yeast two-hybrid screen. The first archaeal GINS ortholog identified was shown to be related to both Psf2 and Psf3 and was thus named Gins23 (Marinsek et al. 2006). Interestingly, the *gins23* gene is encoded within a bi-cistronic operon with *mcm*. Biochemical studies revealed that Gins23 co-purified with another small protein that was revealed to be related to Psf1 and Sld5 and thus named Gins15. The archaeal GINS assembly was shown to be a tetramer, containing two copies each of Gins15 and Gins23 (Marinsek et al. 2006). While the initial work was performed in *Sulfolobus*, the archaeal GINS complex is now known to be conserved across the archaeal domain of life (MacNeill 2010; Oyama et al. 2011; Yoshimochi et al. 2008; Oyama et al. 2016). During the biochemical isolation of *Sulfolobus* GINS, a further polypeptide co-purified over eight steps and was identified as being related to the DNA-binding fold of the RecJ superfamily of proteins, leading to its initial name of RecJdbh (Marinsek et al. 2006). Subsequent work has revealed an unambiguous relationship between RecJ and eukaryotic Cdc45, and so RecJdbh has been renamed as Cdc45 (Sanchez-Pulido and Ponting 2011; Xu et al. 2016). Interestingly, Cdc45-related proteins have been identified across the archaeal domain of life but appear phylogenetically diverse (Makarova et al. 2012). One such protein, termed GAN, has been shown to be capable of association with GINS in the organism *Thermococcus kodakarensis* and, intriguingly, appears to be active as a nuclease (Li et al. 2011; Oyama et al. 2016). Recent structural studies have confirmed the GAN•GINS interaction and revealed the basis of the interaction between the GAN and the C-terminal domain of Gins15 (Oyama et al. 2016). Notably, in eukaryotes, an analogous interaction is observed between Psf1's CTD and Cdc45 (Costa et al. 2011).

In *Sulfolobus*, Cdc45 appears to be very tightly associated with GINS as evidenced by their co-purification over multiple steps (Marinsek et al. 2006). Furthermore, experiments with recombinant GINS and Cdc45 have revealed that the Cdc45•GINS complex (termed CG) is resistant to up to 8 M urea (Xu et al. 2016). Chromatin immunoprecipitation experiments have demonstrated that Cdc45 (and by inference, GINS) associates with MCM at replication origins and proceeds with the helicase during DNA synthesis. At the biochemical level, association of CG with MCM leads to a robust stimulation of helicase activity. Importantly, neither Cdc45 nor GINS when individually added to MCM results in detectable stimulation of helicase activity (Xu et al. 2016). While this latter observation agrees with initial reports that *Sulfolobus* GINS did not stimulate MCM's helicase activity (Marinsek et al. 2006), a report from the Huang laboratory has suggested that *Sulfolobus* GINS alone could stimulate MCM (Lang and Huang 2015).

One important difference between the archaeal and eukaryotic Cdc45 and GINS association lies in the composition of the assembly. While both eukaryotic Psf1 and Sld5 possess the AB domain organization, only Psf1 interacts with Cdc45 (Costa et al. 2011). This enforces a stoichiometry of one Cdc45 per GINS complex. In contrast, in the archaeal GINS, two identical copies of Gins15 are present, thus conferring the potential to interact with two Cdc45 molecules per GINS complex. Native electrospray ionization mass spectrometry experiments on the reconstituted *Sulfolobus* CG complex revealed that this was indeed the case, revealing a mass compatible with two copies each of Cdc45, Gins15, and Gins23 (Xu et al. 2016). While it has not been directly determined, it seems likely that this organization will also apply to the euryarchaeal *Thermococcus* GINS•GAN assembly (Oyama et al. 2016). Although this observation suggests a distinct difference between archaeal and eukaryotic CMG, hidden Markov modeling of the predicted structure of *Sulfolobus* Cdc45 revealed a hitherto undocumented similarity with an unanticipated region of eukaryotic Cdc45 (Xu et al. 2016). More specifically, the RecJ fold of eukaryotic Cdc45 is interrupted by a so-called CID domain (Simon et al. 2016). Surprisingly, *Sulfolobus* Cdc45 was predicted to form a structure related to this CID domain. As it had already been documented that *Sulfolobus* Cdc45 has similarities to the RecJ fold, this observation suggests that eukaryotic Cdc45 may have arisen via a gene duplication and internal fusion event, yielding a Russian doll-like organization (Fig. 5.6a). Thus, eukaryotic Cdc45 can be viewed as a pseudodimer when compared to its archaeal antecedents.

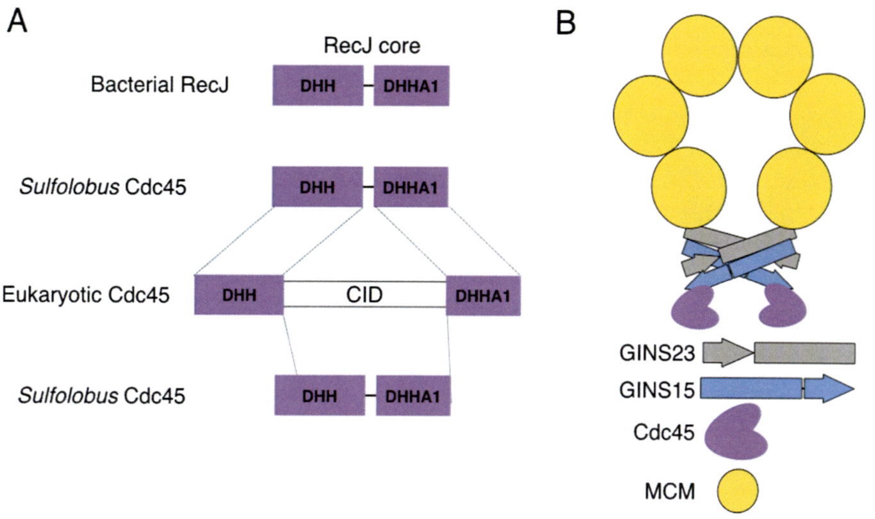

Fig. 5.6 The archaeal CMG complex. (**a**) Relationship between bacterial RecJ and archaeal and eukaryotic Cdc45. The *Sulfolobus* Cdc45 corresponds to the core fold of RecJ – comprised of the DHH and DHHA1 domains. Eukaryotic Cdc45 has these two domains separated by the "CID" domain. Hidden Markov modeling revealed that the CID may have evolved from a partial copy of a core RecJ fold. See Xu et al. (2016) for details. (**b**) Speculative model for the architecture of the archaeal CMG complex. Gins23 and Gins15 are shown in gray and blue, respectively. Their beta-strand-rich domains are shown as arrows and their alpha-helical domains as rectangles. Gins15 and Gins23 form a 2:2 complex. Further, Gins15 interacts with Cdc45, and Gins 23 interacts with MCM. An open-ring form of MCM, such as that loaded on the replication origins, is depicted

Electron microscopy studies of the eukaryotic CMG complex reveal that GINS and Cdc45 interact over the interface between MCM2 and MCM5 subunits (Costa et al. 2011). This interface serves as a gate in the MCM ring, and elegant cross-linking studies have revealed that the ability of this gate to open is key to loading eukaryotic MCM(2-7) at replication origins (Samel et al. 2014). The innate asymmetry of the eukaryotic heterohexameric MCM(2-7) makes it easy to understand how the location and stoichiometry of Cdc45 and GINS association are imposed. This contrasts with the situation in archaea where the MCM is composed of six identical subunits. However, the available data indicate that MCM is recruited to origins in an open-ring form (Samson et al. 2016). It is possible that the nature of the opening between MCM subunits is such that it favors association of CG with that locus on the MCM complex (Fig. 5.6b). It may be significant that CG interacts with MCM's N-terminal domains via the Gins23 subunit (Marinsek et al. 2006). It is conceivable that the presence of two identical MCM-interaction interfaces on archaeal CG favors interactions between MCM N-terminal domains juxtaposed across the opening in the MCM ring.

In eukaryotes, the sequential and regulated associations of first Cdc45 and then GINS with loaded MCM are pivotal events in the control of the initiation of DNA replication (Siddiqui et al. 2013; Tanaka and Araki 2013; Bell and Labib 2016).

Interestingly, the so-called firing factors that facilitate this process (e.g., Sld2, Sld3, Sld7, Dpb11) are eukaryotic innovations with no discernable homologs in the archaea. Furthermore, the CDK and DDK kinases that in turn govern the behavior of the firing factors are also absent from archaea. The tight association of Cdc45 and GINS in archaeal cell extracts might imply that these factors interact en bloc with origin-associated MCM, leading to activation of MCM's helicase activity. Whether this step in archaeal DNA replication initiation is subject to regulatory control is currently unknown. However, in species such as *Sulfolobus* where multiple replication origins are coordinately regulated to trigger a single initiation event per cell cycle, it is very tempting to speculate that MCM activation by CG could be a key and committing step in regulating replication initiation.

Acknowledgments I would like to thank Rachel Samson for helpful discussions of this material.

References

Barry ER, Bell SD (2006) DNA replication in the archaea. Microbiol Mol Biol Rev 70(4):876–887. https://doi.org/10.1128/mmbr.00029-06

Bell SD (2012) Archaeal orc1/cdc6 proteins. Subcell Biochem 62:59–69. https://doi.org/10.1007/978-94-007-4572-8_4

Bell SP, Labib K (2016) Chromosome duplication in Saccharomyces cerevisiae. Genetics 203(3):1027–1067. https://doi.org/10.1534/genetics.115.186452

Bleichert F, Botchan MR, Berger JM (2017) Mechanisms for initiating cellular DNA replication. Science 355:eaah6317. https://doi.org/10.1126/science.aah6317

Bochman ML, Schwacha A (2009) The Mcm complex: unwinding the mechanism of a replicative helicase. Microbiol Mol Biol Rev 73(4):652–683. https://doi.org/10.1128/mmbr.00019-09

Breuert S, Allers T, Spohn G, Soppa J (2006) Regulated polyploidy in halophilic archaea. PLoS One 1(1):e92. https://doi.org/10.1371/journal.pone.0000092

Brochier-Armanet C, Boussau B, Gribaldo S, Forterre P (2008) Mesophilic crenarchaeota: proposal for a third archaeal phylum, the Thaumarchaeota. Nat Rev Microbiol 6(3):245–252. https://doi.org/10.1038/nrmicro1852

Chen YJ, Yu XO, Kasiviswanathan R, Shin JH, Kelman Z, Egelman EH (2005) Structural polymorphism of Methanothermobacter thermautotrophicus MCM. J Mol Biol 346(2):389–394. https://doi.org/10.1016/j.jmb.2004.11.076

Costa A, Onesti S (2009) Structural biology of MCM helicases. Crit Rev Biochem Mol Biol 44(5):326–342. https://doi.org/10.1080/10409230903186012

Costa A, Ilves I, Tamberg N, Petojevic T, Nogales E, Botchan MR, Berger JM (2011) The structural basis for MCM2-7 helicase activation by GINS and Cdc45. Nat Struct Mol Biol 18(4):471-U110. https://doi.org/10.1038/nsmb.2004

Dueber ELC, Corn JE, Bell SD, Berger JM (2007) Replication origin recognition and deformation by a heterodimeric archaeal Orc1 complex. Science 317(5842):1210–1213. https://doi.org/10.1126/science.1143690

Dueber EC, Costa A, Corn JE, Bell SD, Berger JM (2011) Molecular determinants of origin discrimination by Orc1 initiators in archaea. Nucleic Acids Res 39(9):3621–3631. https://doi.org/10.1093/nar/gkq1308

Duggin IG, McCallum SA, Bell SD (2008) Chromosome replication dynamics in the archaeon Sulfolobus acidocaldarius. Proc Natl Acad Sci U S A 105(43):16737–16742. https://doi.org/10.1073/pnas.0806414105

Edgell DR, Doolittle WF (1997) Archaea and the origin(s) of DNA replication proteins. Cell 89(7):995–998

Forterre P (2015) The universal tree of life: an update. Front Microbiol 6:717. https://doi.org/10.3389/fmicb.2015.00717

Frols S, Gordon PM, Panlilio MA, Duggin IG, Bell SD, Sensen CW, Schleper C (2007) Response of the hyperthermophilic archaeon Sulfolobus solfataricus to UV damage. J Bacteriol 189(23):8708–8718. https://doi.org/10.1128/JB.01016-07

Gaudier M, Schuwirth BS, Westcott SL, Wigley DB (2007) Structural basis of DNA replication origin recognition by an ORC protein. Science 317(5842):1213–1216. https://doi.org/10.1126/science.1143664

Gotz D, Paytubi S, Munro S, Lundgren M, Bernander R, White MF (2007) Responses of hyperthermophilic crenarchaea to UV irradiation. Genome Biol 8(10):R220. https://doi.org/10.1186/gb-2007-8-10-r220

Grainge I, Gaudier M, Schuwirth BS, Westcott SL, Sandall J, Atanassova N, Wigley DB (2006) Biochemical analysis of a DNA replication origin in the archaeon Aeropyrum pernix. J Mol Biol 363(2):355–369. https://doi.org/10.1016/jmb.2006.07.076

Guy L, Ettema TJ (2011) The archaeal 'TACK' superphylum and the origin of eukaryotes. Trends Microbiol 19(12):580–587. https://doi.org/10.1016/j.tim.2011.09.002

Hawkins M, Malla S, Blythe MJ, Nieduszynski CA, Allers T (2013) Accelerated growth in the absence of DNA replication origins. Nature 503(7477):544–547. https://doi.org/10.1038/nature12650

Hildenbrand C, Stock T, Lange C, Rother M, Soppa J (2011) Genome copy numbers and gene conversion in methanogenic archaea. J Bacteriol 193:734–743

Kelman LM, Kelman Z (2014) Archaeal DNA replication. Annu Rev Genet 48:71–97. https://doi.org/10.1146/annurev-genet-120213-092148

Labib K, Gambus A (2007) A key role for the GINS complex at DNA replication forks. Trends Cell Biol 17(6):271–278. https://doi.org/10.1016/j.tcb.2007.04.002

Lang S, Huang L (2015) The Sulfolobus solfataricus GINS complex stimulates DNA binding and processive DNA unwinding by minichromosome maintenance helicase. J Bacteriol 197(21):3409–3420. https://doi.org/10.1128/JB.00496-15

Li Z, Pan M, Santangelo TJ, Chemnitz W, Yuan W, Edwards JL, Hurwitz J, Reeve JN, Kelman Z (2011) A novel DNA nuclease is stimulated by association with the GINS complex. Nucleic Acids Res 39(14):6114–6123. https://doi.org/10.1093/nar/gkr181

Lundgren M, Andersson A, Chen LM, Nilsson P, Bernander R (2004) Three replication origins in Sulfolobus species: synchronous initiation of chromosome replication and asynchronous termination. Proc Natl Acad Sci U S A 101(18):7046–7051. https://doi.org/10.1073/pnas.0400656101

Lundgren M, Malandrin L, Eriksson S, Huber H, Bernander R (2008) Cell cycle characteristics of Crenarchaeota: unity among diversity. J Bacteriol 190(15):5362–5367. https://doi.org/10.1128/jb.00330-08

Maaty WS, Wiedenheft B, Tarlykov P, Schaff N, Heinemann J, Robison-Cox J, Valenzuela J, Dougherty A, Blum P, Lawrence CM, Douglas T, Young MJ, Bothner B (2009) Something old, something new, something borrowed; how the thermoacidophilic archaeon Sulfolobus solfataricus responds to oxidative stress. PLoS One 4(9):e6964. https://doi.org/10.1371/journal.pone.0006964

MacNeill SA (2010) Structure and function of the GINS complex, a key component of the eukaryotic replisome. Biochem J 425(3):489–500. https://doi.org/10.1042/BJ20091531

Makarova KS, Koonin EV, Kelman Z (2012) The CMG (CDC45/RecJ, MCM, GINS) complex is a conserved component of the DNA replication system in all archaea and eukaryotes. Biol Direct 7:7. https://doi.org/10.1186/1745-6150-7-7

Majernik AI, Chong JP (2008) A conserved mechanism for replication origin recognition and binding in archaea. Biochem J 409(2):511–518

Marinsek N, Barry ER, Makarova KS, Dionne I, Koonin EV, Bell SD (2006) GINS, a central nexus in the archaeal DNA replication fork. EMBO Rep 7(5):539–545. https://doi.org/10.1038/sj.embor.7400649

McGeoch AT, Bell SD (2008) Extra-chromosomal elements and the evolution of cellular DNA replication machineries. Nat Rev Mol Cell Biol 9(7):569–574. https://doi.org/10.1038/nrm2426

Michel B, Bernander R (2014) Chromosome replication origins: do we really need them? BioEssays 36(6):585–590. https://doi.org/10.1002/bies.201400003

Myllykallio H, Lopez P, Lopez-Garcia P, Heilig R, Saurin W, Zivanovic Y, Philippe H, Forterre P (2000) Bacterial mode of replication with eukaryotic-like machinery in a hyperthermophilic archaeon. Science 288(5474):2212–2215

Naor A, Lapierre P, Mevarech M, Papke RT, Gophna U (2012) Low species barriers in halophilic archaea and the formation of recombinant hybrids. Curr Biol 22(15):1444–1448. https://doi.org/10.1016/j.cub.2012.05.056

Norais C, Hawkins M, Hartman AL, Eisen JA, Myllykallio H, Allers T (2007) Genetic and physical mapping of DNA replication origins in Haloferax volcanii. PLoS Genet 3:729–743. https://doi.org/10.1371/journal.pgen.0030077

Oyama T, Ishino S, Fujino S, Ogino H, Shirai T, Mayanagi K, Saito M, Nagasawa N, Ishino Y, Morikawa K (2011) Architectures of archaeal GINS complexes, essential DNA replication initiation factors. BMC Biol 9:28. https://doi.org/10.1186/1741-7007-9-28

Oyama T, Ishino S, Shirai T, Yamagami T, Nagata M, Ogino H, Kusunoki M, Ishino Y (2016) Atomic structure of an archaeal GAN suggests its dual roles as an exonuclease in DNA repair and a CMG component in DNA replication. Nucleic Acids Res 44(19):9505–9517. https://doi.org/10.1093/nar/gkw789

Pape T, Meka H, Chen SX, Vicentini G, van Heel M, Onesti S (2003) Hexameric ring structure of the full-length archaeal MCM protein complex. EMBO Rep 4(11):1079–1083. https://doi.org/10.1038/sj.embor.7400010

Pelve EA, Lindas AC, Knoppel A, Mira A, Bernander R (2012) Four chromosome replication origins in the archaeon Pyrobaculum calidifontis. Mol Microbiol 85(5):986–995. https://doi.org/10.1111/j.1365-2958.2012.08155.x

Pelve EA, Martens-Habbena W, Stahl DA, Bernander R (2013) Mapping of active replication origins in vivo in thaum- and euryarchaeal replicons. Mol Microbiol 90(3):538–550. https://doi.org/10.1111/mmi.12382

Rivera MC, Lake JA (2004) The ring of life provides evidence for a genome fusion origin of eukaryotes. Nature 431(7005):152–155. https://doi.org/10.1038/nature02848

Robinson NP, Bell SD (2007) Extrachromosomal element capture and the evolution of multiple replication origins in archaeal chromosomes. Proc Natl Acad Sci U S A 104(14):5806–5811. https://doi.org/10.1073/pnas.0700206104

Robinson NP, Dionne I, Lundgren M, Marsh VL, Bernander R, Bell SD (2004) Identification of two origins of replication in the single chromosome of the Archaeon Sulfolobus solfataricus. Cell 116(1):25–38

Robinson NP, Blood KA, McCallum SA, Edwards PAW, Bell SD (2007) Sister chromatid junctions in the hyperthermophilic archaeon Sulfolobus solfataricus. EMBO J 26(3):816–824. https://doi.org/10.1038/sj.emboj.7601529

Sakakibara N, Kelman LM, Kelman Z (2009) How is the archaeal MCM helicase assembled at the origin? Possible mechanisms. Biochem Soc Trans 37(Pt 1):7–11. https://doi.org/10.1042/BST0370007

Samel SA, Fernandez-Cid A, Sun J, Riera A, Tognetti S, Herrera MC, Li H, Speck C (2014) A unique DNA entry gate serves for regulated loading of the eukaryotic replicative helicase MCM2-7 onto DNA. Genes Dev 28(15):1653–1666. https://doi.org/10.1101/gad.242404.114

Samson RY, Bell SD (2014) Archaeal chromosome biology. J Mol Microbiol Biotechnol 24(5–6):420–427. https://doi.org/10.1159/000368854

Samson RY, Bell SD (2016) Archaeal DNA replication origins and recruitment of the MCM replicative helicase. Enzyme 39:169–190. https://doi.org/10.1016/bs.enz.2016.03.002

Samson RY, Xu Y, Gadelha C, Stone TA, Faqiri JN, Li D, Qin N, Pu F, Liang YX, She Q, Bell SD (2013) Specificity and function of archaeal DNA replication initiator proteins. Cell Rep 3(2):485–496. https://doi.org/10.1016/j.celrep.2013.01.002

Samson RY, Abeyrathne PD, Bell SD (2016) Mechanism of archaeal MCM helicase recruitment to DNA replication origins. Mol Cell 61(2):287–296. https://doi.org/10.1016/j.molcel.2015.12.005

Sanchez-Pulido L, Ponting CP (2011) Cdc45: the missing RecJ ortholog in eukaryotes? Bioinformatics 27(14):1885–1888. https://doi.org/10.1093/bioinformatics/btr332

Sherratt DJ (2003) Bacterial chromosome dynamics. Science 301(5634):780–785

Siddiqui K, On KF, Diffley JF (2013) Regulating DNA replication in eukarya. Cold Spring Harb Perspect Biol 5(9):a010371. https://doi.org/10.1101/cshperspect.a012930

Simon AC, Sannino V, Costanzo V, Pellegrini L (2016) Structure of human Cdc45 and implications for CMG helicase function. Nat Commun 7:11638. https://doi.org/10.1038/ncomms11638

Slaymaker IM, Fu Y, Toso DB, Ranatunga N, Brewster A, Forsburg SL, Zhou ZH, Chen XS (2013) Mini-chromosome maintenance complexes form a filament to remodel DNA structure and topology. Nucleic Acids Res 41(5):3446–3456. https://doi.org/10.1093/nar/gkt022

Tanaka S, Araki H (2013) Helicase activation and establishment of replication forks at chromosomal origins of replication. Cold Spring Harb Perspect Biol 5(12):a010371. https://doi.org/10.1101/cshperspect.a010371

Tiengwe C, Marcello L, Farr H, Gadelha C, Burchmore R, Barry JD, Bell SD, McCulloch R (2012) Identification of ORC1/CDC6-interacting factors in Trypanosoma brucei reveals critical features of origin recognition complex architecture. PLoS One 7(3):e32674. https://doi.org/10.1371/journal.pone.0032674

Williams TA, Foster PG, Cox CJ, Embley TM (2013) An archaeal origin of eukaryotes supports only two primary domains of life. Nature 504(7479):231–236. https://doi.org/10.1038/nature12779

Woese CR, Fox GE (1977) Phylogenetic structure of the prokaryotic domain: the primary kingdoms. Proc Natl Acad Sci U S A 74(11):5088–5090

Xu Y, Gristwood T, Hodgson B, Trinidad JC, Albers SV, Bell SD (2016) Archaeal orthologs of Cdc45 and GINS form a stable complex that stimulates the helicase activity of MCM. Proc Natl Acad Sci U S A 113(47):13390–13395. https://doi.org/10.1073/pnas.1613825113

Yang H, Wu Z, Liu J, Liu X, Wang L, Cai S, Xiang H (2015) Activation of a dormant replication origin is essential for Haloferax mediterranei lacking the primary origins. Nat Commun 6:8321. https://doi.org/10.1038/ncomms9321

Yao NY, O'Donnell ME (2016) Evolution of replication machines. Crit Rev Biochem Mol Biol 51(3):135–149. https://doi.org/10.3109/10409238.2015.1125845

Yardimci H, Walter JC (2014) Prereplication-complex formation: a molecular double take? Nat Struct Mol Biol 21(1):20–25. https://doi.org/10.1038/nsmb.2738

Yoshimochi T, Fujikane R, Kawanami M, Matsunaga F, Ishino Y (2008) The GINS complex from Pyrococcus furiosus stimulates the MCM helicase activity. J Biol Chem 283(3):1601–1609. https://doi.org/10.1074/jbc.M707654200

Zerulla K, Soppa J (2014) Polyploidy in haloarchaea: advantages for growth and survival. Front Microbiol 5:274. https://doi.org/10.3389/fmicb.2014.00274

Zerulla K, Chimileski S, Nather D, Gophna U, Papke RT, Soppa J (2014) DNA as a phosphate storage polymer and the alternative advantages of polyploidy for growth or survival. PLoS One 9(4):e94819. https://doi.org/10.1371/journal.pone.0094819

Chapter 6
Mechanism of Lagging-Strand DNA Replication in Eukaryotes

Joseph L. Stodola and Peter M. Burgers

Abstract This chapter focuses on the enzymes and mechanisms involved in lagging-strand DNA replication in eukaryotic cells. Recent structural and biochemical progress with DNA polymerase α-primase (Pol α) provides insights how each of the millions of Okazaki fragments in a mammalian cell is primed by the primase subunit and further extended by its polymerase subunit. Rapid kinetic studies of Okazaki fragment elongation by Pol δ illuminate events when the polymerase encounters the double-stranded RNA-DNA block of the preceding Okazaki fragment. This block acts as a progressive molecular break that provides both time and opportunity for the flap endonuclease 1 (FEN1) to access the nascent flap and cut it. The iterative action of Pol δ and FEN1 is coordinated by the replication clamp PCNA and produces a regulated degradation of the RNA primer, thereby preventing the formation of long-strand displacement flaps. Occasional long flaps are further processed by backup nucleases including Dna2.

Keywords DNA replication • Lagging strand • Okazaki fragment maturation • DNA polymerase α-primase • DNA polymerase δ • Flap endonuclease 1 • Dna2

6.1 Introduction

Three DNA polymerases are responsible for the bulk of genomic DNA replication, Pol α, Pol δ, and Pol ε. A preponderance of evidence supports the following division of labor at the replication fork: The Pol α-primase complex primes synthesis on both the leading and lagging strands, with Pol ε synthesizing the leading strand and Pol δ synthesizing the discontinuous Okazaki fragments that make up the lagging strand (Burgers 2009). This model has been supported by analysis of replication errors (Pursell et al. 2007; Nick McElhinny et al. 2008; Larrea et al. 2010), studies of

J.L. Stodola • P.M. Burgers (✉)
Department of Biochemistry and Molecular Biophysics, Washington University School of Medicine, Saint Louis, MO, USA
e-mail: burgers@biochem.wustl.edu

H. Masai, M. Foiani (eds.), *DNA Replication*, Advances in Experimental Medicine and Biology 1042, https://doi.org/10.1007/978-981-10-6955-0_6

polymerase localization on replication forks (Yu et al. 2014), and genomic rNMP incorporation studies (Nick McElhinny et al. 2010a; Miyabe et al. 2011; Reijns et al. 2015; Daigaku et al. 2015; Koh et al. 2015; Clausen et al. 2015). Biochemical studies have shown that Pols ε and δ replicate their respective strands spontaneously in the presence of purified CMG helicase (Cdc45-Mcm_{2-7}-GINS) complex (Georgescu et al. 2014a, 2015) and are excluded from the incorrect strand (Schauer and O'Donnell 2017). For these reasons, the model placing Pol ε on the leading strand and Pol δ on the lagging strand has become widely accepted.

A recent study has suggested an alternate arrangement of polymerases at the replication fork (Johnson et al. 2015), concluding that Pol δ replicates both strands of the replication fork. These conclusions have become a matter of debate in the field (Stillman 2015; Johnson et al. 2016; Burgers et al. 2016), and some very recent biochemical data support a very limited engagement of Pol δ during the initiation of leading-strand DNA replication (Yeeles et al. 2017). However, no study disputes the current model of lagging-strand DNA replication involving the synthetic activities of Pol α-primase and Pol δ, which will be the primary focus of this review.

6.2 Priming by Pol α-Primase

DNA synthesis on both strands of the fork is initiated by the synthesis of RNA primers by the Pol α-primase complex. Pol α and its associated primase each contain one accessory subunit, forming a hetero-tetrameric complex overall, often designated as the eukaryotic primosome. The polymerase catalytic and accessory subunits are Pol1 and Pol12, respectively, in budding yeast and p180 and p70 in human cells (Johansson and Dixon 2013). The catalytic subunit is one of the four, eukaryotic B-family polymerases, which comprises a conserved polymerase domain and a separate C-terminal domain that is connected to the polymerase domain by a flexible linker (Klinge et al. 2009; Suwa et al. 2015; Kilkenny et al. 2012; Baranovskiy et al. 2016a). Interactions between the catalytic and the accessory subunit are made through this C-terminal domain (denoted $p180_C$ below). Similarly, the primase contains a catalytic and an accessory subunit: Pri1 and Pri2, respectively, in yeast, and p49 and p58 in humans. Integral to the mechanism described below, the primase accessory subunit contains two domains (N-terminal and C-terminal, denoted $p58_N$ and $p58_C$ below) connected by a flexible linker (Baranovskiy et al. 2015, 2016b).

The primase initiates RNA synthesis de novo, synthesizing an 8–10-nucleotide primer that is transferred to the polymerase subunit of the Pol α-primase complex for extension with dNTPs (Baranovskiy et al. 2016b; Singh et al. 1986; Kuchta et al. 1990; Kuchta and Stengel 2010) and then creating an ~30-nucleotide hybrid primer that becomes the substrate for Pol δ (Bullock et al. 1991; Murakami and Hurwitz 1993). The mechanism by which Pol α-primase makes uniformly sized RNA primers has long been unclear. Recent structural and biochemical studies with the human and yeast primosome have contributed to the proposal of a new model for primer synthesis (Klinge et al. 2009; Baranovskiy et al. 2015; Vaithiyalingam et al. 2014;

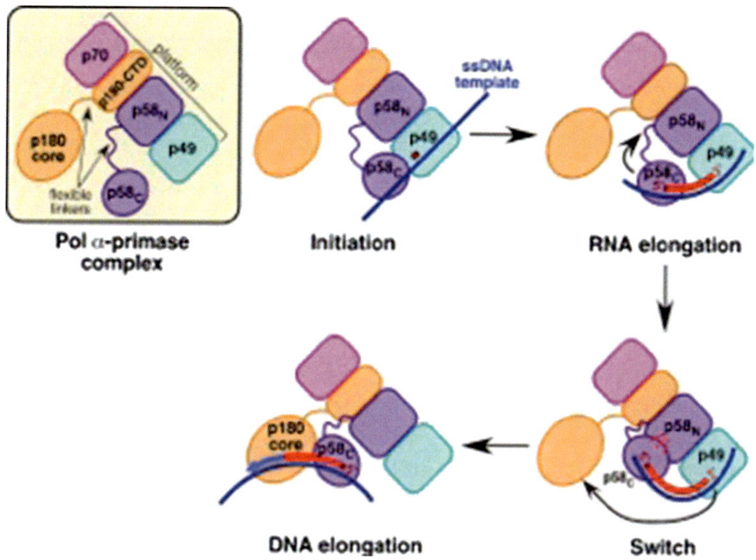

Fig. 6.1 Priming of DNA synthesis by Pol α-primase. The sequential steps in the initiation of RNA priming, the elongation of the RNA primer, and the switch to DNA synthesis are shown. The model of Pol α-primase is based on Baranovskiy et al. (2016a)

Agarkar et al. 2011; Nunez-Ramirez et al. 2011; Sauguet et al. 2010; Kilkenny et al. 2013; Perera et al. 2013). This model is outlined by Baranovskiy et al. and described below (Baranovskiy et al. 2016a).

The crystal structure of the apo form of the primosome (not bound to DNA) indicates that the entire complex is built upon a stable platform with a p49-p58$_N$-p180$_C$-p70 arrangement (human subunit designations). Flexible linkers connect the polymerase (p180$_{core}$) and the C-terminal half of the primase accessory subunit (p58$_C$) to this platform (Fig. 6.1). Large conformational changes of p180$_{core}$ and p58$_C$ with respect to the primosome platform enable the substrate exchanges necessary for priming and extension. De novo RNA synthesis occurs at the interface of the primase catalytic and accessory subunits (p49-p58$_C$ interface) (Zerbe and Kuchta 2002). As the new primer grows, p58$_C$ retains interactions with its 5′-terminus and rotates away from p49. Eventually, this rotation brings on steric clashes between p58$_C$ and p58$_N$ (Fig. 6.1). Molecular modeling predicts that these clashes would occur when the RNA primer had reached about ten nucleotides in length, providing an explanation for why RNA primers longer than ten nucleotides are rarely produced (Baranovskiy et al. 2016a).

After further RNA synthesis is inhibited, the DNA/RNA duplex is intramolecularly transferred from the primase to the polymerase subunit (Fig. 6.1). Since p58$_C$ makes extensive contacts with the duplex and p49 is only weakly bound, it is predicted that p58$_C$ delivers the primer terminus to the polymerase. Molecular modeling of the potential transfer complex predicts that the polymerase is only able to access the 3′-primer terminus when the primer is at least nine nucleotides in length, consistent with biochemical data (Baranovskiy et al. 2016a). Pol α extends the RNA primers for an additional 20–30 nucleotides with dNTPs, yielding a 30–40-nucleotide-long primer. This estimate dates back to early in vitro SV40 replication studies (Bullock et al. 1991; Murakami and Hurwitz 1993).

6.3 Polymerase Exchanges at Pol α-Synthesized Primers

After primer synthesis, Pol α is exchanged for Pol ε or Pol δ for further high-fidelity DNA replication. It is still unclear how DNA synthesis by Pol α remains so precisely limited and how polymerase exchange occurs. Several mechanisms have been proposed. First, it has been hypothesized that the different helical characteristics of the RNA/DNA duplex and double-stranded DNA may be sensed by the Pol α active site. Pol α has been shown to bind more tightly to RNA/DNA duplexes, which adopt an A-form helix, than to the B-form DNA helix (Perera et al. 2013). As Pol α extends the RNA primer with dNTPs, the A-form helix initially present will be converted to a B-form helix. It has been proposed that the formation of the B-form structure inhibits further synthesis by Pol α (Perera et al. 2013). However, the biochemical experiments supporting this hypothesis were performed using poly(dT) templates, where formation of triplex structures (dT-dA-dT) after limited replication causes inhibition of DNA synthesis by most DNA polymerases, and not just Pol α (Mikhailov and Bogenhagen 1996; Zhang et al. 2016). As a result, the extent to which Pol α extended these homopolymeric templates was artificially low.

Alternately, it has been suggested that the switch from Pol α to Pol δ is mediated by loading of PCNA onto 3′-primer termini by the RFC complex (Schauer and O'Donnell 2017; Tsurimoto and Stillman 1991; Eki et al. 1992; Yuzhakov et al. 1999; Maga et al. 2000; Mossi et al. 2000). In the absence of PCNA, the RFC complex has been shown to inhibit Pol α activity when present at high concentrations (Yuzhakov et al. 1999; Maga et al. 2000). However, Pol α inhibition is greatly enhanced when both RFC and PCNA are both present, suggesting that clamp loading is integral to polymerase switching (Schauer and O'Donnell 2017; Tsurimoto and Stillman 1991). Polymerase switching has also been shown to be stimulated by the presence of the single-stranded binding protein RPA at the template-primer junction. RPA directly binds RFC, providing specificity of PCNA loading and the displacement of Pol α (Yuzhakov et al. 1999; Gomes and Burgers 2001). Regardless of the exact details of the mechanism of Pol α ejection, the preponderance of evidence points to PCNA loading by RFC as essential to the recruitment of Pol δ, which prevents rebinding of Pol α.

CMG helicase-dependent leading- and lagging-strand synthesis has recently been reconstituted in vitro using the budding yeast replication system (Georgescu et al. 2015, 2014b; Yeeles et al. 2017; Devbhandari et al. 2017). These studies have provided biochemical support for the current model of the eukaryotic replication fork, with Pol ε replicating the leading strand and Pol δ the lagging strand, and Pol α priming synthesis on both strands. It appears from these studies that replicating a bidirectional fork in the presence of the CMG helicase complex enforces the division of labor of the replication machinery; i.e., Pol ε is suppressed on the lagging strand, and Pol δ is suppressed on the leading strand (Schauer and O'Donnell 2017). Interestingly, Diffley and coworkers found that leading-strand replication proceeded more efficiently if the initial elongation of the leading-strand primer was carried out by Pol δ, followed by a second exchange from Pol δ to Pol ε (Yeeles et al. 2017). Presumably, the latter polymerase exchange occurs when the elongating PCNA-Pol δ complex collides with the leading CMG complex ahead of it and GINS enforces the exchange to Pol ε.

6.4 Strand Displacement Synthesis and Nick Translation

Pol δ extends primers on the lagging strand until it reaches the 5′-end of the preceding Okazaki fragment. Before ligation, however, the initiator RNA at the 5′-terminus of the primer must be removed. Biochemical and genetic studies support a model in which the initiator RNA is predominantly removed through the joint action of Pol δ and flap endonuclease 1 (FEN1), a structure-specific nuclease (Grasby et al. 2012; Balakrishnan and Bambara 2013). When Pol δ collides with the previous Okazaki fragment, it continues replicating through limited displacement of the RNA primer, forming a short 5′-flap. This flap is the substrate for FEN1; repetition of strand displacement synthesis by Pol δ followed by FEN1 cleavage removes the initiator RNA (Garg et al. 2004; Rossi and Bambara 2006; Stodola and Burgers 2016). Most frequently, one- or two-nucleotide products are liberated by FEN1 (Stodola and Burgers 2016; Stith et al. 2008). After removal of the RNA through these iterative actions of Pol δ and FEN1, a process termed "nick translation," the nick can be sealed by DNA ligase. These basic steps are sufficient to process the vast majority of Okazaki fragments. On rare occasions, strand displacement synthesis may become decoupled from flap cutting, and flaps can grow to lengths that cannot be processed by FEN1 (Murante et al. 1995; Bae et al. 2001a). Backup mechanisms, described below, are required to cleave these flaps so that they do not lead to DNA damage.

Pol δ possesses two enzymatic activities, a DNA polymerase activity and a 3–5′-exonuclease activity. The exonuclease is required for proofreading of misincorporated nucleotides during DNA replication but also plays an important role in Okazaki fragment maturation (Jin et al. 2001). After Pol δ initiates strand displacement synthesis, the forward, flap-generating movement of the polymerase is countered by the exonucleolytic activity of the polymerase. After formation of a short flap, the exonuclease activity of Pol δ cuts out the nucleotides that the polymerase had inserted, with the release of dNMPs. Repeated short-flap formation followed by exonucleolytic cleavage back to the nick position has been termed "polymerase idling." This activity appears to be unique to the lagging-strand polymerase; Pol ε exhibits very weak strand displacement and idling activities (Garg et al. 2004; Ganai et al. 2016). In most sequence contexts, idling is sufficient to restrict forward movement of Pol δ to within three nucleotides of the nick position, and most frequently, one- or two-nucleotide products are liberated by FEN1 (Garg et al. 2004; Stodola and Burgers 2016; Stith et al. 2008).

Polymerase idling is not the only restraint placed on strand displacement. As Pol δ initiates strand displacement, the rate of forward polymerase movement slows down progressively as the 5′-flap grows longer (Stodola and Burgers 2016), i.e., the growing flap inhibits further synthesis in a length-dependent manner. Thus, the nascent flap acts as a "molecular brake" on the polymerase. Idling and flap inhibition allow Pol δ to produce a substrate with a short 5′-flap for FEN1 while simultaneously limiting extensive strand displacement synthesis. This cooperation is necessary. If the rate of strand displacement remained constant irrespective of flap length, idling alone would be insufficient to constrain the polymerase near the nick.

Surprisingly, the ability of Pol δ to displace the duplex region of the preceding Okazaki fragment is not dependent on the nature of the block, i.e., RNA versus DNA, but solely on the stability of the duplex (Stodola and Burgers 2016; Stith et al. 2008). Extensive strand displacement synthesis is favored in sequence contexts with low duplex stability such as AT-rich sequences. Furthermore, when flaps reach a critical length, Pol δ continues strand displacement synthesis in a manner that is decoupled from its regulatory mechanisms, generating long flaps (Ayyagari et al. 2003). This "critical length" remains poorly defined. It is possible that the "molecular brake" exerted on Pol δ only applies in situations where flaps are very short, perhaps due to interactions between the polymerase and the 5′-end of the flap (Koc et al. 2015). Perhaps, the failure of the flap-controlling mechanisms could be caused by a failure of very long flaps to interact with the enzyme. Further investigation is required to more fully examine this phenomenon.

6.5 Short-Flap Processing by FEN1

The iterative action of Pol δ and FEN1 removes initiator RNA so that nick ligation can occur. In vitro, these enzymes together comprise an efficient maturation machine, rapidly degrading either RNA or DNA annealed downstream of Pol δ (Stodola and Burgers 2016; Stith et al. 2008; Lin et al. 2013). In the absence of DNA ligase, nick translation can continue indefinitely unless it is blocked by other DNA-binding proteins, as observed in yeast (Smith and Whitehouse 2012). Much effort has been dedicated to determine the structure of FEN1's optimal substrate. The consensus model is that FEN1 most efficiently cuts double-flap structures with a single-nucleotide 3′-flap and a variable length 5′-flap (Kao et al. 2002; Tsutakawa et al. 2011, 2014). Irrespective of the length of the 5′-flap, FEN1 cuts a single base into the 5′-duplex region, yielding a ligatable nick when the single-nt

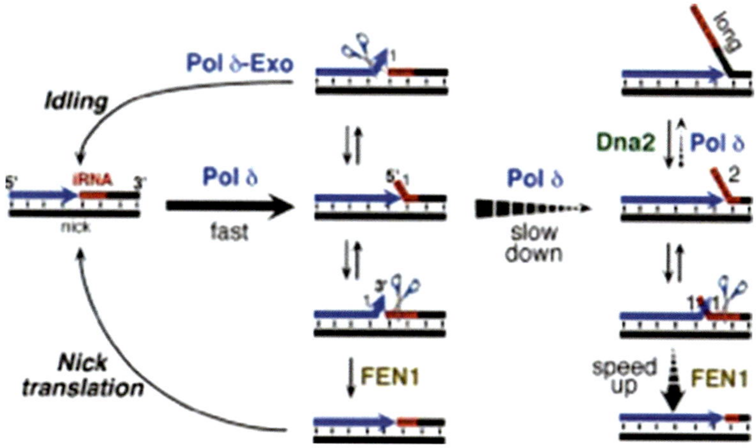

Fig. 6.2 Regulatory steps that limit strand displacement synthesis by Pol δ. The formation of long flaps is restricted by the 3′-exonuclease activity of Pol δ (*idling*), by a progressive slowdown of strand displacement synthesis as the flap grows, and by cutting of the nascent flap by FEN1. FEN1 cleavage may be accelerated when double-flap structures are formed as flaps grow in size, thereby further limiting their length. Long flaps that still do occur are trimmed by Dna2

3′-flap reanneals to the template (Fig. 6.2) (Kao et al. 2002; Kaiser et al. 1999; Xie et al. 2001).

Whether this optimal substrate, requiring at least two unpaired nucleotides, represents the substrate that is *most often* cut during nick translation has recently been addressed. Data suggest that the major FEN1 substrate in nick translation results from a single-nucleotide flap (Stodola and Burgers 2016). Since a 3′-flap is required for FEN1 activity, we hypothesize that the single-nucleotide 5′-flap formed by Pol δ strand displacement must re-equilibrate into a single-nucleotide 3′-flap and a fully base-paired 5′-junction before cutting (Fig. 6.2). Although shown to be efficiently cut by FEN1, this single-flap structure is not processed as avidly as double-flap structures. Thus, these data suggests that in most contexts, the major FEN1 substrate is not actually the optimal substrate (Stodola and Burgers 2016; Kao et al. 2002). These observations could be interpreted as a contradiction, but it may in fact represent another layer of regulation limiting the formation of long flaps. It is likely that during nick translation, FEN1 binds and cuts double-flap structures more avidly than single 3′-flap structures. The higher-affinity binding of FEN1 to these double-flap intermediates would aid in the preferential recruitment of FEN1 to longer flaps in the case that the enzyme was not associated with PCNA-Pol δ at the start of strand displacement synthesis. Such a mechanism would ensure that flaps longer than a single nucleotide are processed before Pol δ strand displacement extends too far.

6.6 Alternative and Long-Flap Processing

The occurrence of long flaps in the cell was initially inferred from genetic studies in *S. cerevisiae*. Deletion of *RAD27*, which encodes FEN1, is associated with a dramatic increase in the occurrence of duplications between direct repeats, up to ~100 nt in length, as it could result from slippage mispairing of long 5′-flaps (Tishkoff et al. 1997a). The related exonuclease 1 (Exo1) also shows flap processing activity and can process nascent flaps generated during strand displacement synthesis by Pol δ, although less efficiently than FEN1 (Tran et al. 2002; Sparks et al. 2012). However, the spectrum of mutations observed in an *exo1Δ* strain is most consistent with a defect in mismatch repair rather than in Okazaki fragment maturation (Tran et al. 2001). Because *rad27 exo1* double mutants are lethal, the model has been proposed that Exo1 serves as a backup nuclease for FEN1, and in the absence of both enzymes, the burden of long flaps overwhelms the ability of the cell to process them (Stith et al. 2008; Tishkoff et al. 1997b). Further genetic studies have highlighted Dna2 as the principal enzyme responsible for processing long flaps. For instance, the conditional lethality of *DNA2* mutations is suppressed by overexpression of *RAD27*, and the temperature sensitivity of *rad27Δ* is suppressed by *DNA2* overexpression (Budd and Campbell 1997).

Based on biochemical studies, FEN1 has been apportioned the task of processing short flaps and Dna2 that of long flaps (reviewed in Burgers 2009; Kang et al. 2010; Balakrishnan and Bambara 2010; Burgers and Kunkel 2017). Long flaps are operationally defined as those longer than ~20 nucleotides, the length at which RPA stably binds flaps (Kumaran et al. 2006). FEN1 itself can cleave long flaps in vitro, but when the 5′-flap is coated with RPA or assumes a secondary structure, FEN1 cutting is abrogated (Murante et al. 1995). In wild-type cells, long flaps could be formed in certain sequence environments, such as AT-rich sequences, where strand displacement synthesis by Pol δ is predicted to be very rapid (Stodola and Burgers 2016). Alternatively, Pol δ strand displacement could become decoupled from flap cutting for other reasons, e.g., if FEN1 and Exo1 were absent from the replisome. In addition, the generation of long flaps is enhanced by the action of Pif1 helicase (Budd et al. 2006; Rossi et al. 2008) or by a defect in the proofreading activity of Pol δ (Jin et al. 2003). Therefore, backup mechanisms are required to process long flaps and rescue replication forks (Stith et al. 2008; Jin et al. 2001, 2005).

S. cerevisiae Dna2 is a multifunctional enzyme with nuclease, helicase, and cell-cycle checkpoint activities (Lee et al. 2000; Budd et al. 2000; Bae et al. 2001b; Kumar and Burgers 2013). Of these activities, the nuclease is most critical to Okazaki fragment maturation. Dna2 nuclease threads onto the 5′-end of flaps, displacing RPA before cutting DNA (Stewart et al. 2010; Zhou et al. 2015). In several reports, Dna2 was observed to cleave flaps several nucleotides away from their base, leaving behind a ~5–8-nucleotide 5′-flap (Bae et al. 2001a, 1998; Kao et al. 2004). Additionally, in one report, cutting at the base of the flap was also observed (Levikova and Cejka 2015). However, regardless of the exact cleavage accuracy of Dna2, efficient Okazaki fragment maturation of long-flap intermediates requires additional nucleolytic processing beyond that by Dna2. Either additional 5′-flap cutting by FEN1 or 3′-exonucleolytic processing by the proofreading activity of Pol δ is required to produce ligatable nicks with high efficiency. When in biochemical studies, Dna2 was the only 5′-nuclease provided, the maturation of long flaps carried out with a proofreading-defective form of Pol δ produced ligatable nicks very inefficiently (Jin et al. 2003; Levikova and Cejka 2015). Consistent with these biochemical results is the observation that yeast mutants defective for the Pol δ 3′-exonuclease activity are exquisitely sensitive to additional defects in FEN1 (Jin et al. 2001).

A recent electron-microscopic study of isolated fission yeast replication forks provides structural support for the existence of the long-flap pathway (Liu et al. 2017). In wild-type cells, 10% of the isolated forks had associated with it a 40–50-nt-long flap. Often, these long flaps were detected kilobases distant from the fork. Because the EM methodology cannot detect very short flaps and nicks that are normally generated during short-flap processing, one conclusion from these data is that long flaps are rare. The frequency of long flaps increased in *rad2* (*S. pombe* FEN1) mutants as well as in *dna2* mutants. These results are consistent with the model, because FEN1 defects are expected to generate more long flaps while Dna2 defects are expected to abrogate their resolution. Accordingly, quantification of the frequency of long flaps in the *dna2*⁻ mutant should give a good estimate of their normal occurrence during Okazaki fragment maturation. In *dna2*⁻ cells, 32% of the forks showed long flaps, and the average distance between long flaps was about 6.5 kb. If one assumes that an Okazaki fragment is ~150 nt in length, one can estimate that long flaps are generated at a frequency of 1–2%. With about 50,000 Okazaki fragments being generated per fission yeast cell cycle, this amounts to as many as 500–1000 long flaps, which makes it unsurprising that *dna2* is essential for cell growth in *S. pombe*, as it is in *S. cerevisiae* (Kang et al. 2000; Budd et al. 1995).

The same EM study also determined the role of fission yeast RNase H2 and Exo1 in Okazaki fragment maturation (Liu et al. 2017). Defects in RNase H2 (*rnh201Δ)* did not result in a significant increase in the frequency of long flaps, suggesting that this enzyme does not participate in the degradation of the RNA primers during Okazaki fragment maturation. However, an *exo1Δ* mutant showed a clear increase in the frequency of long flaps, suggesting that Exo1 participates in Okazaki fragment maturation in wild-type cells. When compared with the known phenotypes of *S. cerevisiae exo1Δ* (see above), it appears that in *S. pombe*, Exo1 plays a more prominent role in Okazaki fragment maturation.

While there is strong evidence in both yeasts for the processing of long flaps by Dna2, the situation is less clear in human cells. Human Dna2 has been shown to play a role in nuclear genome maintenance, specifically promoting the rescue of stalled replication forks (Thangavel et al. 2015). However, currently there is no strong evidence for a role for Dna2 in Okazaki fragment maturation analogous to its role in both yeasts (Duxin et al. 2009, 2012). It is unknown whether human Okazaki fragment maturation can be accomplished by just FEN1 and Exo1 or whether long flaps are processed by additional nucleases redundant with Dna2, or different nucleases.

6.7 DNA Ligation

Following the removal of initiator RNA, nicks are sealed by DNA ligase I (*cdc9* in budding yeast and *LIG1* in human cells) (Howes and Tomkinson 2012). The eukaryotic ligase contains a conserved PCNA-interacting protein motif that binds PCNA in the interdomain connection loop (Vijayakumar et al. 2007). This interaction is important for localizing ligase to replication foci and for completing Okazaki fragment maturation in mammalian cells (Montecucco et al. 1998; Levin et al. 2000). PCNA has also been shown to stimulate ligase activity on nicked DNA substrates (Tom et al. 2001). Despite these effects, it is unclear whether the ligase is actually a stable component of the PCNA-mediated maturation complex like Pol δ and FEN1. During in vitro Okazaki fragment maturation, yeast ligase acts distributively, with the position of ligation following RNA removal dependent on the ligase concentration (Ayyagari et al. 2003). The cause of this observation remains unclear, but it is possible that when Pol δ and FEN1 are bound to PCNA, ligase cannot gain access to the PCNA ring, resulting in distributive ligation.

6.8 Limits to Nick Translation and the Size of Okazaki Fragments

The transient nature of Okazaki fragments has made the study of their properties in vivo difficult. However, advances have been made in recent years in isolating and examining Okazaki fragments in vivo and also in reconstituting lagging-strand replication in vitro. Both approaches have yielded new insights into the controls placed on Okazaki fragment synthesis and maturation.

Recent in vitro replication studies showed that, when lagging-strand replication was coupled to leading-strand synthesis by CMG-Pol ε, Pol α spontaneously primed on the lagging strand (Georgescu et al. 2015; Yeeles et al. 2017). The distance between priming events decreased as the concentration of Pol α in the assay was raised, indicating that priming itself is stochastic (Yeeles et al. 2017). PCNA-Pol δ spontaneously extended these Pol α-synthesized primers, producing Okazaki fragments that ranged from 100 to 500 nucleotides (Georgescu et al. 2015; Yeeles et al. 2017). Chromatin structure further modulated the size distribution of Okazaki fragments (Devbhandari et al. 2017; Kurat et al. 2017).

Maturation of these synthesized Okazaki fragments minimally requires FEN1 and DNA ligase as well as PCNA-Pol δ. In the absence of ligase, Pol δ and FEN1 could perform nick translation indefinitely (Ayyagari et al. 2003), although this would represent a major inefficiency in lagging-strand DNA replication. There is strong evidence that the chromatin context of the cell places a limit on the amount of nick translation synthesis that can be performed by Pol δ FEN1 (Smith and Whitehouse 2012). By purifying Okazaki fragments from a budding yeast strain deficient for DNA ligase, the Whitehouse Group found that the size distribution of Okazaki fragments was strongly influenced by the placement of nucleosomes, with Okazaki fragment termini preferentially located at nucleosome dyads (Smith and Whitehouse 2012). Thus, it appears that a bound nucleosome upstream of the nick translation machinery is enough to block its further movement. This phenomenon has been extended to other DNA-binding proteins that bind the double-stranded DNA downstream of the migrating nick; transcription factor binding sites have been shown to be correlated with Okazaki fragment termination sites (Reijns et al. 2015; Smith and Whitehouse 2012). The lagging-strand replication machinery has also been shown to be blocked by double-stranded DNA-binding proteins in vitro (Koc et al. 2016).

It is currently unknown whether nick translation regularly extends to where nucleosomes or protein blocks are positioned or whether DNA is ligated before PCNA-Pol δ and FEN1 reach these blocks. Since the observations discussed above were generated in a yeast strain deficient for ligase, this data may report more on the limits placed on the maturation machinery rather than representing true Okazaki fragments (Smith and Whitehouse 2012; Smith et al. 2015). Since ligase acts distributively, in some situations ligation could leave some Pol α-synthesized DNA in the mature genome, despite the fact that more extensive nick translation would replace the lower-fidelity DNA produced by Pol α with that of the higher-fidelity Pol δ (Kadyrov et al. 2009; Liu et al. 2015). Indeed, several studies have shown that a significant amount of Pol α-synthesized DNA remains in the mature yeast genome (Reijns et al. 2015; Nick McElhinny et al. 2010b; Lujan et al. 2014). It remains to be determined to what extent nucleosome positioning directly influences the retention of Pol α-synthesized DNA.

Acknowledgments The research in the authors' laboratory is supported by grants from the US National Institutes of Health (GM032431, GM083970, and GM118129 to P.B). J.S. was supported in part by a grant from the USA-Israel Binational Science Foundation (2013358).

References

Agarkar VB, Babayeva ND, Pavlov YI, Tahirov TH (2011) Crystal structure of the C-terminal domain of human DNA primase large subunit: implications for the mechanism of the primase-polymerase alpha switch. Cell Cycle 10(6):926–931. https://doi.org/10.4161/cc.10.6.15010

Ayyagari R, Gomes XV, Gordenin DA, Burgers PM (2003) Okazaki fragment maturation in yeast. I. Distribution of functions between FEN1 AND DNA2. J Biol Chem 278(3):1618–1625

Bae SH, Choi E, Lee KH, Park JS, Lee SH, Seo YS (1998) Dna2 of Saccharomyces cerevisiae possesses a single-stranded DNA-specific endonuclease activity that is able to act on double-stranded DNA in the presence of ATP. J Biol Chem 273(41):26880–26890

Bae SH, Bae KH, Kim JA, Seo YS (2001a) RPA governs endonuclease switching during processing of Okazaki fragments in eukaryotes. Nature 412(6845):456–461

Bae SH, Kim JA, Choi E, Lee KH, Kang HY, Kim HD, Kim JH, Bae KH, Cho Y, Park C, Seo YS (2001b) Tripartite structure of Saccharomyces cerevisiae Dna2 helicase/endonuclease. Nucleic Acids Res 29(14):3069–3079

Balakrishnan L, Bambara RA (2010) Eukaryotic lagging strand DNA replication employs a multi-pathway mechanism that protects genome integrity. J Biol Chem 286(9):6865–6870. R110.209502 [pii] 10.1074/jbc.R110.209502

Balakrishnan L, Bambara RA (2013) Flap endonuclease 1. Annu Rev Biochem 82:119–138. https://doi.org/10.1146/annurev-biochem-072511-122603

Baranovskiy AG, Zhang Y, Suwa Y, Babayeva ND, Gu J, Pavlov YI, Tahirov TH (2015) Crystal structure of the human primase. J Biol Chem 290(9):5635–5646. https://doi.org/10.1074/jbc.M114.624742

Baranovskiy AG, Babayeva ND, Zhang Y, Gu J, Suwa Y, Pavlov YI, Tahirov TH (2016a) Mechanism of concerted Rna-DNA primer synthesis by the human primosome. J Biol Chem. https://doi.org/10.1074/jbc.M116.717405

Baranovskiy AG, Zhang Y, Suwa Y, Gu J, Babayeva ND, Pavlov YI, Tahirov TH (2016b) Insight into the human DNA primase interaction with template-primer. J Biol Chem 291(9):4793–4802. https://doi.org/10.1074/jbc.M115.704064

Budd ME, Campbell JL (1997) A yeast replicative helicase, Dna2 helicase, interacts with yeast FEN-1 nuclease in carrying out its essential function. Mol Cell Biol 17(4):2136–2142

Budd ME, Choe W-C, Campbell J (1995) *DNA2* encodes a DNA helicase essential for replication of eukaryotic chromosomes. J Biol Chem 270(45):26766–26769

Budd ME, Choe W, Campbell JL (2000) The nuclease activity of the yeast DNA2 protein, which is related to the RecB-like nucleases, is essential in vivo. J Biol Chem 275(22):16518–16529

Budd ME, Reis CC, Smith S, Myung K, Campbell JL (2006) Evidence suggesting that Pif1 helicase functions in DNA replication with the Dna2 helicase/nuclease and DNA polymerase delta. Mol Cell Biol 26(7):2490–2500

Bullock PA, Seo YS, Hurwitz J (1991) Initiation of simian virus 40 DNA synthesis in vitro. Mol Cell Biol 11(5):2350–2361

Burgers PM (2009) Polymerase dynamics at the eukaryotic DNA replication fork. J Biol Chem 284(7):4041–4045

Burgers PM, Kunkel TA (2017) Eukaryotic DNA replication fork. Annu Rev Biochem. https://doi.org/10.1146/annurev-biochem-061516-044709

Burgers PM, Gordenin D, Kunkel TA (2016) Who is leading the replication fork, Pol epsilon or Pol delta? Mol Cell 61(4):492–493. https://doi.org/10.1016/j.molcel.2016.01.017

Clausen AR, Lujan SA, Burkholder AB, Orebaugh CD, Williams JS, Clausen MF, Malc EP, Mieczkowski PA, Fargo DC, Smith DJ, Kunkel TA (2015) Tracking replication enzymology in vivo by genome-wide mapping of ribonucleotide incorporation. Nat Struct Mol Biol 22(3):185–191. nsmb.2957 [pii] 10.1038/nsmb.2957

Daigaku Y, Keszthelyi A, Muller CA, Miyabe I, Brooks T, Retkute R, Hubank M, Nieduszynski CA, Carr AM (2015) A global profile of replicative polymerase usage. Nat Struct Mol Biol 22(3):192–198. https://doi.org/10.1038/nsmb.2962

Devbhandari S, Jiang J, Kumar C, Whitehouse I, Remus D (2017) Chromatin constrains the initiation and elongation of DNA replication. Mol Cell 65(1):131–141. https://doi.org/10.1016/j.molcel.2016.10.035

Duxin JP, Dao B, Martinsson P, Rajala N, Guittat L, Campbell JL, Spelbrink JN, Stewart SA (2009) Human Dna2 is a nuclear and mitochondrial DNA maintenance protein. Mol Cell Biol 29(15):4274–4282. MCB.01834-08 [pii] 10.1128/MCB.01834-08

Duxin JP, Moore HR, Sidorova J, Karanja K, Honaker Y, Dao B, Piwnica-Worms H, Campbell JL, Monnat RJ Jr, Stewart SA (2012) Okazaki fragment processing-independent role for human Dna2 enzyme during DNA replication. J Biol Chem 287(26):21980–21991. https://doi.org/10.1074/jbc.M112.359018

Eki T, Matsumoto T, Murakami Y, Hurwitz J (1992) The replication of DNA containing the simian virus 40 origin by the monopolymerase and dipolymerase systems. J Biol Chem 267(11):7284–7294

Ganai RA, Zhang XP, Heyer WD, Johansson E (2016) Strand displacement synthesis by yeast DNA polymerase epsilon. Nucleic Acids Res. https://doi.org/10.1093/nar/gkw556

Garg P, Stith CM, Sabouri N, Johansson E, Burgers PM (2004) Idling by DNA polymerase delta maintains a ligatable nick during lagging-strand DNA replication. Genes Dev 18(22):2764–2773

Georgescu RE, Langston L, Yao NY, Yurieva O, Zhang D, Finkelstein J, Agarwal T, O'Donnell ME (2014a) Mechanism of asymmetric polymerase assembly at the eukaryotic replication fork. Nat Struct Mol Biol 21(8):664–670. https://doi.org/10.1038/nsmb.2851

Georgescu RE, Yao N, Indiani C, Yurieva O, O'Donnell ME (2014b) Replisome mechanics: lagging strand events that influence speed and processivity. Nucleic Acids Res 42(10):6497–6510. https://doi.org/10.1093/nar/gku257

Georgescu RE, Schauer GD, Yao NY, Langston LD, Yurieva O, Zhang D, Finkelstein J, O'Donnell ME (2015) Reconstitution of a eukaryotic replisome reveals suppression mechanisms that define leading/lagging strand operation. Elife 4:e04988. https://doi.org/10.7554/eLife.04988

Gomes XV, Burgers PM (2001) ATP utilization by yeast replication factor C. I. ATP-mediated interaction with DNA and with proliferating cell nuclear antigen. J Biol Chem 276(37):34768–34775

Grasby JA, Finger LD, Tsutakawa SE, Atack JM, Tainer JA (2012) Unpairing and gating: sequence-independent substrate recognition by FEN superfamily nucleases. Trends Biochem Sci 37(2):74–84. https://doi.org/10.1016/j.tibs.2011.10.003

Howes TR, Tomkinson AE (2012) DNA ligase I, the replicative DNA ligase. Subcell Biochem 62:327–341. https://doi.org/10.1007/978-94-007-4572-8_17

Jin YH, Obert R, Burgers PM, Kunkel TA, Resnick MA, Gordenin DA (2001) The 3′-->5′ exonuclease of DNA polymerase delta can substitute for the 5′ flap endonuclease Rad27/Fen1 in processing Okazaki fragments and preventing genome instability. Proc Natl Acad Sci U S A 98(9):5122–5127

Jin YH, Ayyagari R, Resnick MA, Gordenin DA, Burgers PM (2003) Okazaki fragment maturation in yeast. II. Cooperation between the polymerase and 3′-5′-exonuclease activities of Pol delta in the creation of a ligatable nick. J Biol Chem 278(3):1626–1633

Jin YH, Garg P, Stith CM, Al Refai H, Sterling J, Weston L, Kunkel T, Resnick MA, Burgers PM, Gordenin DA (2005) The multiple biological roles for the 3′-5′-exonuclease of DNA polymerase d require switching between the polymerase and exonuclease domains. Mol Cell Biol 25:461–471

Johansson E, Dixon N (2013) Replicative DNA polymerases. Cold Spring Harb Perspect Biol 5(6):1–14. 5/6/a012799 [pii] 10.1101/cshperspect.a012799

Johnson RE, Klassen R, Prakash L, Prakash S (2015) A major role of DNA polymerase delta in replication of both the leading and lagging DNA strands. Mol Cell 59(2):163–175. S1097-2765(15)00439-6 [pii] 10.1016/j.molcel.2015.05.038

Johnson RE, Klassen R, Prakash L, Prakash S (2016) Response to Burgers et al. Mol Cell 61(4):494–495. https://doi.org/10.1016/j.molcel.2016.01.018

Kadyrov FA, Genschel J, Fang Y, Penland E, Edelmann W, Modrich P (2009) A possible mechanism for exonuclease 1-independent eukaryotic mismatch repair. Proc Natl Acad Sci U S A 106(21):8495–8500. https://doi.org/10.1073/pnas.0903654106

Kaiser MW, Lyamicheva N, Ma W, Miller C, Neri B, Fors L, Lyamichev VI (1999) A comparison of eubacterial and archaeal structure-specific 5′-exonucleases. J Biol Chem 274(30):21387–21394

Kang HY, Choi E, Bae SH, Lee KH, Gim BS, Kim HD, Park C, MacNeill SA, Seo YS (2000) Genetic analyses of Schizosaccharomyces pombe dna2(+) reveal that dna2 plays an essential role in Okazaki fragment metabolism. Genetics 155(3):1055–1067

Kang YH, Lee CH, Seo YS (2010) Dna2 on the road to Okazaki fragment processing and genome stability in eukaryotes. Crit Rev Biochem Mol Biol 45(2):71–96. https://doi.org/10.3109/10409230903578593

Kao HI, Henricksen LA, Liu Y, Bambara RA (2002) Cleavage specificity of Saccharomyces cerevisiae flap endonuclease 1 suggests a double-flap structure as the cellular substrate. J Biol Chem 277(17):14379–14389

Kao HI, Campbell JL, Bambara RA (2004) Dna2p helicase/nuclease is a tracking protein, like FEN1, for flap cleavage during Okazaki fragment maturation. J Biol Chem 279(49):50840–50849. https://doi.org/10.1074/jbc.M409231200

Kilkenny ML, De Piccoli G, Perera RL, Labib K, Pellegrini L (2012) A conserved motif in the C-terminal tail of DNA polymerase alpha tethers primase to the eukaryotic replisome. J Biol Chem 287(28):23740–23747. M112.368951 [pii] 10.1074/jbc.M112.368951

Kilkenny ML, Longo MA, Perera RL, Pellegrini L (2013) Structures of human primase reveal design of nucleotide elongation site and mode of Pol alpha tethering. Proc Natl Acad Sci U S A 110(40):15961–15966. https://doi.org/10.1073/pnas.1311185110

Klinge S, Nunez-Ramirez R, Llorca O, Pellegrini L (2009) 3D architecture of DNA Pol alpha reveals the functional core of multi-subunit replicative polymerases. EMBO J 28(13):1978–1987. emboj2009150 [pii] 10.1038/emboj.2009.150

Koc KN, Stodola JL, Burgers PM, Galletto R (2015) Regulation of yeast DNA polymerase delta-mediated strand displacement synthesis by 5′-flaps. Nucleic Acids Res 43(8):4179–4190. gkv260 [pii] 10.1093/nar/gkv260

Koc KN, Singh SP, Stodola JL, Burgers PM, Galletto R (2016) Pif1 removes a Rap1-dependent barrier to the strand displacement activity of DNA polymerase delta. Nucleic Acids Res 44(8):3811–3819. https://doi.org/10.1093/nar/gkw181

Koh KD, Balachander S, Hesselberth JR, Storici F (2015) Ribose-seq: global mapping of ribonucleotides embedded in genomic DNA. Nat Methods 12(3):251–257. nmeth.3259 [pii] 10.1038/nmeth.3259

Kuchta RD, Stengel G (2010) Mechanism and evolution of DNA primases. Biochim Biophys Acta 1804(5):1180–1189. https://doi.org/10.1016/j.bbapap.2009.06.011

Kuchta RD, Reid B, Chang LM (1990) DNA primase. Processivity and the primase to polymerase alpha activity switch. J Biol Chem 265(27):16158–16165

Kumar S, Burgers PM (2013) Lagging strand maturation factor Dna2 is a component of the replication checkpoint initiation machinery. Genes Dev 27(3):313–321. gad.204750.112 [pii] 10.1101/gad.204750.112

Kumaran S, Kozlov AG, Lohman TM (2006) Saccharomyces cerevisiae replication protein A binds to single-stranded DNA in multiple salt-dependent modes. Biochemistry 45(39):11958–11973. https://doi.org/10.1021/bi060994r

Kurat CF, Yeeles JT, Patel H, Early A, Diffley JF (2017) Chromatin controls DNA replication origin selection, lagging-strand synthesis, and replication fork rates. Mol Cell 65(1):117–130. https://doi.org/10.1016/j.molcel.2016.11.016

Larrea AA, Lujan SA, Nick McElhinny SA, Mieczkowski PA, Resnick MA, Gordenin DA, Kunkel TA (2010) Genome-wide model for the normal eukaryotic DNA replication fork. Proc Natl Acad Sci U S A 107(41):17674–17679. 1010178107 [pii] 10.1073/pnas.1010178107

Lee KH, Kim DW, Bae SH, Kim JA, Ryu GH, Kwon YN, Kim KA, Koo HS, Seo YS (2000) The endonuclease activity of the yeast Dna2 enzyme is essential in vivo. Nucleic Acids Res 28(15):2873–2881

Levikova M, Cejka P (2015) The Saccharomyces cerevisiae Dna2 can function as a sole nuclease in the processing of Okazaki fragments in DNA replication. Nucleic Acids Res 43(16):7888–7897. https://doi.org/10.1093/nar/gkv710

Levin DS, McKenna AE, Motycka TA, Matsumoto Y, Tomkinson AE (2000) Interaction between PCNA and DNA ligase I is critical for joining of Okazaki fragments and long-patch base-excision repair. Curr Biol 10(15):919–922

Lin SH, Wang X, Zhang S, Zhang Z, Lee EY, Lee MY (2013) Dynamics of enzymatic interactions during short flap human Okazaki fragment processing by two forms of human DNA polymerase delta. DNA Repair (Amst) 12(11):922–935. https://doi.org/10.1016/j.dnarep.2013.08.008

Liu S, Lu G, Ali S, Liu W, Zheng L, Dai H, Li H, Xu H, Hua Y, Zhou Y, Ortega J, Li GM, Kunkel TA, Shen B (2015) Okazaki fragment maturation involves alpha-segment error editing by the mammalian FEN1/MutS alpha functional complex. EMBO J 34(13):1829–1843. 10.15252/embj.201489865

Liu B, Hu J, Wang J, Kong D (2017) Direct visualization of RNA-DNA primer removal from Okazaki fragments provides support for flap cleavage and exonucleolytic pathways in eukaryotic cells. J Biol Chem. https://doi.org/10.1074/jbc.M116.758599

Lujan SA, Clausen AR, Clark AB, MacAlpine HK, MacAlpine DM, Malc EP, Mieczkowski PA, Burkholder AB, Fargo DC, Gordenin DA, Kunkel TA (2014) Heterogeneous polymerase fidelity and mismatch repair bias genome variation and composition. Genome Res 24(11):1751–1764. https://doi.org/10.1101/gr.178335.114

Maga G, Stucki M, Spadari S, Hubscher U (2000) DNA polymerase switching: I. Replication factor C displaces DNA polymerase alpha prior to PCNA loading. J Mol Biol 295(4):791–801

Mikhailov VS, Bogenhagen DF (1996) Termination within oligo(dT) tracts in template DNA by DNA polymerase gamma occurs with formation of a DNA triplex structure and is relieved by mitochondrial single-stranded DNA-binding protein. J Biol Chem 271(48):30774–30780

Miyabe I, Kunkel TA, Carr AM (2011) The major roles of DNA polymerases epsilon and delta at the eukaryotic replication fork are evolutionarily conserved. PLoS Genet 7(12):e1002407. https://doi.org/10.1371/journal.pgen.1002407. PGENETICS-D-11-01459 [pii]

Montecucco A, Rossi R, Levin DS, Gary R, Park MS, Motycka TA, Ciarrocchi G, Villa A, Biamonti G, Tomkinson AE (1998) DNA ligase I is recruited to sites of DNA replication by an interaction with proliferating cell nuclear antigen: identification of a common targeting mechanism for the assembly of replication factories. EMBO J 17(13):3786–3795

Mossi R, Keller RC, Ferrari E, Hubscher U (2000) DNA polymerase switching: II. Replication factor C abrogates primer synthesis by DNA polymerase alpha at a critical length. J Mol Biol 295(4):803–814

Murakami Y, Hurwitz J (1993) Functional interactions between SV40 T antigen and other replication proteins at the replication fork. J Biol Chem 268(15):11008–11017

Murante RS, Rust L, Bambara RA (1995) Calf 5′ to 3′ exo/endonuclease must slide from a 5′ end of the substrate to perform structure-specific cleavage. J Biol Chem 270(51):30377–30383

Nick McElhinny SA, Gordenin DA, Stith CM, Burgers PM, Kunkel TA (2008) Division of labor at the eukaryotic replication fork. Mol Cell 30(2):137–144

Nick McElhinny SA, Watts BE, Kumar D, Watt DL, Lundstrom EB, Burgers PM, Johansson E, Chabes A, Kunkel TA (2010a) Abundant ribonucleotide incorporation into DNA by yeast replicative polymerases. Proc Natl Acad Sci U S A 107(11):4949–4954. 0914857107 [pii] 10.1073/pnas.0914857107

Nick McElhinny SA, Kissling GE, Kunkel TA (2010b) Differential correction of lagging-strand replication errors made by DNA polymerases {alpha} and {delta}. Proc Natl Acad Sci U S A 107(49):21070–21075. https://doi.org/10.1073/pnas.1013048107

Nunez-Ramirez R, Klinge S, Sauguet L, Melero R, Recuero-Checa MA, Kilkenny M, Perera RL, Garcia-Alvarez B, Hall RJ, Nogales E, Pellegrini L, Llorca O (2011) Flexible tethering of primase and DNA Pol alpha in the eukaryotic primosome. Nucleic Acids Res 39(18):8187–8199. https://doi.org/10.1093/nar/gkr534

Perera RL, Torella R, Klinge S, Kilkenny ML, Maman JD, Pellegrini L (2013) Mechanism for priming DNA synthesis by yeast DNA polymerase alpha. Elife 2:e00482. https://doi.org/10.7554/eLife.0048200482. [pii]

Pursell ZF, Isoz I, Lundstrom EB, Johansson E, Kunkel TA (2007) Yeast DNA polymerase epsilon participates in leading-strand DNA replication. Science 317(5834):127–130

Reijns MA, Kemp H, Ding J, de Proce SM, Jackson AP, Taylor MS (2015) Lagging-strand replication shapes the mutational landscape of the genome. Nature 518(7540):502–506. nature14183 [pii] 10.1038/nature14183

Rossi ML, Bambara RA (2006) Reconstituted Okazaki fragment processing indicates two pathways of primer removal. J Biol Chem 281(36):26051–26061. https://doi.org/10.1074/jbc.M604805200

Rossi ML, Pike JE, Wang W, Burgers PM, Campbell JL, Bambara RA (2008) Pif1 helicase directs eukaryotic Okazaki fragments toward the two-nuclease cleavage pathway for primer removal. J Biol Chem 283(41):27483–27493. M804550200 [pii] 10.1074/jbc.M804550200

Sauguet L, Klinge S, Perera RL, Maman JD, Pellegrini L (2010) Shared active site architecture between the large subunit of eukaryotic primase and DNA photolyase. PLoS One 5(4):e10083. https://doi.org/10.1371/journal.pone.0010083

Schauer GD, O'Donnell ME (2017) Quality control mechanisms exclude incorrect polymerases from the eukaryotic replication fork. Proc Natl Acad Sci U S A 114(4):675–680. https://doi.org/10.1073/pnas.1619748114

Singh H, Brooke RG, Pausch MH, Williams GT, Trainor C, Dumas LB (1986) Yeast DNA primase and DNA polymerase activities. An analysis of RNA priming and its coupling to DNA synthesis. J Biol Chem 261(18):8564–8569

Smith DJ, Whitehouse I (2012) Intrinsic coupling of lagging-strand synthesis to chromatin assembly. Nature 483(7390):434–438. https://doi.org/10.1038/nature10895

Smith DJ, Yadav T, Whitehouse I (2015) Detection and sequencing of Okazaki fragments in S. cerevisiae. Methods Mol Biol 1300:141–153. https://doi.org/10.1007/978-1-4939-2596-4_10

Sparks JL, Chon H, Cerritelli SM, Kunkel TA, Johansson E, Crouch RJ, Burgers PM (2012) RNase H2-initiated ribonucleotide excision repair. Mol Cell 47(6):980–986. S1097-2765(12)00599-0 [pii] 10.1016/j.molcel.2012.06.035

Stewart JA, Campbell JL, Bambara RA (2010) Dna2 is a structure-specific nuclease, with affinity for 5′-flap intermediates. Nucleic Acids Res 38(3):920–930. https://doi.org/10.1093/nar/gkp1055

Stillman B (2015) Reconsidering DNA polymerases at the replication fork in eukaryotes. Mol Cell 59(2):139–141. S1097-2765(15)00535-3 [pii] 10.1016/j.molcel.2015.07.004

Stith CM, Sterling J, Resnick MA, Gordenin DA, Burgers PM (2008) Flexibility of eukaryotic Okazaki fragment maturation through regulated strand displacement synthesis. J Biol Chem 283(49):34129–34140

Stodola JL, Burgers PM (2016) Resolving individual steps of Okazaki-fragment maturation at a millisecond timescale. Nat Struct Mol Biol 23(5):402–408. https://doi.org/10.1038/nsmb.3207

Suwa Y, Gu J, Baranovskiy AG, Babayeva ND, Pavlov YI, Tahirov TH (2015) Crystal structure of the human pol alpha B subunit in complex with the C-terminal domain of the catalytic subunit. J Biol Chem 290(23):14328–14337. https://doi.org/10.1074/jbc.M115.649954

Thangavel S, Berti M, Levikova M, Pinto C, Gomathinayagam S, Vujanovic M, Zellweger R, Moore H, Lee EH, Hendrickson EA, Cejka P, Stewart S, Lopes M, Vindigni A (2015) DNA2 drives processing and restart of reversed replication forks in human cells. J Cell Biol 208(5):545–562. jcb.201406100 [pii] 10.1083/jcb.201406100

Tishkoff DX, Filosi N, Gaida GM, Kolodner RD (1997a) A novel mutation avoidance mechanism dependent on S. cerevisiae RAD27 is distinct from DNA mismatch repair [see comments]. Cell 88(2):253–263

Tishkoff DX, Boerger AL, Bertrand P, Filosi N, Gaida GM, Kane MF, Kolodner RD (1997b) Identification and characterization of Saccharomyces cerevisiae EXO1, a gene encoding an exonuclease that interacts with MSH2. Proc Natl Acad Sci U S A 94(14):7487–7492

Tom S, Henricksen LA, Park MS, Bambara RA (2001) DNA ligase I and proliferating cell nuclear antigen form a functional complex. J Biol Chem 276(27):24817–24825

Tran PT, Simon JA, Liskay RM (2001) Interactions of Exo1p with components of MutLalpha in Saccharomyces cerevisiae. Proc Natl Acad Sci U S A 98(17):9760–9765. https://doi.org/10.1073/pnas.161175998

Tran PT, Erdeniz N, Dudley S, Liskay RM (2002) Characterization of nuclease-dependent functions of Exo1p in Saccharomyces cerevisiae. DNA Repair (Amst) 1(11):895–912

Tsurimoto T, Stillman B (1991) Replication factors required for SV40 DNA replication in vitro. II. Switching of DNA polymerase alpha and delta during initiation of leading and lagging strand synthesis. J Biol Chem 266(3):1961–1968

Tsutakawa SE, Classen S, Chapados BR, Arvai AS, Finger LD, Guenther G, Tomlinson CG, Thompson P, Sarker AH, Shen B, Cooper PK, Grasby JA, Tainer JA (2011) Human flap endonuclease structures, DNA double-base flipping, and a unified understanding of the FEN1 superfamily. Cell 145(2):198–211. S0092-8674(11)00241-8 [pii] 10.1016/j.cell.2011.03.004

Tsutakawa SE, Lafrance-Vanasse J, Tainer JA (2014) The cutting edges in DNA repair, licensing, and fidelity: DNA and RNA repair nucleases sculpt DNA to measure twice, cut once. DNA Repair (Amst) 19:95–107. https://doi.org/10.1016/j.dnarep.2014.03.022

Vaithiyalingam S, Arnett DR, Aggarwal A, Eichman BF, Fanning E, Chazin WJ (2014) Insights into eukaryotic primer synthesis from structures of the p48 subunit of human DNA primase. J Mol Biol 426(3):558–569. https://doi.org/10.1016/j.jmb.2013.11.007

Vijayakumar S, Chapados BR, Schmidt KH, Kolodner RD, Tainer JA, Tomkinson AE (2007) The C-terminal domain of yeast PCNA is required for physical and functional interactions with Cdc9 DNA ligase. Nucleic Acids Res 35(5):1624–1637. gkm006 [pii] 10.1093/nar/gkm006

Xie Y, Liu Y, Argueso JL, Henricksen LA, Kao HI, Bambara RA, Alani E (2001) Identification of rad27 mutations that confer differential defects in mutation avoidance, repeat tract instability, and flap cleavage. Mol Cell Biol 21(15):4889–4899

Yeeles JT, Janska A, Early A, Diffley JF (2017) How the eukaryotic replisome achieves rapid and efficient DNA replication. Mol Cell 65(1):105–116. https://doi.org/10.1016/j.molcel.2016.11.017

Yu C, Gan H, Han J, Zhou ZX, Jia S, Chabes A, Farrugia G, Ordog T, Zhang Z (2014) Strand-specific analysis shows protein binding at replication forks and PCNA unloading from lagging strands when forks stall. Mol Cell 56(4):551–563. https://doi.org/10.1016/j.molcel.2014.09.017

Yuzhakov A, Kelman Z, Hurwitz J, O'Donnell M (1999) Multiple competition reactions for RPA order the assembly of the DNA polymerase delta holoenzyme. EMBO J 18(21):6189–6199

Zerbe LK, Kuchta RD (2002) The p58 subunit of human DNA primase is important for primer initiation, elongation, and counting. Biochemistry 41(15):4891–4900

Zhang Y, Baranovskiy AG, Tahirov ET, Tahirov TH, Pavlov YI (2016) Divalent ions attenuate DNA synthesis by human DNA polymerase alpha by changing the structure of the template/primer or by perturbing the polymerase reaction. DNA Repair (Amst) 43:24–33. https://doi.org/10.1016/j.dnarep.2016.05.017

Zhou C, Pourmal S, Pavletich NP (2015) Dna2 nuclease-helicase structure, mechanism and regulation by Rpa. Elife 4. https://doi.org/10.7554/eLife.09832

Chapter 7
Functions of Multiple Clamp and Clamp-Loader Complexes in Eukaryotic DNA Replication

Eiji Ohashi and Toshiki Tsurimoto

Abstract Proliferating cell nuclear antigen (PCNA) and replication factor C (RFC) were identified in the late 1980s as essential factors for replication of simian virus 40 DNA in human cells, by reconstitution of the reaction in vitro. Initially, they were only thought to be involved in the elongation stage of DNA replication. Subsequent studies have demonstrated that PCNA functions as more than a replication factor, through its involvement in multiple protein-protein interactions. PCNA appears as a functional hub on replicating and replicated chromosomal DNA and has an essential role in the maintenance genome integrity in proliferating cells.

Eukaryotes have multiple paralogues of sliding clamp, PCNA and its loader, RFC. The PCNA paralogues, RAD9, HUS1, and RAD1 form the heterotrimeric 9-1-1 ring that is similar to the PCNA homotrimeric ring, and the 9-1-1 clamp complex is loaded onto sites of DNA damage by its specific loader RAD17-RFC. This alternative clamp-loader system transmits DNA-damage signals in genomic DNA to the checkpoint-activation network and the DNA-repair apparatus.

Another two alternative loader complexes, CTF18-RFC and ELG1-RFC, have roles that are distinguishable from the role of the canonical loader, RFC. CTF18-RFC interacts with one of the replicative DNA polymerases, Polε, and loads PCNA onto leading-strand DNA, and ELG1-RFC unloads PCNA after ligation of lagging-strand DNA. In the progression of S phase, these alternative PCNA loaders maintain appropriate amounts of PCNA on the replicating sister DNAs to ensure that specific enzymes are tethered at specific chromosomal locations.

Keywords DNA polymerase • PCNA • RFC • PIP box • Leading strand • Lagging strand • Chromatin • Cohesion • Unloading

---

E. Ohashi • T. Tsurimoto (✉)
Department of Biology, Faculty of Science, Kyushu University, Fukuoka, Japan
e-mail: ohashi.eiji.173@m.kyushu-u.ac.jp; tsurimoto.toshiki.549@m.kyushu-u.ac.jp

H. Masai, M. Foiani (eds.), *DNA Replication*, Advances in Experimental Medicine and Biology 1042, https://doi.org/10.1007/978-981-10-6955-0_7

7.1 Overview of the Clamp-Loader System

7.1.1 PCNA and RFC

Proliferating cell nuclear antigen (PCNA) was originally identified as an autoimmune-disease antigen, which specifically accumulated in S-phase nuclei; it was later rediscovered as an essential replication factor in human cells (Prelich et al. 1987). Subsequent studies demonstrated that PCNA functions as the processivity factor for DNA polymerase δ (Polδ) through its activity as a sliding clamp on DNA (Tinker et al. 1994). Sliding clamps are highly conserved in all organisms; examples include β-clamp in *Escherichia coli*, the phage T4 gene 45 product (gp45), and PCNA in archaea and eukaryotes (Georgescu et al. 2015b). Clamps have dimeric or trimeric ring structures, which enable them to bind double-stranded DNA (dsDNA) topologically and slide along it freely. Binding requires temporary opening of the closed ring by the action of specific loader complexes. RFC, the loader for PCNA, consists of one large and four small subunits (RFC1 and RFC2-5, respectively), which are highly related proteins of the AAA+ ATPase family (Majka and Burgers 2004; Kelch et al. 2012; Yao and O'Donnell 2012). PCNA and RFC are involved in the elongation stage of DNA replication and are essential for the coordinated synthesis of leading-strand and lagging-strand DNA (Prelich and Stillman 1988; Tsurimoto and Stillman 1989).

Because of the essential role of sliding clamps in DNA replication, PCNA-loading mechanisms have been intensively studied by biochemical and structural approaches (Indiani and O'Donnell 2006; Yao and O'Donnell 2012). The PCNA ring consists of three identical subunits that connect in a head-to-tail manner. The ring has asymmetric surfaces, known as the N face and C face (or the back face and front face), because of the presence of N termini and C termini, respectively (Fig. 7.1). The PCNA protomer has two repetitive domains, one and two, each of which consists of a framework of β-sheet arrays and two α-helices facing the hole of the ring. These two domains are bridged by the interdomain connecting loop (IDCL), which locates on the C face and provides a structure that interacts with many PCNA-binding proteins (Tsurimoto 1999; Moldovan et al. 2007; Park et al. 2016; Choe and Moldovan 2017).

A sophisticated process has been demonstrated, in which RFC opens the ring structure to enable loading of PCNA on a DNA substrate (Fig. 7.1). In the presence of ATP, RFC attaches to the C face of PCNA, opens one interface between subunits, and binds to DNA at a 3′-terminal primer-template junction. Subsequent ATP hydrolysis induces a structural change in RFC and dissociation from PCNA and DNA. The closed PCNA ring remains on the dsDNA and is "loaded" with its C face directed to the 3′ end of the primer. The loaded PCNA recruits Polδ to the 3′ primer end, using the C face as the docking surface. Polδ is thereby bound to DNA in the correct orientation, enabling processive DNA synthesis by the Polδ-PCNA complex.

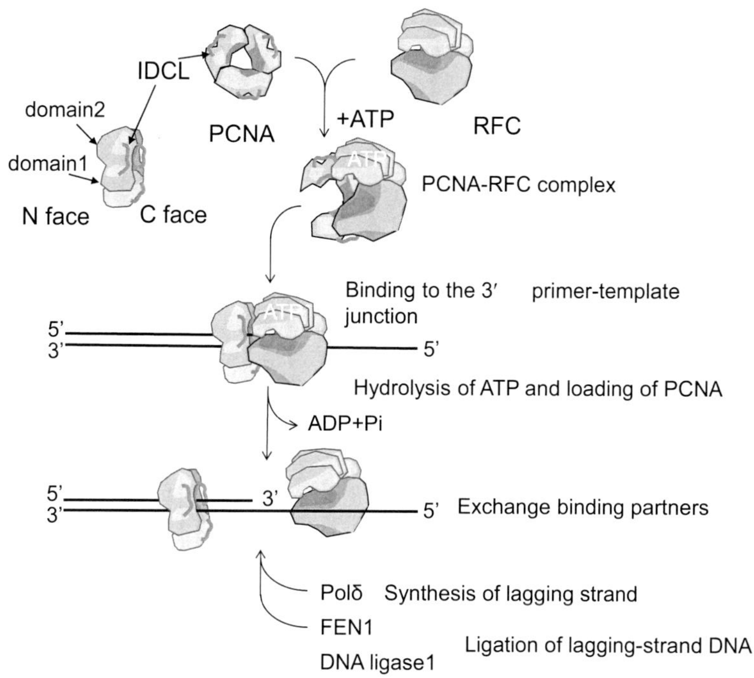

Fig. 7.1 A schematic model of PCNA loading by RFC. The homotrimeric ring of PCNA has a head-to-tail configuration of subunits. The ring has asymmetric side surfaces known as the N face and C face. PCNA protomer has two repetitive domains, 1 and 2 that are bridged by IDCL, which is located on the C face. In the presence of ATP, RFC attaches to the C face, opens one interface between the subunits, and binds to the 3′ primer-template junction. Upon ATP hydrolysis, the structure of RFC changes to dissociate from PCNA and DNA, leaving a closed PCNA ring that is loaded on the duplex DNA with the C face directed to the 3′ end of the primer. Polδ then binds to the 3′ primer end using the C face of PCNA as its docking surface and synthesizes lagging-strand DNA processively. After completion of the DNA elongation, FEN1 and DNA ligase 1 are tethered sequentially to ligate the lagging strands

The dynamic status of PCNA on dsDNA was proposed by structural analyses, single-molecule imaging, and molecular-dynamics simulations (Kochaniak et al. 2009; De March et al. 2017). PCNA moves along dsDNA in a diffusive fashion in both directions. Most of the time, PCNA tracks rotationally the helical pitch of dsDNA by tilting with the DNA axis. This rotational motion of the tilted PCNA on DNA facilitates formation of a large number of electrostatic interactions between DNA backbone and the positively charged residues lining the PCNA inner surface, and it may provide a structure that captures a proper PCNA-binding partner.

7.1.2 PCNA and Replicative DNA Polymerases

Eukaryotes have three distinct replicative DNA polymerases: Polα, Polδ, and Polε (Waga and Stillman 1998, Burgers 2009). Polα DNA synthesis does not depend on PCNA; instead, Polα tightly associates with primase (pri) subunits. The Polα-pri complex synthesizes a 20–30 nucleotides-long RNA-DNA hybrid that acts as a primer. The 3′ end of this primer is a target of PCNA loading and efficient switching to Polδ, resulting in DNA synthesis for lagging strands. The Polδ-PCNA complex has a major role in lagging-strand DNA synthesis in the eukaryotic replication fork, and in repair DNA synthesis, at the late stage of various repair reactions (Prindle and Loeb 2012). The importance of PCNA for Polδ-mediated DNA synthesis is apparent, as Polδ alone can only incorporate several nucleotides at the primer end; whereas in the presence of PCNA, it can produce DNA strands longer than 200–300 nucleotides. The PCNA-RFC-Polδ system can efficiently fill DNA gaps from short patches to lagging-strand sizes.

The largest subunit of Polε consists of a catalytic N-terminal half and a C-terminal half that is involved in the protein-protein interaction necessary for initiation of DNA replication (Dua et al. 1999). The catalytic half is not essential for viability of *Saccharomyces cerevisiae* and *Schizosaccharomyces pombe* (Kesti et al. 1999; Feng and D'Urso 2001), but it has a major role in leading-strand synthesis (Pursell et al. 2007). Polε alone has polymerase activity, but, for its full activity in leading-strands synthesis, other replication-fork components are necessary, such as CMG helicase (see previous chapters) and PCNA (Georgescu et al. 2014). The requirement for PCNA for full activity of Polδ and Polε indicates that the synthesis of both leading and lagging strands in eukaryotes is dependent on PCNA.

7.1.3 Alternative Clamps and Loaders

Three *RFC1* paralogues, *RAD17* (*RAD24* in *S. cerevisiae*), *CTF18* (chromosome transmission fidelity 18), and *ELG1* (enhanced level of genome instability 1; in human, also called as *ATAD5*, ATPase family, AAA domain containing 5 or *FRAG1*, FGF receptor activating protein 1) have been identified in eukaryotes. Proteins from these genes form the alternative clamp-loader-type complexes RAD17-RFC, CTF18-RFC, and ELG1-RFC in association with RFC2-5, (reviewed in Kim and MacNeill 2003, Majka and Burgers 2004, Kubota et al. 2013b, Shiomi and Nishitani 2017) (Fig. 7.2). The functions of these complexes have mainly been analyzed in yeast, where they are involved in checkpoint responses, sister chromatid cohesion, and maintenance of chromosome stability, respectively. Furthermore, three proteins that share significant amino acid sequence similarities with PCNA (RAD9, RAD1 and HUS1 in humans and *S. pombe*, and DDC1, RAD17 and MEC3 in *S. cerevisiae*) are necessary for the checkpoint-response pathway, along with RAD17-RFC (Kondo et al. 1999; Volkmer and Karnitz 1999; St Onge et al. 1999; Caspari et al.

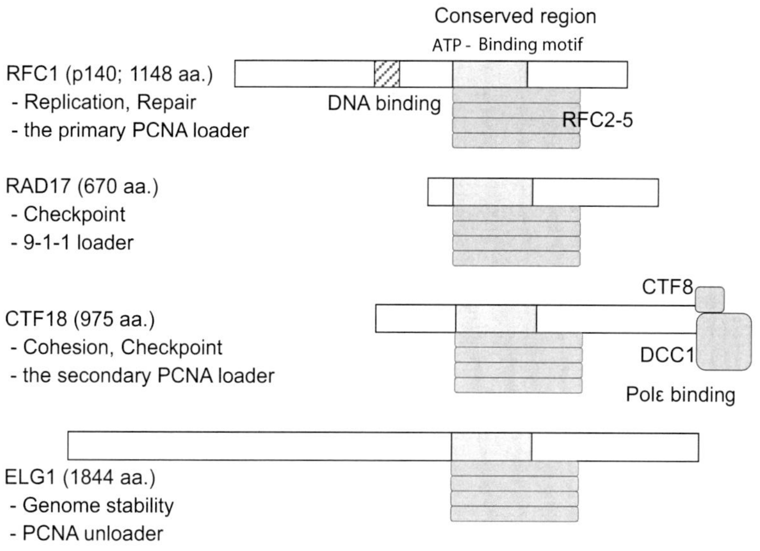

Fig. 7.2 Structures and functions of multiple clamp-loader complexes in humans. Schematic structures and representative functions of four clamp-loader complexes in human cells are shown

2000). The three proteins form a heterotrimeric complex RAD9-HUS1-RAD1 (9-1-1) with a ring structure similar to that of PCNA (Burtelow et al. 2001; Kaur et al. 2001; Lindsey-Boltz et al. 2001; Shiomi et al. 2002; Griffith et al. 2002). Thus, eukaryotes have two clamps and four clamp-loader complexes, which are functionally distinct from each other.

7.2 PCNA Is a Hub Protein That Connects DNA Replication and Peripheral Chromosomal Reactions

7.2.1 During DNA Synthesis

Nearly 50 PCNA-interacting proteins have been identified, and most are involved in DNA replication, repair, and cell cycle control (reviewed in Moldovan et al. 2007, Park et al. 2016). Many of these proteins have a conserved motif, known as the PCNA-interacting protein (PIP) box, at their N-termini or C-termini (Mailand et al. 2013; Boehm and Washington 2016). The PIP box motif is defined by the sequence Q-X-X-Ψ-X-X-Θ-Θ (where Ψ is I/L/M/V, and Θ is F/Y), and it usually protrudes from the main body of a protein and associates with IDCL in PCNA. Studies with PCNA mutants have demonstrated that PCNA-partner interactions have coevolved during evolution, and are not optimized for affinity, but rather may be designed for rapid exchangeability (Zamir et al. 2012; Fridman et al. 2013).

In eukaryotes, the interactions between PCNA, RFC, and Polδ are essential for chromosomal DNA replication. Several other enzymes that interact with DNA, such as FEN1 nuclease and DNA ligase 1, also bind PCNA via PIP boxes and thereby facilitate efficient lagging-strand processing (as described in the previous chapter). Similarly, DNA polymerases, Polη, Polι, Polκ, Polλ, REV1, and Polζ involved in translesion synthesis (TLS) can interact with PCNA via the PIP box or its variants (Mailand et al. 2013). PCNA provides a common platform, enabling rapid exchange of polymerases at 3′ DNA ends and resulting in loading of enzymes that are compatible with DNA structures at stalled forks (Maga and Hübscher 2003).

The clamp-based switching of DNA polymerases and related enzymes is an important feature of DNA replication and is highly conserved among organisms. The *E. coli* β-clamp, like PCNA, interacts with lagging-strand-processing enzymes DNA polymerase I (Pol I) and DNA ligase, and also with TLS polymerases Pol II, PolIV, and Pol V, in addition to the replicative DNA polymerase Pol III (Indiani et al. 2005).

7.2.2 After DNA Synthesis

Even with the proofreading activity of replicative DNA polymerases, a number of mis-incorporated nucleotides occur in replicated DNA. The mismatch repair (MMR) reaction corrects such errors after DNA replication, thereby avoiding mutation. MutS in bacteria (Iyer et al. 2006), and two MutS-related heterodimer complexes (MutSα (MSH2-MSH6) and MutSβ (MSH2-MSH3) in eukaryotes) (Marsischky et al. 1996), function as sliding clamps in MMR and have essential roles in the recognition of unmatched bases. ATP-dependent mismatch recognition by MutS discriminates between DNA strands to specifically repair mis-incorporated nucleotides in newly synthesized DNA. In *E. coli*, the hemi-methylated structures produced by DNA replication of A-methylated GATC sites enable this discrimination. A major mechanism for discrimination in eukaryotes involves the targeting of single-strand breaks in the nascent DNA strands by the MutS complexes (Iyer et al. 2006; Kunkel and Erie 2005). However, although these mechanisms differ, the interactions of MMR proteins with clamp molecules are widely conserved from *E. coli* to humans, indicating the importance of PCNA in the coordination between replication and MMR (López de Saro and O'Donnell 2001). Subunits of eukaryotic MutS complexes, MSH3 and MSH6, interact with PCNA for efficient mismatch recognition (Flores-Rozas et al. 2000; Kleczkowska et al. 2001). After replication, PCNA remaining on the DNA is positioned with a specific orientation in the direction of synthesis, and it functions as the strand-discrimination marker for MutS complexes (Pluciennik et al. 2010; Georgescu et al. 2015a; Kawasoe et al. 2016). In the processing of mismatched DNA, the endonuclease MutL and an exonuclease EXO1 also interact with PCNA (Kadyrov et al. 2006, 2007; Pluciennik et al. 2010; Chen et al. 2013; Goellner et al. 2014). MutL complex that consists of MLH1 and

PMS2 interacts with MutS complexes and incises DNA strands near mismatched bases, and EXO1 excises the mismatched DNA region. Subsequently, PCNA-RFC-Polδ repairs the processed DNA gap. Thus, PCNA functions in the sequential steps of MMR by tethering MMR proteins, discriminating the newly synthesized DNA strands, and facilitating processing of the mismatched DNA region (López de Saro and O'Donnell 2001).

Following DNA synthesis at the replication fork, various chromosomal structures including epigenetic information encoded by covalent modifications of DNA and histones must be reconstituted. DNA methyltransferase 1 (DNMT1) has a PIP box in its N-terminal region and physically interacts with PCNA (Chuang et al. 1997). This interaction is not essential for the DNMT1 action, but it facilitates efficient methylation of hemi-methylated DNA in vitro and may help to maintain the methylation status of replicated DNA in S phase (Iida et al. 2002). Laser micro-irradiation experiments with mammalian cells demonstrated that interaction with PCNA is required for efficient recruitment of DNMT1 to DNA-repair sites, suggesting its importance for restoration of epigenetic information during DNA repair (Mortusewicz et al. 2005).

Chromatin assembly factor-1 is an evolutionarily conserved histone-chaperone complex that is capable of chromatin assembly in newly replicated DNA in vitro. This complex associates histones H3 and H4 with replicated DNA through its interaction with PCNA (Shibahara and Stillman 1999) and is required for the maintenance of epigenetic information in *S. cerevisiae* (Zhang et al. 2000). Similarly, several regulators of chromatin, such as histone deacetylase 1, histone acetyltransferase p300, and the WSTF-SNF2H chromatin remodeling complex, bind to PCNA, indicating its pleiotropic significance for the maintenance of chromatin structures in replicated DNA (Moldovan et al. 2007).

Cohesion of sister chromosomes is essential for precise segregation of replicated chromosomes to daughter cells (Skibbens 2009). This physical coupling of the replicated DNA is mediated by a protein ring, known as cohesin, which consists of SMC1, SMC3, SCC1, and SCC3 proteins. Cohesion is established in S phase simultaneously with the passage of the replication fork. The N-acetyltransferase ECO1 of *S. cerevisiae* (ESCO1 and ESCO2 in humans), which acetylates SMC3, has an essential role in this step. Replication-fork components, FEN1, RFC, and PCNA interact with ECO1-family proteins and couple cohesion establishment with replication (Rudra and Skibbens 2013). Indeed, ECO1-family proteins contain PIP-box motifs that are essential for PCNA binding and establishment of cohesion in S phase (Moldovan et al. 2006).

7.2.3 Repair of Damaged DNA

Genomic DNA is exposed to various damaging events, such as ionizing and ultraviolet radiation, exposures to chemicals, and oxidation. Various repair systems have evolved that utilize PCNA as the assembly target and as a component of the resynthesis machinery after processing of damaged DNA, as in MMR. Base excision repair (BER) proteins correct DNA with damaged bases. This reaction is initiated by the recognition and removal of damaged bases from nucleotides by specific DNA glycosylases. The phosphodiester bond of the 5′ abasic site is then cleaved by apurinic/apyrimidinic (AP) endonucleases (Krokan and Bjørås 2013). PCNA binds to four DNA glycosylases (UNG2, MPG, NTH1, and hMYH) and two AP endonucleases (APE1 and APE2) to colocalize with them and stimulate their activities (Moldovan et al. 2007). In the reaction that follows, short-patch repair reaction using Polβ and XRCC1-DNA ligase 3 complex seals the gap (reviewed in Dianov et al. 2003). Polβ and XRCC1 also interact with PCNA (Kedar et al. 2002; Fan et al. 2004). Thus, PCNA may recruit these BER proteins sequentially to the repair sites.

Nucleotide excision repair (NER) removes bulky DNA lesions, and as with other types of repair, it involves PCNA (Sancar et al. 2004). Seven xeroderma pigmentosum (XP) proteins engage in detection of lesions, unwinding of damaged duplex DNA, and removal of the damaged DNA strand by endonuclease and exonuclease activities. Among XP proteins, the endonuclease XPG has a PIP box in its C-terminal region, which is required for PCNA binding and NER activity. This interaction may facilitate DNA resynthesis after DNA excision by XPG (Gary et al. 1997).

7.2.4 Cell-Cycle Regulation

PCNA was characterized as a marker protein of S-phase cells; it is involved in the activities of these cells and in cell proliferation. PCNA interacts with proteins that regulate cell-cycle progression, such as the cyclin-dependent kinases (CDKs) (Zhang et al. 1993) and the tumor-suppressor protein p21 (Waga et al. 1994), which binds to CDK and inhibits its activity (Xiong et al. 1993). These interactions produce a quaternary complex, PCNA–p21/CDK-cyclin, the function of which is unclear. The interaction of PCNA with CDK-cyclin may locate CDK activity to the replisome and result in phosphorylation of PCNA-binding partners, such as RFC, DNA ligase 1 (Koundrioukoff et al. 2000), and FEN1 (Henneke et al. 2003). p21 has a typical PIP box at the C-terminus and binds stably to PCNA (Warbrick et al. 1995). This binding results in competitive inhibition of PCNA binding to its partners, including Polδ, and inhibits DNA replication in vitro and in vivo (Rousseau et al. 1999; Waga et al. 1994). Thus, p21 may regulate PCNA-stimulated reactions competitively in accordance with cell-cycle progression. p21 is in turn regulated by PCNA-dependent proteolysis (Chuang and Yew 2001; Kim et al. 2008; Nishitani et al. 2008; Abbas et al. 2008). p21 has a specialized PIP box, the "PIP degron," which is characterized by the sequence Q-X-X-Ψ-T-D-Θ-Θ-X-X-X-B (B: K/R) (Havens and Walter 2009). An ubiquitin ligase ($CRL4^{Cdt2}$) is activated by chromatin-loaded PCNA, and poly-ubiquitinates the PCNA-associated PIP degron proteins, such as p21. Because chromatin-loaded PCNA is maintained during DNA replication and repair, this PCNA-dependent proteolysis is responsible for degradation of p21 from G1/S transition to S phase and upon UV irradiation (Abbas et al. 2008).

One of the licensing factors, CDT1 (*cdc10* dependent transcript 1) (Hofmann and Beach 1994), which is required in G1 phase to commit replication origins to initiation of DNA synthesis, is the first protein that is characterized to have a PIP degron (Arias and Walter 2005; Nishitani et al. 2006). CDT1 has a PIP degron at its N-terminus and exhibits similar cellular behavior to p21, accumulating in G1 phase and degrading from G1/S transition to S phase, concomitantly with loading of PCNA to chromatin. CDT1 also degrades upon DNA damage. PIP-degron-dependent CDT1 degradation is the major cause of the temporal alteration of chromosomal CDT1, and it ensures the initiation of DNA replication once per cell cycle. Similar PIP-degron-dependent proteolysis has been reported for the transcription factor E2F, the histone H4K20 mono methyltransferase Set8 and Polη (Shibutani et al. 2008, Kim and Michael 2008, Abbas et al. 2010), which also exhibit S-phase and DNA-damage-dependent degradation.

7.3 PCNA Dynamics

When 3′ DNA ends are produced during DNA replication and repair, PCNA is loaded on the sites mainly by RFC, setting the stage for the binding of other DNA-interacting proteins (Leonhardt et al. 2000; Kim and Lee 2008). The topologically bound PCNA remains on the DNA until a specific unloading process takes place. Important reactions occur sequentially on replicated chromosomes, with PCNA as a platform, and the timing and preference of target sites for PCNA loading and unloading, along with the life-span of loaded PCNA, are essential factors in the maintenance of chromosomal structure and progression of the cell cycle (Fig. 7.3). PCNA and its loading proteins participate in a multilayered regulatory network to maintain genome integrity.

7.3.1 Primary PCNA Loading by RFC

Biochemical analyses showed that RFC recognizes the 3′ end of primer without any preference for the DNA sequence. RFC will load PCNA on any primer end to synthesize lagging strands in a replication fork. Coordination of clamp-loading activity with replication forks is likely to require physical association of the clamp loader with the replisome complex. The γ clamp-loader complex of *E. coli* is physically integrated in the replisome (McInerney et al. 2007). The γ complex consists of five subunits (δ, δ’, and three subunits of γ and/or τ) and associates with Pol III and DnaB DNA helicase via the C-terminal domain of τ subunits. The γ complex is the core protein for assembly of the multiple DNA polymerases and DNA helicase in the *E. coli* replisome. However, unlike the *E. coli* loader, direct connections of RFC and Polδ to replisome components apart from PCNA are currently unknown. Instead, Polε and Polα-pri are integrated in the eukaryotic replisome through direct

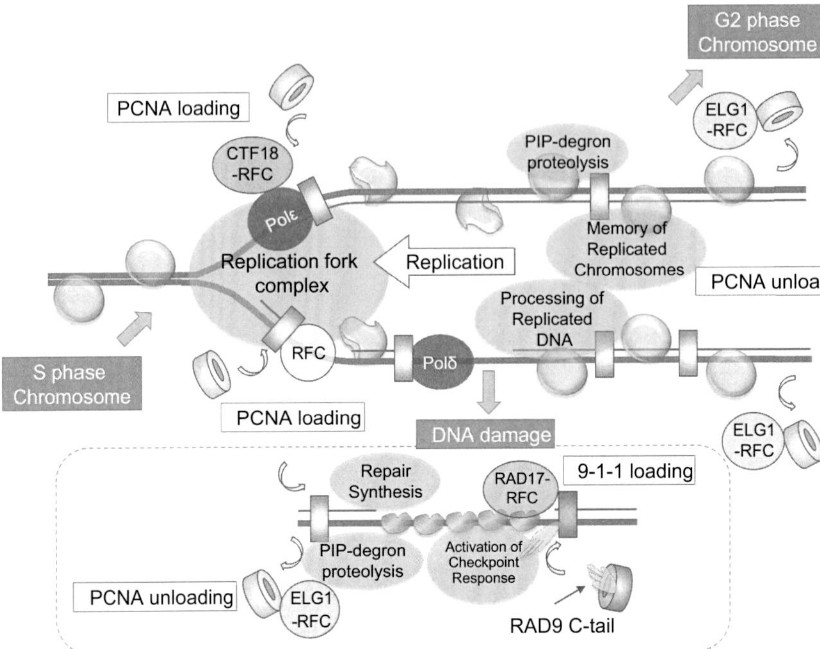

Fig. 7.3 PCNA dynamics (loading and unloading) on replicating and post-replication chromosomes. PCNA is loaded on lagging-strand DNA by RFC and on leading-strand DNA probably by Polε–CTF18-RFC complex. This PCNA loading makes the chromosome region in S-phase mode, in which various DNA processing reactions take place. After completion of these reactions, ELG1-RFC unloads PCNA from chromosomes and switches to the G2-phase mode. Proper switching of the chromosome modes is crucial for precise maintenance of genomic integrity and transmission of epigenetic information. Similar PCNA dynamics, along with ATR checkpoint-signal activation by 9-1-1 and RAD17-RFC, occur at sites of DNA damage. To simplify this model, several essential components, such as CMG helicase, Polα/primase, and most of PCNA-interacting proteins are represented by unlabeled markers. Ubiquitination and SUMOylation that occur on DNA-bound PCNA are also not indicated

or indirect interactions with the replicative helicase CMG complex (O'Donnell and Li 2016). This observation suggests that the action of RFC-PCNA-Polδ during DNA replication will be distributive with CMG complex in the DNA area proximal to the replication fork. RFC-PCNA-Polδ associates distributively with gapped DNA regions between short RNA-DNA hybrid primers and the previously synthesized lagging strands, and fills a few hundred nucleotides in the gaps, as it does in the resynthesis of excised damaged DNAs.

In one particular situation, DNA-sequence-specific PCNA loading occurs via interaction of RFC with a sequence-specific DNA-binding protein. Kaposi's sarcoma-associated herpesvirus (KSHV) latency-associated nuclear antigen mediates viral DNA replication and persistence, binding specifically to the viral terminal-repeat sequence that contains the viral replication origin and interacting with RFC. Through this interaction, RFC loads PCNA to the terminal-repeat DNA efficiently in vitro (Sun et al. 2014). Thus, the viral antigen facilitates efficient virus replication by recruiting host-cell replication machinery, such as the PCNA-loading system, to its replication origin.

7.3.2 Secondary PCNA Loading by CTF18-RFC

The heptameric complex, CTF18-RFC, in addition to CTF18 and RFC2–5, contains the subunits DCC1 and CTF8 (Mayer et al. 2001; Merkle et al. 2003) (Fig. 7.2) and is involved in replication-fork progression, DNA-damage response, and proper establishment of cohesion (Hanna et al. 2001; Naiki et al. 2001; Terret et al. 2009; Crabbé et al. 2010). In DNA replication, CTF18 localizes at the replication fork in *S. cerevisiae* and humans (Lengronne et al. 2006; Sirbu et al. 2013; Alabert et al. 2014). Biochemical studies demonstrated that CTF18-RFC functions as a second PCNA loader, loading PCNA at the 3′ end of a primer-template junction through ATP hydrolysis and stimulating Polδ-mediated DNA synthesis (Bermudez et al. 2003; Shiomi et al. 2004). However, PCNA-unloading activity has also been reported for CTF18-RFC in *S. cerevisiae* (Bylund and Burgers 2005). Purified CTF18-RFC unloads PCNA in the presence of a sufficient amount of replication protein A (RPA), although the physiological relevance of this unloading has not been fully addressed.

CTF18-RFC directly interacts with Polε via a trimeric assembly consisting of CTF18, DCC1, and CTF8 (Murakami et al. 2010; García-Rodríguez et al. 2015). The results of a recent study (Fujisawa et al. 2017) demonstrated that PCNA-loading activity by CTF18-RFC alone is intrinsically weak and almost inactive at a near-physiological salt concentration in the presence of RPA. However, if CTF18-RFC is associated with Polε, its PCNA-loading activity is restored even in these conditions. This result indicates that the active status of Ctf18-RFC has to be in a complex with Polε, and it suggests the involvement of the secondary PCNA loader CTF18-RFC in the leading strand DNA-polymerase complex. PCNA loading by the CTF18-RFC–Polε complex occurs when Polε is in non-synthesis mode, and placing of Polε in the synthesis mode greatly suppressed the loading. PCNA loading forms the novel complex of CTF18-RFC–Polε–PCNA, which synthesizes DNA processively. From these results, two roles of PCNA loading by CTF18-RFC–Polε for DNA replication have been proposed. One role is the maintenance of leading-strand synthesis at template DNA structures that disrupt Polε progression. The second role involves active PCNA loading on a routine basis by CTF18-RFC–Polε during replication, which balances PCNA dosage between the leading and lagging DNA strands, enabling PCNA-binding proteins to function properly on both strands. Accordingly, it was reported that CTF18-RFC has a key role in the PCNA-degron-dependent CDT1 degradation in S phase, whereas RFC is required for CDT1 degradation after UV irradiation (Shiomi et al. 2012). These results suggest that the ubiquitin ligase $CRL4^{Cdt2}$ may be activated by PCNA on the leading strand via an unknown mechanism.

7.3.3 Unloading via ELG1-RFC, and PCNA Memory

Defects of *ELG1* exhibit various genomic instabilities (Kubota et al. 2013b; Ulrich 2013; Shiomi and Nishitani 2017). Control of the life-span of PCNA loaded on replicated chromosome is necessary to adjust enzymatic activity on the chromosomes from S phase to G2 phase (Fig. 7.3). ELG1-RFC unloads PCNA from chromosome in yeast and mammalian cells (Lee et al. 2013; Kubota et al. 2013a; Shiomi and Nishitani 2013). Depletion of ELG1 from *S. cerevisiae* or human cells results in an accumulation of chromatin-bound PCNA, which is removed upon expression of ELG1 in vivo. Similarly, treatment of isolated chromatin with partially purified ELG1-RFC causes PCNA unloading in vitro (Kubota et al. 2013a). Recent analyses by eSPAN (enrichment and sequencing of protein-associated nascent DNA) showed that the ratio of PCNA on the lagging strand to that on the leading strand in *S. cerevisiae* is about two to one, but PCNA becomes preferentially associated with the leading strand in a stalled replication fork (Yu et al. 2014). ELG1-RFC is required for this shift of PCNA distribution and has a role as an unloader of PCNA from the lagging strand in a stalled replication fork. PCNA unloading by ELG1-RFC will shorten the life-span of PCNA on S-phase chromatin.

ELG1 is not essential for cell viability in *S. cerevisiae*, suggesting that other loader complexes can complement its activity during normal DNA replication. Indeed, as RFC and CTF18-RFC unload PCNA from DNA with nicks or gaps in vitro (Yao et al. 1996; Shibahara and Stillman 1999; Bylund and Burgers 2005), their activities may provide the unloading function in the absence of ELG1.

Timing of unloading is another important aspect of PCNA functionality. S-phase-specific PCNA modifications, such as SUMOylation (addition of small ubiquitin-like modifier protein (SUMO)), may be involved in this timing. *S. cerevisiae* ELG1 has putative SUMO-interaction motifs (SIMs) within its N-terminal domain, and ELG1-RFC preferentially interacts with SUMOylated PCNA through these motifs (Parnas et al. 2010). Furthermore, SUMOylated PCNA accumulates on chromatin in an *elg1*Δ mutant, especially in the presence of methyl methanesulfonate. These results suggest that *S. cerevisiae* ELG1-RFC is an unloader that is specific to SUMOylated PCNA. However, SUMOylation is not absolutely required for PCNA unloading by ELG1-RFC (Kubota et al. 2013a), suggesting that accumulation of SUMOylated PCNA in an *elg1*Δ mutant is probably the result of prolonged retention of PCNA on chromatin, caused by the absence of its unloader.

Although a PCNA-unloading reaction with purified ELG1-RFC has not been reconstituted, partially purified yeast ELG1-RFC unloads PCNA from chromatin of an *elg1*Δ mutant in vitro in an ATP-dependent manner (Kubota et al. 2013a). This reaction also demonstrates that PCNA unloading by ELG1-RFC is replication-coupled, but inactive with a ligase-deficient cell extract, although it can be restored by addition of an exogenous ligase (Kubota et al. 2015). Thus, ELG1-RFC acts as a PCNA unloader on dsDNA, after lagging-strand ligation. This property may distinguish the active timing of ELG1-RFC from that of RFC and CTF18-RFC, which require unligated 3′ ends for their activity (Yao et al. 1996, Shibahara and Stillman 1999, Bylund and Burgers 2005).

DNA-bound PCNA will transmit various signals to DNA-transaction enzymes; the presence of PCNA represents the "memory" of specific events related to newly synthesized DNA strands, sister DNAs, de novo replicated DNA regions, and incompletely replicated or repaired regions (Georgescu et al. 2015b). DNA-bound PCNA induces strand-specific MMR in *Xenopus*-egg extracts through the interaction of MutSα with PCNA (Kawasoe et al. 2016). MutSα that recognizes unmatched bases prevents PCNA from being removed from the DNA. This interaction between MutSα and PCNA keeps the memory of the new DNA strand that is contained in the orientation of PCNA, through the inhibition of PCNA unloading, until MMR has taken place. In vivo, PCNA unloading during S phase largely depends on ELG1-RFC, suggesting that the PIP box in MutSα may limit the access of ELG1-RFC to DNA-bound PCNA, thereby blocking the unloading.

If PCNA functions as a memory molecule on replicated chromosomes, ELG1-RFC acts as an eraser of PCNA memory. PCNA-binding proteins may also regulate the period of memory mediated by DNA-bound PCNA. The proper duration of this memory is crucial to the maintenance of genome integrity after DNA replication, and functional defects in the loader and unloader complexes lead to chromosomal abnormalities (Johnson et al. 2016).

7.3.4 Post-translational Modifications

Quantitative regulation of PCNA levels on chromosomes is essential to maintain genome stability in S phase and after DNA damages. In addition, qualitative alterations to PCNA, in the form of several post-translational modifications (PTMs), affect interactions with various partners, and function as "PCNA code" to select these partners and thereby regulate PCNA functions. Among various PTMs, we focus on two modifications, ubiquitination and SUMOylation (Hoege et al. 2002; Stelter and Ulrich 2003), which sort the two DNA-damage-tolerance pathways during S phase.

The highly conserved lysine 164 (K164) of PCNA is the target site of the PTMs (PCNA in *S. cerevisiae* has a secondary SUMOylation site at lysine 127). Mono-ubiquitination of K164 occurs by Rad6-Rad18 E2/E3 ligase upon DNA damage in S phase and promotes the TLS pathway. Mono-ubiquitination of K164 also occurs in human cells, via CRL4^{Cdt2} ubiquitin ligase. Unlike yeast, a temporal-negative regulation of a ubiqutin-specific protease, USP1, results in the accumulation of mono-ubiquitinated PCNA in S phase and upon DNA damage in human cells (Huang et al. 2006; Terai et al. 2010). Most TLS polymerases have ubiquitin-binding domains that are necessary for their activities, in addition to the PCNA binding motifs (Bienko et al. 2005; Guo et al. 2006). At stalled replication forks, mono-ubiquitination of K164 switches the affinity of PCNA from high-fidelity replicative polymerases to low-fidelity TLS polymerases, to bypass the lesions and continue the replication (Masuda et al. 2010; Edmunds et al. 2008).

Mono-ubiquitination of K164 also links with poly-ubiquitination at the same site by MMS2-UBC13 and RAD5, another E2-E3 ligase (Ulrich and Jentsch 2000). The poly-ubiquitinated PCNA leads an alternative, error-free lesion bypass through template switching between de novo synthesized sister DNAs by recombination-like processes. The different ubiquitination modifications can switch between error-prone and error-free pathways at stalled replication forks, to bypass DNA lesions and to restart the forks. However, the mechanism by which poly-ubiquitinated PCNA promotes template switching is not yet understood.

SUMO modification of PCNA is more prominent in *S. cerevisiae* than other species. In *S. cerevisiae*, this modification occurs at K164 in DNA-bound PCNA constitutively during S phase by UBC9-SIZ1 SUMO ligase (Hoege et al. 2002) and results in the recruitment of the anti-recombination helicase SRS2 via its PIP box and SIM. K164 SUMOylation is also detectable in human cells, and a functional orthologue of SRS2, PCNA-associated recombination inhibitor PARI (which contains a PIP box and SIM), has been identified (Gali et al. 2012; Moldovan et al. 2012). These results show that PCNA SUMOylation has a role in the inhibition of homologous recombination, and may suppress recombination-based restart of stalled replication forks, which would cause potential deleterious recombination events between newly synthesized sister chromatids, leading to chromosome rearrangements (Carr and Lambert 2013). SUMOylated PCNA is also a target of ELG1-RFC, and may have a role in regulation of the unloading of DNA-bound PCNA, although an active role for the unloading remains to be demonstrated.

7.4 Checkpoint Clamp and Loader

7.4.1 Loading of the 9-1-1 Clamp onto DNA by RAD17-RFC

The heterotrimer clamp 9-1-1 has the characteristic RAD9 C-terminal extension ("C-tail") that is intrinsically disordered, in addition to its PCNA-like ring structure. 9-1-1 and RAD17-RFC are thought to be the specialized clamp and clamp loader for "emergency" situations. When DNA replication is perturbed by DNA damage, uncoupling of replicative DNA helicase and DNA polymerase occurs, and single-stranded DNA (ssDNA) regions are exposed and covered by the ssDNA-binding protein, RPA (Byun et al. 2005). Nucleolytic degradation after generation DNA double-strand breaks (DSBs), for example, by ionizing radiation, also leads to exposure of ssDNA regions. Whereas RFC loads PCNA at 3′-recessed ends of DNA, human RAD17-RFC and its *S. cerevisiae* orthologue load 9-1-1 onto 5′-recessed ends, in a reaction that is dependent on ATP and RPA (Ellison and Stillman 2003, Majka et al. 2006).

DNA-damaging treatments result in increased chromatin association of 9-1-1 and nuclear-foci formation of its components (Caspari et al. 2000; Burtelow et al. 2000; Zou et al. 2002; You et al. 2002). These responses are dependent on RAD17 and are thought to correspond to the loading of 9-1-1 onto DNA by RAD17-RFC. RPA has an important role in the localization of 9-1-1 to damaged chromatin, probably through the interaction between RPA and the RPA-binding motif in the C-tail of RAD9 (Wu et al. 2005; Xu et al. 2008). RAD17 is also recruited to the site of DNA damage through the interaction with RPA (Lee et al. 2003a; Zou et al. 2003). Phosphorylation of RAD17 at two serine residues (S635 and S645) is induced by DNA-damaging treatments and is required for the checkpoint responses, probably through stimulation of its 9-1-1 loading (Bao et al. 2001; Wang et al. 2006). S635 and S645 are the targets of two major checkpoint kinases, ATM (ataxia telansiectasia mutated) and ATR (ATM and Rad3-related) (Kim et al. 1999, O'Neill et al. 2000, Bao et al. 2001, Matsuoka et al. 2007). Activated ATR phosphorylates RAD17, which stimulates 9-1-1 loading that in turn contributes to ATR activation, implying a positive-feedback loop to maintain activation of ATR until the DNA damage is removed. The 9-1-1 complex is also phosphorylated by ATR or ATM in the presence of DNA damage (Chen et al. 2001; Roos-Mattjus et al. 2002, 2003), but the biological significance of this phosphorylation is unclear.

7.4.2 Activation of the ATR-Dependent DNA-Damage-Checkpoint Pathway

7.4.2.1 From Loading of 9-1-1 to Activation of ATR

ATM and ATR respond mainly to exposure to DSBs and ssDNA, respectively (reviewed in Lavin 2008, Cimprich and Cortez 2008, Maréchal and Zou 2013, Awasthi et al. 2015). The activated kinases phosphorylate and activate their downstream kinases, CHK2 (for ATM) and CHK1 (for ATR). All the activated kinases phosphorylate and regulate the functions of a range of proteins involved in DNA repair, cell-cycle regulation, and apoptosis. RAD17-RFC and 9-1-1 are involved in an early stage of DNA-damage response, in activation of the ATR–CHK1 pathway (Fig. 7.4). 9-1-1 is loaded by RAD17-RFC at the junction of RPA-bound ssDNA and dsDNA. ATR complexed with ATRIP (ATR-interacting protein) is also recruited to the RPA-bound ssDNA via an interaction between ATRIP and RPA. Thus, ATR–ATRIP and 9-1-1 complexes are recruited to the damaged DNA independently (Kondo et al. 2001; Melo et al. 2001; Zou et al. 2002; You et al. 2002). The DNA topoisomerase 2-binding protein 1(TopBP1) binds to 9-1-1 and to ATR-ATRIP complexes, and both of these interactions are involved in the activation of ATR (Kumagai et al. 2006; Mordes et al. 2008; Navadgi-Patil and Burgers 2008).

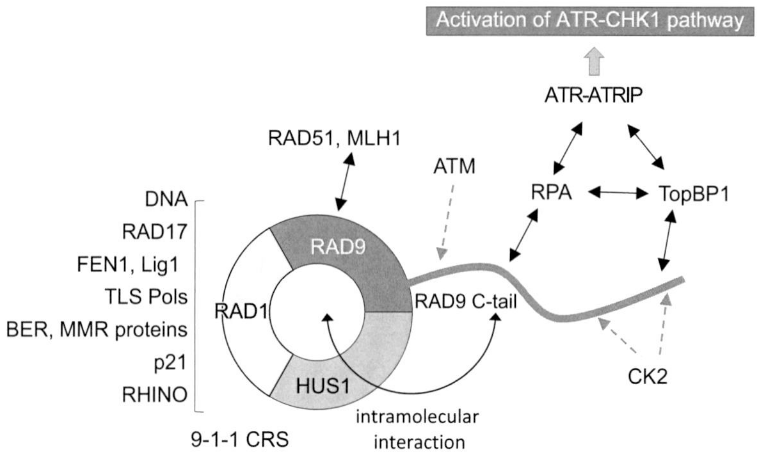

Fig. 7.4 Activation of the ATR–CHK1 pathway and interaction of 9-1-1 with factors involved in DNA-damage responses. Proteins interacting with 9-1-1 and RAD9 described in this review are shown. Double-headed arrows indicate interactions. Single-headed arrows with broken lines indicate phosphorylations on 9-1-1. CRS, core ring structure; Lig1, DNA ligase 1; TLS pols, translesion synthesis polymerases

7.4.2.2 Phosphorylation of the C-Tail and Activation of ATR

Phosphorylation of S387 in the RAD9 C-tail is required for the interaction of 9-1-1 with TopBP1 and the subsequent activation of ATR (St Onge et al. 2003; Delacroix et al. 2007; Lee et al. 2007). The C-tail has multiple phosphorylation sites (10 in humans: S272, S277, T292, S328, S336, S341, T355, S375, S380, and S387). Among these sites, S272 is known to be phosphorylated by ATM in response to DNA damage, although the phosphorylation is not involved in the interaction with TopBP1 (Longhese et al. 1997; Furuya et al. 2004; St Onge et al. 2001; St Onge et al. 2003; Roos-Mattjus et al. 2003). S387 and the internal S341 are typical targets of casein kinase 2 (CK2), a ubiquitous, pleiotropic, and constitutively active kinase (Fig. 7.4). The constitutive phosphorylation of S387 and S341 promotes binding of 9-1-1 to TopBP1 (Takeishi et al. 2010; Rappas et al. 2011) and is required for efficient phosphorylation of CHK1 (Ueda et al. 2012).

As the CK2-dependent interaction between 9-1-1 and TopBP1 is constitutive, a pathway should exist for immediate ATR activation upon DNA damage. Indeed, the interaction of the counterparts of 9-1-1 and TopBP1 in *S. pombe* is damage-inducible, and the phosphorylation sites responsible for their interaction in both *S. cerevisiae* and *S. pombe* are damage-inducible using ATM/ATR kinase activities (Furuya et al. 2004; Puddu et al. 2008; Navadgi-Patil and Burgers 2009). In addition, the C-tail is also able to stimulate ATR (Mec1), in the absence of the TopBP1 (Dpb11) in *S. cerevisiae* (Navadgi-Patil and Burgers 2009). Different models for the binding of 9-1-1 and TopBP1 to damaged chromatin have been reported. One suggestion is that 9-1-1 recruits TopBP1 in the vicinity of ATR-ATRIP complex (Greer et al. 2003; Delacroix et al. 2007; Lee et al. 2007), whereas an alternative is that 9-1-1 is recruited to sites of replication stress via the pre-assigned TopBP1 (Yan and Michael 2009). However, recent studies with *Xenopus*-egg extract and human cells demonstrated that 9-1-1, TopBP1, and ATR-ATRIP can be recruited to sites of stress independently of each other (Duursma et al. 2013; Lee and Dunphy 2013; Gong et al. 2010; Ohashi et al. 2014; Acevedo et al. 2016). The RAD9, HUS1, and RAD1-interacting nuclear orphan protein 1 (RHINO) has been identified in human cells as a co-binding factor to TopBP1 and 9-1-1 (Cotta-Ramusino et al. 2011). Recruitment of this protein to the site of DNA damage is dependent on the 9-1-1 complex and may have a role in damage-dependent activation of the 9-1-1–TopBP1 pathway in vertebrates (Lindsey-Boltz et al. 2015).

7.4.2.3 The Intramolecular Interaction of the C-Tail

The RAD9 C-tail and 9-1-1 core ring structure (CRS) interact in the regulation of 9-1-1 function. The C-tail binds to the 9-1-1 CRS, and this intramolecular interaction interferes with the DNA binding of 9-1-1 (Fig. 7.4) (Takeishi et al. 2015). The region in the C-tail that is necessary for binding to the CRS partially overlaps with the region for binding to TopBP1. When TopBP1 binds to the C-tail, the C-tail unfolds and its binding motif(s) on the CRS are exposed. Many proteins interact with 9-1-1, and some of them may bind through the same binding motif(s) on the CRS as the C-tail does. The intramolecular interaction of the C-tail with the CRS may alter the interaction of 9-1-1 with DNA, TopBP1, and other proteins and consequently may have a role in genome maintenance following DNA damage.

7.4.3 Roles in Other Pathways

7.4.3.1 The Response to DSBs and Homologous Recombination

The results of early studies suggested that 9-1-1 and RAD17-RFC are involved in ATR activation, but are not required for phosphorylation of CHK2, a mediator kinase of the ATM pathway for response to DSBs (Weiss et al. 2002; Roos-Mattjus et al. 2003; Wang et al. 2003). However, results from a number of studies involving knockout of genes encoding 9-1-1 and RAD17-RFC in *S. cerevisiae*, chicken DT40 cells, and mouse cells demonstrated that these complexes have direct roles in DSB repair and homologous recombination. Compared with wild-type cells, the mutants show higher sensitivity to ionizing radiation, as well as inefficient gene targeting and gene conversion (Aylon and Kupiec 2003, Kobayashi et al. 2004, Saberi et al. 2008, Nishino et al. 2008). In humans, RAD9 interacts with RAD51 recombinase and influences the DSB repair activity through homologous recombination (Pandita et al. 2006).

7.4.3.2 TLS and Template Switching

The 9-1-1 complex is involved in the two DNA damage-tolerance pathways (TLS and template switching), as is ubiquitinated PCNA. In *S. cerevisiae*, the 9-1-1 complex interacts with the TLS polymerase Polζ, and enhances efficiency of Polζ-dependent mutagenesis (Paulovich et al. 1998; Sabbioneda et al. 2005). The *S. pombe* checkpoint protein Rad17 is required for the enhanced chromatin association of Polκ (Kai and Wang 2003). Human 9-1-1 complex stimulates the activity of Polβ and Polι (Smith et al. 2012). Evidence suggests that 9-1-1 is also involved in template switching. The DNA-damage-inducible phosphorylation of Rad9 at Thr-225 by ATR in *S. pombe* promotes interaction with Mms2, which is involved in polyubiquitylation of PCNA, and represses the error-prone TLS pathway (Kai et al. 2007). In addition, 9-1-1 and RAD17-RFC promote the template-switching pathway independently of their canonical checkpoint-signaling pathway in *S. cerevisiae* (Karras et al. 2013).

7.4.3.3 DNA Repair Pathways

9-1-1 also interacts with many proteins that are involved in BER and MMR, such as Polβ, FEN1, DNA ligase 1, several DNA glycosylases, AP endonuclease 1, MLH1, MSH2, MSH3, and MSH6, and enhances their activities (reviewed in Li et al. 2016). Thus, 9-1-1 may have a direct role for efficient DNA-repair reactions. Because these BER and MMR proteins have also been reported to interact with PCNA, it is unclear which clamp is more important for the repair pathways (Gembka et al. 2007; Balakrishnan et al. 2009). Notably, because 9-1-1 is involved in the DNA-damage-checkpoint pathway, interaction of these proteins with 9-1-1 could provide signals to activate the ATR pathway. 9-1-1 also interacts with p21 as PCNA does (Doré et al. 2009). p21 binding to 9-1-1 was suggested to competitively block the interaction between 9-1-1 and these repair enzymes.

7.4.3.4 Meiosis

DSBs occur during meiosis and are required for meiotic recombination and segregation of homologous chromosomes. To ensure the appropriate resolution of meiosis, many canonical DNA-damage-response proteins are involved (reviewed in Hochwagen and Amon 2006, Subramanian and Hochwagen 2014). RAD17-RFC and 9-1-1 participate in meiosis through regulation of both meiotic checkpoint and meiotic recombination. 9-1-1 is loaded to the ssDNA region close to the resected meiotic DSB sites, and this loading can induce delays in the entry into meiosis I (Lydall et al. 1996). In addition, RAD17-RFC and 9-1-1 have a direct role in meiotic recombination (Grushcow et al. 1999; Thompson and Stahl 1999; Shinohara et al. 2003).

7.5 Conclusion

PCNA has important roles, both as a component of the replisome and as a memory molecule marking newly synthesized DNA strands to recruit specific DNA transaction enzymes under a defined spatiotemporal regulation. The dynamics of PCNA binding to DNA are involved in switching from G1 phase to S phase and from S phase to G2 phase (Fig. 7.3). PCNA is, therefore, essential for maintenance of genome integrity and epigenetic programming. We have a growing understanding of the molecular mechanisms that determine how, where, and when PCNA loading and unloading occur. However, many questions remain to be answered. The mechanism that determines whether PCNA is loaded by RFC or CTF18-RFC is not yet known. We have not yet discovered how PCNA loading discriminate between leading and lagging strands and how the loading process "determines" the relative frequencies in a certain DNA region. Whether particular DNA sequences, chromatin structures, stages in S phase, and cell-cycle progression events are able to switch PCNA dynamics actively through effects on the interacting proteins and PTMs is also not known. Furthermore, the mechanisms that determine the order and the specificity of the actions of PCNA-binding proteins remain to be determined. Approaches to address these questions should clarify the dynamic features of multiple clamps and loaders in eukaryotes, and how they change from fork to fork. Factors that can influence these events include chromatin structure, DNA sequence, and replication timing. Progress in the development of analytical technologies for single molecules and single cells, along with data mining, should help to elucidate the biological significance of multiple clamps and loaders in replicating chromosomes.

Acknowledgments We would like to thank Dr. T. S. Takahashi (Kyushu University) for the comments on the manuscript. We apologize to those colleagues whose work is not cited because of space restrictions. This work is supported by Grants-in-aid for Scientific Research (KAKENHI) 25131714, 25440011, 26114714, and 16H04743.

References

Abbas T, Sivaprasad U, Terai K, Amador V, Pagano M, Dutta A (2008) PCNA-dependent regulation of p21 ubiquitylation and degradation via the CRL4Cdt2 ubiquitin ligase complex. Genes Dev 22:2496–2506

Abbas T, Shibata E, Park J, Jha S, Karnani N, Dutta A (2010) CRL4 Cdt2 regulates cell proliferation and histone gene expression by targeting PR-Set7/Set8 for degradation. Mol Cell 40:9–21

Acevedo J, Yan S, Michael WM (2016) Direct binding to replication protein a (RPA)-coated single-stranded DNA allows recruitment of the ATR activator TopBP1 to sites of DNA damage. J Biol Chem 291:13124–13131

Alabert C, Bukowski-Wills JC, Lee SB, Kustatscher G, Nakamura K, de Lima AF, Menard P, Mejlvang J, Rappsilber J, Groth A (2014) Nascent chromatin capture proteomics determines chromatin dynamics during DNA replication and identifies unknown fork components. Nat Cell Biol 16:281–293

Arias EE, Walter JC (2005) Replication-dependent destruction of Cdt1 limits DNA replication to a single round per cell cycle in Xenopus egg extracts. Genes Dev 19:114–126
Awasthi P, Foiani M, Kumar A (2015) ATM and ATR signaling at a glance. J Cell Sci 128:4255–4262
Aylon Y, Kupiec M (2003) The checkpoint protein Rad24 of saccharomyces cerevisiae is involved in processing double-strand break ends and in recombination partner choice. Mol Cell Biol 23(18):6585–6596
Balakrishnan L, Brandt PD, Lindsey-Boltz LA, Sancar A, Bambara RA (2009) Long patch base excision repair proceeds via coordinated stimulation of the multienzyme DNA repair complex. J Biol Chem 284:15158–15172
Bao S, Tibbetts RS, Brumbaugh KM, Fang Y, Richardson DA, Ali A, Chen SM, Abraham RT, Wang XF (2001) ATR/ATM-mediated phosphorylation of human Rad17 is required for genotoxic stress responses. Nature 411:969–974
Bermudez VP, Lindsey-Boltz LA, Cesare AJ, Maniwa Y, Griffith JD, Hurwitz J, Sancar A (2003) Loading of the human 9-1-1 checkpoint complex onto DNA by the checkpoint clamp loader hRad17-replication factor C complex in vitro. Proc Natl Acad Sci U S A 100:1633–1638
Bienko M, Green CM, Crosetto N, Rudolf F, Zapart G, Coull B, Kannouche P, Wider G, Peter M, Lehmann AR, Hofmann K, Dikic I (2005) Ubiquitin-binding domains in Y-family polymerases regulate translesion synthesis. Science 310:1821–1824
Boehm EM, Washington MT (2016) R.I.P. to the PIP: PCNA-binding motif no longer considered specific: PIP motifs and other related sequences are not distinct entities and can bind multiple proteins involved in genome maintenance. BioEssays 38:1117–1122
Burgers PM (2009) Polymerase dynamics at the eukaryotic DNA replication fork. J Biol Chem 284:4041–4055
Burtelow MA, Kaufmann SH, Karnitz LM (2000) Retention of the human Rad9 checkpoint complex in extraction-resistant nuclear complexes after DNA damage. J Biol Chem 275:26343–26348
Burtelow MA, Roos-Mattjus PM, Rauen M, Babendure JR, Karnitz LM (2001) Reconstitution and molecular analysis of the hRad9-hHus1-hRad1 (9-1-1) DNA damage responsive checkpoint complex. J Biol Chem 276:25903–25909
Bylund GO, Burgers PM (2005) Replication protein A-directed unloading of PCNA by the Ctf18 cohesion establishment complex. Mol Cell Biol 25:5445–5455
Byun TS, Pacek M, Yee MC, Walter JC, Cimprich KA (2005) Functional uncoupling of MCM helicase and DNA polymerase activities activates the ATR-dependent checkpoint. Genes Dev 19:1040–1052
Carr AM, Lambert S (2013) Replication stress-induced genome instability: the dark side of replication maintenance by homologous recombination. J Mol Biol 425:4733–4744
Caspari T, Dahlen M, Kanter-Smoler G, Lindsay HD, Hofmann K, Papadimitriou K, Sunnerhagen P, Carr AM (2000) Characterization of *Schizosaccharomyces pombe* Hus1: a PCNA-related protein that associates with Rad1 and Rad9. Mol Cell Biol 74:1254–1262
Chen MJ, Lin YT, Lieberman HB, Chen G, Lee EY (2001) ATM-dependent phosphorylation of human Rad9 is required for ionizing radiation-induced checkpoint activation. J Biol Chem 276:16580–16586
Chen X, Paudyal SC, Chin RI, You Z (2013) PCNA promotes processive DNA end resection by Exo1. Nucleic Acids Res 41:9325–9338
Choe KN, Moldovan GL (2017) Forging ahead through darkness: PCNA, still the principal conductor at the replication fork. Mol Cell 65:380–392
Chuang LC, Yew PR (2001) Regulation of nuclear transport and degradation of the Xenopus cyclin-dependent kinase inhibitor, p27Xic1. J Biol Chem 276:1610–1617
Chuang LS, Ian HI, Koh TW, Ng HH, Xu G, Li BF (1997) Human DNA-(cytosine-5) methyltransferase-PCNA complex as a target for p21WAF1. Science 277:1996–2000
Cimprich KA, Cortez D (2008) ATR: an essential regulator of genome integrity. Nat Rev Mol Cell Biol 9:616–627
Cotta-Ramusino C, McDonald ER 3rd, Hurov K, Sowa ME, Harper JW, Elledge SJ (2011) A DNA damage response screen identifies RHINO, a 9-1-1 and TopBP1 interacting protein required for ATR signaling. Science 332:1313–1317

Crabbé L, Thomas A, Pantesco V, De Vos J, Pasero P, Lengronne A (2010) Analysis of replication profiles reveals key role of RFC–Ctf18 in yeast replication stress response. Nat Struct Mol Biol 17:1391–1397

De March M, Merino N, Barrera-Vilarmau S, Crehuet R, Onesti S, Blanco FJ, De Biasio A (2017) Structural basis of human PCNA sliding on DNA. Nat Commun 8:13935

Delacroix S, Wagner JM, Kobayashi M, Yamamoto K, Karnitz LM (2007) The Rad9-Hus1-Rad1 (9-1-1) clamp activates checkpoint signaling via TopBP1. Genes Dev 21:1472–1477

Dianov GL, Sleeth KM, Dianova II, Allinson SL (2003) Repair of abasic sites in DNA. Mutat Res 531:157–163

Doré AS, Kilkenny ML, Rzechorzek NJ, Pearl LH (2009) Crystal structure of the rad9-rad1-hus1 DNA damage checkpoint complex--implications for clamp loading and regulation. Mol Cell 34:735–745

Dua R, Levy DL, Campbell JL (1999) Analysis of the essential functions of the C-terminal protein/protein interaction domain of Saccharomyces cerevisiae pol ε and its unexpected ability to support growth in the absence of the DNA polymerase domain. J Biol Chem 274:22283–22288

Duursma AM, Driscoll R, Elias JE, Cimprich KA (2013) A role for the MRN complex in ATR activation via TOPBP1 recruitment. Mol Cell 50:116–122

Edmunds CE, Simpson LJ, Sale JE (2008) PCNA ubiquitination and REV1 define temporally distinct mechanisms for controlling translesion synthesis in the avian cell line DT40. Mol Cell 30:519–529

Ellison V, Stillman B (2003) Biochemical characterization of DNA damage checkpoint complexes: clamp loader and clamp complexes with specificity for 5′ recessed DNA. PLoS Biol 1:e33

Fan J, Otterlei M, Wong HK, Tomkinson AE, Wilson DM 3rd (2004) XRCC1 co-localizes and physically interacts with PCNA. Nucleic Acids Res 32:2193–2201

Feng W, D'Urso G (2001) Schizosaccharomyces pombe cells lacking the amino-terminal catalytic domains of DNA polymerase ε are viable but require the DNA damage checkpoint control. Mol Cell Biol 21:4495–4504

Flores-Rozas H, Clark D, Kolodner RD (2000) Proliferating cell nuclear antigen and Msh2p-Msh6p interact to form an active mispair recognition complex. Nat Genet 26:375–378

Fridman Y, Gur E, Fleishman SJ, Aharoni A (2013) Computational protein design suggests that human PCNA-partner interactions are not optimized for affinity. Proteins 81:341–348

Fujisawa R, Ohashi E, Hirota K, Tsurimoto T (2017) Human CTF18-RFC clamp-loader complexed with non-synthesising DNA polymerase ε efficiently loads the PCNA sliding clamp. Nucleic Acids Res 45(8):4550–4563

Furuya K, Poitelea M, Guo L, Caspari T, Carr AM (2004) Chk1 activation requires Rad9 S/TQ-site phosphorylation to promote association with C-terminal BRCT domains of Rad4TOPBP1. Genes Dev 18:1154–1164

Gali H, Juhasz S, Morocz M, Hajdu I, Fatyol K, Szukacsov V, Burkovics P, Haracska L (2012) Role of SUMO modification of human PCNA at stalled replication fork. Nucleic Acids Res 40:6049–6059

García-Rodríguez LJ, De Piccoli G, Marchesi V, Jones RC, Edmondson RD, Labib K (2015) A conserved Polε binding module in Ctf18-RFC is required for S-phase checkpoint activation downstream of Mec1. Nucleic Acids Res 43:8830–8838

Gary R, Ludwig DL, Cornelius HL, MacInnes MA, Park MS (1997) The DNA repair endonuclease XPG binds to proliferating cell nuclear antigen (PCNA) and shares sequence elements with the PCNA-binding regions of FEN-1 and cyclin-dependent kinase inhibitor p21. J Biol Chem 272:24522–24529

Gembka A, Toueille M, Smirnova E, Poltz R, Ferrari E, Villani G, Hübscher U (2007) The checkpoint clamp, Rad9-Rad1-Hus1 complex, preferentially stimulates the activity of apurinic/apyrimidinic endonuclease 1 and DNA polymerase beta in long patch base excision repair. Nucleic Acids Res 35:2596–2608

Georgescu RE, Langston L, Yao NY, Yurieva O, Zhang D, Finkelstein J, Agarwal T, O'Donnell ME (2014) Mechanism of asymmetric polymerase assembly at the eukaryotic replication fork. Nat Struct Mol Biol 21:664–670

Georgescu RE, Schauer GD, Yao NY, Langston LD, Yurieva O, Zhang D, Finkelstein J, O'Donnell ME (2015a) Reconstitution of a eukaryotic replisome reveals suppression mechanisms that define leading/lagging strand operation. Elife 4:e04988

Georgescu R, Langston L, O'Donnell M (2015b) A proposal: evolution of PCNA's role as a marker of newly replicated DNA. DNA Repair (Amst) 29:4–15

Goellner EM, Smith CE, Campbell CS, Hombauer H, Desai A, Putnam CD, Kolodner RD (2014) PCNA and Msh2-Msh6 activate an Mlh1-Pms1 endonuclease pathway required for Exo1-independent mismatch repair. Mol Cell 55:291–304

Gong Z, Kim JE, Leung CC, Glover JN, Chen J (2010) BACH1/FANCJ acts with TopBP1 and participates early in DNA replication checkpoint control. Mol Cell 37:438–446

Greer DA, Besley BD, Kennedy KB, Davey S (2003) hRad9 rapidly binds DNA containing double-strand breaks and is required for damage-dependent topoisomerase II beta binding protein 1 focus formation. Cancer Res 63:4829–4835

Griffith JD, Lindsey-Boltz LA, Sancar A (2002) Structures of the human Rad17-replication factor C and checkpoint Rad 9-1-1 complexes visualized by glycerol spray/low voltage microscopy. J Biol Chem 277:15233–15236

Grushcow JM, Holzen TM, Park KJ, Weinert T, Lichten M, Bishop DK (1999) Saccharomyces cerevisiae checkpoint genes MEC1, RAD17 and RAD24 are required for normal meiotic recombination partner choice. Genetics 153:607–620

Guo C, Tang TS, Bienko M, Parker JL, Bielen AB, Sonoda E, Takeda S, Ulrich HD, Dikic I, Friedberg EC (2006) Ubiquitin-binding motifs in REV1 protein are required for its role in the tolerance of DNA damage. Mol Cell Biol 26:8892–8900

Hanna JS, Kroll ES, Lundblad V, Spencer FA (2001) Saccharomyces cerevisiae CTF18 and CTF4 are required for sister chromatid cohesion. Mol Cell Biol 21:3144–3158

Havens CG, Walter JC (2009) Docking of a specialized PIP box onto chromatin-bound PCNA creates a Degron for the ubiquitin ligase $CRL4^{Cdt2}$. Mol Cell 35:93–104

Henneke G, Koundrioukoff S, Hubscher U (2003) Phosphorylation of human Fen1 by cyclin-dependent kinase modulates its role in replication fork regulation. Oncogene 22:4301–4313

Hochwagen A, Amon A (2006) Checking your breaks: surveillance mechanisms of meiotic recombination. Curr Biol 16:R217–R228

Hoege C, Pfander B, Moldovan GL, Pyrowolakis G, Jentsch S (2002) RAD6-dependent DNA repair is linked to modification of PCNA by ubiquitin and SUMO. Nature 419:135–141

Hofmann JFX, Beach D (1994) cdt1 is an essential target of the Cdc10/Sct1 transcription factor: requirement for DNA replication and inhibition of mitosis. EMBO J 13:425–434

Huang TT, Nijman SM, Mirchandani KD, Galardy PJ, Cohn MA, Haas W, Gygi SP, Ploegh HL, Bernards R, D'Andrea AD (2006) Regulation of monoubiquitinated PCNA by DUB autocleavage. Nat Cell Biol 8:339–347

Iida T, Suetake I, Tajima S, Morioka H, Ohta S, Obuse C, Tsurimoto T (2002) PCNA clamp facilitates action of DNA cytosine methyltransferase 1 on hemimethylated DNA. Genes Cells 7:997–1007

Indiani C, O'Donnell M (2006) The replication clamp-loading machine at work in the three domains of life. Nat Rev Mol Cell Biol 7:751–761

Indiani C, McInerney P, Georgescu R, Goodman MF, O'Donnell M. (2005) A sliding-clamp tool belt binds high- and low-fidelity DNA polymerases simultaneously. Mol Cell 19:805–815

Iyer RR, Pluciennik A, Burdett V, Modrich PL (2006) DNA mismatch repair: functions and mechanisms. Chem Rev 106:302–323

Johnson C, Gali VK, Takahashi TS, Kubota T (2016) PCNA retention on DNA into G2/M phase causes genome instability in cells lacking Elg1. Cell Rep 16:684–695

Kadyrov FA, Dzantiev L, Constantin N, Modrich P (2006) Endonucleolytic function of MutLalpha in human mismatch repair. Cell 126:297–308

Kadyrov FA, Holmes SF, Arana ME, Lukianova OA, O'Donnell M, Kunkel TA, Modrich P (2007) Saccharomyces cerevisiae MutLalpha is a mismatch repair endonuclease. J Biol Chem 282:37181–37190

Kai M, Wang TS (2003) Checkpoint activation regulates mutagenic translesion synthesis. Genes Dev 17:64–76
Kai M, Furuya K, Paderi F, Carr AM, Wang TS (2007) Rad3-dependent phosphorylation of the checkpoint clamp regulates repair-pathway choice. Nat Cell Biol 9:691–697
Karras GI, Fumasoni M, Sienski G, Vanoli F, Branzei D, Jentsch S (2013) Noncanonical role of the 9-1-1 clamp in the error-free DNA damage tolerance pathway. Mol Cell 49:536–546
Kaur R, Kostrub CF, Enoch T (2001) Structure-function analysis of fission yeast Hus1-Rad1-Rad9 checkpoint complex. Mol Biol Cell 12:3744–3758
Kawasoe Y, Tsurimoto T, Nakagawa T, Masukata H, Takahashi TS (2016) MutSa maintains the mismatch repair capability by inhibiting PCNA unloading. Elife 5:e15155
Kedar PS, Kim SJ, Robertson A, Hou E, Prasad R, Horton JK, Wilson SH (2002) Direct interaction between mammalian DNA polymerase beta and proliferating cell nuclear antigen. J Biol Chem 277:31115–311123
Kelch BA, Makino DL, O'Donnell M, Kuriyan J (2012) Clamp loader ATPases and the evolution of DNA replication machinery. BMC Biol 10:34
Kesti T, Flick K, Keranen S, Syvaoja JE, Wittenberg C (1999) DNA polymerase ε catalytic domains are dispensable for DNA replication, DNA repair, and cell viability. Mol Cell 3:679–685
Kim BJ, Lee H (2008) Lys-110 is essential for targeting PCNA to replication and repair foci, and the K110A mutant activates apoptosis. Biol Cell 100:675–686
Kim J, MacNeill SA (2003) Genome stability: a new member of the RFC family. Curr Biol 13:R873–R875
Kim SH, Michael WM (2008) Regulated proteolysis of DNA polymerase eta during the DNA-damage response in C. Elegans. Mol Cell 32:757–766
Kim ST, Lim DS, Canman CE, Kastan MB (1999) Substrate specificities and identification of putative substrates of ATM kinase family members. J Biol Chem 274:37538–37543
Kim Y, Starostina NG, Kipreos ET (2008) The CRL4Cdt2 ubiquitin ligase targets the degradation of p21Cip1 to control replication licensing. Genes Dev 22:2507–2519
Kleczkowska HE, Marra G, Lettieri T, Jiricny J (2001) hMSH3 and hMSH6 interact with PCNA and colocalize with it to replication foci. Genes Dev 15:724–736
Kobayashi M, Hirano A, Kumano T, Xiang SL, Mihara K, Haseda Y, Matsui O, Shimizu H, Yamamoto K (2004) Critical role for chicken Rad17 and Rad9 in the cellular response to DNA damage and stalled DNA replication. Genes Cells 9:291–303
Kochaniak AB, Habuchi S, Loparo JJ, Chang DJ, Cimprich KA, Walter JC, van Oijen AM (2009) Proliferating cell nuclear antigen uses two distinct modes to move along DNA. J Biol Chem 284:17700–117710
Kondo T, Matsumoto K, Sugimoto K (1999) Role of a complex containing Rad17, Mec3, and Ddc1 in the yeast DNA damage checkpoint pathway. Mol Cell Biol 19:1136–1143
Kondo T, Wakayama T, Naiki T, Matsumoto K, Sugimoto K (2001) Recruitment of Mec1 and Ddc1 checkpoint proteins to double-strand breaks through distinct mechanisms. Science 294:867–870
Koundrioukoff S, Jónsson ZO, Hasan S, de Jong RN, van der Vliet PC, Hottiger MO, Hübscher U (2000) A direct interaction between proliferating cell nuclear antigen (PCNA) and Cdk2 targets PCNA-interacting proteins for phosphorylation. J Biol Chem 275:22882–22887
Krokan HE, Bjørås M (2013) Base excision repair. Cold Spring Harb Perspect Biol 5:a012583
Kubota T, Nishimura K, Kanemaki MT, Donaldson AD (2013a) The Elg1 replication factor C-like complex functions in PCNA unloading during DNA replication. Mol Cell 50:273–280
Kubota T, Myung K, Donaldson AD (2013b) Is PCNA unloading the central function of the Elg1/ATAD5 replication factor C-like complex? Cell Cycle 12:2570–2579
Kubota T, Katou Y, Nakato R, Shirahige K, Donaldson AD (2015) Replication-coupled PCNA unloading by the Elg1 complex occurs genome-wide and requires Okazaki fragment ligation. Cell Rep 12:774–787
Kumagai A, Lee J, Yoo HY, Dunphy WG (2006) TopBP1 activates the ATR-ATRIP complex. Cell 124:943–955

Kunkel TA, Erie DA (2005) DNA mismatch repair. Annu Rev Biochem 74:681–710
Lavin MF (2008) Ataxia-telangiectasia: from a rare disorder to a paradigm for cell signalling and cancer. Nat Rev Mol Cell Biol 9:759–769
Lee J, Dunphy WG (2013) The Mre11-Rad50-Nbs1 (MRN) complex has a specific role in the activation of Chk1 in response to stalled replication forks. Mol Biol Cell 24:1343–1353
Lee J, Kumagai A, Dunphy WG (2003a) Claspin, a Chk1-regulatory protein, monitors DNA replication on chromatin independently of RPA, ATR, and Rad17. Mol Cell 11:329–340
Lee J, Kumagai A, Dunphy WG (2007) The Rad9-Hus1-Rad1 checkpoint clamp regulates interaction of TopBP1 with ATR. J Biol Chem 282:28036–28044
Lee KY, Fu H, Aladjem MI, Myung K (2013) ATAD5 regulates the lifespan of DNA replication factories by modulating PCNA level on the chromatin. J Cell Biol 200:31–44
Lengronne A, McIntyre J, Katou Y, Kanoh Y, Hopfner KP, Shirahige K, Uhlmann F (2006) Establishment of sister chromatid cohesion at the S. Cerevisiae replication fork. Mol Cell 23:787–799
Leonhardt H, Rahn HP, Weinzierl P, Sporbert A, Cremer T, Zink D, Cardoso MC (2000) Dynamics of DNA replication factories in living cells. J Cell Biol 149:271–280
Li Z, Pearlman AH, Hsieh P (2016) DNA mismatch repair and the DNA damage response. DNA Repair (Amst) 38:94–101
Lindsey-Boltz LA, Bermudez VP, Hurwitz J, Sancar A (2001) Purification and characterization of human DNA damage checkpoint Rad complexes. Proc Natl Acad Sci U S A 98:11236–11241
Lindsey-Boltz LA, Kemp MG, Capp C, Sancar A (2015) RHINO forms a stoichiometric complex with the 9-1-1 checkpoint clamp and mediates ATR-Chk1 signaling. Cell Cycle 14:99–108
Longhese MP, Paciotti V, Fraschini R, Zaccarini R, Plevani P, Lucchini G (1997) The novel DNA damage checkpoint protein ddc1p is phosphorylated periodically during the cell cycle and in response to DNA damage in budding yeast. EMBO J 16:5216–5226
López de Saro FJ, O'Donnell M (2001) Interaction of the beta sliding clamp with MutS, ligase, and DNA polymerase I. Proc Natl Acad Sci U S A 98:8376–8380
Lydall D, Nikolsky Y, Bishop DK, Weinert T (1996) A meiotic recombination checkpoint controlled by mitotic checkpoint genes. Nature 383:840–843
Maga G, Hübscher U (2003) Proliferating cell nuclear antigen (PCNA): a dancer with many partners. J Cell Sci 116:3051–3060
Mailand N, Gibbs-Seymour I, Bekker-Jensen S (2013) Regulation of PCNA-protein interactions for genome stability. Nat Rev Mol Cell Biol 14:269–282
Majka J, Burgers PM (2004) The PCNA-RFC families of DNA clamps and clamp loaders. Prog Nucleic Acid Res Mol Biol 78:227–260
Majka J, Binz SK, Wold MS, Burgers PM (2006) Replication protein a directs loading of the DNA damage checkpoint clamp to 5′-DNA junctions. J Biol Chem 281:27855–27861
Maréchal A, Zou L (2013) DNA damage sensing by the ATM and ATR kinases. Cold Spring Harb Perspect Biol 5:a012716
Marsischky GT, Filosi N, Kane MF, Kolodner R (1996) Redundancy of Saccharomyces cerevisiae MSH3 and MSH6 in MSH2-dependent mismatch repair. Genes Dev 10:407–420
Masuda Y, Piao J, Kamiya K (2010) DNA replication-coupled PCNA mono-ubiquitination and polymerase switching in a human in vitro system. J Mol Biol 396:487–500
Matsuoka S, Ballif BA, Smogorzewska A, McDonald ER, Hurov KE, Luo J, Bakalarski CE, Zhao Z, Solimini N, Lerenthal Y, Shiloh Y, Gygi SP, Elledge SJ (2007) ATM and ATR substrate analysis reveals extensive protein networks responsive to DNA damage. Science 316(5828):1160–1166
Mayer ML, Gygi SP, Aebersold R, Hieter P (2001) Identification of RFC(Ctf18p, Ctf8p, Dcc1p): an alternative RFC complex required for sister chromatid cohesion in S. Cerevisiae. Mol Cell 7:959–970
McInerney P, Johnson A, Katz F, O'Donnell M (2007) Characterization of a triple DNA polymerase replisome. Mol Cell 27:527–538
Melo JA, Cohen J, Toczyski DP (2001) Two checkpoint complexes are independently recruited to sites of DNA damage in vivo. Genes Dev 15:2809–2821

Merkle CJ, Karnitz LM, Henry-Sánchez JT, Chen J (2003) Cloning and characterization of hCTF18, hCTF8, and hDCC1. Human homologs of a Saccharomyces cerevisiae complex involved in sister chromatid cohesion establishment. J Biol Chem 278:30051–30056

Moldovan GL, Pfander B, Jentsch S (2006) PCNA controls establishment of sister chromatid cohesion during S phase. Mol Cell 23:723–732

Moldovan GL, Pfander B, Jentsch S (2007) PCNA, the maestro of the replication fork. Cell 129:665–679

Moldovan GL, Dejsuphong D, Petalcorin MI, Hofmann K, Takeda S, Boulton SJ, D'Andrea AD (2012) Inhibition of homologous recombination by the PCNA-interacting protein PARI. Mol Cell 45:75–86

Mordes DA, Glick GG, Zhao R, Cortez D (2008) TopBP1 activates ATR through ATRIP and a PIKK regulatory domain. Genes Dev 22:1478–1489

Mortusewicz O, Schermelleh L, Walter J, Cardoso MC, Leonhardt H (2005) Recruitment of DNA methyltransferase I to DNA repair sites. Proc Natl Acad Sci U S A 102:8905–8909

Murakami T, Takano R, Takeo S, Taniguchi R, Ogawa K, Ohashi E, Tsurimoto T (2010) Stable interaction between the human proliferating cell nuclear antigen loader complex Ctf18-replication factor C (RFC) and DNA polymerase ε is mediated by the cohesion-specific subunits, Ctf18, Dcc1, and Ctf8. J Biol Chem 285:34608–34615

Naiki T, Kondo T, Nakada D, Matsumoto K, Sugimoto K (2001) Chl12 (Ctf18) forms a novel replication factor C-related complex and functions redundantly with Rad24 in the DNA replication checkpoint pathway. Mol Cell Biol 21:5838–5845

Navadgi-Patil VM, Burgers PM (2008) Yeast DNA replication protein Dpb11 activates the Mec1/ATR checkpoint kinase. J Biol Chem 283:35853–35859

Navadgi-Patil VM, Burgers PM (2009) The unstructured C-terminal tail of the 9-1-1 clamp subunit Ddc1 activates Mec1/ATR via two distinct mechanisms. Mol Cell 36:743–753

Nishino K, Inoue E, Takada S, Abe T, Akita M, Yoshimura A, Tada S, Kobayashi M, Yamamoto K, Seki M, Enomoto T (2008) A novel role for Rad17 in homologous recombination. Genes Genet Syst 83:427–431

Nishitani H, Sugimoto N, Roukos V, Nakanishi Y, Saijo M, Obuse C, Tsurimoto T, Nakayama KI, Nakayama K, Fujita M, Lygerou Z, Nishimoto T (2006) Two E3 ubiquitin ligases, SCF-Skp2 and DDB1-Cul4, target human Cdt1 for proteolysis. EMBO J 25:1126–1136

Nishitani H, Shiomi Y, Iida H, Michishita M, Takami T, Tsurimoto T (2008) CDK inhibitor p21 is degraded by a proliferating cell nuclear antigen-coupled Cul4-DDB1Cdt2 pathway during S phase and after UV irradiation. J Biol Chem 283:29045–29052

O'Donnell M, Li H (2016) The eukaryotic replisome goes under the microscope. Curr Biol 26:R247–R256

Ohashi E, Takeishi Y, Ueda S, Tsurimoto T (2014) Interaction between Rad9-Hus1-Rad1 and TopBP1 activates ATR-ATRIP and promotes TopBP1 recruitment to sites of UV-damage. DNA Repair (Amst) 21:1–11

O'Neill T, Dwyer AJ, Ziv Y, Chan DW, Lees-Miller SP, Abraham RH, Lai JH, Hill D, Shiloh Y, Cantley LC, Rathbun GA (2000) Utilization of oriented peptide libraries to identify substrate motifs selected by ATM. J Biol Chem 275:22719–22727

Pandita RK, Sharma GG, Laszlo A, Hopkins KM, Davey S, Chakhparonian M, Gupta A, Wellinger RJ, Zhang J, Powell SN, Roti Roti JL, Lieberman HB, Pandita TK (2006) Mammalian Rad9 plays a role in telomere stability, S- and G2-phase-specific cell survival, and homologous recombinational repair. Mol Cell Biol 26:1850–1864

Park SY, Jeong MS, Han CW, Yu HS, Jang SB (2016) Structural and functional insight into proliferating cell nuclear antigen. J Microbiol Biotechnol 26:637–647

Parnas O, Zipin-Roitman A, Pfander B, Liefshitz B, Mazor Y, Ben-Aroya S, Jentsch S, Kupiec M (2010) Elg1, an alternative subunit of the RFC clamp loader, preferentially interacts with SUMOylated PCNA. EMBO J 29:2611–2622

Paulovich AG, Armour CD, Hartwell LH (1998) The Saccharomyces cerevisiae RAD9, RAD17, RAD24 and MEC3 genes are required for tolerating irreparable, ultraviolet-induced DNA damage. Genetics 150:75–93

Pluciennik A, Dzantiev L, Iyer RR, Constantin N, Kadyrov FA, Modrich P (2010) PCNA function in the activation and strand direction of MutLα endonuclease in mismatch repair. Proc Natl Acad Sci U S A 107:16066–16071

Prelich G, Stillman B (1988) Coordinated leading and lagging strand synthesis during SV40 DNA replication in vitro requires PCNA. Cell 53:117–126

Prelich G, Kostura M, Marshak DR, Mathews MB, Stillman B (1987) The cell-cycle regulated proliferating cell nuclear antigen is required for SV40 DNA replication in vitro. Nature 326:471–475

Prindle MJ, Loeb LA (2012) DNA polymerase delta in DNA replication and genome maintenance. Environ Mol Mutagen 53:666–682

Puddu F, Granata M, Di Nola L, Balestrini A, Piergiovanni G, Lazzaro F, Giannattasio M, Plevani P, Muzi-Falconi M (2008) Phosphorylation of the budding yeast 9-1-1 complex is required for Dpb11 function in the full activation of the UV-induced DNA damage checkpoint. Mol Cell Biol 28:4782–4793

Pursell ZF, Isoz I, Lundström EB, Johansson E, Kunkel TA (2007) Yeast DNA polymerase epsilon participates in leading-strand DNA replication. Science 317:127–130

Rappas M, Oliver AW, Pearl LH (2011) Structure and function of the Rad9-binding region of the DNA-damage checkpoint adaptor TopBP1. Nucleic Acids Res 39:313–324

Roos-Mattjus P, Vroman BT, Burtelow MA, Rauen M, Eapen AK, Karnitz LM (2002) Genotoxin-induced Rad9-Hus1-Rad1 (9-1-1) chromatin association is an early checkpoint signaling event. J Biol Chem 277:43809–43812

Roos-Mattjus P, Hopkins KM, Oestreich AJ, Vroman BT, Johnson KL, Naylor S, Lieberman HB, Karnitz LM (2003) Phosphorylation of human Rad9 is required for genotoxin-activated checkpoint signaling. J Biol Chem 278:24428–24437

Rousseau D, Cannella D, Boulaire J, Fitzgerald P, Fotedar A, Fotedar R (1999) Growth inhibition by CDK-cyclin and PCNA binding domains of p21 occurs by distinct mechanisms and is regulated by ubiquitin-proteasome pathway. Oncogene 18:4313–4325

Rudra S, Skibbens RV (2013) Cohesin codes - interpreting chromatin architecture and the many facets of cohesin function. J Cell Sci 126:31–41

Sabbioneda S, Minesinger BK, Giannattasio M, Plevani P, Muzi-Falconi M, Jinks-Robertson S (2005) The 9-1-1 checkpoint clamp physically interacts with polzeta and is partially required for spontaneous polzeta-dependent mutagenesis in Saccharomyces cerevisiae. J Biol Chem 280:38657–38665

Saberi A, Nakahara M, Sale JE, Kikuchi K, Arakawa H, Buerstedde JM, Yamamoto K, Takeda S, Sonoda E (2008) The 9-1-1 DNA clamp is required for immunoglobulin gene conversion. Mol Cell Biol 28:6113–6122

Sancar A, Lindsey-Boltz LA, Unsal-Kaçmaz K, Linn S (2004) Molecular mechanisms of mammalian DNA repair and the DNA damage checkpoints. Annu Rev Biochem 73:39–85

Shibahara K, Stillman B (1999) Replication-dependent marking of DNA by PCNA facilitates CAF-1-coupled inheritance of chromatin. Cell 96:575–585

Shibutani ST, de la Cruz AF, Tran V, Turbyfill WJ 3rd, Reis T, Edgar BA, Duronio RJ (2008) Intrinsic negative cell cycle regulation provided by PIP box- and Cul4Cdt2-mediated destruction of E2f1 during S phase. Dev Cell 15:890–900

Shinohara M, Sakai K, Ogawa T, Shinohara A (2003) The mitotic DNA damage checkpoint proteins Rad17 and Rad24 are required for repair of double-strand breaks during meiosis in yeast. Genetics 164:855–865

Shiomi Y, Nishitani H (2013) Alternative replication factor C protein, Elg1, maintains chromosome stability by regulating PCNA levels on chromatin. Genes Cells 18:946–959

Shiomi Y, Nishitani H (2017) Control of genome integrity by RFC complexes; conductors of PCNA loading onto and unloading from chromatin during DNA replication. Genes (Basel) 8., pii:E52

Shiomi Y, Shinozaki A, Nakada D, Sugimoto K, Usukura J, Obuse C, Tsurimoto T (2002) Clamp and clamp loader structures of the human checkpoint protein complexes, Rad9-1-1 and Rad17-RFC. Genes Cells 7:861–868

Shiomi Y, Shinozaki A, Sugimoto K, Usukura J, Obuse C, Tsurimoto T (2004) The reconstituted human Chl12-RFC complex functions as a second PCNA loader. Genes Cells 9:279–290

Shiomi Y, Hayashi A, Ishii T, Shinmyozu K, Nakayama J, Sugasawa K, Nishitani H (2012) Two different replication factor C proteins, Ctf18 and RFC1, separately control PCNA-CRL4Cdt2-mediated Cdt1 proteolysis during S phase and following UV irradiation. Mol Cell Biol 32:2279–2288

Sirbu BM, McDonald WH, Dungrawala H, Badu-Nkansah A, Kavanaugh GM, Chen Y, Tabb DL, Cortez D (2013) Identification of proteins at active, stalled, and collapsed replication forks using isolation of proteins on nascent DNA (iPOND) coupled with mass spectrometry. J Biol Chem 288:31458–31467

Skibbens RV (2009) Establishment of sister chromatid cohesion. Curr Biol 19:R1126–R1132

Smith LA, Makarova AV, Samson L, Thiesen KE, Dhar A, Bessho T (2012) Bypass of a psoralen DNA interstrand cross-link by DNA polymerases β, ι, and κ in vitro. Biochemistry 51:8931–8938

St Onge RP, Udell CM, Casselman R, Davey S (1999) The human G2 checkpoint control protein hRAD9 is a nuclear phosphoprotein that forms complexes with hRAD1 and hHUS1. Mol Biol Cell 10:1985–1995

St Onge RP, Besley BD, Park M, Casselman R, Davey S (2001) DNA damage-dependent and -independent phosphorylation of the hRad9 checkpoint protein. J Biol Chem 276:41898–41905

St Onge RP, Besley BD, Pelley JL, Davey S (2003) A role for the phosphorylation of hRad9 in checkpoint signaling. J Biol Chem 278:26620–26628

Stelter P, Ulrich HD (2003) Control of spontaneous and damage-induced mutagenesis by SUMO and ubiquitin conjugation. Nature 425:188–191

Subramanian VV, Hochwagen A (2014) The meiotic checkpoint network: step-by-step through meiotic prophase. Cold Spring Harb Perspect Biol 6:a016675

Sun Q, Tsurimoto T, Juillard F, Li L, Li S, De León VE, Chen S, Kaye K (2014) Kaposi's sarcoma-associated herpesvirus LANA recruits the DNA polymerase clamp loader to mediate efficient replication and virus persistence. Proc Natl Acad Sci U S A 111:11816–11821

Takeishi Y, Ohashi E, Ogawa K, Masai H, Obuse C, Tsurimoto T (2010) Casein kinase 2-dependent phosphorylation of human Rad9 mediates the interaction between human Rad9-Hus1-Rad1 complex and TopBP1. Genes Cells 15:761–771

Takeishi Y, Iwaya-Omi R, Ohashi E, Tsurimoto T (2015) Intramolecular binding of the Rad9 C terminus in the checkpoint clamp Rad9-Hus1-Rad1 is closely linked with its DNA binding. J Biol Chem 290:19923–19932

Terai K, Abbas T, Jazaeri AA, Dutta A (2010) CRL4(Cdt2) E3 ubiquitin ligase monoubiquitinates PCNA to promote translesion DNA synthesis. Mol Cell 37:143–149

Terret ME, Sherwood R, Rahman S, Qin J, Jallepalli PV (2009) Cohesin acetylation speeds the replication fork. Nature 462:231–234

Thompson DA, Stahl FW (1999) Genetic control of recombination partner preference in yeast meiosis. Isolation and characterization of mutants elevated for meiotic unequal sister-chromatid recombination. Genetics 153:621–641

Tinker RL, Kassavetis GA, Geiduschek EP (1994) Detecting the ability of viral, bacterial and eukaryotic proteins to track along DNA. EMBO J 13:5330–5337

Tsurimoto T (1999) PCNA binding proteins. Front Biosci 4:D849–D858

Tsurimoto T, Stillman B (1989) Purification of a cellular replication factor, RF-C, that is required for coordinated synthesis of leading and lagging strands during simian virus 40 DNA replication in vitro. Mol Cell Biol 9:609–619

Ueda S, Takeishi Y, Ohashi E, Tsurimoto T (2012) Two serine phosphorylation sites in the C-terminus of Rad9 are critical for 9-1-1 binding to TopBP1 and activation of the DNA damage checkpoint response in HeLa cells. Genes Cells 17:807–816

Ulrich HD (2013) New insights into replication clamp unloading. J Mol Biol 425:4727–4732
Ulrich HD, Jentsch S (2000) Two RING finger proteins mediate cooperation between ubiquitin-conjugating enzymes in DNA repair. EMBO J 19:3388–3397
Volkmer E, Karnitz LM (1999) Human homologs of Schizosaccharomyces pombe rad1, hus1, and rad9 form a DNA damage-responsive protein complex. J Biol Chem 274:567–570
Waga S, Stillman B (1998) The DNA replication fork in eukaryotic cells. Annu Rev Biochem 67:721–751
Waga S, Hannon GJ, Beach D, Stillman B (1994) The p21 inhibitor of cyclin-dependent kinases controls DNA replication by interaction with PCNA. Nature 369:574–578
Wang X, Zou L, Zheng H, Wei Q, Elledge SJ, Li L (2003) Genomic instability and endoreduplication triggered by RAD17 deletion. Genes Dev 17:965–970
Wang X, Hu B, Weiss RS, Wang Y (2006) The effect of Hus1 on ionizing radiation sensitivity is associated with homologous recombination repair but is independent of nonhomologous end-joining. Oncogene 25:1980–1983
Warbrick E, Lane DP, Glover DM, Cox LS (1995) A small peptide inhibitor of DNA replication defines the site of interaction between the cyclin-dependent kinase inhibitor p21waf1 and the proliferating cell nuclear antigen. Curr Biol 5:275–282
Weiss RS, Matsuoka S, Elledge SJ, Leder P (2002) Hus1 acts upstream of chk1 in a mammalian DNA damage response pathway. Curr Biol 12:73–77
Wu X, Shell SM, Zou Y (2005) Interaction and colocalization of Rad9/Rad1/Hus1 checkpoint complex with replication protein a in human cells. Oncogene 24:4728–4735
Xiong Y, Hannon GJ, Zhang H, Casso D, Kobayashi R, Beach D (1993) p21 is a universal inhibitor of cyclin kinases. Nature 366:701–704
Xu X, Vaithiyalingam S, Glick GG, Mordes DA, Chazin WJ, Cortez D (2008) The basic cleft of RPA70N binds multiple checkpoint proteins, including RAD9, to regulate ATR signaling. Mol Cell Biol 28:7345–7353
Yan S, Michael WM (2009) TopBP1 and DNA polymerase-alpha directly recruit the 9-1-1 complex to stalled DNA replication forks. J Cell Biol 184:793–804
Yao NY, O'Donnell M (2012) The RFC clamp loader: structure and function. Subcell Biochem 62:259–279
Yao N, Turner J, Kelman Z, Stukenberg PT, Dean F, Shechter D, Pan ZQ, Hurwitz J, O'Donnell M (1996) Clamp loading, unloading and intrinsic stability of the PCNA, beta and gp45 sliding clamps of human, E. Coli and T4 replicases. Genes Cells 1:101–113
You Z, Kong L, Newport J (2002) The role of single-stranded DNA and polymerase alpha in establishing the ATR, Hus1 DNA replication checkpoint. J Biol Chem 277:27088–27093
Yu C, Gan H, Han J, Zhou ZX, Jia S, Chabes A, Farrugia G, Ordog T, Zhang Z (2014) Strand-specific analysis shows protein binding at replication forks and PCNA unloading from lagging strands when forks stall. Mol Cell 56:551–563
Zamir L, Zaretsky M, Fridman Y, Ner-Gaon H, Rubin E, Aharoni A (2012) Tight coevolution of proliferating cell nuclear antigen (PCNA)-partner interaction networks in fungi leads to interspecies network incompatibility. Proc Natl Acad Sci U S A 109:E406–E414
Zhang H, Xiong Y, Beach D (1993) Proliferating cell nuclear antigen and p21 are components of multiple cell cycle kinase complexes. Mol Biol Cell 4:897–906
Zhang Z, Shibahara K, Stillman B (2000) PCNA connects DNA replication to epigenetic inheritance in yeast. Nature 408:221–225
Zou L, Cortez D, Elledge SJ (2002) Regulation of ATR substrate selection by Rad17-dependent loading of Rad9 complexes onto chromatin. Genes Dev 16:198–208
Zou L, Liu D, Elledge SJ (2003) Replication protein A-mediated recruitment and activation of Rad17 complexes. Proc Natl Acad Sci U S A 100:13827–11383

Chapter 8
Termination of Eukaryotic Replication Forks

Agnieszka Gambus

Abstract Termination of DNA replication forks takes place when two replication forks coming from neighbouring origins meet each other usually in the midpoint of the replicon. At this stage, the remaining fragments of DNA have to be unwound, all remaining DNA replicated and newly synthesised strands ligated to produce continuous sister chromatids. Finally, the replication machinery has to be taken off, chromatin re-assembled, and entwisted sister chromatids resolved topologically.

Over the last few decades, we have learned a lot about the assembly of the helicase and replisome and the initiation stage of DNA replication. We also know much more about the ability of forks to cope with replication stress. However, only within recent years we have gained the first glimpse of the mechanism of replication fork termination. In this chapter I will summarise the recent findings on replication termination, weigh this against the past literature and discuss relevant consequences and views for the future.

Keywords Eukaryotic DNA replication • Termination of DNA replication • Ubiquitin • Cdc48 p97 segregase • Cullins

8.1 Introduction

To maintain genomic stability, it is essential that every step of DNA replication is faultlessly executed. Mistakes during eukaryotic replication that are not efficiently repaired can lead to mutations and genome rearrangements that promote changes leading to development of cancer and other disorders.

DNA replication can be divided into three stages: initiation, elongation and termination. Initiation of DNA replication happens when licenced origins of replication fire creating two DNA replication forks, which move in opposite directions. The elongation stage involves the progression of replication forks as they unwind and replicate DNA. Finally, termination happens when two replication forks from

A. Gambus (✉)
Institute of Cancer and Genomic Sciences, University of Birmingham, Birmingham, UK
e-mail: a.gambus@bham.ac.uk

H. Masai, M. Foiani (eds.), *DNA Replication*, Advances in Experimental Medicine and Biology 1042, https://doi.org/10.1007/978-981-10-6955-0_8

neighbouring origins converge and the duplication of remaining fragment of DNA is neatly completed. Over the years we have learnt a lot about the mechanisms of DNA replication initiation and elongation (briefly explained below), but until recently our knowledge of replication termination was very restricted. The last few years have brought a breakthrough in our understanding of mechanisms of replication termination: we have learnt that converging replication forks can pass each other when terminating, and we have also unravelled the workings of disassembly of terminated replisomes.

8.2 Replication Fork Termination Occurs Throughout S-Phase

When does replication termination take place? In our mind replication termination should happen mostly at the end of the whole replication process, so in terms of cell cycle stages – at the end of S-phase. In reality, however, DNA replication forks encounter forks from neighbouring origins throughout the entire S-phase. Forks emanating from origin clusters firing in early S-phase will also terminate in early S-phase; with average replicon size of 31 kbp (Moreno et al. 2016; Picard et al. 2014) and an average fork speed of 1.5 kb/min (Conti et al. 2007), it takes about 10 min for the two neighbouring forks to reach one another. In fact there is likely more termination occurring in mid S-phase than in late S-phase as the strict replication timing programme driving replication in each cell means that only difficult-to-replicate regions are replicated in late S-phase (Gilbert 2010).

8.3 Replication Initiation and Elongation

To ensure that all of the large eukaryotic genomes are duplicated in full before each cell division, eukaryotic DNA replication starts from multiple origins of replication. Human cells have on average about 50,000 of them spread throughout the genome. It is also essential that DNA is replicated just once per cell cycle as re-replication of parts of the genome is a threat to the maintenance of genome integrity. To achieve this, the replicative helicase (protein complex, which unwinds double-stranded DNA during replication) can be loaded onto DNA only before the onset of S-phase when CDK activity is low and can be activated only during S-phase when CDK activity is high. Origins of replication are therefore "licenced" in late M and G1 stages of the cell cycle, by loading of the core of the replicative helicase: Mcm2–7 (Minichromosome maintenance 2,3,4,5,6,7) complexes. Double hexamers of the Mcm2–7 complexes are loaded onto origins through the concerted action of ORC (origin recognition complex), Cdc6 and Cdt1 factors. These double hexamers encircle the double-stranded DNA and are arranged in N-terminus to N-terminus orientation with the C-terminal helicase domains on the outside. There are multiple Mcm2–7 double hexamers loaded around each origin of replication, which may be facilitated by their ability to slide on double-stranded DNA (Evrin et al. 2009; Gambus et al. 2011; Remus et al. 2009).

The initiation of DNA replication requires the activity of two S-phase kinases: Cdc7/Dbf4 (DDK – Dbf4-dependent kinase) and Cdk/cyclin (CDK – Cyclin-dependent kinase). DDK phosphorylates double hexamers of Mcm2–7, while CDK drives association of GINS (Go-Ichi-Ni-San, complex of Sld5, Psf1, Psf2 and Psf3 or GINS1,2,3,4) and Cdc45 with the Mcm2–7 complexes, forming the CMG complex (**C**dc45/**M**cm2–7/**G**INS), which is an active replicative helicase (Ilves et al. 2010; Moyer et al. 2006; Simon et al. 2016). The initiation process leads to rearrangement of Mcm2–7 complexes: the double hexamers split into two CMGs, and each of them now likely encircles just single-stranded DNA (Costa et al. 2011; Gambus et al. 2006; Yardimci et al. 2010).

During the elongation stage of DNA replication, the helicase (CMG complex) travels at the tip of the replication fork, unwinding the double-stranded DNA and exposing single strands that can act as a template for DNA synthesis by DNA replication polymerases. The MCM motor of CMG belongs to the superfamily of AAA+ ATPases and is a 3–5′ DNA translocase, which encircles the leading strand of the replication fork (reviewed in Pellegrini and Costa (2016)). The Polα-primase complex initiates DNA synthesis with a short RNA primer that is then elongated for another 20-nt by Polα polymerase activity. The leading strand is believed to be synthesised mainly by DNA Polε (DNA polymerase epsilon) in a continuous manner, while the lagging strand is thought to be completed by DNA Polδ (DNA polymerase delta) (Daigaku et al. 2015; Georgescu et al. 2015; Pavlov et al. 2006). The latter synthesises short Okazaki fragments in the opposite direction to the movement of the fork, and these fragments need to therefore be processed and ligated to produce the continuous DNA strand (maturation of Okazaki fragments). DNA Polε is therefore following the helicase and indeed has a number of connections linking it directly to the helicase to facilitate the smooth progression of the fork (see below and reviewed in Pellegrini and Costa 2016).

DNA unwinding generates a compensatory increase in the intertwining of parental strands, which can be converted into helical overwinding (positive supercoiling) of the unreplicated portions of the DNA ahead of the forks (Postow et al. 2001; Wang 2002). This mechanical strain can be transmitted to replicated DNA by rotation at the branching point of the replication fork, thus generating intertwining of the daughter duplexes (known as precatenates) (Been and Champoux 1980) (Fig. 8.1). Recent research in budding yeast has shown, however, that during normal progression of replication forks, fork rotation and precatenation are actively inhibited by components of the replisome Timeless/Tof1 and Tipin/Csm3 (Schalbetter et al. 2015). Instead, supercoils generated during replication elongation can be relaxed by both type I and type II topoisomerases (Wang 2002). Indeed, the current view assumes that positive supercoiling is mainly relaxed by type I enzymes (Topo I, *S. cerevisiae* Top1) anywhere in the unreplicated region (Postow et al. 2001). The replisome progression complex (RPC) built around CMG at the tip of the fork

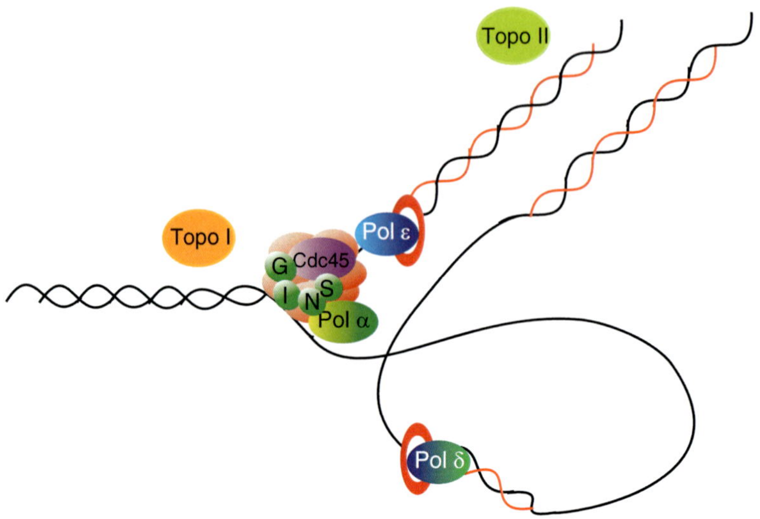

Fig. 8.1 Topoisomerases at the replication fork. Topo I relaxes the positive supercoiling building up ahead of the fork. Sometimes this supercoiling can lead to rotation of the fork and intertwining of the daughter strands of DNA behind the fork (precatenates). These are resolved by Topo II

contains Top1, positioning it perfectly for its function ahead of the fork (Gambus et al. 2006) (Fig. 8.1). Interestingly, yeast cells without Top1 and also *top2* mutants can replicate DNA, but replication is not possible when both proteins are defective (Bermejo et al. 2007; Brill et al. 1987).

Importantly, replication forks do not move through naked DNA but through a chromatin structure. Nucleosomes therefore need to be dismantled ahead of the forks and rebuilt behind the forks. The efficient repositioning of parental histones is essential for full reconstitution of epigenetic markings throughout the replicating genome. Studies of SV40 replication forks provided evidence for the existence of only 200–300 bp of apparently nucleosome-free DNA behind the replication fork (Gasser et al. 1996), and the nucleosomes in yeast were shown recently to be positioned immediately after the fork passage and restrict Okazaki fragments sizes (Smith and Whitehouse 2012). Progressing replication forks need also to remove other proteins attached to DNA, for example, the unfired Mcm2–7 double hexamers, which, loaded in excess, serve as dormant origins ready to rescue collapsed forks. Finally, sister chromatids are topologically embraced and held together until mitosis by cohesin ring complexes. This cohesion is established during DNA replication as forks progress (reviewed by Uhlmann 2009).

8.4 Where Does Termination of Eukaryotic Replication Forks Happen?

The simplest answer to this question is wherever the two neighbouring forks meet each other. Recent analysis of genome-wide replication profiles in budding yeast, both through high-resolution replication profiling (Hawkins et al. 2013) and through deep sequencing of Okazaki fragments (McGuffee et al. 2013), showed that termination generally occurs midway between two adjacent replication origins. The precise position of termination depends on the relative activation time of each of the origins and their variable efficiency. Okazaki fragment mapping in human cells (HeLa and GM06990) also confirmed such midpoint localisation (Petryk et al. 2016).

Eukaryotes not only have specific spatial patterns but also possess temporal patterns of genome replication, which are executed by regulated activation of replication origins throughout S-phase. High-throughput experiments allowed the identification of a genome-wide temporal order of replication (Gilbert 2010). In early S-phase, "active" chromatin is replicated with origins of replication located in general in-between the genes. Not surprisingly, therefore, many termination events in early S-phase were found to overlap with transcribed genes. In late S-phase, however, when heterochromatin is replicated, many termination zones were found in large non-expressed regions of DNA (Petryk et al. 2016).

This sequence independent localisation of termination sites is in sharp contrast to the organisation of termination events in *E. coli* chromosome where termination takes place within a broad region containing several specialised fork barriers, i.e. Tus-TER complexes, which confine fork fusion to a site of 270 kb (reviewed in Dimude et al. (2016)). Due to these defined prokaryotic termination regions (TER), for a number of years, termination of eukaryotic replication forks was studied only at the existing few loci within the eukaryotic genome, which contain specialised replication fork barriers (RFBs). The best characterised of such sites are: the RTS1 site in *S. pombe*, which regulates mating type switching (Brewer and Fangman 1988), and the rDNA locus within ribosomal DNA repeats of metazoa and yeasts (Dalgaard and Klar 2000). The RFB barriers are able to arrest one of the two neighbouring forks and therefore create specific termination sites (Bastia and Zaman 2014; Dalgaard et al. 2009). To minimise fork pausing at RFBs, the protein displacement helicase Rrm3 helps to displace the barriers to allow replication passage and is required for fork termination at these sites. In yeast lacking Rrm3, tenfold accumulation of termination structures ("X"-shaped DNA structures in 2D DNA gels) was observed, while only twofold accumulation of paused forks at the barrier (Ivessa et al. 2000, 2003). Despite this, Rrm3 is not required for bulk replisome unloading during normal termination (Maric et al. 2014), and it is needed only for fork convergence at rare situations when one fork is paused.

In 2010, Fachinetti et al. identified 71 termination regions (TERs) in budding yeast, through a combination of chromatin immunoprecipitation (ChiP) and BrdU incorporation experiments. Their work found that the majority of these regions contain fork-pausing elements, such as transcription clusters, and that efficient termination at the identified sites requires activity of Rrm3 and Top2 (Fachinetti et al. 2010). However, the more recent high-resolution approaches suggest that these TERs actually represent sites with a higher than average probability of termination as they are flanked by early-firing efficient origins. Importantly, changes of origin-firing pattern moved the termination positioning both in non-TER and TER replicons, indicating that it is the timing and efficiency of origin firing and not fork-pausing elements that dictate the precise place of replication fork convergence (Hawkins et al. 2013; McGuffee et al. 2013).

8.5 How Do Replication Forks Converge?

Figure 8.2 summarises our current model of replication fork termination. To allow convergence of two approaching DNA replication forks, all of the proteins bound to DNA between them must be evicted (Fig. 8.2a). Unwinding of final stretches of DNA can present a problem for the forks as the torsional stress created ahead of the fork cannot be easily released due to lack of access for Top I (see below for more details) and has to be translated into precatenates, which accumulate behind the fork (Fig. 8.2b). Two converging forks present two large protein machineries approaching one another and heading for head-on collision while unwinding the remaining DNA between them (Fig. 8.2c). After forks converge, all of the remaining DNA needs to be replicated, and the RNA-DNA primer of the last Okazaki fragment on the lagging strand needs to be processed (Fig. 8.2d, e). Once this is complete, DNA needs to be ligated into a continuous strand, and replisomes need to be disassembled (Fig. 8.2e). Finally, the entangled sister chromatids need to be resolved into two separate strands (Fig. 8.2g).

Recent years have brought a breakthrough in our understanding of the above processes. Beautiful work from Prof. Johannes Walter's lab shed light on the mechanism by which forks converge and termination is resolved (Dewar et al. 2015). To synchronise termination events and facilitate their analysis, they constructed plasmids with an array of *lac* repressors (LacRs) bound to *lac* operators (LacOs), which can be disrupted by IPTG. Such plasmids replicated in cell-free *Xenopus laevis* egg extract accumulated blocked forks at the edges of the array. The blocked forks were then released by addition of IPTG and proceeded to terminate within the DNA fragment comprising the array. Using this system, Dewar et al. could monitor unwinding of DNA as forks approach each other, synthesis of DNA, ligation of the replicated DNA and decatenation of daughter molecules. Strikingly, the rate of DNA synthesis within the array was almost perfectly linear after IPTG addition and resembled the fork progression speed reported in the same extracts. It suggests therefore that converging forks do not slow significantly before they meet; they do not collide with each other or stall but rather pass each other (Dewar et al. 2015) (Fig. 8.2c, d). Such passage can be possible as CMGs encircle the leading strand of the replication fork and therefore approach each other on opposite strands when converging at termination (Ali et al. 2016; Costa et al. 2011; Fu et al. 2011). Interestingly, how-

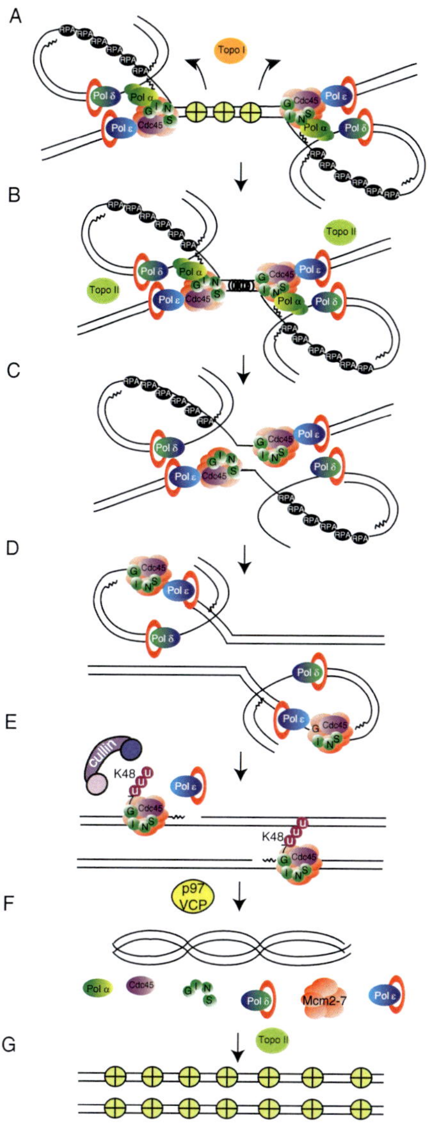

Fig. 8.2 Model of termination of eukaryotic replication forks. When two neighbouring replication forks approach each other from opposite directions, all of the proteins organising DNA in between the forks (nucleosomes and others) have to be removed, while Topo I relaxes the torsional stress (positive supercoiling) (**a**). When two terminating forks converge, the supercoiling of DNA between them cannot be resolved by Topo I due to lack of space for it to act. Instead, terminating forks depend on transmission of this torsional stress behind the forks creating precatenates resolved by Topo II (**b**). During convergence two replisomes approach each other moving on opposite strands of DNA (leading strand of each fork) (**c**). The replisomes can pass each other and most likely CMG slides onto the double-stranded DNA of last Okazaki fragment (**d**). The synthesis of DNA is completed, the last Okazaki fragment matured and the DNA is ligated. Helicase is then ubiquitylated and removed by p97/VCP/Cdc48 segregase (**e**). Intertwined sister chromatids need to be resolved by Topo II (**f**). The final product: two individual sister chromatids with reconstituted chromatin structure (**g**)

ever, recent reports suggest that large protein barriers on the lagging strand can indeed slow down progression of the fork (Duxin et al. 2014; Langston and O'Donnell 2017). Does the approaching neighbour replisome, which is on the lagging strand, not present such a barrier? Is there an active mechanism regulating the smooth passage of the replisomes? Or are the replisomes idling at the edge of the barrier, in the attempt to unwind it, especially prepared to deal with barriers laying ahead and hence better at passing each other smoothly? More work is needed to answer these questions.

The results presented by Dewar et al. also suggest, at least in the context of the plasmid template, that torsional stress building up ahead of the forks does not slow down fork convergence (Dewar et al. 2015) (see also below for role of topoisomerases) (Fig. 8.2b). The removal of proteins (nucleosomes) ahead of the fork could not be directly addressed in this setup due to the artificial "clearing up" of chromatin ahead of the fork due to removal of the *lac* array. Interestingly, the in vitro reconstitution of eukaryotic DNA replication with purified budding yeast proteins revealed that nucleosomal packaging does appear to inhibit replication termination. As the elongation stage of the reaction was efficient, but termination alone was blocked, it suggests that the termination stage may be especially sensitive to the presence of chromatin structure (Devbhandari et al. 2017) (Fig. 8.2a).

8.6 The Completion of DNA Synthesis

Data provided by Dewar et al. suggest that leading-strand DNA is replicated up to a few bases away from the end of the last Okazaki fragment of the encountered lagging strand (Fig. 8.2d, e). There is no evidence for persistent gaps between these strands upon termination (Dewar et al. 2015). These data, however, do not explain which polymerase carries on synthesis of last fragments of DNA and maturation of the last Okazaki fragment. The RNA-DNA primer of each Okazaki fragment on the lagging strand is removed by concerted action of DNA Polδ and Fen1 endonuclease (reviewed in Balakrishnan and Bambara (2013)). DNA Polδ can support strand displacement resynthesis of the DNA previously synthesised by Polα and in doing so can progress until it encounters the nucleosome or another DNA-binding protein, both of which are efficiently repositioned behind the replication fork (Smith and Whitehouse 2012). Interestingly, fragments of DNA synthesised by Polα can be detected in mature genome mostly at the junctions of Okazaki fragments, usually at the nucleosome midpoint (dyad position). In total about 1.5% of mature genome was shown to be synthesised by Polα (Reijns et al. 2015).

Is the last Okazaki fragment matured by DNA Polε? The holoenzyme of DNA Polε is unable to carry on extended strand displacement synthesis in in vitro reconstitution experiments, unless its 3′–5′ exonuclease activity is removed, and it cannot mature Okazaki fragments on lagging strand (Devbhandari et al. 2017; Ganai et al. 2016). However, DNA Polε in the context of the replisome tightly associates with the CMG complex through the Dpb2 subunit of Polε and GINS and forms a functional unit (Langston et al. 2014; Muramatsu et al. 2010; Sengupta et al. 2013). A recent negative stain electron microscopy reconstruction of a CMG-Polε complex visualised the close association of this complex (Pellegrini and Costa 2016; Sun et al. 2015), and we found that the post-replication replisome in both *C. elegans* and *X. laevis* interacts with Polε and not Polδ (Sonneville et al. 2017). This interaction of Polε with the replisome likely acts as additional processivity factor for Polε, in addition to action of PCNA (Kang et al. 2012; Langston et al. 2014; Yeeles et al. 2017). It would be interesting to investigate Polε strand displacement activity in the context of the replisome. In support of the Polε role at termination, analysis of the genome-wide location of ribonucleotides incorporated into DNA by mutants of Polδ and Polε especially prone to such misincorporations discovered a substantial bias towards Polδ proximal to origins which declined towards the centre of the replicons where Polε synthesis was more evident (Daigaku et al. 2015). This would suggest that Polε carries out the replication at sites of termination.

Can Polε mature the last Okazaki fragment? Can it sustain strand displacement synthesis when supported by both PCNA and the CMG? It remains to be unravelled. Importantly, DNA Polε on its own does not interact with Fen1 (Garg et al. 2004); therefore, another processing mechanism would be required to complete maturation of the last Okazaki fragment, unless Fen1 is brought in by a different component of the terminating replisome. Alternatively, Polε can slide along the last Okazaki fragment together with the post-termination replisome, making room for Polδ to displace and mature the last RNA-DNA primer. Much is to be discovered about the ability of the terminated CMG to move away from the termination site especially in the context of re-established nucleosomes. However, Polδ has been shown previously to play a role in leading-strand synthesis in vivo (Daigaku et al. 2015; Johnson et al. 2015; Waga et al. 2001). Moreover, recent data obtained from the budding yeast in vitro reconstitution system of replication revealed that polymerase switching may be more common than expected. Polδ can play an important role in establishing leading-strand synthesis (Yeeles et al. 2017), and Polδ assembled at the leading strand was shown to be displaced if Polε was added after DNA synthesis has initiated (Georgescu et al. 2014). More research is required to show which of the polymerases finishes the replication job.

8.7 Role of Topoisomerases During DNA Replication Termination

The ability of topoisomerases to act ahead of the replication forks becomes very limited as two replication forks converge (Sundin and Varshavsky 1980). In this circumstance, fork rotation and precatenation become the primary pathway of DNA relaxing ahead of the fork. Catenated, double-stranded DNA (intertwined sister chromatids) can only be resolved by type II topoisomerases (Topo II, *S. cerevisiae* Top2) (Fig. 8.2b). Experiments with Topo II inhibitors in *Xenopus* egg extract showed that Topo II can be trapped behind, but not in front of the forks, and resolves replication intermediates in a nonredundant manner with Topo I (Hyrien 2009; Lucas et al. 2001). Interestingly, Top2 depletion in yeast does not stop cells from completing DNA replication, nor passing through mitosis, although they do dramatically mis-segregate and break their chromosomes due to sister chromatid catenation. On the other hand, inhibition of Top2 enzymatic activity in a way that Top2 is still able to bind DNA but unable to catalyse strand breakage causes incomplete DNA replication and induces G2/M cell cycle arrest (Baxter and Diffley 2008). Similarly, inhibition of Topo II activity in higher eukaryotes with the small molecule inhibitor ICRF-193 was shown to block termination of DNA replication in *Xenopus* egg extract and induce G2 arrest in human cells without the high level of DNA strand breaks associated with Topo II poisons (Cuvier et al. 2008; Downes et al. 1994; Skoufias et al. 2004). ICRF-193 traps Topo II on the DNA in the form of a non-covalent intermediate named the closed clamp (Roca et al. 1994). It is unclear therefore whether replication termination defects observed upon addition of ICRF-193 to *Xenopus* egg extracts are due to inhibition of Topo II activity or some other effect of the closed clamps, such as changes to nucleosome spacing and chromatin structure (Gaggioli et al. 2013; Germe and Hyrien 2005).

In agreement with the role of Topo II in replication fork termination, post-termination replisomes from *C. elegans* and *X. laevis* contain Topo II, unlike the budding yeast replisome progression complex, which represents active helicase and contains Top1 (Gambus et al. 2006; Sonneville et al. 2017). Moreover, Dewar et al. reported that site-specific termination plasmids (described above) require Topo II for decatenation of daughter plasmids, but Topo II activity is not needed for fork convergence and DNA ligation (Dewar et al. 2015) (Fig. 8.2b, f).

8.8 Replisome Disassembly

The data presented by Dewar et al. suggest that the dissolution of the replisome in the plasmid-based system is the last stage of replication fork termination, executed after ligation of leading and lagging strands (Dewar et al. 2015) (Fig. 8.2e). Work in budding yeast and *Xenopus laevis* egg extract discovered the first elements of this dissolution mechanism, which was found to be a highly evolutionary conserved process (Maric et al. 2014; Moreno et al. 2014) (Fig. 8.3a, b). In both model organisms, the Mcm7 subunit of the CMG complex becomes polyubiquitylated when forks terminate. The ubiquitin chains attached to Mcm7 are linked through lysine 48 (K48), but ubiquitylated Mcm7 is not degraded directly on chromatin as inhibition of proteasomal activity does not inhibit CMG disassembly. Instead, a protein remodeller Cdc48 (p97, VCP, segregase) recognises the ubiquitylated CMG and through its ATPase activity removes the CMG complexes from chromatin (Maric et al. 2014; Moreno et al. 2014). It is unclear at present whether the ubiquitylated Mcm7 is degraded upon removal from chromatin or de-ubiquitylated. A recent report by Fullbright et al. suggests that during unperturbed DNA replication in *Xenopus* egg

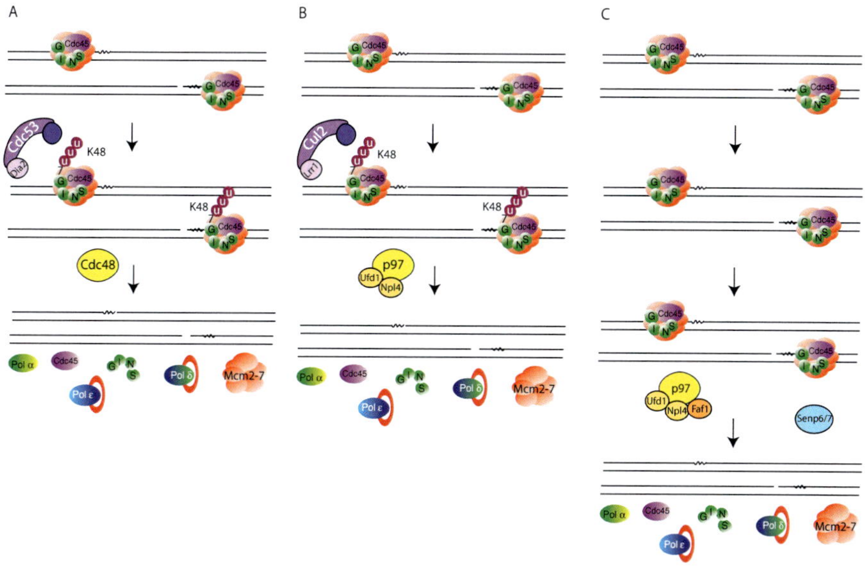

Fig. 8.3 Model of replisome disassembly at the termination of replication forks. In budding yeast *S. cerevisiae*, the Mcm7 subunit of the terminating replisome is ubiquitylated by SCF^{Dia2} and removed from chromatin by Cdc48 segregase (**a**). In *Xenopus* egg extract and *C. elegans* embryos, CRL2^{Lrr1} ubiquitylates Mcm7 during termination of replication forks, and CDC-48/p97 segregase removes it from chromatin with help of Ufd1/Npl4 cofactors (**b**). If the mechanism of removal of the replisome during termination of forks in S-phase does not work, *C. elegans* embryos have a backup mechanism removing replisomes in prophase in mitosis. This mechanism requires CDC-48/p97 and Npl4/Ufd1 but also UBXN-3/FAF1 cofactor and is regulated by ULP-4/Senp6,7 (**c**)

extract, ubiquitylated Mcm7 is likely to be de-ubiquitylated (Fullbright et al. 2016). Interestingly, ubiquitylation of human Mcm7 (both endogenous and exogenously expressed in cells) was reported in the past, but the fate of the ubiquitylated form of Mcm7 and the function of the ubiquitylation were not clear (Buchsbaum et al. 2007; Kuhne and Banks 1998).

8.8.1 SCF^{Dia2} Ubiquitin Ligase in Budding Yeast

In budding yeast the ubiquitin ligase, which ubiquitylates Mcm7, is SCF^{Dia2} (Maric et al. 2014). SCF^{Dia2} is a multisubunit ligase built around a Cdc53 cullin scaffold (homologue of Cullin 1 in higher eukaryotes) (Fig. 8.3a). Dia2 is the substrate-specific receptor, F-box protein, which binds through the substrate adaptor (Skp1) to the N-terminal part of Cdc53. The C-terminus of Cdc53, on the other hand, binds RING domain factor Hrt1, connecting the ligase to the ubiquitin-conjugating enzyme (E2) Cdc34 (SCF = **S**kp1 + **C**ullin1 + **F**-box) (Fig. 8.4b). SCF^{Dia2} was shown to be essential for Mcm7 ubiquitylation, specifically in the context of CMG

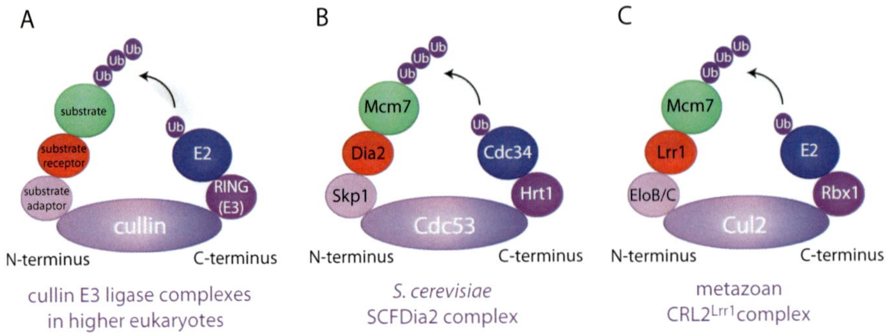

Fig. 8.4 Model of cullin ligases ubiquitylating Mcm7 during termination of replication forks. General model of organisation of cullin family members (**a**). Model of SCF^{Dia2} ubiquitylating Mcm7 in *S. cerevisiae* (**b**). Model of $CRL2^{Lrr1}$ ubiquitylating Mcm7 in *C. elegans* embryos and *Xenopus* egg extract (**c**)

during S-phase – both in vitro and in vivo. Moreover cells lacking Dia2 (*dia2Δ*) retain CMG complexes on chromatin after S-phase until the next G1 stage of the cell cycle (Maric et al. 2014). Not surprisingly budding yeast cells lacking Dia2, although viable, are defective in S-phase progression and present high rates of endogenous DNA damage and genome instability. They are also unable to grow at low temperatures and are sensitive to DNA-damaging agents that affect replication fork progression (Blake et al. 2006; Koepp et al. 2006).

Dia2 contains a protein-protein interaction N-terminal tetratricopeptide repeat (TPR) domain, a nuclear localisation signal (NLS), an F-box that connects it to the rest of the SCF ligase, and a C-terminal substrate recognition domain comprising of leucine-rich repeats (LRR). The TPR domain of Dia2 was shown to interact with Mrc1 and Tof1 components of the replisome progression complex (RPC) built around the CMG helicase (Gambus et al. 2006; Morohashi et al. 2009). As a result, Dia2 was detected interacting with RPC in S-phase, and this interaction was preserved when cells were treated with hydroxyurea (HU) to stall progressing replication forks (Morohashi et al. 2009). Interestingly cells lacking the TPR domain within Dia2 (*dia2-ΔTPR*) do not present the severe phenotype of *dia2Δ* cells – with the exception of synthetic lethality with rrm3Δ (a helicase supporting passage of forks past protein-DNA barriers). Cells lacking the TPR domain in Dia2 were, however, shown consequently to have a partial defect in Mcm7 ubiquitylation and CMG disassembly (Maculins et al. 2015). It seems that attaching SCF^{Dia2} to the replisome via the TPR domain increases the efficiency of CMG ubiquitylation. It may not be essential for normal CMG disassembly as the LRR domain can still recognise its substrate even without the tethering, but there may be situations when this stabilised interaction with the replisome is more vital – for example, when forks struggle to pass DNA-protein barriers in the absence of Rrm3.

8.8.2 CRL2^{Lrr1} Ubiquitin Ligase in Higher Eukaryotes

Recent research from our and two other groups discovered that in higher eukaryotes the ubiquitin ligase ubiquitylating Mcm7 at termination of replication forks is not an SCF but a Cullin2-based ubiquitin ligase with a leucine-rich repeat 1 protein (Lrr1) as a substrate receptor (**C**ullin-**R**ing **L**igase 2 with Lrr1 = CRL2^{Lrr1}) (Dewar et al. 2017; Sonneville et al. 2017) (Fig. 8.4c). Both in *Xenopus* egg extract and in *C. elegans* embryos, inhibition or downregulation of Cullin 1 ligase activity did not influence Mcm7 ubiquitylation nor helicase disassembly during S-phase (Sonneville et al. 2017 and our unpublished data). On the other hand, siRNA downregulation of CUL-2/LRR-1 complex in *C. elegans* embryos and immunodepletion of CRL2^{Lrr1} in egg extract blocked both phenotypes (Dewar et al. 2017; Sonneville et al. 2017) (Fig. 8.3b). CRL2^{Lrr1} was also shown to be the only cullin-type ubiquitin ligase that interacts with post-termination replisomes in *Xenopus* egg extract and *C. elegans* embryos and accumulates at the sites of termination in plasmid-based termination system described above (Dewar et al. 2017; Sonneville et al. 2017). Importantly, both studies found that CRL2^{Lrr1} interacts specifically with terminating CMG and not with actively unwinding helicase nor double Mcm2–7 hexamers of dormant origins. The regulated binding of CRL2^{Lrr1} to post-termination replisome represents therefore the first known step of replisome disassembly. Finally, the ubiquitin ligase activity of CRL2^{Lrr1} is necessary for the Mcm7 ubiquitylation and helicase disassembly, as a mutant of Cul2-Rbx1 complex, which cannot be activated by neddylation, is unable to rescue the CRL2^{Lrr1}-immunodepleted egg extract unlike a wild-type fully functioning complex (Sonneville et al. 2017).

What is CRL2^{Lrr1}? Previous work has shown that *C. elegans* LRR-1 is an essential gene (Piano et al. 2002). LRR-1 is required for embryonic development, but maternal rescue allows analysis of *lrr-1* loss of function in adult tissues. *Lrr-1* mutants are sterile owing to severe defects in germ cell proliferation (Merlet et al. 2010; Starostina et al. 2010). Inactivation of *lrr-1* induces DNA damage, which may arise due to DNA re-replication problems (ssDNA/RPA-1 foci accumulate in *lrr-1* germ cells, which also contain greater than 4 N DNA content). This in turn leads to hyperactivation of ATL-1/CHK-1 pathway (ATR/Chk1 pathway in vertebrates), which delays mitotic entry and results in embryonic lethality. Inactivation of ATL-1/CHK-1 checkpoint components supresses the proliferation defect and fully restores *lrr-1* mutant fertility (Burger et al. 2013; Merlet et al. 2010). How the re-replication/DNA damage is induced in *lrr-1* worms is not as yet determined. Interestingly, an RNAi-based suppressor screen of *lrr-1* and *cul-2* mutants identified two genes encoding components of the GINS complex, as well as CDC-7 and MUS-101, which are needed for CMG activation (Ossareh-Nazari et al. 2016). These data suggest that reducing CMG levels on chromatin can supress the DNA damage created in *lrr-1* mutants and supress their lethality. This is in agreement with LRR-1's role in Mcm7 ubiquitylation as lower levels of CMG on chromatin would compensate for a defect in CMG unloading.

On the other hand, another study found that *C. elegans lrr-1* mutants germ cells arrest with 2C DNA content, which may be due to accumulation of CDK inhibitor CKI-1 as deletion of one copy of CKI-1 or *cki-1* RNAi treatment can rescue *lrr-1* mutant germ cells numbers. In support of the CUL-2/LRR-1 role in targeting CKI-1 for degradation, study in human cells found that overexpressed CKI-1 was degraded faster when LRR-1 was also overexpressed (Starostina et al. 2010). Interestingly, LRR1 or CUL2 knockdown in HeLa cells did not induce a strong cell cycle arrest, and LRR1 was shown to be important to regulate levels of cytoplasmic p21 (human CKI) to control actin cytoskeleton remodelling (Starostina et al. 2010). Further studies are required to analyse in depth the role of LRR1 in human cells and the interplay between different substrates of this ubiquitin ligase.

Several questions remain – what is the signal for polyubiquitylation of Mcm7 and removal of helicase? How are CMGs protected from ubiquitylation during elongation and efficiently ubiquitylated at termination (Fig. 8.2)? Dewar et al. hypothesise that it may be conformational changes within CMG upon transition from encircling single-stranded DNA to double-stranded DNA of last Okazaki fragment that provide this post-termination specificity (Dewar et al. 2015). In support of this hypothesis, it was shown that CMG is indeed able to slide on double-stranded DNA (Kang et al. 2012).

We should also keep in mind that many substrate-specific receptors of CRLs recognise their substrates only when they are post-translationally modified; e.g. F-box receptors of SCF often recognise phosphorylated proteins, and VHL interacting with CRL2 recognises Hif1α upon its hydroxylation. It is possible therefore that terminating CMG is first modified in a yet undiscovered manner before being ubiquitylated. Budding yeast Mcm2–7 complex has been recently shown to be SUMOylated upon loading at origins in G1 stage of cell cycle before Mcm2–7 phosphorylation. The level of Mcm2–6 SUMOylation decreases during S-phase as MCM becomes phosphorylated and activated, with exception of Mcm7, which SUMOylation was retained during S-phase (Wei and Zhao 2016). Additionally, deubiquitylating enzyme Usp7 was described recently as a SUMO-specific DUB, removing ubiquitin from SUMOylated proteins and maintaining high SUMO/low ubiquitin ratio at replication forks (Lecona et al. 2016; Lopez-Contreras et al. 2013). A theory was therefore proposed that SUMO-driven ubiquitylation could act as a signal for the termination of DNA replication (Lecona and Fernandez-Capetillo 2016). Usp7 was also previously shown to interact with MCM-binding protein MCM-BP and to cooperate with it to unload the Mcm2-7 complexes from chromatin at the end of S-phase (Jagannathan et al. 2014; Nishiyama et al. 2011). Is Usp7 DUB activity for SUMOylated proteins linked with its MCM-BP interaction? Is Mcm7 in higher eukaryotes modified by SUMO? Is SUMOylation of Mcm7 regulating its ubiquitylation at termination events? More work is needed to understand fully this complex process.

Another possibility in need of investigation is involvement of priming ubiquitin ligase. Indeed ARIH1, an Ariadne family Ring-Between-Ring (RBR) ubiquitin ligase, was shown recently to interact with a number of CRLs including CRL2s and prime their substrates (Scott et al. 2016). It is probable, therefore, that such a priming ligase recognises the terminating helicase and CRL2^{Lrr1} only acts on primed substrate.

8.8.3 The Role of p97 Segregase in Replisome Disassembly

p97, also known as VCP in metazoans, CDC-48 in *C. elegans*, Cdc48 in yeast and Ter94 in insects, is a ubiquitin-dependent segregase that plays a central role in the regulation of protein homeostasis. Once bound to ubiquitylated substrates, this conserved hexameric AAA+ ATPase utilises the energy released from ATP hydrolysis to undergo a conformational change across its hexamer structure called interprotomer motion transmission (Huang et al. 2012; Li et al. 2012). This movement allows p97 to remove substrates from different cellular locations and complexes, likely by substrate translocation through p97's narrow central pore (Tonddast-Navaei and Stan 2013). The separated or unfolded substrates can then be directed to the proteasome and degraded or de-ubiquitylated and recycled with the help of DUBs associating with p97. p97 carries on this segregase/unfoldase activity on a myriad of substrates participating in a large variety of cellular processes. Not surprisingly, knockdown of both p97 alleles causes early embryonic lethality in mice, and siRNA-depletion of p97 in cells causes apoptosis (Muller et al. 2007; Wojcik et al. 2004).

The interaction of p97 with its many different substrates is mediated by a group of about 30 adaptor proteins that specifically recruit ubiquitylated proteins (Meyer et al. 2012; Yeung et al. 2008). The cofactors usually bind to the N-terminal domain of p97 using p97 interacting motifs. The best characterised major p97 cofactors include Ufd1/Npl4 heterodimer and p47, which bind to the p97 in mutually exclusive manner (Bruderer et al. 2004). Further, minor cofactors such as FAF1 or UBXD7 can then associate to the p97 complex with a major cofactor (Hanzelmann et al. 2011). Some of the cofactors, such as UBXD7, can also interact with various ubiquitin ligases and streamline the process of ubiquitin-dependent substrate removal/degradation (reviewed in Meyer et al. 2012).

The role of p97 during DNA replication was first suggested in *C. elegans* embryos. RNAi-mediated depletion of the CDC-48 complex leads to a defect in cell division: mitotic entry was delayed as a result of the activation of the DNA damage checkpoint. The severe chromatin defects observed in embryos as well as mitotic cells of the gonads included mitotic bridges and accumulated foci of RAD-51 DNA repair protein. Moreover, embryos lacking CDC-48, UFD-1 or NPL-4 are strongly reduced in DNA content (Deichsel et al. 2009; Mouysset et al. 2008). It was subsequently shown that embryos lacking CDC-48 or UFD1/NPL-4 cofactors accumulate origin licencing factor CDT-1 on mitotic chromatin and present persistent chromatin association of CDC-45/GINS after S-phase is completed (Franz et al. 2011). This process involves another p97 cofactor UBXN-3/FAF1 (Franz et al. 2016). Interestingly, inhibition of CDT-1 degradation and its accumulation on chromatin in embryos lacking CDC-48 or UFD1/NPL-4 does not lead to re-replication phenotype in these embryos but rather a strong reduction in their DNA content.

In the case of replisome disassembly, the segregase function was shown to be essential to disassemble ubiquitylated post-termination CMG in budding yeast, *C. elegans* embryos and *Xenopus* egg extract (Maric et al. 2014; Moreno et al. 2014; Sonneville et al. 2017). The ATPase activity of p97 is essential for this disassembly function as the replisome can be blocked on chromatin when two ATPase domains of p97 (D1 and D2) are mutated or the activity of p97 is blocked with a small molecule inhibitor NMS973 (Dewar et al. 2017; Moreno et al. 2014; Sonneville et al. 2017). This replisome disassembly defect phenotype is not driven through Cdt1 deregulation nor represents novel binding of GINS/Cdc45 to mitotic chromosomes (Moreno et al. 2014; Sonneville et al. 2017). In worm embryos, RNAi-directed inactivation of *ufd-1* and *npl-4* leads to a defect in replisome unloading, and the Ufd1/Npl4 heterodimer is found to interact with the post-termination replisome in *Xenopus* egg extracts (Dewar et al. 2017; Sonneville et al. 2017). Moreover, plasmids with accumulated terminating forks contain enriched Ubxn7 and Dvc1/SPRTN bound to them (Dewar et al. 2017). Future work will show whether these additional cofactors play a role in replisome disassembly.

8.8.4 Backup Pathway for Replisome Disassembly

Importantly, work in *C. elegans* embryos revealed that if the removal of CMG complexes is not accomplished during S-phase due to defective $CRL2^{Lrr1}$, then they can be removed from chromatin at the beginning of mitosis, in late prophase (Sonneville et al. 2017) (Fig. 8.3c). This backup mitotic pathway of replisome disassembly also requires p97/Ufd1/Npl4 (worm CDC-48/UFD-1/NPL-4) segregase, but to accomplish it p97 requires yet another cofactor: Fas-associated factor 1 FAF1 (worm UBXN-3) (Sonneville et al. 2017). FAF1 is an evolutionarily conserved proapoptotic factor that contains multiple protein interaction domains: ubiquitin-associated UBA, ubiquitin-like UBL1 and UBL2, Fas-interacting domain FID, death effector domain-interacting domain DEDID, Ubiquitin-associated UAS and ubiquitin regulatory X UBX (Lee et al. 2013; Menges et al. 2009). FAF1 is an essential gene (Adham et al. 2008), an established modulator of apoptosis, which regulates NFκB and is involved in ubiquitin-mediated protein turnover (reviewed in Menges et al. (2009)). FAF1 was also shown to bind to p97-Ufd1-Npl4 complex via the UBX domain and polyubiquitylated proteins via the UBA domain to promote endoplasmic reticulum-associated degradation ERAD (Lee et al. 2013). Finally, recent work from the Thorsten Hoppe lab showed that FAF-1/UBXN-3 is required for cell cycle progression in *C. elegans* embryo due to the problem with CDT-1 degradation and its inappropriate maintenance on chromatin during mitosis, together with CDC-45 and GINS (Franz et al. 2016). Moreover, Franz et al. have shown that downregulation of FAF1 by siRNA in human cells causes a pronounced replication stress phenotype: defective fork progression, fork stalling, dormant origin firing and activation of both S-phase checkpoint (ATR/Chk1) and DNA damage checkpoint (ATM/Chk2) (Franz et al. 2016). It remains to be investigated whether this observed replication stress is the result of Cdt1-induced re-replication, a defect in unloading of the post-termination replisomes or one of the many other FAF1 functions.

Intriguingly, the backup mitotic pathway of replisome disassembly in *C. elegans* embryos is modulated by the activity of the SUMO protease ULP-4: co-depletion of ULP-4 with LRR-1 delayed the release of CMG components from chromatin (Sonneville et al. 2017). ULP-4 is a major mitotic SUMO protease in worms and is present at mitotic chromosomes and at the spindle midzone (Pelisch et al. 2014). The ULP-4 analogous proteases in human cells are SENP6–7. It remains to be unravelled whether SUMO plays a regulatory role in the backup process or whether ULP-4 functions in another way, e.g. by bridging some important interactions and allowing p97 complex recruitment. It would be very interesting to investigate the existence of such a potential backup pathway in human somatic cells.

8.9 The Importance of Faultless Termination

Does deregulation of termination contribute to genomic instability and human disease? Cancer chromosomal instability (CIN) is observed in most solid tumours and is associated with poor prognosis and drug resistance (McGranahan et al. 2012). CIN leads to increased rate of changes in chromosomal numbers and structure and generates intra-tumour heterogeneity. Recent data implicate a central role for replication stress in the generation of CIN (Burrell et al. 2013). Can faulty termination provide a source of replication stress, which then contributes to the generation of genomic instability and CIN? What are the ways in which problems during replication fork termination could lead to genomic instability? At present we have limited experimental data on consequences of problems with replication fork termination, but we can speculate based on what we know.

We know that failure to decatenate newly replicated sister chromatids upon termination of replication forks does not tend to be detected by G2/M checkpoint but leads to dramatic mis-segregation of chromosomes during mitosis (Baxter and Diffley 2008). What about other stages of the termination process?

What would happen if forks cannot converge properly? What if their passing each other at the termination stage is blocked? We can imagine that problems during convergence of replication forks could lead to similar torsional stresses as these created by lack of topoisomerase I during elongation. Inhibition of Topo I activity in human cells, mouse embryonic fibroblasts and *Xenopus laevis* egg extract frequently induces replication fork reversal (reviewed in Neelsen and Lopes 2015). Fork reversal can have physiological roles during replication but can also have pathological consequences, contribute to genome instability in neurodegenerative syndromes and cancer. A small but reproducible number of reversed forks were detected also in various unchallenged human cell lines, while deregulation of poly (ADP-ribose) metabolism, which regulates fork reversal and restart, induces high level of reversed forks even in the absence of genotoxic replication stress (reviewed in Neelsen and Lopes 2015). Fork reversal is also very frequent in mouse embryonic stem cells (Ahuja et al. 2016). Where do these reversed forks come from? Could problems with termination of replication forks be one of the sources of such reversed forks? Interestingly, transient over-replication, fork reversal and end-processing by exonucleases were recently associated with completion of replication termination in *E. coli* (Wendel et al. 2014). More research and visualisation of converging forks either unchallenged or upon termination perturbations are needed to elucidate the possibility of fork reversal at sites of troubled replication fork termination.

Can failure to complete DNA synthesis at termination sites create genome instabilities? It has been shown recently that not all of the DNA is always replicated in human cells during S-phase – unreplicated segments resulting from double fork stalling in large replicons are frequently present in G2. They can be partially resolved during mitosis, create ultrafine bridges during segregation in mitosis and are subsequently recognised in the G1 stage of the cell cycle by DNA repair protein p53-binding protein 1 (53BP1) to be resolved in this new cell cycle (Moreno et al. 2016). Failure to complete DNA synthesis at termination sites would likely lead to a similar scenario.

What about inhibiting disassembly of the replisome? This is the part of the termination process that we understand best at present. If disassembly of the replisome constitutes the last step of replication termination, then the failure to remove it should not leave unligated DNA nor unusual DNA structures (Dewar et al. 2015). It would however leave a DNA helicase on a DNA substrate. Tested on synthetic in vitro substrates CMG can translocate on double-stranded DNA and then start unwinding DNA if a fork structure is present (Kang et al. 2012). One can imagine therefore that the second to the last Okazaki fragment, which may be in a mid-maturation stage with a flap created by Polδ, could be such a substrate for the approaching post-termination CMG to start de novo unwinding. In bacteria, recent data suggest that in termination zones 3′ ssDNA flaps are created that, if not removed by RecG nuclease (in RecG mutants), can provide substrates for de novo replication, leading to re-replication and creating pathological DNA structures. Tus termination sequences limit the extent of such re-replication initiated in termination zones (Rudolph et al. 2013). What about eukaryotic cells? They do not have Tus termination sequences. Can faulty termination of replication forks initiate re-replication?

Moreover, CMG complexes left behind on chromatin would disturb proper chromatin re-establishment and pose a problem to processes for which DNA is a substrate, such as transcription and next replication. As mentioned above, CMGs can translocate on double-stranded DNA (Kang et al. 2012), and by moving along DNA, they could displace other proteins bound to DNA. At present we do not know whether CMG sliding on dsDNA can displace nucleosomes or if they will be trapped by them.

A final potential problem arising from lack of efficient disassembly of the CMG complexes at the termination of replication forks is entrapment of Cdc45 and GINS within these post-termination complexes. Cdc45 was shown to be a rate limiting factor for DNA replication in mammalian cells. It was proposed that regulated expression levels of Cdc45 enforce reutilisation of existing Cdc45 during S-phase, which in turn can limit and stagger origin activation throughout the S-phase (Kohler et al. 2016; Wong et al. 2011). A lack of Cdc45 available for recycling can therefore potentially slow S-phase progression and inhibit DNA synthesis. Primary untransformed human cells with reduced levels of GINS components present all the phenotypes of replication stress and accumulation of DNA damage (Barkley et al. 2009). Future studies of replisome disassembly in human somatic cells are essential to shed light at this possibility as so far this process was investigated only in embryonic systems (*Xenopus laevis* egg extract and *C. elegans* embryos) which have higher levels of Cdc45 and GINS.

Is there experimental evidence that faulty disassembly of the replisome can lead to genome instability? *S. cerevisiae* cells lacking Dia2, which are unable to remove post-termination CMG from chromatin, are viable but present very high levels of genomic instabilities (described above). LRR-1 – the Mcm7-specific substrate receptor in higher eukaryotes – is an essential gene in *C. elegans*. LRR1 may have other than Mcm7 substrates, as CMG becomes unloaded by a backup system in *lrr-1* embryos. However, partial downregulation of LRR-1 together with downregulation of the backup pathway factors FAF-1/UBXN-3 or ULP4 results in synthetic lethality, suggesting that inhibition of CMG removal by partial blocking of both pathways results in non-viable worms (Sonneville et al. 2017). FAF1 itself is a factor often downregulated or mutated in multiple cancers. It may be its proapoptotic function that drives this downregulation, but in consequence these cancers could exhibit higher levels of genomic instability due to their replication fork termination problems. It is crucial therefore that we investigate the process of replisome disassembly in human cells to confirm its analogy.

Factors that drive replication initiation and the assembly of CMG, such as Cdc7 kinase and TopBP1 (Cut5) initiation factor, are currently being explored as potential anticancer therapy targets in tumours that present defects in chromosome replication (Chowdhury et al. 2014; Montagnoli et al. 2010). Can CMG disassembly also serve as a potential target for future therapies? Could we target the S-phase pathway of CMG disassembly in cancers with mutated or downregulated FAF1? For this we need to understand the CMG disassembly process in much more detail and crucially confirm its conservation in human cells. It seems likely that ubiquitylation is rate limiting for CMG disassembly, although it needs to be demonstrated by mapping the ubiquitylation sites and creating an unmodifiable mutant. It is clear, however, that Mcm7 ubiquitylation is regulated in a precise fashion on many levels, both spatially and temporally.

Finally, many factors implicated in DNA replication fork termination and replisome disassembly, such as p97 segregase and Usp7, are also targets of small molecule inhibitors used or being tested for antitumour therapies (Magnaghi et al. 2013; Reverdy et al. 2012). A better understanding of CMG disassembly pathway and replication fork termination in human cells might help us to explain the mode of action of these inhibitors in clinic.

References

Adham IM, Khulan J, Held T, Schmidt B, Meyer BI, Meinhardt A, Engel W (2008) Fas-associated factor (FAF1) is required for the early cleavage-stages of mouse embryo. Mol Hum Reprod 14:207–213

Ahuja AK, Jodkowska K, Teloni F, Bizard AH, Zellweger R, Herrador R, Ortega S, Hickson ID, Altmeyer M, Mendez J et al (2016) A short G1 phase imposes constitutive replication stress and fork remodelling in mouse embryonic stem cells. Nat Commun 7:10660

Ali N, Ismail IM, Khoder M, Shamy M, Alghamdi M, Costa M, Ali LN, Wang W, Eqani SA (2016) Polycyclic aromatic hydrocarbons (PAHs) in indoor dust samples from cities of Jeddah and Kuwait: levels, sources and non-dietary human exposure. Sci Total Environ 573:1607–1614

Balakrishnan L, Bambara RA (2013) Okazaki fragment metabolism. Cold Spring Harb Perspect Biol 5

Barkley LR, Song IY, Zou Y, Vaziri C (2009) Reduced expression of GINS complex members induces hallmarks of pre-malignancy in primary untransformed human cells. Cell Cycle 8:1577–1588

Bastia D, Zaman S (2014) Mechanism and physiological significance of programmed replication termination. Semin Cell Dev Biol 30:165–173

Baxter J, Diffley JF (2008) Topoisomerase II inactivation prevents the completion of DNA replication in budding yeast. Mol Cell 30:790–802

Been MD, Champoux JJ (1980) Breakage of single-stranded DNA by rat liver nicking-closing enzyme with the formation of a DNA-enzyme complex. Nucleic Acids Res 8:6129–6142

Bermejo R, Doksani Y, Capra T, Katou YM, Tanaka H, Shirahige K, Foiani M (2007) Top1- and Top2-mediated topological transitions at replication forks ensure fork progression and stability and prevent DNA damage checkpoint activation. Genes Dev 21:1921–1936

Blake D, Luke B, Kanellis P, Jorgensen P, Goh T, Penfold S, Breitkreutz BJ, Durocher D, Peter M, Tyers M (2006) The F-box protein Dia2 overcomes replication impedance to promote genome stability in Saccharomyces cerevisiae. Genetics 174:1709–1727

Brewer BJ, Fangman WL (1988) A replication fork barrier at the 3′ end of yeast ribosomal RNA genes. Cell 55:637–643

Brill SJ, DiNardo S, Voelkel-Meiman K, Sternglanz R (1987) Need for DNA topoisomerase activity as a swivel for DNA replication for transcription of ribosomal RNA. Nature 326:414–416

Bruderer RM, Brasseur C, Meyer HH (2004) The AAA ATPase p97/VCP interacts with its alternative co-factors, Ufd1-Npl4 and p47, through a common bipartite binding mechanism. J Biol Chem 279:49609–49616

Buchsbaum S, Morris C, Bochard V, Jalinot P (2007) Human INT6 interacts with MCM7 and regulates its stability during S phase of the cell cycle. Oncogene 26:5132–5144

Burger J, Merlet J, Tavernier N, Richaudeau B, Arnold A, Ciosk R, Bowerman B, Pintard L (2013) CRL2(LRR-1) E3-ligase regulates proliferation and progression through meiosis in the Caenorhabditis elegans germline. PLoS Genet 9:e1003375

Burrell RA, McClelland SE, Endesfelder D, Groth P, Weller MC, Shaikh N, Domingo E, Kanu N, Dewhurst SM, Gronroos E et al (2013) Replication stress links structural and numerical cancer chromosomal instability. Nature 494:492–496

Chowdhury P, Lin GE, Liu K, Song Y, Lin FT, Lin WC (2014) Targeting TopBP1 at a convergent point of multiple oncogenic pathways for cancer therapy. Nat Commun 5:5476

Conti C, Sacca B, Herrick J, Lalou C, Pommier Y, Bensimon A (2007) Replication fork velocities at adjacent replication origins are coordinately modified during DNA replication in human cells. Mol Biol Cell 18:3059–3067

Costa A, Ilves I, Tamberg N, Petojevic T, Nogales E, Botchan MR, Berger JM (2011) The structural basis for MCM2-7 helicase activation by GINS and Cdc45. Nat Struct Mol Biol 18:471–477

Cuvier O, Stanojcic S, Lemaitre JM, Mechali M (2008) A topoisomerase II-dependent mechanism for resetting replicons at the S-M-phase transition. Genes Dev 22:860–865

Daigaku Y, Keszthelyi A, Muller CA, Miyabe I, Brooks T, Retkute R, Hubank M, Nieduszynski CA, Carr AM (2015) A global profile of replicative polymerase usage. Nat Struct Mol Biol 22:192–198

Dalgaard JZ, Klar AJ (2000) swi1 and swi3 perform imprinting, pausing, and termination of DNA replication in S. pombe. Cell 102:745–751

Dalgaard JZ, Eydmann T, Koulintchenko M, Sayrac S, Vengrova S, Yamada-Inagawa T (2009) Random and site-specific replication termination. Methods Mol Biol 521:35–53

Deichsel A, Mouysset J, Hoppe T (2009) The ubiquitin-selective chaperone CDC-48/p97, a new player in DNA replication. Cell Cycle 8:185–190

Devbhandari S, Jiang J, Kumar C, Whitehouse I, Remus D (2017) Chromatin constrains the initiation and elongation of DNA replication. Mol Cell 65:131–141

Dewar JM, Budzowska M, Walter JC (2015) The mechanism of DNA replication termination in vertebrates. Nature 525:345–350

Dewar JM, Low E, Mann M, Raschle M, Walter JC (2017) CRL2Lrr1 promotes unloading of the vertebrate replisome from chromatin during replication termination. Genes Dev 31:275–290

Dimude JU, Midgley-Smith SL, Stein M, Rudolph CJ (2016). Replication termination: containing fork fusion-mediated pathologies in Escherichia coli. Genes (Basel) 7

Downes CS, Clarke DJ, Mullinger AM, Gimenez-Abian JF, Creighton AM, Johnson RT (1994) A topoisomerase II-dependent G2 cycle checkpoint in mammalian cells. Nature 372:467–470

Duxin JP, Dewar JM, Yardimci H, Walter JC (2014) Repair of a DNA-protein crosslink by replication-coupled proteolysis. Cell 159:346–357

Evrin C, Clarke P, Zech J, Lurz R, Sun J, Uhle S, Li H, Stillman B, Speck C (2009) A double-hexameric MCM2-7 complex is loaded onto origin DNA during licensing of eukaryotic DNA replication. Proc Natl Acad Sci U S A 106:20240–20245

Fachinetti D, Bermejo R, Cocito A, Minardi S, Katou Y, Kanoh Y, Shirahige K, Azvolinsky A, Zakian VA, Foiani M (2010) Replication termination at eukaryotic chromosomes is mediated by Top2 and occurs at genomic loci containing pausing elements. Mol Cell 39:595–605

Franz A, Orth M, Pirson PA, Sonneville R, Blow JJ, Gartner A, Stemmann O, Hoppe T (2011) CDC-48/p97 coordinates CDT-1 degradation with GINS chromatin dissociation to ensure faithful DNA replication. Mol Cell 44:85–96

Franz A, Pirson PA, Pilger D, Halder S, Achuthankutty D, Kashkar H, Ramadan K, Hoppe T (2016) Chromatin-associated degradation is defined by UBXN-3/FAF1 to safeguard DNA replication fork progression. Nat Commun 7:10612

Fu YV, Yardimci H, Long DT, Ho TV, Guainazzi A, Bermudez VP, Hurwitz J, van Oijen A, Scharer OD, Walter JC (2011) Selective bypass of a lagging strand roadblock by the eukaryotic replicative DNA helicase. Cell 146:931–941

Fullbright G, Rycenga HB, Gruber JD, Long DT (2016) p97 promotes a conserved mechanism of helicase unloading during DNA cross-link repair. Mol Cell Biol 36:2983–2994

Gaggioli V, Le Viet B, Germe T, Hyrien O (2013) DNA topoisomerase II alpha controls replication origin cluster licensing and firing time in Xenopus egg extracts. Nucleic Acids Res 41:7313–7331

Gambus A, Jones RC, Sanchez-Diaz A, Kanemaki M, van Deursen F, Edmondson RD, Labib K (2006) GINS maintains association of Cdc45 with MCM in replisome progression complexes at eukaryotic DNA replication forks. Nat Cell Biol 8:358–366

Gambus A, Khoudoli GA, Jones RC, Blow JJ (2011) MCM2-7 form double hexamers at licensed origins in Xenopus egg extract. J Biol Chem 286:11855–11864

Ganai RA, Zhang XP, Heyer WD, Johansson E (2016) Strand displacement synthesis by yeast DNA polymerase epsilon. Nucleic Acids Res 44:8229–8240

Garg P, Stith CM, Sabouri N, Johansson E, Burgers PM (2004) Idling by DNA polymerase delta maintains a ligatable nick during lagging-strand DNA replication. Genes Dev 18:2764–2773

Gasser R, Koller T, Sogo JM (1996) The stability of nucleosomes at the replication fork. J Mol Biol 258:224–239

Georgescu RE, Langston L, Yao NY, Yurieva O, Zhang D, Finkelstein J, Agarwal T, O'Donnell ME (2014) Mechanism of asymmetric polymerase assembly at the eukaryotic replication fork. Nat Struct Mol Biol 21:664–670

Georgescu RE, Schauer GD, Yao NY, Langston LD, Yurieva O, Zhang D, Finkelstein J, O'Donnell ME (2015) Reconstitution of a eukaryotic replisome reveals suppression mechanisms that define leading/lagging strand operation. Elife 4:e04988

Germe T, Hyrien O (2005) Topoisomerase II-DNA complexes trapped by ICRF-193 perturb chromatin structure. EMBO Rep 6:729–735

Gilbert DM (2010) Cell fate transitions and the replication timing decision point. J Cell Biol 191:899–903

Hanzelmann P, Buchberger A, Schindelin H (2011) Hierarchical binding of cofactors to the AAA ATPase p97. Structure 19:833–843

Hawkins M, Retkute R, Muller CA, Saner N, Tanaka TU, de Moura AP, Nieduszynski CA (2013) High-resolution replication profiles define the stochastic nature of genome replication initiation and termination. Cell Rep 5:1132–1141

Huang C, Li G, Lennarz WJ (2012) Dynamic flexibility of the ATPase p97 is important for its interprotomer motion transmission. Proc Natl Acad Sci U S A 109:9792–9797

Hyrien O (2009) Topological analysis of plasmid DNA replication intermediates using two-dimensional agarose gels. Methods Mol Biol 521:139–167

Ilves I, Petojevic T, Pesavento JJ, Botchan MR (2010) Activation of the MCM2-7 helicase by association with Cdc45 and GINS proteins. Mol Cell 37:247–258

Ivessa AS, Zhou JQ, Zakian VA (2000) The Saccharomyces Pif1p DNA helicase and the highly related Rrm3p have opposite effects on replication fork progression in ribosomal DNA. Cell 100:479–489

Ivessa AS, Lenzmeier BA, Bessler JB, Goudsouzian LK, Schnakenberg SL, Zakian VA (2003) The Saccharomyces cerevisiae helicase Rrm3p facilitates replication past nonhistone protein-DNA complexes. Mol Cell 12:1525–1536

Jagannathan M, Nguyen T, Gallo D, Luthra N, Brown GW, Saridakis V, Frappier L (2014) A role for USP7 in DNA replication. Mol Cell Biol 34:132–145

Johnson RE, Klassen R, Prakash L, Prakash S (2015) A major role of DNA polymerase delta in replication of both the leading and lagging DNA strands. Mol Cell 59:163–175

Kang YH, Galal WC, Farina A, Tappin I, Hurwitz J (2012) Properties of the human Cdc45/Mcm2-7/GINS helicase complex and its action with DNA polymerase epsilon in rolling circle DNA synthesis. Proc Natl Acad Sci U S A 109:6042–6047

Koepp DM, Kile AC, Swaminathan S, Rodriguez-Rivera V (2006) The F-box protein Dia2 regulates DNA replication. Mol Biol Cell 17:1540–1548

Kohler C, Koalick D, Fabricius A, Parplys AC, Borgmann K, Pospiech H, Grosse F (2016) Cdc45 is limiting for replication initiation in humans. Cell Cycle 15:974–985

Kuhne C, Banks L (1998) E3-ubiquitin ligase/E6-AP links multicopy maintenance protein 7 to the ubiquitination pathway by a novel motif, the L2G box. J Biol Chem 273:34302–34309

Langston L, O'Donnell M (2017) Action of CMG with strand-specific DNA blocks supports an internal unwinding mode for the eukaryotic replicative helicase. Elife 6

Langston LD, Zhang D, Yurieva O, Georgescu RE, Finkelstein J, Yao NY, Indiani C, O'Donnell ME (2014) CMG helicase and DNA polymerase epsilon form a functional 15-subunit holoenzyme for eukaryotic leading-strand DNA replication. Proc Natl Acad Sci U S A 111:15390–15395

Lecona E, Fernandez-Capetillo O (2016) A SUMO and ubiquitin code coordinates protein traffic at replication factories. BioEssays 38:1209–1217

Lecona E, Rodriguez-Acebes S, Specks J, Lopez-Contreras AJ, Ruppen I, Murga M, Munoz J, Mendez J, Fernandez-Capetillo O (2016) USP7 is a SUMO deubiquitinase essential for DNA replication. Nat Struct Mol Biol 23:270–277

Lee JJ, Park JK, Jeong J, Jeon H, Yoon JB, Kim EE, Lee KJ (2013) Complex of Fas-associated factor 1 (FAF1) with valosin-containing protein (VCP)-Npl4-Ufd1 and polyubiquitinated proteins promotes endoplasmic reticulum-associated degradation (ERAD). J Biol Chem 288:6998–7011

Li G, Huang C, Zhao G, Lennarz WJ (2012) Interprotomer motion-transmission mechanism for the hexameric AAA ATPase p97. Proc Natl Acad Sci U S A 109:3737–3741

Lopez-Contreras AJ, Ruppen I, Nieto-Soler M, Murga M, Rodriguez-Acebes S, Remeseiro S, Rodrigo-Perez S, Rojas AM, Mendez J, Munoz J et al (2013) A proteomic characterization of factors enriched at nascent DNA molecules. Cell Rep 3:1105–1116

Lucas I, Germe T, Chevrier-Miller M, Hyrien O (2001) Topoisomerase II can unlink replicating DNA by precatenane removal. EMBO J 20:6509–6519

Maculins T, Nkosi PJ, Nishikawa H, Labib K (2015) Tethering of SCF(Dia2) to the replisome promotes efficient ubiquitylation and disassembly of the CMG helicase. Curr Biol 25:2254–2259

Magnaghi P, D'Alessio R, Valsasina B, Avanzi N, Rizzi S, Asa D, Gasparri F, Cozzi L, Cucchi U, Orrenius C et al (2013) Covalent and allosteric inhibitors of the ATPase VCP/p97 induce cancer cell death. Nat Chem Biol 9:548–556

Maric M, Maculins T, De Piccoli G, Labib K (2014) Cdc48 and a ubiquitin ligase drive disassembly of the CMG helicase at the end of DNA replication. Science 346:1253596

McGranahan N, Burrell RA, Endesfelder D, Novelli MR, Swanton C (2012) Cancer chromosomal instability: therapeutic and diagnostic challenges. EMBO Rep 13:528–538

McGuffee SR, Smith DJ, Whitehouse I (2013) Quantitative, genome-wide analysis of eukaryotic replication initiation and termination. Mol Cell 50:123–135

Menges CW, Altomare DA, Testa JR (2009) FAS-associated factor 1 (FAF1): diverse functions and implications for oncogenesis. Cell Cycle 8:2528–2534

Merlet J, Burger J, Tavernier N, Richaudeau B, Gomes JE, Pintard L (2010) The CRL2LRR-1 ubiquitin ligase regulates cell cycle progression during C. elegans development. Development 137:3857–3866

Meyer H, Bug M, Bremer S (2012) Emerging functions of the VCP/p97 AAA-ATPase in the ubiquitin system. Nat Cell Biol 14:117–123

Montagnoli A, Moll J, Colotta F (2010) Targeting cell division cycle 7 kinase: a new approach for cancer therapy. Clin Cancer Res 16:4503–4508

Moreno SP, Bailey R, Campion N, Herron S, Gambus A (2014) Polyubiquitylation drives replisome disassembly at the termination of DNA replication. Science 346:477–481

Moreno A, Carrington JT, Albergante L, Al Mamun M, Haagensen EJ, Komseli ES, Gorgoulis VG, Newman TJ, Blow JJ (2016) Unreplicated DNA remaining from unperturbed S phases passes through mitosis for resolution in daughter cells. Proc Natl Acad Sci U S A 113:E5757–E5764

Morohashi H, Maculins T, Labib K (2009) The amino-terminal TPR domain of Dia2 tethers SCF(Dia2) to the replisome progression complex. Curr Biol 19:1943–1949

Mouysset J, Deichsel A, Moser S, Hoege C, Hyman AA, Gartner A, Hoppe T (2008) Cell cycle progression requires the CDC-48UFD-1/NPL-4 complex for efficient DNA replication. Proc Natl Acad Sci U S A 105:12879–12884

Moyer SE, Lewis PW, Botchan MR (2006) Isolation of the Cdc45/Mcm2-7/GINS (CMG) complex, a candidate for the eukaryotic DNA replication fork helicase. Proc Natl Acad Sci U S A 103:10236–10241

Muller JM, Deinhardt K, Rosewell I, Warren G, Shima DT (2007) Targeted deletion of p97 (VCP/CDC48) in mouse results in early embryonic lethality. Biochem Biophys Res Commun 354:459–465

Muramatsu S, Hirai K, Tak YS, Kamimura Y, Araki H (2010) CDK-dependent complex formation between replication proteins Dpb11, Sld2, Pol (epsilon), and GINS in budding yeast. Genes Dev 24:602–612

Neelsen KJ, Lopes M (2015) Replication fork reversal in eukaryotes: from dead end to dynamic response. Nat Rev Mol Cell Biol 16:207–220

Nishiyama A, Frappier L, Mechali M (2011) MCM-BP regulates unloading of the MCM2-7 helicase in late S phase. Genes Dev 25:165–175

Ossareh-Nazari B, Katsiarimpa A, Merlet J, Pintard L (2016) RNAi-based suppressor screens reveal genetic interactions between the CRL2LRR-1 E3-ligase and the DNA replication Machinery in Caenorhabditis elegans. G3 (Bethesda) 6:3431–3442

Pavlov YI, Frahm C, Nick McElhinny SA, Niimi A, Suzuki M, Kunkel TA (2006) Evidence that errors made by DNA polymerase alpha are corrected by DNA polymerase delta. Curr Biol 16:202–207
Pelisch F, Sonneville R, Pourkarimi E, Agostinho A, Blow JJ, Gartner A, Hay RT (2014) Dynamic SUMO modification regulates mitotic chromosome assembly and cell cycle progression in Caenorhabditis elegans. Nat Commun 5:5485
Pellegrini L, Costa A (2016) New insights into the mechanism of DNA duplication by the eukaryotic replisome. Trends Biochem Sci 41:859–871
Petryk N, Kahli M, d'Aubenton-Carafa Y, Jaszczyszyn Y, Shen Y, Silvain M, Thermes C, Chen CL, Hyrien O (2016) Replication landscape of the human genome. Nat Commun 7:10208
Piano F, Schetter AJ, Morton DG, Gunsalus KC, Reinke V, Kim SK, Kemphues KJ (2002) Gene clustering based on RNAi phenotypes of ovary-enriched genes in C. elegans. Curr Biol 12:1959–1964
Picard F, Cadoret JC, Audit B, Arneodo A, Alberti A, Battail C, Duret L, Prioleau MN (2014) The spatiotemporal program of DNA replication is associated with specific combinations of chromatin marks in human cells. PLoS Genet 10:e1004282
Postow L, Crisona NJ, Peter BJ, Hardy CD, Cozzarelli NR (2001) Topological challenges to DNA replication: conformations at the fork. Proc Natl Acad Sci U S A 98:8219–8226
Reijns MA, Kemp H, Ding J, de Proce SM, Jackson AP, Taylor MS (2015) Lagging-strand replication shapes the mutational landscape of the genome. Nature 518:502–506
Remus D, Beuron F, Tolun G, Griffith JD, Morris EP, Diffley JF (2009) Concerted loading of Mcm2-7 double hexamers around DNA during DNA replication origin licensing. Cell 139:719–730
Reverdy C, Conrath S, Lopez R, Planquette C, Atmanene C, Collura V, Harpon J, Battaglia V, Vivat V, Sippl W et al (2012) Discovery of specific inhibitors of human USP7/HAUSP deubiquitinating enzyme. Chem Biol 19:467–477
Roca J, Ishida R, Berger JM, Andoh T, Wang JC (1994) Antitumor bisdioxopiperazines inhibit yeast DNA topoisomerase II by trapping the enzyme in the form of a closed protein clamp. Proc Natl Acad Sci U S A 91:1781–1785
Rudolph CJ, Upton AL, Stockum A, Nieduszynski CA, Lloyd RG (2013) Avoiding chromosome pathology when replication forks collide. Nature 500:608–611
Schalbetter SA, Mansoubi S, Chambers AL, Downs JA, Baxter J (2015) Fork rotation and DNA precatenation are restricted during DNA replication to prevent chromosomal instability. Proc Natl Acad Sci U S A 112:E4565–E4570
Scott DC, Rhee DY, Duda DM, Kelsall IR, Olszewski JL, Paulo JA, de Jong A, Ovaa H, Alpi AF, Harper JW et al (2016) Two distinct types of E3 ligases work in unison to regulate substrate ubiquitylation. Cell 166(1198–1214):e1124
Sengupta S, van Deursen F, de Piccoli G, Labib K (2013) Dpb2 integrates the leading-strand DNA polymerase into the eukaryotic replisome. Curr Biol 23:543–552
Simon AC, Sannino V, Costanzo V, Pellegrini L (2016) Structure of human Cdc45 and implications for CMG helicase function. Nat Commun 7:11638
Skoufias DA, Lacroix FB, Andreassen PR, Wilson L, Margolis RL (2004) Inhibition of DNA decatenation, but not DNA damage, arrests cells at metaphase. Mol Cell 15:977–990
Smith DJ, Whitehouse I (2012) Intrinsic coupling of lagging-strand synthesis to chromatin assembly. Nature 483:434–438
Sonneville R, Moreno SP, Knebel A, Johnson C, Hastie CJ, Gartner A, Gambus A, Labib K (2017) CUL-2LRR-1 and UBXN-3 drive replisome disassembly during DNA replication termination and mitosis. Nat Cell Biol 19(5):468–479
Starostina NG, Simpliciano JM, McGuirk MA, Kipreos ET (2010) CRL2(LRR-1) targets a CDK inhibitor for cell cycle control in C. elegans and actin-based motility regulation in human cells. Dev Cell 19:753–764
Sun J, Shi Y, Georgescu RE, Yuan Z, Chait BT, Li H, O'Donnell ME (2015) The architecture of a eukaryotic replisome. Nat Struct Mol Biol 22:976–982

Sundin O, Varshavsky A (1980) Terminal stages of SV40 DNA replication proceed via multiply intertwined catenated dimers. Cell 21:103–114
Tonddast-Navaei S, Stan G (2013) Mechanism of transient binding and release of substrate protein during the allosteric cycle of the p97 nanomachine. J Am Chem Soc 135:14627–14636
Uhlmann F (2009) A matter of choice: the establishment of sister chromatid cohesion. EMBO Rep 10:1095–1102
Waga S, Masuda T, Takisawa H, Sugino A (2001) DNA polymerase epsilon is required for coordinated and efficient chromosomal DNA replication in Xenopus egg extracts. Proc Natl Acad Sci U S A 98:4978–4983
Wang JC (2002) Cellular roles of DNA topoisomerases: a molecular perspective. Nat Rev Mol Cell Biol 3:430–440
Wei L, Zhao X (2016) A new MCM modification cycle regulates DNA replication initiation. Nat Struct Mol Biol 23:209–216
Wendel BM, Courcelle CT, Courcelle J (2014) Completion of DNA replication in Escherichia coli. Proc Natl Acad Sci U S A 111:16454–16459
Wojcik C, Yano M, DeMartino GN (2004) RNA interference of valosin-containing protein (VCP/p97) reveals multiple cellular roles linked to ubiquitin/proteasome-dependent proteolysis. J Cell Sci 117:281–292
Wong PG, Winter SL, Zaika E, Cao TV, Oguz U, Koomen JM, Hamlin JL, Alexandrow MG (2011) Cdc45 limits replicon usage from a low density of preRCs in mammalian cells. PLoS One 6:e17533
Yardimci H, Loveland AB, Habuchi S, van Oijen AM, Walter JC (2010) Uncoupling of sister replisomes during eukaryotic DNA replication. Mol Cell 40:834–840
Yeeles JT, Janska A, Early A, Diffley JF (2017) How the eukaryotic replisome achieves rapid and efficient DNA replication. Mol Cell 65:105–116
Yeung HO, Kloppsteck P, Niwa H, Isaacson RL, Matthews S, Zhang X, Freemont PS (2008) Insights into adaptor binding to the AAA protein p97. Biochem Soc Trans 36:62–67

Chapter 9
Structure of the MCM2-7 Double Hexamer and Its Implications for the Mechanistic Functions of the Mcm2-7 Complex

Yuanliang Zhai and Bik-Kwoon Tye

Abstract The eukaryotic minichromosome maintenance 2–7 complex is the core of the inactive MCM replication licensing complex and the catalytic core of the Cdc45-MCM-GINS replicative helicase. The years of effort to determine the structure of parts or the whole of the heterohexameric complex by X-ray crystallography and conventional cryo-EM produced limited success. Modern cryo-EM technology ushered in a new era of structural biology that allowed the determination of the structure of the inactive double hexamer at an unprecedented resolution of 3.8 Å. This review will focus on the fine details observed in the Mcm2-7 double hexameric complex and their implications for the function of the Mcm2-7 hexamer in its different roles during DNA replication.

Keywords DNA replication • Replication licensing • MCM2-7 • Cryo-EM structure

Y. Zhai
Division of Life Science, Hong Kong University of Science and Technology, Hong Kong, China

Institute for Advanced Study, Hong Kong University of Science and Technology, Hong Kong, China

B.-K. Tye (✉)
Department of Molecular Biology and Genetics, College of Agriculture and Life Sciences, Cornell University, Ithaca, NY, USA

Division of Life Science, Hong Kong University of Science and Technology, Hong Kong, China
e-mail: biktye@ust.hk

H. Masai, M. Foiani (eds.), *DNA Replication*, Advances in Experimental Medicine and Biology 1042, https://doi.org/10.1007/978-981-10-6955-0_9

9.1 Introduction

Of the three macromolecule-synthesizing machines that synthesize proteins, mRNAs and DNA, respectively, the crystal structures of the ribosome (Ramakrishnan 2010; Steitz 2010; Yonath 2010) and RNA polymerase (Kornberg 2007) have been determined at atomic resolution, and Nobel prizes were even awarded for these achievements. However, the replisome and many of its integral components have been recalcitrant to crystallization, and structural determination of the replisome has lagged behind. The replisome is a complex machine with multiple engines including a DNA-unwinding helicase at the front end and one leading-strand and two lagging-strand DNA polymerases chugging along at the back end. The size, asymmetry, flexibility, and multipartite nature of the replisome have posed difficult challenges for structural biologists on the mission to unravel its anatomical secrets. However, the prospects for unraveling the atomic structure of the replisome soon will change because of the advent of the cryo-EM revolution that took place in 2013 (Fernandez-Leiro and Scheres 2016). To date, two major structures that are critical for unraveling the anatomy of the helicase component of the replisome have been determined at near-atomic resolution. The first is the inactive MCM double hexamer determined at an overall resolution of 3.8 Å (Li et al. 2015). The unprecedented resolution achieved in this study lays a strong foundation for structural work on the helicase component of the replisome and will serve as a template for subsequent atomic modeling of all MCM-associated DNA replication complexes. The second is the CMG helicase resolved at a slightly lower resolution by two groups (Abid Ali et al. 2016; Yuan et al. 2016). In this chapter, we will focus on the structure and structure-informed functions of the MCM double hexamer. In the following chapter, the structure of the CMG helicase and the architecture of the replisome will be reviewed.

The Mcm2-7 hexameric complex is the core of the inactive MCM replication licensing complex (Tye 1999a; Donovan et al. 1997; Chong et al. 1995; Thommes et al. 1997) and the catalytic core of the Cdc45-MCM-GINS replicative helicase (Labib et al. 2000; Moyer et al. 2006; Gambus et al. 2006; Ilves et al. 2010). Three of the MCM subunits (Mcm2, Mcm3, and Mcm5) were initially identified in a yeast mutant hunt for proteins that regulate the initiation step of DNA replication (Tye 1999a). This screen utilized the maintenance of minichromosomes as an assay and hence the name (Maine et al. 1984; Tye 1999b). Paralogs of these Mcm proteins were subsequently identified in the yeasts and other eukaryotes from screens not necessarily related to DNA replication functions (Moir et al. 1982; Johnston and Thomas 1982; Bae et al. 2009). There is a total of eight Mcm paralogs. Mcm8 and Mcm9 are only found in metazoans, and their functions are less well known (Maiorano et al. 2005; Gambus and Blow 2013; Lutzmann et al. 2005; Nishimura et al. 2012; Traver et al. 2015). Mcm2, Mcm3, Mcm4, Mcm5, Mcm6, and Mcm7 form a hexameric complex that is involved in replication initiation and elongation in all eukaryotes (Bochman and Schwacha 2009). However, purified Mcm2-7 complexes show little or no helicase activity unless provided with the accessory factors, the tetrameric GINS and Cdc45 (Bochman and Schwacha 2008; Ilves et al. 2010).

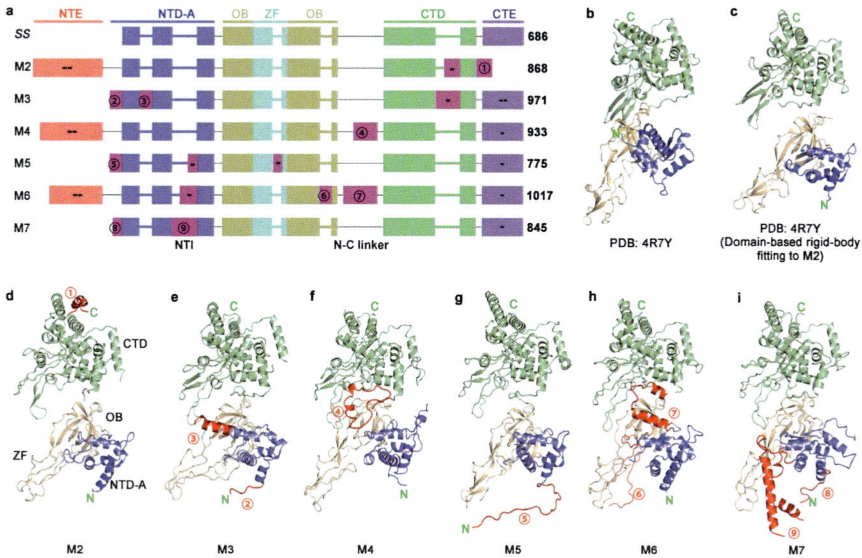

Fig. 9.1 Structural organization of the MCM2-7 subunits. **a** Schematic illustration of domain organization and subunit-specific features of MCM2-7 subunits, with comparison to the archaeal MCM (SS, *Sulfolobus solfataricus*). Numbered regions correspond to numbered extensions and insertions highlighted in (**d–i**). "-" symbols denote corresponding regions with reliable densities to trace the main-chain direction, but not sufficient for atomic modeling. "--" symbols denote sequences with highly disordered densities. (**b, c**) A protomer of the crystal structure of a chimeric archaeal MCM hexamer structure (PDB code 4R7Y) was used as the template for modeling with its subdomains divided and colored. The archaeal MCM was aligned globally (**b**) or domain-based flexibly fitted (**c**) to the atomic model of Mcm2. (**d–i**) Side-by-side structural comparison of MCM2-7 proteins, with Mcm3-7 globally aligned to the atomic model of Mcm2. The well-resolved insertions and extensions of each MCM subunit (**d–i**) are numbered and colored in *red* (Reproduced from Li et al. 2015)

In contrast, in archaebacteria, a single Mcm protein forms a homohexameric ring that displays robust helicase activity (Kelman et al. 1999; Chong et al. 2000). An alignment of the Mcm2, 3, 4, 5, 6, 7 amino acid sequence with their archaeal homologue shows that the CTD catalytic core and NTD core are highly conserved (Fig. 9.1a). Each of the Mcm2-7 subunit has its own characteristic CTE, NTI, and NTE that suggest subunit-specific functions.

The assembly and activation of Mcm2-7 helicases at replication origins are strictly regulated during each cell division cycle, ensuring replication initiation to occur at each origin no more than once (Diffley et al. 1994; Siddiqui et al. 2013). Several laboratories have been able to reconstitute the DNA replication system using purified yeast proteins (Georgescu et al. 2015; Yeeles et al. 2015). These laboratories have taken the approach to look at in vitro assembled replication complexes by negative staining EM and cryo-EM (Remus et al. 2009; Costa et al. 2011, 2014; Sun et al. 2013, 2014, 2015; Yuan et al. 2016; Abid Ali et al. 2016). An alternative approach is

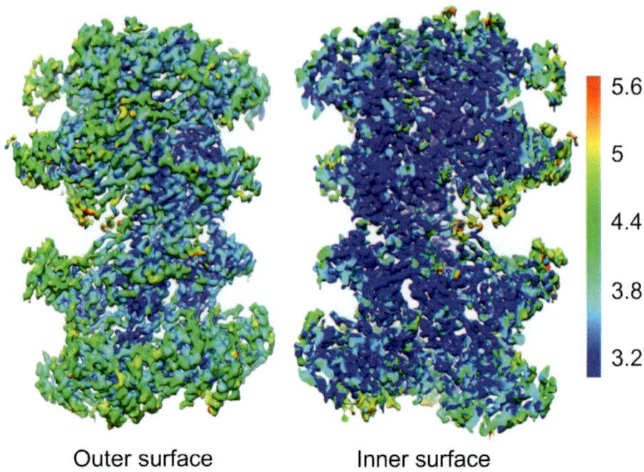

Fig. 9.2 The density map of the MCM2-7 double hexamer (sharpened) is shown in two views for the outer and inner surfaces. The map is color-coded to indicate the range of the local resolution

to purify endogenously assembled DNA replication complexes that may have a better chance of preserving fidelity but may compromise on the yield of such preparations.

In yeast, the soluble form of the Mcm2-7 complex is a single hexamer, and the chromatin-bound form of the complex is a double hexamer. Tye and collaborators purified the soluble single hexamer (SH) from an overexpression strain and the double hexamer (DH) from native chromatin and examined their structures by cryo-EM (Li et al. 2015). After much effort, they were unable to get a respectable 3-D structure of the SH because of its flexibility and instability. In contrast, the double hexamer was clearly much more stable. After only two sessions of data collection, a refined 3-D structure of the DH was obtained with a resolution of about 4.4 Å for the more flexible exterior and a better than 3.2 Å for the interior core (Fig. 9.2). The high quality of this EM density map allowed the assignment of each of the MCM subunits unambiguously without the help of conventional tagging strategy (Figs. 9.1d–i and 9.3c–e). The crystal structure of the conserved core region of the chimera of *Sulfolobus solfataricus* NTD and *Pyrococcus furiosus* CTD (Fig. 9.1b) (Miller et al. 2014) was used for the atomic modeling. Rigid body domains of the chimera were divided into four segments (NTD-A, OB-fold, ZF, and CTD) and fitted onto the electron density map with manual adjustments according to predicted secondary structures (Fig. 9.1b–i). For regions without a known template, the model relied on predicted secondary structure and tracing of the main chain based solely on densities. As a result, an atomic model of the DH for ~80% of its sequence, including most of the subunit-specific extensions and insertions, was built.

9.2 Asymmetry of the Double Hexamer

The double hexamer is a rigid, head-to-head double ring structure that is slightly tilted, twisted, and offset with respect to each other, similar to that described previously for in vitro purified DH (Sun et al. 2014). The relative positions of the subunits

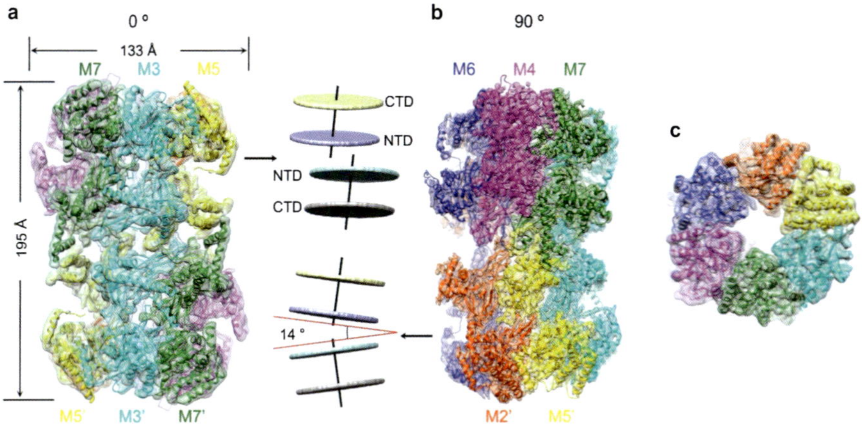

Fig. 9.3 Overall structure of the MCM2-7 double hexamer. (**a–c**) Side (**a, b**) and *top* (**c**) views of the cryo-EM map of the MCM2-7 DH purified from native chromatin (Reproduced from Li et al. 2015). The map is superimposed with the atomic model. Unsharpened map (**a**) is displayed from the twofold axis, and sharpened map (**b**) is displayed with indicated rotations relative to **a** along the cylinder axis. The tilted and twisted arrangements of the two single hexamers are illustrated in the side panels of (**a**, **b**)

were also in the precise predicted order determined by low-resolution cryo-EM except that the assignment of the CTD relative to the NTD was different. In the atomic model, the CTD is almost vertically aligned with respect to the NTD (Fig. 9.3) rather than the sharp anticlockwise displacement in the earlier model. The top view from the CTD shows a compact closed ring with an open channel wide enough for dsDNA to pass through without the obstructions of the CTE reported in Sun et al. (2014). These observed differences could result from in vitro versus in vivo assembly.

The tilted arrangement of the two SHs forms a 14° wedge in between (Fig. 9.3). The Mcm2/Mcm6/Mcm4 subunits from both SHs are all positioned at the thick edge. Although the very long NTEs from Mcm2, Mcm4, and Mcm6 (Fig. 9.1a) appear too disordered to produce observable electron density, their bulkiness is believed to be the cause of the tilted conformation of the DH. The localization of these three subunits at the thick edge of the DH wedged site allows an increased opening between the two hexamers and thus a solvent exposure of their NTEs. Notably, they are side by side on one SH and head-to-head on the DH (Fig. 9.3), and hence, a large surface area on top of six NTEs from two hexamers virtually forms a docking platform for regulatory kinases such as DDK during helicase activation. Consistent with this notion, the NTEs of Mcm2 (Lei et al. 1997), Mcm4, and Mcm6 (Sheu and Stillman 2006; Randell et al. 2010) are known substrates of DDK, and the DH, but not the SH, is the preferred substrate for DDK (Sun et al. 2014). In vitro assembly assays indicate that phosphorylated DH supports initial recruitment of Sld3 and Cdc45 (Deegan et al. 2016), followed by GINS and further phosphorylation by CDK, to form a functional CMG helicase for DNA unwinding (Yeeles et al. 2015; Heller et al. 2011). Low-resolution EM studies showed that DDK phosphorylation neither triggers the uncoupling of the DH nor causes conformational change in its overall structure (On et al. 2014; Sun et al. 2014). However, subtle alterations in DH structure induced by DDK may be critical for the transformation of the inactive DH into the active CMG helicases via Mcm2/Mcm5 gate opening and/or DH uncoupling. High-resolution structure determination of the phosphorylated DH using advanced cryo-EM analysis may provide insight into this intricate pivotal mechanism.

9.3 Tight Junction at the NTDs

Archaeal homohexameric MCM complex can form DH in solution directly through head-to-head interactions between zinc finger motifs from opposite hexamers (Chong et al. 2000; Brewster et al. 2008). In contrast, the eukaryotic heterohexameric Mcm2-7 can only be assembled into a dimer of hexamers on duplex origin DNA (Remus et al. 2009; Evrin et al. 2009). Furthermore, this assembly process is tightly controlled in a stepwise manner. The Mcm2-7 hexamer is recruited to the ORC-Cdc6-binding site during the G1 phase one at a time (Ticau et al. 2015; Sun et al. 2014). Once assembled, the Mcm2-7 hexamers form a topologically stable, inactive head-to-head double hexamer that encircles duplex origin DNA (Remus et al. 2009). The assembly of the double hexamer is an energy-consuming process requiring the hydrolysis of ATP by both ORC-Cdc6 and Mcm2-7 during the loading of the first and the second hexamer (Coster et al. 2014; Kang et al. 2014). Similar to archaeal DH formation, interactions among ZF motifs from Mcm2-7 subunits contribute to the stabilization of DH. However, unlike the simple head-on interactions between ZFs from two archaeal SHs, the inter-ZF interactions of eukaryotic Mcm2-7 appear to be more versatile, forming additional interactions with NTEs and β turns of the opposing hexamer in some cases (Fig. 9.4). This elaborate intertwining and extensions of NTEs into the opposite hexamers suggest that formation of the head-to-head tight junction is an ATP-consuming process that requires partial unfolding of the NTDs before forming the rigid interlocked structure.

Due to the tilted and twisted arrangement of the two SHs, the orientations of ZFs are completely different at two sides of the wedged interface of the DH: vertical at the thick Mcm2/Mcm6/Mcm4 side and horizontal at the thin Mcm5/3/7 side. This unique arrangement ensures intimate contacts among ZFs. In addition, according to the buried surface calculated at the hexamer interface, NTDs, NTEs, and NTIs from Mcm5/3/7 play significant roles in contributing to a stable DH (Fig. 9.4), which is even greater than ZF interactions. The involvement of NTEs of Mcm2/Mcm6/Mcm4 in stabilizing DH remains unknown because of their highly disordered feature in the structure. More importantly, although N-terminal sequences of each MCM subunits are less conserved in higher eukaryotes, most of the NTDs, NTEs, and NTIs (mainly NTI of M7, NTE of M5, β-turn of M6) involved in inter-hexamer interactions are conserved, suggesting the inter-subunit interactions are most likely maintained in metazoans as well.

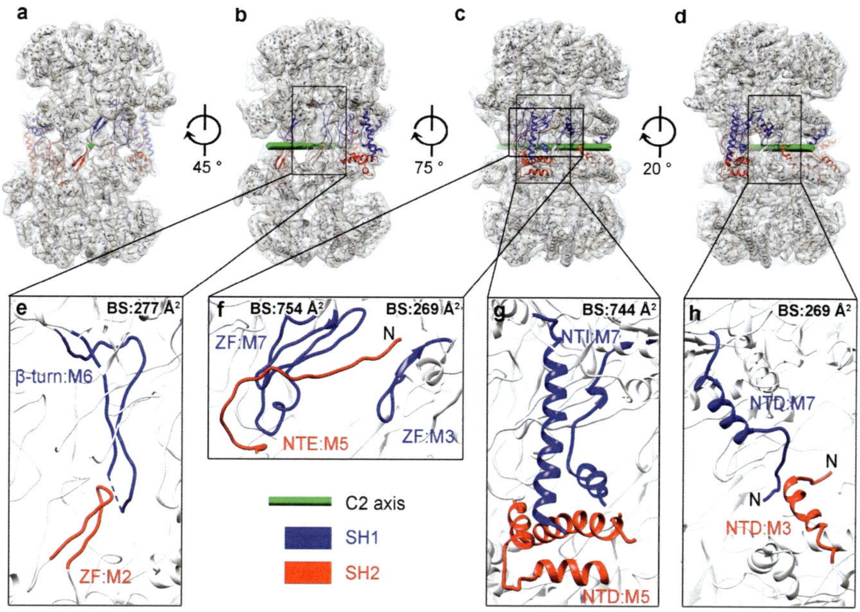

Fig. 9.4 Stabilization of MCM2-7 double hexamer contributed by NTEs and NTIs. (**a–d**) Side views of the MCM2-7 DH, with indicated rotations around the cylinder axis. Atomic structure is superimposed with the unsharpened map. The sequence elements involved in inter-hexamer interactions are highlighted. (**e–h**) Zoomed-in views of the boxed regions from (**b–d**). Buried areas (Å^2) of these interfaces in (**e–h**) are labeled. BS, buried surface (Reproduced from Li et al. 2015)

The tight coupling of the double hexamer at the NTD raises the question of how the two hexamers uncouple upon helicase activation. In vitro reconstitution and genetic analysis show that separation of MCM DH in S phase involves the engagement of other initiation factors such as Cdc45, GINS, and Mcm10 to unravel the entanglement between the NTDs and NTIs of M5, M3, and M7. Consistent with this idea, recent CMG structure showed that two α-helices of Psf2 interact with NTD of Mcm5 (Abid Ali et al. 2016; Yuan et al. 2016) which is contacted by the NTE of the opposite Mcm7 in the DH. Previous study also showed that Cdc45 can interact with Mcm5-N (N-terminus) and Mcm2-N and Psf2 with Mcm5-N and Psf3 with Mcm3-N(Costa et al. 2011; Yuan et al. 2016; Abid Ali et al. 2016); in addition, Mcm10 has been shown to interact physically with MCM6 (van Deursen et al. 2012) and MCM7 (Homesley et al. 2000). Presumably these factors work cooperatively to uncouple the MCM DH. The recruitment of all these factors requires the action of DDK and CDKs during S phase to ensure the sequential coupling and uncoupling of the MCM hexamers at separate stages of each cell division cycle (Heller et al. 2011; Yeeles et al. 2015). Again, the role of DDK phosphorylation in triggering the DH uncoupling is still unclear. A better understanding of this process will rely on the determination of the atomic structures of the intermediate replication complexes during helicase activation and the delineation of conformational changes induced by the uncoupling factors.

9.4 A Unique Side Channel at the Mcm2-Mcm6 Interface

The archaeal MCM helicase is made up of six identical subunits, which form a ring with six identical interfaces (Bochman and Schwacha 2009). The atomic model of the archaeal MCM helicase built from the crystal structure of the near-full-length protomer of the ssoMCM by applying a sixfold symmetry showed six side channels at the neck region of the ring (Brewster et al. 2008). The size of the side channels large enough for ssDNA to pass through raised speculations that dsDNA may pass through the central channel at the CTD and unwound by extrusion of one strand laterally at the side channel. This configuration would favor the strand extrusion model for DNA unwinding. Throughout the years, this model has been disfavored based on in vitro biochemical studies using bulky chemical crosslinks on the leading and lagging strands (Fu et al. 2011). The CMG helicase efficiently bypasses a roadblock embedded on the lagging strand more efficiently than that embedded on the leading strand, arguing against the translocation of CMG helicase on dsDNA. Most recently, the structure of the apo CMG helicase has been determined by cryo-EM (Yuan et al. 2016). The WH domain of Mcm5 restricts the main channel of the CMG helicase to the extent that only ssDNA can be accommodated. The positions of the ssDNA associated with the CMG are consistent with the steric exclusion model in which duplex DNA is unwound by the translocation of the CMG on ssDNA.

The observation that the yeast MCM double hexamer contains a central channel that can accommodate dsDNA and a side channel between Mcm2 and Mcm6 wide enough to accommodate ssDNA (Fig. 9.5a) encourages a revisit of the strand extrusion model. Despite all the evidence against this model, it is worthwhile to reevaluate the evidence. First, the in vitro roadblock bypass experiment showed only a bias in the efficiency of bypass rather than a clear-cut bypass on the lagging strand but not the leading strand. Also, the latching of the Mcm2-Mcm5 gate by GINS and Cdc45 in the CMG helicase structure suggests that the gate is only loosely fastened and may not be closed shut or irreversibly latched on during translocation (Costa et al. 2011, 2014). As for the threading of the artificial Y-shaped DNA substrate through the in vitro assembled CMG helicase, it may not correctly represent how DNA is threaded in a CMG helicase assembled de novo at replication origins. Most interestingly, a recent model for the architecture of the in vitro assembled core replisome suggests that DNA polymerase epsilon is placed ahead of the helicase on the leading strand (Sun et al. 2015). Although the path of the DNA could not be traced, the position of Polε would require a sharp U-turn in the threading of the leading strand from the NTD of the CMG helicase to the DNA polymerase if the steric exclusion model is correct (Fig. 9.5b). Indeed, without the evidence of an atomic structure of an endogenous replisome, one can easily come up with radical models that are more compatible with the positioning of Polε by using the unique side channel (Fig. 9.5c). In this model, the side channel would act like a plowshare to force apart the two strands as the helicase plows ahead. Whether the CMG helicase unwinds DNA by strand extrusion, steric exclusion or some other mechanisms remains to be verified by cryo-EM structures of CMG helicase assembled from replication origins.

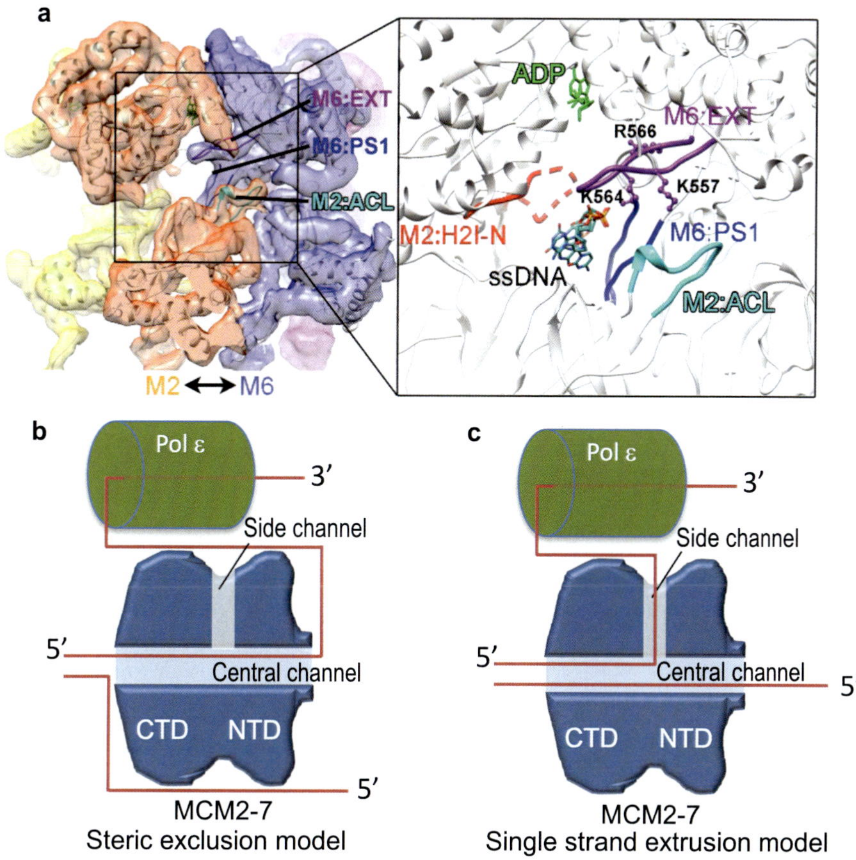

Fig. 9.5 A unique side channel between Mcm2 and Mcm6. **a** Outer surface representation of the Mcm2/Mcm6 interface. The boxed region is shown in a zoomed-in view (*right*) with individual components (H2I-N, EXT, PS1, and ACL) colored individually. A short piece of ssDNA is modeled in the structure to indicate the size of this side channel large enough to act as a pore for strand extrusion from the central channel during DNA unwinding along with basic residues (Arg566, Lys557, and Lys564) of the EXT hairpin from MCM6. The H2I-N is partially disordered. (**b, c**) Threading paths proposed for leading- and lagging-strand DNA based on the architecture of the replisome (Sun et al. 2015) in the steric exclusion model (**b**) versus the strand extrusion model where the 3′-5′ strand extrudes through Mcm2/Mcm6 side channel (**c**)

9.5 ATPase Pockets in the Double Hexamer

Mcm2-7 forms the core of the replication-licensing complex that assembles the replication initiation complex at replication origins and the core of the replicative helicase that unwinds dsDNA at replication forks (Bochman and Schwacha 2009). In both capacities, ATP hydrolysis by the Mcm2-7 catalytic core is required to fuel the energy-consuming functions that it performs.

Heterohexameric Mcm2-7 belongs to the AAA+ family of ATPases, whose ATPase active sites are formed at inter-subunit interfaces. One subunit contributes the Walker A and B motifs, while the other contributes the arginine finger. Thus, each Mcm subunit participates in two ATPase active sites, resulting in six distinct ATPase pockets. At each step of the assembly, starting from the single hexamers to the pre-RC and then from the pre-RC to the pre-IC, a different set of ATPases is called into action. Exhaustive mutant and biochemical analysis combined with in vitro reconstitution studies was able to identify which ATPase pocket is required for activity at each assembly step (Bochman and Schwacha 2008; Coster et al. 2014; Kang et al. 2014). Figure 9.6g (modified from Kang et al. 2014) summarizes the results of these elegant and intricate studies that show which ATPase pocket(s) is required at each stage. During this process, the 2:6 ATPase appears to play a role in the loading of the first SH. The 4:7, 6:4, 5:2, and 3:5 ATPases are required for the loading and head-to-head fusion of the second SH to form the DH. Up to this point, the only ATPase pocket that is not required for activity is the 7:3 pocket as if all of the energy consumed up to this point is to assemble this high-energy structure that is stable throughout G1 phase. Interestingly, in the high-resolution structure of the inactive DH, only the 7:3 ATPase pocket shows extra density indicating the possibility of ATP bound (Fig. 9.6a–f) (Li et al. 2015). It would be interesting to find out what exactly happens in the steps that follow. For example, is ATP hydrolysis by the 7:3 pocket the first responder of DDK or CDK phosphorylation to initiate DNA melting at the G1-S phase transition, which in turn recruits Sld3 and Cdc45? According to the in vitro reconstitution study, all of the ATPases except 6:4 act immediately after the CDK action to recruit GINS and RPA (Kang et al. 2014). What seems remarkable is how the asymmetry of the MCM ATPase pockets directs the sequential assembly of the replication initiation complex to form the helicase engine of the replisome. However, to unravel which ATPase is called into action at which specific step would require the high-resolution structure of each staged assembly that shows the nucleotide-binding state of each ATPase pocket as achieved for the double hexamer (Li et al. 2015). The double hexamer is an unusually stable structure that can withstand washing by 0.5 M salt (Bowers et al. 2004; Remus et al. 2009; Evrin et al. 2009) and therefore more amenable to structural analysis. It may be difficult to find stable conditions for each of the intermediate assembly structures for high-resolution structural analysis.

As a footnote to the importance of high-resolution structures, a previous study using negative staining EM suggested that the inactivation of the ATPase pocket in the DH is due to the uncoupling of the ATP hydrolysis motifs by the staggering of the CTD and NTD of each subunit (Sun et al. 2014). It should be pointed out that the model presented was based on the misassignment of the CTDs of the subunits that led the authors to these conclusions.

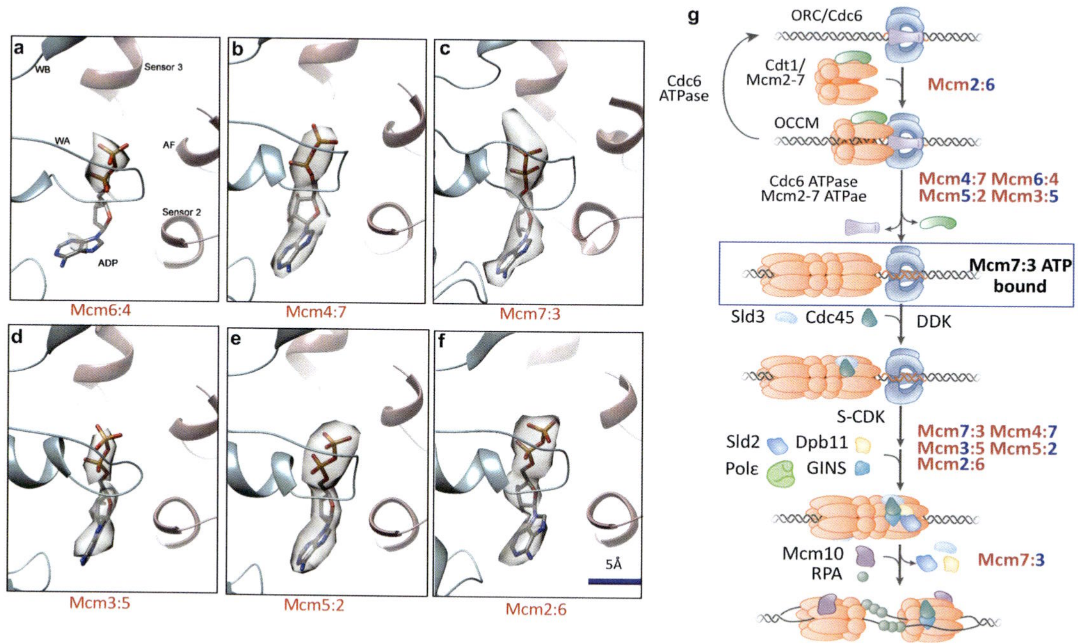

Fig. 9.6 MCM2-7 ATPase pockets involved in different stages of helicase loading and activation. (**a–f**) Zoomed-in views of nucleotide occupancy at the six ATPase centers of MCM2-7 double hexamer. The conserved ATPase elements of the active centers are labeled as in (**a**). Segmented nucleotide densities at a contour level of 5.5σ were superimposed (transparent gray). Nucleotides were modeled using ADP (Reproduced from Li et al. 2015). (**g**) Kang et al. (2014) purified a series of Cdt1-MCM2-7 heptamers to examine the role of each MCM2-7 ATPase active site during DNA replication initiation. Each complex contains a single mutation in either Walker A (K to A) or arginine finger (R to A) motif of a particular Mcm subunit. The identified ATPase active sites are summarized in this cartoon. McmX:X in which number highlighted in blue represents the subunit location of that single mutation: left, Walker A motif, and right, arginine finger motif (Adapted from Kang et al. (2014) Fig. 7 by permission from Elsevier publisher Inc. copyright 2014). The unique ATP-bound state of the 7:3 ATPase in the DH atomic model is highlighted by a *boxed* frame

9.6 Central Channel and Model of Initial Origin Melting

Like many hexameric AAA+ machineries (Brewster et al. 2010; Enemark and Joshua-Tor 2006), Mcm2-7 complex also has a central-pored chamber decorated by multiple layers of hairpin loops formed by the H2Is and β-turn motifs from the OB domains. These motifs are placed in axially staggered positions, resembling archaeal MCM structures in contacting with ssDNA and dsDNA (Miller et al. 2014). The atomic model for the DH suggests that MCM2-7 helicase employs a similar mechanism in controlling the axial displacement of these loops to facilitate DNA translocation and unwinding (Fig. 9.7a, b).

The most striking feature of the MCM2-7 DH structure is that the central channel, formed by these two staggered MCM rings, has four constriction points in the channel and a kink at the interface of the two rings (Fig. 9.7a, b). Modeling with dsDNA shows that these constriction points interact with the major and minor grooves of dsDNA consistent with observations that the DH complexes stably associate with linear dsDNA under physiological condition and only becomes mobile on dsDNA in high-salt buffer (Remus et al. 2009; Kumar and Remus 2016). These fine structures of the MCM2-7 DH provide insight into its function in DNA melting (Fig. 9.8). First, the kinked interface of the two rings would deform dsDNA (Fig. 9.7e), which could serve as a nucleation center for DNA destabilization. It has been shown that DNA bending that distorts hydrogen bonds between base pairs causes local DNA melting and facilitates DNA unwinding during transcription (Tang and Patel 2006). Second, the tight grip of duplex DNA by the two restriction points in the two SH chambers would further deform DNA at the nucleation point if rotated against each other. Third, possible rotations between ring structures formed by subdomains of each hexamer would lower the activation energy for DNA destabilization even fur-

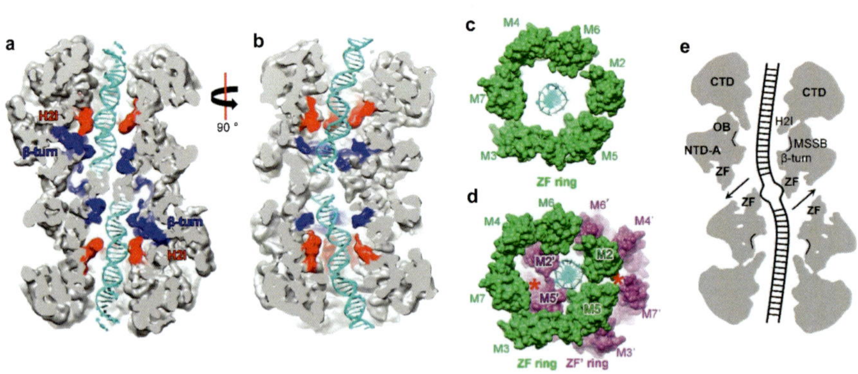

Fig. 9.7 Central channel and its implication in origin melting. (**a, b**) Cutaway views of the density map (unsharpened) with two dsDNA fragments fitted in the central channel. Surface representation of the atomic models of ZFs from one single hexamer (**c**) and ZFs from both single hexamers (**d**). (**e**) A simplified diagram of the cryo-EM map of the MCM2-7 DH. The structural features that may involve in the initial melting step are labeled as indicated (Reproduced from Li et al. 2015)

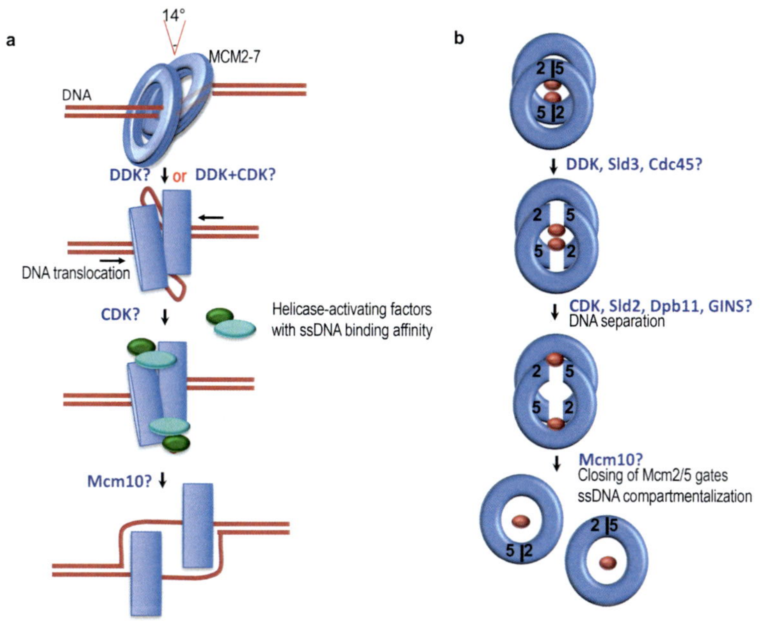

Fig. 9.8 A model for gate opening and initial DNA melting during helicase activation. Although many factors have been shown to be required for this process, it is unclear when and how they are involved in the actual gate opening and DNA melting events. See text for details

ther. Recent single-molecule analysis of *Dm*CMG helicase showed that the CTD and NTD of the DmCMG helicase could rotate in opposite directions, clockwise and anticlockwise, during its engagement with DNA (Abid Ali et al. 2016). These special features of the atomic model of the DH suggest that allosteric conformational changes following the activation of the MCM2-7 complex by cell cycle-regulated kinases might bring about DNA destabilization for initial DNA melting.

Another notable feature about the DH structure is that the overlapping of two central channels at the interface forms a narrowed main channel and two minor or exit channels. Interestingly, the main channel is formed mostly by two ZF pairs from the gate-forming subunits Mcm2 and Mcm5 (Fig. 9.7c, d). The size of the main channel is just wide enough to fit dsDNA. However, once the Mcm2/Mcm5 gates are opened at the interface region, the three channels would join together to form a space large enough for strand separation (Fig. 9.8b). Strand separation is likely assisted by the MSSB motifs conveniently located on opposite sides of the enlarged chamber (Fig. 9.7e) (Miller et al. 2014). During this DNA melting step, if the two MCM hexamers fused at the NTDs translocate along dsDNA in opposite directions, they will effectively pump in dsDNA from both ends. The already deformed DNA at the kink region would be subsequently separated (Fig. 9.8b). The melted single-DNA strands that become accessible to the outside by looping out from the exit channels could be captured by helicase-activating factors to secure their separation (Fig. 9.8a). Several of the helicase-activating factors, such as Sld2,

Sld3, Cdc45, and Mcm10, have been shown to have ssDNA-binding property (Bruck and Kaplan 2011, 2013; Costa et al. 2014; Homesley et al. 2000; Eisenberg et al. 2009). At the final step, which is yet to be understood, the DH is uncoupled, and the lagging-strand DNA is excluded from each of the MCM hexameric ring (Yardimci et al. 2010; Fu et al. 2011).

9.7 Looking Ahead

Despite the many hurdles that have held back fine-detailed structural studies of the DNA replication machinery, there is reason for optimism for unraveling this last macromolecule-synthesizing machine in the near future. Determining the near-atomic resolution structure of the inactive MCM double hexamer is a crucial start of this endeavor. The Mcm2-7 complex is a major player throughout the entire process of DNA replication, from replication licensing to initial melting of origin DNA, progression of bidirectional forks, and finally replication termination (Bell and Labib 2016). In each step, the MCM complex appears to play a somewhat different role first as an inert assembly platform, and then a duplex DNA pump, a translocator on ssDNA, and finally, a disbander of the replisome. The recent advances in cryo-EM promise that the determination of many of these MCM-associated structures is at hand. Combined with single-molecule studies and the in vitro reconstitution of the stepwise-assembled intermediates of the replisome, the next decade will witness the visualization of the workings of the replisome that is so vividly portrayed in textbooks from elegant biochemical and genetic studies of decades before.

References

Abid Ali F, Renault L, Gannon J, Gahlon HL, Kotecha A, Zhou JC, Rueda D, Costa A (2016) Cryo-EM structures of the eukaryotic replicative helicase bound to a translocation substrate. Nat Commun 7:10708. https://doi.org/10.1038/ncomms10708

Bae B, Chen YH, Costa A, Onesti S, Brunzelle JS, Lin Y, Cann IK, Nair SK (2009) Insights into the architecture of the replicative helicase from the structure of an archaeal MCM homolog. Structure 17(2):211–222. https://doi.org/10.1016/j.str.2008.11.010

Bell SP, Labib K (2016) Chromosome duplication in Saccharomyces cerevisiae. Genetics 203(3):1027–1067. https://doi.org/10.1534/genetics.115.186452

Bochman ML, Schwacha A (2008) The Mcm2-7 complex has in vitro helicase activity. Mol Cell 31(2):287–293. https://doi.org/10.1016/j.molcel.2008.05.020

Bochman ML, Schwacha A (2009) The Mcm complex: unwinding the mechanism of a replicative helicase. Microbiol Mol Biol Rev: MMBR 73(4):652–683. https://doi.org/10.1128/MMBR.00019-09

Bowers JL, Randell JC, Chen S, Bell SP (2004) ATP hydrolysis by ORC catalyzes reiterative Mcm2-7 assembly at a defined origin of replication. Mol Cell 16(6):967–978. https://doi.org/10.1016/j.molcel.2004.11.038

Brewster AS, Wang G, Yu X, Greenleaf WB, Carazo JM, Tjajadi M, Klein MG, Chen XS (2008) Crystal structure of a near-full-length archaeal MCM: functional insights for an AAA+ hexameric helicase. Proc Natl Acad Sci U S A 105(51):20191–20196. https://doi.org/10.1073/pnas.0808037105

Brewster AS, Slaymaker IM, Afif SA, Chen XS (2010) Mutational analysis of an archaeal minichromosome maintenance protein exterior hairpin reveals critical residues for helicase activity and DNA binding. BMC Mol Biol 11:62. https://doi.org/10.1186/1471-2199-11-62

Bruck I, Kaplan DL (2011) Origin single-stranded DNA releases Sld3 protein from the Mcm2-7 complex, allowing the GINS tetramer to bind the Mcm2-7 complex. J Biol Chem 286(21):18602–18613. https://doi.org/10.1074/jbc.M111.226332

Bruck I, Kaplan DL (2013) Cdc45 protein-single-stranded DNA interaction is important for stalling the helicase during replication stress. J Biol Chem 288(11):7550–7563. https://doi.org/10.1074/jbc.M112.440941

Chong JP, Mahbubani HM, Khoo CY, Blow JJ (1995) Purification of an MCM-containing complex as a component of the DNA replication licensing system. Nature 375(6530):418–421. https://doi.org/10.1038/375418a0

Chong JP, Hayashi MK, Simon MN, Xu RM, Stillman B (2000) A double-hexamer archaeal minichromosome maintenance protein is an ATP-dependent DNA helicase. Proc Natl Acad Sci U S A 97(4):1530–1535. https://doi.org/10.1073/pnas.030539597

Costa A, Ilves I, Tamberg N, Petojevic T, Nogales E, Botchan MR, Berger JM (2011) The structural basis for MCM2-7 helicase activation by GINS and Cdc45. Nat Struct Mol Biol 18(4):471–477. https://doi.org/10.1038/nsmb.2004

Costa A, Renault L, Swuec P, Petojevic T, Pesavento JJ, Ilves I, MacLellan-Gibson K, Fleck RA, Botchan MR, Berger JM (2014) DNA binding polarity, dimerization, and ATPase ring remodeling in the CMG helicase of the eukaryotic replisome. elife 3:e03273. https://doi.org/10.7554/eLife.03273

Coster G, Frigola J, Beuron F, Morris EP, Diffley JF (2014) Origin licensing requires ATP binding and hydrolysis by the MCM replicative helicase. Mol Cell 55(5):666–677. https://doi.org/10.1016/j.molcel.2014.06.034

Deegan TD, Yeeles JT, Diffley JF (2016) Phosphopeptide binding by Sld3 links Dbf4-dependent kinase to MCM replicative helicase activation. EMBO J 35(9):961–973. 10.15252/embj.201593552

van Deursen F, Sengupta S, De Piccoli G, Sanchez-Diaz A, Labib K (2012) Mcm10 associates with the loaded DNA helicase at replication origins and defines a novel step in its activation. EMBO J 31(9):2195–2206. https://doi.org/10.1038/emboj.2012.69

Diffley JF, Cocker JH, Dowell SJ, Rowley A (1994) Two steps in the assembly of complexes at yeast replication origins in vivo. Cell 78(2):303–316

Donovan S, Harwood J, Drury LS, Diffley JF (1997) Cdc6p-dependent loading of Mcm proteins onto pre-replicative chromatin in budding yeast. Proc Natl Acad Sci U S A 94(11):5611–5616

Eisenberg S, Korza G, Carson J, Liachko I, Tye BK (2009) Novel DNA binding properties of the Mcm10 protein from Saccharomyces cerevisiae. J Biol Chem 284(37):25412–25420. https://doi.org/10.1074/jbc.M109.033175

Enemark EJ, Joshua-Tor L (2006) Mechanism of DNA translocation in a replicative hexameric helicase. Nature 442(7100):270–275. https://doi.org/10.1038/nature04943

Evrin C, Clarke P, Zech J, Lurz R, Sun J, Uhle S, Li H, Stillman B, Speck C (2009) A double-hexameric MCM2-7 complex is loaded onto origin DNA during licensing of eukaryotic DNA replication. Proc Natl Acad Sci U S A 106(48):20240–20245. https://doi.org/10.1073/pnas.0911500106

Fernandez-Leiro R, Scheres SHW (2016) Unravelling biological macromolecules with cryo-electron microscopy. Nature 537(7620):339–346. https://doi.org/10.1038/nature19948

Fu YV, Yardimci H, Long DT, Ho TV, Guainazzi A, Bermudez VP, Hurwitz J, van Oijen A, Scharer OD, Walter JC (2011) Selective bypass of a lagging strand roadblock by the eukaryotic replicative DNA helicase. Cell 146(6):931–941. https://doi.org/10.1016/j.cell.2011.07.045

Gambus A, Blow JJ (2013) Mcm8 and Mcm9 form a dimeric complex in Xenopus laevis egg extract that is not essential for DNA replication initiation. Cell Cycle 12(8):1225–1232. https://doi.org/10.4161/cc.24310

Gambus A, Jones RC, Sanchez-Diaz A, Kanemaki M, van Deursen F, Edmondson RD, Labib K (2006) GINS maintains association of Cdc45 with MCM in replisome progression complexes at eukaryotic DNA replication forks. Nat Cell Biol 8(4):358–366. https://doi.org/10.1038/ncb1382

Georgescu RE, Schauer GD, Yao NY, Langston LD, Yurieva O, Zhang D, Finkelstein J, O'Donnell ME (2015) Reconstitution of a eukaryotic replisome reveals suppression mechanisms that define leading/lagging strand operation. elife 4:e04988. https://doi.org/10.7554/eLife.04988

Heller RC, Kang S, Lam WM, Chen S, Chan CS, Bell SP (2011) Eukaryotic origin-dependent DNA replication in vitro reveals sequential action of DDK and S-CDK kinases. Cell 146(1):80–91. https://doi.org/10.1016/j.cell.2011.06.012

Homesley L, Lei M, Kawasaki Y, Sawyer S, Christensen T, Tye BK (2000) Mcm10 and the MCM2-7 complex interact to initiate DNA synthesis and to release replication factors from origins. Genes Dev 14(8):913–926

Ilves I, Petojevic T, Pesavento JJ, Botchan MR (2010) Activation of the MCM2-7 helicase by association with Cdc45 and GINS proteins. Mol Cell 37(2):247–258. https://doi.org/10.1016/j.molcel.2009.12.030

Johnston LH, Thomas AP (1982) The isolation of new DNA synthesis mutants in the yeast Saccharomyces cerevisiae. Mol Gen Genet 186(3):439–444

Kang S, Warner MD, Bell SP (2014) Multiple functions for Mcm2-7 ATPase motifs during replication initiation. Mol Cell 55(5):655–665. https://doi.org/10.1016/j.molcel.2014.06.033

Kelman Z, Lee J-K, Hurwitz J (1999) The single minichromosome maintenance protein of Methanobacterium thermoautotrophicum ΔH contains DNA helicase activity. Proc Natl Acad Sci 96(26):14783–14788. https://doi.org/10.1073/pnas.96.26.14783

Kornberg R (2007) The molecular basis of eukaryotic transcription (Nobel lecture). Angew Chem 46(37):6956–6965. https://doi.org/10.1002/anie.200701832

Kumar C, Remus D (2016) Eukaryotic replication origins: strength in flexibility. Nucleus 7(3):292–300. https://doi.org/10.1080/19491034.2016.1187353

Labib K, Tercero JA, Diffley JF (2000) Uninterrupted MCM2-7 function required for DNA replication fork progression. Science 288(5471):1643–1647

Lei M, Kawasaki Y, Young MR, Kihara M, Sugino A, Tye BK (1997) Mcm2 is a target of regulation by Cdc7-Dbf4 during the initiation of DNA synthesis. Genes Dev 11(24):3365–3374

Li N, Zhai Y, Zhang Y, Li W, Yang M, Lei J, Tye BK, Gao N (2015) Structure of the eukaryotic MCM complex at 3.8 A. Nature 524(7564):186–191. https://doi.org/10.1038/nature14685

Lutzmann M, Maiorano D, Mechali M (2005) Identification of full genes and proteins of MCM9, a novel, vertebrate-specific member of the MCM2-8 protein family. Gene 362:51–56. https://doi.org/10.1016/j.gene.2005.07.031

Maine GT, Sinha P, Tye BK (1984) Mutants of S. cerevisiae defective in the maintenance of minichromosomes. Genetics 106(3):365–385

Maiorano D, Cuvier O, Danis E, Mechali M (2005) MCM8 is an MCM2-7-related protein that functions as a DNA helicase during replication elongation and not initiation. Cell 120(3):315–328. https://doi.org/10.1016/j.cell.2004.12.010

Miller JM, Arachea BT, Epling LB, Enemark EJ (2014) Analysis of the crystal structure of an active MCM hexamer. elife 3:e03433. https://doi.org/10.7554/eLife.03433

Moir D, Stewart SE, Osmond BC, Botstein D (1982) Cold-sensitive cell-division-cycle mutants of yeast: isolation, properties, and pseudoreversion studies. Genetics 100(4):547–563

Moyer SE, Lewis PW, Botchan MR (2006) Isolation of the Cdc45/Mcm2-7/GINS (CMG) complex, a candidate for the eukaryotic DNA replication fork helicase. Proc Natl Acad Sci U S A 103(27):10236–10241. https://doi.org/10.1073/pnas.0602400103

Nishimura K, Ishiai M, Horikawa K, Fukagawa T, Takata M, Takisawa H, Kanemaki MT (2012) Mcm8 and Mcm9 form a complex that functions in homologous recombination repair induced by DNA interstrand crosslinks. Mol Cell 47(4):511–522. https://doi.org/10.1016/j.molcel.2012.05.047

On KF, Beuron F, Frith D, Snijders AP, Morris EP, Diffley JF (2014) Prereplicative complexes assembled in vitro support origin-dependent and independent DNA replication. EMBO J 33(6):605–620. https://doi.org/10.1002/embj.201387369
Ramakrishnan V (2010) Unraveling the structure of the ribosome (Nobel lecture). Angew Chem 49(26):4355–4380. https://doi.org/10.1002/anie.201001436
Randell JC, Fan A, Chan C, Francis LI, Heller RC, Galani K, Bell SP (2010) Mec1 is one of multiple kinases that prime the Mcm2-7 helicase for phosphorylation by Cdc7. Mol Cell 40(3):353–363. https://doi.org/10.1016/j.molcel.2010.10.017
Remus D, Beuron F, Tolun G, Griffith JD, Morris EP, Diffley JF (2009) Concerted loading of Mcm2-7 double hexamers around DNA during DNA replication origin licensing. Cell 139(4):719–730. https://doi.org/10.1016/j.cell.2009.10.015
Sheu YJ, Stillman B (2006) Cdc7-Dbf4 phosphorylates MCM proteins via a docking site-mediated mechanism to promote S phase progression. Mol Cell 24(1):101–113. https://doi.org/10.1016/j.molcel.2006.07.033
Siddiqui K, On KF, Diffley JF (2013) Regulating DNA replication in eukarya. Cold Spring Harb Perspect Biol 5(9). https://doi.org/10.1101/cshperspect.a012930
Steitz TA (2010) From the structure and function of the ribosome to new antibiotics (Nobel lecture). Angew Chem 49(26):4381–4398. https://doi.org/10.1002/anie.201000708
Sun J, Evrin C, Samel SA, Fernandez-Cid A, Riera A, Kawakami H, Stillman B, Speck C, Li H (2013) Cryo-EM structure of a helicase loading intermediate containing ORC-Cdc6-Cdt1-MCM2-7 bound to DNA. Nat Struct Mol Biol 20(8):944–951. https://doi.org/10.1038/nsmb.2629
Sun J, Fernandez-Cid A, Riera A, Tognetti S, Yuan Z, Stillman B, Speck C, Li H (2014) Structural and mechanistic insights into Mcm2-7 double-hexamer assembly and function. Genes Dev 28(20):2291–2303. https://doi.org/10.1101/gad.242313.114
Sun J, Shi Y, Georgescu RE, Yuan Z, Chait BT, Li H, O'Donnell ME (2015) The architecture of a eukaryotic replisome. Nat Struct Mol Biol 22(12):976–982. https://doi.org/10.1038/nsmb.3113
Tang GQ, Patel SS (2006) T7 RNA polymerase-induced bending of promoter DNA is coupled to DNA opening. Biochemistry 45(15):4936–4946. https://doi.org/10.1021/bi0522910
Thommes P, Kubota Y, Takisawa H, Blow JJ (1997) The RLF-M component of the replication licensing system forms complexes containing all six MCM/P1 polypeptides. EMBO J 16(11):3312–3319. https://doi.org/10.1093/emboj/16.11.3312
Ticau S, Friedman LJ, Ivica NA, Gelles J, Bell SP (2015) Single-molecule studies of origin licensing reveal mechanisms ensuring bidirectional helicase loading. Cell 161(3):513–525. https://doi.org/10.1016/j.cell.2015.03.012
Traver S, Coulombe P, Peiffer I, Hutchins JR, Kitzmann M, Latreille D, Mechali M (2015) MCM9 is required for mammalian DNA mismatch repair. Mol Cell 59(5):831–839. https://doi.org/10.1016/j.molcel.2015.07.010
Tye BK (1999a) MCM proteins in DNA replication. Annu Rev Biochem 68:649–686. https://doi.org/10.1146/annurev.biochem.68.1.649
Tye BK (1999b) Minichromosome maintenance as a genetic assay for defects in DNA replication. Methods 18(3):329–334. https://doi.org/10.1006/meth.1999.0793
Yardimci H, Loveland AB, Habuchi S, van Oijen AM, Walter JC (2010) Uncoupling of sister replisomes during eukaryotic DNA replication. Mol Cell 40(5):834–840. https://doi.org/10.1016/j.molcel.2010.11.027
Yeeles JT, Deegan TD, Janska A, Early A, Diffley JF (2015) Regulated eukaryotic DNA replication origin firing with purified proteins. Nature. https://doi.org/10.1038/nature14285
Yonath A (2010) Polar bears, antibiotics, and the evolving ribosome (Nobel lecture). Angew Chem 49(26):4341–4354. https://doi.org/10.1002/anie.201001297
Yuan Z, Bai L, Sun J, Georgescu R, Liu J, O'Donnell ME, Li H (2016) Structure of the eukaryotic replicative CMG helicase suggests a pumpjack motion for translocation. Nat Struct Mol Biol 23(3):217–224. https://doi.org/10.1038/nsmb.3170

Chapter 10
Architecture of the *Saccharomyces cerevisiae* Replisome

Lin Bai, Zuanning Yuan, Jingchuan Sun, Roxana Georgescu, Michael E. O'Donnell, and Huilin Li

Abstract Eukaryotic replication proteins are highly conserved, and thus study of *Saccharomyces cerevisiae* replication can inform about this central process in higher eukaryotes including humans. The *S. cerevisiae* replisome is a large and dynamic assembly comprised of ~50 proteins. The core of the replisome is composed of 31 different proteins including the 11-subunit CMG helicase; RFC clamp loader pentamer; PCNA clamp; the heteroligomeric DNA polymerases ε, δ, and α-primase; and the RPA heterotrimeric single strand binding protein. Many additional protein factors either travel with or transiently associate with these replisome proteins at particular times during replication. In this chapter, we summarize several recent structural studies on the *S. cerevisiae* replisome and its subassemblies using single particle electron microscopy and X-ray crystallography. These recent structural studies have outlined the overall architecture of a core replisome subassembly and shed new light on the mechanism of eukaryotic replication.

Keywords Eukaryotic DNA replication • Replisome • Replicative helicase • Cryo-EM • Mcm • CMG • DNA polymerase • Structural biology

L. Bai • J. Sun
Cryo-EM Structural Biology Laboratory, Van Andel Research Institute, Grand Rapids, MI, USA

Z. Yuan • H. Li (✉)
Biochemistry and Structural Biology Graduate Program, Stony Brook University, Stony Brook, NY, USA

Cryo-EM Structural Biology Laboratory, Van Andel Research Institute, Grand Rapids, MI, USA
e-mail: Huilin.Li@vai.org

R. Georgescu • M.E. O'Donnell (✉)
Howard Hughes Medical Institute, The Rockefeller University, New York, NY, USA
e-mail: odonnel@rockefeller.edu

H. Masai, M. Foiani (eds.), *DNA Replication*, Advances in Experimental Medicine and Biology 1042, https://doi.org/10.1007/978-981-10-6955-0_10

10.1 Introduction

The DNA genome is duplicated in semiconservative fashion wherein each strand serves as the template for synthesis of its complementary strand (Meselson and Stahl 1958; Watson and Crick 1953a, b). This process may seem simple, but nothing could be further from the truth; it requires >50 different proteins in eukaryotes, and many of their individual functions remain unknown, much less how they work together to accomplish high-fidelity genome duplication (Costa et al. 2013; Li and Araki 2013; MacNeill 2012). Unlike protein synthesis and mRNA synthesis that are well conserved across the three domains of life, DNA synthesis is not well conserved. Many key replication proteins, such as the replicative helicase, DNA polymerases, primase, and single strand binding protein, are evolutionarily unrelated between bacteria and archaea/eukarya (Forterre et al. 2004; Leipe et al. 1999). Thus one cannot rely on the wealth of bacterial studies to understand eukaryotic replication, and detailed studies of eukaryotic replication mechanisms are warranted. Fortunately, the replication machinery within eukaryotes is highly conserved, and thus the relatively simple *Saccharomyces cerevisiae* can be used as a reliable model for replication in higher eukaryotes (Leman and Noguchi 2013).

The "core replisome" encompasses proteins that are necessary to propagate the replication fork (Zhang and O'Donnell 2016). The eukaryotic replisome core is composed of CMG helicase, the leading strand DNA polymerase (Pol) ε, lagging strand Pol δ, Pol α-primase, along with the RFC clamp loader, PCNA clamp, and RPA single strand (ss) DNA-binding protein. Many other components, such as Ctf4, Mcm10, Topo I, the checkpoint response factors Mrc1, Tof1, Csm3, and the FACT nucleosome mobility factor, either travel with the replisome or hop on and off at different points during replication (Bell and Labib 2016).

The eukaryotic helicase is an 11-protein machine composed of *C*dc45, the *M*cm2-7 heterohexamer, and the four-protein *G*INS complex (Psf1-3, and Sld5) referred to as CMG (Bochman et al. 2008; Ilves et al. 2010; Moyer et al. 2006). Eukaryotic CMG helicase is assembled through a series of replication initiation events (reviewed in (Costa et al. 2013; O'Donnell et al. 2013; Tanaka and Araki 2013; Tognetti et al. 2015)). Briefly, the origin recognition complex (ORC) along with Cdc6 loads onto replication origins in the G1 phase of the cell cycle (Bell and Stillman 1992) and then recruits the helicase core Mcm2-7 heterohexamer with the help of Cdt1 (Cocker et al. 1996; Liang et al. 1995; Mimura et al. 2004; Nishitani et al. 2000; Santocanale and Diffley 1996; Sun et al. 2013; Tanaka and Diffley 2002). Subsequently, another Mcm2-7 is recruited onto double-stranded (ds) DNA forming a double hexamer of Mcm2-7 that is inactive (Duzdevich et al. 2015; Evrin et al. 2009; Remus et al. 2009; Sun et al. 2014; Ticau et al. 2015). Finally, at the G1-to-S transition, the double hexamers are converted to two active helicases comprised of Cdc45-Mcm2-7-GINS (CMG) that each encircle opposite strands of single-stranded (ss) DNA to carry out bidirectional separation of the DNA duplex (Bell and Botchan 2013; Duzdevich et al. 2015).

Eukaryotic replication requires at least three DNA polymerases to propagate fork movement (Burgers 2009). These include Pol α-primase, Pol ε, and Pol δ. Pols α, ε, and δ are all members of the B family of DNA polymerases (Steitz 1999). Pol α initiates DNA synthesis on both the leading and lagging strands by synthesizing a RNA/DNA hybrid primer (Conaway and Lehman 1982). Numerous genetic, cell biology, and biochemical studies concur that Pol ε and Pol δ extend primers on the leading and lagging strands, respectively (Burgers et al. 2016; Clausen et al. 2015; Kunkel and Burgers 2008; Miyabe et al. 2011; Nick McElhinny et al. 2008; Pursell et al. 2007), although one report suggests that Pol δ functions on both strands leaving some uncertainty for future studies to resolve (Johnson et al. 2015).

The antiparallel structure of dsDNA, combined with unidirectional synthesis by DNA polymerases, results in continuous synthesis on the leading strand that travels in the direction of DNA unwinding and discontinuous synthesis on the lagging strand that is extended as multiple Okazaki fragments in the direction opposite DNA unwinding. A preformed primed site is required for DNA polymerase function, and both strands are primed by Pol α-primase. However, Pol α-primase is needed repeatedly on the lagging strand, once for each 100–200 bp Okazaki fragment. Pol α-primase is a four-subunit enzyme; the largest subunit is a DNA polymerase that lacks a proofreading 3′–5′ exonuclease, the second largest is the B subunit (unknown function), and the two small subunits, PriL and PriS, contain the priming activity (Banks et al. 1979; Conaway and Lehman 1982; Kaguni et al. 1983a, b).

There have been numerous important advances on the structure of components of the eukaryotic replisome in the last decade. In this chapter, we focus mainly on structures that have been solved during the past 3 years, specifically of proteins directly involved in moving replication forks. We first review the cryo-EM structure of the CMG helicase and compare the structures of CMG components GINS and Cdc45 of yeast and human. We then review the relative positions of Pol ε, Ctf4, and Pol α in the context of the CMG helicase and the possible replisome architecture at the replication fork. We end our chapter by providing a brief perspective on key missing information and what to expect in the coming years.

10.2 The CMG Structure

The first structure of eukaryotic CMG was determined in the *Drosophila melanogaster* system by negative stain EM at low resolution (Costa et al. 2011). The structure revealed a Mcm2-7 ring braced on the side by Cdc45 and GINS that interacted mainly with Mcm2, Mcm3, and Mcm5, apparently forming a secondary pore. However, this secondary pore is essentially filled in by side chains in the higher resolution cryo-EM map of the *S. cerevisiae* CMG at a resolution of 3.7–4.8 Å (Fig. 10.1a–c) (Yuan et al. 2016). The secondary pore also disappears in a medium resolution cryo-EM map of *Drosophila* CMG at 7–10 Å resolution (Abid Ali et al. 2016). The Mcm2-7 core in the CMG forms a two-tiered ring structure: an N-terminal domain (NTD) tier composed of a helical subdomain, a Zn-binding

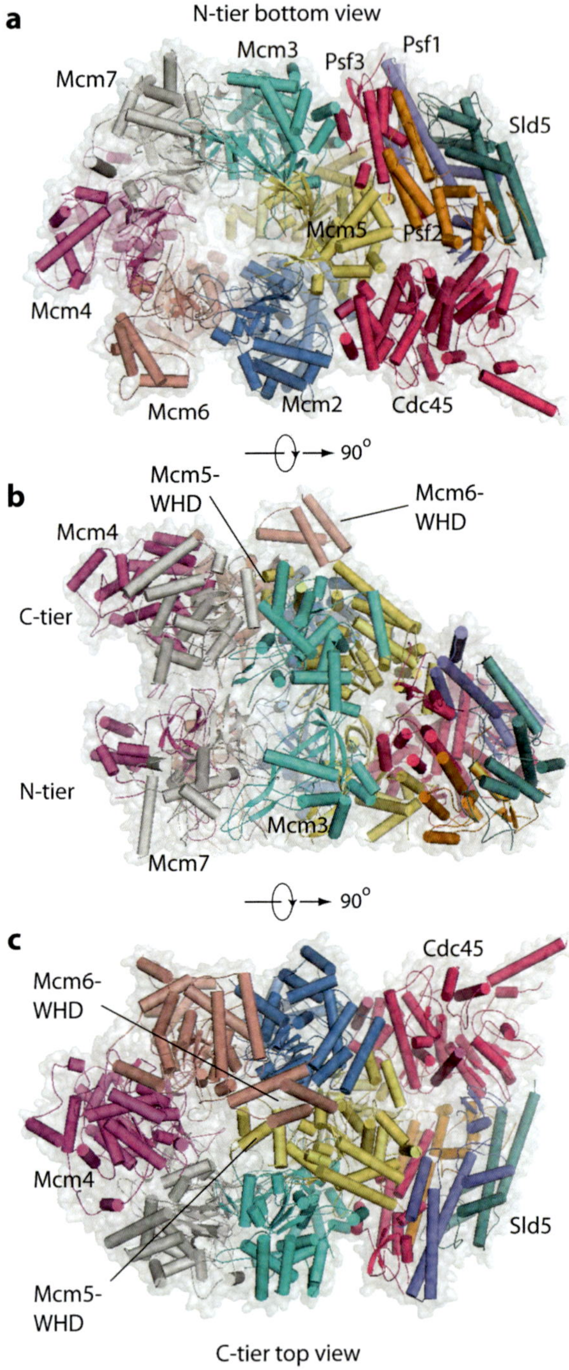

Fig. 10.1 The CMG structure. The CMG structure is shown in cartoon view with each subunit a different color (PDB 3JC7). The surface representation of the structure is superimposed as a dim gray surface view. The structure is shown from the following angles: (**a**) top CTD ring view, (**b**) a side view, and (**c**) a bottom NTD view

motif, and an oligonucleotide/oligosaccharide-binding (OB) motif, while the C-terminal domain (CTD) tier contains the AAA+ motors. This domain architecture is similar to that of other hexameric helicases including the archaeal MCM hexamer and the inactive yeast Mcm2-7 in the double hexamer (Brewster et al. 2008; Enemark and Joshua-Tor 2006; Li et al. 2015; Miller et al. 2014; Singleton et al. 2000; Slaymaker and Chen 2012). Cdc45 and GINS mainly bind the NTD-tier ring of Mcm2-7, forming a rigid unit of Cdc45-GINS-Mcm2-7 NTD. More precisely, Cdc45 contacts only the NTDs of Mcm2 and Mcm5, and GINS contacts only the NTDs of Mcm3 and Mcm5 (GINS and Cdc45 also contact one another). It is the tight interaction between the NTD of Mcm5 and Cdc45-GINS that occludes the secondary pore.

In the medium resolution cryo-EM structures of the *Drosophila* CMG, density corresponding to six nucleotides of ssDNA was observed either inside the CTD-tier motor ring or inside the NTD-tier ring (Abid Ali et al. 2016). This observation is consistent with CMG acting as an ssDNA translocase. Interestingly, in an archaeal MCM hexamer, ssDNA was found to bind the NTD-tier ring as a circle perpendicular to the MCM channel axis and was proposed to be an intermediate in origin initiation (Froelich et al. 2014). How a CMG engages DNA at a replication fork remains an important issue in the field.

10.3 The Mcm2-7 Hexamer Undergoes Large Conformational Changes from the Inactive Double Hexamer to the Active CMG Helicase

Recent studies have provided two yeast Mcm2-7 hexamer structures: one determined in the context of the inactive Mcm2-7 double hexamer and the other in the context of active CMG as described above (Li et al. 2015; Yuan et al. 2016). This allows a comparison of the structures to gain insight into the mechanism by which the helicase is activated. To transition from the inactive Mcm2-7 double hexamer to the single Mcm2-7 in CMG requires large domain movements of up to 20 Å and rotations of up to 30° (Yuan et al. 2016). Movements in the CTD-tier motor domains are more extensive than the NTD-tier ring. As an example, we compare Mcm2-Mcm5 in the two structures (Fig. 10.2a–c). Notably, the interface between the CTDs of Mcm2 and Mcm5 is looser in the CMG structure than in the double hexamer. Activation of CMG requires it to encircle the leading strand and exclude the lagging strand. The looser Mcm2/5 interface at the C-tier ring may have allowed the extrusion of the lagging strand to the outside surface during activation of the CMG helicase. Another major change is the insertion of the Mcm5 C-terminal winged helix domain (WHD) into the interior channel in the active helicase, while the Mcm5 WHD of the inactive Mcm2-7 hexamer is visualized as weak density outside the central channel (Li et al. 2015; Yuan et al. 2016). The Mcm6 C-terminal WHD in the active and inactive Mcm2-7 was also reconfigured. Considering that the WHD is

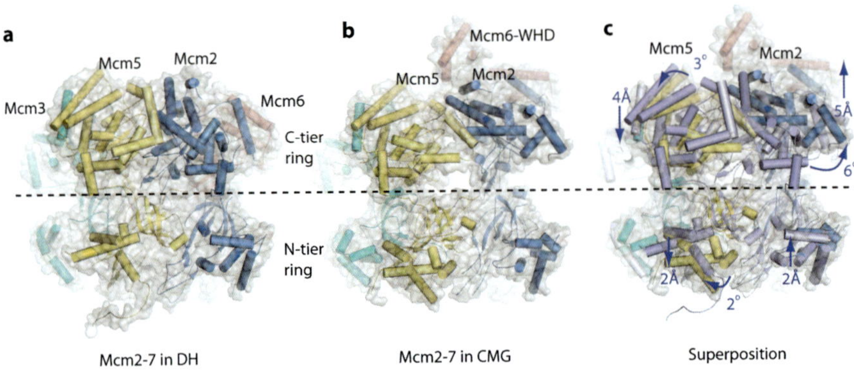

Fig. 10.2 Conformation changes proceeding from the inactive Mcm2-7 at an origin to Mcm2-7 in the active CMG helicase. The figure focuses on a side view of the Mcm2 and Mcm5 subunits in the Mcm2-7 hexamer in which one Mcm2-7 hexamer has been extracted from the inactive double hexamer (PDB 3JA8) (left side) and the Mcm2-7 hexamer has been extracted from the active CMG structure (PDB 3JC7) (middle). The two Mcm2-7 structures are then superimposed (right side). Structures are in cartoon view wrapped in a dim gray surface view. Mcm5 is gold and Mcm2 is blue. The Mcm2/5 interface is nearly opened in Mcm2-7 within CMG

usually involved in DNA binding or interaction with other proteins, the observed movement of these WHDs may have important consequences for helicase activation and replisome function during fork progression.

10.4 GINS Is a Protein-Binding Hub of the Replisome but Does Not Contain a Central Channel for ssDNA

The yeast CMG structure reveals for the first time the complete structure of the GINS heterotetramer (Yuan et al. 2016). The GINS subunits are related in evolution and are composed of two domains each (A and B) (Chang et al. 2007; Choi et al. 2007; Kamada et al. 2007). The B domain of the Psf1 subunit was missing in all previously reported GINS structures (Chang et al. 2007; Choi et al. 2007; Kamada et al. 2007). The Psf1 B domain interacts with Cdc45 by forming an interfacial cross-molecule β-sheet (Yuan et al. 2016). Perhaps the Psf1 B domain is disordered in the absence of Cdc45, explaining why it was not resolved in the crystal structures of the GINS alone.

Earlier low-resolution EM observations as well as the crystal structure of the human GINS suggested a potential hole in the middle of GINS (Boskovic et al. 2007; Chang et al. 2007; Kubota et al. 2003). It was further suggested that ssDNA might thread through the central hole of the GINS. The structure of the full-length yeast GINS contains only a small hole of ~7 Å diameter, too small to thread ssDNA (Fig. 10.3) (Yuan et al. 2016). Importantly, although the GINS structures are very similar between human and yeast, structural features surrounding the central region

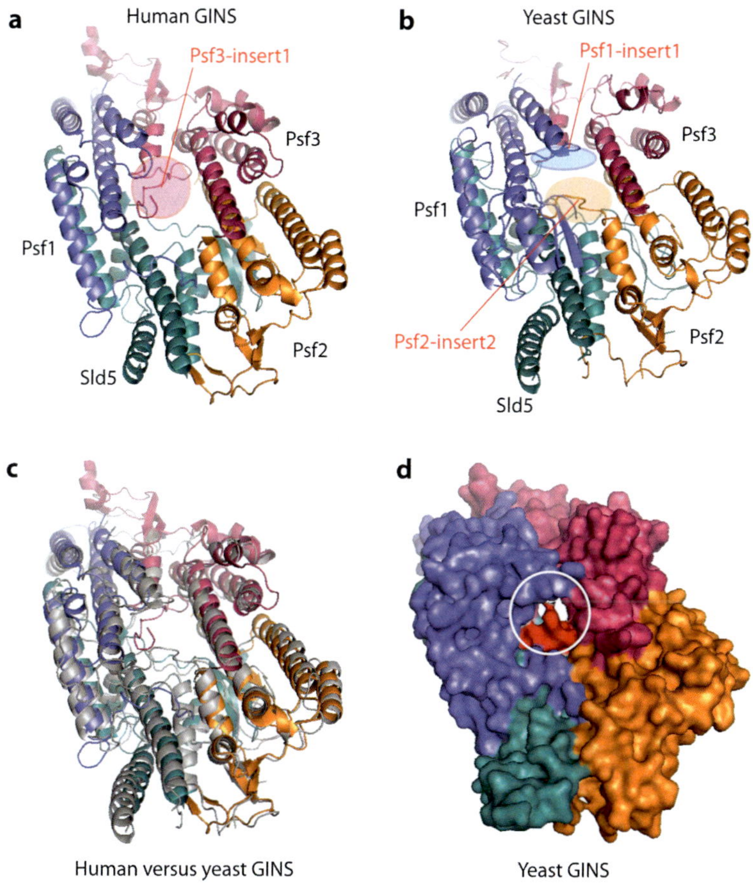

Fig. 10.3 Comparison of the human and the yeast GINS structures. (**a**) Human GINS with each subunit shown in a different color (PDB 2Q9Q). An N-terminal insertion loop in Psf3 is highlighted by a semitransparent circle near the central pore region. (**b**) The yeast GINS structure, extracted from the CMG structure (PDB 3JC6), is shown in the same color scheme as in (**a**). (**c**) Superimposition of the human and the yeast GINS structures, revealing that the structural variation is focused in the central region. (**d**) Surface representation of the yeast GINS structure. The central pore is too small to thread ssDNA

vary (Chang et al. 2007; Choi et al. 2007; Kamada et al. 2007; Yuan et al. 2016). The yeast GINS has two insertions in Psf1 and Psf2 that are absent in human GINS, and the human GINS has a Psf3 insertion loop that is absent in the yeast GINS. The structural variation, together with the small size, argues against ssDNA threading through the central hole. Instead, the GINS functions mainly to scaffold the replisome by recruiting other proteins. The CMG structure appears to bear this out. Thus GINS interacts extensively with Mcm3/5 and Cdc45 (Yuan et al. 2016). The Psf1 B domain forms the main connection to Cdc45, and the Sld5 subunit binds the Ctf4 trimer scaffold protein that in turn binds to Pol α-primase (Simon et al. 2014). An interaction between Psf1 and the B subunit of Pol ε has also been documented by biochemical studies (Gambus et al. 2009), and recently, the direct interaction between GINS and Pol ε was visualized by negative stain EM (Sun et al. 2015).

10.5 Cdc45 Has Two RecJ-Like Domains and a Protruding Helical Motif

Cdc45 has been predicted to have a RecJ-like fold (Krastanova et al. 2012; Onesti and MacNeill 2013; Sanchez-Pulido and Ponting 2011). The structure of yeast and human Cdc45 confirmed this prediction and further revealed that there are actually two RecJ-like α-/β-domains that are separated by a small helical inter-domain (ID) (Fig. 10.4a–c) (Simon et al. 2016; Yuan et al. 2016). Human and yeast Cdc45 are highly conserved at the structure level and superimpose with a root-mean-square deviation of <2 Å (Fig. 10.4c). Cdc45 has a protruding helical motif (PHM) that contains a highly negatively charged and disordered loop (D166 – R217 in yeast). The protrusion may interact and support the largely flexible N-terminal catalytic domain of Pol2 (see below), because a lysine at the tip of the PHM helix (K222) cross-links to the N-terminal catalytic domain of Pol2 (Yuan et al. 2016). The helical ID contacts and stabilizes the NTDs of Mcm2 and Mcm5 (Fig. 10.1a, c).

The fact that Cdc45 faces the neighboring Mcm2 and Mcm5 subunits in the CMG helicase is suggested to poise Cdc45 to capture DNA if the Mcm2/5 gate breaches open (Petojevic et al. 2015). Although Cdc45 does not have a nuclease activity, there is evidence that Cdc45 binds DNA weakly and nonspecifically (Krastanova et al. 2012; Simon et al. 2016; Bruck and Kaplan 2013). RecJ forms an O-like structure with a pore in the middle to serve as a ssDNA-binding groove, but unlike RecJ the yeast and human Cdc45 do not have a central pore. Therefore, Cdc45 does not contain an internal DNA-binding groove. It is possible that the exterior surface of Cdc45 may help to coordinate DNA in the context of CMG, but the nature of the DNA contact is currently unknown.

Meier-Gorlin syndrome (MGS) is a rare autosomal recessive disorder characterized by short stature (dwarfism) (Klingseisen and Jackson 2011). Previous work has found that mutations in genes involved in establishing the pre-replication complex, such as ORC1, ORC4, ORC6, CDT1, and CDC6, can cause MGS (Hossain and Stillman 2016). More recently, eight point mutations in *CDC45* have been identified in MGS patients (Fenwick et al. 2016). The mutations are scattered across the protein structure (Fig. 10.4d). Because Cdc45 is involved in both replication initiation and elongation, it is unclear whether initiation or ongoing fork movement, or both, is the underlying cause of the disease.

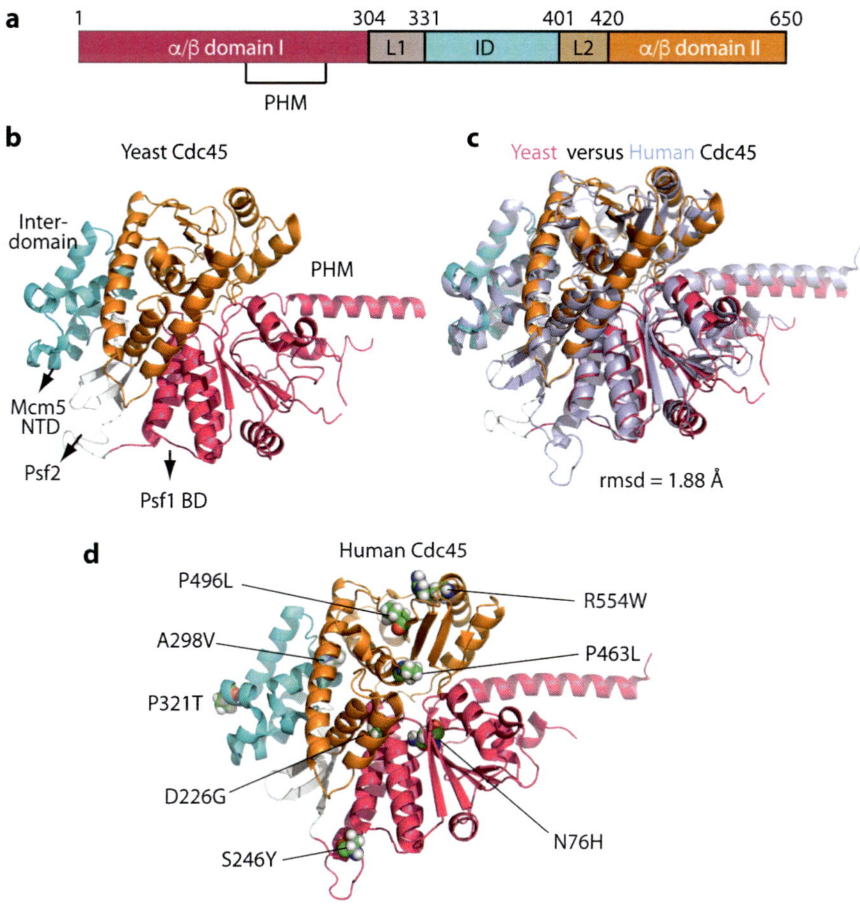

Fig. 10.4 Comparison of the yeast and human Cdc45 structures. (**a**) The domain structure of Cdc45. (**b**) Cartoon view of the yeast Cdc45 structure extracted from the CMG structure (PDB 3JC6), with each domain colored according to the domain sketch in (**a**). PHM refers to protruding helical motif. (**c**) Superimposition of the yeast and human Cdc45 structures. Human Cdc45 is shown in light purple (PDB 5DGO). (**d**) Human Cdc45 point mutations identified in patients with Meier-Gorlin syndrome are highlighted as spheres

10.6 CMG Helicase May Function as an Oil Rig Pump Jack to Inchworm Along DNA

Cryo-EM analysis of CMG particles embedded in vitreous ice revealed two conformations, an extended structure (conformer I) and a compact structure (conformer II) at 4.7 Å and 4.8 Å resolution, respectively (Fig. 10.5a, b) (Yuan et al. 2016). In conformer I, when viewed from the Cdc45-GINS side of the structure, the CTD ring of Mcm2-7 containing the AAA+ motors is tilted by ~10° with respect to the NTD ring, leading to an approximate spiral arrangement of motor domains comprising

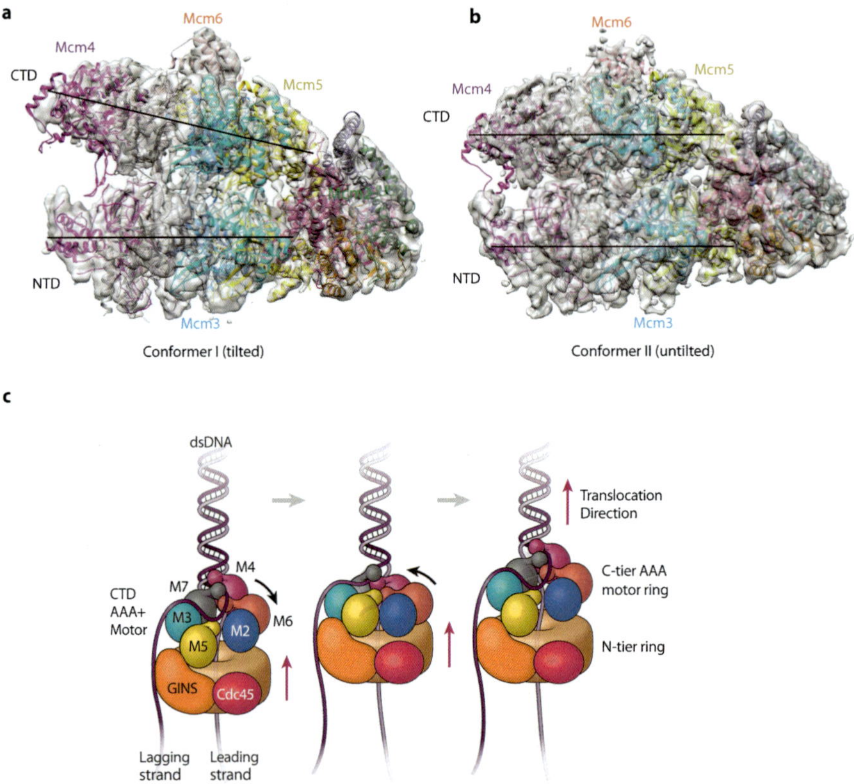

Fig. 10.5 CMG helicase alternates between tilted (extended) and untilted (compact) conformations. (**a**) Side view of CMG conformer I (extended) in which the CTD motor ring is tilted relative to the NTD ring. (**b**) Side view of the CMG in conformer II (compact) with an untilted CTD ring. In panels (**a**, **b**), the cryo-EM density map is shown as a semitransparent surface rendering, and the atomic model is shown in cartoon (EMD-6535, EMD-6536, PDB 3JC5, and 3JC7). (**c**) An oil rig-like pump jack DNA unwinding model. Panel (**c**) is reproduced in part from Figure 7 in Yuan et al. (2016) with permission. Note that the CTD-tier ring pushing on the dsDNA is only for the purpose of illustration. The pump jack model would still function as a translocase if CMG were oriented with the NTD-tier ring pushing on the dsDNA

the CTD ring. In conformer II, the CTD motor ring is approximately parallel to the NTD ring, and CMG is more compact than conformer I. In both conformers, the Mcm2-7 NTD ring-GINS-Cdc45 unit appears to be a rigid platform upon which the CTD AAA+ motor domains switch between extended and compact states during cycles of ATP hydrolysis. These structures suggest that CMG may function like an oil rig pump jack attached to a stable platform, nodding up and down to inchworm along ssDNA and unwind dsDNA at a forked junction (Fig. 10.5c) (Yuan et al. 2016). This linear inchworm ratchet mechanism is distinct from the sequential rotary ATP hydrolysis unwinding mechanism proposed for the homo-hexameric helicases such as *E. coli* DnaB, *E. coli* Rho transcription terminator, and bovine papillomavirus (BPV) E1 (Enemark and Joshua-Tor 2006; Itsathitphaisarn et al. 2012; Thomsen and Berger 2009). Additionally, single point mutants of the ATP sites in the Mcm2-7 subunits of *Drosophila* CMG show major helicase defects in only two of the six mutants, inconsistent with a sequential rotary hydrolysis mechanism that predicts each ATP site would be important to helicase activity (Ilves et al. 2010). Similar conclusions of nonequivalent ATP sites have been made by mutational studies of yeast Mcm2-7 (Schwacha and Bell 2001).

Aside from its main function in unwinding parental DNA, CMG also functions as a scaffold for assembly of the replisome. Indeed, the CMG helicase interacts with many other replicative factors, chief among them are the leading strand Pol ε (Langston et al. 2014) and the Pol α-primase that is required repeatedly on the lagging strand (Gambus et al. 2006).

10.7 Pol ε Binds to the CTD Motor Side of CMG

In 1990 Pol ε was identified as a third DNA polymerase in budding yeast that is essential for cellular replication (Morrison et al. 1990). Subsequent studies revealed that Pol ε is also involved in many other pathways such as the DNA damage checkpoint response, epigenetic silencing, sister chromatid cohesion, and possibly DNA recombination during repair of DNA lesions (Pursell and Kunkel 2008). Nearly two decades after its discovery, Pol ε was assigned as the major leading strand polymerase (Clausen et al. 2015; Kunkel and Burgers 2008; Miyabe et al. 2011; Nick McElhinny et al. 2008; Pursell et al. 2007). In all eukaryotes studied to date, Pol ε consists of four proteins; the largest is the DNA polymerase, followed by the B subunit and two small histone-fold subunits (Pursell and Kunkel 2008). The Pol ε of *Saccharomyces cerevisiae* consists of Pol2, Dpb2 (DNA polymerase-binding protein 2), Dpb3, and Dpb4. Pol2 is composed of two subdomains: a N-terminal catalytic domain that contains the polymerase and exonuclease active sites and an inactive C-terminal domain that shares homology to B-family DNA polymerases (Fig. 10.6a) (Tahirov et al. 2009). The ternary crystal structure of the N-terminal catalytic subdomain of Pol2 in complex with primer-template DNA and incoming dNTP was recently determined, and it shares many features of other B-family DNA polymerases such as Pol δ (Hogg et al. 2014). However, a P domain, new to the B-family polymerases, enables Pol ε to encircle the nascent dsDNA and enhances the processivity of DNA synthesis. This may explain why Pol ε has the highest fidelity among B-family polymerases despite the absence of an extended β-hairpin loop that is required for high-fidelity replication by other B-family polymerases (Fortune et al. 2005; Hogg and Johansson 2012; Hogg et al. 2014).

Cryo-EM analysis of yeast Pol ε revealed a bilobed overall architecture (Asturias et al. 2006). Pol ε has also recently been shown to bind directly to CMG helicase, independent of DNA, and the 15 protein complex is referred to as CMGE (Langston et al. 2014). Negative stain EM analysis of CMGE provided a 16 Å 3D map

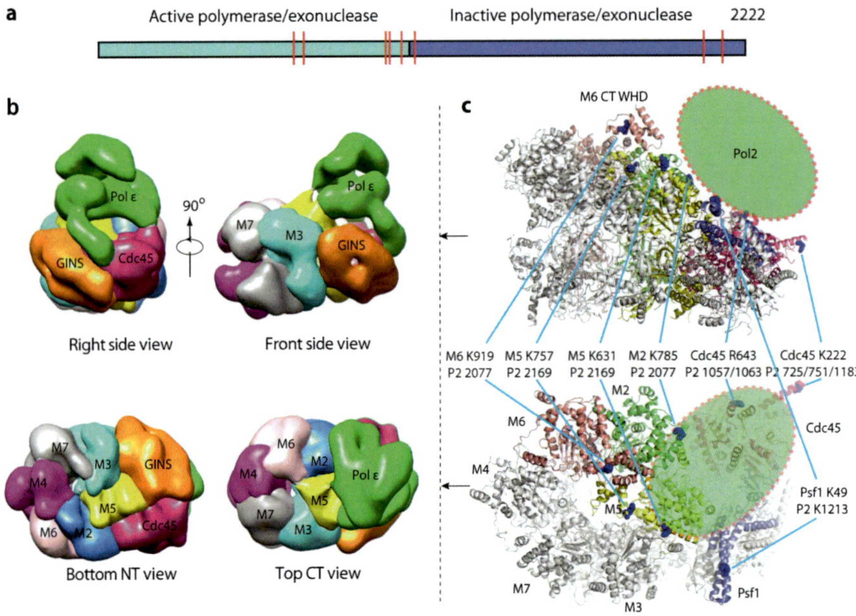

Fig. 10.6 Architecture of the CMGE complex. (**a**) The Pol2 (P2) subunit of Pol ε is comprised of a catalytic N-terminal subdomain and an inactive B-family Pol in the C-terminal subdomain. (**b**) Negative stain 3D EM map of the CMGE in four different views (EMD-6465). The EM density is segmented and colored according to individual subunits except for the Pol ε where the complex is shown in green. Part of the Pol ε density is not visible in the EM and may be the N-terminal subdomain of Pol2 as proposed in Sun et al. (2015). (**c**) Cross-linking mass spectrometry identified multiple contacts between Pol2 and GINS, Cdc45, and the CTD regions of Mcm2, Mcm5, and Mcm6 (This figure is modified with permission from Figure 6 of Yuan et al. 2016)

(Fig. 10.6b) (Sun et al. 2015). This low-resolution structure revealed that Pol ε is positioned on the CTD-tier side of CMG, sitting atop the CTDs of Mcm2 and Mcm5, and GINS and Cdc45. This Pol ε position was confirmed by extensive cross-linking mass spectrometry analysis (Fig. 10.6c). The visible density of Pol ε could only account for about 70 % the mass of Pol ε, and one suggestion for this missing density was that the catalytic Pol ε N-terminal subdomain was too mobile to visualize (Fig. S6 in ref. (Sun et al. 2015)). However, cross-linking mass spectrometry analysis of CMGE demonstrated that the Pol ε N-terminal subdomain was in proximity to the other subunits of Pol ε and thus located nearby.

10.8 Pol α and Ctf4 Are Located on the NTD-Tier Side of CMG, Opposite from Pol ε

DNA synthesis activity of Pol ε and Pol δ requires the primed site synthesized by the four-subunit Pol α-primase, which synthesizes a hybrid RNA/DNA primer (Conaway and Lehman 1982). A 7-9mer RNA is synthesized de novo by the Pri1/2 subunits followed by an internal transfer of the RNA primer to the DNA polymerase which adds a further 10–20 deoxyribonucleotides (Nethanel et al. 1992). Thus Pol α-primase "counts" the length of RNA and DNA and stops synthesis. The underlying mechanism of how this enzyme functions has been revealed by a series of elegant crystal structures and biochemical studies (Coloma et al. 2016; Kilkenny et al. 2013; Klinge et al. 2009; Perera et al. 2013).

Crystal structures of the yeast Pol α catalytic core (349–1258; 910 amino acids) in the apo form, bound to an RNA/DNA primer template (the binary complex), and in a ternary complex with RNA/DNA primer template and incoming dGTP show that Pol α specifically recognizes the A-form RNA/DNA helix for extension with dNTPs (Perera et al. 2013). But once a full turn of double-helix DNA has been synthesized, Pol α is no longer in direct contact with the A-form RNA/DNA helix, which causes its release from DNA. A recent crystal structure of human Pol α-polymerase subunit in complex with a DNA/DNA helix supports this view, but interestingly, the DNA/DNA helix in contact with the polymerase is in a hybrid A-B form (Coloma et al. 2016). It was suggested that the free energy cost of distorting DNA from B- to the A-B hybrid, rather than the loss of specific contacts, drives the termination of primer synthesis.

Pol α-primase is a bilobed structure in which Pol1 binds to the B subunit and Pri1-Pri2 dimer through its CTD, and the CTD of Pol1 is connected to the polymerase region by a flexible stalk (Nunez-Ramirez et al. 2011). Pol α-primase requires CMG for priming activity on both the leading and lagging strands during unwinding of the forked DNA in the presence of RPA (Georgescu et al. 2015b). Pol α-primase is anchored to CMG helicase by the Ctf4 homotrimer (Gambus et al. 2009; Simon et al. 2014; Tanaka et al. 2009). The C-terminal domain of Ctf4 self-associates into a disk-shaped trimer that binds to a peptide sequence in Sld5, a subunit of the GINS complex within CMG (Simon et al. 2014). The same region of Ctf4 also binds to a homologous peptide sequence within the catalytic subunit of Pol α (Simon et al. 2014). Therefore, Ctf4 acts as a scaffold that cross-links CMG and Pol α-primase, underlying the mechanism by which Ctf4 recruits Pol α-primase to the replisome (Gambus et al. 2009). These physical interactions were visualized in a recent EM analysis of combinations of CMG with Pol ε, Pol α-primase, and Ctf4 (Fig. 10.7) (Sun et al. 2015). Unexpectedly, Pol α-primase and Ctf4 were found to bind to the NTD-tier of CMG, opposite from the CTD-tier where Pol ε binds. This organization was novel because all previous illustrations of the replisome placed the polymerases and primase on the same side of the helicase, after DNA unwinding.

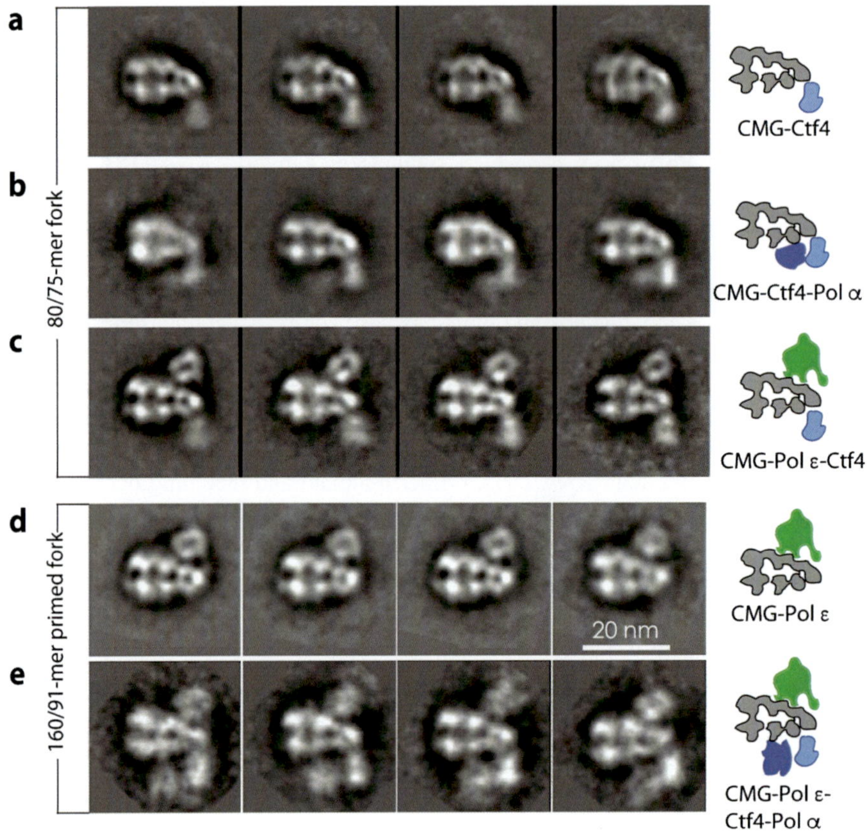

Fig. 10.7 Selected 2D class averages of several replisome complexes. (**a**) The Ctf4 trimer binds to the GINS from the bottom NTD side of CMG. (**b**) Pol α-primase appears as a fuzzy density suggesting a flexible association with the Ctf4 just under the NTD ring of Mcm2-7. (**c**) Simultaneous binding of Pol ε to the CTD side of Mcm2-7 on top and the Ctf4 trimer to the NTD side of CMG. (**d**) Pol ε binds to the CTD side of CMG in the presence of a primed forked DNA. (**e**) Pol ε binds to the CTD side of CMG, while Ctf4 and Pol α bind to the NTD side of CMG (This figure is modified with permission from Figure 5 of Sun et al. 2015)

10.9 Future Perspectives

We have reviewed here just some of the numerous recent structural advances in the field of replication that have revealed the structures of replicative polymerases, the 11 subunit CMG helicase, and the overall architecture of the eukaryotic replisome core. There are numerous important questions that remain to be answered, in fact too many to list. One important question for future studies is to understand the DNA path through the replisome and how CMG translocates along DNA. Available biochemical data indicate that CMG functions by steric exclusion in which both the CTD and NTD rings of Mcm2-7 encircle and translocate along one strand of

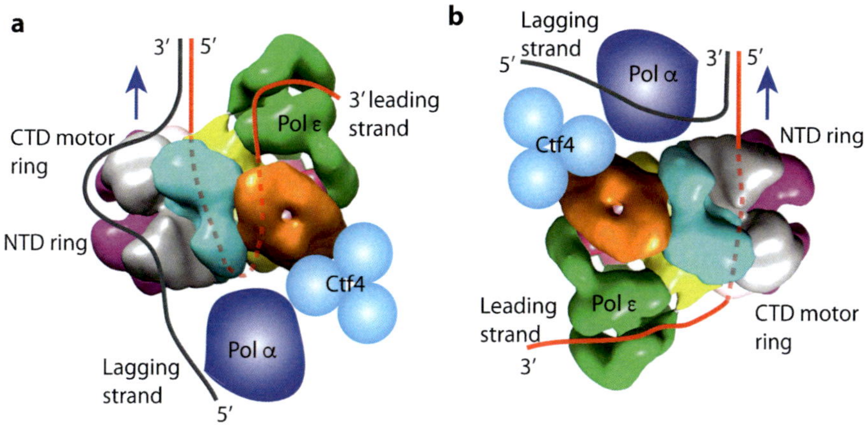

Fig. 10.8 The eukaryotic replisome positions Pols ε and α on opposite sides of CMG. Depending on whether the CTD motor ring or the NTD ring pushes against the forked junction, the polymerases are positioned either: (**a**) with Pol ε in front and Pol α-primase in back of CMG or (**b**) with Pol α-primase in front and Pol ε in back of CMG

ssDNA, excluding the other strand to the outside of the ring and thus separating the strands (Fu et al. 2011). However, the atomic structure of the Mcm2-7 double hexamer reveals a possible side channel for one DNA strand that lies between the CTD and NTD rings, suggesting the DNA split point may be internal to CMG (Li et al. 2015). Also, the proposed inchworm pump jack mechanism of CMG translocation (Yuan et al. 2016) is a new alternative to the rotary sequential ATP hydrolysis mechanism of hexameric helicase translocation (Enemark and Joshua-Tor 2006; Itsathitphaisarn et al. 2012; Thomsen and Berger 2009). However, there is still very little evidence to support either model. Given the knowledge that CMG helicase moves on the leading strand in a 3′–5′ direction (Bochman and Schwacha 2008; Kang et al. 2012; Moyer et al. 2006), that the 3′ terminus of ssDNA enters CMG through the CTD of Mcm2-7 (Costa et al. 2014), and that Pol ε binds to the CTD side of CMG (Sun et al. 2015), the leading strand DNA may need to travel a long and winding path before reaching Pol ε. These facts, taken together, suggest that the leading strand ssDNA threads through the CTD ring of Mcm2-7, exits the NTD ring of Mcm2-7, and then needs to make a U-turn to reach Pol ε at the CTD side of CMG (Fig. 10.8a). This DNA path assumes that the Mcm2-7 CTD motor ring translocates ahead of the NTD ring and thus pushes against the dsDNA/forked junction, as indicated by EM studies using a biotinylated DNA-streptavidin bound to *Drosophila* CMG (Costa et al. 2014). Furthermore, in the steric exclusion model of unwinding, the DNA split point would be just before DNA entry into CMG, and thus the newly unwound lagging strand ssDNA would need to traverse the outside surface of the Mcm2-7 ring to reach Pol α-primase on the NTD-tier side of CMG. Why nature would thread DNA through the replisome in this fashion is not understood, considering the extensive ssDNA exposed in this orientation. However, one may propose that this orientation enables Pol ε to be the first to interact with nucleosomes, as Pol ε is known to interact with nucleosomes (Foltman et al. 2013; Tackett et al. 2005) and to function in heterochromatin maintenance (Iida and Araki 2004; Tackett et al. 2005). On the other hand, it is interesting to note that BPV E1 helicase translocates on DNA such that the NTD ring faces the forked junction (Enemark and Joshua-Tor 2008; Lee et al. 2014). If the NTD ring of Mcm2-7 within CMG were to push against the forked junction, this orientation would place Pol α-primase at the top of the fork adjacent to the DNA split point, where Pol α-primase could prime the DNA template immediately after strand separation, and Pol ε would be behind CMG to extend the leading strand immediately after the DNA exits from the Mcm2-7 channel (Fig. 10.8b). We note that CMG helicase and the replisome architectures determined so far are based on in vitro assembled complexes on artificial substrates. It remains to be investigated if the in vivo assembled structures starting from origin DNA are the same especially in the threading of the DNA.

The DNA replication machinery must perform many additional tasks beyond DNA synthesis. For example, the replisome must deal with DNA lesions, transcribing RNA polymerase, DNA-bound proteins, recombination intermediates, nucleosomes in both euchromatin and heterochromatin, and cohesion rings that hold the sister chromosomes together after replication. Thus it may come as no surprise that there are numerous additional proteins that travel with or transiently interact with the replisome, and it is very likely that many new replisome interactive proteins will be identified in future studies. A first glimpse at the multitude of proteins that bind the replisome is provided by pull-outs using antibodies directed against CMG components followed by mass spectrometry to identify CMG-binding proteins (Gambus et al. 2006). This methodology identified a group of proteins referred to as the replisome progression complex (RPC) that contains CMG along with Ctf4, Mcm10, Tof1, Csm3, checkpoint mediator Mrc1, histone chaperone FACT (Spt16 and Pob3), and topoisomerase I (Gambus et al. 2006). The function of FACT in binding H2A/H2B and stimulating transcription through chromatin and of Topo I in relief of supercoiling is understood (Belotserkovskaya et al. 2003; Brill et al. 1987; Kim and Wang 1989; Orphanides et al. 1998; Orphanides et al. 1999; Schlesinger and Formosa 2000). However, most proteins of the RPC have nearly unknown functions. The Ctf4 trimer recognizes a peptide sequence to bind Pol α and Sld5 suggesting that Ctf4 acts as a platform for recruitment and exchange of different replication factors (Villa et al. 2016), in loose analogy to the many proteins that traffic on and off PCNA (Georgescu et al. 2015a). The Mrc1 checkpoint mediator is a stably attached replisome component that appears to aid normal replisome progression in addition to its checkpoint function (Lou et al. 2008), but how Mrc1 facilitates these activities and the spatial orientation of Mrc1 within the replisome are unknown. Tof1 and Csm3 are two proteins that ride with the replisome and appear to pause replication forks at protein-DNA barriers, but how they carry out this function is not understood (Calzada et al. 2005; Mohanty et al. 2006; Tourriere et al. 2005). Mcm10 is an essential replication factor that is reported to interact with ssDNA, dsDNA, Pol α-primase, the CMG helicase, and other factors (reviewed in Bielinsky 2016; Thu and Bielinsky 2013). Mcm10 appears to be the last initiation protein to act at an origin, because two complete CMGs form at an origin yet cannot produce ssDNA until Mcm10 is present (Heller et al. 2011; Kanke et al. 2012; Watase et al. 2012; Yeeles et al. 2015). How Mcm10 fulfills this initiation role and whether Mcm10 moves with the replisome are presently unknown. Besides the RPC proteins, the flap endonuclease 1 (FEN1) and DNA ligase both interact with PCNA and function with Pol δ/PCNA on the lagging strand to remove the RNA primer and join the Okazaki fragments, respectively. How the Pol δ lagging strand polymerase and the RFC clamp loader associate with the replisome, if at all, is another ongoing mystery. Finally, the replisome must replicate through every nucleosome in the genome, including tightly packed heterochromatin without disrupting the epigenetic state of the cell. The CMG, Pol ε, and Pol α all bind nucleosomes or histone subassemblies (Foltman et al. 2013; Huang et al. 2015; Tackett et al. 2005; Wang et al. 2015), yet how the replisome handles nucleosomes and passes them onto the daughter strands is largely unexplored territory. Answers to these many important questions will require a multitude of experimental approaches, but one can be quite certain that cryo-EM structure determination will be essential to understanding the complete picture of these very large and complicated multiprotein assemblies.

References

Abid Ali F, Renault L, Gannon J, Gahlon HL, Kotecha A, Zhou JC, Rueda D, Costa A (2016) Cryo-EM structures of the eukaryotic replicative helicase bound to a translocation substrate. Nat Commun 7:10708

Asturias FJ, Cheung IK, Sabouri N, Chilkova O, Wepplo D, Johansson E (2006) Structure of Saccharomyces cerevisiae DNA polymerase epsilon by cryo-electron microscopy. Nat Struct Mol Biol 13:35–43

Banks GR, Boezi JA, Lehman IR (1979) A high molecular weight DNA polymerase from Drosophila melanogaster embryos. Purification, structure, and partial characterization. J Biol Chem 254:9886–9892

Bell SD, Botchan MR (2013) The minichromosome maintenance replicative helicase. Cold Spring Harb Perspect Biol 5:a012807

Bell SP, Labib K (2016) Chromosome duplication in Saccharomyces cerevisiae. Genetics 203:1027–1067

Bell SP, Stillman B (1992) ATP-dependent recognition of eukaryotic origins of DNA replication by a multiprotein complex. Nature 357:128–134

Belotserkovskaya R, Oh S, Bondarenko VA, Orphanides G, Studitsky VM, Reinberg D (2003) FACT facilitates transcription-dependent nucleosome alteration. Science 301:1090–1093

Bielinsky AK (2016) Mcm10: the glue at replication forks. Cell Cycle 15:1–2

Bochman ML, Schwacha A (2008) The Mcm2-7 complex has in vitro helicase activity. Mol Cell 31:287–293

Bochman ML, Bell SP, Schwacha A (2008) Subunit organization of Mcm2-7 and the unequal role of active sites in ATP hydrolysis and viability. Mol Cell Biol 28:5865–5873

Boskovic J, Coloma J, Aparicio T, Zhou M, Robinson CV, Mendez J, Montoya G (2007) Molecular architecture of the human GINS complex. EMBO Rep 8:678–684

Brewster AS, Wang G, Yu X, Greenleaf WB, Carazo JM, Tjajadi M, Klein MG, Chen XS (2008) Crystal structure of a near-full-length archaeal MCM: functional insights for an AAA+ hexameric helicase. Proc Natl Acad Sci U S A 105:20191–20196

Brill SJ, DiNardo S, Voelkel-Meiman K, Sternglanz R (1987) Need for DNA topoisomerase activity as a swivel for DNA replication for transcription of ribosomal RNA. Nature 326:414–416
Bruck I, Kaplan DL (2013) Cdc45 protein-single-stranded DNA interaction is important for stalling the helicase during replication stress. J Biol Chem 288:7550–7563
Burgers PM (2009) Polymerase dynamics at the eukaryotic DNA replication fork. J Biol Chem 284:4041–4045
Burgers PM, Gordenin D, Kunkel TA (2016) Who is leading the replication fork, Pol epsilon or Pol delta? Mol Cell 61:492–493
Calzada A, Hodgson B, Kanemaki M, Bueno A, Labib K (2005) Molecular anatomy and regulation of a stable replisome at a paused eukaryotic DNA replication fork. Genes Dev 19:1905–1919
Chang YP, Wang G, Bermudez V, Hurwitz J, Chen XS (2007) Crystal structure of the GINS complex and functional insights into its role in DNA replication. Proc Natl Acad Sci U S A 104:12685–12690
Choi JM, Lim HS, Kim JJ, Song OK, Cho Y (2007) Crystal structure of the human GINS complex. Genes Dev 21:1316–1321
Clausen AR, Lujan SA, Burkholder AB, Orebaugh CD, Williams JS, Clausen MF, Malc EP, Mieczkowski PA, Fargo DC, Smith DJ et al (2015) Tracking replication enzymology in vivo by genome-wide mapping of ribonucleotide incorporation. Nat Struct Mol Biol 22:185–191
Cocker JH, Piatti S, Santocanale C, Nasmyth K, Diffley JF (1996) An essential role for the Cdc6 protein in forming the pre-replicative complexes of budding yeast. Nature 379:180–182
Coloma J, Johnson RE, Prakash L, Prakash S, Aggarwal AK (2016) Human DNA polymerase alpha in binary complex with a DNA:DNA template-primer. Sci Rep 6:23784
Conaway RC, Lehman IR (1982) A DNA primase activity associated with DNA polymerase alpha from Drosophila melanogaster embryos. Proc Natl Acad Sci U S A 79:2523–2527
Costa A, Ilves I, Tamberg N, Petojevic T, Nogales E, Botchan MR, Berger JM (2011) The structural basis for MCM2-7 helicase activation by GINS and Cdc45. Nat Struct Mol Biol 18:471–477
Costa A, Hood IV, Berger JM (2013) Mechanisms for initiating cellular DNA replication. Annu Rev Biochem 82:25–54
Costa A, Renault L, Swuec P, Petojevic T, Pesavento JJ, Ilves I, MacLellan-Gibson K, Fleck RA, Botchan MR, Berger JM (2014) DNA binding polarity, dimerization, and ATPase ring remodeling in the CMG helicase of the eukaryotic replisome. Elife 3:e03273
Duzdevich D, Warner MD, Ticau S, Ivica NA, Bell SP, Greene EC (2015) The dynamics of eukaryotic replication initiation: origin specificity, licensing, and firing at the single-molecule level. Mol Cell 58:483–494
Enemark EJ, Joshua-Tor L (2006) Mechanism of DNA translocation in a replicative hexameric helicase. Nature 442:270–275
Enemark EJ, Joshua-Tor L (2008) On helicases and other motor proteins. Curr Opin Struct Biol 18:243–257
Evrin C, Clarke P, Zech J, Lurz R, Sun J, Uhle S, Li H, Stillman B, Speck C (2009) A double-hexameric MCM2-7 complex is loaded onto origin DNA during licensing of eukaryotic DNA replication. Proc Natl Acad Sci U S A 106:20240–20245
Fenwick AL, Kliszczak M, Cooper F, Murray J, Sanchez-Pulido L, Twigg SR, Goriely A, McGowan SJ, Miller KA, Taylor IB et al (2016) Mutations in CDC45, encoding an essential component of the pre-initiation complex, cause Meier-Gorlin syndrome and Craniosynostosis. Am J Hum Genet 99:125–138
Foltman M, Evrin C, De Piccoli G, Jones RC, Edmondson RD, Katou Y, Nakato R, Shirahige K, Labib K (2013) Eukaryotic replisome components cooperate to process histones during chromosome replication. Cell Rep 3:892–904
Forterre P, Filee J, Myllykallio H (2004) Origin and evolution of DNA and DNA replication. In: Landes Bioscience, p 24
Fortune JM, Pavlov YI, Welch CM, Johansson E, Burgers PM, Kunkel TA (2005) Saccharomyces cerevisiae DNA polymerase delta: high fidelity for base substitutions but lower fidelity for single- and multi-base deletions. J Biol Chem 280:29980–29987

Froelich CA, Kang S, Epling LB, Bell SP, Enemark EJ (2014) A conserved MCM single-stranded DNA binding element is essential for replication initiation. Elife 3:e01993
Fu YV, Yardimci H, Long DT, Ho TV, Guainazzi A, Bermudez VP, Hurwitz J, van Oijen A, Scharer OD, Walter JC (2011) Selective bypass of a lagging strand roadblock by the eukaryotic replicative DNA helicase. Cell 146:931–941
Gambus A, Jones RC, Sanchez-Diaz A, Kanemaki M, van Deursen F, Edmondson RD, Labib K (2006) GINS maintains association of Cdc45 with MCM in replisome progression complexes at eukaryotic DNA replication forks. Nat Cell Biol 8:358–366
Gambus A, van Deursen F, Polychronopoulos D, Foltman M, Jones RC, Edmondson RD, Calzada A, Labib K (2009) A key role for Ctf4 in coupling the MCM2-7 helicase to DNA polymerase alpha within the eukaryotic replisome. EMBO J 28:2992–3004
Georgescu R, Langston L, O'Donnell M (2015a) A proposal: evolution of PCNA's role as a marker of newly replicated DNA. DNA Repair (Amst) 29:4–15
Georgescu RE, Schauer GD, Yao NY, Langston LD, Yurieva O, Zhang D, Finkelstein J, O'Donnell ME (2015b) Reconstitution of a eukaryotic replisome reveals suppression mechanisms that define leading/lagging strand operation. Elife 4:e04988
Heller RC, Kang S, Lam WM, Chen S, Chan CS, Bell SP (2011) Eukaryotic origin-dependent DNA replication in vitro reveals sequential action of DDK and S-CDK kinases. Cell 146:80–91
Hogg M, Johansson E (2012) DNA polymerase epsilon. Subcell Biochem 62:237–257
Hogg M, Osterman P, Bylund GO, Ganai RA, Lundstrom EB, Sauer-Eriksson AE, Johansson E (2014) Structural basis for processive DNA synthesis by yeast DNA polymerase varepsilon. Nat Struct Mol Biol 21:49–55
Hossain M, Stillman B (2016) Meier-Gorlin syndrome. In: Kaplan DL (ed) The initiation of DNA replication in eukaryotes. Springer International Publishing, Cham, p 563
Huang H, Stromme CB, Saredi G, Hodl M, Strandsby A, Gonzalez-Aguilera C, Chen S, Groth A, Patel DJ (2015) A unique binding mode enables MCM2 to chaperone histones H3-H4 at replication forks. Nat Struct Mol Biol 22:618–626
Iida T, Araki H (2004) Noncompetitive counteractions of DNA polymerase epsilon and ISW2/yCHRAC for epigenetic inheritance of telomere position effect in Saccharomyces cerevisiae. Mol Cell Biol 24:217–227
Ilves I, Petojevic T, Pesavento JJ, Botchan MR (2010) Activation of the MCM2-7 helicase by association with Cdc45 and GINS proteins. Mol Cell 37:247–258
Itsathitphaisarn O, Wing RA, Eliason WK, Wang J, Steitz TA (2012) The hexameric helicase DnaB adopts a nonplanar conformation during translocation. Cell 151:267–277
Johnson RE, Klassen R, Prakash L, Prakash S (2015) A major role of DNA polymerase delta in replication of both the leading and lagging DNA strands. Mol Cell 59:163–175
Kaguni LS, Rossignol JM, Conaway RC, Banks GR, Lehman IR (1983a) Association of DNA primase with the beta/gamma subunits of DNA polymerase alpha from Drosophila melanogaster embryos. J Biol Chem 258:9037–9039
Kaguni LS, Rossignol JM, Conaway RC, Lehman IR (1983b) Isolation of an intact DNA polymerase-primase from embryos of Drosophila melanogaster. Proc Natl Acad Sci U S A 80:2221–2225
Kamada K, Kubota Y, Arata T, Shindo Y, Hanaoka F (2007) Structure of the human GINS complex and its assembly and functional interface in replication initiation. Nat Struct Mol Biol 14:388–396
Kang YH, Galal WC, Farina A, Tappin I, Hurwitz J (2012) Properties of the human Cdc45/Mcm2-7/GINS helicase complex and its action with DNA polymerase epsilon in rolling circle DNA synthesis. Proc Natl Acad Sci U S A 109:6042–6047
Kanke M, Kodama Y, Takahashi TS, Nakagawa T, Masukata H (2012) Mcm10 plays an essential role in origin DNA unwinding after loading of the CMG components. EMBO J 31:2182–2194
Kilkenny ML, Longo MA, Perera RL, Pellegrini L (2013) Structures of human primase reveal design of nucleotide elongation site and mode of Pol alpha tethering. Proc Natl Acad Sci U S A 110:15961–15966

Kim RA, Wang JC (1989) Function of DNA topoisomerases as replication swivels in Saccharomyces cerevisiae. J Mol Biol 208:257–267
Klinge S, Nunez-Ramirez R, Llorca O, Pellegrini L (2009) 3D architecture of DNA Pol alpha reveals the functional core of multi-subunit replicative polymerases. EMBO J 28:1978–1987
Klingseisen A, Jackson AP (2011) Mechanisms and pathways of growth failure in primordial dwarfism. Genes Dev 25:2011–2024
Krastanova I, Sannino V, Amenitsch H, Gileadi O, Pisani FM, Onesti S (2012) Structural and functional insights into the DNA replication factor Cdc45 reveal an evolutionary relationship to the DHH family of phosphoesterases. J Biol Chem 287:4121–4128
Kubota Y, Takase Y, Komori Y, Hashimoto Y, Arata T, Kamimura Y, Araki H, Takisawa H (2003) A novel ring-like complex of Xenopus proteins essential for the initiation of DNA replication. Genes Dev 17:1141–1152
Kunkel TA, Burgers PM (2008) Dividing the workload at a eukaryotic replication fork. Trends Cell Biol 18:521–527
Langston LD, Zhang D, Yurieva O, Georgescu RE, Finkelstein J, Yao NY, Indiani C, O'Donnell ME (2014) CMG helicase and DNA polymerase epsilon form a functional 15-subunit holoenzyme for eukaryotic leading-strand DNA replication. Proc Natl Acad Sci U S A 111:15390–15395
Lee SJ, Syed S, Enemark EJ, Schuck S, Stenlund A, Ha T, Joshua-Tor L (2014) Dynamic look at DNA unwinding by a replicative helicase. Proc Natl Acad Sci U S A 111:E827–E835
Leipe DD, Aravind L, Koonin EV (1999) Did DNA replication evolve twice independently? Nucleic Acids Res 27:3389–3401
Leman AR, Noguchi E (2013) The replication fork: understanding the eukaryotic replication machinery and the challenges to genome duplication. Genes (Basel) 4:1–32
Li Y, Araki H (2013) Loading and activation of DNA replicative helicases: the key step of initiation of DNA replication. Genes Cells 18:266–277
Li N, Zhai Y, Zhang Y, Li W, Yang M, Lei J, Tye BK, Gao N (2015) Structure of the eukaryotic MCM complex at 3.8 A. Nature 524:186–191
Liang C, Weinreich M, Stillman B (1995) ORC and Cdc6p interact and determine the frequency of initiation of DNA replication in the genome. Cell 81:667–676
Lou H, Komata M, Katou Y, Guan Z, Reis CC, Budd M, Shirahige K, Campbell JL (2008) Mrc1 and DNA polymerase epsilon function together in linking DNA replication and the S phase checkpoint. Mol Cell 32:106–117
MacNeill S (2012) Composition and dynamics of the eukaryotic replisome: a brief overview. Subcell Biochem 62:1–17
Meselson M, Stahl FW (1958) The replication of DNA in Escherichia Coli. Proc Natl Acad Sci U S A 44:671–682
Miller JM, Arachea BT, Epling LB, Enemark EJ (2014) Analysis of the crystal structure of an active MCM hexamer. Elife 3:e03433
Mimura S, Seki T, Tanaka S, Diffley JF (2004) Phosphorylation-dependent binding of mitotic cyclins to Cdc6 contributes to DNA replication control. Nature 431:1118–1123
Miyabe I, Kunkel TA, Carr AM (2011) The major roles of DNA polymerases epsilon and delta at the eukaryotic replication fork are evolutionarily conserved. PLoS Genet 7:e1002407
Mohanty BK, Bairwa NK, Bastia D (2006) The Tof1p-Csm3p protein complex counteracts the Rrm3p helicase to control replication termination of Saccharomyces cerevisiae. Proc Natl Acad Sci U S A 103:897–902
Morrison A, Araki H, Clark AB, Hamatake RK, Sugino A (1990) A third essential DNA polymerase in S. cerevisiae. Cell 62:1143–1151
Moyer SE, Lewis PW, Botchan MR (2006) Isolation of the Cdc45/Mcm2-7/GINS (CMG) complex, a candidate for the eukaryotic DNA replication fork helicase. Proc Natl Acad Sci U S A 103:10236–10241
Nethanel T, Zlotkin T, Kaufmann G (1992) Assembly of simian virus 40 Okazaki pieces from DNA primers is reversibly arrested by ATP depletion. J Virol 66:6634–6640

Nick McElhinny SA, Gordenin DA, Stith CM, Burgers PM, Kunkel TA (2008) Division of labor at the eukaryotic replication fork. Mol Cell 30:137–144

Nishitani H, Lygerou Z, Nishimoto T, Nurse P (2000) The Cdt1 protein is required to license DNA for replication in fission yeast. Nature 404:625–628

Nunez-Ramirez R, Klinge S, Sauguet L, Melero R, Recuero-Checa MA, Kilkenny M, Perera RL, Garcia-Alvarez B, Hall RJ, Nogales E et al (2011) Flexible tethering of primase and DNA Pol alpha in the eukaryotic primosome. Nucleic Acids Res 39:8187–8199

O'Donnell M, Langston L, Stillman B (2013) Principles and concepts of DNA replication in bacteria, archaea, and eukarya. Cold Spring Harb Perspect Biol 5

Onesti S, MacNeill SA (2013) Structure and evolutionary origins of the CMG complex. Chromosoma 122:47–53

Orphanides G, LeRoy G, Chang CH, Luse DS, Reinberg D (1998) FACT, a factor that facilitates transcript elongation through nucleosomes. Cell 92:105–116

Orphanides G, Wu WH, Lane WS, Hampsey M, Reinberg D (1999) The chromatin-specific transcription elongation factor FACT comprises human SPT16 and SSRP1 proteins. Nature 400:284–288

Perera RL, Torella R, Klinge S, Kilkenny ML, Maman JD, Pellegrini L (2013) Mechanism for priming DNA synthesis by yeast DNA polymerase alpha. Elife 2:e00482

Petojevic T, Pesavento JJ, Costa A, Liang J, Wang Z, Berger JM, Botchan MR (2015) Cdc45 (cell division cycle protein 45) guards the gate of the Eukaryote replisome helicase stabilizing leading strand engagement. Proc Natl Acad Sci U S A 112:E249–E258

Pursell ZF, Kunkel TA (2008) DNA polymerase epsilon: a polymerase of unusual size (and complexity). Prog Nucleic Acid Res Mol Biol 82:101–145

Pursell ZF, Isoz I, Lundstrom EB, Johansson E, Kunkel TA (2007) Yeast DNA polymerase epsilon participates in leading-strand DNA replication. Science 317:127–130

Remus D, Beuron F, Tolun G, Griffith JD, Morris EP, Diffley JF (2009) Concerted loading of Mcm2-7 double hexamers around DNA during DNA replication origin licensing. Cell 139:719–730

Sanchez-Pulido L, Ponting CP (2011) Cdc45: the missing RecJ ortholog in eukaryotes? Bioinformatics 27:1885–1888

Santocanale C, Diffley JF (1996) ORC- and Cdc6-dependent complexes at active and inactive chromosomal replication origins in Saccharomyces cerevisiae. EMBO J 15:6671–6679

Schlesinger MB, Formosa T (2000) POB3 is required for both transcription and replication in the yeast Saccharomyces cerevisiae. Genetics 155:1593–1606

Schwacha A, Bell SP (2001) Interactions between two catalytically distinct MCM subgroups are essential for coordinated ATP hydrolysis and DNA replication. Mol Cell 8:1093–1104

Simon AC, Zhou JC, Perera RL, van Deursen F, Evrin C, Ivanova ME, Kilkenny ML, Renault L, Kjaer S, Matak-Vinkovic D et al (2014) A Ctf4 trimer couples the CMG helicase to DNA polymerase alpha in the eukaryotic replisome. Nature 510:293–297

Simon AC, Sannino V, Costanzo V, Pellegrini L (2016) Structure of human Cdc45 and implications for CMG helicase function. Nat Commun 7:11638

Singleton MR, Sawaya MR, Ellenberger T, Wigley DB (2000) Crystal structure of T7 gene 4 ring helicase indicates a mechanism for sequential hydrolysis of nucleotides. Cell 101:589–600

Slaymaker IM, Chen XS (2012) MCM structure and mechanics: what we have learned from archaeal MCM. Subcell Biochem 62:89–111

Steitz TA (1999) DNA polymerases: structural diversity and common mechanisms. J Biol Chem 274:17395–17398

Sun J, Evrin C, Samel SA, Fernandez-Cid A, Riera A, Kawakami H, Stillman B, Speck C, Li H (2013) Cryo-EM structure of a helicase loading intermediate containing ORC-Cdc6-Cdt1-MCM2-7 bound to DNA. Nat Struct Mol Biol 20:944–951

Sun J, Fernandez-Cid A, Riera A, Tognetti S, Yuan Z, Stillman B, Speck C, Li H (2014) Structural and mechanistic insights into Mcm2-7 double-hexamer assembly and function. Genes Dev 28:2291–2303

Sun J, Shi Y, Georgescu RE, Yuan Z, Chait BT, Li H, O'Donnell ME (2015) The architecture of a eukaryotic replisome. Nat Struct Mol Biol 22:976–982

Tackett AJ, Dilworth DJ, Davey MJ, O'Donnell M, Aitchison JD, Rout MP, Chait BT (2005) Proteomic and genomic characterization of chromatin complexes at a boundary. J Cell Biol 169:35–47

Tahirov TH, Makarova KS, Rogozin IB, Pavlov YI, Koonin EV (2009) Evolution of DNA polymerases: an inactivated polymerase-exonuclease module in Pol epsilon and a chimeric origin of eukaryotic polymerases from two classes of archaeal ancestors. Biol Direct 4:11

Tanaka S, Araki H (2013) Helicase activation and establishment of replication forks at chromosomal origins of replication. Cold Spring Harb Perspect Biol 5:a010371

Tanaka S, Diffley JF (2002) Interdependent nuclear accumulation of budding yeast Cdt1 and Mcm2-7 during G1 phase. Nat Cell Biol 4:198–207

Tanaka H, Katou Y, Yagura M, Saitoh K, Itoh T, Araki H, Bando M, Shirahige K (2009) Ctf4 coordinates the progression of helicase and DNA polymerase alpha. Genes Cells 14:807–820

Thomsen ND, Berger JM (2009) Running in reverse: the structural basis for translocation polarity in hexameric helicases. Cell 139:523–534

Thu YM, Bielinsky AK (2013) Enigmatic roles of Mcm10 in DNA replication. Trends Biochem Sci 38:184–194

Ticau S, Friedman LJ, Ivica NA, Gelles J, Bell SP (2015) Single-molecule studies of origin licensing reveal mechanisms ensuring bidirectional helicase loading. Cell 161:513–525

Tognetti S, Riera A, Speck C (2015) Switch on the engine: how the eukaryotic replicative helicase MCM2-7 becomes activated. Chromosoma 124:13–26

Tourriere H, Versini G, Cordon-Preciado V, Alabert C, Pasero P (2005) Mrc1 and Tof1 promote replication fork progression and recovery independently of Rad53. Mol Cell 19:699–706

Villa F, Simon AC, Ortiz Bazan MA, Kilkenny ML, Wirthensohn D, Wightman M, Matak-Vinkovic D, Pellegrini L, Labib K (2016) Ctf4 is a hub in the eukaryotic replisome that links multiple CIP-box proteins to the CMG helicase. Mol Cell 63:385–396

Wang H, Wang M, Yang N, Xu RM (2015) Structure of the quaternary complex of histone H3-H4 heterodimer with chaperone ASF1 and the replicative helicase subunit MCM2. Protein Cell 6:693–697

Watase G, Takisawa H, Kanemaki MT (2012) Mcm10 plays a role in functioning of the eukaryotic replicative DNA helicase, Cdc45-Mcm-GINS. Curr Biol 22:343–349

Watson JD, Crick FH (1953a) Genetical implications of the structure of deoxyribonucleic acid. Nature 171:964–967

Watson JD, Crick FH (1953b) Molecular structure of nucleic acids; a structure for deoxyribose nucleic acid. Nature 171:737–738

Yeeles JT, Deegan TD, Janska A, Early A, Diffley JF (2015) Regulated eukaryotic DNA replication origin firing with purified proteins. Nature 519:431–435

Yuan Z, Bai L, Sun J, Georgescu R, Liu J, O'Donnell ME, Li H (2016) Structure of the eukaryotic replicative CMG helicase suggests a pumpjack motion for translocation. Nat Struct Mol Biol 23:217–224

Zhang D, O'Donnell M (2016) The eukaryotic replication machine. Enzyme 39:191–229

Chapter 11
Replication Domains: Genome Compartmentalization into Functional Replication Units

Peiyao A. Zhao, Juan Carlos Rivera-Mulia, and David M. Gilbert

Abstract DNA replication occurs in a defined temporal order during S phase, known as the replication timing programme, which is regulated not only during the cell cycle but also during the process of development and differentiation. The units of replication timing regulation, known as replication domains (RDs), frequently comprise several nearly synchronously firing replication origins. Replication domains correspond to topologically associating domains (TADs) mapped by chromatin conformation capture methods and are likely to be the molecular equivalents of replication foci observed using cytogenetic methods. Both TAD and replication foci are considered to be stable structural units of chromosomes, conserved through the cell cycle and development, and accordingly, the boundaries of RDs also appear to be stable in different cell types. During both normal development and progression of disease, distinct cell states are characterized by unique replication timing signatures, with approximately half of genomic RDs switching replication timing between these cell states. Advances in functional genomics provide hope that we can soon gain an understanding of the cause and consequence of the replication timing programme and its myriad correlations with chromatin context and gene regulation.

Keywords Replication timing (RT) • Replication foci • Replication domain (RD) • Topologically associating domain (TAD) • Chromatin • Nucleus • Cell cycle

---

P.A. Zhao • J.C. Rivera-Mulia
Department of Biological Science, Florida State University, Tallahassee, FL, USA

D.M. Gilbert (✉)
Department of Biological Science, Florida State University, Tallahassee, FL, USA

Center for Genomics and Personalized Medicine, Florida State University, Tallahassee, FL, USA
e-mail: gilbert@bio.fsu.edu

H. Masai, M. Foiani (eds.), *DNA Replication*, Advances in Experimental Medicine and Biology 1042, https://doi.org/10.1007/978-981-10-6955-0_11

11.1 Introduction

The spatial organization of the genome is critical for proper cellular physiology; multiple nuclear processes and genome functions are linked to chromosome architecture (Cavalli and Misteli 2013; Cremer and Cremer 2010; Pombo and Dillon 2015). Initial cytogenetic characterization of genome organization unveiled the non-random arrangement of chromosomes into territories and revealed replication foci as stable chromosomal structural units or domains that arrange into distinct early and late replicating subnuclear compartments (Cremer et al. 2006; Dimitrova and Gilbert 1999; Ferreira et al. 1997; Jackson and Pombo 1998; Sparvoli et al. 1994). The advent of genome-wide methods enabled replication timing to be measured comprehensively in many different cell types and the identification of units of replication timing regulation, termed 'replication domains' (RDs), many of which could be seen to change replication timing during cell fate transitions. Recently developed chromatin conformation capture methods (e.g. Hi-C) that map long-range chromatin interactions confirmed the partitioning of the genome into distinct compartments within the nucleus that correspond to early and late replication timing compartments and also confirmed the existence of stable units of chromosome structure termed topologically associating domains (TADs) that correspond to RDs (Dixon et al. 2012; Lieberman-Aiden et al. 2009; Pope et al. 2014). Distinct genomic features are associated with early and late replicating compartments, such as GC content, gene density and transcriptional activity. Moreover, differences in chromosomal interactions have been observed between distinct cell types, and individual domains can be repositioned during development between the nuclear interior and periphery correlated with dynamic changes in their replication timing and transcriptional activation/repression (Hiratani et al. 2010; Takebayashi et al. 2012). Here we discuss historical and emerging evidences placing RDs in the context of global chromatin organization and providing a better understanding of the structural and functional properties and biological significance of nuclear organization.

11.2 Replication Foci and the Spatial Organization of the Genome

Duplication of the genome occurs in a defined temporal order known as the replication timing (RT) programme (Rhind and Gilbert 2013). Early cytogenetic studies of DNA replication that visualized DNA synthesis by incubating S phase mammalian cells with nucleotide homologues such as BrdU revealed distinct punctate structures in the nucleus termed replication foci that were found in different spatial patterns at distinct times during S phase (Nakamura et al. 1986; Nakayasu and Berezney 1989). Early in S phase, many small replication foci are associated with euchromatic regions interspersed throughout the nucleoplasm, while as S phase progresses, DNA synthesis moves to the heterochromatin-rich regions at nuclear and nucleolar

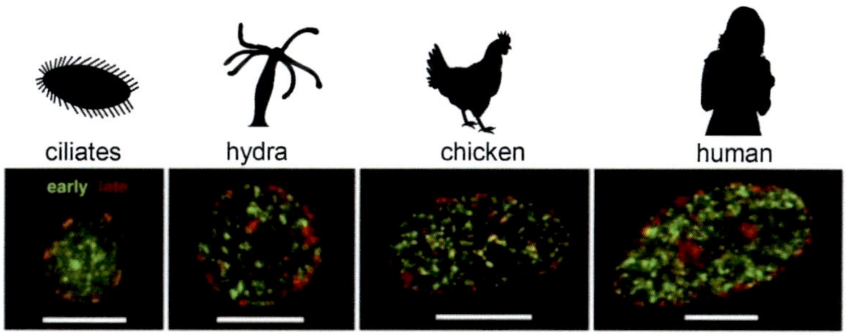

Fig. 11.1 Evolutionary conservation of the spatio-temporal RT programme (Adapted from Solovei et al. 2016). Cells pulse labelled with CldU, chased for approximately half of S phase, and then pulse labelled with IdU, followed by fixation and immunostaining for CldU (*green*) and IdU (*red*). Patterns are remarkably similar in a wide variety of species

peripheries forming what appeared under conventional light microscopy to be larger foci. Remarkably, this spatial segregation of chromatin into early and late replication foci is conserved in all eukaryotes (Fig. 11.1). Replication foci were estimated to contain 0.5–1 Mb of DNA and consist of clusters of replicons that replicate synchronously (Ma et al. 1998). Replication foci also co-localize with replication proteins, either by immunofluorescence or when visualized in living cells using GFP-tagged replication proteins, such as proliferating cell nuclear antigen (PCNA) and replication protein A (RPA) (Sporbert et al. 2002). In living cells, replication proteins can be seen to move from recently completed to spatially adjacent foci (Maya-Mendoza et al. 2010; Sporbert et al. 2002). Pulse-chase-pulse experiments in which cells were labelled with two different nucleotides separated by a variable chase time allowed for the study of the kinetics of replication foci patterns in different stages of S phase. With short chase times, the number of replication foci was found to remain constant and the two labels co-localized, while chase times of 60 minutes or more resulted in complete separation of the two pulse labels, suggesting that the replication of each focus was completed within that time (Dimitrova and Gilbert 1999; Jackson and Pombo 1998). Chase times of several hours resulted in differential labeling of the spatial patterns of replication (Fig. 11.1). Replication foci were further studied by chasing labelled cells for many generations, revealing that each focus retained its shape, size and intensity of label, demonstrating that the DNA that is synthesized together stays together as a structural and functional unit of chromosomes (Berezney et al. 2000; Dimitrova and Gilbert 1999; Ferreira et al. 1997; Jackson and Pombo 1998; Ma et al. 1998; Sadoni et al. 2004; Sparvoli et al. 1994). When spatially adjacent foci were labelled sequentially with different nucleotide analogues separated by an ~60 min chase and then further chased for multiple generations, spatially linked sequentially replicating foci remained adjacent demonstrating that they were genetically linked (Maya-Mendoza et al. 2012; Sporbert et al. 2002). Thus, DNA replication in large genomes is regulated both temporally and spatially.

11.3 Replication Origin Control at a Glance

Several mechanisms and cell cycle checkpoints are in place to ensure proper regulation of replication origin firing under normal vs. replication stress conditions (Fig. 11.2a). First, in telophase/early G1 phase, the pre-replication complex (pre-RC) is assembled at many potential origins under conditions that strictly prevent initiation in a process known as 'origin licencing' (Fragkos et al. 2015). Origin licencing initiates with the recruitment of the origin recognition complex (ORC; consisting of six subunits, ORC1–ORC6) followed by cell division cycle 6 (CDC6) and cdc10-dependent transcript 1 (CDT1), which assist the loading of the MCM helicase complex (MCM2–MCM7) (Deegan and Diffley 2016; Hyrien 2015). Then, at the G1-S phase transition, formation and activation of the pre-initiation complex (pre-IC) are triggered by DBF4-dependent kinase (DDK) and cyclin-dependent kinases (CDKs) (Fragkos et al. 2015). Upon activation, the MCM2-7 double hexamers separate and nucleate the replisome at each replication fork (Deegan and Diffley 2016). Activation of helicases to initiate DNA replication can only occur under conditions of high CDK and DDK, which are also conditions that strictly prevent the assembly of pre-RCs (Blow and Dutta 2005). This two-cycle engine, the temporal uncoupling of origin licencing and activation, is a crucial feature of eukaryotic replication, restricting replication to once-per-cell-cycle regardless of where or when replication initiates.

Only a small fraction of the pre-RCs that were assembled in telophase/early G1 get activated during S phase. The origins that are licenced without firing during one cell cycle are known as 'dormant' origins (Fragkos et al. 2015). It is speculated that assembly of an excess of pre-RCs and activation of a small fraction of potential origins might be a mechanism to ensure that genome is fully replicated, in such a way that in the case of fork stalling during conditions of replication stress, dormant origins fire to complete genome duplication within the immediate vicinity of the clusters of activated origins to rapidly complete replication in already initiated domains (Rivera-Mulia and Gilbert 2016b). At the same time, DNA damage response kinases, e.g. ATR and CHK1, suppress the formation of initiation in unreplicated RDs, restricting new initiation to the firing of only the dormant origins in the vicinity of the existing replication forks (Donzelli and Draetta 2003). In this manner, the cells are able to reduce the number of active replication forks rapidly to minimize reactive dangerous replication forks under conditions of replication stress (Rivera-Mulia and Gilbert 2016a, b).

11.4 Replication Origin Activation Within Replication Foci

The mechanism by which origins fire in clusters within foci remains unclear (Fig. 11.2b). The global RT programme can be seen as being determined at both the level of chromosomal structure when the RT programme is set up through elaborate

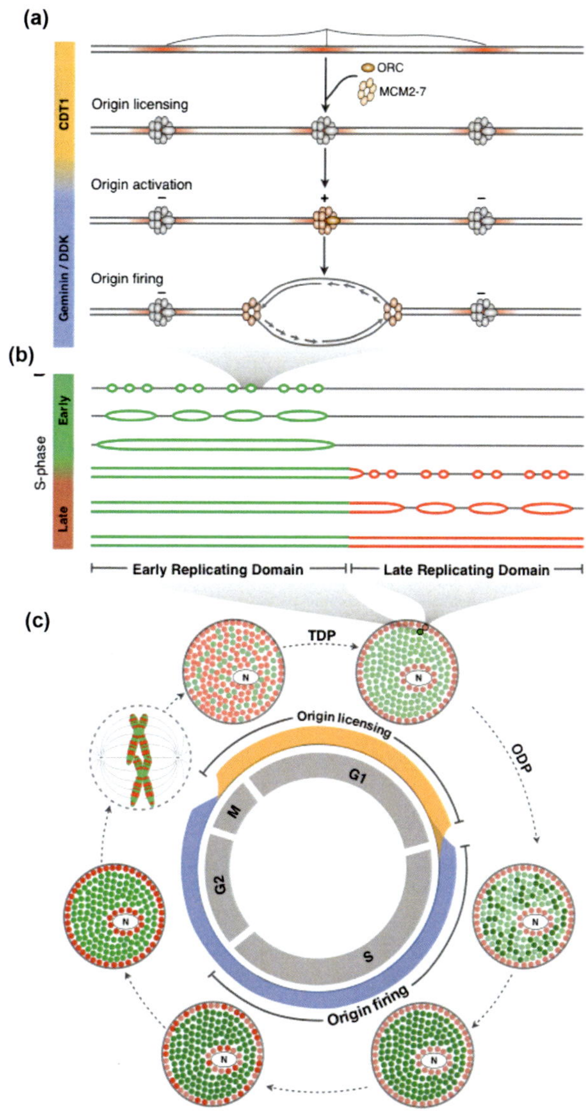

Fig. 11.2 Replication is regulated at scales from origins to domains (Adapted from Rivera-Mulia and Gilbert 2016a). (**a**) Replication origin regulation. The ORC and other contributing proteins load the MCM2-7 double hexamer at origins in a step called origin licencing. A subset of licenced origins are subsequently activated, and origin firing is initiated by the bidirectional unwinding of the DNA strands by MCM2-7 double hexamers (see chapter by Araki in this issue). (**b**) Synchronized firing of clustered origins to form early and late replicating domains. (**c**) DNA replication and its regulation through the cell cycle. Origin licencing occurs from M through most of G1. The timing decision point (TDP) occurs early during G1 and precedes the origin decision point (ODP). RDs are compartmentalized in such a way that early RDs are located at the interior of the nucleus and late RDs are located at the nuclear and nucleolar periphery. Timing determinants are lost in G2 phase and regained at the TDP

mechanisms to commit origins to their characteristic firing times and the activation of individual origins at such times (Rivera-Mulia and Gilbert 2016b). Genome-wide determination of the RT programme will be discussed below. Here we discuss two alternative high-level views of origin regulation (Rhind et al. 2010). A deterministic model assumes a predetermined RT programme in which each origin is programmed to fire at specific times. The model is supported by some studies carried out in *S. cerevisiae*, which feature well-defined, site-specific origins not typically observed in metazoans (Ferguson et al. 1991; Newlon et al. 1991). A stochastic model posits that different origins have different probabilities of firing, giving rise to heterogeneous firing times for individual origins in a population of cells (Kaykov and Nurse 2015; Patel et al. 2006; Saner et al. 2013). Mathematical modelling of replication kinetics has shown that a stochastic firing model fits well with current experimental observations in various systems (Goldar et al. 2008, 2016; Lob et al. 2016; Norio et al. 2005; Yang et al. 2010). Moreover, DNA combing, a technique that allows the investigation of origin firing at the level of single DNA molecules (Kaykov and Nurse 2015) has experimentally demonstrated genome-wide stochastic origin firing in yeast. In other words, normally, early-firing origins have a chance to fire late or not at all, while normally late-firing origins occasionally fire early.

Similar single DNA molecule experiments on a whole genome are presently lacking for mammalian cells. Nonetheless, DNA fibre analyses carried out on the mouse *Igh locus* show different origins utilized on different DNA fibres, consistent with the stochastic usage vs. non-usage of origins in any given cell cycle but not sufficient evidence to demonstrate stochastic temporal firing (Demczuk et al. 2012). Additionally, studies in *Xenopus* egg extracts and mammalian cells have shown that the number of dormant origins firing can be modulated by manipulating the levels of cell cycle checkpoint proteins or depleting the concentration of MCM2-7 (Ge and Blow 2010; Woodward et al. 2006). It is important to emphasize that stochastic firing of origins is premised on a number of criteria regarding origin firing probability. One such criteria is an increase in firing probability as S phase progresses invoked in stochastic firing models to ensure that origins with relatively low firing probability can eventually fire later in S phase (Rhind et al. 2010). Several mechanisms accounting for the increased firing probability have been proposed (Rhind and Gilbert 2013), but all current models leave us with many unanswered questions: do active replication forks induce the firing of origins in their proximity? To what extent is the induction necessary for the timely completion of S phase? What factors regulate the initiation rate in various model systems? More data such as high-resolution genome-wide RT profiles and reliable maps of all potential origins and their frequencies of firing are needed to further refine the model.

There has been a long-standing correlation between transcription and origin selection and activation. However, the studies carried out so far are characterized by contradicting evidence, with some showing an enrichment of origins in active transcription start sites (TSSs) and CpG islands and others observing no significant overlap (Besnard et al. 2012; Karnani et al. 2010; Prioleau 2009; Sequeira-Mendes et al. 2009). The disparity between studies is primarily due to the lack of agreement in origin features and locations between the origin mapping techniques used or the cell types under study (Hyrien 2015). Using nascent strand purification followed by microarray to map replication initiation sites in mouse embryonic stem cells (ESCs) (Sequeira-Mendes et al. 2009), a strong correlation between transcription initiation activity and origin firing efficiency was found with origins associated with CpG islands firing the most efficiently. On the other hand, replication origin sites identified through bubble-seq had surprisingly little overlap between the majority of origins and transcribed genes (Mesner et al. 2013). The lack of agreement between techniques used for origin mapping could potentially be due to heterogeneity in the molecular properties of the origin population that is detected using different methods. For this reason, the most parsimonious conclusion is that a positive correlation between transcription activity and early origin firing is restricted to a subset of origins.

Numerous additional genetic and epigenetic features correlate with origin licencing and firing. Attempts have been made to put origins into distinct classes based on their chromatin association and underlying genetic sequences (Cayrou et al. 2015). The structure known as a G-quadruplex (G4), which is a four-stranded nucleic acid structure with bonds formed between guanines, is a conserved feature that is associated with highly efficient origins in metazoan (Besnard et al. 2012; Cayrou et al. 2012; Valton et al. 2014). There is also evidence suggesting overrepresentation of the association of G4 structures with origins due to the inherent biases of the methods used for origin mapping (Foulk et al. 2015). The exact mechanisms of how G4s regulate origin firing remain unknown. Moreover, despite the strong link, the overlap between G4 structures and activated origins is far from perfect, and G4 structures cannot predict the locations of origin firing, suggesting the involvement of other origin specificity factors. In addition to underlying sequence features, nucleosomal organization near origins has been shown to influence both origin licencing and helicase activation steps of initiation (Azmi et al. 2017). Certain histone modifications and histone modifiers have been shown to be enriched near origins and be implicated in origin selection and activation (Feng et al. 2016; Liu et al. 2012; Pourkarimi et al. 2016; Rondinelli et al. 2015; Tardat et al. 2010; Wu et al. 2017). Acetylation of histones H3 and H4, the modification that is generally associated with open chromatin, has been shown in *Drosophila* to associate with active origins, and surprisingly acetylation of certain lysines is also dependent on ORC binding to the origins (Liu et al. 2012). Dynamic changes in H4K20me1 through the cell cycle have been shown to regulate origin licencing (Tardat et al. 2010). Early and late origins were also found to associate with different histone marks. Demethylation of H3K4me3 was shown to promote early origin firing, whereas the repressive chromatin mark H3K9me3 was found associated with late replicating origins (Rondinelli et al. 2015; Wang et al. 2016).

Unfortunately, the universal abundance of histone modifications across the genome makes any one modification unlikely to be sufficient for dictating origin selection and activation. Moreover, the intimate link between epigenetic landscape and transcription potential of the region further complicates investigation into the role of histone marks in origin regulation. There has been evidence suggesting cell cycle-dependent changes of histone modifications at one origin resulting in coordinated regulation of origin firing and transcription (Kylie et al. 2016). Several transcription factors have also been shown to have the function as origin regulators with effects that can be either origin firing promoting or inhibiting (Bellelli et al. 2014; Knott et al. 2012; Ostrow et al. 2017). Altogether, evidence argues against a single genetic or epigenetic factor being necessary or sufficient for origin regulation. More likely complex combinations of genetic and chromatin features dictate the probability of origin firing. Genome-wide studies correlating dynamic changes in epigenetic landscape with cell cycle-dependent changes in transcription and origin regulation are needed to further evaluate the complex relationship between chromatin modifications, transcription and origin regulation. Most importantly, as discussed above, comprehensive and reliable lists of all origin sites and their probabilities of firing would be transformative for this field. Unfortunately, probabilities are only obtained through single-molecule studies, which to date have been low throughput.

11.5 Replication Factories

The spatial organization of replicons within replication foci has been the subject of much debate. An early electron microscopic study (Hozak et al. 1993) described ovoid-shaped bodies observed under an electron microscope that are ~150 nm in diameter. DNA synthesis first appeared within these 'replication factories' and after a brief chase was seen to extrude from them, leading to a model of a fixed and pre-existing protein complex, through which DNA replication forks are spooled during DNA synthesis. It had been proposed that neighbouring replicons are organized in close physical proximity in these factories with intermediate DNA sequences looped out forming rosette-like synthetic centres to facilitate synchronous firing (Vogelstein et al. 1980). Using time-lapse microscopy, Kitamura et al. (2006) and Meister et al. (2007) have visualized clusters of replication forks being co-ordinately processed during replication in live *S. cerevisiae* and *S. pombe*, respectively, supporting the model of replication factory. More recently, super resolution methods such as 3D-structured illumination microscopy (3D-SIM) and stimulated emission depletion (STED) can visualize structures 30 nm and smaller, providing the optical resolution to distinguish single replicons within foci in mammalian cells (Baddeley et al. 2010; Chagin et al. 2016). A recent study (Chagin et al. 2016) observed ~5000 simultaneously active replicons in human cells, finding that sites of labelling observed within the lower-resolution replication foci were clearly spatially separated, casting doubt on the 'replication factory' model of replicon clustering (Chagin et al. 2016; Lob et al. 2016). By simultaneously imaging PCNA replication foci and the chromosomal regions undergoing replication (Deng et al. 2016), DNA was observed to exit replication foci during replication and return without disturbing higher-order chromatin organization. Taken together, current evidence does not support the existence of fixed pre-existing replication factories but rather supports a model in which independent replicons within each focus are spatially clustered by virtue of their chromatin loop and higher-order chromatin organization.

11.6 Replication Domains and Large-Scale Chromosome Organization

Further investigation into the molecular properties of replication foci has been greatly facilitated by the development of methods to map RT on a genome-wide scale permitting genomic map coordinates to be placed into the context of cytogenetic findings. Several related methods to measure global RT have been developed (Gilbert 1986, 2010; Gilbert and Cohen 1987; Hiratani et al. 2008; Schubeler et al. 2002; Woodfine et al. 2004). RT profiles partition the genome into chromosome segments of uniform replication timing, known as constant timing regions (CTRs), punctuated by timing transition regions (TTRs). CTRs correspond to several tandem RDs that replicate at similar times, and so their boundaries cannot be detected because forks emanating from each RD fuse quickly. RDs have the potential to be independently regulated during development, and in cell types in which they replicate at different times, RD boundaries become revealed as TTRs (Pope et al. 2014; Rhind and Gilbert 2013). Genome-wide studies showed that these dynamic changes in RT occur at the level of sub-chromosomal units of 400–800 kb and that the boundaries of these units were at the same chromosomal map units regardless of the cell type in which they were observed to replicate discordantly with their neighbours. In fact, during cell fate commitment, a process called domain consolidation was found to occur where an increasing number of adjacent RDs replicate synchronously to form larger CTRs that are also consolidated spatially into the same subnuclear compartment coinciding with their re-localization within the nucleus (Fig. 11.3a) (Hiratani et al. 2008; Rivera-Mulia et al. 2015; Yue et al. 2014). From these observations emerged the concept of the replication domain model, that is, independent structural and functional units of chromosomes (RDs) that coalesce to form the variably sized CTRs observed in replication timing datasets (Fig. 11.3b). Although it has not been formally demonstrated that RDs are the molecular equivalents of the replication foci discussed above, their size and replication kinetics account for the observed behaviour of replication foci, and the fact that the boundaries of RDs are conserved in different cell types attests to their stability. Hence, genome-wide RT mapping has revealed RDs as a basic chromatin unit and provides the opportunity to correlate their structure to other chromosomal properties.

The advent of chromatin conformation capture-based methods such as Hi-C provided the opportunity to relate RDs to the chromatin context and genome coordinates. Hi-C identified stable structural units of chromosomes termed topologically self-associating domains (TADs) and detected the folding of TADs into multi-TAD compartments of active/open (A-compartments) or inactive/closed (B-compartments) chromatin. The comparison of RT profiles and Hi-C A/B compartments demonstrated that Hi-C confirms the physical compartmentalization of the genome revealed by prior replication labelling findings (discussed above), with A-compartments representing the early replicating internally localized foci and B-compartments replicating the late replicating and peripherally localized foci (Fig. 11.4a) (Dixon et al. 2012; Moindrot et al. 2012; Ryba et al. 2010; Yaffe et al. 2010). By mapping RT in

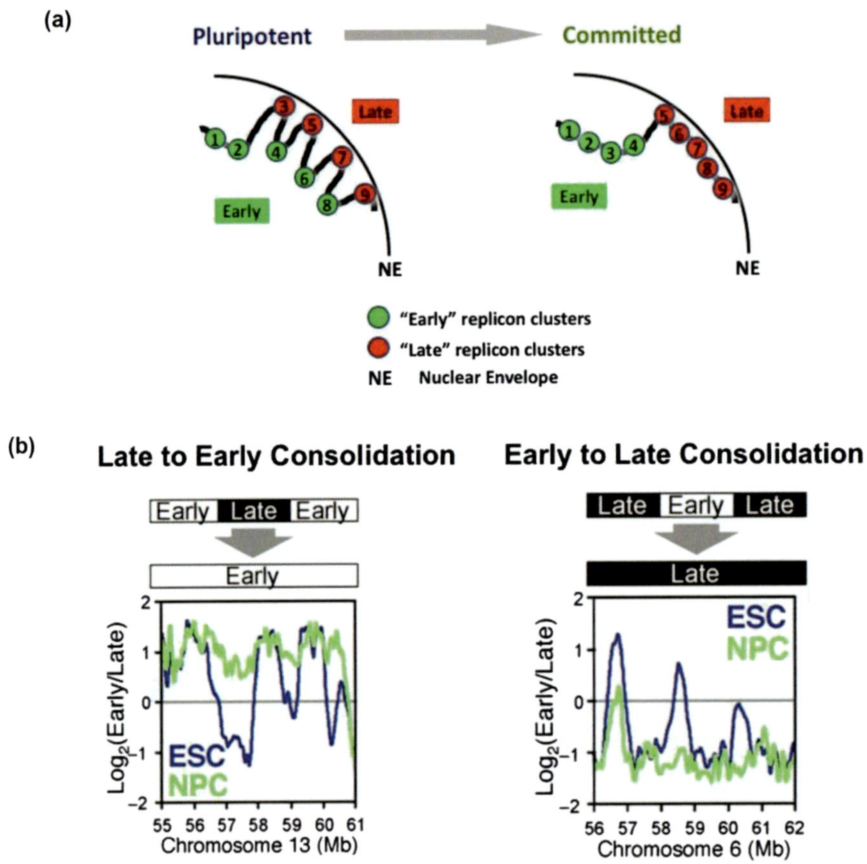

Fig. 11.3 Consolidation of RDs upon differentiation from ESCs to neural progenitor cells (NPCs) (Adapted from Hiratani et al. 2008). (**a**) A model of domain reorganization during neural differentiation. The replicon clusters that replicate differentially in ESCs undergo spatial rearrangement and consolidate into larger co-ordinately replicating domains. (**b**) Exemplary late to early and early to late domain consolidation during differentiation

multiple cell types, the boundaries between RDs were determined and found to align to TAD boundaries, demonstrating that the units of RT regulation correspond to TADs (Fig. 11.4b) (Pope et al. 2014). Additionally, changes in RT during cellular differentiation were found to coincide with changes in inter-TAD interactions, with long-range chromatin interactions between RD/TAD shifting their RT such that they interact with other domains that replicate at similar times (Takebayashi et al. 2012). Together, evidence supports a unified model for large-scale chromosome organization known as the 'Replication Domain model' (Fig. 11.4b).

To better understand causal linkages in this intriguing relationship between genome organization (structure) and RT (function), new approaches to study large-scale chromosome architecture are necessary. One way to investigate causality has

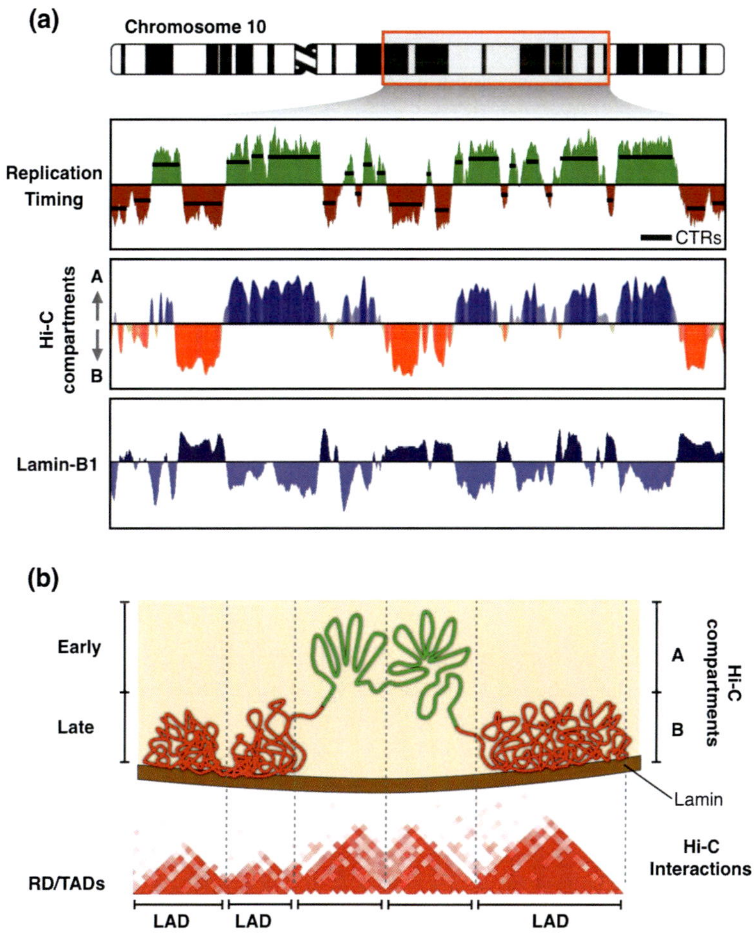

Fig. 11.4 Replication timing reflects genome compartmentalization (Adapted from Rivera-Mulia and Gilbert 2016b). (**a**) Exemplary RT profile of 50 Mb of human Chr10 from IMR90 fibroblasts. The RT profile aligns with Hi-C A/B interaction compartments, and late replicating domains align with lamina-associated domains (LADs). Segments of chromosomes consisting of multiple RDs that replicate at similar times are marked with *horizontal lines* and are known as CTRs. (**b**) The Replication Domain Model: RDs are the equivalent of TADs, which fold in 3D space such that TADs in close proximity replicate at similar times

been to relate the formation of TADs and 3D compartment through the cell cycle to the establishment of RT. It has been shown that 3D structure is dismantled during mitosis, but both TAD structure and interaction compartments are re-established coincidentally with the re-establishment of RT in a very narrow time window during G1 referred to as the timing decision point (TDP) (Fig. 11.2c) (Dileep et al. 2015b). Interestingly, this same study found that, during G2 phase when chromatin has lost properties necessary to dictate a proper RT programme (Lu et al. 2010), TAD structure and inter-TAD interactions are retained, demonstrating that this 3D organization is not sufficient for RT but may provide a scaffold on which other factors (such as Rif1; see chapter by Buonomo in this issue) can operate to define RT. Thanks to genome engineering tools such as CRISPR, we are now in a position to manipulate domain structure and determine the cis-elements necessary for autonomous regulation of TAD RT and eventually the contribution of the 3D chromosome scaffold to the establishment of RT.

In addition to TADs, RT also correlates with lamina-associated domains (LADs), which range in size from 10 kb to 100Mb and form physical contact with the nuclear periphery (Fig. 11.4a) (Amendola and van Steensel 2014; Pope et al. 2014). Consistent with the sequence dependence of RT with late RDs being mostly A/T rich and located at the nuclear periphery, LADs, which are also characterized by their high A/T content and high isochore concordance, are found to coincide with late RDs and timing transition regions (TTRs) that are located between early and late RDs (Meuleman et al. 2013; Pope et al. 2014). Hence, late domains and TTRs both contain origins with a low probability of firing and are enriched in LADs, and LADs tend to extend right to the boundaries of the locally earliest replicating RDs, which project into the nuclear interior. A subset of LADs known as facultative LADs exhibit cell type-specific lamina association, and their relocation towards or away from the nuclear lamina is often correlated with the transcription activation and repression of the genes contained within the domain (Kosak et al. 2002; Meuleman et al. 2013; Peric-Hupkes et al. 2010; Robson et al. 2016). Given the close correspondence between LADs and late RDs, the developmental regulation that facultative LADs are subjected to is often reflected in a switch in RT of the domain during cell differentiation in that the detachment of the domain from the nuclear lamina is often accompanied by a late to early switch in RT (Hiratani et al. 2008). Taken together the evidences suggest that RD and LAD organization are closely linked and RT is a genomic feature that reflects changes in chromatin organization through development.

RDs also share some properties with the sequence-based features of chromosomes known as isochores. DNA isochores are defined as homogenous DNA stretches of similar sequence composition. By assessing the segments' GC content, the human genome was found to be composed of ~3200 isochores divided into five families of increasing GC content (Costantini et al. 2007). Replication origins have been shown to enrich around GC-rich regions such as CpG islands and exhibit early firing in high origin density regions (Cayrou et al. 2011). RT programmes are found to correlate with isochore GC content, LINE density and gene density (Fig. 11.5a) (Costantini and Bernardi 2008; Woodfine et al. 2004). The sequence-based isochore features, however, are not sufficient to dictate RT as developmentally controlled regions residing in AT-rich/LINE-rich isochores have been shown to change their RT during ESC to NPC differentiation (Hiratani et al. 2004, 2008). Developmentally regulated RDs often exhibit an inverse correlation between GC content and gene density. We speculate that isochores that exhibit intermediate or mixed sequence features are more prone to change RT during development due to opposing effects from the underlying isochore sequence and gene density, possibly due to transcription or the regulatory features associated with transcription. It is clear that combina-

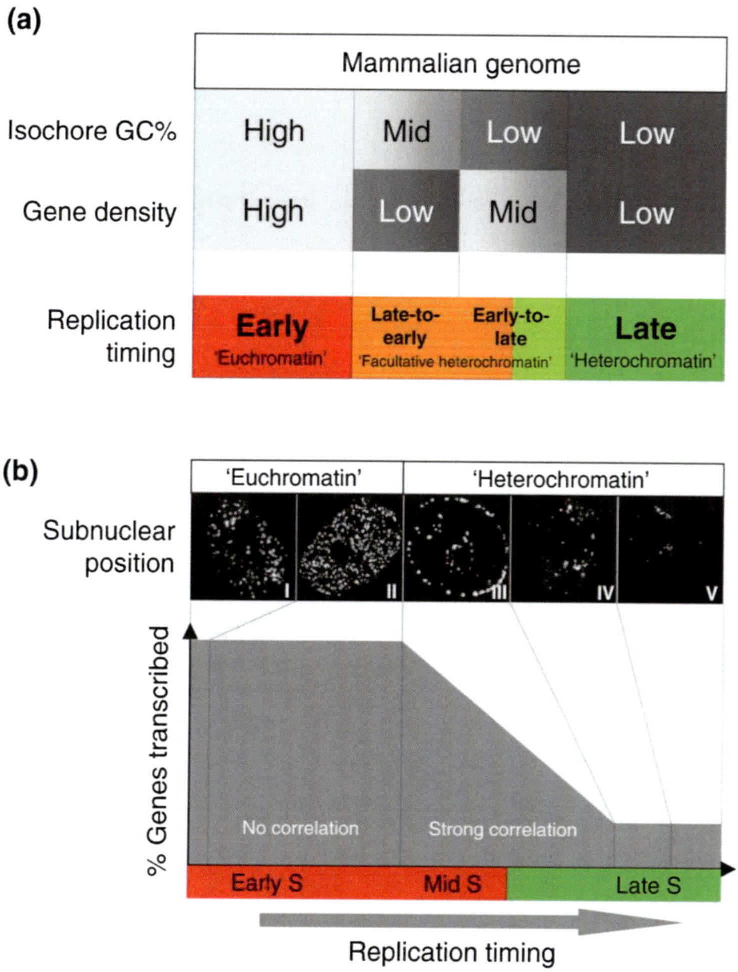

Fig. 11.5 Relationship between RT regulation and isochore properties, subnuclear position and transcription (Adapted from Hiratani et al. 2009). (**a**) RT is correlated with properties of isochores. Mammalian genomes are composed of isochores of different GC content and gene density. Isochores that are high in both GC content and gene density are replicated early in S phase. Isochores that are low in both GC content and gene density are replicated late in S phase. Isochores that have intermediate GC content and gene density are subjected to developmentally regulated RT switches. (**b**) Changes in RT that traverse the middle to late S phase accompany changes in subnuclear position and transcriptional potential. Replication early in S phase (I and II) takes place at the euchromatin compartment located at the interior of the nucleus. Replication in mid- and late S phase (III, IV and V) takes place at the heterochromatin compartment located at the nuclear periphery (III), the nucleolar periphery (III) and interior heterochromatin (IV and V). Genes that are replicated in early S phase (I and II) have equally high probability of being expressed ('no correlation'); therefore, changes in RT for genes in this category are not necessarily accompanied by changes in nuclear position or transcriptional potential. Genes that are replicated in mid-late S phase exhibit a strong correlation between RT and transcription activity

tions of several factors mediate domain-level regulation of RT (discussed below). Isochores implicate the genetic sequence composition, LINE and gene density in RT determination.

11.7 Replication Domains, Gene Regulation and Chromatin Context

A correlation between RT, gene expression and nuclear positioning has been known for decades (Gilbert 2002; Goldman et al. 1984; Hatton et al. 1988). However, mechanisms and causal relationships remain elusive, and recent results suggest that there are likely to be many layers of context-specific complexity. Early studies found that transcriptionally silent heterochromatin generally replicates late during S phase and locates near the nuclear periphery, while active genes tend to replicate early and are preferentially located at the nuclear interior (Baer 1965; Himes 1967; Klevecz and Stubblefield 1967; Lima-De-Faria 1959; Pfeiffer 1968; Taylor 1960). Dynamic changes in RT are accompanied by subnuclear repositioning, further highlighting the importance of nuclear localization to RT (Hiratani et al. 2008). For instance, in the case of *Mash1*, a proneural regulator gene, it was shown that in the process of mouse ESC differentiation to neural cells, the gene switches from late replicating to early replicating accompanied by transcription activation and relocation from the periphery to the interior of the nucleus (Williams et al. 2006). Classical genetic characterization of position-effect variegation in fly demonstrated that changing the chromatin context of a gene influences its transcriptional activity, while studies in mammals showed that the insertion site of a transgene can influence the replication time of the inserted sequence (Elgin and Reuter 2013; Gilbert and Cohen 1990). The most recent development of the high-throughput approach known as thousands of reporters integrated in parallel (TRiP) in chromosomal loci with various chromatin context has further elucidated position effects on a genome-wide scale, and when performed in combination with RT profiling, the technique will provide us with a unique opportunity to investigate the relationship between transcription, chromatin context and RT in a high-throughput manner (Akhtar et al. 2013). Another example that illustrates the relationship between RT and chromatin context is provided by early studies on genomic imprinting in mammalian cells. Genomic imprinting is associated with silencing and delayed replication of the imprinted allele, and early RT is often seen on the monoallelically expressed genes (Farago et al. 2012; Kitsberg et al. 1993; Mostoslavsky et al. 2001; Simon et al. 1999). X chromosome inactivation in female mammals provides another example. The inactive chromosome X (Xi) in female mammals is tightly organized at the nuclear periphery in a nuclear compartment known as the Barr body and replicates late, while the expressed chromosome (Xa) is more interiorly located and replicates earlier than the Xi (Avner and Heard 2001; Koren et al. 2014). Moreover, when autosomes are translocated to the Xi, they become late replicating and genetically inactive suggesting that late replication might lead to stable gene silencing (Hiratani et al. 2010; Keohane et al. 1996). Together these results suggest that the temporal order of replication of chromosomal segments is reflective of transcriptional potential, chromatin compaction and subnuclear position.

If replication timing is causally linked to transcriptional potential, it should also be subject to developmental control. Indeed, the temporal order of replication of chromosomal segments is reflective of cell type-specific chromatin organization and changes co-ordinately with differentiation state during development. Early studies of individual gene RT concluded that constitutively active genes replicate early while tissue-specific genes replicate early only in the cell type in which they are expressed (Ermakova et al. 1999; Gilbert 1986; Goldman et al. 1984; Guinta and Korn 1986; Holmquist 1987). These observations suggested that there might be coordinated changes in gene expression and the temporal order of replication of tissue-specific genes, but as these studies had compared established and often genetically unstable cell lines, they could not distinguish developmental control from genetic aberrations or epigenetic adaptations to cell culture. More recent studies with embryonic stem cells identified dynamically regulated RT switches with late to early switches coordinated with transcriptional activation (Hiratani et al. 2004; Perry et al. 2004). Later, genome-wide studies revealed that dynamic late to early or early to late RT changes are coordinated across hundreds of kb, often closely coincident with transcriptional activation or silencing of genes within the affected domain and the repositioning to the nuclear interior or periphery, respectively (Hiratani et al. 2008). Consistent with the global spatial patterns of replication discussed above, changes in subnuclear position only occurred in cases where the RT change traversed mid-late S phase, which was also the period when RT and transcription have the strongest correlation (Fig. 11.5b) (Hiratani et al. 2008). Ironically, these studies also revealed the existence of a small subset of genes that were transcribed when late replicating or even some that were activated when switching from early to late replication. These genes may be equivalent to genes identified in *Drosophila* that require heterochromatin to be expressed (Yasuhara et al. 2005). A more recent study showed that a strong transactivator targeted to an endogenous promoter to induce transcription resulted in both repositioning to the nuclear interior and a shift from late to early RT (Therizols et al. 2014). This observation led these authors to conclude that transcription was sufficient to advance the RT of a chromosomal domain. However, since there are natural examples of transcription within late replicating domains, this observation remains to be independently confirmed and the conclusions substantiated in a generalized manner.

Taken together, the experimental evidence linking RT, gene expression and genome organization has led to the construction of two alternative but not mutually exclusive models as to how they might be related. In chromatin biology, there is rarely a unidirectional cause and effect but more typically a series of positive feedback loops that reinforce a transcriptional state, and the links between transcription and RT appear to be no exception. First, based on studies in yeasts (see chapter by Masukata in this issue), it is clear that chromatin can influence the timing of replication. Unfortunately, homologues of most of these regulators have little effect on RT in mammalian cells (Dileep et al. 2015a). Moreover, replication initiation zones are enriched for DNase I hypersensitive sites (Petryk et al. 2016), and computational model building has revealed DNase I hypersensitive sites to be the best predictor of RT (Gindin et al. 2014). Since changes in chromatin states are associated with transcriptional activity and the act of transcription can cause distinct components of chromatin to be disassembled, changes in transcriptional regulatory networks associated with cell fate changes and transcription itself are both consistent with an indirect relationship, with perhaps chromatin structure regulating RT (Belotserkovskaya et al. 2003; Bender and Fitzgerald 2002; Chakalova et al. 2005; Hogga and Karch 2002; Kireeva et al. 2002; Schmitt et al. 2005; Schwartz and Ahmad 2005). The alternative model is that RT controls chromatin structure. Since chromatin assembly takes place at the replication fork, it represents a window of opportunity to remodel chromatin structure across the entire chromosome domains by replicating DNA in a different time and place in the nucleus (Gilbert 2002, 2010). In fact, an artificial system in which bovine papilloma virus vectors replicate at different times in subsequent S phases has demonstrated that changes in RT are accompanied by the assembly of different chromatin composition, with earlier replication associated with a higher level of histone acetylation (Lande-Diner et al. 2009; Zhang et al. 2002). While direct evidence for both of these models is still wanting, the two models together would predict a strong positive feedback loop to maintain stable epigenetic states within a cell type and to rapidly alter those states under the conditions of changing transcriptional regulatory networks that occur during cell fate changes.

The results discussed above summarized decades of literature leading to a long-established dogma that early RT is correlated to transcriptional activity. However, recent studies have cast doubt even on this seemingly incontrovertible relationship (Mesner et al. 2013; Rivera-Mulia et al. 2015). In Rivera-Mulia et al. (2015), the authors generated RT profiles for 26 human cell types representing various intermediate stages of development and correlated RT with transcriptome data. The domains with developmentally regulated RT often lacked or exhibited weaker associations of RT to chromatin structure and transcription as compared to RT constitutive domains. Moreover, approximately two thirds of the genes that switch RT were found to be strongly expressed when late replicating in one or more cell types investigated. Despite their ability to be transcribed at high levels when late replicating, transcription of these genes was still coordinated with RT, being upregulated just prior to a late to early switch and downregulated just after an early to late switch. Such observations argue against a model in which active transcription is sufficient for early replication. Moreover, studies in *S. cerevisiae* have found that Fkh1/2-dependent origin and RT regulation are independent from the proteins' role as transcription factors but rather dependent on their structural roles in origin clustering and spatial organization of chromosomal elements, which is likely mediated through dimerization of Fkh1 and Fkh2 (Knott et al. 2012; Ostrow et al. 2017). Collectively, the evidence suggests that RT is related to features of chromatin, including modifications and nuclear architecture, rather than transcription per se. The complexity of behaviour and chromatin marks present on these domains suggests that different domains are likely to be regulated by distinct combinations of mechanisms resulting in an intricate context-dependent interplay between factors such as chromatin modifications, accessibility and 3D conformation.

The connection between RT and chromatin modifications has been characterized extensively, and despite many correlations, no single modification has been found to significantly correlate it with RT on a genome-wide scale in mammalian cells (Petryk et al. 2016; Pope et al. 2014; Ryba et al. 2010). In addition, knockdown of histone-modifying enzymes resulted in no significant differences in global RT compared to wild-type control, suggesting the lack of causality between any single type of chromatin modification and global RT establishment (Pope et al. 2014; Ryba et al. 2010). In *Drosophila* cells the chromatin modification H3K16ac and the histone modifier heterochromatin protein 1 (HP1) have been shown to be implicated in genome-wide RT regulation suggesting potential fundamental differences in mechanisms underlying RT regulation by chromatin modifications between *Drosophila* and mammalian cells (Lubelsky et al. 2014; Schwaiger et al. 2009, 2010). However, at local level, there have been studies suggesting causal links between histone modifications and RT highlighting the possibility of isolated mechanisms at play in specific regions in addition to (or somehow translating into) domain-level regulation in mammalian system (Casas-Delucchi et al. 2011; Goren et al. 2008). These evidences taken together further highlight the context-dependent correlation between RT and chromatin modifications.

The replication domain model posits that replication begins within TADs that reside in the nuclear interior that are characterized by generally high transcriptional potential. Meanwhile, replication gradually advances into adjacent late replicating TADs that are located at the nuclear periphery (LADs) or other repressive compartments and contain features associated with repressed transcription. Upon differentiation, TADs that switch RT acquire features associated with their new nuclear localization, while new compartment boundaries are established with new interactions. Despite the many aspects of the model that require further experimental data to validate, it fits well with existing data that largely argues for a domain-wide RT regulation that is not solely reliant on transcription potential of the loci.

There is some evidence that proteins involved in controlling higher-order chromosome architecture may be important for RT as well. For instance, cohesin, a ring-shaped protein complex, has been shown to be necessary for the formation of large chromatin loops and may be able to limit the number of active origins and slow down S phase progression as a result (Guillou et al. 2010). However, it is difficult to evaluate the involvement of cohesion in DNA replication since cohesion depletion leads to severe cell cycle defects that confound the interpretation of results collected from cycling cells. CTCF is also implicated in the formation of chromatin loops, but loss of CTCF has no effect on RT (Sima and Gilbert, unpublished observations). One protein that may control chromatin architecture and RT is the Rap1-interacting factor (Rif1). (see chapter by Buonomo in this issue) Rif1 is a conserved telomere-binding factor that co-localizes with the mid-S replication foci and may participate in the formation of chromatin loop structures (Foti et al. 2016). Depletion of Rif1 results in loss of mid-S replication foci and enhancement of initiation events in early S phase and disruption of RT across a large portion of the genome (Cornacchia et al. 2012; Yamazaki et al. 2012). Together with the results showing that 3D conformation of chromatin is established coincidentally with the establishment of an RT programme (Dimitrova and Gilbert 1999; Dileep et al. 2015a, b), results suggest RT is regulated by 3D chromatin conformation, but direct evidence and mechanistic insight are lacking.

11.8 Reshaping the Epigenome

DNA methylation, histone modifications and chromatin accessibility constitute the epigenomic landscape, which is critical for cell identity maintenance, gene regulation and proper cellular function. Hence, epigenetic information must be also faithfully transmitted after each cell division to maintain cell identity (Alabert and Groth 2012). Recently, genomic studies have witnessed great progress in the characterization of the epigenome in multiple cell types and its dynamic changes during cell fate specification (Gifford et al. 2013; Roadmap Epigenomics et al. 2015; Xie et al. 2013). However, how the epigenome is inherited and the remodelling of chromatin structure at replication forks has been a long-standing mystery (see chapter by Groth in this issue). During DNA replication, chromatin must be disassembled to allow the replication fork to progress and needs to be reassembled onto the duplicated strands (Ramachandran and Henikoff 2015). Moreover, essential replication proteins at the replication fork are directly involved in recycling of parental histones (Huang et al. 2015). Indeed, distinct types of chromatin are assembled at different times during S phase (Alabert et al. 2015; Lande-Diner et al. 2009), supporting the model that RT can shape the epigenome. By characterizing proteins associated with newly replicated DNA, it has been shown that levels of histone modifications decrease by ~50% on newly synthesized strands, due to a combination of recycled and new unmodified histones. These modifications are restored at different rates thereafter, with some requiring continuous remodelling of both parental and new histones across multiple cell cycles (Alabert et al. 2015). A question going forward will be how the epigenome landscape is remodelled as a result of DNA replication during the process of differentiation.

11.9 Altered RD Organization and Disease

RT is an exceptionally stable epigenetic property that is not easily disturbed by loss or misexpression of single master regulators of cell fate or chromatin regulatory factors (Dileep et al. 2015a; Pope et al. 2011) Nonetheless, every cell fate change has been associated with specific changes in RT. Since the RT profile is such a stable epigenetic characteristic of cell type, it stands to reason that features of the RT programme (i.e. RT of specific domains) would be altered in diseases including cancer. Indeed, a range of genetic disorders have been shown to be associated with aberrant

RT programmes. For instance, in fragile X syndrome, a genetic disorder commonly associated with an affected *FMR1* allele containing a hypermethylated CpG island and expanded CGG repeats, the mutant *FMR1* allele was found to replicate later the unaffected allele (Hansen et al. 1993). On the other hand, ICF immunodeficiency syndrome results from abnormal escape of X chromosome inactivation via hypomethylation and was shown to be dependent on a temporally advanced RT (Hansen et al. 2000). In addition, abnormal regulation of RT has been observed in every cancer investigated to date (Amiel et al. 1998a, b, 1999; Korenstein-Ilan et al. 2002). Several studies indicate malignancies are accompanied by disruption of the temporal order of allelic replication; most regions of the genome normally replicate synchronously, but some have been found to replicate asynchronously in malignant cells and associated with poor prognosis, suggesting that RT might be an attractive biomarker for diagnostic purposes (Donley and Thayer 2013). Moreover, a series of studies linking dysregulated RT with increased mutagenesis and chromosomal instability led to speculation that RT programmes may function as the driving force of mutagenesis during tumorigenesis and late replicating domains have a higher mutagenesis rate than early replicating domains (Lang and Murray 2011; Stamatoyannopoulos et al. 2009). The different mutation rates and patterns associated with early and late replicating regions are likely due to the differential DNA mismatch repair (MMR) efficiency at the regions (Koren et al. 2012; Liu et al. 2013; Supek and Lehner 2015). A chromosome-wide replication delay, which has been observed in many cancer types, was also shown to be associated with a range of tumorigenic abnormalities including delays in mitotic condensation, gain or loss of chromosomes and increased rate of chromosomal rearrangements (Breger et al. 2005; Chang et al. 2007; Smith et al. 2001). Alternatively, RT changes can also be induced by chromosomal rearrangements as a result of juxtaposing regions of differential RT (Breger et al. 2004; Pope et al. 2012; Ryba et al. 2012). Naturally, the interdependent nature of RT changes and chromosomal instability during tumorigenesis has made it a formidable task to delineate cause and consequence. However, these results strongly suggest that alterations in RT have implications in tumorigenesis and the development of various diseases.

The only systematic genome-wide analysis of RT in cancers to date explored paediatric B-cell precursor acute lymphoblastic leukaemia (BCP-ALL) (Ryba et al. 2012). This is an excellent model cancer due to the well-studied genetic subtypes linked to prognosis. The RT profiles of leukaemia exhibited more heterogeneity than non-leukaemic B-cell controls, suggesting either variable arrested/altered differentiation states of the different samples or a generalized epigenetic dysregulation. Several lines of evidence suggested that this heterogeneity was due to altered developmental programmes. First, using a previously developed algorithm to identify the statistically most significant RT differences between samples, the authors identified 'RT fingerprints' unique to genetic subtypes of BCP-ALL, suggesting that at least some of the fingerprints were causally linked to the particular subtype. Secondly, the differences that distinguished leukaemia aligned to the boundaries of developmentally regulated RT domains that distinguish normal cell types further supporting abnormal developmental control. Finally, in a follow-up study, BCP-

ALL samples were passaged serially in immunodeficient mice as patient-derived xenografts and the RT profiles were found to be stable characteristics of patient samples, demonstrating that RT profiling can reveal stable epigenetic features of specific cancer clones (Sasaki et al. 2017). Further research will be needed to reveal whether such features can function as prognostic markers and their mechanistic links to cancer phenotypes. Since different types of chromatin are assembled at replication forks that are active at different stages during S phase (discussed above), aberrations in RT likely contribute to large-scale changes in chromatin structure and epigenetic states that persist in malignant cells thereby having profound effects on their transcriptional programmes. These findings provide strong impetus for further exploration of the mechanisms by which RT aberrations arise and its role in determining the biological features of human cancer.

11.10 Conclusion and Perspectives

From cytogenetic observations of replication foci to genome-wide RT mapping and the discovery of an uncanny alignment between RD and TAD boundaries, the last decade has witnessed a conceptual leap in the linkage between RT and chromatin 3D structure. In what has come to be known as the 'Replication Domain model', RT is regulated at the level of TADs, within which 1–4 replication origins are selected stochastically from many potential sites to replicate the RD/TAD within 45–60 min, while at a higher scale, TADs are organized during nuclear reassembly in such a way that RDs/TADs in close proximity replicate at similar times to form distinct spatially separated nuclear compartments. RDs/TADs are stable chromatin units with profound implications on important aspects of chromosomal properties including transcription regulation (Fig. 11.6). Still, we are a long way from a complete understanding of how RT is regulated and its biological significance, particularly in organisms with large genomes. 3D organization is tightly associated with RT but is not sufficient to dictate the RT programme (Dileep et al. 2015b). Although most individual chromatin readers, writers and erasers are not necessary for the RT programme, there are certainly critical players such as Rif1 (Buonomo chapter in this issue) that interact with this scaffold and contribute to the creation of subnuclear domains of differing initiation potential, and these players must be identified. We need better and more reliable origin-mapping tools, particularly those that can measure single-molecule initiation events and single-cell RT so that the stochastic probabilities of origin firing and domain-scale RT can be related to the single-cell transcription and chromatin state measurements being currently developed (Liu and Trapnell 2016; Ramani et al. 2017). We need better means to unravel the enigmatic relationship between RT and transcription, which seems to correlate well in some contexts and not at all in others. Finally, we need to understand the mechanisms linking higher-order chromosome architecture and RT. The advent of novel robust gene editing tools should permit a genetic dissection of the structure and RT of chromosome domains.

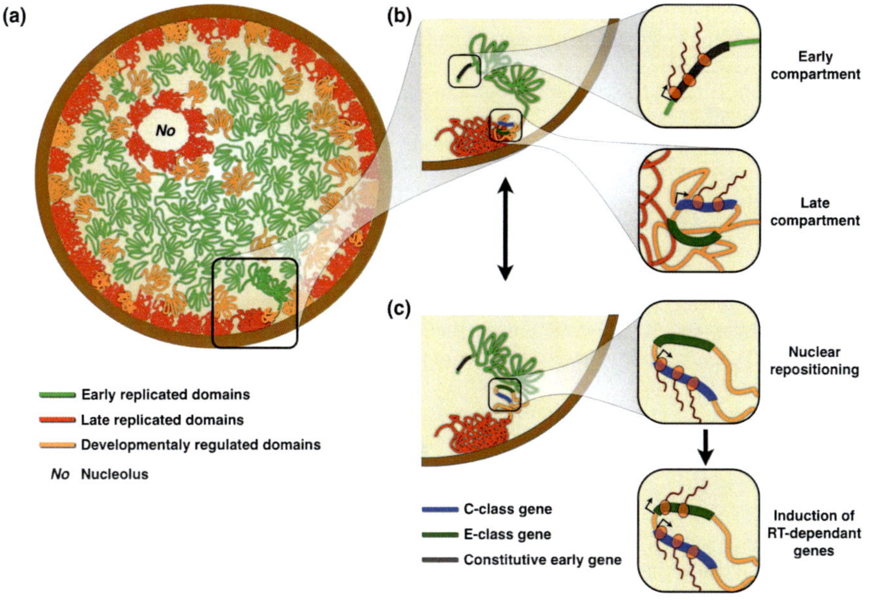

Fig. 11.6 RD organization, RT and gene regulation in the nucleus during development (Adapted from Rivera-Mulia and Gilbert 2016b). Early RDs contain open chromatin at the nuclear interior, whereas late RDs are located at the lamina and nucleoli. During development, certain domains undergo a switch in RT from late to early replication while detaching from nuclear lamina. The RT switching domain harbours genes belonging to different classes based on their transcription potential during RT changes. There are genes (blue) that are induced only after the detachment from lamina and the RT change. There are also genes (green) that are expressed despite being at the periphery and late replicating

References

Akhtar W, de Jong J, Pindyurin AV, Pagie L, Meuleman W, de Ridder J, Berns A, Wessels LF, van Lohuizen M, van Steensel B (2013) Chromatin position effects assayed by thousands of reporters integrated in parallel. Cell 154:914–927

Alabert C, Groth A (2012) Chromatin replication and epigenome maintenance. Nat Rev Mol Cell Biol 13:153–167

Alabert C, Barth TK, Reveron-Gomez N, Sidoli S, Schmidt A, Jensen ON, Imhof A, Groth A (2015) Two distinct modes for propagation of histone PTMs across the cell cycle. Genes Dev 29:585–590

Amendola M, van Steensel B (2014) Mechanisms and dynamics of nuclear lamina-genome interactions. Curr Opin Cell Biol 28:61–68

Amiel A, Kolodizner T, Fishman A, Gaber E, Klein Z, Beyth Y, Fejgin MD (1998a) Replication pattern of the p53 and 21q22 loci in the premalignant and malignant stages of carcinoma of the cervix. Cancer 83:1966–1971

Amiel A, Litmanovitch T, Lishner M, Mor A, Gaber E, Tangi I, Fejgin M, Avivi L (1998b) Temporal differences in replication timing of homologous loci in malignant cells derived from CML and lymphoma patients. Genes Chromosom Cancer 22:225–231

Amiel A, Kirgner I, Gaber E, Manor Y, Fejgin M, Lishner M (1999) Replication pattern in cancer: asynchronous replication in multiple myeloma and in monoclonal gammopathy. Cancer Genet Cytogenet 108:32–37

Avner P, Heard E (2001) X-chromosome inactivation: counting, choice and initiation. Nat Rev Genet 2:59–67

Azmi IF, Watanabe S, Maloney MF, Kang S, Belsky JA, MacAlpine DM, Peterson CL, Bell SP (2017) Nucleosomes influence multiple steps during replication initiation. Elife 6

Baddeley D, Chagin VO, Schermelleh L, Martin S, Pombo A, Carlton PM, Gahl A, Domaing P, Birk U, Leonhardt H et al (2010) Measurement of replication structures at the nanometer scale using super-resolution light microscopy. Nucleic Acids Res 38:–e8

Baer D (1965) Asynchronous replication of DNA in a heterochromatic set of chromosomes in Pseudococcus obscurus. Genetics 52:275–285

Bellelli R, Castellone MD, Guida T, Limongello R, Dathan NA, Merolla F, Cirafici AM, Affuso A, Masai H, Costanzo V et al (2014) NCOA4 transcriptional coactivator inhibits activation of DNA replication origins. Mol Cell 55:123–137

Belotserkovskaya R, Oh S, Bondarenko VA, Orphanides G, Studitsky VM, Reinberg D (2003) FACT facilitates transcription-dependent nucleosome alteration. Science 301:1090–1093

Bender W, Fitzgerald DP (2002) Transcription activates repressed domains in the Drosophila bithorax complex. Development 129:4923–4930

Berezney R, Dubey DD, Huberman JA (2000) Heterogeneity of eukaryotic replicons, replicon clusters, and replication foci. Chromosoma 108:471–484

Besnard E, Babled A, Lapasset L, Milhavet O, Parrinello H, Dantec C, Marin JM, Lemaitre JM (2012) Unraveling cell type-specific and reprogrammable human replication origin signatures associated with G-quadruplex consensus motifs. Nat Struct Mol Biol 19:837–844

Blow JJ, Dutta A (2005) Preventing re-replication of chromosomal DNA. Nat Rev Mol Cell Biol 6:476–486

Breger KS, Smith L, Turker MS, Thayer MJ (2004) Ionizing radiation induces frequent translocations with delayed replication and condensation. Cancer Res 64:8231–8238

Breger KS, Smith L, Thayer MJ (2005) Engineering translocations with delayed replication: evidence for cis control of chromosome replication timing. Hum Mol Genet 14:2813–2827

Casas-Delucchi CS, Brero A, Rahn HP, Solovei I, Wutz A, Cremer T, Leonhardt H, Cardoso MC (2011) Histone acetylation controls the inactive X chromosome replication dynamics. Nat Commun 2:222

Cavalli G, Misteli T (2013) Functional implications of genome topology. Nat Struct Mol Biol 20:290–299

Cayrou C, Coulombe P, Vigneron A, Stanojcic S, Ganier O, Peiffer I, Rivals E, Puy A, Laurent-Chabalier S, Desprat R et al (2011) Genome-scale analysis of metazoan replication origins reveals their organization in specific but flexible sites defined by conserved features. Genome Res 21:1438–1449

Cayrou C, Coulombe P, Puy A, Rialle S, Kaplan N, Segal E, Mechali M (2012) New insights into replication origin characteristics in metazoans. Cell Cycle 11:658–667

Cayrou C, Ballester B, Peiffer I, Fenouil R, Coulombe P, Andrau JC, van Helden J, Mechali M (2015) The chromatin environment shapes DNA replication origin organization and defines origin classes. Genome Res 25:1873–1885

Chagin VO, Casas-Delucchi CS, Reinhart M, Schermelleh L, Markaki Y, Maiser A, Bolius JJ, Bensimon A, Fillies M, Domaing P et al (2016) 4D visualization of replication foci in mammalian cells corresponding to individual replicons. Nat Commun 7:11231

Chakalova L, Debrand E, Mitchell JA, Osborne CS, Fraser P (2005) Replication and transcription: shaping the landscape of the genome. Nat Rev Genet 6:669–677

Chang BH, Smith L, Huang J, Thayer M (2007) Chromosomes with delayed replication timing lead to checkpoint activation, delayed recruitment of Aurora B and chromosome instability. Oncogene 26:1852–1861

Cornacchia D, Dileep V, Quivy JP, Foti R, Tili F, Santarella-Mellwig R, Antony C, Almouzni G, Gilbert DM, Buonomo SB (2012) Mouse Rif1 is a key regulator of the replication-timing programme in mammalian cells. EMBO J 31:3678–3690
Costantini M, Bernardi G (2008) Replication timing, chromosomal bands, and isochores. Proc Natl Acad Sci U S A 105:3433–3437
Costantini M, Clay O, Federico C, Saccone S, Auletta F, Bernardi G (2007) Human chromosomal bands: nested structure, high-definition map and molecular basis. Chromosoma 116:29–40
Cremer T, Cremer M (2010) Chromosome territories. Cold Spring Harb Perspect Biol 2:a003889
Cremer T, Cremer M, Dietzel S, Muller S, Solovei I, Fakan S (2006) Chromosome territories – a functional nuclear landscape. Curr Opin Cell Biol 18:307–316
Deegan TD, Diffley JF (2016) MCM: one ring to rule them all. Curr Opin Struct Biol 37:145–151
Demczuk A, Gauthier MG, Veras I, Kosiyatrakul S, Schildkraut CL, Busslinger M, Bechhoefer J, Norio P (2012) Regulation of DNA replication within the immunoglobulin heavy-chain locus during B cell commitment. PLoS Biol 10:e1001360
Deng X, Zhironkina OA, Cherepanynets VD, Strelkova OS, Kireev II, Belmont AS (2016) Cytology of DNA replication reveals dynamic plasticity of large-scale chromatin fibers. Curr Biol 26:2527–2534
Dileep V, Rivera-Mulia JC, Sima J, Gilbert DM (2015a) Large-scale chromatin structure-function relationships during the cell cycle and development: insights from replication timing. Cold Spring Harb Symp Quant Biol 80:53–63
Dileep V, Ay F, Sima J, Vera DL, Noble WS, Gilbert DM (2015b) Topologically associating domains and their long-range contacts are established during early G1 coincident with the establishment of the replication-timing program. Genome Res 25:1104–1113
Dimitrova DS, Gilbert DM (1999) The spatial position and replication timing of chromosomal domains are both established in early G1 phase. Mol Cell 4:983–993
Dixon JR, Selvaraj S, Yue F, Kim A, Li Y, Shen Y, Hu M, Liu JS, Ren B (2012) Topological domains in mammalian genomes identified by analysis of chromatin interactions. Nature 485:376–380
Donley N, Thayer MJ (2013) DNA replication timing, genome stability and cancer: late and/or delayed DNA replication timing is associated with increased genomic instability. Semin Cancer Biol 23:80–89
Donzelli M, Draetta GF (2003) Regulating mammalian checkpoints through Cdc25 inactivation. EMBO Rep 4:671–677
Elgin SC, Reuter G (2013) Position-effect variegation, heterochromatin formation, and gene silencing in drosophila. Cold Spring Harb Perspect Biol 5:a017780
Ermakova OV, Nguyen LH, Little RD, Chevillard C, Riblet R, Ashouian N, Birshtein BK, Schildkraut CL (1999) Evidence that a single replication fork proceeds from early to late replicating domains in the IgH locus in a non-B cell line. Mol Cell 3:321–330
Farago M, Rosenbluh C, Tevlin M, Fraenkel S, Schlesinger S, Masika H, Gouzman M, Teng G, Schatz D, Rais Y et al (2012) Clonal allelic predetermination of immunoglobulin-kappa rearrangement. Nature 490:561–565
Feng Y, Vlassis A, Roques C, Lalonde ME, Gonzalez-Aguilera C, Lambert JP, Lee SB, Zhao X, Alabert C, Johansen JV et al (2016) BRPF3-HBO1 regulates replication origin activation and histone H3K14 acetylation. EMBO J 35:176–192
Ferguson BM, Brewer BJ, Reynolds AE, Fangman WL (1991) A yeast origin of replication is activated late in S phase. Cell 65:507–515
Ferreira J, Paolella G, Ramos C, Lamond AI (1997) Spatial organization of large-scale chromatin domains in the nucleus: a magnified view of single chromosome territories. J Cell Biol 139:1597–1610
Foti R, Gnan S, Cornacchia D, Dileep V, Bulut-Karslioglu A, Diehl S, Buness A, Klein FA, Huber W, Johnstone E et al (2016) Nuclear architecture organized by Rif1 underpins the replication-timing program. Mol Cell 61:260–273

Foulk MS, Urban JM, Casella C, Gerbi SA (2015) Characterizing and controlling intrinsic biases of lambda exonuclease in nascent strand sequencing reveals phasing between nucleosomes and G-quadruplex motifs around a subset of human replication origins. Genome Res 25:725–735

Fragkos M, Ganier O, Coulombe P, Mechali M (2015) DNA replication origin activation in space and time. Nat Rev Mol Cell Biol 16:360–374

Ge XQ, Blow JJ (2010) Chk1 inhibits replication factory activation but allows dormant origin firing in existing factories. J Cell Biol 191:1285–1297

Gifford CA, Ziller MJ, Gu H, Trapnell C, Donaghey J, Tsankov A, Shalek AK, Kelley DR, Shishkin AA, Issner R et al (2013) Transcriptional and epigenetic dynamics during specification of human embryonic stem cells. Cell 153:1149–1163

Gilbert DM (1986) Temporal order of replication of Xenopus laevis 5S ribosomal RNA genes in somatic cells. Proc Natl Acad Sci U S A 83:2924–2928

Gilbert DM (2002) Replication timing and metazoan evolution. Nat Genet 32:336–337

Gilbert DM (2010) Evaluating genome-scale approaches to eukaryotic DNA replication. Nat Rev Genet 11:673–684

Gilbert DM, Cohen SN (1987) Bovine papilloma virus plasmids replicate randomly in mouse fibroblasts throughout S phase of the cell cycle. Cell 50:59–68

Gilbert DM, Cohen SN (1990) Position effects on the timing of replication of chromosomally integrated simian virus 40 molecules in Chinese hamster cells. Mol Cell Biol 10:4345–4355

Gindin Y, Valenzuela MS, Aladjem MI, Meltzer PS, Bilke S (2014) A chromatin structure-based model accurately predicts DNA replication timing in human cells. Mol Syst Biol 10:722

Goldar A, Labit H, Marheineke K, Hyrien O (2008) A dynamic stochastic model for DNA replication initiation in early embryos. PLoS One 3:e2919

Goldar A, Arneodo A, Audit B, Argoul F, Rappailles A, Guilbaud G, Petryk N, Kahli M, Hyrien O (2016) Deciphering DNA replication dynamics in eukaryotic cell populations in relation with their averaged chromatin conformations. Sci Rep 6:22469

Goldman MA, Holmquist GP, Gray MC, Caston LA, Nag A (1984) Replication timing of genes and middle repetitive sequences. Science 224:686–692

Goren A, Tabib A, Hecht M, Cedar H (2008) DNA replication timing of the human beta-globin domain is controlled by histone modification at the origin. Genes Dev 22:1319–1324

Guillou E, Ibarra A, Coulon V, Casado-Vela J, Rico D, Casal I, Schwob E, Losada A, Mendez J (2010) Cohesin organizes chromatin loops at DNA replication factories. Genes Dev 24:2812–2822

Guinta DR, Korn LJ (1986) Differential order of replication of Xenopus laevis 5S RNA genes. Mol Cell Biol 6:2536–2542

Hansen RS, Canfield TK, Lamb MM, Gartler SM, Laird CD (1993) Association of fragile X syndrome with delayed replication of the FMR1 gene. Cell 73:1403–1409

Hansen RS, Stoger R, Wijmenga C, Stanek AM, Canfield TK, Luo P, Matarazzo MR, D'Esposito M, Feil R, Gimelli G et al (2000) Escape from gene silencing in ICF syndrome: evidence for advanced replication time as a major determinant. Hum Mol Genet 9:2575–2587

Hatton KS, Dhar V, Brown EH, Iqbal MA, Stuart S, Didamo VT, Schildkraut CL (1988) Replication program of active and inactive multigene families in mammalian cells. Mol Cell Biol 8:2149–2158

Himes M (1967) An analysis of heterochromatin in maize root tips. J Cell Biol 35:175–181

Hiratani I, Leskovar A, Gilbert DM (2004) Differentiation-induced replication-timing changes are restricted to AT-rich/long interspersed nuclear element (LINE)-rich isochores. Proc Natl Acad Sci U S A 101:16861–16866

Hiratani I, Ryba T, Itoh M, Yokochi T, Schwaiger M, Chang CW, Lyou Y, Townes TM, Schubeler D, Gilbert DM (2008) Global reorganization of replication domains during embryonic stem cell differentiation. PLoS Biol 6:e245

Hiratani I, Takebayashi S, Lu J, Gilbert DM (2009) Replication timing and transcriptional control: beyond cause and effect – part II. Curr Opin Genet Dev 19:142–149

Hiratani I, Ryba T, Itoh M, Rathjen J, Kulik M, Papp B, Fussner E, Bazett-Jones DP, Plath K, Dalton S et al (2010) Genome-wide dynamics of replication timing revealed by in vitro models of mouse embryogenesis. Genome Res 20:155–169

Hogga I, Karch F (2002) Transcription through the iab-7 cis-regulatory domain of the bithorax complex interferes with maintenance of Polycomb-mediated silencing. Development 129:4915–4922

Holmquist GP (1987) Role of replication time in the control of tissue-specific gene expression. Am J Hum Genet 40:151–173

Hozak P, Hassan AB, Jackson DA, Cook PR (1993) Visualization of replication factories attached to nucleoskeleton. Cell 73:361–373

Huang H, Stromme CB, Saredi G, Hodl M, Strandsby A, Gonzalez-Aguilera C, Chen S, Groth A, Patel DJ (2015) A unique binding mode enables MCM2 to chaperone histones H3-H4 at replication forks. Nat Struct Mol Biol 22:618–626

Hyrien O (2015) Peaks cloaked in the mist: the landscape of mammalian replication origins. J Cell Biol 208:147–160

Jackson DA, Pombo A (1998) Replicon clusters are stable units of chromosome structure: evidence that nuclear organization contributes to the efficient activation and propagation of S phase in human cells. J Cell Biol 140:1285–1295

Karnani N, Taylor CM, Malhotra A, Dutta A (2010) Genomic study of replication initiation in human chromosomes reveals the influence of transcription regulation and chromatin structure on origin selection. Mol Biol Cell 21:393–404

Kaykov A, Nurse P (2015) The spatial and temporal organization of origin firing during the S-phase of fission yeast. Genome Res 25:391–401

Keohane AM, O'Neill LP, Belyaev ND, Lavender JS, Turner BM (1996) X-inactivation and histone H4 acetylation in embryonic stem cells. Dev Biol 180:618–630

Kireeva ML, Walter W, Tchernajenko V, Bondarenko V, Kashlev M, Studitsky VM (2002) Nucleosome remodeling induced by RNA polymerase II: loss of the H2A/H2B dimer during transcription. Mol Cell 9:541–552

Kitamura E, Blow JJ, Tanaka TU (2006) Live-cell imaging reveals replication of individual replicons in eukaryotic replication factories. Cell 125:1297–1308

Kitsberg D, Selig S, Brandeis M, Simon I, Keshet I, Driscoll DJ, Nicholls RD, Cedar H (1993) Allele-specific replication timing of imprinted gene regions. Nature 364:459–463

Klevecz RR, Stubblefield E (1967) RNA synthesis in relation to DNA replication in synchronized Chinese hamster cell cultures. J Exp Zool 165:259–268

Knott SR, Peace JM, Ostrow AZ, Gan Y, Rex AE, Viggiani CJ, Tavare S, Aparicio OM (2012) Forkhead transcription factors establish origin timing and long-range clustering in S. cerevisiae. Cell 148:99–111

Koren A, Polak P, Nemesh J, Michaelson JJ, Sebat J, Sunyaev SR, McCarroll SA (2012) Differential relationship of DNA replication timing to different forms of human mutation and variation. Am J Hum Genet 91:1033–1040

Koren A, Handsaker RE, Kamitaki N, Karlic R, Ghosh S, Polak P, Eggan K, McCarroll SA (2014) Genetic variation in human DNA replication timing. Cell 159:1015–1026

Korenstein-Ilan A, Amiel A, Lalezari S, Lishner M, Avivi L (2002) Allele-specific replication associated with aneuploidy in blood cells of patients with hematologic malignancies. Cancer Genet Cytogenet 139:97–103

Kosak ST, Skok JA, Medina KL, Riblet R, Le Beau MM, Fisher AG, Singh H (2002) Subnuclear compartmentalization of immunoglobulin loci during lymphocyte development. Science 296:158–162

Kylie K, Romero J, Lindamulage IK, Knockleby J, Lee H (2016) Dynamic regulation of histone H3K9 is linked to the switch between replication and transcription at the Dbf4 origin-promoter locus. Cell Cycle 15:2321–2335

Lande-Diner L, Zhang J, Cedar H (2009) Shifts in replication timing actively affect histone acetylation during nucleosome reassembly. Mol Cell 34:767–774

Lang GI, Murray AW (2011) Mutation rates across budding yeast chromosome VI are correlated with replication timing. Genome Biol Evol 3:799–811
Lieberman-Aiden E, van Berkum NL, Williams L, Imakaev M, Ragoczy T, Telling A, Amit I, Lajoie BR, Sabo PJ, Dorschner MO et al (2009) Comprehensive mapping of long-range interactions reveals folding principles of the human genome. Science 326:289–293
Lima-De-Faria A (1959) Differential uptake of tritiated thymidine into hetero- and euchromatin in Melanoplus and Secale. J Biophys Biochem Cytol 6:457–466
Liu S, Trapnell C (2016) Single-cell transcriptome sequencing: recent advances and remaining challenges. F1000Res 5
Liu J, McConnell K, Dixon M, Calvi BR (2012) Analysis of model replication origins in Drosophila reveals new aspects of the chromatin landscape and its relationship to origin activity and the prereplicative complex. Mol Biol Cell 23:200–212
Liu L, De S, Michor F (2013) DNA replication timing and higher-order nuclear organization determine single-nucleotide substitution patterns in cancer genomes. Nat Commun 4:1502
Lob D, Lengert N, Chagin VO, Reinhart M, Casas-Delucchi CS, Cardoso MC, Drossel B (2016) 3D replicon distributions arise from stochastic initiation and domino-like DNA replication progression. Nat Commun 7:11207
Lu J, Li F, Murphy CS, Davidson MW, Gilbert DM (2010) G2 phase chromatin lacks determinants of replication timing. J Cell Biol 189:967–980
Lubelsky Y, Prinz JA, DeNapoli L, Li Y, Belsky JA, MacAlpine DM (2014) DNA replication and transcription programs respond to the same chromatin cues. Genome Res 24:1102–1114
Ma H, Samarabandu J, Devdhar RS, Acharya R, Cheng PC, Meng C, Berezney R (1998) Spatial and temporal dynamics of DNA replication sites in mammalian cells. J Cell Biol 143:1415–1425
Maya-Mendoza A, Olivares-Chauvet P, Shaw A, Jackson DA (2010) S phase progression in human cells is dictated by the genetic continuity of DNA foci. PLoS Genet 6:e1000900
Maya-Mendoza A, Olivares-Chauvet P, Kohlmeier F, Jackson DA (2012) Visualising chromosomal replication sites and replicons in mammalian cells. Methods 57:140–148
Meister P, Taddei A, Ponti A, Baldacci G, Gasser SM (2007) Replication foci dynamics: replication patterns are modulated by S-phase checkpoint kinases in fission yeast. EMBO J 26:1315–1326
Mesner LD, Valsakumar V, Cieslik M, Pickin R, Hamlin JL, Bekiranov S (2013) Bubble-seq analysis of the human genome reveals distinct chromatin-mediated mechanisms for regulating early- and late-firing origins. Genome Res 23:1774–1788
Meuleman W, Peric-Hupkes D, Kind J, Beaudry JB, Pagie L, Kellis M, Reinders M, Wessels L, van Steensel B (2013) Constitutive nuclear lamina-genome interactions are highly conserved and associated with A/T-rich sequence. Genome Res 23:270–280
Moindrot B, Audit B, Klous P, Baker A, Thermes C, de Laat W, Bouvet P, Mongelard F, Arneodo A (2012) 3D chromatin conformation correlates with replication timing and is conserved in resting cells. Nucleic Acids Res 40:9470–9481
Mostoslavsky R, Singh N, Tenzen T, Goldmit M, Gabay C, Elizur S, Qi P, Reubinoff BE, Chess A, Cedar H et al (2001) Asynchronous replication and allelic exclusion in the immune system. Nature 414:221–225
Nakamura H, Morita T, Sato C (1986) Structural organizations of replicon domains during DNA synthetic phase in the mammalian nucleus. Exp Cell Res 165:291–297
Nakayasu H, Berezney R (1989) Mapping replicational sites in the eucaryotic cell nucleus. J Cell Biol 108:1–11
Naumova N, Imakaev M, Fudenberg G, Zhan Y, Lajoie BR, Mirny LA, Dekker J (2013) Organization of the mitotic chromosome. Science 342:948–953
Newlon CS, Lipchitz LR, Collins I, Deshpande A, Devenish RJ, Green RP, Klein HL, Palzkill TG, Ren RB, Synn S et al (1991) Analysis of a circular derivative of Saccharomyces cerevisiae chromosome III: a physical map and identification and location of ARS elements. Genetics 129:343–357
Norio P, Kosiyatrakul S, Yang Q, Guan Z, Brown NM, Thomas S, Riblet R, Schildkraut CL (2005) Progressive activation of DNA replication initiation in large domains of the immunoglobulin heavy chain locus during B cell development. Mol Cell 20:575–587

Ostrow AZ, Kalhor R, Gan Y, Villwock SK, Linke C, Barberis M, Chen L, Aparicio OM (2017) Conserved forkhead dimerization motif controls DNA replication timing and spatial organization of chromosomes in S. cerevisiae. Proc Natl Acad Sci U S A 114:E2411–E2419
Patel PK, Arcangioli B, Baker SP, Bensimon A, Rhind N (2006) DNA replication origins fire stochastically in fission yeast. Mol Biol Cell 17:308–316
Peric-Hupkes D, Meuleman W, Pagie L, Bruggeman SW, Solovei I, Brugman W, Graf S, Flicek P, Kerkhoven RM, van Lohuizen M et al (2010) Molecular maps of the reorganization of genome-nuclear lamina interactions during differentiation. Mol Cell 38:603–613
Perry P, Sauer S, Billon N, Richardson WD, Spivakov M, Warnes G, Livesey FJ, Merkenschlager M, Fisher AG, Azuara V (2004) A dynamic switch in the replication timing of key regulator genes in embryonic stem cells upon neural induction. Cell Cycle 3:1645–1650
Petryk N, Kahli M, d'Aubenton-Carafa Y, Jaszczyszyn Y, Shen Y, Silvain M, Thermes C, Chen CL, Hyrien O (2016) Replication landscape of the human genome. Nat Commun 7:10208
Pfeiffer SE (1968) RNA synthesis in synchronously growing populations of HeLa S3 cells. II. Rate of synthesis of individual RNA fractions. J Cell Physiol 71:95–104
Pombo A, Dillon N (2015) Three-dimensional genome architecture: players and mechanisms. Nat Rev Mol Cell Biol 16:245–257
Pope BD, Tsumagari K, Battaglia D, Ryba T, Hiratani I, Ehrlich M, Gilbert DM (2011) DNA replication timing is maintained genome-wide in primary human myoblasts independent of D4Z4 contraction in FSH muscular dystrophy. PLoS One 6:e27413
Pope BD, Chandra T, Buckley Q, Hoare M, Ryba T, Wiseman FK, Kuta A, Wilson MD, Odom DT, Gilbert DM (2012) Replication-timing boundaries facilitate cell-type and species-specific regulation of a rearranged human chromosome in mouse. Hum Mol Genet 21:4162–4170
Pope BD, Ryba T, Dileep V, Yue F, Wu W, Denas O, Vera DL, Wang Y, Hansen RS, Canfield TK et al (2014) Topologically associating domains are stable units of replication-timing regulation. Nature 515:402–405
Pourkarimi E, Bellush JM, Whitehouse I (2016) Spatiotemporal coupling and decoupling of gene transcription with DNA replication origins during embryogenesis in C. elegans. Elife 5:e21728
Prioleau MN (2009) CpG islands: starting blocks for replication and transcription. PLoS Genet 5:e1000454
Ramachandran S, Henikoff S (2015) Replicating nucleosomes. Sci Adv 1
Ramani V, Deng X, Qiu R, Gunderson KL, Steemers FJ, Disteche CM, Noble WS, Duan Z, Shendure J (2017) Massively multiplex single-cell Hi-C. Nat Methods 14:263–266
Rhind N, Gilbert DM (2013) DNA replication timing. Cold Spring Harb Perspect Biol 5:a010132
Rhind N, Yang SC, Bechhoefer J (2010) Reconciling stochastic origin firing with defined replication timing. Chromosom Res 18:35–43
Rivera-Mulia JC, Gilbert DM (2016a) Replicating large genomes: divide and conquer. Mol Cell 62:756–765. https://doi.org/10.1016/j.molcel.2016.05.007
Rivera-Mulia JC, Gilbert DM (2016b) Replication timing and transcriptional control: beyond cause and effect-part III. Curr Opin Cell Biol 40:168–178. https://doi.org/10.1016/j.ceb.2016.03.022
Rivera-Mulia JC, Buckley Q, Sasaki T, Zimmerman J, Didier RA, Nazor K, Loring JF, Lian Z, Weissman S, Robins AJ et al (2015) Dynamic changes in replication timing and gene expression during lineage specification of human pluripotent stem cells. Genome Res 25:1091–1103
Roadmap Epigenomics C, Kundaje A, Meuleman W, Ernst J, Bilenky M, Yen A, Heravi-Moussavi A, Kheradpour P, Zhang Z, Wang J et al (2015) Integrative analysis of 111 reference human epigenomes. Nature 518:317–330
Robson MI, de Las Heras JI, Czapiewski R, Le Thanh P, Booth DG, Kelly DA, Webb S, Kerr AR, Schirmer EC (2016) Tissue-specific gene repositioning by muscle nuclear membrane proteins enhances repression of critical developmental genes during myogenesis. Mol Cell 62:834–847
Rondinelli B, Schwerer H, Antonini E, Gaviraghi M, Lupi A, Frenquelli M, Cittaro D, Segalla S, Lemaitre JM, Tonon G (2015) H3K4me3 demethylation by the histone demethylase KDM5C/JARID1C promotes DNA replication origin firing. Nucleic Acids Res 43:2560–2574

Ryba T, Hiratani I, Lu J, Itoh M, Kulik M, Zhang J, Schulz TC, Robins AJ, Dalton S, Gilbert DM (2010) Evolutionarily conserved replication timing profiles predict long-range chromatin interactions and distinguish closely related cell types. Genome Res 20:761–770

Ryba T, Battaglia D, Chang BH, Shirley JW, Buckley Q, Pope BD, Devidas M, Druker BJ, Gilbert DM (2012) Abnormal developmental control of replication-timing domains in pediatric acute lymphoblastic leukemia. Genome Res 22:1833–1844

Sadoni N, Cardoso MC, Stelzer EH, Leonhardt H, Zink D (2004) Stable chromosomal units determine the spatial and temporal organization of DNA replication. J Cell Sci 117:5353–5365

Saner N, Karschau J, Natsume T, Gierlinski M, Retkute R, Hawkins M, Nieduszynski CA, Blow JJ, de Moura AP, Tanaka TU (2013) Stochastic association of neighboring replicons creates replication factories in budding yeast. J Cell Biol 202:1001–1012

Sasaki T, Rivera-Mulia JC, Vera D, Zimmerman J, Das S, Padget M, Nakamichi N, Chang BH, Tyner J, Druker BJ, Weng AP, Civin CI, Eaves CJ, Gilbert DM (2017) Stability of patient-specific features of altered DNA replication timing in xenografts of primary human acute lymphoblastic leukemia. Exp Hematol 51:71–82.e3

Schmitt S, Prestel M, Paro R (2005) Intergenic transcription through a polycomb group response element counteracts silencing. Genes Dev 19:697–708

Schubeler D, Scalzo D, Kooperberg C, van Steensel B, Delrow J, Groudine M (2002) Genome-wide DNA replication profile for Drosophila melanogaster: a link between transcription and replication timing. Nat Genet 32:438–442

Schwaiger M, Stadler MB, Bell O, Kohler H, Oakeley EJ, Schubeler D (2009) Chromatin state marks cell-type- and gender-specific replication of the Drosophila genome. Genes Dev 23:589–601

Schwaiger M, Kohler H, Oakeley EJ, Stadler MB, Schubeler D (2010) Heterochromatin protein 1 (HP1) modulates replication timing of the Drosophila genome. Genome Res 20:771–780

Schwartz BE, Ahmad K (2005) Transcriptional activation triggers deposition and removal of the histone variant H3.3. Genes Dev 19:804–814

Sequeira-Mendes J, Diaz-Uriarte R, Apedaile A, Huntley D, Brockdorff N, Gomez M (2009) Transcription initiation activity sets replication origin efficiency in mammalian cells. PLoS Genet 5:e1000446

Simon I, Tenzen T, Reubinoff BE, Hillman D, McCarrey JR, Cedar H (1999) Asynchronous replication of imprinted genes is established in the gametes and maintained during development. Nature 401:929–932

Smith L, Plug A, Thayer M (2001) Delayed replication timing leads to delayed mitotic chromosome condensation and chromosomal instability of chromosome translocations. Proc Natl Acad Sci U S A 98:13300–13305

Solovei I, Thanisch K, Feodorova Y (2016) How to rule the nucleus: divide et impera. Curr Opin Cell Biol 40:47–59. https://doi.org/10.1016/j.ceb.2016.02.014

Sparvoli E, Levi M, Rossi E (1994) Replicon clusters may form structurally stable complexes of chromatin and chromosomes. J Cell Sci 107(Pt 11):3097–3103

Sporbert A, Gahl A, Ankerhold R, Leonhardt H, Cardoso MC (2002) DNA polymerase clamp shows little turnover at established replication sites but sequential de novo assembly at adjacent origin clusters. Mol Cell 10:1355–1365

Stamatoyannopoulos JA, Adzhubei I, Thurman RE, Kryukov GV, Mirkin SM, Sunyaev SR (2009) Human mutation rate associated with DNA replication timing. Nat Genet 41:393–395

Supek F, Lehner B (2015) Differential DNA mismatch repair underlies mutation rate variation across the human genome. Nature 521:81–84

Takebayashi S, Dileep V, Ryba T, Dennis JH, Gilbert DM (2012) Chromatin-interaction compartment switch at developmentally regulated chromosomal domains reveals an unusual principle of chromatin folding. Proc Natl Acad Sci U S A 109:12574–12579

Tardat M, Brustel J, Kirsh O, Lefevbre C, Callanan M, Sardet C, Julien E (2010) The histone H4 Lys 20 methyltransferase PR-Set7 regulates replication origins in mammalian cells. Nat Cell Biol 12:1086–1093

Taylor JH (1960) Asynchronous duplication of chromosomes in cultured cells of Chinese hamster. J Biophys Biochem Cytol 7:455–464
Therizols P, Illingworth RS, Courilleau C, Boyle S, Wood AJ, Bickmore WA (2014) Chromatin decondensation is sufficient to alter nuclear organization in embryonic stem cells. Science 346:1238–1242
Valton AL, Hassan-Zadeh V, Lema I, Boggetto N, Alberti P, Saintome C, Riou JF, Prioleau MN (2014) G4 motifs affect origin positioning and efficiency in two vertebrate replicators. EMBO J 33:732–746
Vogelstein B, Pardoll DM, Coffey DS (1980) Supercoiled loops and eukaryotic DNA replication. Cell 22:79–85
Wang Y, Khan A, Marks AB, Smith OK, Giri S, Lin YC, Creager R, MacAlpine DM, Prasanth KV, Aladjem MI et al (2016) Temporal association of ORCA/LRWD1 to late-firing origins during G1 dictates heterochromatin replication and organization. Nucleic Acids Res 45:2490–2502
Williams RR, Azuara V, Perry P, Sauer S, Dvorkina M, Jorgensen H, Roix J, McQueen P, Misteli T, Merkenschlager M et al (2006) Neural induction promotes large-scale chromatin reorganisation of the Mash1 locus. J Cell Sci 119:132–140
Woodfine K, Fiegler H, Beare DM, Collins JE, McCann OT, Young BD, Debernardi S, Mott R, Dunham I, Carter NP (2004) Replication timing of the human genome. Hum Mol Genet 13:191–202
Woodward AM, Gohler T, Luciani MG, Oehlmann M, Ge X, Gartner A, Jackson DA, Blow JJ (2006) Excess Mcm2-7 license dormant origins of replication that can be used under conditions of replicative stress. J Cell Biol 173:673–683
Wu R, Wang Z, Zhang H, Gan H, Zhang Z (2017) H3K9me3 demethylase Kdm4d facilitates the formation of pre-initiative complex and regulates DNA replication. Nucleic Acids Res 45:169–180
Xie W, Schultz MD, Lister R, Hou Z, Rajagopal N, Ray P, Whitaker JW, Tian S, Hawkins RD, Leung D et al (2013) Epigenomic analysis of multilineage differentiation of human embryonic stem cells. Cell 153:1134–1148
Yaffe E, Farkash-Amar S, Polten A, Yakhini Z, Tanay A, Simon I (2010) Comparative analysis of DNA replication timing reveals conserved large-scale chromosomal architecture. PLoS Genet 6:e1001011
Yamazaki S, Ishii A, Kanoh Y, Oda M, Nishito Y, Masai H (2012) Rif1 regulates the replication timing domains on the human genome. EMBO J 31:3667–3677
Yang SC, Rhind N, Bechhoefer J (2010) Modeling genome-wide replication kinetics reveals a mechanism for regulation of replication timing. Mol Syst Biol 6:404
Yasuhara JC, DeCrease CH, Wakimoto BT (2005) Evolution of heterochromatic genes of Drosophila. Proc Natl Acad Sci U S A 102:10958–10963
Yue F, Cheng Y, Breschi A, Vierstra J, Wu W, Ryba T, Sandstrom R, Ma Z, Davis C, Pope BD et al (2014) A comparative encyclopedia of DNA elements in the mouse genome. Nature 515:355–364
Zhang J, Xu F, Hashimshony T, Keshet I, Cedar H (2002) Establishment of transcriptional competence in early and late S phase. Nature 420:198–202

Chapter 12
Rif1-Dependent Regulation of Genome Replication in Mammals

Sara B.C. Buonomo

Abstract Eukaryotic genomes are replicated starting from multiple origins of replication. Their usage is tightly regulated, and not all the potential origins are activated during a single cell cycle. In addition, the ones that are activated are activated in a sequential order. Why don't origins of replication normally all fire together? Is this important? And if so, why? Would any order of firing do, or does the specific sequence matter? How is this process regulated? These questions concern all eukaryotes but have proven extremely hard to address because replication timing is a process intricately connected with multiple aspects of nuclear function.

Keywords DNA replication timing • Nuclear architecture • Rif1 • Nuclear lamina • Nuclear organization • PP1 • Origin firing • Telomere replication

In 2012 the identification of the first genome-wide regulator of replication timing across evolution, a protein called Rif1 (Cornacchia et al. 2012; Hayano et al. 2012; Yamazaki et al. 2012), provided a long-awaited tool to start addressing these questions. However, Rif1 itself has also proven a very complex protein, confusingly involved in telomere length control (Gallardo et al. 2011; Hardy et al. 1992; Teixeira et al. 2004), DNA repair (Wang et al. 2009; Buonomo et al. 2009; Chapman et al. 2013; Daley and Sung 2013; Feng et al. 2013; Martina et al. 2014), DNA replication timing (Cornacchia et al. 2012; Dave et al. 2014; Hayano et al. 2012; Hiraga et al. 2014; Mattarocci et al. 2014; Peace et al. 2014; Sreesankar et al. 2015; Yamazaki et al. 2012), and nuclear architecture organization (Foti et al. 2016). Thus, the complexity of Rif1 biology reflects the intricacies of replication-timing control at a molecular level.

S.B.C. Buonomo (✉)
School of Biological Sciences, Institute of Cell Biology, University of Edinburgh, Edinburgh, UK
e-mail: sara.buonomo@ed.ac.uk

H. Masai, M. Foiani (eds.), *DNA Replication*, Advances in Experimental Medicine and Biology 1042, https://doi.org/10.1007/978-981-10-6955-0_12

12.1 Rif1 and DNA Replication Timing

Rif1 was originally discovered in budding yeast as a negative regulator of telomere length, where it is brought to the telomeric repeats through its interaction with Rap1 (Hardy et al. 1992), a sequence-specific DNA-binding protein that recognizes and directly binds telomeric repeats in yeast (Conrad et al. 1990; Lustig et al. 1990). Hence, the amount of Rif1 at telomeres is proportional to telomere length (Levy and Blackburn 2004). Interestingly, both sudden telomere shortening (Bianchi and Shore 2007) and Rif1 deletion-induced telomere elongation (Lian et al. 2011) correlate with a shift of telomere replication from late to early S phase. While it was originally hypothesized that telomere length per se might have affected the timing of firing of subtelomeric origins, these data suggested instead that Rif1 presence at budding yeast telomeres was responsible for their late replication (Hiraga et al. 2014). In mammalian cells, Rif1 is not detectable at normal telomeres (Silverman et al. 2004; Xu and Blackburn 2004) nor does it seem to be involved in telomere length regulation (Buonomo et al. 2009). This is consistent with the fact that mammalian telomeres do not specifically replicate in late S phase (Arnoult et al. 2010; Hultdin et al. 2001; Wright et al. 1999; Zou et al. 2004). Rif1 deficiency induces genome-wide changes in replication timing in *Schizosaccharomyces pombe* (Hayano et al. 2012), *Saccharomyces cerevisiae* (Peace et al. 2014), *Drosophila* (Sreesankar et al. 2015), mouse and human cells (Cornacchia et al. 2012; Yamazaki et al. 2012). Hence, Rif1-dependent control of replication timing is highly conserved throughout evolution.

In yeast, the control of replication timing via Rif1 requires its interaction with the Ser/Thr phosphatase PP1 (Dave et al. 2014; Hiraga et al. 2014; Mattarocci et al. 2014). Disruption of this interaction increases the amount of phosphorylated MCM4, thereby potentially promoting origin firing (Fu et al. 2011). According to the model drawn on the basis of these results, during the G1/S transition and the initial part of S phase, Rif1 recruits PP1 to the origins destined to fire later in S phase, acting locally to counteract the activating phosphorylation of MCM4 by Dbf4-dependent kinase (DDK). This hypothetical pathway places Rif1-dependent regulation of origin firing at the stage of execution of the replication-timing program. Recent data obtained also from human cell lines (Alver et al. 2017; Hiraga et al. 2017) support this hypothesis. However, there are also experimental results suggesting that there could be more to Rif1 function. The overlap between Rif1-bound origins and those whose timing of firing is dependent upon Rif1 is modest in yeast (Hayano et al. 2012; Peace et al. 2014), and in mouse embryonic stem cells (ESCs), there is no clear enrichment of Rif1 at the limited number of replication origins analyzed (Foti et al. 2016). This is puzzling in light of the model that sees Rif1 functioning locally at late origins as a PP1 adaptor, suggesting the possibility that Rif1 could control origin activation through additional mechanisms, acting on a genome-wide scale. In mammalian cells, there is evidence that this could be the case. Below, I am going to discuss some of the relevant data and what it tells us about Rif1's role and the relationship between nuclear organization and regulation of replication timing.

12.2 Rif1, Nuclear Organization, and Replication Timing

The idea that spatial nuclear organization and sequential firing of replication origins are connected relies on evidence accumulated over the past several years. For example, based on the frequency of intra-domain chromatin interactions, the mammalian nucleus has been subdivided into two large compartments, A and B (Lieberman-Aiden et al. 2009), displaying a striking correlation with early and late replicating fractions of the genome (Ryba et al. 2010; Yaffe et al. 2010). More recently, it has been shown that the units that constitute the building blocks of these large nuclear compartments, the topologically associated domains (TADs) (Dixon et al. 2012), coincide with the units of replication-timing regulation, the developmentally defined replication domains (Pope et al. 2014). At the molecular level, it has recently been shown in budding yeast that Fkh1 and Fkh2 promote early firing of origins by mediating their clustering (Knott et al. 2012). On the other hand, the relationship between DNA localization at the nuclear periphery and its late replication is well established, with developmentally regulated loci switching replication from late to early S phase (LtoE) while also relocating toward the internal part of the nucleus and vice versa (Hiratani et al. 2008, 2010; Williams et al. 2006; Zhou et al. 2002). Similarly, the time at which replication timing is reestablished after mitosis (timing decision point, TDP), in early G1, coincides with chromosome repositioning, when domains destined to be late replicating occupy the nuclear periphery and peri-nucleolar areas (Dileep et al. 2015; Dimitrova and Gilbert 1999; Raghuraman et al. 1997). The peripheral position of late replicating domains is conserved throughout evolution.

Rif1 localizes to the nuclear periphery in a variety of cells, from yeast to mammals. In mammals roughly half of the pool of Rif1 is associated with the nuclear periphery and interacts with the lamina (Cornacchia et al. 2012; Foti et al. 2016; Roux et al. 2012; Yamazaki et al. 2012). In addition, Rif1 is distributed in large domains (Rif1-associated domains or RADs) strikingly overlapping with the late replicating genome (Foti et al. 2016). This distribution is highly reminiscent of the genome-binding profile of one of the major components of the nuclear lamina, Lamin B1. The nuclear lamina plays an important role in the organization of nuclear architecture (reviewed in Shimi et al. 2010). Lamin B1 covers large domains called Lamin B1-associated domains (LADs) (Peric-Hupkes et al. 2010) that also overlap with the late replicating genome (Peric-Hupkes et al. 2010), although not as extensively as RADs (our unpublished results). Finally, LADs and RADs display a substantial degree of overlap. Overall, these data suggest that Rif1 could have a role in three-dimensional (3D) organization of the genome in the nucleus. In agreement with this hypothesis, Rif1 deficiency in mammalian cells has been shown to alter chromatin loop size (Yamazaki et al. 2012) and chromocenter compaction (Cornacchia et al. 2012) and induce loss of temporal coordination of the replication of neighboring as well as more distantly interacting replicons (Cornacchia et al. 2012; Foti et al. 2016). Nevertheless, these effects could be consequential to the changes of replication timing. However, we have recently shown that deletion of Rif1 in primary mouse embryonic fibroblasts arrested in G0 impacts the 3D organization of replication domains as early as the first G1 after Rif1 deletion (Foti et al. 2016). Such early alterations of chromatin architecture might be responsible for the changes in replication timing that appear in the following S phase. In fact, the constrains that normally limit domain interactions within either early or late replicating domains (Takebayashi et al. 2012) are lost in the absence of Rif1, with late domains promiscuously establishing numerous, low-frequency contacts with early domains and vice versa (Foti et al. 2016) (Fig. 12.1a, b). In agreement with this, we have observed by locus-specific 3D FISH that LtoE switches induced by Rif1 deletion are also accompanied by a tendency to relocate the corresponding portions of the genome from the periphery to the interior of the nucleus (our unpublished data). These observations indicate that in G1, Rif1 plays a role in the organization of nuclear architecture, possibly at two different levels: it orchestrates and spatially constrains contacts between different replication domains and ensures peripheral positioning of at least some late chromosomal domains. We suggest that these functions of Rif1 in nuclear architecture could be a first tier at which Rif1 instructs replication timing, as both of these aspects of nuclear organization have been related to replication timing.

12.3 Rif1 and PP1

In line with the identification of yeast Rif1 as a mediator of PP1 function at replication origins, mammalian and *Drosophila* Rif1s were identified in PP1-associated complexes (Guruharsha et al. 2011; Moorhead et al. 2008; Trinkle-Mulcahy et al. 2006). In addition, we have obtained structural and functional data classifying mouse Rif1 as a *bona fide* PP1 regulatory subunit (Sukackaite et al. 2017). It is therefore clear that one Rif1 function conserved throughout evolution is to act as a regulatory subunit of PP1 (Sreesankar et al. 2012). However, the idea that Rif1 could play a role in G1 during the establishment of replication timing is not mutually exclusive with its proposed function at late origins in G1/S. The spatial sequestration of late origins away from early ones (and from DNA replication's limiting factors? (Mantiero et al. 2011; Patel et al. 2006; Tanaka et al. 2011; Wu and Nurse 2009)) through the formation of RADs would also result in creation of large areas of high PP1 concentration which could facilitate PP1 function on late origins later in the cell cycle. In this scenario, Rif1 would act both in early G1 at the level of establishment (Fig. 12.2a, b) and later, at G1/S transition, at the execution step of the replication-timing program (Fig. 12.2c).

Alternatively, it is formally possible that the architectural function of Rif1 has no bearing on replication timing per se and that the common denominator between nuclear 3D organization and replication timing is Rif1 as a molecule, independently

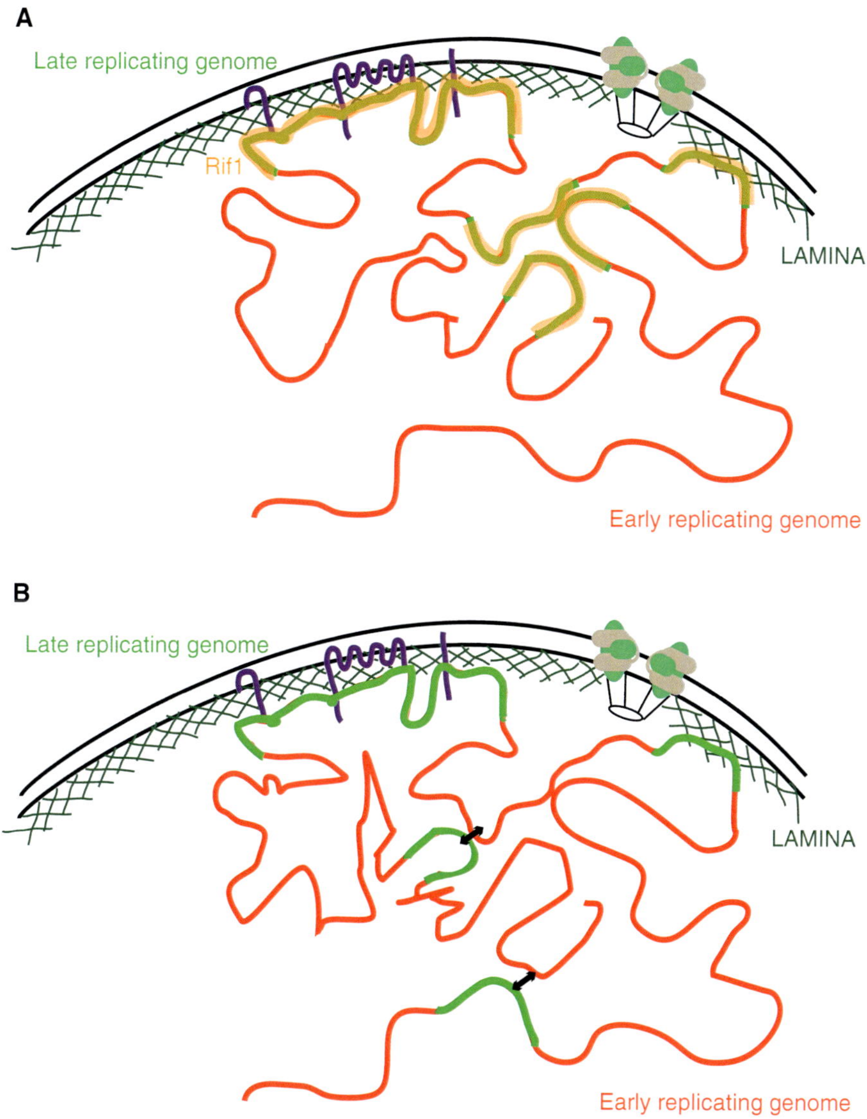

Fig. 12.1 (**a**) Rif1 (orange) coats late replicating genome (green) and limits the interactions between regions with the same replication timing. (**b**) In absence of Rif1 late replicating regions of the genome not associated with Lamin B1 change replication timing as well as lose replication-timing specificity of interactions

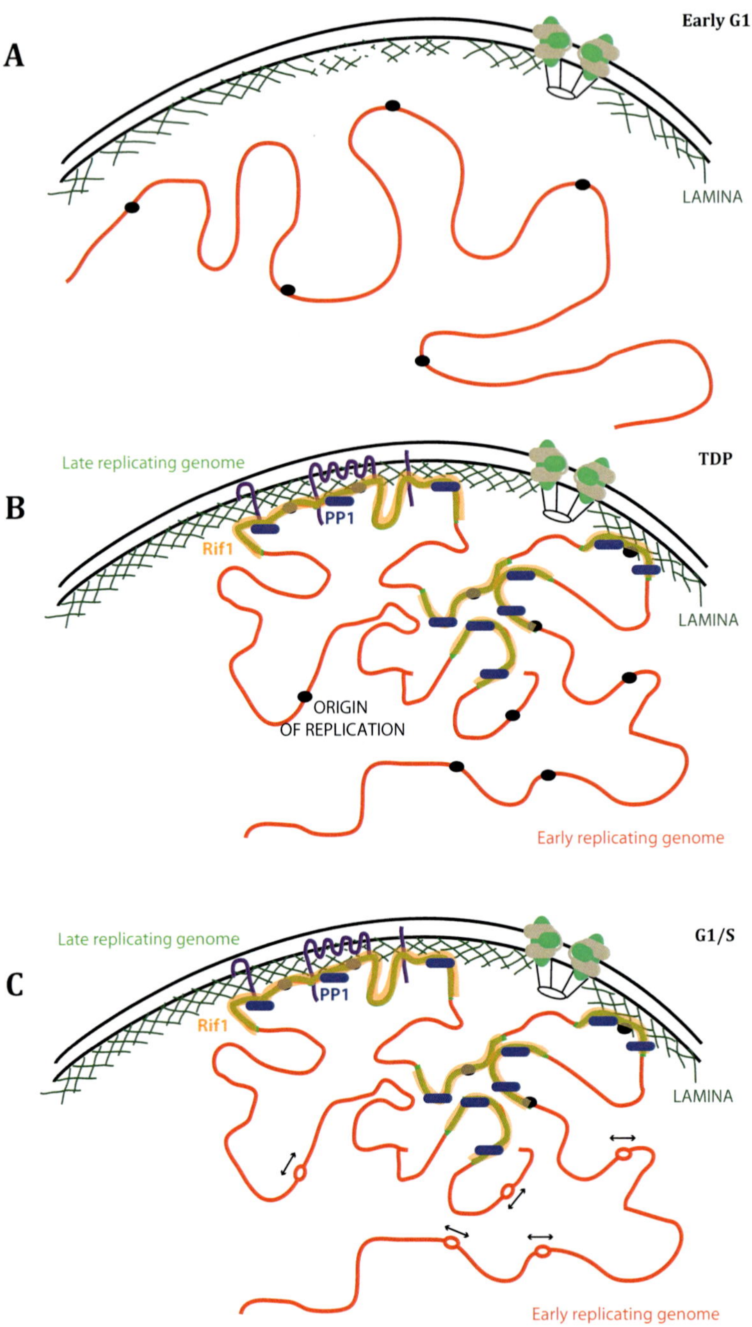

Fig. 12.2 (**a**) At the end of mitosis, replication-timing determinants are lost together with chromatin-organized positions in the nucleus. (**b**) Reestablishment of replication timing takes place in G1. Rif1 associates with chromatin at the end of telophase-early G1 (Yamazaki et al. 2012), in time to enforce the restriction of 3D contacts between genomic regions within the same replication timing at the TDP. The result is the architectural compartmentalization of the late replicating genome and the consequent creation of domains of high PP1 density. (**c**) When DDK activity increases at G1/S, late origins are embedded in RADs and surrounded by high concentrations of PP1. Possible physical sequestration and high concentrations of phosphatase inhibit their firing

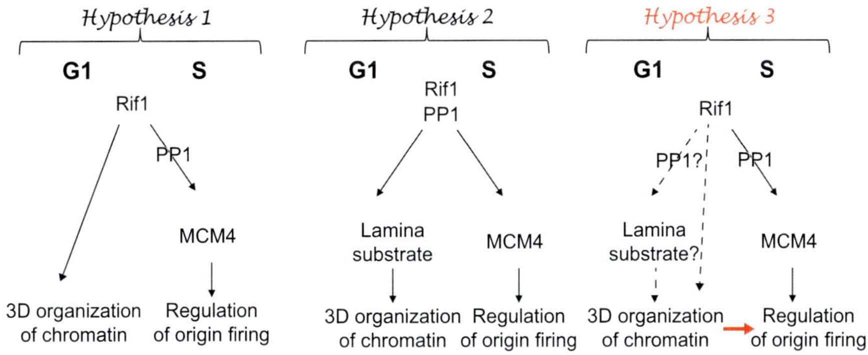

Fig. 12.3 Nuclear architecture and DNA replication timing could be two independent processes connected only by their respective dependence upon Rif1 (Hypothesis 1 and 2). Alternatively, Rif1-dependent organization of nuclear architecture could also influence the firing of replication origins independently of Rif1 (Hypothesis 3)

involved in both processes. However, the loss of replication-timing-specific domain contacts observed in G1 in Rif1 null cells seems to contradict this hypothesis (Fig. 12.3).

Finally, direct transposition of functional data between distant organisms can be misleading. It cannot be excluded that in Metazoa, Rif1 has evolved an architectural function that is absent or not essential in yeast. This change could indeed have taken place concomitantly with the transition from a closed to an open mitosis, the evolution of the lamina, and the need to re-create an ordered nucleus after each mitosis (Sazer et al. 2014). In fact, it is not known whether the lamina or other nuclear architectural components also play a role in either the establishment or execution of the DNA replication-timing program. However, some preliminary observations suggest that they could. We have found that replication-timing regulation of the late replicating genome is indeed differentially sensitive to Rif1 function, depending on the concomitant stable presence of Lamin B1. The regions of the late replicating genome concurrently associated with Rif1 and Lamin B1 (RADs-LB$^+$) do not change their replication timing upon Rif1 deletion. In contrast, late replication of genomic regions associated with Rif1 only (RADs-LB$^-$) is entirely Rif1 dependent (Foti et al. 2016) (Fig. 12.1a, b). We have therefore unexpectedly revealed a further level of complexity in the regulation of replication timing, once again coinciding with some aspects of nuclear architecture organization. This finding also reinforces the question of the role of PP1 in Rif1-mediated control of replication timing. If PP1 is indeed the sole, ultimate effector of Rif1 function, another protein has to be responsible for recruiting PP1 at RADs-LB$^+$ in Rif1 null cells. This is in principle possible, as the PP1 phosphatase has been identified as a partner of different proteins associated with the nuclear envelope, such as AKAP149 (Steen et al. 2000) and LAP1β (Santos et al. 2013).

12.4 Telomere Length Regulation, DNA Repair, Replication Timing, and Nuclear Organization: Where Is the Connection?

Rif1 has been involved in processes other than replication timing and organization of nuclear architecture, namely, DNA repair (Wang et al. 2009; Buonomo et al. 2009; Chapman et al. 2013; Daley and Sung 2013; Feng et al. 2013; Martina et al. 2014) and telomere length regulation (Gallardo et al. 2011; Hardy et al. 1992; Teixeira et al. 2004). Such diversity of roles is puzzling and could either reflect multiple independent functions of different parts of this very large protein, multiple outcomes of a single molecular property of the protein applied in different contexts, or both.

The understanding of the functional organization of Rif1 domains is still rudimentary. The N terminus is highly conserved, composed of numerous HEAT repeats (Silverman et al. 2004; Sreesankar et al. 2012) required for localizing the protein to double-strand breaks in mouse cells, through binding of phosphorylated 53BP1 (Escribano-Diaz et al. 2013). The large middle region is non-conserved and predicted to be a non-structured polypeptide. Interestingly, in both mouse and human Rif1 genes, this region is encoded by a single, ~3000-bp-long exon, accounting for almost half of the entire coding region. This unusual gene organization might suggest acquisition of a mammalian-specific function. The C terminus of Rif1 contains three recognizable domains, indicated as conserved regions C, of which CI and CII are present from yeast to humans, while CIII is only conserved within vertebrates. In Metazoa, CI harbors PP1-interacting motifs, which are located at the N terminus of yeast Rif1. CII is a very intriguing portion of the protein. In Metazoa it contains a DNA-binding domain, preferentially recognizing cruciform structures (Sukackaite et al. 2014; Xu et al. 2010). However, the relevance of this interaction with DNA in vivo is yet to be determined, as Rif1 association with insoluble nuclear fractions renders this technically challenging to assess. Moreover, the in vivo distribution of cruciform DNA is unclear, having been associated with promoter melting, activated origins of replication, and DNA recombination intermediates (Brazda et al. 2011). In vivo DNA binding has been shown for fission yeast Rif1, which recognizes a consensus sequence capable of forming G-quadruplex (G4) (Kanoh et al. 2015). However, the specific domain mediating this interaction is yet to be mapped. A putative DNA-binding domain seems to be present in the budding yeast Rif1 but not in CII (Sreesankar et al. 2012), which is instead required for interaction with Rap1 (Shi et al. 2013), DDK (Hiraga et al. 2014), and for Rif1's tetramerization (Shi et al. 2013). The residues involved in the tetramerization of *S. cerevisiae* Rif1 largely overlap with the positions implicated in DNA binding in the human and mouse homologs, suggesting the possibility that the two functions are related. Multimerization has also been shown for the mammalian protein (Xu et al. 2010), supporting the idea that Rif1 could form patches of a lamina-associated protein network anchoring RADs to the nuclear periphery.

Clearly one of the conserved properties of Rif1 is its interaction with PP1. It is therefore reasonable to hypothesize that PP1 could also mediate Rif1 functions at telomeres (Mattarocci et al. 2016) and/or during the DNA damage response (DDR), by ensuring the dephosphorylation of telomere or DDR-specific substrates. However, at least for budding yeast telomeres, there is an alternative hypothesis that Rif1-dependent control of late replication could be part of the mechanism directing telomerase-dependent telomere elongation. As telomerase-dependent telomere lengthening occurs after conventional replication (Diede and Gottschling 1999), telomere replication in late S phase would in turn limit the time available for telomerase-dependent telomere elongation (Bianchi and Shore 2007). This mechanism could explain how telomere length equilibrium is maintained via differential replication of longer and shorter telomeres: longer telomere length would translate into higher Rif1 occupancy at that specific telomere, and higher Rif1 occupancy would impose later subtelomeric origin firing and telomere replication. Late telomere replication will result in a shorter window of opportunity for telomerase to extend that telomere within a cell cycle. Therefore, longer telomeres are replicated later and are less likely to be extended by telomerase.

12.5 The Intricate Politics of Nuclear Function

The identification of Rif1 has provided an important gateway into understanding the genetic and molecular control of replication timing. However, it has also raised numerous questions, some unexpected, and uncovered a confusing network that interconnects several fundamental nuclear functions. For example, studying the impact of Rif1 deficiency on replication timing, gene expression, and nuclear organization in different cell types has revealed that nuclear architecture could be the common denominator between regulation of gene expression and replication timing. The relationship between these two aspects of nuclear function has been widely debated, as there are general correlations between genomic regions that are early replicating and those that are expressed and, reciprocally, between late replicating regions and heterochromatic, transcriptionally repressed domains. Changes of gene expression and replication timing during development often move in the same direction, with switches of replication from early to late S phase (EtoL) coinciding with reduced gene expression and vice versa. These data have prompted the idea that one aspect of nuclear function could determine the other. However, what controls what has been challenging to determine due to contradicting data (reviewed in Rivera-Mulia and Gilbert 2016). Recently, a detailed analysis of replication timing and gene expression changes at different stages of human ESC differentiation has in fact revealed that the temporal relationship between gene expression and replication-timing changes is different in different regions of the genome (Rivera-Mulia et al. 2015), indicating clearly that at least one important variable is still missing from the picture. Our data indicate that the missing component could be nuclear positioning. We found that "simply" changing replication timing is not sufficient to affect gene expression. Different cell types are either permissive or not for proliferation in the absence of Rif1, probably depending on the status of the DNA damage checkpoints.

For cells like primary mouse embryonic fibroblasts (pMEFs), where Rif1 deletion impairs proliferation, no gene expression changes can be detected as a consequence of the deletion, although chromatin organization and replication timing are affected (Cornacchia et al. 2012; Foti et al. 2016). However, in cells that proliferate upon Rif1 deletion, for example, immortalized MEFs or ESCs, a progressive deregulation of gene expression can be observed in time (Foti et al. 2016). These data suggest that an increasingly amplified deregulation of nuclear architecture could turn into transcriptional changes, in agreement with the evidence linking subnuclear positioning and gene expression (Andrulis et al. 1998; Finlan et al. 2008; Mattout et al. 2011; Peric-Hupkes et al. 2010; Reddy et al. 2008; Zullo et al. 2012). It remains unclear if the effect of nuclear architecture on gene expression is mediated through modification of the epigenetic landscape: we could not detect any major changes in the epigenome after Rif1 deletion, at least for the limited number of histone modifications we have examined (Foti et al. 2016).

12.6 Conclusions

The discovery of Rif1 and its multifaceted functions has represented an entry point into the molecular labyrinth of nuclear functions. Future work will have to span across different fields in order to unravel how specific to Rif1 this functional network is or if it is a general feature of nuclear structural components. Importantly, understanding to what extent PP1 mediates different aspects of Rif1's role bears the potential of creating separation of function mutants. Integrating different aspects of nuclear function is fundamental to dissect the complexity of cellular transitions such as aging, differentiation, and transformation.

Acknowledgments I would like to thank past and present members of the Buonomo group for their work that is the base of this review and the invaluable scientific contribution. I am also grateful to Svetlana Makovets and Elizabeth Bayne for critical reading of the manuscript and helpful suggestions. I apologize for the references omitted due to general nature of this review.

References

Alver RC, Chadha GS, Gillespie PJ, Blow JJ (2017) Reversal of DDK-mediated MCM phosphorylation by Rif1-PP1 regulates replication initiation and replisome stability independently of ATR/Chk1. Cell Rep 18:2508–2520

Andrulis ED, Neiman AM, Zappulla DC, Sternglanz R (1998) Perinuclear localization of chromatin facilitates transcriptional silencing. Nature 394:592–595

Arnoult N, Schluth-Bolard C, Letessier A, Drascovic I, Bouarich-Bourimi R, Campisi J, Kim SH, Boussouar A, Ottaviani A, Magdinier F et al (2010) Replication timing of human telomeres is chromosome arm-specific, influenced by subtelomeric structures and connected to nuclear localization. PLoS Genet 6:e1000920

Bianchi A, Shore D (2007) Early replication of short telomeres in budding yeast. Cell 128:1051–1062

Brazda V, Laister RC, Jagelska EB, Arrowsmith C (2011) Cruciform structures are a common DNA feature important for regulating biological processes. BMC Mol Biol 12:33

Buonomo SB, Wu Y, Ferguson D, de Lange T (2009) Mammalian Rif1 contributes to replication stress survival and homology-directed repair. J Cell Biol 187:385–398

Chapman JR, Barral P, Vannier JB, Borel V, Steger M, Tomas-Loba A, Sartori AA, Adams IR, Batista FD, Boulton SJ (2013) RIF1 is essential for 53BP1-dependent nonhomologous end joining and suppression of DNA double-strand break resection. Mol Cell 49:858–871

Conrad MN, Wright JH, Wolf AJ, Zakian VA (1990) RAP1 protein interacts with yeast telomeres in vivo: overproduction alters telomere structure and decreases chromosome stability. Cell 63:739–750

Cornacchia D, Dileep V, Quivy JP, Foti R, Tili F, Santarella-Mellwig R, Antony C, Almouzni G, Gilbert DM, Buonomo SB (2012) Mouse Rif1 is a key regulator of the replication-timing programme in mammalian cells. EMBO J 31:3678–3690

Daley JM, Sung P (2013) RIF1 in DNA break repair pathway choice. Mol Cell 49:840–841

Dave A, Cooley C, Garg M, Bianchi A (2014) Protein phosphatase 1 recruitment by Rif1 regulates DNA replication origin firing by counteracting DDK activity. Cell Rep 7:53–61

Diede SJ, Gottschling DE (1999) Telomerase-mediated telomere addition in vivo requires DNA primase and DNA polymerases alpha and delta. Cell 99:723–733

Dileep V, Ay F, Sima J, Vera DL, Noble WS, Gilbert DM (2015) Topologically associating domains and their long-range contacts are established during early G1 coincident with the establishment of the replication-timing program. Genome Res 25(8):1104–1113

Dimitrova DS, Gilbert DM (1999) The spatial position and replication timing of chromosomal domains are both established in early G1 phase. Mol Cell 4:983–993

Dixon JR, Selvaraj S, Yue F, Kim A, Li Y, Shen Y, Hu M, Liu JS, Ren B (2012) Topological domains in mammalian genomes identified by analysis of chromatin interactions. Nature 485:376–380

Escribano-Diaz C, Orthwein A, Fradet-Turcotte A, Xing M, Young JT, Tkac J, Cook MA, Rosebrock AP, Munro M, Canny MD et al (2013) A cell cycle-dependent regulatory circuit composed of 53BP1-RIF1 and BRCA1-CtIP controls DNA repair pathway choice. Mol Cell 49:872–883

Feng L, Fong KW, Wang J, Wang W, Chen J (2013) RIF1 counteracts BRCA1-mediated end resection during DNA repair. J Biol Chem 288:11135–11143

Finlan LE, Sproul D, Thomson I, Boyle S, Kerr E, Perry P, Ylstra B, Chubb JR, Bickmore WA (2008) Recruitment to the nuclear periphery can alter expression of genes in human cells. PLoS Genet 4:e1000039

Foti R, Gnan S, Cornacchia D, Dileep V, Bulut-Karslioglu A, Diehl S, Buness A, Klein FA, Huber W, Johnstone E et al (2016) Nuclear architecture organized by Rif1 underpins the replication-timing program. Mol Cell 61:260–273

Fu YV, Yardimci H, Long DT, Ho TV, Guainazzi A, Bermudez VP, Hurwitz J, van Oijen A, Scharer OD, Walter JC (2011) Selective bypass of a lagging strand roadblock by the eukaryotic replicative DNA helicase. Cell 146:931–941

Gallardo F, Laterreur N, Cusanelli E, Ouenzar F, Querido E, Wellinger RJ, Chartrand P (2011) Live cell imaging of telomerase RNA dynamics reveals cell cycle-dependent clustering of telomerase at elongating telomeres. Mol Cell 44:819–827

Guruharsha KG, Rual JF, Zhai B, Mintseris J, Vaidya P, Vaidya N, Beekman C, Wong C, Rhee DY, Cenaj O et al (2011) A protein complex network of *Drosophila melanogaster*. Cell 147:690–703

Hardy CF, Sussel L, Shore D (1992) A RAP1-interacting protein involved in transcriptional silencing and telomere length regulation. Genes Dev 6:801–814

Hayano M, Kanoh Y, Matsumoto S, Renard-Guillet C, Shirahige K, Masai H (2012) Rif1 is a global regulator of timing of replication origin firing in fission yeast. Genes Dev 26:137–150

Hiraga S, Alvino GM, Chang F, Lian HY, Sridhar A, Kubota T, Brewer BJ, Weinreich M, Raghuraman MK, Donaldson AD (2014) Rif1 controls DNA replication by directing Protein

Phosphatase 1 to reverse Cdc7-mediated phosphorylation of the MCM complex. Genes Dev 28:372–383

Hiraga SI, Ly T, Garzon J, Horejsi Z, Ohkubo YN, Endo A, Obuse C, Boulton SJ, Lamond AI, Donaldson AD (2017) Human RIF1 and protein phosphatase 1 stimulate DNA replication origin licensing but suppress origin activation. EMBO Rep 18:403–419

Hiratani I, Ryba T, Itoh M, Yokochi T, Schwaiger M, Chang CW, Lyou Y, Townes TM, Schubeler D, Gilbert DM (2008) Global reorganization of replication domains during embryonic stem cell differentiation. PLoS Biol 6:e245

Hiratani I, Ryba T, Itoh M, Rathjen J, Kulik M, Papp B, Fussner E, Bazett-Jones DP, Plath K, Dalton S et al (2010) Genome-wide dynamics of replication timing revealed by in vitro models of mouse embryogenesis. Genome Res 20:155–169

Hultdin M, Gronlund E, Norrback KF, Just T, Taneja K, Roos G (2001) Replication timing of human telomeric DNA and other repetitive sequences analyzed by fluorescence in situ hybridization and flow cytometry. Exp Cell Res 271:223–229

Kanoh Y, Matsumoto S, Fukatsu R, Kakusho N, Kono N, Renard-Guillet C, Masuda K, Iida K, Nagasawa K, Shirahige K et al (2015) Rif1 binds to G quadruplexes and suppresses replication over long distances. Nat Struct Mol Biol 22:889–897

Knott SR, Peace JM, Ostrow AZ, Gan Y, Rex AE, Viggiani CJ, Tavare S, Aparicio OM (2012) Forkhead transcription factors establish origin timing and long-range clustering in S. cerevisiae. Cell 148:99–111

Levy DL, Blackburn EH (2004) Counting of Rif1p and Rif2p on *Saccharomyces cerevisiae* telomeres regulates telomere length. Mol Cell Biol 24:10857–10867

Lian HY, Robertson ED, Hiraga S, Alvino GM, Collingwood D, McCune HJ, Sridhar A, Brewer BJ, Raghuraman MK, Donaldson AD (2011) The effect of Ku on telomere replication time is mediated by telomere length but is independent of histone tail acetylation. Mol Biol Cell 22:1753–1765

Lieberman-Aiden E, van Berkum NL, Williams L, Imakaev M, Ragoczy T, Telling A, Amit I, Lajoie BR, Sabo PJ, Dorschner MO et al (2009) Comprehensive mapping of long-range interactions reveals folding principles of the human genome. Science (New York, NY) 326:289–293

Lustig AJ, Kurtz S, Shore D (1990) Involvement of the silencer and UAS binding protein RAP1 in regulation of telomere length. Science (New York, NY) 250:549–553

Mantiero D, Mackenzie A, Donaldson A, Zegerman P (2011) Limiting replication initiation factors execute the temporal programme of origin firing in budding yeast. EMBO J 30:4805–4814

Martina M, Bonetti D, Villa M, Lucchini G, Longhese MP (2014) Saccharomyces cerevisiae Rif1 cooperates with MRX-Sae2 in promoting DNA-end resection. EMBO Rep 15:695–704

Mattarocci S, Shyian M, Lemmens L, Damay P, Altintas DM, Shi T, Bartholomew CR, Thoma NH, Hardy CF, Shore D (2014) Rif1 controls DNA replication timing in yeast through the PP1 phosphatase Glc7. Cell Rep 7:62–69

Mattarocci S, Hafner L, Lezaja A, Shyian M, Shore D (2016) Rif1: a conserved regulator of DNA replication and repair hijacked by telomeres in yeasts. Front Genet 7(e1002024):45

Mattout A, Pike BL, Towbin BD, Bank EM, Gonzalez-Sandoval A, Stadler MB, Meister P, Gruenbaum Y, Gasser SM (2011) An EDMD mutation in C. elegans lamin blocks muscle-specific gene relocation and compromises muscle integrity. Curr Biol 21:1603–1614

Moorhead GB, Trinkle-Mulcahy L, Nimick M, De Wever V, Campbell DG, Gourlay R, Lam YW, Lamond AI (2008) Displacement affinity chromatography of protein phosphatase one (PP1) complexes. BMC Biochem 9:28

Patel PK, Arcangioli B, Baker SP, Bensimon A, Rhind N (2006) DNA replication origins fire stochastically in fission yeast. Mol Biol Cell 17:308–316

Peace JM, Ter-Zakarian A, Aparicio OM (2014) Rif1 regulates initiation timing of late replication origins throughout the S. cerevisiae genome. PLoS One 9:e98501

Peric-Hupkes D, Meuleman W, Pagie L, Bruggeman SW, Solovei I, Brugman W, Graf S, Flicek P, Kerkhoven RM, van Lohuizen M et al (2010) Molecular maps of the reorganization of genome-nuclear lamina interactions during differentiation. Mol Cell 38:603–613

Pope BD, Ryba T, Dileep V, Yue F, Wu W, Denas O, Vera DL, Wang Y, Hansen RS, Canfield TK et al (2014) Topologically associating domains are stable units of replication-timing regulation. Nature 515:402–405

Raghuraman MK, Brewer BJ, Fangman WL (1997) Cell cycle-dependent establishment of a late replication program. Science (New York, NY) 276:806–809

Reddy KL, Zullo JM, Bertolino E, Singh H (2008) Transcriptional repression mediated by repositioning of genes to the nuclear lamina. Nature 452:243–247

Rivera-Mulia JC, Gilbert DM (2016) Replication timing and transcriptional control: beyond cause and effect-part III. Curr Opin Cell Biol 40:168–178

Rivera-Mulia JC, Buckley Q, Sasaki T, Zimmerman J, Didier RA, Nazor K, Loring JF, Lian Z, Weissman S, Robins AJ et al (2015) Dynamic changes in replication timing and gene expression during lineage specification of human pluripotent stem cells. Genome Res 25:1091–1103

Roux KJ, Kim DI, Raida M, Burke B (2012) A promiscuous biotin ligase fusion protein identifies proximal and interacting proteins in mammalian cells. J Cell Biol 196:801–810

Ryba T, Hiratani I, Lu J, Itoh M, Kulik M, Zhang J, Schulz TC, Robins AJ, Dalton S, Gilbert DM (2010) Evolutionarily conserved replication timing profiles predict long-range chromatin interactions and distinguish closely related cell types. Genome Res 20:761–770

Santos M, Rebelo S, Van Kleeff PJ, Kim CE, Dauer WT, Fardilha M, da Cruz ESOA, da Cruz ESEF (2013) The nuclear envelope protein, LAP1B, is a novel protein phosphatase 1 substrate. PLoS One 8:e76788

Sazer S, Lynch M, Needleman D (2014) Deciphering the evolutionary history of open and closed mitosis. Curr Biol 24:R1099–R1103

Shi T, Bunker RD, Mattarocci S, Ribeyre C, Faty M, Gut H, Scrima A, Rass U, Rubin SM, Shore D et al (2013) Rif1 and Rif2 shape telomere function and architecture through multivalent Rap1 interactions. Cell 153:1340–1353

Shimi T, Butin-Israeli V, Adam SA, Goldman RD (2010) Nuclear lamins in cell regulation and disease. Cold Spring Harb Symp Quant Biol 75:525–531

Silverman J, Takai H, Buonomo SB, Eisenhaber F, de Lange T (2004) Human Rif1, ortholog of a yeast telomeric protein, is regulated by ATM and 53BP1 and functions in the S-phase checkpoint. Genes Dev 18:2108–2119

Sreesankar E, Senthilkumar R, Bharathi V, Mishra RK, Mishra K (2012) Functional diversification of yeast telomere associated protein, Rif1, in higher eukaryotes. BMC Genomics 13:255

Sreesankar E, Bharathi V, Mishra RK, Mishra K (2015) Drosophila Rif1 is an essential gene and controls late developmental events by direct interaction with PP1-87B. Sci Rep 5:10679

Steen RL, Martins SB, Tasken K, Collas P (2000) Recruitment of protein phosphatase 1 to the nuclear envelope by A-kinase anchoring protein AKAP149 is a prerequisite for nuclear lamina assembly. J Cell Biol 150:1251–1262

Sukackaite R, Jensen MR, Mas PJ, Blackledge M, Buonomo SB, Hart DJ (2014) Structural and biophysical characterization of murine Rif1 C terminus reveals high specificity for DNA cruciform structures. J Biol Chem 289:13903–13911

Sukackaite R, Cornacchia D, Jensen MR, Mas PJ, Blackledge M, Enervald E, Duan G, Auchynnikava T, Kohn M, Hart DJ et al (2017) Mouse Rif1 is a regulatory subunit of protein phosphatase 1 (PP1). Sci Rep 7:2119

Takebayashi S, Dileep V, Ryba T, Dennis JH, Gilbert DM (2012) Chromatin-interaction compartment switch at developmentally regulated chromosomal domains reveals an unusual principle of chromatin folding. Proc Natl Acad Sci U S A 109:12574–12579

Tanaka S, Nakato R, Katou Y, Shirahige K, Araki H (2011) Origin association of Sld3, Sld7, and Cdc45 proteins is a key step for determination of origin-firing timing. Curr Biol 21:2055–2063

Teixeira MT, Arneric M, Sperisen P, Lingner J (2004) Telomere length homeostasis is achieved via a switch between telomerase- extendible and -nonextendible states. Cell 117:323–335

Trinkle-Mulcahy L, Andersen J, Lam YW, Moorhead G, Mann M, Lamond AI (2006) Repo-Man recruits PP1 gamma to chromatin and is essential for cell viability. J Cell Biol 172:679–692

Wang et al (2009) Human RIF1 encodes an anti-apoptotic factor required for DNA repair. Carcinogenesis 8:1314–1319
Williams RR, Azuara V, Perry P, Sauer S, Dvorkina M, Jorgensen H, Roix J, McQueen P, Misteli T, Merkenschlager M et al (2006) Neural induction promotes large-scale chromatin reorganisation of the Mash1 locus. J Cell Sci 119:132–140
Wright WE, Tesmer VM, Liao ML, Shay JW (1999) Normal human telomeres are not late replicating. Exp Cell Res 251:492–499
Wu PY, Nurse P (2009) Establishing the program of origin firing during S phase in fission yeast. Cell 136:852–864
Xu L, Blackburn EH (2004) Human Rif1 protein binds aberrant telomeres and aligns along anaphase midzone microtubules. J Cell Biol 167:819–830
Xu D, Muniandy P, Leo E, Yin J, Thangavel S, Shen X, Ii M, Agama K, Guo R, Fox D 3rd et al (2010) Rif1 provides a new DNA-binding interface for the Bloom syndrome complex to maintain normal replication. EMBO J 29:3140–3155
Yaffe E, Farkash-Amar S, Polten A, Yakhini Z, Tanay A, Simon I (2010) Comparative analysis of DNA replication timing reveals conserved large-scale chromosomal architecture. PLoS Genet 6:e1001011
Yamazaki S, Ishii A, Kanoh Y, Oda M, Nishito Y, Masai H (2012) Rif1 regulates the replication timing domains on the human genome. EMBO J 31:3667–3677
Zhou J, Ermakova OV, Riblet R, Birshtein BK, Schildkraut CL (2002) Replication and subnuclear location dynamics of the immunoglobulin heavy-chain locus in B-lineage cells. Mol Cell Biol 22:4876–4889
Zou Y, Gryaznov SM, Shay JW, Wright WE, Cornforth MN (2004) Asynchronous replication timing of telomeres at opposite arms of mammalian chromosomes. Proc Natl Acad Sci U S A 101:12928–12933
Zullo JM, Demarco IA, Pique-Regi R, Gaffney DJ, Epstein CB, Spooner CJ, Luperchio TR, Bernstein BE, Pritchard JK, Reddy KL et al (2012) DNA sequence-dependent compartmentalization and silencing of chromatin at the nuclear lamina. Cell 149:1474–1487

Chapter 13
G-Quadruplexes and DNA Replication Origins

Marie-Noëlle Prioleau

Abstract DNA replication ensures the accurate duplication of the genome at each cell cycle. During S phase, tens of thousands of replication origins throughout the vertebrate genome are activated according to a spatiotemporal program. The genome-wide mapping of origins in several model systems has identified G-quadruplexes—higher-order DNA structures formed from G-rich sequences—as potential key regulators of origin activity. Here, I describe genetic experiments demonstrating the role of G-quadruplexes in origin function. I discuss the different means by which G-quadruplexes might regulate origin function. Finally, comparisons of replicon organization in the three domains of life suggest that G-quadruplexes may have retained a conserved role in origin function during evolution.

Keywords Replication origin • G-quadruplex • Chromatin • Evolution

13.1 Main Text

In each cell cycle, tens of thousands of start sites for DNA replication are activated to ensure the faithful duplication of complex vertebrate genomes. The DNA replication program is responsible for guaranteeing the correct inheritance of genetic information, but replication origin selection and the timing of activation are flexible. This plasticity facilitates adaptation of the replication program to changes associated with the establishment of cell type specificity, which mostly involves transcriptional modifications. DNA sequence and chromosomal environment contribute to the selection and activation of mammalian replication origins through complex processes including epigenetic modifications, the binding of specific *trans*-acting factors and nuclear localization. These aspects are described in Chaps. 2 and 3 of this book. This more tightly focused section concentrates on the recent identification of a DNA motif common to most strong replication origins in vertebrates.

M.-N. Prioleau (✉)
Institut Jacques Monod, CNRS UMR7592, Université Paris Diderot, Equipe Labellisée Association pour la Recherche sur le Cancer, Paris, France
e-mail: marie-noelle.prioleau@ijm.fr

H. Masai, M. Foiani (eds.), *DNA Replication*, Advances in Experimental Medicine and Biology 1042, https://doi.org/10.1007/978-981-10-6955-0_13

This motif may be considered a "consensus" element within replication origins. We will discuss its potential role in controlling replication initiation in vertebrates and other species.

The distribution of replication origins throughout the genome defines the replicons: regions copied by the bidirectional progression of two replication forks arising from a single origin. Pioneering pulse radio-labeling experiments have estimated mean replicon size at about 150 kb in human cells (Berezney et al. 2000). The replicon model proposed by Jacob, Brenner, and Cuzin describes the regulation of bacterial chromosome replication. According to this model, a positive *trans*-acting factor known as the initiator binds to a specific *cis*-element, the replicator. This binding induces the bidirectional synthesis of the bacterial chromosome in response to a signal during S phase (Jacob et al. 1963). This model was validated by the identification of the DnaA protein as the initiator and of oriC as the replicator for replication of the *E. coli* chromosome. The large size of the genomes of eukaryotic cells results in a need for hundreds (yeast) to tens of thousands (vertebrates) of replicons. A spatiotemporal program of DNA replication is then established by the coordinated activation of these replicons.

Replication is initiated in a two-phase process, to ensure that each replicon is replicated once, and only once, in each cell cycle. During G1 phase, two copies of the MCM2-7 ATPases are loaded onto replication origins in a head-to-head orientation, to form the "pre-replicative complex" (pre-RC). This step requires three "licensing" factors, the hexameric ORC, Cdc6, and Cdt1. Pre-RC formation can occur only in late mitosis and G1 phase, because CDK (cyclin-dependent kinase), a potent inhibitor of pre-RC formation, is inhibited by APC/C and cyclin-dependent kinase inhibitors (CKIs) during these phases of the cell cycle (Diffley 2004). Many more origins than are actually used during S phase are licensed during this period, to ensure that the DNA is loaded with sufficiently large numbers of pre-RCs before the start of S phase. These "dormant" origins provide a backup if the fork stalls and may render the replication program more flexible and robust to perturbation (Ge et al. 2007; Ibarra et al. 2008). At the transition between the G1 and S phases, the APC/C complex and CKIs are inactivated, allowing S phase to be initiated through the cumulative effects of two kinases essential for pre-RC activation: S-CDK and DDK (a Dbf4-dependent kinase composed of Cdc7 kinase and its activator). However, not all replication origins fire at the same time. Instead, each cell type has its own precise temporal program, regulating the coordinated firing of groups of origins and defining replication timing domains with similar replication timing patterns (Hiratani et al. 2008). The activation of replication origins during S phase is regulated in part by limiting concentrations of key activating factors (Mantiero et al. 2011; Tanaka et al. 2011). These important findings led to the development of the hypothesis that there is a general mechanism for the correct temporal activation of origins during S phase. In early S phase, only early-firing origins can be activated, because these origins are the most "accessible" to the limiting firing factors. These limiting factors are released after firing and can then bind to and activate less accessible origins and so on. This discovery led to studies aiming to identify the factors regulating the accessibility of specific pre-RCs to these limiting factors.

13.2 Vertebrate Replication Origins Are Enriched in Potential G4 Motifs

The highest-resolution method currently available for origin mapping in vertebrate cells is based on the purification and quantification of short nascent strands (SNS) (Prioleau and MacAlpine 2016). In this method, 1.5–2.5 kb nascent strands specific to replication origins are purified on the basis of the resistance of these strands to λ-exonuclease digestion due to the incorporation, by the primase, of small RNA primers at their 5′ ends (Bielinsky and Gerbi 1998) (Fig. 13.1a). This step is necessary as it results in digestion of the large excess of broken genomic DNA, which would generate a nonspecific signal if not correctly removed. The initial large-scale mapping of replication origins over a fraction of the genome (1%) in vertebrate cells provided important new insight into the potential mechanisms underlying origin firing (Cadoret et al. 2008; Sequeira-Mendes et al. 2009). These analyses revealed that replication origins were nonrandomly distributed, being particularly abundant in transcriptional regulatory elements, such as TSS and enhancer regions. CpG islands (CGI), which are associated with ~50% of TSS, were identified as the most efficient origins, and about 35% of all origins are of this type (Cadoret et al. 2008; Sequeira-Mendes et al. 2009). These analyses paved the way for genome-wide studies, which confirmed the observations made at a smaller scale. Moreover, these new studies showed that potential G4 motifs (pG4s) were particularly abundant within replication origins in mouse and human cells (Cayrou et al. 2011, 2012; Besnard et al. 2012).

PG4s consist of four tracts of at least three guanine residues separated by other bases (1–7), and they can fold to form a secondary structure known as a G-quadruplex (G4). The G4 is a four-stranded helical DNA structure constructed from G-quartet building blocks, square planar assemblies of four Hoogsteen-bonded guanine bases. These planar G-quartets stack on top of each another, forming four-stranded helical structures. Computational analyses have identified about 370,000 pG4s in the human genome (Huppert and Balasubramanian 2005). The distribution of pG4s is not random; they tend to accumulate in promoter regions and in the 5′ or 3′-untranslated regions of mRNA. There therefore seems to be a selection pressure driving the retention of these motifs at specific sites (Rhodes and Lipps 2015). They also cluster at telomeres, which, in humans, consist of tandemly arranged TTAGGG repeats. Several lines of indirect evidence have convincingly demonstrated the formation of G4 structures in vivo (Valton and Prioleau 2016). The development of specific G-quadruplex-interacting drugs has led to the identification of perturbed genomic regions in vivo that are prone to the formation of G4 structures. Pyridostatin, a specific G4 ligand, has been shown to induce transcription- and replication-dependent DNA damage, resulting in arrest of the cell cycle in the G2/M phase and activation

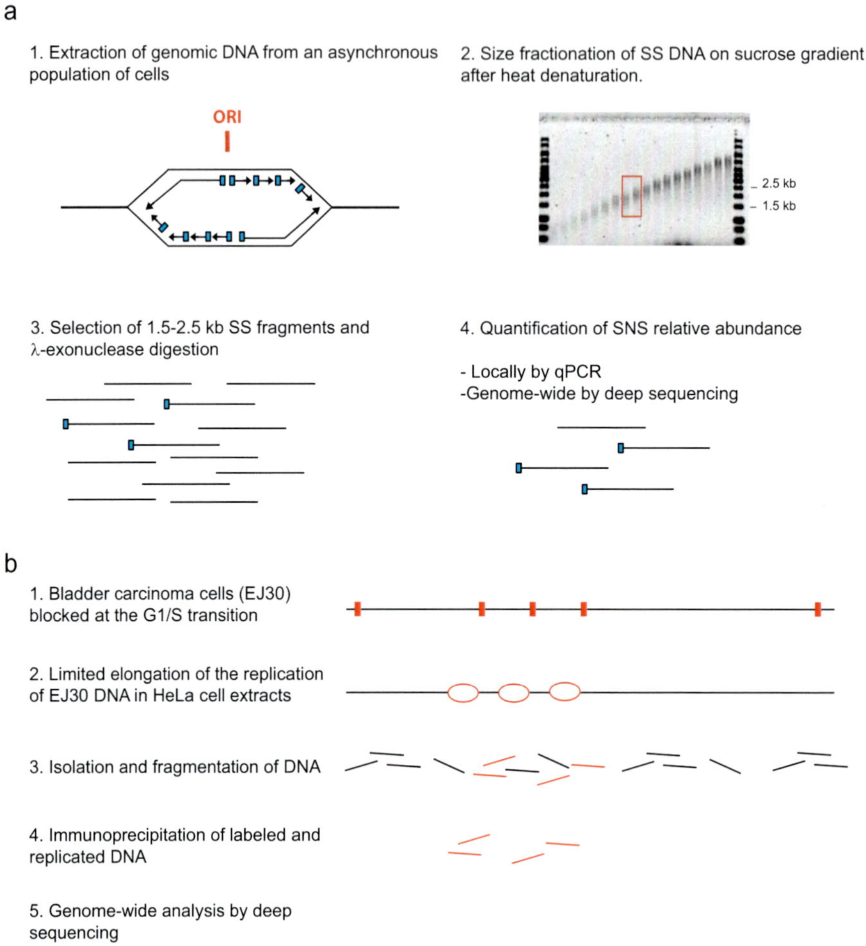

Fig. 13.1 Methods for the genome-wide mapping of replication origins. (**a**) Description of the main steps in the purification of short nascent strands. A bubble located at a replication origin is represented, leading strands and Okazaki fragments synthesized on the lagging strands contain small RNA primers (blue rectangles). Genomic DNA is first gently extracted from an asynchronous population of cells. The purified DNA is then denatured by heating and run on a 5–30% linear sucrose gradient. Fractions containing SS DNA of 1.5–2.5 kb in size are selected. The DNA is phosphorylated and digested with a large excess of λ-exonuclease to remove excess nonspecific broken DNA; the SNSs are protected by their small RNA primers. Analyses may be carried out locally, by qPCR, or genome-wide, by deep sequencing the SNS-enriched preparation. (**b**) Genome-wide mapping of early-firing origins by ini-seq. Cells are synchronized by treatment with mimosine at the G1/S transition to allow origin licensing. Licensed origins are represented by red rectangles. The purified nuclei are then transferred into a HeLa S phase extract, to ensure limited elongation and the labeling of early-firing origins. These origins are indicated by red bubbles. DNA is then isolated and fragmented. Labeled DNA is purified and subjected to deep sequencing (Adapted form Langley et al. 2016)

of the ATR/Chk1 pathway (Rodriguez et al. 2012). Genome-wide mapping of γH2AX, a DNA damage marker, after the addition of pyridostatin showed that this drug mostly targeted gene-rich regions containing clusters of sequences capable of adopting a G-quadruplex conformation. These data suggested that pyridostatin interacted predominantly with non-telomeric regions at low concentration and demonstrated the formation of structured G4s throughout the human genome. The formation of G4 structures was recently captured directly in vivo (Hänsel-Hertsch et al. 2016). In this study, a G4 ChIP-seq protocol was developed with an engineered antibody specific for G4 structures, to map the location of these structures genome-wide in human epidermal keratinocyte HaCaT cells. About 10,000 G4 peaks were mapped, mostly in regulatory regions depleted of nucleosomes. G4 structures were particularly abundant in the promoters and 5′UTRs of highly transcribed genes. Moreover, enhanced G4 formation was found to be associated with stronger transcriptional activity, suggesting links between chromatin structure, gene expression, and transcription in the formation of G4 structures.

13.3 Does the Enrichment of SNSs in pG4 Reflect a Functional Role of G4 in Replication Origin Firing?

Studies in human and mouse cell lines have shown that ~80% of the origins mapped by the SNS method overlap with a pG4 (Cayrou et al. 2011, 2012; Besnard et al. 2012). However, not all the pG4s overlapped with a replication origin, suggesting that pG4s may be important determinants of origin specification but that they are not sufficient on their own (Picard et al. 2014). The discovery that pG4s might play a positive role in replication initiation came as a surprise, as pG4s were generally considered to impede the progression of replication forks (Valton and Prioleau 2016). It was suggested that the strong enrichment in pG4 of origins detected by SNS was a bias induced by the use of λ-exonuclease (Foulk et al. 2015). The authors of this study observed that the digestion of genomic DNA with low concentration of λ-exonuclease (one thousandth those used for SNS purification) tended to result in an enrichment in GC-rich and pG4-containing sequences, due to the lower efficiency of digestion for such sequences with this enzyme. However, the digestion of DNA from nonproliferating cells with a large excess of λ-exonuclease, as in classical SNS preparations, does not lead to specific peaks at pG4-rich sequences (Cayrou et al. 2011, 2015). It is therefore important to follow a very strict protocol for λ-exonuclease digestion. An orthogonal method with a resolution similar to that of SNS was recently developed, to exclude definitively any possible bias due to λ-exonuclease treatment (Langley et al. 2016). In this new method, initiation site sequencing (ini-seq), nuclei were synchronized at the G1/S border by treatment with mimosine and were then transferred to S phase extract (Fig. 13.1b). Newly synthesized DNA was directly labeled with digoxigenin-dUTP, with a short pulse corresponding to elongation by 1.5–2.5 kb. The labeled DNA was purified and analyzed by deep sequencing. This method, which detects only early-firing origins, identified ~25,000 discrete peaks. About 50% of the activated origins identified by ini-seq were at TSS and contained pG4s. There was also a high level of concordance between the sites identified in this way and those identified by SNS-seq, validating the λ-exonuclease approach. Thus, λ-exonuclease, when used properly, is not subject to strong biases, contrary to the suggestions of Foulk et al. (2015).

Many potential initiation sites may be used stochastically within zones of initiation, but SNS preparations may be artificially enriched in sites containing pG4, due to transient pausing at G4 structures. This hypothesis is supported by the observation that SNS enrichment peaks are systematically located 3′ to a pG4, with a sharp decrease observed at the position of the pG4. This peak shape suggests that there is, indeed, a very transient pause at pG4s (Valton et al. 2014; Comoglio et al. 2015). However, the ini-seq method, which is based on very limited elongation, revealed a strong enrichment of TSS containing pG4s, suggesting that site-specific initiation occurs at these positions, with other regions not overrepresented.

A recent genetic study in the chicken DT40 cell line used a new functional origin assay to investigate the role of pG4s in replication initiation (Valton et al. 2014). This test is based on the ability of active origins to advance the replication timing of regions that are naturally replicated late (Fig. 13.2a). The well-characterized chicken β^A origin, which contains a pG4, was inserted, by homologous recombination, into a region devoid of strong initiation events that normally replicates in mid-late S phase. The insertion of the chicken β^A origin was sufficient to cause local SNS enrichment, demonstrating the ability of this origin to function in an ectopic position. However, this origin was not sufficient to advance replication timing. The flanking of this origin with binding sites for USF (upstream stimulating factor), a transcription factor, modified local histone H3 acetylation and H3K4me2, which advances replication timing in a β^A origin-dependent manner (Hassan-Zadeh et al. 2012) (Fig. 13.2b, WT). This functional assay demonstrated that the β^A promoter contained a bona fide origin. The role of the pG4 within the β^A origin was also investigated in detail. Point mutations decreasing the stability of the G4 structure in vitro also reduced origin efficiency, as shown by SNS and timing assays (Fig. 13.2b). Thus, G-quadruplex structures play a critical positive role in replication initiation, independent of fork pausing (Valton et al. 2014). Furthermore, the orientation of the pG4 within the β^A origin determines the position at which replication is initiated, consistent with genome-wide data showing a SNS enrichment peak ~220 bp 3′ to pG4 (Cayrou et al. 2012). Finally, cooperation between the pG4 of the β^A model origin and a 250 bp module located 3′ to it was required for origin function (Fig. 13.2b, Δ3′). This finding is not consistent with SNS enrichment being due to a replication fork barrier induced by the pG4, because the pG4 is maintained in this construct. Deletion of the pG4 in a second model origin also abolished origin function. Overall, these results provide a formal demonstration that, at least in two model origins, pG4 is required for origin function.

a

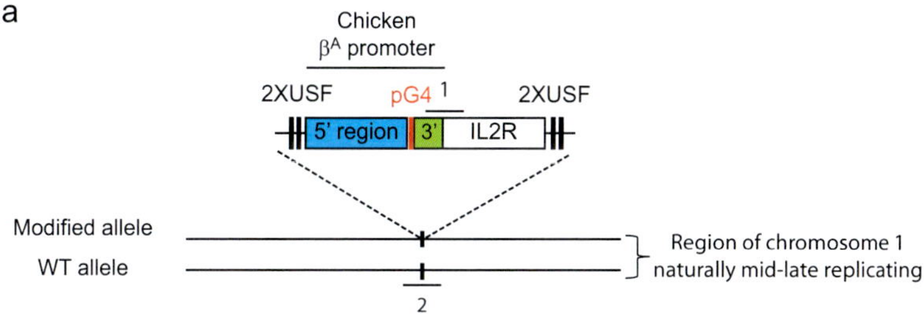

b

Construct	Relative SNS enrichment	Replication timing shift	PG4 probability
WT GGGGGGGGGGGGCGGG	100%	+	1
Δ5’	62%	ND	1
Δ3’	13%	-	1
ΔpG4	11%	-	0
m14 GGGGGGGGGGGGAGGG	95%	+	1
m4 GGGAGGGGGGGGCGGG	56%	ND	0.6
m9 GGGGGGGGAGGGGCGGG	34%	ND	0.6
m6 GGGGGAGGGGGGGCGGG	18%	-	0.3
m16 GGGGGGGGGGGGCGAG	25%	-	0
m12 GGGGGGGGGGAGCGGG	11%	-	0.1

Fig. 13.2 A G4 and a cooperating *cis*-module are required for the activity of the chicken β^A globin origin. (**a**) Strategy for dissecting the role of *cis*-elements in origin activity. A construct containing a 1.1 kb fragment overlapping the chicken β^A globin promoter fused to the reporter IL2R gene and flanked by two copies of binding sites for the transcription factor USF was inserted by homologous recombination. The β^A promoter is represented by three fragments: a blue rectangle corresponding to an 850 bp region upstream from the pG4, a red rectangle corresponding to the pG4, and a green rectangle corresponding to a 250 bp region located 3′ to the pG4. Site-specific insertion was achieved in a 50 kb region of chromosome 1 devoid of strong initiation sites and replicated in mid-late S phase. Origin activity was tested in two independent assays. Relative SNS enrichment was quantified with primer pair 1 (black horizontal bar), systematically compared with a background signal located 5 kb away from the insertion site. SNS were also quantified at an endogenous origin (positive control). In the second method, the capacity of the transgene to advance replication timing was investigated with a pair of primers specific for the transgene (primer pair 1), with comparison to the WT allele analyzed with a primer pair overlapping the insertion site (primer pair 2). (**b**) Analyses of origin activity in different constructs (Adapted from Valton et al. 2014). As a reference, the WT ectopic β^A origin was considered to display 100% relative enrichment. Constructs containing either deletions of part of the β^A promoter (Δ5′, Δ3′ and ΔpG4) or single point mutations affecting the pG4 (indicated in *red*) were tested for origin function. Relative SNS enrichment is indicated for these mutants. The capacity of these constructs to advance replication timing was also analyzed for a fraction of constructs (data for Δ3′ is not published yet). + and – indicate transgenes that do and do not advance replication timing, respectively. The potential G4 probability is also shown in the table. The 17-mer β^A G-rich motif can, in principle, fold into ten intramolecular G4 structures. For each mutant, we defined the PG4 probability as the number of possible G4 structures divided by the number of possible G4 structures for the WT sequence

13.4 To What Extent Have the Mechanisms Involved in Origin Selection Been Conserved During Evolution?

Studies of the nature and mode of operation of replicators and initiators in different model systems may reveal conserved mechanisms involved in origin selection during evolution, shedding light on the role of pG4 in origin selection in vertebrates. The replicator has two main functions, as a site for initiator recognition and as a region favoring the melting of DNA. Replication of the bacterial chromosome is initiated at a single origin of replication, oriC, and is regulated by the coordinated action of numerous *trans*-factors including DnaA, the initiator. All the replicators identified in bacteria to date consist of three functional modules: a cluster of binding sites for the initiator DnaA (DnaA boxes), the DNA unwinding element (DUE), and binding sites for regulatory proteins (Wolański et al. 2014). These elements control the opening up of the DNA double helix and regulate the initiation of bacterial chromosome replication. The DUE is a typical AT-rich stretch. DnaA boxes from different bacteria have a common core sequence motif (TTATCCACA) recognized by the replicator, DnaA. A similar organization is also found in many archaea, despite the use of multiple origins to replicate the genomes of these organisms. Studies in several strains have revealed that the basic structure of replication origins is conserved among archaea, with an AT-rich unwinding element and several conserved repeats (origin recognition box, ORB) (Kelman and Kelman 2014). Biochemical and structural approaches showed that the ORB elements act as recognition sites for the Orc/Cdc6 initiator protein. Interestingly, in one of the two origins (oriC1) found on the main chromosome of several haloarchaea, the conserved "G-string" motif (GGGG) found in most archaea origins is followed by another specific G-rich sequence (TGGGGGGG) (Norais et al. 2007). In *H. hispanica*, this long G-rich motif was shown to be necessary for efficient replication initiation (Wu et al. 2014). The capacity of this G-rich motif to form a G4 structure in vitro was not tested, but it will be important to determine whether or not the regulation of oriC1 in *H. hispanica* involves the formation of a G4 structure. The common G-string element found in most archaeal ORBs is recognized by the AAA+ domain of the replicator Orc/Cdc6 (Dueber et al. 2007; Gaudier et al. 2007). It has therefore been suggested that the extended halophile-specific G-stretches in haloarchaeal ORBs play an essential role in the interaction between the Orc/Cdc6 protein and its origin at high intracellular salt concentrations (Wu et al. 2014). Eukaryotic DNA replication initiates at multiple sites, from several hundred in yeast to tens of thousands in vertebrates. Replication origins have been extensively analyzed in the budding yeast *Saccharomyces cerevisiae*, in which small origin fragments can act as autonomously replicating sequences (ARSs), mediating the propagation of episomal plasmids. The signal includes a 17 bp ARS consensus sequence (ACS) required for recruitment of the ORC complex in a chromatin environment. In addition to the ACS, origin function requires flanking elements, including transcription factor binding sites and/or sequences known to exclude nucleosomes. It is therefore thought that the ACS increases the affinity of a subset of nucleosome-depleted regions (NDRs) for ORC, thereby conferring origin activity on these loci (Eaton et al. 2010). A recent study on an industrially important methylotrophic budding yeast, *Pichia pastoris*, showed that this yeast used at least two types of replication origin: a GC-rich type associated with TSS and an AT-rich type resembling the canonical budding yeast origin (Liachko et al. 2014). GC-ARSs are mostly associated with early-replicating regions, whereas AT-ARSs tend to be associated with regions that replicate later. Moreover, a specific constrained motif was identified (TYGAAC) at GC-ARSs. This motif matches a binding site for human Hsf1 (heat shock factor 1) and could be recognized by the yeast HSF protein. It is possible that this GC-motif is involved in both origin activity and transcriptional regulation. The regions encompassing these origins also have a NDR region, although three nucleosomes are excluded, rather than just one, as for classical AT-rich yeast origins. Thus, this yeast, despite not having pG4 in its origins, is essentially similar to vertebrate origins, some of which have been shown to be GC-rich and associated with TSS. In the fission yeast *Schizosaccharomyces pombe*, origin positioning depends principally on the presence of sequences with a high adenine and thymine content (A+T-rich sequences), but no consensus elements have been identified (Segurado et al. 2003; Heichinger et al. 2006; Hayashi et al. 2007). The information is recognized by the Orc4 subunit of the ORC, which is unique among eukaryotes in that it contains a large domain harboring AT-hook subdomains that target the ORC to A+T-rich sequences (Chuang and Kelly 1999). A combined experimental and bioinformatics approach was developed for the characterization of origins in three distantly related fission yeasts: *S. pombe*, *S. octosporus*, *and S. japonicus* (Xu et al. 2012). A search for motifs predictive of origin function confirmed the key role of poly-A motifs in *S. pombe* and *S. octosporus*. However, *S. japonicus* origins were characterized by three highly abundant motifs, a 17 bp polyG motif, a head to tail repeat of CTCGCT, and the binding site for the Sap1 protein, a *trans*-acting factor known to exclude nucleosomes. It has been suggested that the polyG motif identified favors nucleosome exclusion, although this has not been definitively demonstrated. Origins mapped in a protozoan human parasite (*Leishmania major*) and in *Drosophila* (Cayrou et al. 2012; Comoglio et al. 2015; Lombraña et al. 2016) have also been shown to contain G-rich motifs. Thus, comparisons of the nature of replicators in different species have indicated that polyG motifs have been selected at replication origins several times during evolution. As suggested above, one potential role of polyGs is nucleosome exclusion. This motif has also been shown to be involved in initiator binding in haloarchaea. It would be of particular interest to determine whether G4 structures are involved in the regulation of origins in these species, as shown for the model origin from chicken.

13.5 How Do pG4s Regulate Origin Function?

Two phases of the cell cycle are particularly important for the regulation of DNA replication in eukaryotes. During G1, the chromatin is loaded with prereplication complexes (pre-RCs), but the absence of specific kinases prevents their activation at this time (Zegerman 2015). This phase is known as the licensing period. On entry into S phase, the activation of specific S phase kinases (CDKs and DDK) leads to the firing of pre-RCs, which are then released from the chromatin with no possibility of refolding onto it, thereby preventing re-replication. pG4s may be involved in licensing and/or firing. In one model of origin specification derived from the results of studies in *S. cerevisiae*, pre-RC formation is favored in nucleosome-depleted regions (NDRs) (Eaton et al. 2010; Berbenetz et al. 2010). NDRs containing ACSs have been shown to be the best substrates for pre-RC formation in vivo. It has been suggested that pG4s tend to exclude nucleosomes, thereby potentially favoring pre-RC formation (Fenouil et al. 2012) (Fig. 13.3). Alternatively, G4 formation may facilitate DNA melting and, therefore, the initiation of replication. Finally, pG4s may be recognized by specific factors involved in the formation of a functional origin (Fig. 13.3). Support for this hypothesis has recently been provided by the observation that the ORC binds preferentially to G4 structures formed on RNA or single-stranded DNA (Hoshina et al. 2013). The N-terminal part of human RecQL4, the ortholog of the essential firing factor Sld2 of *S. cerevisiae*, also binds G4 structures with high affinity (Keller et al. 2014). The observation that pG4 determines the location of the initiation site 3′ to the G4 structure suggests the involvement of a *trans*-acting factor able both to recognize the orientation of the G4 and to affect the position of the replication start site. Helicases, such as FANCJ and Pif1, which can interact with and unwind G4 structures in the 5–3′ direction, may control the direction of the process (Fig. 13.3). Genetic and genomic studies have shown that pG4s are not sufficient to define an origin of replication. A 250 bp *cis*-element located 3′ to the pG4 has been shown to cooperate with the pG4 in the chicken β^A origin. This cooperating module may facilitate NDR formation, favor the formation of G4 structures, or facilitate the recruitment of licensing and/or firing factors. The identification of key motifs within the cooperating *cis*-module should improve our understanding of the functioning of pG4.

13.6 Concluding Remarks

The genome-wide mapping of replication origins in vertebrates has led to the identification of pG4s as motifs common to the most efficient, earliest, and constitutive origins. It has been demonstrated that a subset of pG4s adopt a G-quadruplex structure in vivo, particularly at actively transcribed promoters. The dynamics of G4 folding and unfolding may be so rapid at certain sites that G4 structures cannot be detected by existing methodologies. G4 structures are controlled

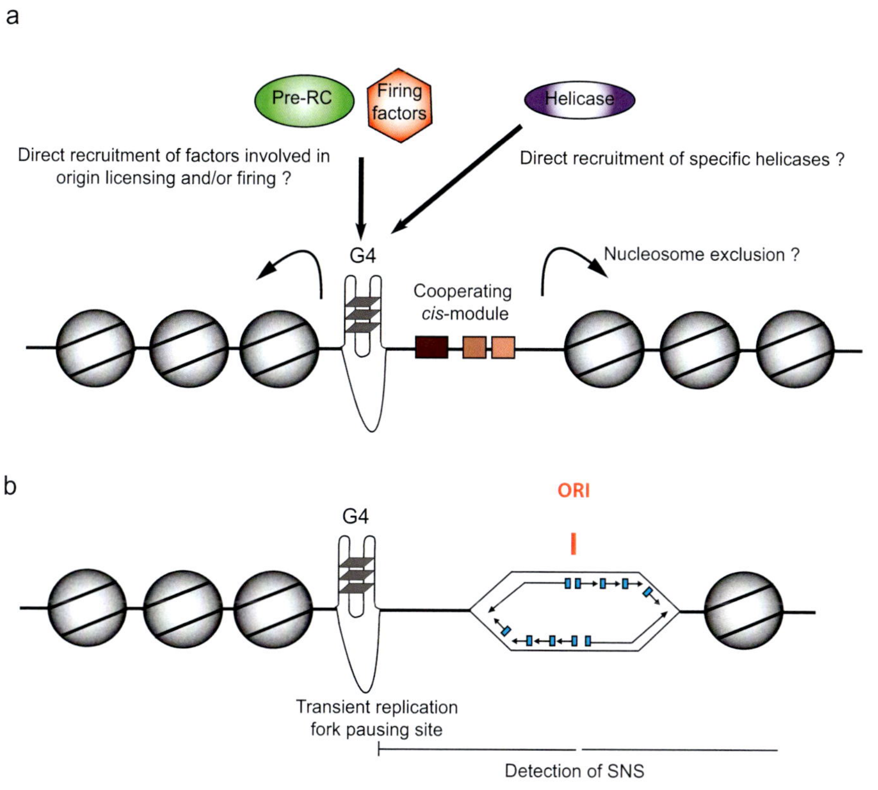

Fig. 13.3 Potential mechanisms involving the action of G4 structures in origin selection and SNS profiles. (**a**) G4 structures can recruit pre-RC, firing components, and/or specific helicases. Alternatively, G4 may be involved in nucleosome exclusion and/or origin melting. A cooperative *cis*-module is required for origin function. The function of this module remains to be determined. (**b**) G4 structures at replication origins transiently block the progression of the leading strand initiated about 250 bp 3′ to the origin

by several mechanisms, including transcription, replication, and the action of specific helicases. Other mechanisms such as molecular crowding and the action of *trans*-acting factors capable of deforming DNA may also be involved. Surprisingly, these motifs have also been identified in replication origins in several other model systems (yeasts, archaea, protozoans, and *Drosophila*) suggesting a common mode of regulation. The development of the powerful CRISPR/Cas9 genome editing system should make it easier to dissect the nature of the *cis*-elements involved in origin selection. These studies should provide new clues to the molecular mechanisms involved in the regulation of origin firing and, more specifically, the role of pG4 in origin selection.

Acknowledgments This work was supported by grants from the Association pour la Recherche sur le Cancer (Equipe labellisée), the Agence Nationale pour la Recherche (ANR-15-CE12-0004-01), and the IdEx Paris Sorbonne to M-N.P. M-N.P is supported by the Inserm.

References

Berbenetz NM, Nislow C, Brown GW (2010) Diversity of eukaryotic DNA replication origins revealed by genome-wide analysis of chromatin structure. PLoS Genet 6:e1001092. https://doi.org/10.1371/journal.pgen.1001092

Berezney R, Dubey DD, Huberman JA (2000) Heterogeneity of eukaryotic replicons, replicon clusters, and replication foci. Chromosoma 108:471–484

Besnard E, Babled A, Lapasset L, Milhavet O, Parrinello H, Dantec C, Marin J-M, Lemaitre J-M (2012) Unraveling cell type-specific and reprogrammable human replication origin signatures associated with G-quadruplex consensus motifs. Nat Struct Mol Biol 19:837–844. https://doi.org/10.1038/nsmb.2339

Bielinsky AK, Gerbi SA (1998) Discrete start sites for DNA synthesis in the yeast ARS1 origin. Science 279:95–98

Cadoret J-C, Meisch F, Hassan-Zadeh V, Luyten I, Guillet C, Duret L, Quesneville H, Prioleau M-N (2008) Genome-wide studies highlight indirect links between human replication origins and gene regulation. Proc Natl Acad Sci U S A 105:15837–15842. https://doi.org/10.1073/pnas.0805208105

Cayrou C, Coulombe P, Vigneron A, Stanojcic S, Ganier O, Peiffer I, Rivals E, Puy A, Laurent-Chabalier S, Desprat R, Méchali M (2011) Genome-scale analysis of metazoan replication origins reveals their organization in specific but flexible sites defined by conserved features. Genome Res 21:1438–1449. https://doi.org/10.1101/gr.121830.111

Cayrou C, Coulombe P, Puy A, Rialle S, Kaplan N, Segal E, Méchali M (2012) New insights into replication origin characteristics in metazoans. Cell Cycle 11:658–667. https://doi.org/10.4161/cc.11.4.19097

Cayrou C, Ballester B, Peiffer I, Fenouil R, Coulombe P, Andrau J-C, van Helden J, Méchali M (2015) The chromatin environment shapes DNA replication origin organization and defines origin classes. Genome Res 25:1873–1885. https://doi.org/10.1101/gr.192799.115

Chuang RY, Kelly TJ (1999) The fission yeast homologue of Orc4p binds to replication origin DNA via multiple AT-hooks. Proc Natl Acad Sci U S A 96:2656–2661

Comoglio F, Schlumpf T, Schmid V, Rohs R, Beisel C, Paro R (2015) High-resolution profiling of *Drosophila* replication start sites reveals a DNA shape and chromatin signature of metazoan origins. Cell Rep 11:821–834. https://doi.org/10.1016/j.celrep.2015.03.070

Diffley JFX (2004) Regulation of early events in chromosome replication. Curr Biol 14:R778–R786. https://doi.org/10.1016/j.cub.2004.09.019

Dueber ELC, Corn JE, Bell SD, Berger JM (2007) Replication origin recognition and deformation by a heterodimeric archaeal Orc1 complex. Science 317:1210–1213. https://doi.org/10.1126/science.1143690

Eaton ML, Galani K, Kang S, Bell SP, MacAlpine DM (2010) Conserved nucleosome positioning defines replication origins. Genes Dev 24:748–753. https://doi.org/10.1101/gad.1913210

Fenouil R, Cauchy P, Koch F, Descostes N, Cabeza JZ, Innocenti C, Ferrier P, Spicuglia S, Gut M, Gut I, Andrau J-C (2012) CpG islands and GC content dictate nucleosome depletion in a transcription-independent manner at mammalian promoters. Genome Res 22:2399–2408. https://doi.org/10.1101/gr.138776.112

Foulk MS, Urban JM, Casella C, Gerbi SA (2015) Characterizing and controlling intrinsic biases of lambda exonuclease in nascent strand sequencing reveals phasing between nucleosomes and

G-quadruplex motifs around a subset of human replication origins. Genome Res. https://doi.org/10.1101/gr.183848.114
Gaudier M, Schuwirth BS, Westcott SL, Wigley DB (2007) Structural basis of DNA replication origin recognition by an ORC protein. Science 317:1213–1216. https://doi.org/10.1126/science.1143664
Ge XQ, Jackson DA, Blow JJ (2007) Dormant origins licensed by excess Mcm2-7 are required for human cells to survive replicative stress. Genes Dev 21:3331–3341. https://doi.org/10.1101/gad.457807
Hänsel-Hertsch R, Beraldi D, Lensing SV, Marsico G, Zyner K, Parry A, Di Antonio M, Pike J, Kimura H, Narita M, Tannahill D, Balasubramanian S (2016) G-quadruplex structures mark human regulatory chromatin. Nat Genet 48:1267–1272. https://doi.org/10.1038/ng.3662
Hassan-Zadeh V, Chilaka S, Cadoret J-C, Ma MK-W, Boggetto N, West AG, Prioleau M-N (2012) USF binding sequences from the HS4 insulator element impose early replication timing on a vertebrate replicator. PLoS Biol 10:e1001277. https://doi.org/10.1371/journal.pbio.1001277
Hayashi M, Katou Y, Itoh T, Tazumi A, Tazumi M, Yamada Y, Takahashi T, Nakagawa T, Shirahige K, Masukata H (2007) Genome-wide localization of pre-RC sites and identification of replication origins in fission yeast. EMBO J 26:1327–1339. https://doi.org/10.1038/sj.emboj.7601585
Heichinger C, Penkett CJ, Bähler J, Nurse P (2006) Genome-wide characterization of fission yeast DNA replication origins. EMBO J 25:5171–5179. https://doi.org/10.1038/sj.emboj.7601390
Hiratani I, Ryba T, Itoh M, Yokochi T, Schwaiger M, Chang C-W, Lyou Y, Townes TM, Schübeler D, Gilbert DM (2008) Global reorganization of replication domains during embryonic stem cell differentiation. PLoS Biol 6:e245. https://doi.org/10.1371/journal.pbio.0060245
Hoshina S, Yura K, Teranishi H, Kiyasu N, Tominaga A, Kadoma H, Nakatsuka A, Kunichika T, Obuse C, Waga S (2013) Human origin recognition complex binds preferentially to G-quadruplex-preferable RNA and single-stranded DNA. J Biol Chem. https://doi.org/10.1074/jbc.M113.492504
Huppert JL, Balasubramanian S (2005) Prevalence of quadruplexes in the human genome. Nucleic Acids Res 33:2908–2916. https://doi.org/10.1093/nar/gki609
Ibarra A, Schwob E, Méndez J (2008) Excess MCM proteins protect human cells from replicative stress by licensing backup origins of replication. Proc Natl Acad Sci U S A 105:8956–8961. https://doi.org/10.1073/pnas.0803978105
Jacob F, Brenner S, Cuzin F (1963) On the regulation of DNA replication in bacteria. Cold Spring Harb Symp Quant Biol 28:329–348. https://doi.org/10.1101/SQB.1963.028.01.048
Keller H, Kiosze K, Sachsenweger J, Haumann S, Ohlenschläger O, Nuutinen T, Syväoja JE, Görlach M, Grosse F, Pospiech H (2014) The intrinsically disordered amino-terminal region of human RecQL4: multiple DNA-binding domains confer annealing, strand exchange and G4 DNA binding. Nucleic Acids Res 42:12614–12627. https://doi.org/10.1093/nar/gku993
Kelman LM, Kelman Z (2014) Archaeal DNA replication. Annu Rev Genet 48:71–97. https://doi.org/10.1146/annurev-genet-120213-092148
Langley AR, Gräf S, Smith JC, Krude T (2016) Genome-wide identification and characterisation of human DNA replication origins by initiation site sequencing (ini-seq). Nucleic Acids Res 44:10230–10247. https://doi.org/10.1093/nar/gkw760
Liachko I, Youngblood RA, Tsui K, Bubb KL, Queitsch C, Raghuraman MK, Nislow C, Brewer BJ, Dunham MJ (2014) GC-rich DNA elements enable replication origin activity in the methylotrophic yeast *Pichia pastoris*. PLoS Genet 10:e1004169. https://doi.org/10.1371/journal.pgen.1004169
Lombraña R, Álvarez A, Fernández-Justel JM, Almeida R, Poza-Carrión C, Gomes F, Calzada A, Requena JM, Gómez M (2016) Transcriptionally driven DNA replication program of the human parasite *Leishmania major*. Cell Rep 16:1774–1786. https://doi.org/10.1016/j.celrep.2016.07.007
Mantiero D, Mackenzie A, Donaldson A, Zegerman P (2011) Limiting replication initiation factors execute the temporal programme of origin firing in budding yeast. EMBO J 30:4805–4814. https://doi.org/10.1038/emboj.2011.404

Norais C, Hawkins M, Hartman AL, Eisen JA, Myllykallio H, Allers T (2007) Genetic and physical mapping of DNA replication origins in *Haloferax volcanii*. PLoS Genet 3:e77. https://doi.org/10.1371/journal.pgen.0030077

Picard F, Cadoret J-C, Audit B, Arneodo A, Alberti A, Battail C, Duret L, Prioleau M-N (2014) The spatiotemporal program of DNA replication is associated with specific combinations of chromatin marks in human cells. PLoS Genet 10:e1004282. https://doi.org/10.1371/journal.pgen.1004282

Prioleau M-N, MacAlpine DM (2016) DNA replication origins-where do we begin? Genes Dev 30:1683–1697. https://doi.org/10.1101/gad.285114.116

Rhodes D, Lipps HJ (2015) G-quadruplexes and their regulatory roles in biology. Nucleic Acids Res 43:8627–8637. https://doi.org/10.1093/nar/gkv862

Rodriguez R, Miller KM, Forment JV, Bradshaw CR, Nikan M, Britton S, Oelschlaegel T, Xhemalce B, Balasubramanian S, Jackson SP (2012) Small-molecule–induced DNA damage identifies alternative DNA structures in human genes. Nat Chem Biol 8:301–310. https://doi.org/10.1038/nchembio.780

Segurado M, de Luis A, Antequera F (2003) Genome-wide distribution of DNA replication origins at A+T-rich islands in *Schizosaccharomyces pombe*. EMBO Rep 4:1048–1053. https://doi.org/10.1038/sj.embor.embor7400008

Sequeira-Mendes J, Díaz-Uriarte R, Apedaile A, Huntley D, Brockdorff N, Gómez M (2009) Transcription initiation activity sets replication origin efficiency in mammalian cells. PLoS Genet 5:e1000446. https://doi.org/10.1371/journal.pgen.1000446

Tanaka S, Nakato R, Katou Y, Shirahige K, Araki H (2011) Origin association of Sld3, Sld7, and Cdc45 proteins is a key step for determination of origin-firing timing. Curr Biol 21:2055–2063. https://doi.org/10.1016/j.cub.2011.11.038

Valton A-L, Prioleau M-N (2016) G-Quadruplexes in DNA replication: a problem or a necessity? Trends Genet 32:697–706. https://doi.org/10.1016/j.tig.2016.09.004

Valton A-L, Hassan-Zadeh V, Lema I, Boggetto N, Alberti P, Saintomé C, Riou J-F, Prioleau M-N (2014) G4 motifs affect origin positioning and efficiency in two vertebrate replicators. EMBO J. https://doi.org/10.1002/embj.201387506

Wolański M, Donczew R, Zawilak-Pawlik A, Zakrzewska-Czerwińska J (2014) oriC-encoded instructions for the initiation of bacterial chromosome replication. Front Microbiol 5:735. https://doi.org/10.3389/fmicb.2014.00735

Wu Z, Liu J, Yang H, Liu H, Xiang H (2014) Multiple replication origins with diverse control mechanisms in *Haloarcula hispanica*. Nucleic Acids Res 42:2282–2294. https://doi.org/10.1093/nar/gkt1214

Xu J, Yanagisawa Y, Tsankov AM, Hart C, Aoki K, Kommajosyula N, Steinmann KE, Bochicchio J, Russ C, Regev A, Rando OJ, Nusbaum C, Niki H, Milos P, Weng Z, Rhind N (2012) Genome-wide identification and characterization of replication origins by deep sequencing. Genome Biol 13:R27. https://doi.org/10.1186/gb-2012-13-4-r27

Zegerman P (2015) Evolutionary conservation of the CDK targets in eukaryotic DNA replication initiation. Chromosoma 124:309–321. https://doi.org/10.1007/s00412-014-0500-y

Chapter 14
Interaction of Rif1 Protein with G-Quadruplex in Control of Chromosome Transactions

Kenji Moriyama, Mong Sing Lai, and Hisao Masai

Abstract Recent studies on G-quadruplex (G4) revealed crucial and conserved functions of G4 in various biological systems. We recently showed that Rif1, a conserved nuclear factor, binds to G4 present in the intergenic regions and plays a major role in spatiotemporal regulation of DNA replication. Rif1 may tether chromatin fibers through binding to G4, generating specific chromatin domains that dictate the replication timing. G4 and its various binding partners are now implicated in many other chromosome regulations, including transcription, replication initiation, recombination, gene rearrangement, and transposition.

Keywords Replication timing • G-quadruplex • Rif1 • Telomere • DSB repair • Nonhomologous end joining • Homologous recombination

14.1 Introduction

DNA with a specific sequence feature or base composition can adopt non-B-type forms of DNA under certain chemical conditions. For example, guanine-rich DNA has been known to take part in structures not only in the standard B-type DNA but also in other nonstandard forms of DNA including Z-DNA, hairpins, cruciforms, triplexes, and G-quadruplexes (Doluca et al. 2013). The presence of these unusual forms of DNA has been known for decades. It is only during the recent years that potential physiological importance of the non-B-type DNA starts to be realized. Among these structures, G-quadruplex (G4) has been a focus of intense research in the past several years. G4 can be formed under physiological conditions, and genetic and physical evidence points to its existence in the living organism. A number of

Kenji Moriyama and Mong Sing Lai contributed equally to this work.

K. Moriyama • M.S. Lai • H. Masai (✉)
Department of Genome Medicine, Tokyo Metropolitan Institute of Medical Science, 2-1-6 Kamikitazawa, Setagaya-ku, Tokyo 156-8506, Japan
e-mail: masai-hs@igakuken.or.jp

H. Masai, M. Foiani (eds.), *DNA Replication*, Advances in Experimental Medicine and Biology 1042, https://doi.org/10.1007/978-981-10-6955-0_14

recent studies strongly suggest that G4 or its related structures play crucial roles in various chromosome metabolism (reviewed in Maizels and Gray 2013; Murat and Balasubramanian 2014; Maizels 2015; Rhodes and Lipps 2015).

We have studied mechanisms of how DNA replication timing during S phase is regulated and have identified a crucial factor, Rif1, which plays a major, conserved role in regulation of spatiotemporal program of DNA replication (Hayano et al. 2012; Yamazaki et al. 2012; Cornacchia et al. 2012). Subsequent studies revealed that Rif1 regulates replication timing by binding to intergenic G4 structures (Kanoh et al. 2015). Other studies have suggested the potential role of G4 in the site selection for initiation of DNA replication in eukaryotic genomes (Besnard et al. 2012; Cayrou et al. 2011, 2012; Valton et al. 2014; Langley et al. 2016).

In this short article, we would like to first discuss Rif1 and its biological roles and then describe biochemical functions of Rif1 with particular emphasis on its ability to bind to G4. We will show a model on how Rif1 may regulate replication timing. Finally, we will introduce latest findings on the physiological functions of G4 structures in regulation of diverse chromosome transactions.

14.2 Rif1 Protein and Chromosome Regulation

14.2.1 Rif1 in Telomere Regulation

Rif1 (Rap1-interacting factor 1) was first identified in budding yeast *Saccharomyces cerevisiae* (Sc) as a cofactor or mediator for Rap1 at silencers and telomeres (Hardy et al. 1992). Rap1 (repressor-activator protein 1) encodes an essential protein that binds extensively to telomeric regions, where its function is related to both transcriptional silencing and telomere maintenance (Marcand et al. 1997). Rif1 interacts with Rap1 at its carboxy-terminal domain (Fig. 14.1a). Cells lacking Rif1 are defective in transcriptional silencing and display an increase in the telomere length. Rap1 also recruits a second interacting factor Rif2 (Rap1-interacting factor 2) via its C-terminus. Rif1 and Rif2 form a functional complex that is essential for telomere length regulation (Wotton and Shore 1997). Recent studies using X-ray crystallography had revealed the architecture of Rap1-Rif1-Rif2 protein backbone at telomeres (Shi et al. 2013). Rif1 assists the essential telomere protection function of the CST (Cdc13-Stn1-Tel1) complex and inhibits the nucleolytic processing at telomeres (Anbalagan et al. 2011; Bonetti et al. 2010). Rif1 also possesses a telomeric anti-checkpoint activity and is required for protection of telomere-free chromosome ends which mimic double-strand breaks (Xue et al. 2011, 2016; Ribeyre and Shore 2012).

Rif1 was also identified in another yeast, fission yeast *Schizosaccharomyces pombe* (Sp) (Kanoh and Ishikawa 2001). However, unlike ScRif1, SpRif1 does not bind to SpRap1. It is recruited to telomeres through a different telomeric protein called Taz1, where it regulates the telomere length and telomere position effect (TPE). SpRap1 and SpRif1 have been shown to exert opposite effects in telomere

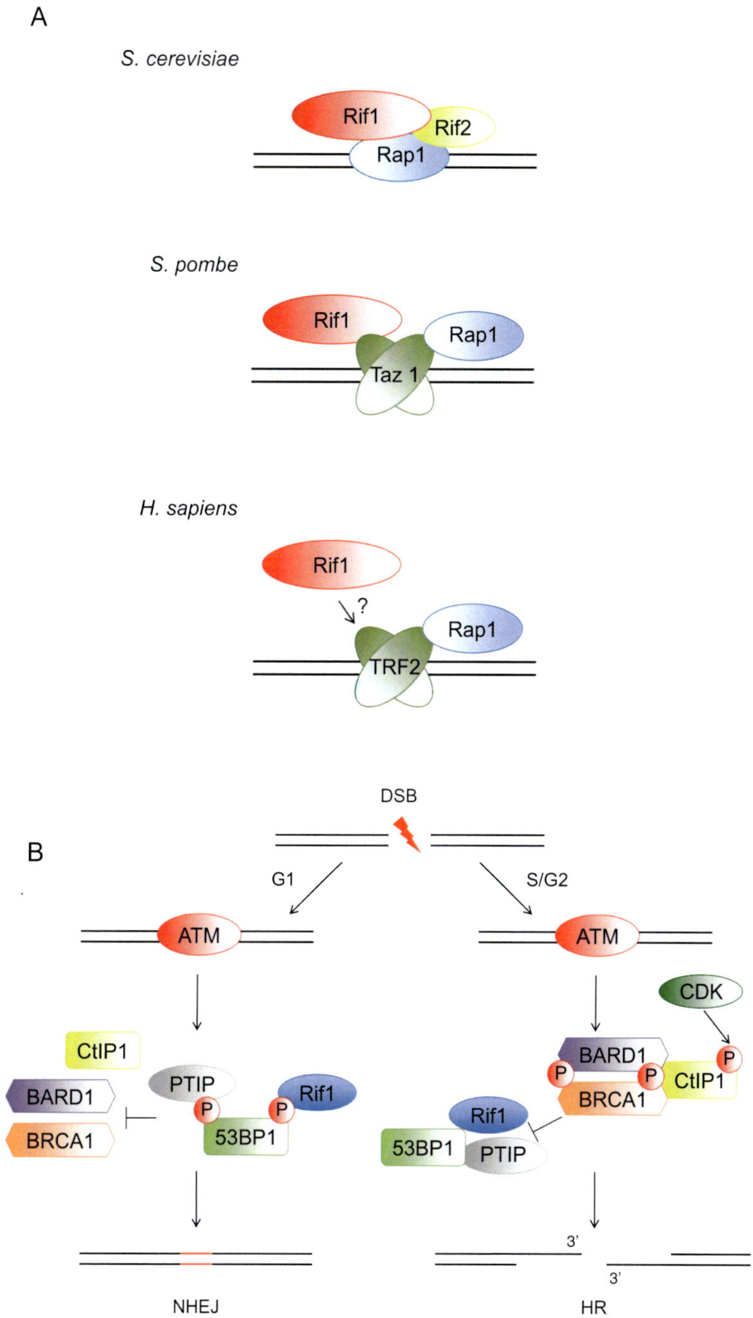

Fig. 14.1 Roles of Rif1 protein in telomere, DSB repair, replication, and transcription. (**a**) Schematic representation of telomeric complexes involving Rif1 in budding yeast (*S. cerevisiae*), fission yeast (*S. pombe*), and human (*H. sapiens*). (**b**) A proposed role for Rif1 in regulation of repair of double-stranded DNA breaks (*DSBs*). In G1, 53BP1 is phosphorylated by ATM and localizes at the site of DSBs. Rif1 and PTIP are then recruited by the phosphorylated 53BP1 to the break

C

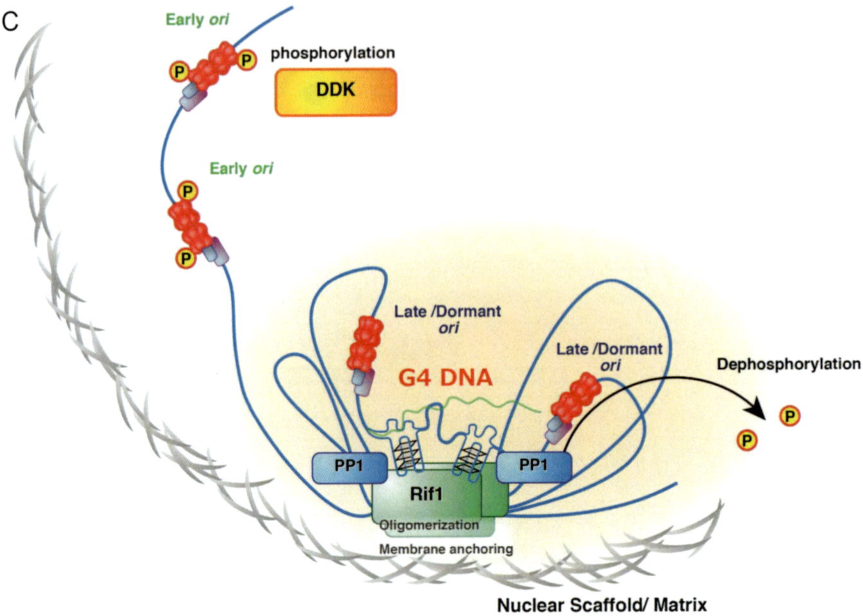

D

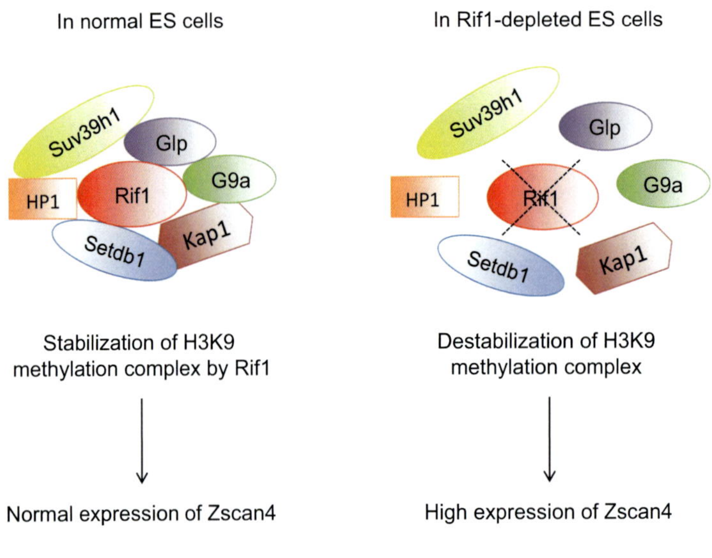

Fig. 14.1 (continued) sites and inhibit the 5'–3' resection of the DNA ends. This process facilitates the repair of the DSBs by nonhomologous end joining (NHEJ). In S/G2, BRCA1 and BARD1 are phosphorylated by ATM, whereas CtIP is phosphorylated by CDK. The phosphorylated complex displaces 53BP1/Rif1/PTIP from the sites of DSBs and recruits the MRX complex to initiate 5'–3' resection and thus promote HR repair. (**c**) Rif1 binds to intergenic G4 structure to generate replication-suppressive chromatin domain near nuclear periphery by facilitating the chromatin loop formation through its G4 DNA-binding and oligomerization activities. (**d**) A model for Rif1-mediated transcriptional regulation of Zscan4 gene

homeostasis. SpRap1, but not SpRif1, is required to regulate 3′ overhang formation and prevents telomere fusions. Whereas *rap1*Δ exacerbates the cold sensitivity of *taz1*Δ cells, *rif1*Δ restores full viability of *taz1*Δ at 25 °C (Miller et al. 2005). In another yeast, *Candida glabrata*, Rif1 is required for correct telomere length regulation and also plays a positive role in subtelomeric silencing (Castaño et al. 2005; Rosas-Hernández et al. 2008).

The presence of Rif1 in higher eukaryotes suggests that the functions of this protein are conserved. Mouse ortholog of Rif1 (mRif1) is highly expressed in pluripotent stem cells, early embryonic cells, and germ cells. In an early report, mRif1 was also reported to be associated with telomeric DNA and to physically interact with a mouse telomere-associated protein, mTrf2 (Adams and McLaren 2004). However, hRif1 may not be involved in telomere regulation, as it can be found at non-telomeric locations and does not colocalize with human telomeric proteins Trf1, Trf2, or human Rap1 at normal telomeres, although overexpression of human Rif1 (hRif1) in *S. cerevisiae* interfered with telomere length control (Silverman et al. 2004; Xu and Blackburn 2004). Consistent with this prediction, Rif1 knockout mice or cells show no obvious telomere defects (Buonomo et al. 2009). On the other hand, Dan et al. (2014) reported that mRif1 controls telomere length through transcriptional silencing of Zscan4, a factor promoting recombination-mediated telomere elongation in ES cells. Thus, Rif1 appears to play an indirect role in telomere length homeostasis, possibly in a cell type-specific manner.

14.2.2 Rif1 in Regulation of Repair of Double-Stranded DNA Breaks

The first indication of Rif1 involvement in DNA damage response came from the observation that human Rif1 only localized to damaged telomeres and accumulated at double-stranded DNA breaks (DSBs) in an ATM- and 53BP1-dependent manner (Silverman et al. 2004; Xu and Blackburn 2004). In vertebrates, Rif1 also participates in the intra-S-phase checkpoint to slow down DNA synthesis in response to DNA damage and also contributes to replication stress survival (Silverman et al. 2004; Buonomo et al. 2009; Xu et al. 2010). In this process, Rif1 accumulates at stalled replication forks in a manner dependent on ATR and 53BP1 (Buonomo et al. 2009). Rif1 interacts directly with BLM, a DNA helicase deficient in Bloom syndrome, and recruitment of Rif1 to stalled forks is delayed in BLM-deficient cells (Xu et al. 2010).

Subsequent studies have further uncovered the potential functions of vertebrate Rif1 in DNA damage responses. Cells respond to DSBs in two major pathways in order to repair the damages. One pathway is called nonhomologous end joining (NHEJ), which involves the religation of DNA ends. Another pathway is called homologous recombination (HR), which utilizes homologous sequence, usually the sister chromatids, to guide error-free repair. How cells select the repair pathways has remained unclear. Earlier studies showed that Rif1 participates in HR-mediated repair of subsets of DSB (Buonomo et al. 2009; Wang et al. 2009). However, Xu et al. (2010) showed that Rif1 plays roles in the recovery of stalled forks generated by replication stress rather than in HR-dependent repair of DSB. More recent studies showed that Rif1 is a key player in DSB repair pathway choice and facilitates NHEJ (Fig. 14.1b). In G1 phase, when DSB occurs, Rif1 is recruited by the phosphorylated 53BP1 at the DSB sites, which in turn inhibits 5'–3' DNA end resection and thus promotes NHEJ. In S/G2 phase, CDK phosphorylation of CtIP facilitates its interaction with BRCA1. The CtIP-BRCA1 complexes displace Rif1 and 53BP1 at the DSB sites and recruit MRN complex to initiate resection that is required for HR repair (Chapman et al. 2013; Di Virgilio et al. 2013; Escribano-Díaz et al. 2013; Zimmermann et al. 2013). The recruitment of Rif1 upon DNA damage depends not only on 53BP1 but also on ATM in both human and mouse cells. Interactions among Rif1, ATM, and MRN complex are observed in *Xenopus* egg (Kumar et al. 2012).

It was previously reported that budding yeast ScRif1 does not significantly associate with DSB (Xue et al. 2011). However, more recent reports show that ScRif1 can also localize near DSB (Martina et al. 2014; Xue et al. 2016). Notably, ScRif1 does not inhibit end resection upon DSB. Instead, ScRif1 promotes DSB resection by cooperating with the MRX-Sae2 complex and thus is important for DSB repair by HR (Martina et al. 2014). Whether and how vertebrate Rif1 participates in HR repair is still under a debate and needs further studies.

14.2.3 Rif1 in Regulation of Temporal and Spatial Regulation of DNA Replication

Besides its functions at telomere regulation and DNA damage response, Rif1 plays conserved and crucial roles in global regulation of replication timing program. DNA replication occurs at different timing and probably at different nuclear locations in different genome segments. This temporal and spatial program of DNA replication is determined at a discrete point during early G1 phase (known as "timing decision point"), where major chromatin repositioning and anchoring take place (Dimitrova and Gilbert 1999) to generate chromosome structures that dictate spatial and temporal organization of replication.

In fission yeast, *hsk1* (ortholog of budding yeast Cdc7) is not essential under certain physiological conditions or genetic backgrounds. In fact, *mrc1*Δ, in which late origins are fired in the presence of hydroxyurea, could suppress *hsk1*Δ and restore growth. Therefore, search for a novel suppressor of *hsk1*Δ cells was expected to lead to discovery of new factors that may regulate replication timing. This screening led to identification of $rif1^+$ (Hayano et al. 2012). In the absence of Rif1, subtelomeric regions as well as many late or dormant origins on the chromosome arms are vigorously fired in the early S phase. Similar results were later reported in budding yeast, although the replication timing effect seems to be more restricted to telomere-proximal regions (Lian et al. 2011; Peace et al. 2014). The striking effect of Rif1 in global regulation of DNA replication was also observed in mouse and human cells (Cornacchia et al. 2012; Yamazaki et al. 2012; Masai et al. 2017). In Rif1-depleted cells, the mid-S-phase-specific pattern of replication foci was lost, and the replication timing program proceeded from early to mid/late S phase with early-replicating foci pattern, whereas the very late S-phase pattern with heterochromatin foci remained.

How does Rif1 influence replication timing on a genome-wide basis? It was initially discovered that phosphorylation of Mcm is significantly enhanced in Rif1-depleted human cells and that this hyper-phosphorylation depends on Cdc7 kinase (Yamazaki et al. 2012). One clue came from the findings that Rif1 possesses a conserved SILK/RVxF motif, which is a docking motif essential for the protein phosphatase 1 (PP1) interaction (Sreesankar et al. 2012; Davé et al. 2014; Hiraga et al. 2014; Mattarocci et al. 2014). In fact, yeast Rif1 interacts with PP1 through the SILK/RVxF motif, and Rif1/PP1 counteracts the DDK-dependent phosphorylation of Mcm complex, thus preventing DNA replication initiation (Fig. 14.1c; Davé et al. 2014; Hiraga et al. 2014; Mattarocci et al. 2014). Genetic and functional interactions between Rif1 and PP1 were also observed in *Drosophila* (Sreesankar et al. 2015). It was recently reported in human cells and *Xenopus* eggs that PP1 interacts with Rif1, and this interaction is required for suppression of Mcm phosphorylation (Hiraga et al. 2017; Alver et al. 2017). Considering the conserved Rif1-PP1 interaction, it is likely that the role of Rif1 in negatively regulating initiation through the recruitment of PP1 may be conserved across species.

14.2.4 Rif1 in Regulation of Transcription

In budding yeast *Saccharomyces cerevisiae*, chromatin immediately adjacent to the telomeric repeats is transcriptionally silenced or heterochromatic (Gottschling et al. 1990). This epigenetic phenomenon, which is known as the telomere position effect (TPE), was proposed to link telomere structure to transcriptional regulation. TPE is mediated by the silent information regulator (Sir) proteins. Rap1 recruits the Sir proteins as well as the Ku70 complex to telomere to promote the spreading of heterochromatin in the subtelomeric region (Kueng et al. 2013). It was shown that *rif1* mutant cells are defective in silencing at *HMR* silencer, a phenotype that is similar to *rap1* mutant cells (Hardy et al. 1992), suggesting a role of Rif1 in transcriptional silencing. In fact, Rif1, together with the Sir proteins (Sir3 and Sir4), interacts with Rap1 to maintain the transcriptional silencing ability (Moretti et al. 1994). In fission yeast, SpRif1 also affects TPE in the subtelomeric region (Greenwood and Cooper 2012).

In HeLa cells, >600 genes are either upregulated or downregulated by depletion of Rif1, suggesting that Rif1 may affect transcription activity (Yamazaki et al. 2012). Notably, a megabase segment containing gene clusters are coordinately deregulated in the absence of Rif1 (Yamazaki et al. unpublished data). Zinc-finger and SCAN domain-containing protein 4 (Zscan4) is upregulated in ES cells depleted of Rif1 (Dan et al. 2014). Rif1 associates with a histone methylation complex in ES cells (Fig. 14.1d), and Rif1 depletion leads to reduction of H3K9me3 in 33% of the genome including the pericentromeric segment and the subtelomeric region where the Zscan4 gene cluster is present (Dan et al. 2014). Coordinated regulation of transcription over several hundred kb by Rif1 may be consistent with the proposal that Rif1 regulates chromosome events through chromatin architecture. Recent report showed that Rif1 expression level in mouse ES cells is also tightly controlled by Smad3 and Oct4 (Li et al. 2015). Developmental control of Rif1 binding on chromatin and their relationship to long-range transcriptional regulation needs to be further studied.

14.3 Rif1 and G-Quadruplex

14.3.1 Interaction of Rif1 with Chromatin Through G-Rich Conserved Sequences

Although an essential role of Rif1 in controlling global replication timing has become evident, how Rif1 is targeted to or selects its binding sites on chromosomes and how it affects origin firing still remain elusive. Recent study carried out in *Schizosaccharomyces pombe* provides evidence that Rif1 in fact can recognize and bind to G4-like structures at selected intergenic regions, thereby generating chromatin environment that can suppress local origin firing (Kanoh et al. 2015).

Using a chromatin immunoprecipitation-sequencing method, Kanoh et al. identified 35 high-affinity Rif1-binding sites (Rif1BSs) on the fission yeast chromosomes. Rif1BSs are associated with the following characteristics. (1) They are located in the intergenic regions. (2) They do not exactly overlap with the locations of pre-RC (Mcm binding sites) or promoter sequences. (3) The maximum chromatin binding is observed during G1, and the binding gradually decreases during S phase. (4) They are more enriched in the vicinity of the late-firing or dormant origins that are converted to early in *rif1*Δ cells.

Detailed analyses of Rif1BS sequences revealed the following features. (1) Rif1BSs contain a conserved motif, CNWWGTGGGGG (N, any nucleotide; W, A, or T), which is named the Rif1 consensus sequence (Rif1CS). (2) The strong Rif1BS almost always contain two or more Rif1CS, and they are arranged in head-to-tail orientation in over 75% of the strong Rif1BS. (Weak Rif1BS often contained only one Rif1CS.) (3) Mutations of the Rif1CSs resulted in the loss of chromatin binding of Rif1 protein specifically at the mutated site. (4) The mutations affected DNA replication initiation over ~100 kb segment encompassing the mutated Rif1BS.

Given the features of Rif1CS that it contains a stretch of five or six guanines and generally contains several other G-tracts, the authors examined a possibility that Rif1BSs adopt G4-like structures in vitro. Single-stranded DNA derived from Rif1BS as well as the double-stranded Rif1BS DNA that has been heat-denatured and reannealed can adopt G4-like structures in a manner dependent on the presence of G-tracts in the Rif1CS as well as at other locations. Purified Rif1 protein binds to G4 DNA with affinity higher than other structured DNAs (see below).

These results strongly suggest that Rif1 binds to the G4-like structures that are generated in the intergenic region and contributes to the formation of the chromatin compartments that are related to the coordinated regulation of replication and possibly to that of transcription (Figs. 14.1c and 14.2a).

14.3.2 Structural and Functional Dissection of Rif1 Protein

The N-terminal 340 amino acid segment of human Rif1 was initially predicted to fold into eight HEAT (huntingtin, elongation factor 3, a subunit of protein phosphatase 2A, and TOR) or armadillo-type helical repeats (Silverman et al. 2004), and this region is most conserved among Rif1 orthologs across the species. Bioinformatics analyses (Xu et al. 2010) predicted that the N-terminal domain of Rif1 consists of 14–21 tandem HEAT-like repeats, whose lengths depend on species (Fig. 14.2b). The HEAT unit consists of 35–45 amino acids bearing loosely conserved Asp and Arg at 19th and 25th residues, respectively, and forms a pair of antiparallel α-helices and a turn surrounding a virtual axis (Andrade et al. 2001). HEAT repeats are present in a variety of proteins and function as a scaffold that bridges different proteins in some cases. The HEAT repeat segment of condensin was reported to serve as a DNA-binding module (Piazza et al. 2014). We have recently found that the isolated HEAT repeat domain of mouse Rif1 not only binds to various G4 DNA but also forms multimers (K.M. and H.M. manuscript submitted). A nutrient-responsive kinase TOR also forms a dimer through its HEAT repeat domain-mediated self-association (Takahara et al. 2006; Baretić et al. 2016). Recently, it was reported that the Rif1 HEAT repeat domain is required for 53BP1-dependent foci formation of MAD2L2 (also known as MAD2B or REV7) following irradiation (Boersma et al. 2015). Thus, the Rif1 HEAT repeat may serve also as a scaffold to which 53BP1 and MAD2L2 are recruited.

Mishra and colleagues analyzed amino acid sequences of Rif1 homologues from 92 organisms (Sreesankar et al. 2012). They identified a highly conserved, Rif1-core region of 101–149 amino acids within the HEAT repeat domain. They also found a docking motif for type I protein phosphatase (PP1), consisting of SILK-RVxF sequences in multicellular organisms. The PP1-binding sequences can be found also in unicellular eukaryotes near the N-terminus in RVxF-SILK arrangement. The authors also pointed out that plant Rif1 homologues lack the PP1-interaction motifs.

In the context of Rif1-mediated regulation of replication timing, Rif1 recruits PP1 via this motif to dephosphorylate Mcm complex and Sld3, thereby repressing origin firing both in budding and fission yeasts (Hiraga et al. 2014; Davé et al. 2014; Mattarocci et al. 2014). Such repression is supposed to be reversed by phosphorylation of Rif1 near this motif probably via CDK/DDK, because this motif is neighbored by nine or more consensus Ser/Thr residues for CDK/DDK, and Asp

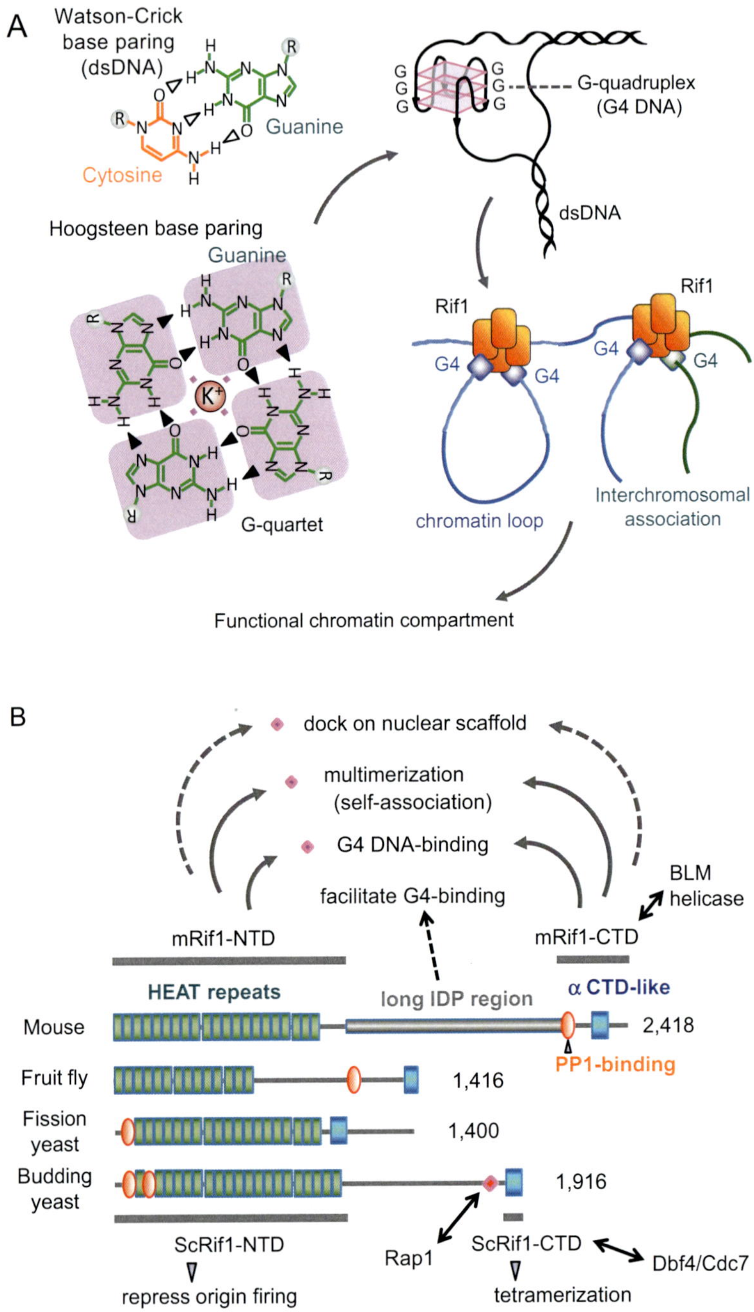

Fig. 14.2 Rif1 protein and G4. (**a**) Schematic drawings of G-quadruplex DNA and its potential roles in Rif1-mediated formation of chromatin loops and association between distant chromosomes. Watson-Crick hydrogen bonds are indicated by open arrowheads, and Hoogsteen hydrogen

substitution of these residues abrogated PP1 binding (Hiraga et al. 2014; Davé et al. 2014; Mattarocci et al. 2014). Additionally, PP1 is involved in Rif1 function in suppression of telomere length in S phase as well as in resolution of non-telomeric chromosomal entanglement in M phase in fission yeast (Zaaijer et al. 2016).

The PP1 docking site exists in the C-terminal conserved domain of vertebrate Rif1 and is designated as CR-I (conserved region I). The CR-I of mouse Rif1 binds to PP1 more tightly than PP1-inhibitor I2 (Sukackaite et al. 2017). There are two additional conserved segments in the C-terminal domain (Rif1-CTD), designated as CR-II and CR-III (Xu et al. 2010). Xu et al. (2010) noticed sequence similarity between CR-II and αCTD domain of bacterial RNA polymerase and thus examined possible DNA binding by Rif1-CTD. They found that recombinant human Rif1-CTD bound branched DNAs, including cruciform and forked DNA, more tightly than dsDNA or ssDNA and that this binding was diminished by mutations in the CR-II region. They also identified Rif1-CTD as BLM-binding domain and suggested that Rif1 may contribute to the process of localizing stalled replication forks or broken DNAs through its ability to selectively recognize branched DNA. Similar DNA-binding properties of mouse Rif1-CTD were reported. It was reported that CR-II is sufficient to bind to cruciform and forked DNA (Sukackaite et al. 2014). An X-ray crystallographic study of the CR-II-related region of *S. cerevisiae* Rif1 (ScRif1) revealed that a short segment of the C-terminal 60 amino acids (residues 1857–1916) forms tetramer, composed of a pair of oppositely oriented dimers (Shi et al. 2013). They argued that the Rif1-CTD domain is conserved from yeast to humans based on their structure-based sequence alignment, and their notion is supported by the phylogenetic sequence analysis (Sreesankar et al. 2012). Another conserved sequence was noticed previously in Rif1-CTD (Silverman et al. 2004), although it appears to be distinct from the above tetramer-forming segment. Multiangle light scattering verified that ScRif1-CTD is also tetrameric in solution. L1905 (in CR-II-related region), highly conserved across the eukaryotic orthologs, is involved in intradimer packing. An arginine mutation of this residue not only led to loss of tetramer formation but also partially inhibited ScRif1 function at telomere (Shi et al. 2014). It was reported that the C-terminal 127 amino acids of ScRif1 associates with an N-terminal BRCT-like domain of Dbf4. However, deletion of C-terminal half of ScRif1 failed to suppress temperature-sensitive growth of *cdc7-1* strain, suggesting that CTD is dispensable for repression of premature origin firing by Rif1 in budding yeast (Hiraga et al. 2014; Mattarocci et al. 2014). We have recently found that isolated CTD of mouse Rif1 binds to various G4 DNAs more tightly than to branched DNAs and also exists as multimers. Thus, it appears to be a

Fig. 14.2 (continuted) bonds by closed arrowheads. Rif1 is present as an oligomer and is proposed to anchor multiple G4 DNAs to nuclear scaffold. (**b**) Domain organization of Rif1 orthologs along their polypeptides. The numbers indicate the length (amino acids) of each protein. PP1-binding motifs are indicated by ovals. Protein-protein interaction is shown by double-headed arrows. The repeat number of HEAT unit may not be accurate. αCTD-like segments are not well conserved among different species, and their assignments in species other than mouse may not be correct. See text for further functional dissections conserved feature of Rif1 that the C-terminal segment is associated with DNA binding and oligomerization.

PP1 forms holoenzymes with more than 200 regulatory proteins, and most of them (≥70%) belong to the class of IDPs (intrinsically disordered proteins) (Choy et al. 2012). Computational disorder prediction indicates the presence of extremely long IDP region between HEAT repeat domain and CTD in vertebrate Rif1 (Sukackaite et al. 2014). As the term indicates, IDP region does not fold into stable tertiary structures under physiological conditions and is also called "naturally unstructured region" (reviewed in Oldfield and Dunker 2014; Shammas et al. 2016; Bah and Forman-Kay 2016). The isolated IDP region of mouse Rif1 does not show DNA-binding activity, but internal deletion of the IDP region resulted in substantial decrease of G4 binding in spite of the presence of DNA binding-proficient C- and N-terminal domains, suggesting that IDP may contribute to the DNA binding of Rif1 (K.M. and H.M. manuscript submitted). The functional and structural roles of the IDP in Rif1 functions will be an important issue to be addressed in the future.

14.3.3 Interaction of Rif1 with G-Quadruplex

As described above, we have recently found that both SpRif1 and mRif1 bind to G4 DNA and presented evidence showing that this interaction is crucial for organization of replication timing domain in fission yeast (Kanoh et al. 2015; K.M. and H.M. manuscript submitted).

SpRif1 not only binds to short oligonucleotides adopting G4 conformation but also to longer dsDNA (derived from SpRif1BS) fragments containing Rif1BS that have been heat denatured/renatured (Kanoh et al. 2015). A classical consensus for G4-forming sequence is represented as $G_{\geq 3}N_{1-7}$ $G_{\geq 3}N_{1-7}G_{\geq 3}N_{1-7}G_{\geq 3}$ (Huppert and Balasubramanian 2005). However, fission yeast Rif1BS sequences deviate greatly from this formula and do not necessarily bear four or more G-tracts aligned on the same strand. In fact, 31 out of the 35 SpRif1BSs do not colocalize with the 446 putative G4 motifs predicted on the fission yeast genome (Sabouri et al. 2014). In spite of that, G4-like stretches in SpRif1BS sequences bind efficiently to a G4-specific antibody, and heat-denatured duplex Rif1BS generates specific higher-order structures, which are stabilized by a chemical compound 7OTD, a derivative of telomestatin, known to bind and stabilize G4 structures (Iida et al. 2013). These structures are not generated by DNA template in which guanine has been substituted by 7-deaza guanine. DNA polymerase stop assays and DMS footprinting analyses also indicated the generation of a specific higher-order structure at the G-tracts of Rif1BS that depends on Hoogsteen base paring. These results strongly point to the ability of Rif1BS to form G4-like structures.

The C-terminal domain of mammalian Rif1 has been reported to bind to branched DNAs including forked, flap, and cruciform DNAs (Xu et al. 2010; Sukackaite et al. 2014). We prepared various branched DNAs and used them in competition assays with the dG_{24} probe, which is known to form a parallel-type G4 structure (Sengar et al. 2014). In EMSA (electrophoretic mobility shift assay), even a tenfold excess of duplex, Y-fork, flap, forked, or cruciform DNA could not outcompete the binding of SpRif1 (Kanoh et al. 2015) or mRif1 (K.M. and H.M. submitted) to the G4 probe. Thus, Rif1 binds to G4 DNA with higher affinity than it does to other structured DNAs. Alternatively, it is also conceivable that G4 DNA and other branched DNAs bind to distinct sites on Rif1. As described in a previous section, both N-terminal and C-terminal domains of mRif1 bind independently to G4 DNA, and they exhibit similar structural preferences for parallel-type G4 DNA. It would be an interesting possibility that N-terminal and C-terminal domains may coordinate the high-affinity binding of Rif1 to G4 structures (Fig. 14.2b).

14.3.4 Potential Modes of Action of Rif1 Protein

In mammals, Rif1 is localized at nuclease-insoluble structures, and Rif1 depletion causes an increase in chromatin loop sizes, suggesting a role in the regulation of higher-order chromatin structures (Cornacchia et al. 2012; Yamazaki et al. 2012). A more recent study in mouse embryonic stem cells (ESCs) further demonstrates that Rif1 is bound extensively with late-replicating regions which are associated with nuclear periphery and regulates chromatin architecture. The inter-replication domain interactions increase in the absence of Rif1 (Foti et al. 2016). Thus, Rif1 creates special nuclear architecture or chromatin domain that influences the regulation of replication timing. In budding yeast, Rif1 undergoes palmitoylation, a post-translational addition of fatty acid that anchors Rif1 to the nuclear periphery (Park et al. 2011).

Since fission yeast Rif1 can exert its inhibitory effect of origin firing as far as up to 50 kb away and biochemical characterization of Rif1 indicated that it forms oligomers (our unpublished data), we proposed that Rif1 potentially anchors to nuclear periphery through lipid modification and creates a specific chromatin architecture through recognition of the G4-like structures generated at specific intergenic regions (Kanoh et al. 2015). Rif1, in conjunction with its ability to recruit phosphatase, may tether multiple chromatin fibers through its ability to oligomerize and generate multiple chromatin loops and can negatively regulate the replication initiation over a long range (Fig. 14.1c). The ability of SpRif1 to interact with G4-DNA structures may provide an important clue to understand the link between nuclear architecture and replication timing regulation.

14.4 Emerging Biological Roles of G-Quadruplex

G-quadruplexes arise in G-rich sequences where four guanine bases make Hoogsteen base paring and thereby form planar G-quartets that stack with three looping strands connecting the G-quartet layers (Fig. 14.2a). G4 formation is kinetically fast, and these structures are stable under physiological molecular crowding conditions, particularly in the presence of K^+ (Bochman et al. 2012). In double-stranded DNA, the opportunity for forming G4 structure may arise during DNA replication, transcription, and repair when DNA is rendered transiently single stranded. However, it was recently reported that Hoogsteen base pairs transiently form in canonical duplex DNA (Nikolova et al. 2011), suggesting that G4 formation may not necessarily require the prior unwinding or melting of DNA duplex. Since there have been a number of excellent reviews on G4 and its biological roles in the past several years (e.g, see Maizels and Gray 2013; Murat and Balasubramanian 2014; Maizels 2015; Rhodes and Lipps 2015; Hänsel-Hertsch et al. 2017), we will focus on very recent progresses on G4 biology that may serve to future study.

A single-chain antibody hf2 was used to map the locations of G4 structures on the human genome by sequencing the immunoprecipitated DNA fragments (Lam et al. 2013). Another recent high-resolution sequencing by using another single-polypeptide G4 antibody, BG4, identified 716,310 distinct G4 structures in the human genome (Chambers et al. 2015). A high G4 density was seen in functional regions, such as 5′ untranslated regions and splicing sites. The putative G4 sites were significantly associated with oncogenes, tumor suppressors, and somatic copy-number alterations related to cancer development. G4 motifs are highly enriched near promoters and transcription start sites. They are more frequently found on the non-template than on the template strand where they either enhance or inhibit transcription, respectively. A genome-wide analysis of the nearly 600,000 regulatory *cis* elements for all known genes in the human genome has shown that DNase hypersensitive (DHS) *cis*-regulatory elements are also enriched in guanines (Hegyi 2015). A possibility was proposed that these distal *cis* regions may form intermolecular G4 structure with the G-rich promoter regions, generating a G4-mediated looping between the non-template strands of the promoter and the enhancer, and this could lead to the generation of DHS segments (Fig. 14.3a).

It has been reported that 67–90% of mammalian replication origins have flanking G4 motifs (Besnard et al. 2012; Cayrou et al. 2012; Valton et al. 2014) and in higher densities near those origins that are used frequently (Besnard et al. 2012). As stated above, G4 motifs were shown to be necessary for origin function in two model origins in chicken DT40 cells, and G4 stability correlates with origin efficiency. In addition, G4 orientation determines the precise position of the replication start site (Cayrou et al. 2012; Valton et al. 2014). However, these studies used G4-sensitive lambda exonuclease (λ exo) to enrich short nascent strands (SNS) for origin mapping, and therefore, concerns about λ exo-associated biases have been pointed out frequently (Foulk et al. 2015; Urban et al. 2015; Prioleau and MacAlpine 2016). A bubble-seq study demonstrated that the majority (59%) of bubble-containing fragments did not bear G4 motifs (Mesner et al. 2013) and consistent with a more recent Okazaki fragment (OK)-seq analysis in HeLa and GM06990 cells (Petryk et al. 2016). Although a nascent strand capture and release (NSCR) approach indicated that, among 5% of the largest SNS peaks, 49.4% harbored associated G4 motifs within 2 kb, only <6% of SNS peaks exhibit a close and orientation-specific localization with G4 motifs in MEF cells (Kunnev et al. 2015). Another BrdU-NS-seq

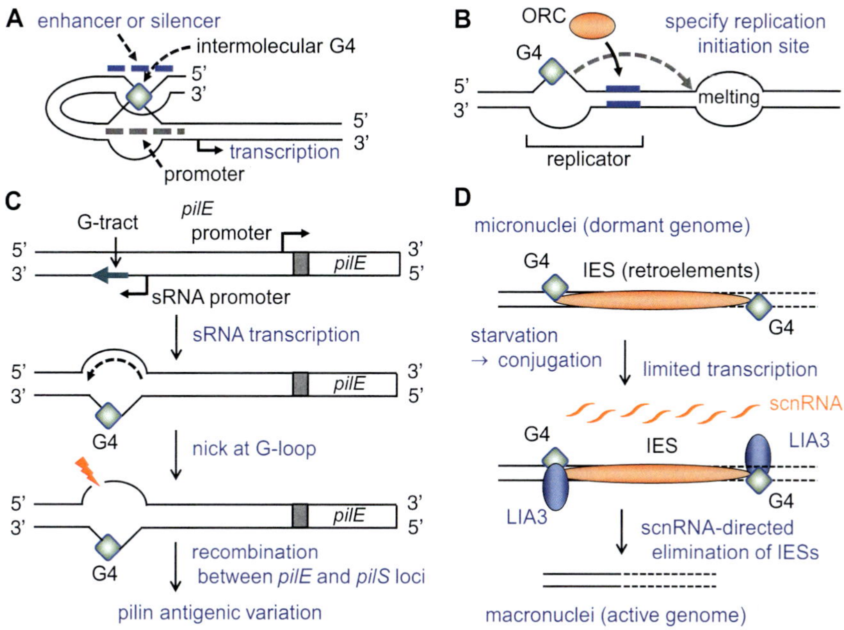

Fig. 14.3 Various potential biological functions of G4 DNA. (**a**) A possible model for transcriptional regulation by G4-mediated DNA looping between gene promoter and its *cis*-regulatory element (Hegyi 2015). (**b**) A potential role of G4 DNA in specification of replication origin in vertebrate. (**c**) Recombination initiation pathway that generates pilin antigenic variation in *Neisseria*. (**d**) Programmed DNA elimination during development of macronucleus from micronucleus in *Tetrahymena*. See text for more details

study also revealed a limited overlap of G4 motifs (~37.5% of mapped origins; Mukhopadhyay et al. 2014). ChIP-seq analysis indicated that only 34.1% of Orc1-binding sites have flanking G4 motifs in HeLa cells (Dellino et al. 2013). A very recent Orc2-mapping study by Miotto et al. (2016) supports this result (31% overlap in K562 cells) and also demonstrated that ORC is enriched in open DNase I-accessible chromatin marked by active histone modifications such as H3K27ac, H3K4me2, and H3K4me3. The apparently discrepant locations between ORC sites and initiation sites (SNS peaks) may be in part explained by potential redistribution or sliding of Mcm2-7 complex after loading. However, it is noteworthy that most approaches found that a significant fraction (>30%) of identified origins (or initiation sites) overlapped with G4 motifs (Fig. 14.3b). The most recent mapping that directly analyzes the initiation site (Ini-seq) and does not involve the SNS methods showed 70 % colocalization of mapped TSS (transcription start site) origins with G4-forming sequences (44% of the total origins; Langley et al. 2016). Thus, it is likely that G4 is at least one of the determinants for origin specification.

Besides replication and transcription, G4 is involved in DNA recombination as well. Certain pathogens use DNA recombination to modify the proteins on their outer surfaces by rearranging their genes and avoid repeated detection by the immune system. Cahoon and Seifert (2009) identified a *cis*-acting G4 motif near the variable pilin genes of the human pathogen *Neisseria gonorrhoeae* that controls recombination of the antigenic locus (Fig. 14.3c). Disruption of the G4 motif either by mutagenesis or with a G4 ligand, N-methyl mesoporphyrin IX (NMM), prevented nicks required for recombination from occurring within the G4 region and thereby inhibited pilin antigenic variation. Interestingly, folding into G4 structure of this site depends on transcription of sRNA, a small noncoding RNA (Cahoon and Seifert 2013). The nicks generated in the G4 DNA are subsequently processed by RecJ exonuclease and RecQ and Rep helicases. Most of the RecQ helicase family members encode one HRDC (helicase and RNase D C-terminal) domain, whereas the *N. gonorrhoeae* RecQ helicase bears three HRDC domains. Deletion of RecQ HRDC domains 2 and 3 causes a decrease in the frequency of pilin antigenic variation, consistent with a decrease in its G4-binding and unwinding activities (Cahoon et al. 2013).

BLM helicase, a Rif1-interacter, is one of the five human RecQ family helicases (Croteau et al. 2014) and is involved in class switch recombination (CSR) of immunoglobulin heavy-chain genes. CSR occurs within S regions, 1- to 8-kb segments containing repetitive G4 motifs, and joins a new constant region to the expressed variable region, thereby conferring new effector functions on the encoded antibody without affecting antigen specificity (Matthews et al. 2014). Recombination to a particular S region requires its prior transcription. The active transcription of S regions produces noncoding transcripts generating R loops that contain a stable RNA/DNA hybrid on the template strand and G-rich ssDNA on the non-template strand harboring repetitive G4 motifs (Zhang et al. 2014). The R-loop region is targeted by factors that promote CSR, including BLM helicase and MutSα (MSH2/MSH6 heterodimer). MutSα functions in telomere maintenance as well as in mismatch repair. Human MutSα binds to G4 in R loops formed by transcribed S regions and promotes synapsis between distinct S regions (Larson et al. 2005).

Another experimental evidence that G4 structure and its associated protein regulate chromosome reorganization emerged recently from the study of programmed DNA elimination events of *Tetrahymena thermophila* (Carle et al. 2016). This ciliate has a whole set of its genome in micronuclei (germline nuclei), but eliminates nearly one-third of its germline genome from each developing somatic nucleus (macronucleus). The excised chromosomal ends either form new telomeres by the action of telomerase or reconnected in a manner similar to DSB repair excluding the internal eliminated sequences (IESs) (Chalker and Yao 2011). Thus, in a sense, these programmed DNA elimination events seem reminiscent of formation of genomic domains or chromosomal boundaries in other eukaryotes (that would be contributed by Rif1; Zofall et al. 2016; Toteva et al. 2017), though these do not accompany genomic deletion. The IESs were targeted for removal by small RNA-directed heterochromatin formation. In cells lacking *LIA3* (*ΔLIA3*), the excision of IESs precisely at specific G-rich sequence (5′-AAAAGGGGG-3′) was impaired and imprecise, whereas the removal of IESs without such controlling sequences was unaffected. The G-rich boundary controlling sequences form parallel G4 structures in vitro that are specifically bound by Lia3 protein (Carle et al. 2016). Thus, G4 DNA and Lia3 crucially participate in global genomic rearrangement in a developmental stage-specific manner (Fig. 14.3d).

Most of IESs appeared to harbor retroelements in *Tetrahymena*. Other eukaryotes also carry huge amounts of retrotransposons which are often present on their genomes as clusters, but do not physically eliminate them during development. As ectopic transcription of retrotransposons is toxic to cells (Hatanaka et al. 2015), they must be silenced during the entire lifespan. Interestingly, knockdown of Rif1 leads to derepression of MuERV-L, a third class of LTR-type retrotransposons as well as other two-cell embryonic stage-specific genes in mouse ES cells (Dan et al. 2014; Yoshizawa et al. manuscript in preparation). LTRs of most LTR-type retrotransposons bear multiple G4 motifs (Lexa et al. 2014; Kejnovsky et al. 2015). G4 motifs are also present in non-LTR retrotransposons, such as LINE or SVA. G4 structure is proposed to affect not only transcription and translation but also reverse transcription and integration of retrotransposons (Lexa et al. 2014; Kejnovsky et al. 2015; Sahakyan et al. 2017). Although it is not known yet whether G4 structure is involved in silencing of retrotransposons or not, the fact that reduced expression of G4-binding proteins, such as Rif1 and chromatin remodeler ATRX (Sadic et al. 2015; Voon and Wong 2016), results in derepression of subsets of endoretroviruses implicates G4 structure as a critical *cis*-element for the gene silencing of retrotransposons. In support of this notion, recent study of immuno-electron microscopy showed exquisite specificity of a G4-specific antibody for heterochromatin (Hoffmann et al. 2016).

G4 structure is also involved in efficient infection and/or transcription of some pathogenic retroviruses including HIV as well as suppression of proviral expansion (Tosoni et al. 2015; Perrone et al. 2016; Métifiot et al. 2014; Harris and Merrick 2015). In addition, small molecule G4 stabilizers can reduce the copy number of episomal genomes of KSHV (Kaposi's sarcoma-associated herpesvirus) and EBV (Epstein-Barr virus) during their latent infection. Their latent-phase replication depends on human pre-RC and may proceeds in a manner similar to that of host DNA replication (Lieberman et al. 2007; Tempera and Lieberman 2010). TR (terminal repeat) of KSHV genome functions as a latent replication origin and harbors G4 motifs. Small molecule G4 stabilizers, such as PhenDC3, slow down fork speed due to fork stalling at the G4 structure and thereby trigger advanced firing of dormant replication origins (Madireddy et al. 2016). Slowing down the latent EBV replication with hydroxyurea also advances replication timing of *oriP*, a latent replication origin of EBV episome, and reduces its copy number (Zhou et al. 2009). BRACO-19, another small molecule G4-stabilizing agent, decreases the copy number of EBV latent genome by disrupting the interaction between EBNA1 and ORC and the ability of EBNA1 to tether to metaphase chromosomes (Norseen et al. 2009). Thus, in these γ-herpesviruses, late-to-early transition of replication timing seems to be intimately correlated with reduction of copy number of latent episomes, and affected by the interference or stabilization of G-quadruplex. These viral studies may provide important insight into spatiotemporal regulation of origin firing and of inter-chromatin associations in eukaryotic genomes.

14.5 Concluding Remarks

G4 was originally discovered from analyses of telomere-derived DNA. In spite of extensive analyses of chemical and physical nature of G4 structure, its biological roles have been rather elusive. However, recent development of new tools and genome-wide analyses led to the consensus that G4 is widespread on the genome. G4, due to their unusual structures, was considered to be inhibitory for transcription or replication. Indeed, numerous studies presented compelling evidence that it is the case. Thus, G4 has been regarded as a "hazardous" genome element. However, recent accumulating evidence points to crucial and unexpected roles of G4 in various aspects of chromosome regulation, including regulation of transcription, replication timing program, replication initiation, recombination, gene rearrangement, viral chromosome attachment, and immune evasion of pathogens.

In each case, G4 is recognized by a specific protein to execute the transaction. Rif1, a conserved nuclear factor, specifically recognizes intergenic G4 sequences to regulate DNA replication timing probably through modulation of chromatin architecture as well as by recruiting a phosphatase. Rif1 is involved also in regulation of telomere functions, DNA repair, and transcription. It would be interesting to examine whether G4 recognition is involved in these processes as well.

More than 370,000 G4-forming sequences are predicted to be present on the human genome, and recent genome-wide search has indeed confirmed this prediction, although it appears that the current algorithm does not precisely predict the in vivo G4 structures. In fact, in cells, Rif1, which shows strong preference for G4 structures in vitro, appears to bind to sequences that do not conform to the canonical consensus sequences. Although various probes or methods have been utilized to determine the genomic locations of G4 in the cells, it still appears to be difficult to obtain the consensus for the fully accurate profile of G4 structures.

Formation of G4 in the cells would be affected not only by sequence itself but also by physiological conditions, epigenetic features, and chromosome microenvironment. Thus, it would be also important to monitor dynamic local changes of DNA shapes in different environmental settings as well as in different cell types.

Acknowledgments We thank Naoko Yoshizawa-Sugata for critical reading of the manuscript. We also thank the past and present members of our laboratory for their contribution to our Rif1 studies. This work was supported by JSPS KAKENHI (Grant-in-Aid for Scientific Research (A)) [Grant Numbers 23247031 and 26251004], Grant-in-Aid for Scientific Research on Priority Areas ["non-coding RNA"; Grant Numbers 24114520], and the Uehara Memorial Foundation Research Support to H.M.

References

Adams IR, McLaren A (2004) Identification and characterisation of mRif1: a mouse telomere-associated protein highly expressed in germ cells and embryo-derived pluripotent stem cells. Dev Dyn 229:733–744

Alver RC, Chadha GS, Gillespie PJ, Blow JJ (2017) Reversal of DDK-mediated MCM phosphorylation by Rif1-PP1 regulates replication initiation and replisome stability independently of ATR/Chk1. Cell Rep 18:2508–2520

Anbalagan S, Bonetti D, Lucchini G, Longhese MP (2011) Rif1 supports the function of the CST complex in yeast telomere capping. PLoS Genet 7:e1002024

Andrade MA, Petosa C, O'Donoghue SI, Müller CW, Bork P (2001) Comparison of ARM and HEAT protein repeats. J Mol Biol 309:1–18

Bah A, Forman-Kay JD (2016) Modulation of intrinsically disordered protein function by post-translational modifications. J Biol Chem 291:6696–6705

Baretić D, Berndt A, Ohashi Y, Johnson CM, Williams RL (2016) Tor forms a dimer through an N-terminal helical solenoid with a complex topology. Nat Commun 7:11016

Besnard E, Babled A, Lapasset L, Milhavet O, Parrinello H, Dantec C, Marin JM, Lemaitre JM (2012) Unraveling cell type-specific and reprogrammable human replication origin signatures associated with G-quadruplex consensus motifs. Nat Struct Mol Biol 19:837–844

Bochman ML, Paeschke K, Zakian VA (2012) DNA secondary structures: stability and function of G-quadruplex structures. Nat Rev Genet 13:770–780

Boersma V, Moatti N, Segura-Bayona S, Peuscher MH, van der Torre J, Wevers BA, Orthwein A, Durocher D, Jacobs JJ (2015) MAD2L2 controls DNA repair at telomeres and DNA breaks by inhibiting 5′ end resection. Nature 521:537–540

Bonetti D, Clerici M, Anbalagan S, Martina M, Lucchini G, Longhese MP (2010) Shelterin-like proteins and Yku inhibit nucleolytic processing of Saccharomyces cerevisiae telomeres. PLoS Genet 6:e1000966

Buonomo SB, Wu Y, Ferguson D, de Lange T (2009) Mammalian Rif1 contributes to replication stress survival and homology-directed repair. J Cell Biol 187:385–398

Cahoon LA, Seifert HS (2009) An alternative DNA structure is necessary for pilin antigenic variation in Neisseria gonorrhoeae. Science 325:764–767

Cahoon LA, Seifert HS (2013) Transcription of a cis-acting, noncoding, small RNA is required for pilin antigenic variation in Neisseria gonorrhoeae. PLoS Pathog 9:e1003074

Cahoon LA, Manthei KA, Rotman E, Keck JL, Seifert HS (2013) Neisseria gonorrhoeae RecQ helicase HRDC domains are essential for efficient binding and unwinding of the pilE guanine quartet structure required for pilin antigenic variation. J Bacteriol 195:2255–2261

Carle CM, Zaher HS, Chalker DL (2016) A parallel G quadruplex-binding protein regulates the boundaries of DNA elimination events of Tetrahymena thermophila. PLoS Genet 12:e1005842

Castaño I, Pan SJ, Zupancic M, Hennequin C, Dujon B, Cormack BP (2005) Telomere length control and transcriptional regulation of subtelomeric adhesins in Candida glabrata. Mol Microbiol 55:1246–1258

Cayrou C, Coulombe P, Vigneron A, Stanojcic S, Ganier O, Peiffer I, Rivals E, Puy A, Laurent-Chabalier S, Desprat R, Méchali M (2011) Genome-scale analysis of metazoan replication origins reveals their organization in specific but flexible sites defined by conserved features. Genome Res 21:1438–1449

Cayrou C, Coulombe P, Puy A, Rialle S, Kaplan N, Segal E, Méchali M (2012) New insights into replication origin characteristics in metazoans. Cell Cycle 11:658–667

Chalker DL, Yao MC (2011) DNA elimination in ciliates: transposon domestication and genome surveillance. Annu Rev Genet 45:227–246

Chambers VS, Marsico G, Boutell JM, Di Antonio M, Smith GP, Balasubramanian S (2015) High-throughput sequencing of DNA G-quadruplex structures in the human genome. Nat Biotechnol 33:877–881

Chapman JR, Barral P, Vannier JB, Borel V, Steger M, Tomas-Loba A, Sartori AA, Adams IR, Batista FD, Boulton SJ (2013) RIF1 is essential for 53BP1-dependent nonhomologous end joining and suppression of DNA double-strand break resection. Mol Cell 49:858–871

Choy MS, Page R, Peti W (2012) Regulation of protein phosphatase 1 by intrinsically disordered proteins. Biochem Soc Trans 40:969–974

Cornacchia D, Dileep V, Quivy JP, Foti R, Tili F, Santarella-Mellwig R, Antony C, Almouzni G, Gilbert DM, Buonomo SB (2012) Mouse Rif1 is a key regulator of the replication-timing programme in mammalian cells. EMBO J 31:3678–3690

Croteau DL, Popuri V, Opresko PL, Bohr VA (2014) Human RecQ helicases in DNA repair, recombination, and replication. Annu Rev Biochem 83:519–552

Dan J, Liu Y, Liu N, Chiourea M, Okuka M, Wu T, Ye X, Mou C, Wang L, Wang L, Yin Y, Yuan J, Zuo B, Wang F, Li Z, Pan X, Yin Z, Chen L, Keefe DL, Gagos S, Xiao A, Liu L (2014) Rif1 maintains telomere length homeostasis of ESCs by mediating heterochromatin silencing. Dev Cell 29:7–19

Davé A, Cooley C, Garg M, Bianchi A (2014) Protein phosphatase 1 recruitment by Rif1 regulates DNA replication origin firing by counteracting DDK activity. Cell Rep 7:53–61

Dellino GI, Cittaro D, Piccioni R, Luzi L, Banfi S, Segalla S, Cesaroni M, Mendoza-Maldonado R, Giacca M, Pelicci PG (2013) Genome-wide mapping of human DNA-replication origins: levels of transcription at ORC1 sites regulate origin selection and replication timing. Genome Res 23:1–11

Di Virgilio M, Callen E, Yamane A, Zhang W, Jankovic M, Gitlin AD, Feldhahn N, Resch W, Oliveira TY, Chait BT, Nussenzweig A, Casellas R, Robbiani DF, Nussenzweig MC (2013) Rif1 prevents resection of DNA breaks and promotes immunoglobulin class switching. Science 339:711–715

Dimitrova DS, Gilbert DM (1999) The spatial position and replication timing of chromosomal domains are both established in early G1 phase. Mol Cell 4:983–993

Doluca O, Withers JM, Filichev VV (2013) Molecular engineering of guanine-rich sequences: Z-DNA, DNA triplexes, and G-quadruplexes. Chem Rev 113(5):3044–3083

Escribano-Díaz C, Orthwein A, Fradet-Turcotte A, Xing M, Young JT, Tkáč J, Cook MA, Rosebrock AP, Munro M, Canny MD, Xu D, Durocher D (2013) A cell cycle-dependent regulatory circuit composed of 53BP1-RIF1 and BRCA1-CtIP controls DNA repair pathway choice. Mol Cell 49:872–883

Foti R, Gnan S, Cornacchia D, Dileep V, Bulut-Karslioglu A, Diehl S, Buness A, Klein FA, Huber W, Johnstone E, Loos R, Bertone P, Gilbert DM, Manke T, Jenuwein T, Buonomo SC (2016) Nuclear architecture organized by Rif1 underpins the replication-timing program. Mol Cell 61:260–273

Foulk MS, Urban JM, Casella C, Gerbi SA (2015) Characterizing and controlling intrinsic biases of lambda exonuclease in nascent strand sequencing reveals phasing between nucleosomes and G-quadruplex motifs around a subset of human replication origins. Genome Res 25:725–735

Greenwood J, Cooper JP (2012) Non-coding telomeric and subtelomeric transcripts are differentially regulated by telomeric and heterochromatin assembly factors in fission yeast. Nucleic Acids Res 40:2956–2963

Gottschling DE, Aparicio OM, Billington BL, Zakian VA (1990) Position effect at S. cerevisiae telomeres: reversible repression of Pol II transcription. Cell 63:751–762

Hänsel-Hertsch R, Antonio MD, Balasubramanian S (2017) DNA G-quadruplexes in the human genome: detection, functions and therapeutic potential. Nat Rev Mol Cell Biol 18:279–284

Hardy CFJ, Sussel L, Shore D (1992) A RAP1-interacting protein involved in silencing and telomere length regulation. Genes Dev 6:801–814

Harris LM, Merrick CJ (2015) G-quadruplexes in pathogens: a common route to virulence control? PLoS Pathog 11:e1004562

Hatanaka Y, Inoue K, Oikawa M, Kamimura S, Ogonuki N, Kodama EN, Ohkawa Y, Tsukada Y, Ogura A (2015) Histone chaperone CAF-1 mediates repressive histone modifications to protect preimplantation mouse embryos from endogenous retrotransposons. Proc Natl Acad Sci U S A 112:14641–14646

Hayano M, Kanoh Y, Matsumoto S, Renard-Guillet C, Shirahige K, Masai H (2012) Rif1 is a global regulator of timing of replication origin firing in fission yeast. Genes Dev 26:137–150
Hegyi H (2015) Enhancer-promoter interaction facilitated by transiently forming G-quadruplexes. Sci Rep 5:9165–9170
Hiraga S, Alvino GM, Chang F, Lian HY, Sridhar A, Kubota T, Brewer BJ, Weinreich M, Raghuraman MK, Donaldson AD (2014) Rif1 controls DNA replication by directing Protein Phosphatase 1 to reverse Cdc7-mediated phosphorylation of the MCM complex. Genes Dev 28:372–383
Hiraga SI, Ly T, Garzón J, Hořejší Z, Ohkubo YN, Endo A, Obuse C, Boulton SJ, Lamond AI, Donaldson AD (2017) Human RIF1 and protein phosphatase 1 stimulate DNA replication origin licensing but suppress origin activation. EMBO Rep 18:403–419
Hoffmann RF, Moshkin YM, Mouton S, Grzeschik NA, Kalicharan RD, Kuipers J, Wolters AH, Nishida K, Romashchenko AV, Postberg J, Lipps H, Berezikov E, Sibon OC, Giepmans BN, Lansdorp PM (2016) Guanine quadruplex structures localize to heterochromatin. Nucleic Acids Res 44:152–163
Huppert JL, Balasubramanian S (2005) Prevalence of quadruplexes in the human genome. Nucleic Acids Res 33:2908–2916
Iida K, Nakamura T, Yoshida W, Tera M, Nakabayashi K, Hata K, Ikebukuro K, Nagasawa K (2013) Fluorescent-ligand-mediated screening of G-quadruplex structures using a DNA microarray. Angew Chem Int Ed Eng 52:12052–12055
Kanoh J, Ishikawa F (2001) spRap1 and spRif1, recruited to telomeres by Taz1, are essential for telomere function in fission yeast. Curr Biol 11:1624–1630
Kanoh Y, Matsumoto S, Fukatsu R, Kakusho N, Kono N, Renard-Guillet C, Masuda K, Iida K, Nagasawa K, Shirahige K, Masai H (2015) Rif1 binds to G quadruplexes and suppresses replication over long distances. Nat Struct Mol Biol 22:889–897
Kejnovsky E, Tokan V, Lexa M (2015) Transposable elements and G-quadruplexes. Chromosome Res 23:615–623
Kueng S, Oppikofer M, Gasser SM (2013) SIR proteins and the assembly of silent chromatin in budding yeast. Annu Rev Genet 47:275–306
Kumar S, Yoo HY, Kumagai A, Shevchenko A, Shevchenko A, Dunphy WG (2012) Role for Rif1 in the checkpoint response to damaged DNA in Xenopus egg extracts. Cell Cycle 11:1183–1194
Kunnev D, Freeland A, Qin M, Leach RW, Wang J, Shenoy RM, Pruitt SC (2015) Effect of minichromosome maintenance protein 2 deficiency on the locations of DNA replication origins. Genome Res 25:558–569
Lam EY, Beraldi D, Tannahill D, Balasubramanian S (2013) G-quadruplex structures are stable and detectable in human genomic DNA. Nat Commun 4:1796
Langley AR, Gräf S, Smith JC, Krude T (2016) Genome-wide identification and characterisation of human DNA replication origins by initiation site sequencing (ini-seq). Nucleic Acids Res 44:10230–10247
Larson ED, Duquette ML, Cummings WJ, Streiff RJ, Maizels N (2005) MutSα binds to and promotes synapsis of transcriptionally activated immunoglobulin switch regions. Curr Biol 15:470–474
Lexa M, Steflova P, Martinek T, Vorlickova M, Vyskot B, Kejnovsky E (2014) Guanine quadruplexes are formed by specific regions of human transposable elements. BMC Genomics 15:1032
Li P, Ma X, Adams IR, Yuan P (2015) A tight control of Rif1 by Oct4 and Smad3 is critical for mouse embryonic stem cell stability. Cell Death Dis 6:e1588
Lian HY, Robertson ED, Hiraga S, Alvino GM, Collingwood D, McCune HJ, Sridhar A, Brewer BJ, Raghuraman MK, Donaldson AD (2011) The effect of Ku on telomere replication time is mediated by telomere length but is independent of histone tail acetylation. Mol Biol Cell 22:1753–1765
Lieberman PM, Hu J, Renne R (2007) Maintenance and replication during latency. In: Human herpesviruses: biology, therapy, and immunoprophylaxis. Cambridge University Press, Cambridge, Chapter 24

Madireddy A, Purushothaman P, Loosbroock CP, Robertson ES, Schildkraut CL, Verma SC (2016) G-quadruplex-interacting compounds alter latent DNA replication and episomal persistence of KSHV. Nucleic Acids Res 44:3675–3694

Maizels N (2015) G4-associated human diseases. EMBO Rep 16:910–922

Maizels N, Gray LT (2013) The G4 genome. PLoS Genet 9:e1003468

Marcand S, Wotton D, Gilson E, Shore D (1997) Rap1p and telomere length regulation in yeast. CIBA Found Symp 211:93–103

Martina M, Bonetti D, Villa M, Lucchini G, Longhese MP (2014) Saccharomyces cerevisiae Rif1 cooperates with MRX-Sae2 in promoting DNA-end resection. EMBO Rep 15:695–704

Masai H, Kanoh Y, Moriyama K, Yamazaki S, Yoshizawa N, Matsumoto S (2017) Telomere-binding factors in the regulation of DNA replication. Genes Genet Syst

Mattarocci S, Shyian M, Lemmens L, Damay P, Altintas DM, Shi T, Bartholomew CR, Thomä NH, Hardy CF, Shore D (2014) Rif1 controls DNA replication timing in yeast through the PP1 phosphatase Glc7. Cell Rep 7:62–69

Matthews AJ, Zheng S, DiMenna LJ, Chaudhuri J (2014) Regulation of immunoglobulin class-switch recombination: choreography of noncoding transcription, targeted DNA deamination, and long-range DNA repair. Adv Immunol 122:1–57

Mesner LD, Valsakumar V, Cieslik M, Pickin R, Hamlin JL, Bekiranov S (2013) Bubble-seq analysis of the human genome reveals distinct chromatin-mediated mechanisms for regulating early- and late-firing origins. Genome Res 23:1774–1788

Métifiot M, Amrane S, Litvak S, Andreola ML (2014) G-quadruplexes in viruses: function and potential therapeutic applications. Nucleic Acids Res 42:12352–12366

Miller KM, Ferreira MG, Cooper JP (2005) Taz1, Rap1 and Rif1 act both interdependently and independently to maintain telomeres. EMBO J 24:3128–3135

Miotto B, Ji Z, Struhl K (2016) Selectivity of ORC binding sites and the relation to replication timing, fragile sites, and deletions in cancers. Proc Natl Acad Sci U S A 113:E4810–E4819

Moretti P, Freeman K, Coodly L, Shore D (1994) Evidence that a complex of SIR proteins interacts with the silencer and telomere-binding protein RAP1. Genes Dev 8:2257–2269

Mukhopadhyay R, Lajugie J, Fourel N, Selzer A, Schizas M, Bartholdy B, Mar J, Lin CM, Martin MM, Ryan M, Aladjem MI, Bouhassira EE (2014) Allele-specific genome-wide profiling in human primary erythroblasts reveal replication program organization. PLoS Genet 10:e1004319

Murat P, Balasubramanian S (2014) Existence and consequences of G-quadruplex structures in DNA. Curr Opin Genet Dev 25:22–29

Nikolova EN, Kim E, Wise AA, O'Brien PJ, Andricioaei I, Al-Hashimi HM (2011) Transient Hoogsteen base pairs in canonical duplex DNA. Nature 470:498–502

Norseen J, Johnson FB, Lieberman PM (2009) Role for G-quadruplex RNA binding by Epstein-Barr virus nuclear antigen 1 in DNA replication and metaphase chromosome attachment. J Virol 83:10336–10346

Oldfield CJ, Dunker AK (2014) Intrinsically disordered proteins and intrinsically disordered protein regions. Annu Rev Biochem 83:553–584

Park S, Patterson EE, Cobb J, Audhya A, Gartenberg MR, Fox CA (2011) Palmitoylation controls the dynamics of budding-yeast heterochromatin via the telomere-binding protein Rif1. Proc Natl Acad Sci U S A 108:14572–14577

Peace JM, Ter-Zakarian A, Aparicio OM (2014) Rif1 regulates initiation timing of late replication origins throughout the S. cerevisiae genome. PLoS One 9:e98501

Perrone R, Butovskaya E, Lago S, Garzino-Demo A, Pannecouque C, Palù G, Richter SN (2016) The G-quadruplex-forming aptamer AS1411 potently inhibits HIV-1 attachment to the host cell. Int J Antimicrob Agents 47:311–316

Petryk N, Kahli M, d'Aubenton-Carafa Y, Jaszczyszyn Y, Shen Y, Silvain M, Thermes C, Chen CL, Hyrien O (2016) Replication landscape of the human genome. Nat Commun 7:10208

Piazza I, Rutkowska A, Ori A, Walczak M, Metz J, Pelechano V, Beck M, Haering CH (2014) Association of condensin with chromosomes depends on DNA binding by its HEAT-repeat subunits. Nat Struct Mol Biol 21:560–568

Prioleau MN, MacAlpine DM (2016) DNA replication origins-where do we begin? Genes Dev 30:1683–1697

Rhodes D, Lipps HJ (2015) G-quadruplexes and their regulatory roles in biology. Nucleic Acids Res 43:8627–8637

Ribeyre C, Shore D (2012) Anticheckpoint pathways at telomeres in yeast. Nat Struct Mol Biol 19:307–313

Rosas-Hernández LL, Juárez-Reyes A, Arroyo-Helguera OE, De Las PA, Pan SJ, Cormack BP, Castaño I (2008) yKu70/yKu80 and Rif1 regulate silencing differentially at telomeres in Candida glabrata. Eukaryot Cell 7:2168–2178

Sabouri N, Capra JA, Zakian VA (2014) The essential Schizosaccharomyces pombe Pfh1 DNA helicase promotes fork movement past G-quadruplex motifs to prevent DNA damage. BMC Biol 12:101

Sadic D, Schmidt K, Groh S, Kondofersky I, Ellwart J, Fuchs C, Theis FJ, Schotta G (2015) Atrx promotes heterochromatin formation at retrotransposons. EMBO Rep 16:836–850

Sahakyan AB, Murat P, Mayer C, Balasubramanian S (2017) G-quadruplex structures within the 3′ UTR of LINE-1 elements stimulate retrotransposition. Nat Struct Mol Biol 24:243–247

Sengar A, Hiddi B, Phan AT (2014) Formation of G-quadruplexes in poly-G sequences: structure of a propeller-type parallel-stranded G-quadruplex formed by a G_{15} stretch. Biochemistry 53:7718–7723

Shammas SL, Crabtree MD, Dahal L, Wicky BI, Clarke J (2016) Insights into coupled folding and binding mechanisms from kinetic studies. J Biol Chem 291:6689–6695

Shi T, Bunker RD, Mattarocci S, Ribeyre C, Faty M, Gut H, Scrima A, Rass U, Rubin SM, Shore D, Thomä NH (2013) Rif1 and Rif2 shape telomere function and architecture through multivalent Rap1 interactions. Cell 153:1340–1353

Silverman J, Takai H, Buonomo SB, Eisenhaber F, de Lange T (2004) Human Rif1, ortholog of a yeast telomeric protein, is regulated by ATM and 53BP1 and functions in the S-phase checkpoint. Genes Dev 18:2108–2119

Sreesankar E, Senthilkumar R, Bharathi V, Mishra RK, Mishra K (2012) Functional diversification of yeast telomere associated protein, Rif1, in higher eukaryotes. BMC Genomics 13:255

Sreesankar E, Bharathi V, Mishra RK, Mishra K (2015) Drosophila Rif1 is an essential gene and controls late developmental events by direct interaction with PP1-87B. Sci Rep 5:10679

Sukackaite R, Jensen MR, Mas PJ, Blackledge M, Buonomo SB, Hart DJ (2014) Structural and biophysical characterization of murine rif1 C terminus reveals high specificity for DNA cruciform structures. J Biol Chem 289:13903–13911

Sukackaite R, Cornacchia D, Jensen MR, Mas PJ, Blackledge M, Enervald E, Duan G, Auchynnikava T, Köhn M, Hart DJ, Buonomo SBC (2017) Mouse Rif1 is a regulatory subunit of protein phosphatase 1 (PP1). Sci Rep 7:2119

Takahara T, Hara K, Yonezawa K, Sorimachi H, Maeda T (2006) Nutrient-dependent multimerization of the mammalian target of rapamycin through the N-terminal HEAT repeat region. J Biol Chem 281:28605–28614

Tempera I, Lieberman PM (2010) Chromatin organization of gammaherpesvirus latent genomes. Biochim Biophys Acta 1799:236–245

Tosoni E, Frasson I, Scalabrin M, Perrone R, Butovskaya E, Nadai M, Palù G, Fabris D, Richter SN (2015) Nucleolin stabilizes G-quadruplex structures folded by the LTR promoter and silences HIV-1 viral transcription. Nucleic Acids Res 43:8884–8897

Toteva T, Mason B, Kanoh Y, Brøgger P, Green D, Verhein-Hansen J, Masai H, Thon G (2017) Establishment of expression-state boundaries by Rif1 and Taz1 in fission yeast. Proc Natl Acad Sci 114(5):1093–1098

Urban JM, Foulk MS, Casella C, Gerbi SA (2015) The hunt for origins of DNA replication in multicellular eukaryotes. F1000Prime Rep 7:30

Valton AL, Hassan-Zadeh V, Lema I, Boggetto N, Alberti P, Saintomé C, Riou JF, Prioleau MN (2014) G4 motifs affect origin positioning and efficiency in two vertebrate replicators. EMBO J 33:732–746

Voon HP, Wong LH (2016) New players in heterochromatin silencing: histone variant H3.3 and the ATRX/DAXX chaperone. Nucleic Acids Res 44:1496–1501

Wang H, Zhao A, Chen L, Zhong X, Liao J, Gao M, Cai M, Lee DH, Li J, Chowdhury D, Yang YG, Pfeifer GP, Yen Y, Xu X (2009) Human RIF1 encodes an anti-apoptotic factor required for DNA repair. Carcinogenesis 30:1314–1319

Wotton D, Shore D (1997) A novel Rap1p-interacting factor, Rif2p, cooperates with Rif1p to regulate telomere length in Saccharomyces cerevisiae. Genes Dev 11:748–760

Xu L, Blackburn EH (2004) Human Rif1 protein binds aberrant telomeres and aligns along anaphase midzone microtubules. J Cell Biol 167:819–830

Xu D, Muniandy P, Leo E, Yin J, Thangavel S, Shen X, Ii M, Agama K, Guo R, Fox D 3rd, Meetei AR, Wilson L, Nguyen H, Weng NP, Brill SJ, Li L, Vindigni A, Pommier Y, Seidman M, Wang W (2010) Rif1 provides a new DNA-binding interface for the Bloom syndrome complex to maintain normal replication. EMBO J 29:3140–3155

Xue Y, Rushton MD, Maringele L (2011) A novel checkpoint and RPA inhibitory pathway regulated by Rif1. PLoS Genet 7:e1002417

Xue Y, Marvin ME, Ivanova IG, Lydall D, Louis EJ, Maringele L (2016) Rif1 and Exo1 regulate the genomic instability following telomere losses. Aging Cell 15:553–562

Yamazaki S, Ishii A, Kanoh Y, Oda M, Nishito Y, Masai H (2012) Rif1 regulates the replication timing domains on the human genome. EMBO J 31:3667–3677

Zaaijer S, Shaikh N, Nageshan RK, Cooper JP (2016) Rif1 regulates the fate of DNA entanglements during mitosis. Cell Rep 16:148–160

Zhang ZZ, Pannunzio NR, Han L, Hsieh CL, Yu K, Lieber MR (2014) The strength of an Ig switch region is determined by its ability to drive R loop formation and its number of WGCW sites. Cell Rep 8:557–569

Zofall M, Smith DR, Mizuguchi T, Dhakshnamoorthy J, Grewal SIS (2016) Taz1-shelterin promotes facultative heterochromatin assembly at chromosome-internal sites containing late replication origins. Mol Cell 62(6):862–874

Zhou J, Snyder AR, Lieberman PM (2009) Epstein-Barr virus episome stability is coupled to a delay in replication timing. J Virol 83:2154–2162

Zimmermann M, Lottersberger F, Buonomo SB, Sfeir A, de Lange T (2013) 53BP1 regulates DSB repair using Rif1 to control 5′ end resection. Science 339:700–704

Chapter 15
Chromatin Replication and Histone Dynamics

Constance Alabert, Zuzana Jasencakova, and Anja Groth

Abstract Inheritance of the DNA sequence and its proper organization into chromatin is fundamental for genome stability and function. Therefore, how specific chromatin structures are restored on newly synthesized DNA and transmitted through cell division remains a central question to understand cell fate choices and self-renewal. Propagation of genetic information and chromatin-based information in cycling cells entails genome-wide disruption and restoration of chromatin, coupled with faithful replication of DNA. In this chapter, we describe how cells duplicate the genome while maintaining its proper organization into chromatin. We reveal how specialized replication-coupled mechanisms rapidly assemble newly synthesized DNA into nucleosomes, while the complete restoration of chromatin organization including histone marks is a continuous process taking place throughout the cell cycle. Because failure to reassemble nucleosomes at replication forks blocks DNA replication progression in higher eukaryotes and leads to genomic instability, we further underline the importance of the mechanistic link between DNA replication and chromatin duplication.

Keywords DNA replication • Nucleosome assembly • Histone chaperone • Histone recycling • Chromatin • Epigenetics

15.1 Introduction

Genome function in eukaryotes is regulated by chromatin, a complex structure consisting of DNA, histones, RNA and a large number of structural and regulatory proteins. Chromatin compacts the genome, restricting access to the DNA template

C. Alabert
Centre for Gene Regulation and Expression, School of Life Sciences, University of Dundee, Dundee, UK

Z. Jasencakova • A. Groth (✉)
Biotech Research and Innovation Centre (BRIC), Health and Medical Faculty, University of Copenhagen, Copenhagen, Denmark
e-mail: anja.groth@bric.ku.dk

H. Masai, M. Foiani (eds.), *DNA Replication*, Advances in Experimental Medicine and Biology 1042, https://doi.org/10.1007/978-981-10-6955-0_15

in a manner that is dependent on chromatin composition and chemical modifications on histones and DNA. Distinct chromatin states can be inherited through mitotic cell division, and this contributes to the correct execution of developmental programmes by establishment and maintenance of gene expression patterns. Duplication of chromatin structures in dividing cells thus impinges on the maintenance of epigenetic states and cell identity. However, while our knowledge about DNA replication is comprehensive, the process of chromatin duplication remains poorly understood. In this chapter we will discuss how chromatin is replicated with the emphasis on histone dynamics during DNA replication and maintenance of histone-based information.

The nucleosome consists of 147 base pairs of DNA wrapped around a histone octamer. Each octamer is composed of one centrally located H3-H4 tetramer, flanked by two H2A-H2B dimers. Histones come in different flavours, so-called variants, which along with the large array of modifications in the histone tails holds information important for the chromatin state. The current view is that the preservation of such histone-based information contributes to inheritance of chromatin states. Pioneering electron microscopy studies of replicating chromatin in *Drosophila* embryos revealed that nucleosomes are formed very rapidly after on the daughter DNA strands (McKnight and Miller 1977). Since then, we have learned that chromatin replication can be divided into five steps (Fig. 15.1): (1) disassembly

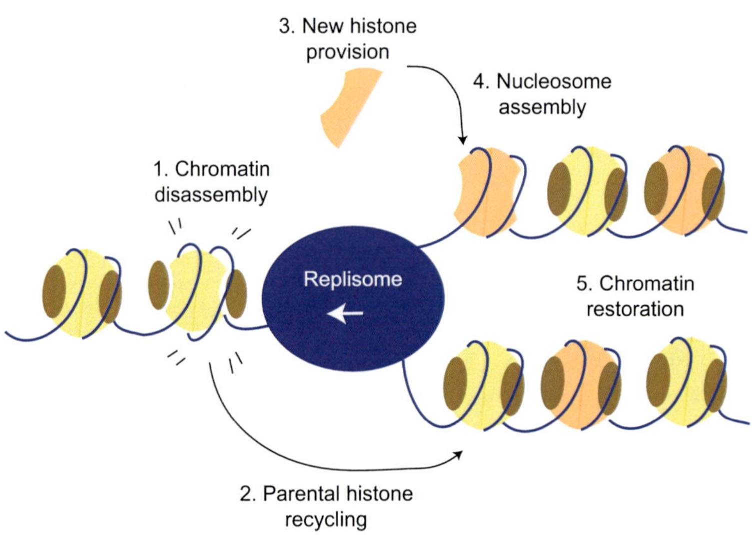

Fig. 15.1 DNA replication in the context of chromatin. The process can be divided into five steps: (*1*) disassembly of chromatin ahead of the replisome, (*2*) recycling of parental histones in a presumably random fashion onto the two daughter strands, (*3*) supply of newly synthesized histones to sites of ongoing DNA replication, (*4*) nucleosome assembly on leading and lagging strands from 50% newly synthesized histones to 50% old recycled histones, (*5*) restoration of marks on DNA and histones and reassociation of other chromatin components including structural and regulatory proteins

of chromatin immediately ahead of the replication machinery, (2) recycling of evicted parental histones onto newly replicated DNA, (3) supply of newly synthesized histones to the sites of ongoing DNA replication, (4) nucleosome assembly from recycled parental and new histones to maintain nucleosome density on the two new daughter DNA strands and (5) restoration of marks on DNA and histones and association of additional chromatin components to restore the chromatin structure. Here we cover mainly steps 1–4 as step 5 is separated from the replication process. Further, we discuss current challenges to understand the dynamic and complex processes of chromatin replication and epigenome maintenance.

15.2 Replication Through Nucleosomes

The DNA molecule is wrapped in a left-handed manner around histone octamers to form nucleosomes (Fig. 15.2). Nucleosomes represent a barrier for DNA-based processes such as transcription, DNA replication and DNA repair. This section focuses

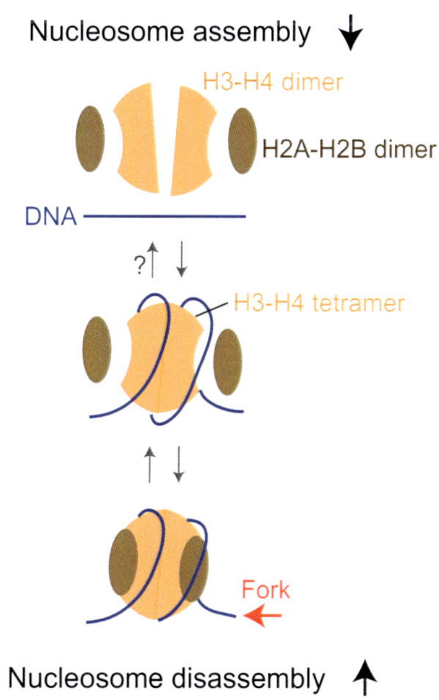

Fig. 15.2 Nucleosome assembly and disassembly processes. In the nucleosome, a histone octamer is wrapped by 147 bp of DNA in a left-handed manner. Nucleosome disassembly (*bottom* to *top*) probably relies on unwinding of DNA by the helicase arriving at the proximal H2A-H2B dimer. This will first release a H2A-H2B dimer and then lead to full disruption of the octamer upon release of the H3-H4 tetramer. Parental histone H3-H4 do not mix with new histone H3-H4, suggesting that separate deposition pathways may exist (see main text). Newly synthesized histones H3-H4 are transported as dimer, which assemble into tetramers in the process of nucleosome assembly

on recent advances in understanding of how nucleosomes are transiently disassembled ahead of replication fork and how evicted histones are recycled on the newly replicated daughter strands.

15.2.1 Chromatin Disassembly Ahead of the Replisome

To sustain an elongation rate of 1.5–2 kb per minute, one nucleosome has to be evicted every 7 second ahead of the replisome. The eviction mechanism involves local destabilization of one (maximally 2) nucleosomes ahead of each replication fork (Gasser et al. 1996; Sogo et al. 1986). Unzipping of the DNA duplex using optical tweezers is sufficient to provoke octamer eviction in vitro (Shundrovsky et al. 2006). However, in vivo studies and recent reconstituted replication systems suggest that a coordinated effort of histone chaperones and chromatin remodellers along with other events allow fork progression through chromatin.

15.2.1.1 Histone Chaperones

Histone chaperones are defined proteins that handle non-nucleosomal histones in vivo and mediate the assembly of nucleosomes from histones in vitro. FACT, consisting of a heterodimer of Spt16 and SSRP1 in humans, is currently viewed as the key chaperone involved in chromatin disruption. FACT binds H2A-H2B dimers (Belotserkovskaya et al. 2003; Orphanides et al. 1999) and H3-H4 tetramer (Tsunaka et al. 2016) and serves a key role in transcription where it aids nucleosome disruption ahead of the RNA polymerase and restores the chromatin template behind (Hammond et al. 2017). Evidence from several organisms have linked FACT to DNA replication, and multiple interactions between FACT and replisome components have been reported (Hammond et al. 2017). However, the exact function of FACT in replication has been difficult to dissect as genetic analysis is hampered by its role in transcription. However, a recent study reconstituting DNA replication on a nucleosomal template from recombinant proteins revealed that FACT is required and sufficient to allow replisome progression in this system (Kurat et al. 2017). How FACT permits the progression of replisomes through nucleosome arrays remains to be determined. As FACT interacts with the replisome (Foltman et al. 2013), one possibility is that FACT promotes nucleosome disassembly as it collides with nucleosomes. As FACT is an extremely abundant protein (one per five nucleosomes in yeast McCullough et al. 2015), another possibility is that FACT destabilizes nucleosomes ahead of the replisome. Since FACT has also been proposed to deposit parental (Foltman et al. 2013) and new histones (Yang et al. 2016) behind the fork, it will be important to dissect these functions from disassembly. Indeed, both nucleosome disassembly ahead and reassembly behind replication fork may control elongation speed, as discussed in Sect. 15.3.

15.2.1.2 ATP-Dependent Chromatin Remodelling Complexes

Chromatin remodellers are large multi-protein complexes that come in different flavours and allow the access to DNA by altering the structure, composition and/or position of nucleosomes. They were first found to play a key role in transcriptional regulation and are now recognized to be part of most chromatin-based processes including DNA replication (Narlikar et al. 2013). Several chromatin remodelling complexes have been suggested to destabilize or remove nucleosomes ahead of replisomes. Recent studies of reconstituted replication on a chromatin template did not find a requirement for ATP-dependent remodelling for fork progression, but the activity of INO80 and ISWIA enhanced the elongation rate (Kurat et al. 2017). Consistent with this, members of the INO80 and ISW2 complex promote efficient fork progression in *S. cerevisiae* (Iida and Araki 2004; Vincent et al. 2008). In mammals, members of the ISWI family (WICH and ACF) and of the INO80 complex have been shown to promote fork progression (Collins et al. 2002; Lee et al. 2014; Poot et al. 2004). However, chromatin remodellers also play an important role behind the fork in establishing nucleosome spacing and position, as discussed in Sect. 15.4. Thus, whether slow fork speed in vivo reflects a function of these remodellers ahead or behind the fork remains unclear.

15.2.1.3 Other Mechanisms Promoting Nucleosome Disassembly

Progression of the replisome creates positive torsional stress of the DNA molecule ahead of the fork. In vitro studies show that nucleosome assembly stalls when DNA is under positive torsional stress (Gupta et al. 2009) and that H2A/H2B dimers could dissociate from the H3-H4 tetramer (Bancaud et al. 2006, 2007). Positive torsional stress could thus induce a structural change in nucleosome that aids disassembly and fork progression (reviewed in Teves and Henikoff 2014). Another mechanism that may aid chromatin disruption is phosphorylation of the linked histone H1. Phosphorylation of histone H1 by cyclin A-CDK2 can decompact chromatin fibres (Contreras et al. 2003) by increasing the dynamic exchange of histone H1. Given that Cdk2 may travel with the fork, it could trigger decompaction of replication domains to facilitate replisome progression (reviewed in Alabert and Groth 2012).

15.2.2 Histone Recycling at Replication Forks

Nucleosomal histones carry modifications that are important for regulation of genome function; thus how they are handled during DNA replication impinges not only on fork progression but also on maintenance of epigenetic states. Histone octamers dissociate upon nucleosome disruption, releasing two H2A-H2B dimers and two H3-H4 dimers under physiological salt conditions (Hammond et al. 2017).

Most evidence suggest that the H3-H4 tetramer remains intact in the process of being recycled onto newly replicated DNA (Fig. 15.2, see below), in which case additional factors like histone chaperones must be involved to maintain the tetrameric state (Hammond et al. 2017). H2A-H2B dimers are also recycled (Alabert et al. 2015), but this process remains poorly understood.

15.2.2.1 H3-H4 Transfer

Upon release from nucleosome, H3-H4 could be transferred as a tetramer or further split into dimers. This has been a major question in the field because of the implicates it has on inheritance of histone-based information; if new and old histone dimers mix after DNA replication, histone modification could be evenly distributed onto the two daughter stands and copied from old histones to new ones within one nucleosome. However, despite the appealing nature of this model, most evidence argue that there is no or only little mixing of new and old histones H3-H4 dimers in the process of DNA replication (Annunziato 2005; Xu et al. 2010). Two complementary technologies have been used to analyse the fate of H3-H4 tetramers upon recycling during DNA replication, differential metabolic labelling of pre-existing and newly synthesized histones (Jackson 1990; Yamasu and Senshu 1990) and differential tagging on newly synthesized and pre-existing histones (Katan-Khaykovich and Struhl 2011; Prior et al. 1980; Xu et al. 2010). Results obtained by both approaches suggest that parental histone H3-H4 do not mix with new histone H3-H4 dimer, which support the idea that the H3-H4 tetramer is transferred as an entity. However, it remains possible that the H3-H4 tetramers split into dimers transiently and reassociate as nucleosomes are assembled on newly replicated DNA. In either case, the absence of new and old H3-H4 dimer mixing in nucleosomes suggests that there are distinct pathways for replication-coupled deposition of new and old histones H3-H4. In relation to epigenetic cell memory, this argues that a potential copy-paste mechanism to propagate histones must operate between nucleosomes rather than internally between tails in the same nucleosome.

15.2.2.2 Mechanisms of Histone Recycling

Pioneer studies of SV40 in vitro replication systems have suggested that histones remain in close proximity to the replisome during recycling (reviewed in Annunziato 2013); however it remains unclear how accurate the recycling process is. Will old histones reassociate with the same DNA sequence or will there be some displacement relative to their original position? A study in yeast used mathematical modelling to address this question, and they estimated that old histones would be recycled roughly within 400 bp of their original position (Radman-Livaja et al. 2011). Recent insights into histone-binding activities within the replisome provide some mechanistic insight into this process of histone segregation (Fig. 15.3).

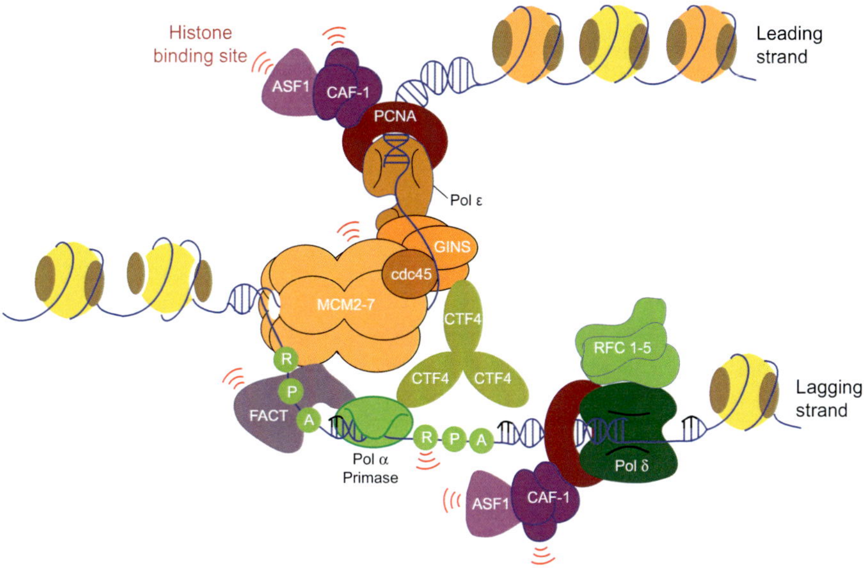

Fig. 15.3 Contact between histones and the replisome. Contact points between histones and replication proteins are indicated in red

MCM2, part of the replicative helicase (MCM2-7), has histone chaperone activity (Ishimi et al. 2001) and binds histone H3-H4 in human (Groth et al. 2007) and yeast (Foltman et al. 2013). Recent crystal structures revealed that MCM2 can chaperone either a H3-H4 tetramer (Richet et al. 2015; Huang et al. 2015) or a H3-H4 dimer in complex with the ASF1 chaperone (Huang et al. 2015; Wang et al. 2015). The histone-binding domain of MCM2 binds H3-H4 tetramers by mimicking DNA in the nucleosome (Huang et al. 2015), which provide a very attractive binding mode for a chaperone involved in recycling. Indeed, MCM2 can handle all H3 variants including H3.3 and CENPA (Huang et al. 2015), and modification of the histone tails should not affect binding. Thus, MCM2 could act as an acceptor of evicted histone H3-H4 genome wide. Consistent with this, mutations of MCM2 histone-binding domain reduce silencing in yeast (Foltman et al. 2013) and slow down cell proliferation in cancer cell lines (Huang et al. 2015). Whether MCM2 operates alone or in conjunction with additional chaperones remains unclear. MCM2 form a co-chaperone complex together with ASF1 in which both chaperones make contact with a H3-H4 dimer (Groth et al. 2007; Huang et al. 2015; Wang et al. 2015). This co-chaperone interaction occurs at replication sites in mammalian cells (Huang et al. 2015), but it remains unclear whether it contributes to normal recycling of H3-H4, through a transient dimeric state, or mainly operates to catch evicted histone upon replication stress as this leads to accumulation of the complex with parental histones (Huang et al. 2015; Jasencakova et al. 2010).

MCM2 also forms a co-chaperone complex with FACT (Foltman et al. 2013). This chaperone travels with the replisome (Alabert et al. 2014; Foltman et al. 2013; Gambus et al. 2006) and can make contact to several replisome components including MCM4 (Gambus et al. 2006; Tan et al. 2006), Polα and RPA1 (VanDemark et al. 2006; Wittmeyer et al. 1999; Zhou and Wang 2004). FACT can bind H3-H4 tetramers (Stuwe et al. 2008; Tsunaka et al. 2016) as well as H2A-H2B dimers (Belotserkovskaya et al. 2003; Orphanides et al. 1999). Recently structural work showed that the binding of H2A-H2B and H3-H4 is not mutually exclusive, and therefore, in theory, FACT could transfer a H3-H4 tetramer together with at least one H2A-H2B dimer. This may happen through a co-chaperone interaction with MCM2, as the structural and biochemical data supports that they can bind H3-H4 simultaneously. FACT is required for replisome progression (Abe et al. 2011; Kurat et al. 2017; Tan et al. 2006); however, as mentioned in Sect. 15.1, genetic studies on FACT are complicated by its multiple functions in transcription and chromatin maintenance.

15.3 New Histone Provision

During replication, pre-existing parental histones are recycled as described above, and in addition newly synthesized histones are delivered and deposited onto the newly synthesized daughter DNA strands to maintain nucleosome density. In human cells about 33 million new nucleosomes have to be assembled in each S phase. To match this high demand for nucleosome assembly, the biosynthesis of new histones and their transfer to replicating DNA are regulated in multiple ways. The replication-dependent canonical histones (H3.1, H3.2, H4, H2A, H2B and H1) are multicopy genes, which are induced at the onset of DNA replication and tightly regulated at both the transcription and post-transcription levels (reviewed in Marzluff et al. 2008). The shortage or excess of histones can block DNA replication and trigger genomic instability in yeast and mammals (Groth et al. 2007; Gunjan and Verreault 2003; Han et al. 1987; Kim et al. 1988; Meeks-Wagner and Hartwell 1986; Mejlvang et al. 2014; Nelson et al. 2002). The massive production of canonical histones also represents a logistic challenge: newly synthesized histones need to be efficiently guided to sites of ongoing replication, a process that is carried out by histone chaperones (reviewed in Hammond et al. 2017). Histone chaperones are a broad class of proteins that handle non-nucleosomal histones in vivo and in vitro can mediate the assembly of nucleosomes when mixed with histones and DNA.

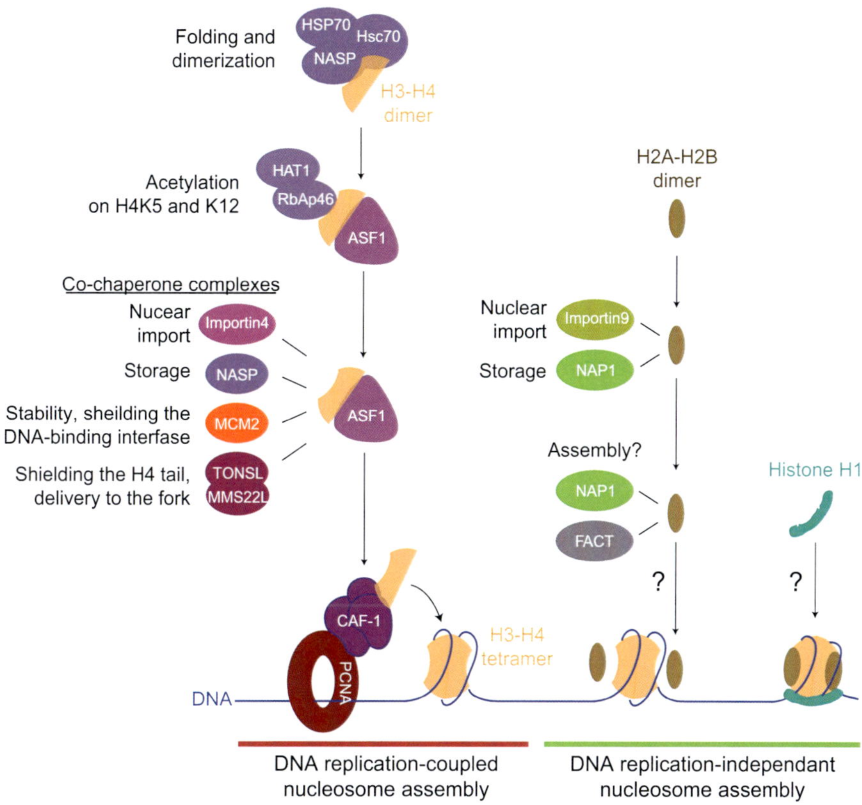

Fig. 15.4 Network of histone chaperones, handling histones from their site of synthesis to their deposition onto newly replicated DNA (See text for details)

15.3.1 Delivery of Newly Synthesized H3.1/H3.2-H4

Soon after their synthesis, histones H3.1/H3.2 and H4 engage with histone chaperones and are transported to sites of DNA replication as dimers (Benson et al. 2006; Tagami et al. 2004). The histone supply chain involves a network of specialized chaperones that are connected via histone-dependent and histone-independent interactions that allow 'on-the-go' modification of the histone, handover of histones between different chaperone complexes and culminates with CAF-1 that assembles H3.1-H4 dimers and newly replicated DNA into tetrasomes, which are completed into nucleosomes upon the addition of H2A-H2B (reviewed in Hammond et al. 2017) (Fig. 15.4).

15.3.1.1 Histone H3.1-H4 Chaperone Network

Biochemical and proteomic characterization of protein complexes interacting with soluble histones has uncovered a large number of histone chaperones, and the list is still growing. While it remains unclear exactly how all these chaperones are organized with respect to each other in the supply chain, the position and role of several key players have been identified based on both biochemical and functional studies. Initially, rapidly after histone synthesis in the cytoplasm, histone H3.1/H3.2/H3.3-H4 can interact with HSC70, HSP90 and NASP that promote the folding and formation of histone dimers (Alvarez et al. 2011; Bowman et al. 2017; Campos and Reinberg 2010). NASP is found in co-chaperone complexes with ASF1 and RBAP46-HAT1 (catalyses H4 acetylation, discussed below) and thus presumably works upstream of these in the network (Bowman et al. 2017; Haigney et al. 2015; Jasencakova et al. 2010).

ASF1 is a central histone H3-H4 chaperone that then engages with several additional histone chaperones, histone modifiers and other accessory partners to form dynamic multi-protein complexes that carry out specialized tasks in histone supply. These tasks include nuclear import (IMPORTIN-4), acetylation of histone tails (RBAP46-HAT1) and regulating the storage pool of soluble histones (NASP or MCM2) (reviewed in Hammond et al. 2017) (Fig. 15.4). These interactions represent co-chaperone relationships where two or more chaperones bind histones concomitantly. ASF1 interacts with the H3 alpha2-alpha3 helices that constitute the H3-H4 tetramerization interface (English et al. 2006; Natsume et al. 2007). Thus, a major consequence of ASF1 binding is to maintain H3-H4 dimeric, while leaving other histone interfaces available for other chaperones like MCM2 (Groth et al. 2007; Huang et al. 2015), RBAP46 (Haigney et al., 2015; Jasencakova et al. 2010), VPS75 (Hammond et al. 2016), NASP (Bowman et al. 2017; Jasencakova et al. 2010) and TONSL (Saredi et al. 2016). It remains to be understood how this dynamic network of multi-chaperone interactions is regulated. However, ASF1 is subject to phosphorylation by the S phase active kinases TLK1 and TLK2 (Sillje and Nigg 2001; Sillje et al. 1999), which are targets of the checkpoint kinase Chk1 (Groth et al. 2003; Krause et al. 2003). TLKs specifically target histone-free ASF1, and phosphorylation in turn promotes histone binding and interaction with downstream chaperones like CAF-1 (Klimovskaia et al. 2014). In contrast, Codanin-1 can act as a negative regulator of histone supply by sequestering ASF1 in the cytoplasm (Ask et al. 2012). Once ASF1 translocate with H3-H4 dimers into the nucleus, it makes direct contact with the two downstream chaperones CAF-1 and HIRA (De Koning et al. 2007; Mello et al. 2002; Tang et al. 2006; Tyler et al. 2001) and, somehow, hand over histone H3-H4 dimers to these chromatin-bound complexes. Whereas CAF-1 deposits canonical histones H3.1/H3.2-H4 in a replication-coupled manner (Sect. 15.3.3) (Smith and Stillman 1989; Tagami et al. 2004), HIRA deposit H3.3-H4 in a replication-independent manner (Ray-Gallet et al. 2002; Tagami et al. 2004) and may serve a gap-filling role under conditions where replication-coupled nucleosome assembly is impaired (Ray-Gallet et al. 2011).

15.3.1.2 Post-translational Modifications of Newly Synthesized H3.1-H4

Newly synthesized histones H4 are acetylated on lysine 5 and 12 by the histone acetyl transferase HAT1 in complex with RBAP46 (Kleff et al. 1995; Parthun et al. 1996; Sobel et al. 1995). This histone H4 di-acetylation is a highly conserved hallmark of new histones, yet its exact function remains unclear. It may contribute to a wide range of events including import of histones into the nucleus (Blackwell et al. 2007; Glowczewski et al. 2004), repair of replication forks (Barman et al. 2006) and CAF-1-dependent chromatin assembly in vivo (Ejlassi-Lassallette et al. 2011; Zhang et al. 2013). Acetylation of new histones generates an open and accessible chromatin organization behind the replication fork (Annunziato and Seale 1983), in part by delaying binding of the linker histone H1 until the acetylations are removed by histone deacetylases 10–20 min after fork passage (Alabert et al. 2014; Perry and Annunziato 1989; Sirbu et al. 2011). Histone H3 is also acetylated, but the sites and abundance differ between species. In yeast, ASF1-bound histone H3 is subject to acetylation on lysine 56 by Rtt109 (Driscoll et al. 2007; Han et al. 2007; Tsubota et al. 2007), and this is a very abundant modification which has major implications for genome stability (Masumoto et al. 2005). H3 K56ac promotes binding to the downstream chaperones CAF-1 and RTT106 and thus facilitates nucleosome assembly (Burgess et al. 2010; Li et al. 2008). The function of H3K56ac does not appear to be conserved in mammals, as H3K56ac is not very abundant (Jasencakova et al. 2010; Ray-Gallet et al. 2011; Xu et al. 2011). In human, soluble H3 is acetylated on lysine 14 and 18 (Jasencakova et al. 2010; Loyola et al. 2006), but their functions remain unclear. Soluble histone H3.1/H3.2 can also be monomethylated on lysine 9 by SETDB1 (Loyola et al. 2006) both early on in the process of translation (Rivera et al. 2015) and late in the assembly pathway as part of a heterochromatin-associated CAF-1 complex containing HP1a and SetDB1 (Loyola et al. 2009). Pre-deposition mono-methylation of H3K9 can prime for the establishment of H3K9me3, which is associated with repressive chromatin (Loyola et al. 2009; Pinheiro et al. 2012); however H3K9me1 may also serve additional roles in the assembly process itself.

15.3.2 Delivery of Newly Synthesized H2A-H2B

The supply chain that delivers histone H2A-H2B to newly replicated DNA appears to be less complex (reviewed in Hammond et al. 2017), although this may reflect that fewer studies have focused on H2A-H2B chaperones. NAP1, the main chaperone for soluble H2A-H2B, stabilizes the H2A-H2B dimer (D'Arcy et al. 2013) and together with Importin-9 facilitates nuclear import (Straube et al. 2010). The binding of NAP1 prevents unscheduled accumulation of H2A-H2B on DNA (Andrews et al. 2010; D'Arcy et al. 2013), and NAP1 is a prime candidate to deliver H2A-H2B to newly formed tetrasomes to complete nucleosome assembly. However, given that FACT is also a histone H2A-H2B (Belotserkovskaya and Reinberg 2004; Hondele et al. 2013) and it is present at replication forks, it might also contribute to H2A-H2B deposition (reviewed in Hammond et al. 2017).

15.3.3 De Novo Deposition

Based on the average replication rate of 1.5–2 kb/min, about one nucleosome assembles on leading and lagging strands every 7 second. Electron microscopy data of replication intermediates suggest that the first nucleosome is assembled about 225 and 285 bp behind the fork for leading and lagging strands, respectively (Gasser et al. 1996). The nucleosome assembly is a stepwise process: the H3-H4 tetramer assembles first followed by two H2A-H2B dimers (Hammond et al. 2017; Smith and Stillman 1991) (Fig. 15.4). While new and old H3-H4 dimers are segregated into separate nucleosomes, new and old H2A-H2B dimers can associate with both new and old H3.1-H4 tetramers (Jackson 1987; Xu et al. 2010). Deposition of histone H1 occurs later as part of the chromatin maturation process, probably depending on deacetylation of the new histones and nucleosome remodelling to establish order nucleosome arrays (Alabert et al. 2014; Perry and Annunziato 1989).

15.3.3.1 Mechanisms of Nucleosome Assembly

Under physiological salt concentrations, histones bind nonspecifically to naked DNA and do not form nucleosomes, and efficient nucleosome assembly thus relies on the concerted action of histone chaperones (reviewed in Hammond et al. 2017). The key chaperone responsible for histone H3.1-H4 deposition onto newly replicated DNA is CAF-1 (Smith and Stillman 1989), which receives histone dimers from ASF1 (Tyler et al. 1999). CAF-1 is composed of three subunits: p150, p60 and RBAP48 (reviewed in Hammond et al. 2017). CAF-1 p150 binds directly to PCNA via a dual PIP-box motif (Moggs et al. 2000; Shibahara and Stillman 1999), elegantly coupling CAF-1 activity to DNA synthesis. ASF1 binds directly to the B domain of the CAF-1 p60 subunit (Mello et al. 2002; Tang et al. 2006; Tyler et al. 2001) through a surface that is an opposite site to the H3-H4-binding site on ASF1 (English et al. 2006; Natsume et al. 2007) (Fig. 15.4). While CAF-1 potentially could bind H3-H4 through RBAP48 (Verreault et al. 1996), recent data argue that it is mainly the p150 subunit that interacts with histones (Kim et al. 2016; Liu et al. 2016). CAF-1 has been found to mainly bind H3.1-H4 dimers (Tagami et al. 2004), but biophysical studies of CAF-1 function show that two CAF-1 complexes may work together to assembly a tetramer during assembly (Mattiroli et al. 2017); Sauer et al. 2017).

In addition to CAF-1, additional pathways for replication-coupled nucleosome assembly are emerging. However, their relative importance and integration with each other remain unclear. RTT106, a yeast chaperone with no clear mammalian homolog, works together with CAF-1 in replication-coupled nucleosome assembly

(Clemente-Ruiz et al. 2011; Li et al. 2008), but it is not clear how it is recruited to sites of DNA replication. Yeast RPA, a critical constituent of DNA replication machinery that binds ssDNA, was recently found to bind histone H3-H4 and facilitate replication-coupled deposition of new histones H3-H4 (Liu et al. 2017). In yeast, FACT has also been implicated in new histone H3-H4 deposition (Yang et al. 2016), and there is evidence that yeast ASF1 interacts with RFC3, part of the clamp loader (Franco et al. 2005). In addition, MCM2 is able to bind both new and old histones at the replication fork and could thus play a role in new histone deposition together with ASF1 (Huang et al. 2015; Jasencakova et al. 2010), although interaction with MCM2 is not a prerequisite for H3.1-H4 incorporation in mammalian cells (Huang et al. 2015). A critical question for future research is whether any of these deposition pathways are specific to either new or old histone H3-H4 and how their activities are coordinated to ensure balanced deposition of new and old H3-H4 on both leading and lagging strands as predicted in the current models.

15.4 Coordination of Nucleosome Assembly and Fork Progression

In yeast, cells can complete one round of DNA replication in the absence of newly synthesized histones (Kim et al. 1988; Prado and Aguilera 2005). In mammals, inhibition of histone production rapidly blocks replisome progression (Mejlvang et al. 2014; Seale and Simpson 1975; Weintraub 1972), reflecting a requirement to coordinate DNA replication and nucleosome assembly. The pool of newly synthesized histones represents only about 1% of the total number of histones in the cells, which explain the need for high rates of histone biosynthesis throughout S phase. The prediction is that replication rates will be sensitive to small changes in new histone availability, as it is the case for deoxyribonucleotides (Earp et al. 2015). However, whereas lack of dNTPs rapidly induces a checkpoint response in part due to excessive DNA unwinding at replication forks (Zeman and Cimprich 2014), lack of histones and impaired nucleosome assembly slows down replication forks without exposing ssDNA and triggering a strong checkpoint response (Groth et al. 2007; Mejlvang et al. 2014). Nevertheless, forks arrested due to lack of nucleosome assembly do become unstable over time giving rise to DNA damage and genomic instability (Hoek and Stillman 2003; Mejlvang et al. 2014; Ye et al. 2003). While replication-coupled nucleosome assembly is not required for fork progression in yeast, it is necessary to maintain integrity of advancing forks. Indeed, mutants defective in replication-coupled chromatin assembly show large chromosomal rearrangements, elevated sister chromatid exchange and loss of replication fork integrity (Clemente-Ruiz et al. 2011; Clemente-Ruiz and Prado 2009; Myung et al. 2003; Prado et al. 2004).

One mechanism proposed to directly link nucleosome assembly and replisome progression involves nucleosome assembly on the lagging strand. Here, histones can be in principle be deposited onto the growing Okazaki fragment once sufficient DNA has been synthesized. Consistent with this idea, the two enzymes important for Okazaki fragment processing, FEN1 and DNA ligase I, can operate efficiently on a nucleosomal substrate (Chafin et al. 2000; Huggins et al. 2002). Intriguingly, Okazaki fragment size is close to the size of a nucleosome repeat (about 200 bp in human cells), suggesting a link between the two processes. This link has now been supported by in vitro and in vivo work in yeast (Devbhandari et al. 2017; Kurat et al. 2017; Smith and Whitehouse 2012). In vivo mapping of Okazaki fragments in yeast revealed that the ligation sites between Okazaki fragments are found in close proximity to the nucleosome dyads and that Okazaki fragment length is increased in CAF-1 mutants deficient in nucleosome assembly (Smith and Whitehouse 2012). More recently, elegant systems for reconstituting DNA replication in vitro demonstrated that nucleosome assembly determines Okazaki fragment length by restricting Pol delta (Polδ) progression (Devbhandari et al. 2017). The evidence thus support a model in which newly assembled nucleosomes will block Polδ progression as the polymerases reaches the nucleosome, which in turn is the signal to release the polymerase and ligate the fragment to the lagging strand. Consistent with this, PCNA unloading, which requires ligation of Okazaki fragments (Kubota et al. 2015), is also impaired when nucleosome assembly is impaired (Mejlvang et al. 2014). Nucleosome assembly on the lagging strand would therefore prime for Polδ release, Okazaki fragment maturation and PCNA unloading, potentially tethering replisome progression to chromatin assembly. How the interplay and potential interdependency between DNA synthesis by Pol epsilon and PCNA-dependent nucleosome assembly operate on the leading strand remains unclear.

15.5 Epigenome Maintenance

DNA replication leads to a genome-wide disruption of chromatin along with two-fold dilution of histone modification as new histones are deposited to double the nucleosome content (Alabert et al. 2015; Alabert and Groth 2012). How chromatin-based information is restored on the two new daughter DNA strands remains a major question. The mechanisms governing restoration of DNA methylation have been extensively studied, and here the maintenance DNA methyltransferase DNMT1 is recruited to hemi-methylated sites in new DNA shortly after fork passage through a mechanism that involves the UHFR1/2 cofactors directly recognizing hemi-methylated CpG sites (Almouzni and Cedar 2016). This section focuses on the restoration of nucleosome positioning and histone modifications. Reincorporation of histone variants and restoration of high-order structures are reviewed elsewhere (Dileep et al. 2015; Talbert and Henikoff 2017).

15.5.1 Nucleosome Positioning

The position of nucleosomes contributes to gene regulation by controlling access to DNA. Accordingly, active promoters and enhancers consist of well-defined nucleosome-free regions (NFR). At these loci, newly assembled nucleosomes thus have to be repositioned to match the pre-replicative state. Nucleosome positioning behind replication forks has recently been analysed genome-wide in yeast and *Drosophila*. In yeast, nucleosomes realign according to their initial profile within minutes (Fennessy and Owen-Hughes 2016; Vasseur et al. 2016). The positioning occurs more rapidly in gene bodies of highly transcribed genes, suggesting a role for transcription elongation (Vasseur et al. 2016). However, transcription elongation-independent mechanisms are also involved in nucleosome positioning, and transcription factors have been suggested to work as so-called rulers that define upstream and downstream nucleosome occupancy patterns (Fennessy and Owen-Hughes 2016; Vasseur et al. 2016). In *Drosophila*, nucleosome positioning is blurred after DNA replication with nucleosome gain at NFR and weakening of strong promoter nucleosome positioning (Ramachandran and Henikoff 2016). One hour later, nucleosomes largely regain patterns, an occupancy pattern resembling the pre-replicative state. These observations suggest that in *Drosophila*, but not in yeast, nucleosomes may outcompete transcription factor during chromatin assembly, potentially creating a window for switching expression states by forming a post-replicative chromatin state characterized by a more uniform nucleosome occupancy pattern.

15.5.2 Histone Post-translational Modifications

A 1:1 mix of newly synthesized and recycled histones is assembled into nucleosomes on the daughter DNA strands (Alabert et al. 2015) (Fig. 15.1), leading to a twofold dilution of histone modifications as the new histones carry mainly pre-deposition-specific acetylation marks. How the pre-replication level of histone modifications is restored after DNA replication is a central question in epigenetics, given that histone modifications sustain transcriptional programmes and must be stably maintained across cellular division (Pengelly et al. 2013). Mass spectrometry data in human cells and genetic analysis in *C. elegans* and fission yeast support that histones retain their modifications during recycling (Alabert et al. 2015; Audergon et al. 2015; Gaydos et al. 2014; Ragunathan et al. 2015; Scharf et al. 2009; Xu et al. 2011). Once incorporated, newly synthesized histones must acquire modification similar to the old ones in the locus where they are incorporated. Several studies have followed this process (Alabert et al. 2015; Pesavento et al. 2008; Scharf et al. 2009; Xu et al. 2011; Zee et al. 2012) and conclude that restoration of the pre-replicative histone modification level is a slow process continuing even in G1 after cell division. This argues that the level of histone modification at a given site oscillates during the cell cycle. It remains unclear how these fluctuations affect chromatin-based processes such as transcription. However, new histones were recently shown to provide a signature of post-replicative chromatin that is recognized by a protein complex involved in homologous recombination (Saredi et al. 2016). This signature relies on the unmethylated state of histone H4 K20, which remains unmodified on new histones until late G2 (Alabert et al. 2015; Pesavento et al. 2008), at which time almost all new histones acquire mono- and di-methylation at this site in a stepwise fashion. In this way, unmethylated H4K20 marks regions of chromatin that have been replicated as good substrates for homologous recombination (Saredi et al. 2016).

How newly assembled histones acquire modifications to restore the pre-replicative state remains unclear. Several histone modifiers bind to PCNA (reviewed in Alabert and Groth 2012) which could serve as a landing platform and provide a direct link between DNA replication and histone modification. However, the slow rates with which marks are established on the new histones question the idea of a tight coupling to the replication process. One possibility is that the replication machinery and parental histones recruit activities that prime histones towards a certain modification state (e.g. H3K9me1 and H3K27me1 are imposed very rapidly, Alabert et al. 2015) and that general chromatin maintenance mechanisms known from studies of transcriptional regulation ensure that the final modification state is restored prior to the next round of cell division. These processes are under intense studies since they are instrumental for maintenance of epigenetic states in cycling cells and likely to play an important role in both organismal development and disease.

15.6 Concluding Remarks

Our understanding of how histones are handled at replication forks is continuing to expand. In the future it will also be important to understand how non-nucleosomal proteins, such as regulatory factors, structural proteins and transcription factors, reassociate on daughter strands and whether their occupancy pattern is affected by DNA replication. Another exciting area of research is to understand how replication stress impinges on the ability of cells to maintain chromatin states (reviewed in Dabin et al. 2016; Svikovic and Sale 2016). As shown in this chapter, nucleosome dynamics are tightly linked to replisome progression, and it remains unclear how fork arrest, processing and collapse will affect the process of chromatin duplication.

References

Abe T, Sugimura K, Hosono Y, Takami Y, Akita M, Yoshimura A, Tada S, Nakayama T, Murofushi H, Okumura K et al (2011) The histone chaperone facilitates chromatin transcription (FACT) protein maintains normal replication fork rates. J Biol Chem 286:30504–30512

Alabert C, Groth A (2012) Chromatin replication and epigenome maintenance. Nat Rev Mol Cell Biol 13:153–167

Alabert C, Bukowski-Wills JC, Lee SB, Kustatscher G, Nakamura K, de Lima Alves F, Menard P, Mejlvang J, Rappsilber J, Groth A (2014) Nascent chromatin capture proteomics determines chromatin dynamics during DNA replication and identifies unknown fork components. Nat Cell Biol 16:281–293

Alabert C, Barth TK, Reveron-Gomez N, Sidoli S, Schmidt A, Jensen ON, Imhof A, Groth A (2015) Two distinct modes for propagation of histone PTMs across the cell cycle. Genes Dev 29:585–590

Almouzni G, Cedar H (2016) Maintenance of epigenetic information. Cold Spring Harb Perspect Biol 8:a019372

Alvarez F, Munoz F, Schilcher P, Imhof A, Almouzni G, Loyola A (2011) Sequential establishment of marks on soluble histones H3 and H4. J Biol Chem 286:17714–17721

Andrews AJ, Chen X, Zevin A, Stargell LA, Luger K (2010) The histone chaperone Nap1 promotes nucleosome assembly by eliminating nonnucleosomal histone DNA interactions. Mol Cell 37:834–842

Annunziato AT (2005) Split decision: what happens to nucleosomes during DNA replication? J Biol Chem 280:12065–12068

Annunziato AT (2013) Assembling chromatin: the long and winding road. Biochim Biophys Acta 1819:196–210

Annunziato AT, Seale RL (1983) Histone deacetylation is required for the maturation of newly replicated chromatin. J Biol Chem 258:12675–12684

Ask K, Jasencakova Z, Menard P, Feng Y, Almouzni G, Groth A (2012) Codanin-1, mutated in the anaemic disease CDAI, regulates Asf1 function in S-phase histone supply. EMBO J 31:2013–2023

Audergon PN, Catania S, Kagansky A, Tong P, Shukla M, Pidoux AL, Allshire RC (2015) Epigenetics. Restricted epigenetic inheritance of H3K9 methylation. Science 348:132–135

Bancaud A, Conde e Silva N, Barbi M, Wagner G, Allemand JF, Mozziconacci J, Lavelle C, Croquette V, Victor JM, Prunell A et al (2006) Structural plasticity of single chromatin fibers revealed by torsional manipulation. Nat Struct Mol Biol 13:444–450

Bancaud A, Wagner G, Conde ESN, Lavelle C, Wong H, Mozziconacci J, Barbi M, Sivolob A, Le Cam E, Mouawad L et al (2007) Nucleosome chiral transition under positive torsional stress in single chromatin fibers. Mol Cell 27:135–147

Barman HK, Takami Y, Ono T, Nishijima H, Sanematsu F, Shibahara K, Nakayama T (2006) Histone acetyltransferase 1 is dispensable for replication-coupled chromatin assembly but contributes to recover DNA damages created following replication blockage in vertebrate cells. Biochem Biophys Res Commun 345:1547–1557

Belotserkovskaya R, Reinberg D (2004) Facts about FACT and transcript elongation through chromatin. Curr Opin Genet Dev 14:139–146

Belotserkovskaya R, Oh S, Bondarenko VA, Orphanides G, Studitsky VM, Reinberg D (2003) FACT facilitates transcription-dependent nucleosome alteration. Science 301:1090–1093

Benson LJ, Gu Y, Yakovleva T, Tong K, Barrows C, Strack CL, Cook RG, Mizzen CA, Annunziato AT (2006) Modifications of H3 and H4 during chromatin replication, nucleosome assembly, and histone exchange. J Biol Chem 281:9287–9296

Blackwell JS Jr, Wilkinson ST, Mosammaparast N, Pemberton LF (2007) Mutational analysis of H3 and H4 N termini reveals distinct roles in nuclear import. J Biol Chem 282:20142–20150

Bowman A, Koide A, Goodman JS, Colling ME, Zinne D, Koide S, Ladurner AG (2017) sNASP and ASF1A function through both competitive and compatible modes of histone binding. Nucleic Acids Res 45:643–656

Burgess RJ, Zhou H, Han J, Zhang Z (2010) A role for Gcn5 in replication-coupled nucleosome assembly. Mol Cell 37:469–480

Campos EI, Reinberg D (2010) New chaps in the histone chaperone arena. Genes Dev 24:1334–1338

Chafin DR, Vitolo JM, Henricksen LA, Bambara RA, Hayes JJ (2000) Human DNA ligase I efficiently seals nicks in nucleosomes. EMBO J 19:5492–5501

Clemente-Ruiz M, Prado F (2009) Chromatin assembly controls replication fork stability. EMBO Rep 10:790–796

Clemente-Ruiz M, Gonzalez-Prieto R, Prado F (2011) Histone H3K56 acetylation, CAF1, and Rtt106 coordinate nucleosome assembly and stability of advancing replication forks. PLoS Genet 7:e1002376

Collins N, Poot RA, Kukimoto I, Garcia-Jimenez C, Dellaire G, Varga-Weisz PD (2002) An ACF1-ISWI chromatin-remodeling complex is required for DNA replication through heterochromatin. Nat Genet 32:627–632

Contreras A, Hale TK, Stenoien DL, Rosen JM, Mancini MA, Herrera RE (2003) The dynamic mobility of histone H1 is regulated by cyclin/CDK phosphorylation. Mol Cell Biol 23:8626–8636

D'Arcy S, Martin KW, Panchenko T, Chen X, Bergeron S, Stargell LA, Black BE, Luger K (2013) Chaperone Nap1 shields histone surfaces used in a nucleosome and can put H2A-H2B in an unconventional tetrameric form. Mol Cell 51:662–677

Dabin J, Fortuny A, Polo SE (2016) Epigenome maintenance in response to DNA damage. Mol Cell 62:712–727

De Koning L, Corpet A, Haber JE, Almouzni G (2007) Histone chaperones: an escort network regulating histone traffic. Nat Struct Mol Biol 14:997–1007

Devbhandari S, Jiang J, Kumar C, Whitehouse I, Remus D (2017) Chromatin constrains the initiation and elongation of DNA replication. Mol Cell 65:131–141

Dileep V, Rivera-Mulia JC, Sima J, Gilbert DM (2015) Large-scale chromatin structure-function relationships during the cell cycle and development: insights from replication timing. Cold Spring Harb Symp Quant Biol 80:53–63

Driscoll R, Hudson A, Jackson SP (2007) Yeast Rtt109 promotes genome stability by acetylating histone H3 on lysine 56. Science 315:649–652

Earp C, Rowbotham S, Merenyi G, Chabes A, Cha RS (2015) S phase block following MEC1ATR inactivation occurs without severe dNTP depletion. Biol Open 4:1739–1743

Ejlassi-Lassallette A, Mocquard E, Arnaud MC, Thiriet C (2011) H4 replication-dependent diacetylation and Hat1 promote S-phase chromatin assembly in vivo. Mol Biol Cell 22:245–255

English CM, Adkins MW, Carson JJ, Churchill ME, Tyler JK (2006) Structural basis for the histone chaperone activity of Asf1. Cell 127:495–508

Fennessy RT, Owen-Hughes T (2016) Establishment of a promoter-based chromatin architecture on recently replicated DNA can accommodate variable inter-nucleosome spacing. Nucleic Acids Res 44:7189–7203

Foltman M, Evrin C, De Piccoli G, Jones RC, Edmondson RD, Katou Y, Nakato R, Shirahige K, Labib K (2013) Eukaryotic replisome components cooperate to process histones during chromosome replication. Cell Rep 3:892–904

Franco AA, Lam WM, Burgers PM, Kaufman PD (2005) Histone deposition protein Asf1 maintains DNA replisome integrity and interacts with replication factor C. Genes Dev 19:1365–1375

Gambus A, Jones RC, Sanchez-Diaz A, Kanemaki M, van Deursen F, Edmondson RD, Labib K (2006) GINS maintains association of Cdc45 with MCM in replisome progression complexes at eukaryotic DNA replication forks. Nat Cell Biol 8:358–366

Gasser R, Koller T, Sogo JM (1996) The stability of nucleosomes at the replication fork. J Mol Biol 258:224–239

Gaydos LJ, Wang W, Strome S (2014) Gene repression. H3K27me and PRC2 transmit a memory of repression across generations and during development. Science 345:1515–1518

Glowczewski L, Waterborg JH, Berman JG (2004) Yeast chromatin assembly complex 1 protein excludes nonacetylatable forms of histone H4 from chromatin and the nucleus. Mol Cell Biol 24:10180–10192

Groth A, Lukas J, Nigg EA, Sillje HH, Wernstedt C, Bartek J, Hansen K (2003) Human tousled like kinases are targeted by an ATM- and Chk1-dependent DNA damage checkpoint. EMBO J 22:1676–1687

Groth A, Corpet A, Cook AJ, Roche D, Bartek J, Lukas J, Almouzni G (2007) Regulation of replication fork progression through histone supply and demand. Science 318:1928–1931

Gunjan A, Verreault A (2003) A Rad53 kinase-dependent surveillance mechanism that regulates histone protein levels in S. cerevisiae. Cell 115:537–549

Gupta P, Zlatanova J, Tomschik M (2009) Nucleosome assembly depends on the torsion in the DNA molecule: a magnetic tweezers study. Biophys J 97:3150–3157

Haigney A, Ricketts MD, Marmorstein R (2015) Dissecting the molecular roles of histone chaperones in histone acetylation by type B histone acetyltransferases (HAT-B). J Biol Chem 290:30648–30657

Hammond CM, Sundaramoorthy R, Larance M, Lamond A, Stevens MA, El-Mkami H, Norman DG, Owen-Hughes T (2016) The histone chaperone Vps75 forms multiple oligomeric assemblies capable of mediating exchange between histone H3-H4 tetramers and Asf1-H3-H4 complexes. Nucleic Acids Res 44:6157–6172

Hammond CM, Stromme CB, Huang H, Patel DJ, Groth A (2017) Histone chaperone networks shaping chromatin function. Nat Rev Mol Cell Biol 18:141–158

Han M, Chang M, Kim UJ, Grunstein M (1987) Histone H2B repression causes cell-cycle-specific arrest in yeast: effects on chromosomal segregation, replication, and transcription. Cell 48:589–597

Han J, Zhou H, Horazdovsky B, Zhang K, Xu RM, Zhang Z (2007) Rtt109 acetylates histone H3 lysine 56 and functions in DNA replication. Science 315:653–655

Hoek M, Stillman B (2003) Chromatin assembly factor 1 is essential and couples chromatin assembly to DNA replication in vivo. Proc Natl Acad Sci U S A 100:12183–12188

Hondele M, Stuwe T, Hassler M, Halbach F, Bowman A, Zhang ET, Nijmeijer B, Kotthoff C, Rybin V, Amlacher S et al (2013) Structural basis of histone H2A-H2B recognition by the essential chaperone FACT. Nature 499:111–114

Huang H, Stromme CB, Saredi G, Hodl M, Strandsby A, Gonzalez-Aguilera C, Chen S, Groth A, Patel DJ (2015) A unique binding mode enables MCM2 to chaperone histones H3-H4 at replication forks. Nat Struct Mol Biol 22:618–626

Huggins CF, Chafin DR, Aoyagi S, Henricksen LA, Bambara RA, Hayes JJ (2002) Flap endonuclease 1 efficiently cleaves base excision repair and DNA replication intermediates assembled into nucleosomes. Mol Cell 10:1201–1211

Iida T, Araki H (2004) Noncompetitive counteractions of DNA polymerase epsilon and ISW2/yCHRAC for epigenetic inheritance of telomere position effect in Saccharomyces cerevisiae. Mol Cell Biol 24:217–227

Ishimi Y, Komamura-Kohno Y, Arai K, Masai H (2001) Biochemical activities associated with mouse Mcm2 protein. J Biol Chem 276:42744–42752

Jackson V (1987) Deposition of newly synthesized histones: new histones H2A and H2B do not deposit in the same nucleosome with new histones H3 and H4. Biochemistry 26:2315–2325

Jackson V (1990) In vivo studies on the dynamics of histone-DNA interaction: evidence for nucleosome dissolution during replication and transcription and a low level of dissolution independent of both. Biochemistry 29:719–731

Jasencakova Z, Scharf AN, Ask K, Corpet A, Imhof A, Almouzni G, Groth A (2010) Replication stress interferes with histone recycling and predeposition marking of new histones. Mol Cell 37:736–743

Katan-Khaykovich Y, Struhl K (2011) Splitting of H3-H4 tetramers at transcriptionally active genes undergoing dynamic histone exchange. Proc Natl Acad Sci U S A 108:1296–1301

Kim UJ, Han M, Kayne P, Grunstein M (1988) Effects of histone H4 depletion on the cell cycle and transcription of Saccharomyces cerevisiae. EMBO J 7:2211–2219

Kim D, Setiaputra D, Jung T, Chung J, Leitner A, Yoon J, Aebersold R, Hebert H, Yip CK, Song JJ (2016) Molecular architecture of yeast chromatin assembly factor 1. Sci Rep 6:26702
Kleff S, Andrulis ED, Anderson CW, Sternglanz R (1995) Identification of a gene encoding a yeast histone H4 acetyltransferase. J Biol Chem 270:24674–24677
Klimovskaia IM, Young C, Stromme CB, Menard P, Jasencakova Z, Mejlvang J, Ask K, Ploug M, Nielsen ML, Jensen ON et al (2014) Tousled-like kinases phosphorylate Asf1 to promote histone supply during DNA replication. Nat Commun 5:3394
Krause DR, Jonnalagadda JC, Gatei MH, Sillje HH, Zhou BB, Nigg EA, Khanna K (2003) Suppression of Tousled-like kinase activity after DNA damage or replication block requires ATM, NBS1 and Chk1. Oncogene 22:5927–5937
Kubota T, Katou Y, Nakato R, Shirahige K, Donaldson AD (2015) Replication-coupled PCNA unloading by the Elg1 complex occurs genome-wide and requires Okazaki fragment ligation. Cell Rep 12:774–787
Kurat CF, Yeeles JT, Patel H, Early A, Diffley JF (2017) Chromatin controls DNA replication origin selection, lagging-strand synthesis, and replication fork rates. Mol Cell 65:117–130
Lee HS, Lee SA, Hur SK, Seo JW, Kwon J (2014) Stabilization and targeting of INO80 to replication forks by BAP1 during normal DNA synthesis. Nat Commun 5:5128
Li Q, Zhou H, Wurtele H, Davies B, Horazdovsky B, Verreault A, Zhang Z (2008) Acetylation of histone H3 lysine 56 regulates replication-coupled nucleosome assembly. Cell 134:244–255
Liu WH, Roemer SC, Zhou Y, Shen ZJ, Dennehey BK, Balsbaugh JL, Liddle JC, Nemkov T, Ahn NG, Hansen KC, et al (2016) The Cac1 subunit of histone chaperone CAF-1 organizes CAF-1-H3/H4 architecture and tetramerizes histones. Elife 5:e18023
Liu S, Xu Z, Leng H, Zheng P, Yang J, Chen K, Feng J, Li Q (2017) RPA binds histone H3-H4 and functions in DNA replication-coupled nucleosome assembly. Science 355:415–420
Loyola A, Bonaldi T, Roche D, Imhof A, Almouzni G (2006) PTMs on H3 variants before chromatin assembly potentiate their final epigenetic state. Mol Cell 24:309–316
Loyola A, Tagami H, Bonaldi T, Roche D, Quivy JP, Imhof A, Nakatani Y, Dent SY, Almouzni G (2009) The HP1alpha-CAF1-SetDB1-containing complex provides H3K9me1 for Suv39-mediated K9me3 in pericentric heterochromatin. EMBO Rep 10:769–775
Marzluff WF, Wagner EJ, Duronio RJ (2008) Metabolism and regulation of canonical histone mRNAs: life without a poly(A) tail. Nat Rev Genet 9:843–854
Masumoto H, Hawke D, Kobayashi R, Verreault A (2005) A role for cell-cycle-regulated histone H3 lysine 56 acetylation in the DNA damage response. Nature 436:294–298
Mattiroli F, Gu Y, Yadav T, Balsbaugh JL, Harris MR, Findlay ES, Liu Y, Radebaugh CA, Stargell LA, Ahn NG, Whitehouse I, Luger K (2017) DNA-mediated association of two histone-bound complexes of yeast Chromatin Assembly Factor-1 (CAF-1) drives tetrasome assembly in the wake of DNA replication. elife 6:e22799
McCullough L, Connell Z, Petersen C, Formosa T (2015) The abundant histone chaperones Spt6 and FACT collaborate to assemble, inspect, and maintain chromatin structure in Saccharomyces cerevisiae. Genetics 201:1031–1045
McKnight SL, Miller OL Jr (1977) Electron microscopic analysis of chromatin replication in the cellular blastoderm Drosophila melanogaster embryo. Cell 12:795–804
Meeks-Wagner D, Hartwell LH (1986) Normal stoichiometry of histone dimer sets is necessary for high fidelity of mitotic chromosome transmission. Cell 44:43–52
Mejlvang J, Feng Y, Alabert C, Neelsen KJ, Jasencakova Z, Zhao X, Lees M, Sandelin A, Pasero P, Lopes M et al (2014) New histone supply regulates replication fork speed and PCNA unloading. J Cell Biol 204:29–43
Mello JA, Sillje HH, Roche DM, Kirschner DB, Nigg EA, Almouzni G (2002) Human Asf1 and CAF-1 interact and synergize in a repair-coupled nucleosome assembly pathway. EMBO Rep 3:329–334
Moggs JG, Grandi P, Quivy JP, Jonsson ZO, Hubscher U, Becker PB, Almouzni G (2000) A CAF-1-PCNA-mediated chromatin assembly pathway triggered by sensing DNA damage. Mol Cell Biol 20:1206–1218

Myung K, Pennaneach V, Kats ES, Kolodner RD (2003) Saccharomyces cerevisiae chromatin-assembly factors that act during DNA replication function in the maintenance of genome stability. Proc Natl Acad Sci U S A 100:6640–6645

Narlikar GJ, Sundaramoorthy R, Owen-Hughes T (2013) Mechanisms and functions of ATP-dependent chromatin-remodeling enzymes. Cell 154:490–503

Natsume R, Eitoku M, Akai Y, Sano N, Horikoshi M, Senda T (2007) Structure and function of the histone chaperone CIA/ASF1 complexed with histones H3 and H4. Nature 446:338–341

Nelson DM, Ye X, Hall C, Santos H, Ma T, Kao GD, Yen TJ, Harper JW, Adams PD (2002) Coupling of DNA synthesis and histone synthesis in S phase independent of cyclin/cdk2 activity. Mol Cell Biol 22:7459–7472

Orphanides G, Wu WH, Lane WS, Hampsey M, Reinberg D (1999) The chromatin-specific transcription elongation factor FACT comprises human SPT16 and SSRP1 proteins. Nature 400:284–288

Parthun MR, Widom J, Gottschling DE (1996) The major cytoplasmic histone acetyltransferase in yeast: links to chromatin replication and histone metabolism. Cell 87:85–94

Pengelly AR, Copur O, Jackle H, Herzig A, Muller J (2013) A histone mutant reproduces the phenotype caused by loss of histone-modifying factor Polycomb. Science 339:698–699

Perry CA, Annunziato AT (1989) Influence of histone acetylation on the solubility, H1 content and DNase I sensitivity of newly assembled chromatin. Nucleic Acids Res 17:4275–4291

Pesavento JJ, Bullock CR, LeDuc RD, Mizzen CA, Kelleher NL (2008) Combinatorial modification of human histone H4 quantitated by two-dimensional liquid chromatography coupled with top down mass spectrometry. J Biol Chem 283:14927–14937

Pinheiro I, Margueron R, Shukeir N, Eisold M, Fritzsch C, Richter FM, Mittler G, Genoud C, Goyama S, Kurokawa M et al (2012) Prdm3 and Prdm16 are H3K9me1 methyltransferases required for mammalian heterochromatin integrity. Cell 150:948–960

Poot RA, Bozhenok L, van den Berg DL, Steffensen S, Ferreira F, Grimaldi M, Gilbert N, Ferreira J, Varga-Weisz PD (2004) The Williams syndrome transcription factor interacts with PCNA to target chromatin remodelling by ISWI to replication foci. Nat Cell Biol 6:1236–1244

Prado F, Aguilera A (2005) Partial depletion of histone H4 increases homologous recombination-mediated genetic instability. Mol Cell Biol 25:1526–1536

Prado F, Cortes-Ledesma F, Aguilera A (2004) The absence of the yeast chromatin assembly factor Asf1 increases genomic instability and sister chromatid exchange. EMBO Rep 5:497–502

Prior CP, Cantor CR, Johnson EM, Allfrey VG (1980) Incorporation of exogenous pyrene-labeled histone into Physarum chromatin: a system for studying changes in nucleosomes assembled in vivo. Cell 20:597–608

Radman-Livaja M, Verzijlbergen KF, Weiner A, van Welsem T, Friedman N, Rando OJ, van Leeuwen F (2011) Patterns and mechanisms of ancestral histone protein inheritance in budding yeast. PLoS Biol 9:e1001075

Ragunathan K, Jih G, Moazed D (2015) Epigenetics. Epigenetic inheritance uncoupled from sequence-specific recruitment. Science 348:1258699

Ramachandran S, Henikoff S (2016) Transcriptional regulators compete with nucleosomes post-replication. Cell 165:580–592

Ray-Gallet D, Quivy JP, Scamps C, Martini EM, Lipinski M, Almouzni G (2002) HIRA is critical for a nucleosome assembly pathway independent of DNA synthesis. Mol Cell 9:1091–1100

Ray-Gallet D, Woolfe A, Vassias I, Pellentz C, Lacoste N, Puri A, Schultz DC, Pchelintsev NA, Adams PD, Jansen LE et al (2011) Dynamics of histone H3 deposition in vivo reveal a nucleosome gap-filling mechanism for H3.3 to maintain chromatin integrity. Mol Cell 44:928–941

Richet N, Liu D, Legrand P, Velours C, Corpet A, Gaubert A, Bakail M, Moal-Raisin G, Guerois R, Compper C et al (2015) Structural insight into how the human helicase subunit MCM2 may act as a histone chaperone together with ASF1 at the replication fork. Nucleic Acids Res 43:1905–1917

Rivera C, Saavedra F, Alvarez F, Diaz-Celis C, Ugalde V, Li J, Forne I, Gurard-Levin ZA, Almouzni G, Imhof A et al (2015) Methylation of histone H3 lysine 9 occurs during translation. Nucleic Acids Res 43:9097–9106

Saredi G, Huang H, Hammond CM, Alabert C, Bekker-Jensen S, Forne I, Reveron-Gomez N, Foster BM, Mlejnkova L, Bartke T et al (2016) H4K20me0 marks post-replicative chromatin and recruits the TONSL-MMS22L DNA repair complex. Nature 534:714–718

Sauer PV, Timm J, Liu D, Sitbon D, Boeri-Erba E, Velours C, Mucke N, Langowski J, Ochsenbein F, Almouzni G, Panne D (2017) Insights into the molecular architecture and histone H3-H4 deposition mechanism of yeast Chromatin assembly factor 1. elife 6:e23474

Scharf AN, Barth TK, Imhof A (2009) Establishment of histone modifications after chromatin assembly. Nucleic Acids Res 37:5032–5040

Seale RL, Simpson RT (1975) Effects of cycloheximide on chromatin biosynthesis. J Mol Biol 94:479–501

Shibahara K, Stillman B (1999) Replication-dependent marking of DNA by PCNA facilitates CAF-1-coupled inheritance of chromatin. Cell 96:575–585

Shundrovsky A, Smith CL, Lis JT, Peterson CL, Wang MD (2006) Probing SWI/SNF remodeling of the nucleosome by unzipping single DNA molecules. Nat Struct Mol Biol 13:549–554

Sillje HH, Nigg EA (2001) Identification of human Asf1 chromatin assembly factors as substrates of Tousled-like kinases. Curr Biol 11:1068–1073

Sillje HH, Takahashi K, Tanaka K, Van Houwe G, Nigg EA (1999) Mammalian homologues of the plant Tousled gene code for cell-cycle-regulated kinases with maximal activities linked to ongoing DNA replication. EMBO J 18:5691–5702

Sirbu BM, Couch FB, Feigerle JT, Bhaskara S, Hiebert SW, Cortez D (2011) Analysis of protein dynamics at active, stalled, and collapsed replication forks. Genes Dev 25:1320–1327

Smith S, Stillman B (1989) Purification and characterization of CAF-I, a human cell factor required for chromatin assembly during DNA replication in vitro. Cell 58:15–25

Smith S, Stillman B (1991) Stepwise assembly of chromatin during DNA replication in vitro. EMBO J 10:971–980

Smith DJ, Whitehouse I (2012) Intrinsic coupling of lagging-strand synthesis to chromatin assembly. Nature 483:434–438

Sobel RE, Cook RG, Perry CA, Annunziato AT, Allis CD (1995) Conservation of deposition-related acetylation sites in newly synthesized histones H3 and H4. Proc Natl Acad Sci U S A 92:1237–1241

Sogo JM, Stahl H, Koller T, Knippers R (1986) Structure of replicating simian virus 40 minichromosomes. The replication fork, core histone segregation and terminal structures. J Mol Biol 189:189–204

Straube K, Blackwell JS Jr, Pemberton LF (2010) Nap1 and Chz1 have separate Htz1 nuclear import and assembly functions. Traffic 11:185–197

Stuwe T, Hothorn M, Lejeune E, Rybin V, Bortfeld M, Scheffzek K, Ladurner AG (2008) The FACT Spt16 "peptidase" domain is a histone H3-H4 binding module. Proc Natl Acad Sci U S A 105:8884–8889

Svikovic S, Sale JE (2016) The effects of replication stress on S phase histone management and epigenetic memory. J Mol Biol 429:2011–2029

Tagami H, Ray-Gallet D, Almouzni G, Nakatani Y (2004) Histone H3.1 and H3.3 complexes mediate nucleosome assembly pathways dependent or independent of DNA synthesis. Cell 116:51–61

Talbert PB, Henikoff S (2017) Histone variants on the move: substrates for chromatin dynamics. Nat Rev Mol Cell Biol 18:115–126

Tan BC, Chien CT, Hirose S, Lee SC (2006) Functional cooperation between FACT and MCM helicase facilitates initiation of chromatin DNA replication. EMBO J 25:3975–3985

Tang Y, Poustovoitov MV, Zhao K, Garfinkel M, Canutescu A, Dunbrack R, Adams PD, Marmorstein R (2006) Structure of a human ASF1a-HIRA complex and insights into specificity of histone chaperone complex assembly. Nat Struct Mol Biol 13:921–929

Teves SS, Henikoff S (2014) DNA torsion as a feedback mediator of transcription and chromatin dynamics. Nucleus 5:211–218
Tsubota T, Berndsen CE, Erkmann JA, Smith CL, Yang L, Freitas MA, Denu JM, Kaufman PD (2007) Histone H3-K56 acetylation is catalyzed by histone chaperone-dependent complexes. Mol Cell 25:703–712
Tsunaka Y, Fujiwara Y, Oyama T, Hirose S, Morikawa K (2016) Integrated molecular mechanism directing nucleosome reorganization by human FACT. Genes Dev 30:673–686
Tyler JK, Adams CR, Chen SR, Kobayashi R, Kamakaka RT, Kadonaga JT (1999) The RCAF complex mediates chromatin assembly during DNA replication and repair. Nature 402:555–560
Tyler JK, Collins KA, Prasad-Sinha J, Amiott E, Bulger M, Harte PJ, Kobayashi R, Kadonaga JT (2001) Interaction between the Drosophila CAF-1 and ASF1 chromatin assembly factors. Mol Cell Biol 21:6574–6584
VanDemark AP, Blanksma M, Ferris E, Heroux A, Hill CP, Formosa T (2006) The structure of the yFACT Pob3-M domain, its interaction with the DNA replication factor RPA, and a potential role in nucleosome deposition. Mol Cell 22:363–374
Vasseur P, Tonazzini S, Ziane R, Camasses A, Rando OJ, Radman-Livaja M (2016) Dynamics of nucleosome positioning maturation following genomic replication. Cell Rep 16:2651–2665
Verreault A, Kaufman PD, Kobayashi R, Stillman B (1996) Nucleosome assembly by a complex of CAF-1 and acetylated histones H3/H4. Cell 87:95–104
Vincent JA, Kwong TJ, Tsukiyama T (2008) ATP-dependent chromatin remodeling shapes the DNA replication landscape. Nat Struct Mol Biol 15:477–484
Wang H, Wang M, Yang N, Xu RM (2015) Structure of the quaternary complex of histone H3-H4 heterodimer with chaperone ASF1 and the replicative helicase subunit MCM2. Protein Cell 6:693–697
Weintraub H (1972) A possible role for histone in the synthesis of DNA. Nature 240:449–453
Wittmeyer J, Joss L, Formosa T (1999) Spt16 and Pob3 of Saccharomyces cerevisiae form an essential, abundant heterodimer that is nuclear, chromatin-associated, and copurifies with DNA polymerase alpha. Biochemistry 38:8961–8971
Xu M, Long C, Chen X, Huang C, Chen S, Zhu B (2010) Partitioning of histone H3-H4 tetramers during DNA replication-dependent chromatin assembly. Science 328:94–98
Xu M, Wang W, Chen S, Zhu B (2011) A model for mitotic inheritance of histone lysine methylation. EMBO Rep 13:60–67
Yamasu K, Senshu T (1990) Conservative segregation of tetrameric units of H3 and H4 histones during nucleosome replication. J Biochem 107:15–20
Yang J, Zhang X, Feng J, Leng H, Li S, Xiao J, Liu S, Xu Z, Xu J, Li D et al (2016) The histone chaperone FACT contributes to DNA replication-coupled nucleosome assembly. Cell Rep 16:3414
Ye X, Franco AA, Santos H, Nelson DM, Kaufman PD, Adams PD (2003) Defective S phase chromatin assembly causes DNA damage, activation of the S phase checkpoint, and S phase arrest. Mol Cell 11:341–351
Zee BM, Britton LM, Wolle D, Haberman DM, Garcia BA (2012) Origins and formation of histone methylation across the human cell cycle. Mol Cell Biol 32:2503–2514
Zeman MK, Cimprich KA (2014) Causes and consequences of replication stress. Nat Cell Biol 16:2–9
Zhang W, Tyl M, Ward R, Sobott F, Maman J, Murthy AS, Watson AA, Fedorov O, Bowman A, Owen-Hughes T et al (2013) Structural plasticity of histones H3-H4 facilitates their allosteric exchange between RbAp48 and ASF1. Nat Struct Mol Biol 20:29–35
Zhou Y, Wang TS (2004) A coordinated temporal interplay of nucleosome reorganization factor, sister chromatin cohesion factor, and DNA polymerase alpha facilitates DNA replication. Mol Cell Biol 24:9568–9579

Chapter 16
The Temporal Regulation of S Phase Proteins During G_1

Gavin D. Grant and Jeanette G. Cook

Abstract Successful DNA replication requires intimate coordination with cell-cycle progression. Prior to DNA replication initiation in S phase, a series of essential preparatory events in G_1 phase ensures timely, complete, and precise genome duplication. Among the essential molecular processes are regulated transcriptional upregulation of genes that encode replication proteins, appropriate post-transcriptional control of replication factor abundance and activity, and assembly of DNA-loaded protein complexes to license replication origins. In this chapter we describe these critical G_1 events necessary for DNA replication and their regulation in the context of both cell-cycle entry and cell-cycle progression.

Keywords Cell cycle • Origin licensing • RB • E2F • Cyclin • CDK • APC/C • Quiescence • DREAM complex

16.1 Introduction

Proper progression from the G_1 cell-cycle phase into S phase and the accurate duplication of chromosomal DNA are key for successful completion of the cell division cycle. In this chapter we describe the molecular events of G_1 that are critical for successful initiation and completion of DNA replication and, thus, S phase. We primarily focus on events that have been studied in human cell lines, but when appropriate, we expand our focus to other model systems. While the ultimate goal of G_1 is to prepare for S phase, there are many sequential G_1 processes required for the proper transition into S phase including appropriate control of gene expression,

G.D. Grant • J.G. Cook (✉)
Department of Biochemistry and Biophysics, University of North Carolina at Chapel Hill, Chapel Hill, NC, USA

Lineberger Comprehensive Cancer Center, University of North Carolina at Chapel Hill, Chapel Hill, NC, USA
e-mail: jean_cook@med.unc.edu

H. Masai, M. Foiani (eds.), *DNA Replication*, Advances in Experimental Medicine and Biology 1042, https://doi.org/10.1007/978-981-10-6955-0_16

protein accumulation, and protein-DNA complex assembly. We have divided our discussion of G_1 into three temporal segments: G_0, early G_1, and late G_1.

Because error-free DNA replication is so critical, many different cell-cycle checkpoints and levels of regulation ensure that this process is precise. Many of these checkpoints happen in G_1 before DNA replication begins. Given that DNA replication is essentially irreversible, it makes sense that multiple inputs are integrated into the decision to begin replication. If cells receive appropriate inputs in G_1, and the external signals and environment are compatible with another round of the cell cycle, then an intracellular signaling cascade ultimately results in irrevocable commitment to S phase entry; this commitment is known as passage through the "restriction point" (Pardee 1974). The restriction point, as canonically known, is the point at which cells transition from requiring mitogens (extracellular growth factors) for S phase entry to no longer requiring sustained mitogenic signaling for progressing into S phase and through the rest of the cell cycle (Blagosklonny and Pardee 2002). If mitogenic stimuli are removed before cells pass the restriction point, they may enter G_0, or quiescence, a temporary withdrawal from the cell cycle. Quiescence is a normal part of organismal development and homeostasis and is important for regenerative processes and the long-term maintenance of stem cell populations (Rumman et al. 2015).

When reduced to its simplest form, progression through G_1 is relatively straightforward: there is a period of low cyclin-dependent kinase (CDK) activity after mitosis or during G_0. Then mitogenic signals trigger intracellular signal transduction networks that ultimately activate the heterodimeric protein kinase cyclin D/CDK4 or cyclin D/CDK6; CDK4 is the catalytic subunit responsible for phosphotransfer but is inactive in the absence of cyclin binding (Choi and Anders 2014). (Current evidence indicates that cyclin-dependent kinases 4 and 6 are functionally redundant in G_1; thus, to simplify, we will only designate CDK4.) Cyclin D/CDK4, in turn, upregulates cyclin E/CDK2, which both drives expression of genes required for DNA replication and triggers the initiation of DNA synthesis (Fig. 16.1). These two major cyclin/CDK pairs drive cells through early then late G_1 and into S phase. There are three different human genes for cyclin D (1, 2, and 3) and two different genes that encode cyclin E (1 and 2); we refer to the class of genes (e.g., cyclin D) instead of specifying each gene product.

Regulatory events or genetic perturbations can prevent, delay, speed up, or enhance any of the essential G_1 processes. Both positive and negative regulation during G_1 influences the rate of cell-cycle progression. For instance, cyclins fluctuate with peak abundance at different cell-cycle phases, and these peaks are driven by accumulation (largely from gene induction) and induced protein degradation. The catalytic subunits of CDKs are present at relatively stable levels throughout the cell cycle but are not active unless bound to a cyclin. Once bound by cyclin, CDK undergoes a series of modifications (protein domain remodeling, phosphorylations, dephosphorylations, etc.) that either activate or inhibit kinase activity. The importance of controlling these events in the cell cycle is illustrated by the fact that most of the proteins involved are misregulated in some way during tumorigenesis.

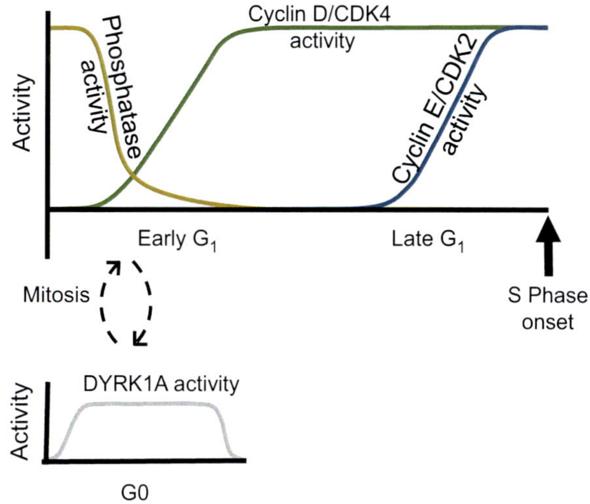

Fig. 16.1 Activities of cell cycle-dependent kinases and phosphatases during G_1 and G_0. Dashed lines indicate entry and exit from G_0

Complete and precise genome duplication in organisms with large genomes divided among many discrete chromosomes requires a complex network of pathways and molecules. Decades of research in both the cell cycle and the DNA replication fields has identified a great many proteins that act at key points in G_1 entry and progression. To aid readers, we provide a list of a subset of principle G_1 proteins and their general functions relevant to our discussion in Table 16.1.

16.2 Quiescence (G_0)

Before passing through the restriction point, cells can enter a nonproliferative phase, called G_0 or quiescence, which is considered to be outside of the cell cycle. This noncycling state is fundamentally different from other cell-cycle exits such as senescence or terminal differentiation, which are irreversible; cells can re-enter the cell cycle from G_0 by transitioning into and progressing through G_1. G_0 is critically important to organismal health and maintenance because, when needed, populations of stem cells in G_0 can re-enter the cell cycle to replenish the overall cell population (Rumman et al. 2015). The availability of G_0 cells is especially important in wound healing, the immune response, the routine regeneration of the intestinal epithelium, and in bone marrow (where hematopoietic stem cells produce blood cells). Similarly, populations of stem cells in G_0 provide protection against acute toxic events such as radiation or chemotherapy. Many types of chemotherapy are designed to induce toxic damage in cycling cells, often by forming DNA adducts that prevent cells from successfully replicating their DNA. It has been proposed recently, however, that some populations of cancer cells are able to escape genotoxic drugs and thus the effects of chemotherapy by entering a G_0 state. After treatment has ceased, these

Table 16.1 A subset of important G_1 proteins and their functions

Factor	Function
Kinases	
Cyclin D/CDK4	Rb phosphorylation
Cyclin E/CDK2	Rb phosphorylation, initiating DNA synthesis at origins
DYRK1A	DREAM complex assembly, Lin52 phosphorylation
GSK3β	p130 stabilization and cyclin D destabilization in G_0
Protein kinase inhibitors	
p21	CDK2 inhibition (cyclin D/CDK4 assembly)
p27	CDK2 inhibition
Phosphatases	
PP1	Rb dephosphorylation
PP2A	Rb dephosphorylation
Transcription factors	
E2F1,2,3	Cell-cycle gene activation
E2F4,5,6,7,8	Cell-cycle gene repression
DP1	Obligate binding partner of activating E2F1,2,3
Transcriptional corepressors	
Rb	Activator E2F repression in G_1
p107	Activator E2F repression in S phase
p130	Cell-cycle gene repression in G0 (DREAM complex)
HDAC	Histone deacetylase
DREAM complex	Repression in G_0 (E2F4/DP1/p130 and Lin52-containing MuvB complex)
Ubiquitylation	
APC/C	E3 ubiquitin ligase complex in mitosis, G_1, and G_0
Cdc20	Substrate-recruiting subunit of APC/C in mitosis
Cdh1	Substrate-recruiting subunit of APC/C in mitosis, G_1, and G_0
Emi1	Pseudosubstrate inhibitor of APC/C^{Cdh1} in G_1
SCF	E3 ubiquitin ligase complex throughout the cell cycle
Skp2	Substrate-recruiting subunit of SCF
26S proteasome	Degradation of ubiquitylated proteins
Origin licensing	
MCM	DNA helicase (Mcm2-Mcm7)
ORC	DNA-binding MCM loading enzyme (Orc1-Orc6)
Cdc6	MCM loading enzyme
Cdt1	MCM recruitment and loading
Geminin	Cdt1 inhibitor
Set8 (PR-Set7)	Histone H4 K20 monomethylase

cells can re-enter the cell cycle and form a new tumor, causing a cancer relapse (O'Connor et al. 2014). Thus, quiescence provides a reservoir of proliferative capacity that maintains adult tissues but also presents a challenge for cancer treatment.

16.2.1 Global Transcriptional Control in G_0

One characteristic of quiescent cells is the active repression of transcriptional activity, especially the repression of genes involved in cell-cycle progression. The molecular mechanisms that occur during the transition from the cell cycle into G_0 are largely unknown, but investigators have identified some key molecular characteristics of cells that are already in G_0 (Litovchick et al. 2007, 2011). Whole-genome transcriptomics studies revealed unique quiescence-associated transcriptional programs (Coller et al. 2006). This microarray based approached catalogued groups of up- or downregulated genes in primary human fibroblasts driven into G_0 by three different methods (serum starvation, contact inhibition, or "loss of adhesion"). While the investigators found one set of genes that were similarly regulated in each of the three methods, indicating there is a "core" transcriptional program in quiescent cells, the regulation of some genes were synchronization method specific. The differences among these three datasets suggest that there are different transcriptional programs active in quiescence depending on the mechanism of cell-cycle exit (Coller et al. 2006). Many of the genes that were downregulated in at least one condition are central to DNA replication and cell-cycle control such as the mitotic cyclin B1, enzymes required to create DNA synthetic precursors, and subunits of the major E3 ubiquitin ligases that act in S phase or mitosis. Among the commonly upregulated genes were transcriptional repressors such as *E2F4* and *E2F5*, which cooperate with the so-called pocket proteins to suppress the expression of DNA replication and cell-cycle genes during G_0.

16.2.2 E2F and Pocket Proteins

The majority of genes that encode proteins needed for cell-cycle progression, DNA synthesis, and mitosis are regulated by a class of transcription factors known as E2Fs. The activating subclass of E2F transcription factors (E2F1, E2F2, and E2F3) is one of the main drivers for the cell-cycle transcriptional program. We will refer to the class of activating transcription factors and their binding partners, DP1 or DP2, as E2F, instead of the individual gene products. A second subclass of E2F transcription factors, known as the repressor E2Fs (E2F4, E2F5, E2F6, E2F7, and E2F8), generally inhibit expression of target genes (Frolov and Dyson 2004). Each E2F protein has a DNA-binding domain, an RB-binding domain, and a DP1-binding domain that consists of a leucine-repeat sequence and the marked-box domain (Fig. 16.2a). E2F is primarily functional during late G_1 and early S phase as cells accumulate proteins

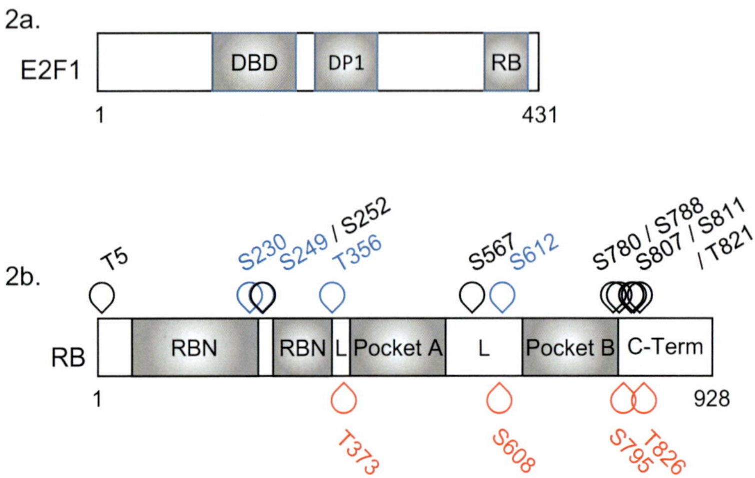

Fig. 16.2 E2F and RB proteins. (**a**) Domain structure of the transcription factor E2F1 as an representative activator E2F. *DBD* DNA-binding domain, *DP1* DP-binding domain, *RB* RB-binding domain. (**b**) RB phosphorylation sites. Blue sites have been implicated in increased binding affinity to E2F transcription factors; red sites decrease affinity for E2F, and black indicates no change in E2F affinity. *RBN* N-terminal domains of RB, *L* linker domains

in preparation for DNA replication. E2F transcriptionally activates the expression of genes encoding proteins involved in positive feedback loops, such as E2F itself, and cyclin E that drive the cell into S phase. During mid-S phase through mitosis and early G_1, E2F is bound to and repressed by the pocket protein RB which is encoded by the human retinoblastoma gene, the first tumor suppressor identified (Knudson 1971; Benedict et al. 1983; Murphree and Benedict 1984). E2F-regulated genes are actively repressed during G_0 by the action of E2F-associated corepressors and that repression is mediated by E2F interaction with the p130 and RB-pocket proteins.

Each member of the activating E2F subclass can largely compensate for the loss of the other two members; however, loss of all three E2F proteins results in severely impaired cell proliferation (Wu et al. 2001). *E2f1* null mice are viable and reproduce normally, indicating that the other two activating E2F transcription factors are able to compensate for the loss of E2F1 (Field et al. 1996). Likewise, *E2f1*$^{-/-}$ *E2f2*$^{-/-}$ mice also survive to adulthood. Interestingly, *E2f3*$^{-/-}$ mice die early in development, indicating E2F3 plays a critical role in mouse development, but *E2f3*$^{-/-}$ cells proliferate nonetheless (Wu et al. 2001). Using conditional knockout mice, Wu et al. generated mouse fibroblast lines with combinations of two or three null alleles in the three activating *E2f* genes. The cells deficient for two of the three E2F transcription factors proliferated slowly, whereas cells lacking all three had no measurable proliferation (Wu et al. 2001).

There are three E2F-binding "pocket proteins" encoded by mammalian genomes: the retinoblastoma protein (RB), p107 (RBL1), and p130 (or RBL2). They are named for a structural domain containing a protein-binding cleft. Though they do

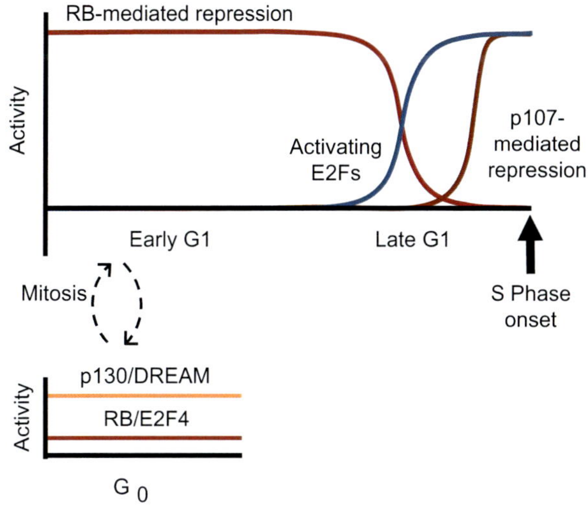

Fig. 16.3 Activities of the pocket proteins RB, p107, and p130 and activating E2F transcription factors during G_1 and G_0. Dashed lines indicate entry and exit from G_0

not bind DNA directly, these proteins mediate transcriptional repression of many cell-cycle genes and are key regulators of cell-cycle progression. Each pocket protein consists of five major domains: the N- and C- terminal domains, the A- and B-pocket domains, and a linker region between the two pocket domains (Fig. 16.2b).

The three pocket proteins have both overlapping and distinct roles, patterns of expression, regulation, and binding partners during the cell cycle (Classon and Harlow 2002). RB is primarily active in G_1 by binding to and inactivating the activating family of E2F transcription factors (E2F1, E2F2, and E2F3). p130 is highly expressed in G_0, whereas p107 expression increases after mitogenic stimulation to peak in S phase (Fig. 16.3) (Litovchick et al. 2007; Devoto et al. 1992; Cobrinik et al. 1993; Burkhart et al. 2010). If one pocket protein is lost through mutation, the others can functionally compensate in some but not all settings.

16.2.3 *E2F and Pocket Proteins in G_0*

During G_0, p130 is abundant and bound to DNA through E2F4 as a member of a multisubunit complex, known as the DREAM complex (Litovchick et al. 2007; Guiley et al. 2015). In contrast, RB is much less abundant, and p107 is almost undetectable in quiescent cells (Hurford et al. 1997; Moberg et al. 1996). The p130-containing DREAM complex only assembles during G_0 and is named for its components: DP, RB-like (e.g., p130), E2F, and MuvB subcomplex, which itself consists of subunits LIN9, LIN37, LIN52, LIN54, and RBAP48. The DREAM complex represses both E2F target genes and genes that are expressed in late S phase and G_2. This repression prevents aberrant expression that could lead to improper cell-cycle reentry (Litovchick et al. 2007; Muller et al. 2012, 2016).

DREAM complex assembly requires phosphorylation of the LIN52 subunit of MuvB at serine 28. In G_0, Lin52 Ser28 is phosphorylated by DYRK1A kinase, a protein kinase with multiple roles in cell proliferation and neuronal development (Fig. 16.1) (Chen et al. 2013; Soppa et al. 2014). Phosphorylation of p130 also contributes to DREAM assembly in G_0 by increasing p130 stability. Glycogen synthase kinase 3β (GSK3β), a protein kinase that is highly active during G_0, phosphorylates three sites in the loop region in the B-pocket domain of p130. These three phosphorylations stabilize p130 to promote DREAM assembly but are removed during cell-cycle reentry to facilitate DREAM disassembly (Litovchick et al. 2004). Cyclin D/CDK4 phosphorylates p130 to disrupt its incorporation into the DREAM complex; cyclin D/CKD4 activity is low in G_0, allowing the DREAM complex to fully assemble (Sandoval et al. 2009). These p130 phosphorylations occur in the p130 interdomain linker region and disrupt the association between p130 and LIN52 to derepress DREAM-regulated genes (Guiley et al. 2015).

The MuvB subcomplex associates with DP, RB-like, and E2F proteins to form the DREAM complex in G_0, but it also binds to the transcription factor BMYB during S phase to form the MMB complex. The MMB complex is important for transactivation of genes involved in G_2 and M phase progression in tandem with the transcription factors BMYB and FOXM1 (Sadasivam et al. 2012). Somewhat paradoxically, the MuvB subcomplex is critical for the maintenance of quiescence as part of the transcriptionally repressive DREAM complex but is also critical for the proper transcriptional expression of G_2/M genes.

Consistent with the idea that cells are sometimes considered to be "resting" during quiescence, overall mRNA abundance is reduced, as are global levels of translation (Williams and Penman 1975; Levine et al. 1965; Degen et al. 1983). These reductions in both mRNA and protein may be due to the ability of RB and p130 to downregulate rRNA gene expression by inhibiting RNA polymerase I (Hannan et al. 2000), or the unique ability of RB to reduce tRNA levels by inhibition of RNA polymerase complex TFIIIB (Scott et al. 2001; White et al. 1996). The resulting reduction in ribosomal RNAs and tRNAs slows protein translation globally.

16.3 Early G_1

Despite its temporal distance from S phase, many events that occur in early G_1 influence S phase progression. These early G_1 events may ultimately impact when a segment of DNA is replicated in S phase, help determine the length of G_1, or ensure that proteins required for S phase do not accumulate prematurely. As described briefly in the introduction, G_1 progression requires the function of two heterodimeric kinases, cyclin D/CDK4 and cyclin E/CDK2. Cyclin E/CDK2 is also essential to initiate DNA synthesis in S phase. All forms of cell cycle-driving CDKs are inactivated at mitosis or kept inactive during G_0 through multiple mechanisms targeting both the cyclin and CDK subunits individually and the complexes themselves. These inactivating mechanisms are reversed in early G_1.

16.3.1 Cyclin D/CDK4 in Early G_1

During G_0 and also between anaphase in mitosis and mid-G_1 in actively proliferating cells, CDK activity is minimal due to low transcriptional activation and targeted destruction of cyclins. Immediately after cell division, cyclin D protein levels are very low due to a combination of inactive *cyclin d* gene expression and active cyclin D protein degradation. Unlike cyclins E, A, or B, the protein levels of cyclin D do not rise and fall in a cell cycle-dependent fashion but instead are regulated by the presence of growth factors in the environment (Reviewed in (Choi and Anders 2014)). There are three primary mechanisms of cyclin D accumulation. The first is transcriptional activation by transcription factors controlled by mitogenic signaling pathways (e.g. c-Jun, c-Fos, or MYC) activated by cytokines (e.g., NF-κB) or other signaling factors (e.g., Notch or CREB). The second method of cyclin D accumulation is through the PI3K-AKT-mTOR-S6 K1 signaling pathway, which downregulates glycogen synthase kinase β and upregulates cyclin D translation (Diehl et al. 1998) (Koziczak and Hynes 2004; Fornoni et al. 2008; Katoh and Katoh 2006). The third method is stabilization and nuclear localization of cyclin D, which is regulated by phosphorylation of cyclin D itself. Cyclin D phosphorylation by GSK3β creates a phospho-dependent binding site for an E3 ubiquitin ligase complex, SCF (Alt et al. 2000; Barbash et al. 2009; Diehl et al. 1998). SCF family enzymes are multi-subunit ubiquitylases that use a variety of substrate-targeting subunits and mechanisms. Substrate-binding stimulates polyubiquitination which typically targets proteins to a degradative protein complex called the 26S proteasome. Since GSK3β is inhibited in G_1, cyclin D1 is not phosphorylated and therefore not targeted for ubiquitylation and destruction; hence cyclin D1 accumulates in G_1.

Many different cofactors associate with cyclin D/CDK4 complexes to regulate kinase activity either positively or negatively depending on context. These cofactors include the Cip/Kip family of proteins (p21, p27, and p57). Although Cip/Kip proteins are strictly inhibitors of cyclin E/CDK2 and mitotic CDKs, they can act as positive regulators of cyclin D/CDK4 (Bao et al. 2006; Cheng et al. 1999; LaBaer et al. 1997; Chen et al. 1995; Fotedar et al. 1996; Luo et al. 1995). Specifically, phosphorylated forms of these proteins play integral roles in cyclin D/CDK4 assembly, activation, and nuclear localization. The Cip/Kip family of proteins is required for cyclin D/CDK4 complex assembly, which is a precursor for activation. After the cyclin D-CDK4-Cip/Kip tertiary complex is formed, a substrate-blocking loop in CDK4 is exposed for an essential activating phosphorylation by the CDK7/cyclin H/MAT1 complex, known as CDK-activating kinase (CAK) (Schachter et al. 2013). This phosphorylation event physically relocates the T-loop from the active site of the CDK subunit, allowing substrates access to the active kinase site. Finally, cyclin D/CDK4 complexes do not have nuclear localization sequences themselves, whereas Cip/Kip proteins have bipartite nuclear localization sequences. Thus, the association between cyclin D/CDK four complexes and Cip/Kip proteins promotes the assembly and nuclear localization of the cyclin D/CDK4 complex, and therefore access to nuclear substrates, most notably RB.

16.3.2 E2F and Pocket Proteins in Early G_1

Perhaps the most consequential event leading to the transition from G_1 into S phase is passing through the restriction point, a cell-cycle checkpoint between early G_1 and S (Pardee 1974; Blagosklonny and Pardee 2002). Cells that have not reached the restriction point can either progress through the cell cycle or can transition into quiescence, G_0. Cells that have progressed past the restriction point are destined to initiate DNA replication and complete the cell cycle. Progression past the restriction point is generally correlated with the phosphorylation state of RB. RB phosphorylation is dependent on cyclin-dependent kinases, in particular cyclin D/CDK4 in early G_1 and cyclin E/CDK2 in late G_1. Prior to the restriction point, RB is not fully phosphorylated and binds activating E2F transcription factors (E2F1–3) to prevent activated E2F target gene expression. Once RB is fully phosphorylated, it releases E2F, allowing activated transcription of most of the genes that encode DNA replication proteins.

RB activity effectively represses the transcription of E2F target genes, particularly those necessary for DNA replication, during early G_1, and prevents cells from prematurely passing the restriction point and irreversibly committing to S phase entry. The mechanism of repression is primarily through local chromatin changes at E2F-dependent promoters. RB recruits HDAC1 or HDAC2, two histone deacetylases that repress gene transcription by maintaining a closed chromatin state at the promoters of E2F target genes (Brehm et al. 1998; Takaki et al. 2004). RB is also linked to the mSin·HDAC complex (containing RBAP46, RBAP48, SAP18, and SAP45) via the RBP1 and SAP30 linker proteins (Fig. 16.4) (Suryadinata et al. 2011). The ultimate effect of these interactions is not only the inability of E2F to stimulate transcription but also active gene repression.

RB is phosphorylated by the G_1 cyclin/CDK complexes, and these phosphorylation events induce structural changes in RB that, in turn, weaken (but do not fully disrupt) the interaction of RB with E2F. The human retinoblastoma protein has 16 potential CDK phosphorylation sites (S/T-P sites) (Fig. 16.2b). Twelve of the 16 RB phosphorylation sites are optimal CDK consensus sites (S/T-P-X-K/R motifs) and can be readily phosphorylated by cyclin/CDK complexes (Ubersax and Ferrell 2007). Each of the 16 phosphorylation sites is also detectably phosphorylated in vivo (Brown et al. 1999; Lents et al. 2006; Zarkowska and Mittnacht 1997; Lees et al. 1991; Harbour et al. 1999; Huttlin et al. 2010; Dephoure et al. 2008; Connell-Crowley et al. 1997). Using a combination of in vitro kinase assays and two-dimensional tryptic phosphopeptide maps, Zarkowska and Mittnacht found that RB is preferentially phosphorylated by different cyclin/CDK pairs (Zarkowska and Mittnacht 1997; Zarkowska et al. 1997). Cyclin D/CDK4 preferentially phosphorylates T5, S249, T252, T356, T373, S608, S788, S795, S807, S811, and T826. Cyclin E/CDK2 phosphorylates T5, T373, S612, S795, and T821. Finally, cyclin A/CDK2 best phosphorylates RB positions T5, S608, S612, S795, and T821.While some positions are only phosphorylated by a single cyclin/CDK pair (for instance, S811 by cyclin D/CDK4), many can be phosphorylated by two different pairs or even, like T5, all three pairs (Zarkowska and Mittnacht 1997).

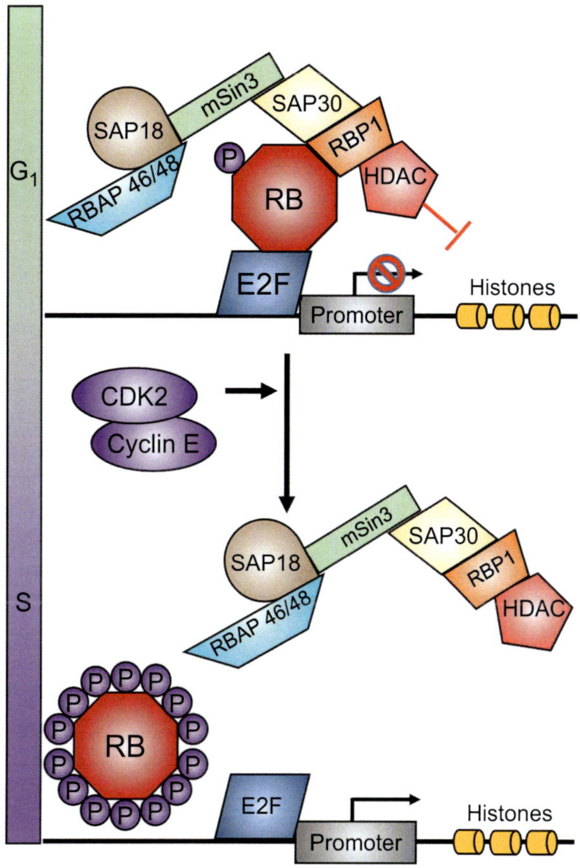

Fig. 16.4 Inactivation of E2F target genes by the RB-associated SAP30-mSin3-HDAC complex in G_1. Monophosphorylated RB binds activating E2F transcription factors at promoters of E2F target genes. The SAP30-mSin3-HDAC complex binds to RB, and the associated HDACs deacetylate the E2F target gene, repressing transcription. Hyperphosphorylated RB releases E2F and dissociates from the SAP30-mSin3-HDAC complex, relieving the transcriptional repression of E2F target genes

These cyclin/CDK RB phosphorylation site preferences, combined with the temporal regulation of cyclin abundance (Fig. 16.1), suggest a temporal order for when individual sites are likely phosphorylated in vivo: In early G_1, RB is phosphorylated by cyclin D/CDK4 at eight sites (S249, T252, T356, S608, S788, S807, S811, and S826) (Zarkowska and Mittnacht 1997). After these initial eight sites are phosphorylated, a next round of phosphorylation occurs via either cyclin D/CDK4 complexes or cyclin E/CDK2 complexes at threonine 373. Finally, cyclin E/CDK2 phosphorylates serine 612 (Zarkowska and Mittnacht 1997). However, to date there has been no biochemical proof that these phosphorylations occur in the temporal order suggested, nor are there biochemical data precisely defining what functionally constitutes hypophosphorylated vs. hyperphosphorylated RB (Narasimha et al. 2014). Using two-dimensional isoelectric focusing (2D IEF), Narasimha et al. showed that during early G_1, RB is monophosphorylated by cyclin D/CDK4 at any one of 14 different sites, but there was no evidence for individual RB molecules with more than one phosphate until very late in G_1. (The T5 and S567 sites were not analyzed in that study.) Later in G_1, upon upregulation of cyclin E expression and activation

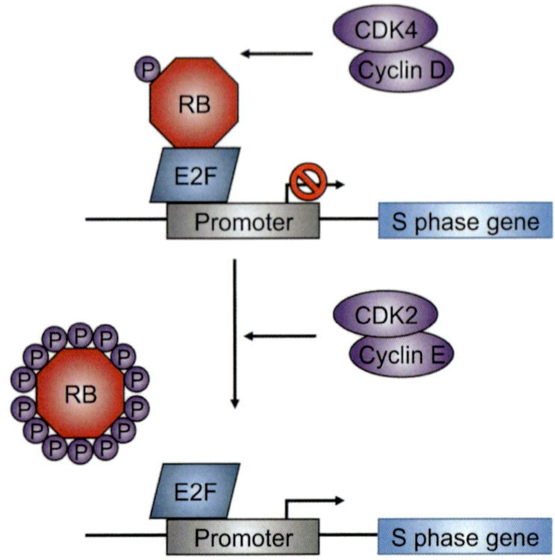

Fig. 16.5 Inactivation of E2F by monophosphorylation of RB. In early G_1, E2F is bound and inactivated by RB, which is monophosphorylated by cyclin D/CDK4. Later in G_1, cyclin E/CDK2 becomes active and hyperphosphorylates RB, releasing E2F which then activates transcription of target genes

of cyclin E/CDK2, the vast majority of RB molecules were hyperphosphorylated on at least 14 sites and no longer bound to E2F (Fig. 16.5). Reducing RB hyperphosphorylation with different concentrations of the CDK2 inhibitor roscovitine, a compound that has little effect on cyclin D/CDK4, suggested that cyclin E/CDK2 is the principle processive RB kinase that rapidly converts monophosphorylated RB molecules to hyperphosphorylated RB molecules in a single binding step.

While phosphorylation of either S788 or S795 somewhat reduces RB affinity for E2F, simultaneous phosphorylation of both residues is additive. Characterizing the effects of multiple phosphorylation events on RB, Burke et al. showed that phosphorylation of S788/S795 destabilizes the interaction of RB and the E2F1-DP1 complex as well as inducing a conformational shift in RB, allowing a disordered linker, RB amino acids 787–816, to bind to the pocket domain. This phosphorylated linker region competes with E2F to bind the pocket domain of RB (Burke et al. 2014). These phosphorylation events weaken the overall interactions between RB and the E2F-DP1 complex as cells progress to late G_1, thus stimulating the production of gene products necessary for DNA replication.

Why is RB phosphorylation so complex (Fig. 16.2b)? For example, monophosphorylation at four sites (S230, S249, T356, or S612) may in fact increase the affinity of RB for E2F1, while phosphorylation at three other sites (T373, S608, or S795) decrease RB-E2F1 affinity (Burke et al. 2010, 2012, 2014; Narasimha et al. 2014). Currently, the biological significance of the various individual phosphorylation sites on RB interactions affinity is unclear. These different phosphorylation sites may modulate binding affinities for the different members of the activating E2F transcription factor family. Another possibility is that the differences in RB affinity for E2F1 allow for activation of some subset of E2F1 target genes, but not other subsets. Furthermore, CDK-dependent phosphorylation affects not only RB-E2F binding but also induces conformational changes in RB that disrupt interactions with histone deacetylases and associated proteins (Takaki et al. 2004). The abundance of phospho-sites on RB may also facilitate integration of different signaling pathways and prevent aberrant E2F release and activation.

During early G_1 the other two pocket proteins, p107 and p130, are complexed with E2F4 at promoters contributing to gene repression, though they are less abundant than RB. In situations where RB has been lost, such as in an *Rb* null mouse, transcriptional repression and G_1 restraint is partially covered by the action of p130 and p107 (Cobrinik et al. 1996). Because RB phosphorylation and the subsequent E2F release stimulate progression from G_1 into S phase, this process is susceptible to oncogenic perturbation. Interestingly, there are only a few cancer types, such as retinoblastoma, osteosarcoma, and small cell lung cancer, which are associated with inactivating *RB* mutations, possibly reflecting the partial redundancy with p107 and p130 (Wikenheiser-Brokamp 2006; Kaye and Harbour 2004). In contrast, many tumors have mutations in upstream regulators of pocket proteins that render them constitutively inactive in cell-cycle control (Wikenheiser-Brokamp 2006; Paternot et al. 2010).

16.3.3 APC^{CDH1}-Mediated Protein Degradation in Early G_1

In addition to the complex network of transcriptional control described above, post-transcriptional regulation ensures that early G_1 is free from both mitotic and S phase activities, to provide essential protected time to properly prepare for S phase. Prior to the cascade of RB phosphorylation events, both during cell-cycle reentry from G_0 and in early G_1 after mitosis, somatic cells experience a period of low cyclin-dependent kinase activity and high phosphatase activity (Fig. 16.1). This period removes mitosis-associated phosphorylations and introduces a cell-cycle pause to be receptive to inputs from the environment, such as growth factors. Activities that ensure this low CDK period, which ultimately determines the length of G_1, are the APC/C and protein phosphatases.

The anaphase-promoting complex/cyclosome, APC/C, is a multisubunit E3 ubiquitin ligase named for its essential role in mitotic progression, but APC/C also has critical functions throughout G_1. The substrate-targeting subunit of APC/C varies such that substrates in anaphase are recognized by the CDC20-targeting subunit, whereas substrates in G_1 are recognized by the CDH1 subunit (Manchado et al. 2010). As cells exit mitosis, APC/C^{CDC20} targets the mitotic cyclins for degradation via the 26S proteasome, reducing the cellular kinase activity and allowing the activation of mitotic exit phosphatases, such as PP1 or PP2A (Wu et al. 2009; van Leuken et al. 2008). The reduced mitotic kinase activity promotes the assembly of APC/C^{CDH1}. Like APC/C^{CDC20}, the newly formed APC/C^{CDH1} targets mitotic cyclins and kinases for degradation. APC/C^{CDH1} also targets CDC20 for destruction, allowing for the complete switch from APC/C^{CDC20} to APC/C^{CDH1}. Once cells have progressed through telophase and have completed the division into two daughter cells, APC/C^{CDH1} continues to target mitotic proteins for destruction throughout G_1. One of the mitotic cyclins is cyclin A, which is active from mid-S phase until early mitosis (Fig. 16.1), and since cyclin A/CDK2 can trigger DNA synthesis, its removal before and during G_1 helps prevent premature DNA replication (Coverley et al. 2002; Erlandsson et al. 2000).

Alongside the elimination of cyclin/CDK activity, phosphatases reverse mitotic (or G_0) phosphorylations. During mitosis, protein phosphatase 1 (PP1) is inactivated by CDK-mediated phosphorylation at threonine 320. In anaphase, PP1 removes the inhibitory phosphorylation on threonine 320 by auto-dephosphorylation as CDK activity drops. PP1 also inactivates its own inhibitor, inhibitor-1, a protein uniquely expressed in G_2/M (Wu et al. 2009). Activated PP1 then removes the phosphates from hyperphosphorylated RB until it is in the hypophosphorylated state which reestablishes RB-mediated repression. This dephosphorylation activity lasts from anaphase to mid-G_1 when the kinase activity toward substrates outpaces the phosphatase. Another phosphatase, the PP2A holoenzyme, also dephosphorylates the pocket proteins. The PP2A holoenzyme can also dephosphorylate the three pocket proteins after a variety of signals that, if ignored, can result in genome instability. These signals include oxidative stress, UV radiation, and DNA damage (Avni et al. 2003; Cicchillitti et al. 2003; Magenta et al. 2008; Voorhoeve et al. 1999). The dephosphorylation of RB delays activating E2F activity, allowing time to recover from the genomic insult.

In early G_1, APC/C^{CDH1} also targets the SKP2 substrate receptor of the SCF (SKP1-Cullin-F-box) E3 ubiquitin ligase (Bassermann et al. 2014; Bashir et al. 2004; Wei et al. 2004). The resulting loss of SCF activity leads to accumulation of SCF substrates that include the cyclin-dependent kinase inhibitors p21^{Cip1} and p27^{Kip1}. High levels of p21 and p27 help to maintain low CDK activity in early G_1. In human cells and mouse models lacking the CDH1 substrate-targeting subunit for APC/C, the exit from mitosis occurs normally, but cells begin G_1 with aberrantly high CDK activity. High CDK activity shortens G_1, i.e., causes premature entry into S phase. Early entry into S phase, in turn, can result in increased endogenous DNA damage, presumably from inadequate G_1 preparation or an uncoordinated G_1/S transition.

APC/C^{CDH1} substrates include other proteins involved in controlling the length of G_1. ETS2, one such target, induces cyclin D expression. APC/C^{CDH1} directed destruction of ETS2 delays *cyclin D* gene expression and prolongs G_1. Two of the repressor E2F transcription factors, E2F7 and E2F8, are also targets of APC/C^{CDH1} and repress transcription of the CDH1 inhibitor, the EMI1 pseudosubstrate. The APC/C^{CDH1} directed destruction of these two E2F repressors and subsequent de-repression of EMI1 contribute to CDH1 inactivation in late G_1. Throughout G_1 APC/C^{CDH1} targets cyclin F, a substrate receptor of SCF; the targeting of cyclin F prevents the SCF$^{cyclin\ F}$ from reciprocally degrading its target,CDH1, and shortening G1 phase (Choudhury et al. 2016). APC/C^{CDH1} also targets CDC6 and ORC1, two proteins involved in origin licensing for DNA replication which itself contributes to CDK2 activation in late G_1 (described below). Altogether, APC/C targets a cohort of proteins in early G_1 including both cyclins and CDK regulators, to ultimately set the timing of the onset of DNA replication.

16.3.4 Nuclear and Chromatin Architecture Changes in Early G_1

Transcriptional and post-transcriptional control are not the sole determinants of DNA replication parameters that are established during G_1. One special aspect of DNA replication in eukaryotes is the phenomenon of replication timing. Genomic DNA replication does not initiate simultaneously at all sites at the beginning of S phase, but rather some loci are replicated early in S phase and others much later in S phase. Much of this replication timing pattern is, in fact, established during early G_1. Local chromatin structure is an important contributor to replication timing, and progress toward understanding this phenomenon is described in chapters by Zhao, Rivera-Mulia and Gilbert and by Alabert, Jasencakova, and Groth in this book.

As a test of when replication timing is determined, Lu et al. (2010) introduced nuclei isolated from cells in specific cell-cycle phases (e.g., early or late G_1) into a cell-free DNA replication initiation system that triggers initiation immediately. They found that nuclei from cells in mid-to-late G_1 displayed a pattern of temporal regulation of DNA in replication domains similar to their pattern in intact cells, but nuclei from cells in early G_1 or G_2 lacked the temporal replication (Lu et al. 2010). They proposed that DNA replication timing is determined in a ~1 h window during early G_1, a model strengthened by more recent live cell imaging analysis by Wilson et al. (2016).

Wilson et al. (2016) used live cell imaging to show that as cells differentiate, the replication domains coalesce into larger regions of dense chromatin that are replicated in a coordinated fashion (Wilson et al. 2016; Dixon et al. 2015). It is well known that the nucleus has a three-dimensional architecture and that this architecture is modified as animal cells progress through development (Hiratani et al. 2010). The interior of the nucleus generally contains highly transcribed genes that are replicated early in S phase, while portions of chromosomes that are closer to the nuclear periphery are likely to be transcriptionally silent and replicated later in S phase. For example, in mammalian females, the inactivated X chromosome is compacted and localized to the nuclear periphery and is replicated much later in S phase than the active X chromosome (Lyon 1961; Morishima et al. 1962; Barr and Bertram 1949). This architectural compaction, DNA replication delay, and localization to the periphery also apply to pluripotency genes when they are silenced during differentiation (Meshorer et al. 2006; Hiratani et al. 2010). Thus, both chromatin and nuclear architecture that are established in early G_1 influence subsequent S phase events.

16.4 Late G_1

The transition from "early G_1" to "late G_1" is not readily defined by any discrete molecular markers; nonetheless, a collection of molecular events are typically confined to a period of G_1 just before the onset of S phase. Many of these late G_1 events depend on successful completion of early G_1 steps, such as initial RB phosphorylation to begin the accumulation of essential DNA replication proteins. In particular, a burst of cyclin E/CDK2 activity coincides with both induced E2F-dependent gene expression and with the completion of DNA replication origin licensing in late G_1. A sequence of events in late G_1 reflects passage through the restriction point and leads inexorably to the initiation of DNA synthesis in S phase. Disruptions to this sequence, particularly after cells are already committed to S phase initiation, increase the likelihood of replication failure or genome instability. Recent advances in live cell imaging have begun to elucidate stereotypical orders of molecular events in a window of 1–2 h (in somatic human cells) at the G_1/S transition; these tools will ultimately support investigations of how that order promotes normal S phase completion (Cappell et al. 2016; Coleman et al. 2015).

16.4.1 Transcriptional Activation via E2F

As cells pass the restriction point, commit to S phase, and therefore commit to progression through the cell cycle, they must accumulate the proteins required for DNA replication. Full release of E2F does not occur until late G_1 upon full RB hyperphosphorylation and is one indicator that the requirements of the restriction point have been fulfilled. Once fully released by RB hyperphosphorylation, activating E2F family members stimulate transcription of key genes involved in DNA replication and S phase onset (Grant et al. 2013; Muller et al. 2016; Helin 1998; Johnson et al. 1994; Slansky and Farnham 1996). Additionally, the dissociation of RB from the activating family of E2F transcription factors allows for the association of other transcriptional coactivators with E2F, such as p300 or CBP (Ait-Si-Ali et al. 2000; Ferreira et al. 1998).

E2F1 specifically is part of two positive feedforward loops that help to drive the cell into S phase with little possibility of turning back to a G_1-like state. The first of these loops involves *cyclin E* and was described in the previous section: cyclin E/CDK2-mediated RB hyperphosphorylation stimulates E2F1-mediated transcription of the *cyclin E* gene to further increase cyclin E/CDK2. The second feedforward loop is autoregulatory: E2F1 binds to and activates its own promoter, driving expression of more E2F1. Additionally, E2F1 is part of a negative feedback loop by activating transcription of the gene encoding the E3 ubiquitin ligase substrate receptor, SKP2. SKP2 is active from late G_1 through G_2 (Marti et al. 1999) and is responsible not only for the eventual ubiquitin-mediated degradation of E2F1 in S phase but also for ubiquitylating the Cip/Kip family of CDK inhibitors. p27 destruction leads to higher cyclin E/CDK2 activity and yet more robust RB hyperphosphorylation creating another feedforward loop (Sheaff et al. 1997; Yung et al. 2007). The result of these reinforcing relationships is maximal cyclin E/CDK2 activity and E2F protein levels at the G_1-to-S phase transition.

The E2F-mediated induction of genes involved in the positive feedback loops occurs before the transcriptional activation of genes involved in the negative feedback loops (Grant et al. 2013; Whitfield et al. 2002; Eser et al. 2011). This temporal arrangement provides time for cells to accumulate sufficient DNA replication proteins while still limiting that time, so replication protein levels do not increase to unregulatable levels. The offset in expression timing also prevents G_1-S phase genes from being expressed in late S phase or during G_2 when they are not needed or could be disruptive.

Nearly every protein required for DNA replication preparation in G_1 or DNA synthesis in S phase is the product of an E2F-regulated gene. Using a combination of microarray expression analysis of cell cycle-regulated transcription in synchronous cells and genome-wide chromatin precipitation followed by high-throughput sequencing (ChIP-seq), E2F1 has been detected at a majority of genes involved in DNA replication in cancer cells and at the promoters of over 2,500 genes across the genome (Grant et al. 2013). Since E2F transcription factors regulate the expression of many genes and are active during late G_1 and S phase, it is not surprising that they regulate a large number of genes involved in all stages of DNA replication. E2F-dependent genes encoding proteins critical for DNA replication include DNA polymerase delta 3 (*POLD3*), replication factor C subunit 4 (*RFC4*), proliferating cell nuclear antigen (*PCNA*), and histone mRNA stem-loop binding protein (*SLBP*). RFC4 and PCNA are accessory proteins for leading strand DNA replication by polymerase delta. SLBP binds to stem loops at the 3′ end of histone RNAs; this binding stabilizes the RNAs and promotes histone protein production (Townley-Tilson et al. 2006; Wang et al. 1996; Whitfield et al. 2000). Many genes whose products are involved in nucleotide metabolism are E2F targets including thymidine kinase 1 (*TK1*), thymidylate synthetase (*TYMS*), dihydrofolate reductase (*DHFR*), and ribonucleotide reductase subunit M2 (*RRM2*) (DeGregori et al. 1995, 1997). Genes whose products are essential for origin licensing that are regulated by E2F transcription factors include *CDC6*, *CDT1*, *ORC1*, and *MCM2–7* (Yoshida and Inoue 2004b; Ohtani et al. 1996, 1998, 1999; Tsuruga et al. 1997; Leone et al. 1998; Yan et al. 1998b).

16.4.2 Origin Licensing

Creating a single exact copy of the genome is, of course, an S phase event and is a critical process for somatic cell viability and genome stability. To replicate the vast quantity of DNA in each eukaryotic cell in a timely fashion, cells initiate DNA replication at thousands of DNA replication origins located throughout the genome. Eukaryotic origins are not strictly defined by specific DNA sequences and are strongly influenced by local chromatin structure. For example, histone H4 lysine 20 monomethylation shows a strong positive correlation with origin activity (Tardat et al. 2010; Jorgensen et al. 2013). Discussions of the molecular features associated with eukaryotic origins, and ongoing efforts to identify molecular determinants of mammalian origins, can be found in other chapters in this book, and therefore are not addressed here.

Prior to replication initiation DNA replication origins must have been pre-loaded in G1 phase with mini-chromosome maintenance (MCM) complexes. MCM complexes are the core component of the replicative DNA helicase (Bell and Kaguni 2013; Remus and Diffley 2009), and the process of loading MCM complexes onto DNA is known as "origin licensing." MCM complexes are stable heterohexameric ring-shaped complexes of subunits MCM2-MCM7, and their sequences are conserved not only throughout eukarya but also in archaeal species (Bell 2012). The majority of knowledge to date about the mechanism of MCM loading comes from pioneering studies with purified *Saccharomyces cerevisiae* or *Xenopus laevis* proteins to reconstitute the MCM loading reaction (Remus and Diffley 2009; Gillespie et al. 2001), but the strong evolutionary conservation among licensing proteins gives confidence that insights into the functions of licensing components are readily extrapolatable to mammalian licensing systems. Despite this strong functional conservation, the regulation of origin-licensing proteins themselves varies among different eukaryotic species. In addition, some mammalian licensing proteins have non-replication functions including roles in transcriptional control, centrosome duplication, CDK regulation, chromosome segregation, and cell division (Kawasaki et al. 2006; Varma et al. 2012; Prasanth et al. 2002, 2004, 2010; Hemerly et al. 2009; Tachibana and Nigg 2006; Hossain and Stillman 2012).

During G_1, MCM complexes are loaded such that double-stranded DNA passes through their central channels (Fig. 16.6) (Bell and Botchan 2013; Gambus et al. 2011; Remus and Diffley 2009; Evrin et al. 2009); a detailed discussion of MCM structure can be found in the chapter by Zhai and Tye (Chapter 9). Functionally licensed origins have at least two MCM complexes loaded in anticipation of their lead roles at bidirectional replication forks in S phase (Sun et al. 2014; Li et al. 2015). MCM complexes are loaded to create these double hexamers by the concerted action of three essential loading factors: the heterohexameric origin recognition complex (ORC), the CDC6 protein, and the CDT1 protein (Siddiqui et al. 2013; Nishitani and Lygerou 2004). ORC and CDC6 are each members of the AAA+ family of ATPases (Duncker et al. 2009; Lee and Bell 2000; Speck et al. 2005; Clarey et al. 2006) and bear some sequence similarity to replication factor C (RFC) which is responsible for the DNA loading of another ring-shaped complex, the PCNA sliding clamp (Schepers and Diffley 2001). By analogy to RFC function, ORC and CDC6 are thought to bind and open MCM rings to allow double-stranded DNA to pass into their central channels (Samel et al. 2014; Bochman and Schwacha 2008). Non-productive MCM hexamer loading attempts are removed through the ATPase activity of CDC6 (Cocker et al. 1996; Frigola et al. 2013). The CDT1 protein binds MCM and is required for MCM loading, but CDT1 has no known enzymatic activity (Xouri et al. 2007; Zhang et al. 2010; Jee et al. 2010; Khayrutdinov et al. 2009). At least one role for CDT1 in origin licensing is to recruit MCM complexes to ORC and CDC6, which are resident at origins in G_1 (Xouri et al. 2007).

Based on the reconstituted yeast MCM loading reaction, once a first MCM hexamer is properly loaded onto DNA, CDT1 and CDC6 are released and can presumably recruit additional MCM complexes (Fig. 16.6) (Randell et al. 2006; Chen et al. 2007; Duzdevich et al. 2015; Ticau et al. 2015). This dynamic behavior of CDT1

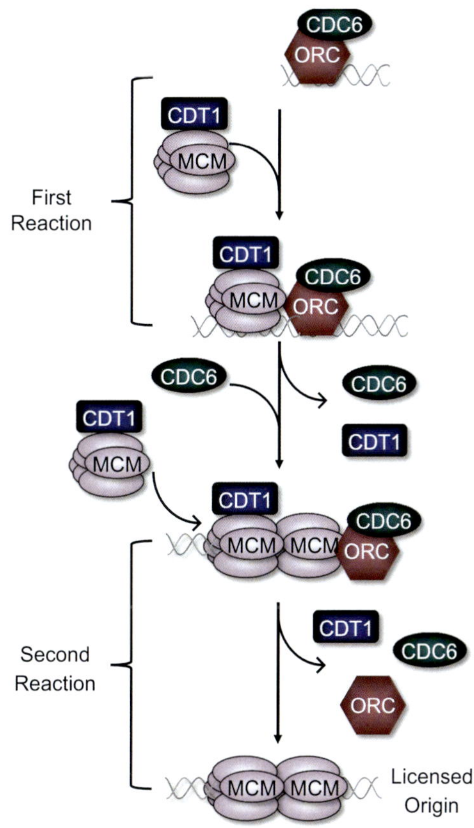

Fig. 16.6 Origin licensing. DNA replication origins are licensed by loading MCM complexes. The first CDT1-associated MCM complex is loaded onto DNA through the action of the DNA bound origin recognition complex (ORC) and CDC6. Once the first MCM complex is successfully loaded by ORC, CDC6, and CDT1, CDC6 and CDT1 dissociate from the DNA bound MCM and ORC complexes. A second CDC6 molecule then associates with the DNA-bound complexes. Finally, a second CDT1 associated MCM complex is loaded onto the DNA in a head to head fashion, and CDC6 and CDT1 are released

relative to ORC implies that CDT1 and CDC6 can readily participate in MCM loading at many origins over the course of G_1 (Xouri et al. 2007). Perhaps for this reason, human CDT1 in particular is the most highly regulated of the origin-licensing proteins, and this regulation is critical to restrict licensing to only G_1. CDT1 is degraded during S phase, bound by the geminin inhibitor protein during S phase and G_2, and inhibited by phosphorylation during G_2 and M phase (Wohlschlegel et al. 2000; Xouri et al. 2007; Chandrasekaran et al. 2011; Coulombe et al. 2013; Nishitani et al. 2006). A second molecule of CDC6 assists with loading a second MCM hexamer. During this process, the second molecules of CDC6 and CDT1 are removed from the complex, completing origin licensing (Ticau et al. 2015; Duzdevich et al. 2015).

An important feature of origin licensing is that once MCM complexes are loaded, none of the loading factors are required for MCM to maintain stable DNA associations or replication competence (Rowles et al. 1999; Yeeles et al. 2015; Tsakraklides and Bell 2010; Ticau et al. 2015; Bowers et al. 2004). Each replication origin that does fire should fire at most once per S phase; only a subset of licensed origins fire in a typical S phase (Woodward et al. 2006; Ge et al. 2007). The complex mechanisms and regulations for converting a licensed origin to an active replication fork are described in several chapters of this book, including Parts IV, V, VI, and XVIII. If an origin fires twice during a single-cell cycle, it results in DNA re-replication, a form of endogenous DNA damage that increases genome instability (Green and Li 2005; Vaziri et al. 2003; Davidson et al. 2006; Arentson et al. 2002; Liontos et al. 2007). Origin licensing can only occur during a period beginning in late mitosis (telophase) throughout all of G_1 until the onset of S phase; telophase and G_1 are the only times when all licensing proteins are abundant and active. At all other times from the beginning of S phase until the end of mitosis, origin licensing is prevented by an extensive series of overlapping molecular mechanisms that inactivate or destroy MCM loading proteins. More extensive descriptions of the mechanisms preventing re-replication after G_1 may be found in several other chapters, such as those contributed Li, by Abbas and Dutta, and by Teixeira and Reed.

16.4.3 Origin-Licensing Regulation During G_1

Origin licensing is blocked during G_2 and mitosis by several mechanisms; one of these involves tight binding of the licensing inhibitor, geminin, to CDT1 (Wohlschlegel et al. 2000). Geminin accumulates throughout S phase and reaches peak levels in late G_2. Geminin binding to CDT1 interferes with the CDT1-MCM interaction, and since that interaction is essential for origin licensing, high levels of geminin block origin licensing during G_2 and mitosis (Cook et al. 2004; Yanagi et al. 2002). Geminin is subject to ubiquitin-mediated proteolysis during anaphase, so its reduction by the beginning of G_1 releases CDT1 to once again bind MCM (McGarry and Kirschner 1998). CDT1 is also hyperphosphorylated during G_2 and M phase, and at least some of these phosphorylations interfere with CDT1 function, though the mechanism of that interference is not yet known. CDT1 is dephosphorylated in early G_1, which presumably increases CDT1 licensing activity (Chandrasekaran et al. 2011; Coulombe et al. 2013).

Each protein in the licensing system is the product of an E2F-regulated gene. As a consequence, each of the genes for licensing proteins is induced during cell-cycle reentry from quiescence, and at least some of them are also subject to cell cycle-dependent fluctuations during active proliferation (Ohtani et al. 1996; DeGregori et al. 1995; Yoshida and Inoue 2004a; Yan et al. 1998a). For example, five of the six subunits of the origin recognition complex (*ORC2–6*) are constitutively expressed throughout the cell cycle, while *ORC1* is cell cycle regulated with peak expression at G_1/S (Whitfield et al. 2002; Grant et al. 2013).

In addition to the genes encoding proteins that act at origins, E2F-dependent *cyclin E* expression impacts origin licensing during late G_1. Like geminin, the CDC6 protein is targeted for degradation during mitosis. Both geminin and CDC6 are substrates of the APC/C^{Cdh1} E3 ubiquitin ligase, which is activated in anaphase and remains active until the onset of S phase (McGarry and Kirschner 1998; Petersen et al. 2000). CDC6 escapes APC/C ^{Cdh1}-mediated degradation in late G_1 because cyclin E/CDK2 phosphorylates CDC6 near the APC/C^{Cdh1}-binding site, and this phosphorylation blocks CDC6 ubiquitylation (Mailand and Diffley 2005). (CDK-mediated CDC6 phosphorylation at a second site also stimulates cytoplasmic translocation later in S phase (Yim et al. 2013).) As a result, cyclin E accumulation and cyclin E/CDK2 activation in late G_1 stabilize CDC6; the subsequent increase in CDC6 stimulates MCM loading. Cyclin E has also been ascribed non-kinase roles in origin licensing, though the mechanism of these roles remain to be determined (Geng et al. 2007). The relationship between CDK activity and licensing is complex, however, because very high levels of cyclin E/CDK2 activity can block origin licensing through multiple routes that are independent of CDC6 stabilization. These routes include premature S phase onset, phosphorylation-stimulated CDT1 and ORC1 degradation, and phosphorylation-mediated inhibition of interactions among licensing proteins or between licensing proteins and chromatin (Wheeler et al. 2008; Ekholm-Reed et al. 2004; Mendez and Stillman 2000; Takeda et al. 2005). Thus, the licensing inhibitory CDK function helps block origin relicensing and the consequent genome instability.

16.4.4 The Origin-Licensing Checkpoint

Due to both the enormity of the task and the potential dire consequences of even small errors, there are checks and balances built into replication control, particularly at the transition from G_1 to S phase. Mechanisms that prevent origin relicensing during S phase set up a separate challenge when considering the need for complete genome duplication. Entering S phase with too few licensed origins to fully duplicate each chromosome results in under-replication which, like re-replication, is also a form of endogenous DNA damage. Reducing the number of licensed origins in G_1 can lead to sections of unreplicated DNA segregating during mitosis which then require DNA repair in the subsequent G_1 (Moreno et al. 2016). In addition, a destabilizing mutation in an MCM subunit causes reduced origin licensing, chromosomal instability, and development of cancer in mice (Shima et al. 2007; Pruitt et al. 2007). Moreover, replication forks can stall if they encounter bulky lesions or interstrand crosslinks, but as long as a fork converges from the other side of the lesion, repair and replication can be completed (Raschle et al. 2008; Moreno et al. 2016). In response to fork stalling, nearby licensed origins are activated to generate such converging forks, but those origins must have been licensed in the previous G_1 (Woodward et al. 2006; Ge et al. 2007; Ge and Blow 2010; Ibarra et al. 2008).

The tight regulation of licensing means that there is no opportunity to license additional origins after G_1, but cells typically license many more origins than are strictly required to accommodate replication fork stalling and to ensure complete replication. Moreover, individual origins can be loaded with more than one MCM double hexamer. At least in vitro, loaded MCM complexes can slide away from their initial loading site, freeing origins to receive additional MCM complexes (Remus and Diffley 2009; Gros et al. 2015; Evrin et al. 2009). As long as the MCM complexes remain DNA-loaded as double hexamers, they can be activated in S phase with no need for ORC, CDC6, or CDT1. The additional licensed origins may never be utilized during S phase, but these dormant origins can be activated if needed.

Though it is imperative that enough origins are licensed to support complete replication before S phase begins, it is not clear how cells couple the completion of origin licensing to the timing of S phase onset. In the early years of origin-licensing investigations, the consequences of licensing failure were evident, but the existence of an origin-licensing checkpoint that can prevent S phase onset before some critical threshold of licensing was reached and sensed was questioned. Budding and fission yeasts with null alleles in essential licensing components do not execute a cell-cycle arrest in G_1 but rather proceed through cell division without replication (Piatti et al. 1995; Kelly et al. 1993; Hofmann and Beach 1994). Depleting origin-licensing proteins by RNAi in the most commonly used cancer-derived human cell lines results in the predicted DNA synthesis defects, but no G_1 arrest (Nevis et al. 2009; Shreeram et al. 2002). These early results argued against any origin-licensing sensing mechanism.

Subsequently, however, several groups found that depleting origin-licensing proteins in non-transformed human cells did indeed cause delayed S phase onset. The delay could only be detected in untransformed cell lines, suggesting that growth control pathways disrupted in cancer-derived cell lines are normally required to link origin-licensing status in G_1 to S phase initiation (Nevis et al. 2009; Shreeram et al. 2002; Teer et al. 2006). In each of these studies, reducing origin licensing caused defects in the activation of cyclin-dependent kinases, especially cyclin E/CDK2.

Cyclin E/CDK2 is a major driver in the transition from G_1 into S phase; constitutive or ectopic expression of cyclin E leads to a shortened G_1 (Resnitzky and Reed 1995) (reviewed in (Clurman et al. 1996). CDK activity is required for origin firing, and early S phase origin firing is driven by CDK2 complexes with cyclin E or cyclin A (Fig. 16.1); chapters on replication initiation describe the molecular mechanisms of replication initiation. Delayed CDK2 activation from impaired origin-licensing consequently delays origin firing and, by definition, S phase entry. No single molecular mechanism to link origin licensing to CDK activity has yet emerged. In different studies, elevated CDK inhibitor proteins, CDK localization or phosphorylation, or expression of G_1 cyclins were each implicated in the S phase delay caused by insufficient origin licensing (Nevis et al. 2009; Shreeram et al. 2002; Liu et al. 2009; Teer et al. 2006; Lunn et al. 2010). It may be that each of these mechanisms can operate in all non-transformed cells, but different cell types are more or less dependent on specific ones.

16.4.5 APC/C^{CDH1} Inactivation and S Phase Entry

The CDH1 substrate-targeting subunit of APC/C must be degraded for proper entry into S phase to allow full CDK activation and cyclin A accumulation. There are three main mechanisms for inactivating APC/C^{CDH1}. The first mechanism is the E2F-dependent accumulation of the APC/C^{CDH1} pseudosubstrate early mitotic inhibitor-1 (EMI1). EMI1 binds to CDH1 complexed with APC/C but is not degraded, thus acting as a competitive inhibitor. Once EMI1 binds APC/C^{CDH1}, an irreversible cascade toward APC/C inactivation is triggered (Cappell et al. 2016). The EMI1-driven partial reduction in APC/C^{CDH1} activity stabilizes cyclin A. Cyclin A/CDK2 phosphorylates CDH1, preventing it from binding APC/C. Cyclin A/CDK2 phosphorylates CDH1 at four different amino acids (S40, S151, S163, and T121) (Lukas et al. 1999). The phosphorylation of serine 40 and threonine 121 create two binding domains for the protein kinase, PLK1. PLK1 can then bind and phosphorylate CDH1 at serines 138 and 146 (Fukushima et al. 2013). Phosphorylated CDH1 molecules are then targeted for degradation by the $SCF^{cyclin\ F}$ complex (Fig. 16.7) (Fukushima et al. 2013; Choudhury et al. 2016).

Cells that lack CDH1 enter S phase early but also experience increased rates of DNA damage. This increased damage could be from insufficient origin licensing if origin firing begins before chromosomes are fully licensed (Ayuda-Duran et al. 2014). In addition CDH1-deficient cells show enhanced mutagenesis. This increased mutation rate could be due to the accumulation of a pair of proteins involved in dTTP formation: thymidine kinase 1 (TK1) and thymidylate kinase (TMPK) (Ke et al. 2005, 2007). The over-accumulation of these two proteins leads to an imbalance in the dNTP pool. This imbalance then leads to an increased rate of dNTP misincorporation, reducing the fidelity of DNA replication and increasing the mutation rate (Ke et al. 2005). Conversely, human osteosarcoma cells that overexpress CDH1 using inducible expression have a delayed S phase entry as well as a slower rate of DNA replication (Sorensen et al. 2000). Thus, the proper regulation of CDH1 is important for proper cell-cycle progression and S phase entry.

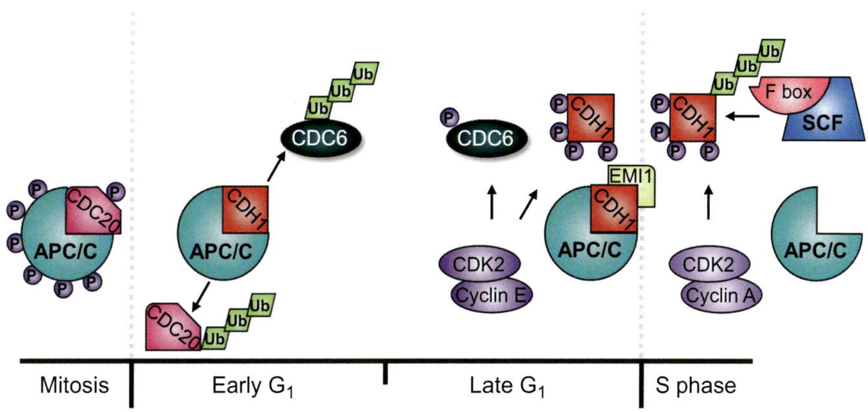

Fig. 16.7 Modifications and activation of APC/C^{CDH1} from mitosis to the onset of S phase. In late mitosis, CDH1 replaces CDC20 as the substrate receptor of the APC/C E3 ligase and is active during early G_1. Two APC/C^{CDH1} substrates in early G1 are CDC20 and the origin-licensing protein, CDC6. In late G_1 cyclin E/CDK2 phosphorylates and protects CDC6. Also in late G1, the pseudosubstrate, EMI1, binds to CDH1, inactivating the complex. In late G_1 and early S phase, CDH1 is phosphorylated by cyclin E/CDK2 and cyclin A/CDK2, respectively; this phosphorylation dissociates CDH1 from APC/C, thus inactivating the complex (From S phase through mitosis, the E3 ligase CDH1 is polyubiquitylated by the $SCF^{cyclin\ F}$ complex and degraded by the 26S proteasome)

16.5 Conclusion

Successful completion of S phase relies on proper progression through G_1. In this chapter we have discussed some of the many different G_1 events and checkpoints that contribute to the proper initiation and completion of DNA replication including quiescence, the restriction point, origin licensing, and cyclin-dependent kinase regulation. Many of the topics covered in this chapter are regulated to occur normally in a stereotypical order and were initially studied using assays of bulk cell populations such as immunoblots, microarray analysis, or ChIP-seq. As technology progresses, researchers will be able to analyze cell-cycle kinetics of individual cells using single-cell assays such as single-cell RNA-seq or high-content live cell imaging with cell-cycle biosensors (Kaida et al. 2011; Pauklin and Vallier 2013; Sugiyama et al. 2009; Wilson et al. 2016; Cappell et al. 2016; Spencer et al. 2013; Coleman et al. 2015; Purvis et al. 2012). These next-generation technologies to study cell-cycle progression with timescales of mere minutes have begun to reveal new timelines of G_1 progression (Cappell et al. 2016; Spencer et al. 2013; Coleman et al. 2015; Matson and Cook 2017).

Acknowledgements We thank members of the Cook lab for feedback and discussion about the manuscript and DBG for critical reading of the manuscript. This work was supported by funding to G.D.G. from the ITCMS training grant (T32CA009156) and to J.G.C. from the National Institutes of Health NIGMS (R01GM102413) and the W.M. Keck Foundation.

Conflicts of Interest The authors have no conflicts of interest.

References

Ait-Si-Ali S, Polesskaya A, Filleur S, Ferreira R, Duquet A, Robin P, Vervish A, Trouche D, Cabon F, Harel-Bellan A (2000) CBP/p300 histone acetyl-transferase activity is important for the G1/S transition. Oncogene 19(20):2430–2437. https://doi.org/10.1038/sj.onc.1203562

Alt JR, Cleveland JL, Hannink M, Diehl JA (2000) Phosphorylation-dependent regulation of cyclin D1 nuclear export and cyclin D1-dependent cellular transformation. Genes Dev 14(24):3102–3114

Arentson E, Faloon P, Seo J, Moon E, Studts JM, Fremont DH, Choi K (2002) Oncogenic potential of the DNA replication licensing protein CDT1. Oncogene 21(8):1150–1158. https://doi.org/10.1038/sj.onc.1205175

Avni D, Yang H, Martelli F, Hofmann F, ElShamy WM, Ganesan S, Scully R, Livingston DM (2003) Active localization of the retinoblastoma protein in chromatin and its response to S phase DNA damage. Mol Cell 12(3):735–746

Ayuda-Duran P, Devesa F, Gomes F, Sequeira-Mendes J, Avila-Zarza C, Gomez M, Calzada A (2014) The CDK regulators Cdh1 and Sic1 promote efficient usage of DNA replication origins to prevent chromosomal instability at a chromosome arm. Nucleic Acids Res 42(11):7057–7068. https://doi.org/10.1093/nar/gku313

Bao GC, Wang JG, Jong A (2006) Increased p21 expression and complex formation with cyclin E/CDK2 in retinoid-induced pre-B lymphoma cell apoptosis. FEBS Lett 580(15):3687–3693. https://doi.org/10.1016/j.febslet.2006.05.052

Barbash O, Egan E, Pontano LL, Kosak J, Diehl JA (2009) Lysine 269 is essential for cyclin D1 ubiquitylation by the SCF(Fbx4/alphaB-crystallin) ligase and subsequent proteasome-dependent degradation. Oncogene 28(49):4317–4325. https://doi.org/10.1038/onc.2009.287
Barr ML, Bertram EG (1949) A morphological distinction between neurones of the male and female, and the behaviour of the nucleolar satellite during accelerated nucleoprotein synthesis. Nature 163(4148):676
Bashir T, Dorrello NV, Amador V, Guardavaccaro D, Pagano M (2004) Control of the SCF(Skp2-Cks1) ubiquitin ligase by the APC/C(Cdh1) ubiquitin ligase. Nature 428(6979):190–193. https://doi.org/10.1038/nature02330
Bassermann F, Eichner R, Pagano M (2014) The ubiquitin proteasome system – implications for cell cycle control and the targeted treatment of cancer. Biochim Biophys Acta 1843(1):150–162. https://doi.org/10.1016/j.bbamcr.2013.02.028
Bell SD (2012) Archaeal orc1/cdc6 proteins. Subcell Biochem 62:59–69. https://doi.org/10.1007/978-94-007-4572-8_4
Bell SD, Botchan MR (2013) The minichromosome maintenance replicative helicase. Cold Spring Harb Perspect Biol 5(11):a012807. https://doi.org/10.1101/cshperspect.a012807
Bell SP, Kaguni JM (2013) Helicase loading at chromosomal origins of replication. Cold Spring Harb Perspect Biol 5(6). https://doi.org/10.1101/cshperspect.a010124
Benedict WF, Murphree AL, Banerjee A, Spina CA, Sparkes MC, Sparkes RS (1983) Patient with 13 chromosome deletion: evidence that the retinoblastoma gene is a recessive cancer gene. Science 219(4587):973–975
Blagosklonny MV, Pardee AB (2002) The restriction point of the cell cycle. Cell Cycle 1(2):103–110
Bochman ML, Schwacha A (2008) The Mcm2-7 complex has in vitro helicase activity. Mol Cell 31(2):287–293. https://doi.org/10.1016/j.molcel.2008.05.020
Bowers JL, Randell JC, Chen S, Bell SP (2004) ATP hydrolysis by ORC catalyzes reiterative Mcm2-7 assembly at a defined origin of replication. Mol Cell 16(6):967–978. https://doi.org/10.1016/j.molcel.2004.11.038
Brehm A, Miska EA, McCance DJ, Reid JL, Bannister AJ, Kouzarides T (1998) Retinoblastoma protein recruits histone deacetylase to repress transcription. Nature 391(6667):597–601. https://doi.org/10.1038/35404
Brown VD, Phillips RA, Gallie BL (1999) Cumulative effect of phosphorylation of pRB on regulation of E2F activity. Mol Cell Biol 19(5):3246–3256
Burke JR, Deshong AJ, Pelton JG, Rubin SM (2010) Phosphorylation-induced conformational changes in the retinoblastoma protein inhibit E2F transactivation domain binding. J Biol Chem 285(21):16286–16293. https://doi.org/10.1074/jbc.M110.108167
Burke JR, Hura GL, Rubin SM (2012) Structures of inactive retinoblastoma protein reveal multiple mechanisms for cell cycle control. Genes Dev 26(11):1156–1166. https://doi.org/10.1101/gad.189837.112
Burke JR, Liban TJ, Restrepo T, Lee HW, Rubin SM (2014) Multiple mechanisms for E2F binding inhibition by phosphorylation of the retinoblastoma protein C-terminal domain. J Mol Biol 426(1):245–255. https://doi.org/10.1016/j.jmb.2013.09.031
Burkhart DL, Wirt SE, Zmoos AF, Kareta MS, Sage J (2010) Tandem E2F binding sites in the promoter of the p107 cell cycle regulator control p107 expression and its cellular functions. PLoS Genet 6(6):e1001003. https://doi.org/10.1371/journal.pgen.1001003
Cappell SD, Chung M, Jaimovich A, Spencer SL, Meyer T (2016) Irreversible APC(Cdh1) inactivation underlies the point of no return for cell-cycle entry. Cell 166(1):167–180. https://doi.org/10.1016/j.cell.2016.05.077
Chandrasekaran S, Tan TX, Hall JR, Cook JG (2011) Stress-stimulated mitogen-activated protein kinases control the stability and activity of the Cdt1 DNA replication licensing factor. Mol Cell Biol 31(22):4405–4416. https://doi.org/10.1128/MCB.06163-11
Chen J, Jackson PK, Kirschner MW, Dutta A (1995) Separate domains of p21 involved in the inhibition of Cdk kinase and PCNA. Nature 374(6520):386–388. https://doi.org/10.1038/374386a0

Chen S, de Vries MA, Bell SP (2007) Orc6 is required for dynamic recruitment of Cdt1 during repeated Mcm2-7 loading. Genes Dev 21(22):2897–2907. https://doi.org/10.1101/gad.1596807
Chen JY, Lin JR, Tsai FC, Meyer T (2013) Dosage of Dyrk1a shifts cells within a p21-cyclin D1 signaling map to control the decision to enter the cell cycle. Mol Cell 52(1):87–100. https://doi.org/10.1016/j.molcel.2013.09.009
Cheng M, Olivier P, Diehl JA, Fero M, Roussel MF, Roberts JM, Sherr CJ (1999) The p21(Cip1) and p27(Kip1) CDK 'inhibitors' are essential activators of cyclin D-dependent kinases in murine fibroblasts. EMBO J 18(6):1571–1583. https://doi.org/10.1093/emboj/18.6.1571
Choi YJ, Anders L (2014) Signaling through cyclin D-dependent kinases. Oncogene 33(15):1890–1903. https://doi.org/10.1038/onc.2013.137
Choudhury R, Bonacci T, Arceci A, Lahiri D, Mills CA, Kernan JL, Branigan TB, DeCaprio JA, Burke DJ, Emanuele MJ (2016) APC/C and SCF(cyclin F) constitute a reciprocal feedback circuit controlling S-phase entry. Cell Rep 16(12):3359–3372. https://doi.org/10.1016/j.celrep.2016.08.058
Cicchillitti L, Fasanaro P, Biglioli P, Capogrossi MC, Martelli F (2003) Oxidative stress induces protein phosphatase 2A-dependent dephosphorylation of the pocket proteins pRb, p107, and p130. J Biol Chem 278(21):19509–19517. https://doi.org/10.1074/jbc.M300511200
Clarey MG, Erzberger JP, Grob P, Leschziner AE, Berger JM, Nogales E, Botchan M (2006) Nucleotide-dependent conformational changes in the DnaA-like core of the origin recognition complex. Nat Struct Mol Biol 13(8):684–690. https://doi.org/10.1038/nsmb1121
Classon M, Harlow E (2002) The retinoblastoma tumour suppressor in development and cancer. Nat Rev Cancer 2(12):910–917. https://doi.org/10.1038/nrc950
Clurman BE, Sheaff RJ, Thress K, Groudine M, Roberts JM (1996) Turnover of cyclin E by the ubiquitin-proteasome pathway is regulated by cdk2 binding and cyclin phosphorylation. Genes Dev 10(16):1979–1990
Cobrinik D, Whyte P, Peeper DS, Jacks T, Weinberg RA (1993) Cell cycle-specific association of E2F with the p130 E1A-binding protein. Genes Dev 7(12A):2392–2404
Cobrinik D, Lee MH, Hannon G, Mulligan G, Bronson RT, Dyson N, Harlow E, Beach D, Weinberg RA, Jacks T (1996) Shared role of the pRB-related p130 and p107 proteins in limb development. Genes Dev 10(13):1633–1644
Cocker JH, Piatti S, Santocanale C, Nasmyth K, Diffley JF (1996) An essential role for the Cdc6 protein in forming the pre-replicative complexes of budding yeast. Nature 379(6561):180–182. https://doi.org/10.1038/379180a0
Coleman KE, Grant GD, Haggerty RA, Brantley K, Shibata E, Workman BD, Dutta A, Varma D, Purvis JE, Cook JG (2015) Sequential replication-coupled destruction at G1/S ensures genome stability. Genes Dev 29(16):1734–1746. https://doi.org/10.1101/gad.263731.115
Coller HA, Sang L, Roberts JM (2006) A new description of cellular quiescence. PLoS Biol 4(3):e83. https://doi.org/10.1371/journal.pbio.0040083
Connell-Crowley L, Harper JW, Goodrich DW (1997) Cyclin D1/Cdk4 regulates retinoblastoma protein-mediated cell cycle arrest by site-specific phosphorylation. Mol Biol Cell 8(2):287–301
Cook JG, Chasse DA, Nevins JR (2004) The regulated association of Cdt1 with minichromosome maintenance proteins and Cdc6 in mammalian cells. J Biol Chem 279(10):9625–9633. https://doi.org/10.1074/jbc.M311933200
Coulombe P, Gregoire D, Tsanov N, Mechali M (2013) A spontaneous Cdt1 mutation in 129 mouse strains reveals a regulatory domain restraining replication licensing. Nat Commun 4:2065. https://doi.org/10.1038/ncomms3065
Coverley D, Laman H, Laskey RA (2002) Distinct roles for cyclins E and A during DNA replication complex assembly and activation. Nat Cell Biol 4(7):523–528. https://doi.org/10.1038/ncb813
Davidson IF, Li A, Blow JJ (2006) Deregulated replication licensing causes DNA fragmentation consistent with head-to-tail fork collision. Mol Cell 24(3):433–443. https://doi.org/10.1016/j.molcel.2006.09.010

Degen JL, Neubauer MG, Degen SJ, Seyfried CE, Morris DR (1983) Regulation of protein synthesis in mitogen-activated bovine lymphocytes. Analysis of actin-specific and total mRNA accumulation and utilization. J Biol Chem 258(20):12153–12162
DeGregori J, Kowalik T, Nevins JR (1995) Cellular targets for activation by the E2F1 transcription factor include DNA synthesis- and G1/S-regulatory genes. Mol Cell Biol 15(8):4215–4224
DeGregori J, Leone G, Miron A, Jakoi L, Nevins JR (1997) Distinct roles for E2F proteins in cell growth control and apoptosis. Proc Natl Acad Sci U S A 94(14):7245–7250
Dephoure N, Zhou C, Villen J, Beausoleil SA, Bakalarski CE, Elledge SJ, Gygi SP (2008) A quantitative atlas of mitotic phosphorylation. Proc Natl Acad Sci U S A 105(31):10762–10767. https://doi.org/10.1073/pnas.0805139105
Devoto SH, Mudryj M, Pines J, Hunter T, Nevins JR (1992) A cyclin A-protein kinase complex possesses sequence-specific DNA binding activity: p33cdk2 is a component of the E2F-cyclin A complex. Cell 68(1):167–176
Diehl JA, Cheng M, Roussel MF, Sherr CJ (1998) Glycogen synthase kinase-3beta regulates cyclin D1 proteolysis and subcellular localization. Genes Dev 12(22):3499–3511
Dixon JR, Jung I, Selvaraj S, Shen Y, Antosiewicz-Bourget JE, Lee AY, Ye Z, Kim A, Rajagopal N, Xie W, Diao Y, Liang J, Zhao H, Lobanenkov VV, Ecker JR, Thomson JA, Ren B (2015) Chromatin architecture reorganization during stem cell differentiation. Nature 518(7539):331–336. https://doi.org/10.1038/nature14222
Duncker BP, Chesnokov IN, McConkey BJ (2009) The origin recognition complex protein family. Genome Biol 10(3):214. https://doi.org/10.1186/gb-2009-10-3-214
Duzdevich D, Warner MD, Ticau S, Ivica NA, Bell SP, Greene EC (2015) The dynamics of eukaryotic replication initiation: origin specificity, licensing, and firing at the single-molecule level. Mol Cell 58(3):483–494. https://doi.org/10.1016/j.molcel.2015.03.017
Ekholm-Reed S, Mendez J, Tedesco D, Zetterberg A, Stillman B, Reed SI (2004) Deregulation of cyclin E in human cells interferes with prereplication complex assembly. J Cell Biol 165(6):789–800. https://doi.org/10.1083/jcb.200404092
Erlandsson F, Linnman C, Ekholm S, Bengtsson E, Zetterberg A (2000) A detailed analysis of cyclin A accumulation at the G(1)/S border in normal and transformed cells. Exp Cell Res 259(1):86–95. https://doi.org/10.1006/excr.2000.4889
Eser U, Falleur-Fettig M, Johnson A, Skotheim JM (2011) Commitment to a cellular transition precedes genome-wide transcriptional change. Mol Cell 43(4):515–527. https://doi.org/10.1016/j.molcel.2011.06.024
Evrin C, Clarke P, Zech J, Lurz R, Sun J, Uhle S, Li H, Stillman B, Speck C (2009) A double-hexameric MCM2-7 complex is loaded onto origin DNA during licensing of eukaryotic DNA replication. Proc Natl Acad Sci U S A 106(48):20240–20245. https://doi.org/10.1073/pnas.0911500106
Ferreira R, Magnaghi-Jaulin L, Robin P, Harel-Bellan A, Trouche D (1998) The three members of the pocket proteins family share the ability to repress E2F activity through recruitment of a histone deacetylase. Proc Natl Acad Sci U S A 95(18):10493–10498
Field SJ, Tsai FY, Kuo F, Zubiaga AM, Kaelin WG Jr, Livingston DM, Orkin SH, Greenberg ME (1996) E2F-1 functions in mice to promote apoptosis and suppress proliferation. Cell 85(4):549–561
Fornoni A, Pileggi A, Molano RD, Sanabria NY, Tejada T, Gonzalez-Quintana J, Ichii H, Inverardi L, Ricordi C, Pastori RL (2008) Inhibition of c-jun N terminal kinase (JNK) improves functional beta cell mass in human islets and leads to AKT and glycogen synthase kinase-3 (GSK-3) phosphorylation. Diabetologia 51(2):298–308. https://doi.org/10.1007/s00125-007-0889-4
Fotedar R, Fitzgerald P, Rousselle T, Cannella D, Doree M, Messier H, Fotedar A (1996) p21 contains independent binding sites for cyclin and cdk2: both sites are required to inhibit cdk2 kinase activity. Oncogene 12(10):2155–2164
Frigola J, Remus D, Mehanna A, Diffley JF (2013) ATPase-dependent quality control of DNA replication origin licensing. Nature 495(7441):339–343. https://doi.org/10.1038/nature11920

Frolov MV, Dyson NJ (2004) Molecular mechanisms of E2F-dependent activation and pRB-mediated repression. J Cell Sci 117(Pt 11):2173–2181. https://doi.org/10.1242/jcs.01227
Fukushima H, Ogura K, Wan L, Lu Y, Li V, Gao D, Liu P, Lau AW, Wu T, Kirschner MW, Inuzuka H, Wei W (2013) SCF-mediated Cdh1 degradation defines a negative feedback system that coordinates cell-cycle progression. Cell Rep 4(4):803–816. https://doi.org/10.1016/j.celrep.2013.07.031
Gambus A, Khoudoli GA, Jones RC, Blow JJ (2011) MCM2-7 form double hexamers at licensed origins in Xenopus egg extract. J Biol Chem 286(13):11855–11864. https://doi.org/10.1074/jbc.M110.199521
Ge XQ, Blow JJ (2010) Chk1 inhibits replication factory activation but allows dormant origin firing in existing factories. J Cell Biol 191(7):1285–1297. https://doi.org/10.1083/jcb.201007074
Ge XQ, Jackson DA, Blow JJ (2007) Dormant origins licensed by excess Mcm2-7 are required for human cells to survive replicative stress. Genes Dev 21(24):3331–3341. https://doi.org/10.1101/gad.457807
Geng Y, Lee YM, Welcker M, Swanger J, Zagozdzon A, Winer JD, Roberts JM, Kaldis P, Clurman BE, Sicinski P (2007) Kinase-independent function of cyclin E. Mol Cell 25(1):127–139. https://doi.org/10.1016/j.molcel.2006.11.029
Gillespie PJ, Li A, Blow JJ (2001) Reconstitution of licensed replication origins on Xenopus sperm nuclei using purified proteins. BMC Biochem 2:15
Grant GD, Brooks L 3rd, Zhang X, Mahoney JM, Martyanov V, Wood TA, Sherlock G, Cheng C, Whitfield ML (2013) Identification of cell cycle-regulated genes periodically expressed in U2OS cells and their regulation by FOXM1 and E2F transcription factors. Mol Biol Cell 24(23):3634–3650. https://doi.org/10.1091/mbc.E13-05-0264
Green BM, Li JJ (2005) Loss of rereplication control in Saccharomyces cerevisiae results in extensive DNA damage. Mol Biol Cell 16(1):421–432. https://doi.org/10.1091/mbc.E04-09-0833
Gros J, Kumar C, Lynch G, Yadav T, Whitehouse I, Remus D (2015) Post-licensing specification of eukaryotic replication origins by facilitated Mcm2-7 sliding along DNA. Mol Cell 60(5):797–807. https://doi.org/10.1016/j.molcel.2015.10.022
Guiley KZ, Liban TJ, Felthousen JG, Ramanan P, Litovchick L, Rubin SM (2015) Structural mechanisms of DREAM complex assembly and regulation. Genes Dev 29(9):961–974. https://doi.org/10.1101/gad.257568.114
Hannan KM, Hannan RD, Smith SD, Jefferson LS, Lun M, Rothblum LI (2000) Rb and p130 regulate RNA polymerase I transcription: Rb disrupts the interaction between UBF and SL-1. Oncogene 19(43):4988–4999. https://doi.org/10.1038/sj.onc.1203875
Harbour JW, Luo RX, Dei Santi A, Postigo AA, Dean DC (1999) Cdk phosphorylation triggers sequential intramolecular interactions that progressively block Rb functions as cells move through G1. Cell 98(6):859–869
Helin K (1998) Regulation of cell proliferation by the E2F transcription factors. Curr Opin Genet Dev 8(1):28–35
Hemerly AS, Prasanth SG, Siddiqui K, Stillman B (2009) Orc1 controls centriole and centrosome copy number in human cells. Science 323(5915):789–793. https://doi.org/10.1126/science.1166745
Hiratani I, Ryba T, Itoh M, Rathjen J, Kulik M, Papp B, Fussner E, Bazett-Jones DP, Plath K, Dalton S, Rathjen PD, Gilbert DM (2010) Genome-wide dynamics of replication timing revealed by in vitro models of mouse embryogenesis. Genome Res 20(2):155–169. https://doi.org/10.1101/gr.099796.109
Hofmann JF, Beach D (1994) cdt1 is an essential target of the Cdc10/Sct1 transcription factor: requirement for DNA replication and inhibition of mitosis. EMBO J 13(2):425–434
Hossain M, Stillman B (2012) Meier-Gorlin syndrome mutations disrupt an Orc1 CDK inhibitory domain and cause centrosome reduplication. Genes Dev 26(16):1797–1810. https://doi.org/10.1101/gad.197178.112
Hurford RK Jr, Cobrinik D, Lee MH, Dyson N (1997) pRB and p107/p130 are required for the regulated expression of different sets of E2F responsive genes. Genes Dev 11(11):1447–1463

Huttlin EL, Jedrychowski MP, Elias JE, Goswami T, Rad R, Beausoleil SA, Villen J, Haas W, Sowa ME, Gygi SP (2010) A tissue-specific atlas of mouse protein phosphorylation and expression. Cell 143(7):1174–1189. https://doi.org/10.1016/j.cell.2010.12.001
Ibarra A, Schwob E, Mendez J (2008) Excess MCM proteins protect human cells from replicative stress by licensing backup origins of replication. Proc Natl Acad Sci U S A 105(26):8956–8961. https://doi.org/10.1073/pnas.0803978105
Jee J, Mizuno T, Kamada K, Tochio H, Chiba Y, Yanagi K, Yasuda G, Hiroaki H, Hanaoka F, Shirakawa M (2010) Structure and mutagenesis studies of the C-terminal region of licensing factor Cdt1 enable the identification of key residues for binding to replicative helicase Mcm proteins. J Biol Chem 285(21):15931–15940. https://doi.org/10.1074/jbc.M109.075333
Johnson DG, Ohtani K, Nevins JR (1994) Autoregulatory control of E2F1 expression in response to positive and negative regulators of cell cycle progression. Genes Dev 8(13):1514–1525
Jorgensen S, Schotta G, Sorensen CS (2013) Histone H4 lysine 20 methylation: key player in epigenetic regulation of genomic integrity. Nucleic Acids Res 41(5):2797–2806. https://doi.org/10.1093/nar/gkt012
Kaida A, Sawai N, Sakaguchi K, Miura M (2011) Fluorescence kinetics in HeLa cells after treatment with cell cycle arrest inducers visualized with Fucci (fluorescent ubiquitination-based cell cycle indicator). Cell Biol Int 35(4):359–363. https://doi.org/10.1042/CBI20100643
Katoh M, Katoh M (2006) Cross-talk of WNT and FGF signaling pathways at GSK3beta to regulate beta-catenin and SNAIL signaling cascades. Cancer Biol Ther 5(9):1059–1064
Kawasaki Y, Kim HD, Kojima A, Seki T, Sugino A (2006) Reconstitution of Saccharomyces cerevisiae prereplicative complex assembly in vitro. Genes Cells 11(7):745–756. https://doi.org/10.1111/j.1365-2443.2006.00975.x
Kaye FJ, Harbour JW (2004) For whom the bell tolls: susceptibility to common adult cancers in retinoblastoma survivors. J Natl Cancer Inst 96(5):342–343
Ke PY, Kuo YY, Hu CM, Chang ZF (2005) Control of dTTP pool size by anaphase promoting complex/cyclosome is essential for the maintenance of genetic stability. Genes Dev 19(16):1920–1933. https://doi.org/10.1101/gad.1322905
Ke PY, Hu CM, Chang YC, Chang ZF (2007) Hiding human thymidine kinase 1 from APC/C-mediated destruction by thymidine binding. FASEB J 21(4):1276–1284. https://doi.org/10.1096/fj.06-7272com
Kelly TJ, Martin GS, Forsburg SL, Stephen RJ, Russo A, Nurse P (1993) The fission yeast cdc18+ gene product couples S phase to START and mitosis. Cell 74(2):371–382
Khayrutdinov BI, Bae WJ, Yun YM, Lee JH, Tsuyama T, Kim JJ, Hwang E, Ryu KS, Cheong HK, Cheong C, Ko JS, Enomoto T, Karplus PA, Guntert P, Tada S, Jeon YH, Cho Y (2009) Structure of the Cdt1 C-terminal domain: conservation of the winged helix fold in replication licensing factors. Protein Sci 18(11):2252–2264. https://doi.org/10.1002/pro.236
Knudson AG Jr (1971) Mutation and cancer: statistical study of retinoblastoma. Proc Natl Acad Sci U S A 68(4):820–823
Koziczak M, Hynes NE (2004) Cooperation between fibroblast growth factor receptor-4 and ErbB2 in regulation of cyclin D1 translation. J Biol Chem 279(48):50004–50011. https://doi.org/10.1074/jbc.M404252200
LaBaer J, Garrett MD, Stevenson LF, Slingerland JM, Sandhu C, Chou HS, Fattaey A, Harlow E (1997) New functional activities for the p21 family of CDK inhibitors. Genes Dev 11(7):847–862
Lee DG, Bell SP (2000) ATPase switches controlling DNA replication initiation. Curr Opin Cell Biol 12(3):280–285
Lees JA, Buchkovich KJ, Marshak DR, Anderson CW, Harlow E (1991) The retinoblastoma protein is phosphorylated on multiple sites by human cdc2. EMBO J 10(13):4279–4290
Lents NH, Gorges LL, Baldassare JJ (2006) Reverse mutational analysis reveals threonine-373 as a potentially sufficient phosphorylation site for inactivation of the retinoblastoma tumor suppressor protein (pRB). Cell Cycle 5(15):1699–1707. https://doi.org/10.4161/cc.5.15.3126

Leone G, DeGregori J, Yan Z, Jakoi L, Ishida S, Williams RS, Nevins JR (1998) E2F3 activity is regulated during the cell cycle and is required for the induction of S phase. Genes Dev 12(14):2120–2130
Levine EM, Becker Y, Boone CW, Eagle H (1965) Contact inhibition, macromolecular synthesis, and polyribosomes in cultured human diploid fibroblasts. Proc Natl Acad Sci U S A 53:350–356
Li N, Zhai Y, Zhang Y, Li W, Yang M, Lei J, Tye BK, Gao N (2015) Structure of the eukaryotic MCM complex at 3.8 A. Nature 524(7564):186–191. https://doi.org/10.1038/nature14685
Liontos M, Koutsami M, Sideridou M, Evangelou K, Kletsas D, Levy B, Kotsinas A, Nahum O, Zoumpourlis V, Kouloukoussa M, Lygerou Z, Taraviras S, Kittas C, Bartkova J, Papavassiliou AG, Bartek J, Halazonetis TD, Gorgoulis VG (2007) Deregulated overexpression of hCdt1 and hCdc6 promotes malignant behavior. Cancer Res 67(22):10899–10909. https://doi.org/10.1158/0008-5472.CAN-07-2837
Litovchick L, Chestukhin A, DeCaprio JA (2004) Glycogen synthase kinase 3 phosphorylates RBL2/p130 during quiescence. Mol Cell Biol 24(20):8970–8980. https://doi.org/10.1128/MCB.24.20.8970-8980.2004
Litovchick L, Sadasivam S, Florens L, Zhu X, Swanson SK, Velmurugan S, Chen R, Washburn MP, Liu XS, DeCaprio JA (2007) Evolutionarily conserved multisubunit RBL2/p130 and E2F4 protein complex represses human cell cycle-dependent genes in quiescence. Mol Cell 26(4):539–551. https://doi.org/10.1016/j.molcel.2007.04.015
Litovchick L, Florens LA, Swanson SK, Washburn MP, DeCaprio JA (2011) DYRK1A protein kinase promotes quiescence and senescence through DREAM complex assembly. Genes Dev 25(8):801–813. https://doi.org/10.1101/gad.2034211
Liu P, Slater DM, Lenburg M, Nevis K, Cook JG, Vaziri C (2009) Replication licensing promotes cyclin D1 expression and G1 progression in untransformed human cells. Cell Cycle 8(1):125–136. https://doi.org/10.4161/cc.8.1.7528
Lu J, Li F, Murphy CS, Davidson MW, Gilbert DM (2010) G2 phase chromatin lacks determinants of replication timing. J Cell Biol 189(6):967–980. https://doi.org/10.1083/jcb.201002002
Lukas C, Sorensen CS, Kramer E, Santoni-Rugiu E, Lindeneg C, Peters JM, Bartek J, Lukas J (1999) Accumulation of cyclin B1 requires E2F and cyclin-A-dependent rearrangement of the anaphase-promoting complex. Nature 401(6755):815–818. https://doi.org/10.1038/44611
Lunn CL, Chrivia JC, Baldassare JJ (2010) Activation of Cdk2/cyclin E complexes is dependent on the origin of replication licensing factor Cdc6 in mammalian cells. Cell Cycle 9(22):4533–4541. https://doi.org/10.4161/cc.9.22.13789
Luo Y, Hurwitz J, Massague J (1995) Cell-cycle inhibition by independent CDK and PCNA binding domains in p21Cip1. Nature 375(6527):159–161. https://doi.org/10.1038/375159a0
Lyon MF (1961) Gene action in the X-chromosome of the mouse (Mus Musculus L.) Nature 190:372–373
Magenta A, Fasanaro P, Romani S, Di Stefano V, Capogrossi MC, Martelli F (2008) Protein phosphatase 2A subunit PR70 interacts with pRb and mediates its dephosphorylation. Mol Cell Biol 28(2):873–882. https://doi.org/10.1128/MCB.00480-07
Mailand N, Diffley JF (2005) CDKs promote DNA replication origin licensing in human cells by protecting Cdc6 from APC/C-dependent proteolysis. Cell 122(6):915–926. https://doi.org/10.1016/j.cell.2005.08.013
Manchado E, Eguren M, Malumbres M (2010) The anaphase-promoting complex/cyclosome (APC/C): cell-cycle-dependent and -independent functions. Biochem Soc Trans 38(Pt 1):65–71. https://doi.org/10.1042/BST0380065
Marti A, Wirbelauer C, Scheffner M, Krek W (1999) Interaction between ubiquitin-protein ligase SCFSKP2 and E2F-1 underlies the regulation of E2F-1 degradation. Nat Cell Biol 1(1):14–19. https://doi.org/10.1038/8984
Matson JP, Cook JG (2017) Cell cycle proliferation decisions: the impact of single cell analyses. FEBS J 284(3):362–375. https://doi.org/10.1111/febs.13898
McGarry TJ, Kirschner MW (1998) Geminin, an inhibitor of DNA replication, is degraded during mitosis. Cell 93(6):1043–1053

Mendez J, Stillman B (2000) Chromatin association of human origin recognition complex, cdc6, and minichromosome maintenance proteins during the cell cycle: assembly of prereplication complexes in late mitosis. Mol Cell Biol 20(22):8602–8612
Meshorer E, Yellajoshula D, George E, Scambler PJ, Brown DT, Misteli T (2006) Hyperdynamic plasticity of chromatin proteins in pluripotent embryonic stem cells. Dev Cell 10(1):105–116. https://doi.org/10.1016/j.devcel.2005.10.017
Moberg K, Starz MA, Lees JA (1996) E2F-4 switches from p130 to p107 and pRB in response to cell cycle reentry. Mol Cell Biol 16(4):1436–1449
Moreno A, Carrington JT, Albergante L, Al Mamun M, Haagensen EJ, Komseli ES, Gorgoulis VG, Newman TJ, Blow JJ (2016) Unreplicated DNA remaining from unperturbed S phases passes through mitosis for resolution in daughter cells. Proc Natl Acad Sci U S A. https://doi.org/10.1073/pnas.1603252113
Morishima A, Grumbach MM, Taylor JH (1962) Asynchronous duplication of human chromosomes and the origin of sex chromatin. Proc Natl Acad Sci U S A 48:756–763
Muller GA, Quaas M, Schumann M, Krause E, Padi M, Fischer M, Litovchick L, DeCaprio JA, Engeland K (2012) The CHR promoter element controls cell cycle-dependent gene transcription and binds the DREAM and MMB complexes. Nucleic Acids Res 40(4):1561–1578. https://doi.org/10.1093/nar/gkr793
Muller GA, Stangner K, Schmitt T, Wintsche A, Engeland K (2016) Timing of transcription during the cell cycle: protein complexes binding to E2F, E2F/CLE, CDE/CHR, or CHR promoter elements define early and late cell cycle gene expression. Oncotarget. 10.18632/oncotarget.10888
Murphree AL, Benedict WF (1984) Retinoblastoma: clues to human oncogenesis. Science 223(4640):1028–1033
Narasimha AM, Kaulich M, Shapiro GS, Choi YJ, Sicinski P, Dowdy SF (2014) Cyclin D activates the Rb tumor suppressor by mono-phosphorylation. elife 3. doi:https://doi.org/10.7554/eLife.02872
Nevis KR, Cordeiro-Stone M, Cook JG (2009) Origin licensing and p53 status regulate Cdk2 activity during G(1). Cell Cycle 8(12):1952–1963. https://doi.org/10.4161/cc.8.12.8811
Nishitani H, Lygerou Z (2004) DNA replication licensing. Front Biosci 9:2115–2132
Nishitani H, Sugimoto N, Roukos V, Nakanishi Y, Saijo M, Obuse C, Tsurimoto T, Nakayama KI, Nakayama K, Fujita M, Lygerou Z, Nishimoto T (2006) Two E3 ubiquitin ligases, SCF-Skp2 and DDB1-Cul4, target human Cdt1 for proteolysis. EMBO J 25(5):1126–1136. https://doi.org/10.1038/sj.emboj.7601002
O'Connor ML, Xiang D, Shigdar S, Macdonald J, Li Y, Wang T, Pu C, Wang Z, Qiao L, Duan W (2014) Cancer stem cells: a contentious hypothesis now moving forward. Cancer Lett 344(2):180–187. https://doi.org/10.1016/j.canlet.2013.11.012
Ohtani K, DeGregori J, Leone G, Herendeen DR, Kelly TJ, Nevins JR (1996) Expression of the HsOrc1 gene, a human ORC1 homolog, is regulated by cell proliferation via the E2F transcription factor. Mol Cell Biol 16(12):6977–6984
Ohtani K, Tsujimoto A, Ikeda M, Nakamura M (1998) Regulation of cell growth-dependent expression of mammalian CDC6 gene by the cell cycle transcription factor E2F. Oncogene 17(14):1777–1785. https://doi.org/10.1038/sj.onc.1202105
Ohtani K, Iwanaga R, Nakamura M, Ikeda M, Yabuta N, Tsuruga H, Nojima H (1999) Cell growth-regulated expression of mammalian MCM5 and MCM6 genes mediated by the transcription factor E2F. Oncogene 18(14):2299–2309. https://doi.org/10.1038/sj.onc.1202544
Pardee AB (1974) A restriction point for control of normal animal cell proliferation. Proc Natl Acad Sci U S A 71(4):1286–1290
Paternot S, Bockstaele L, Bisteau X, Kooken H, Coulonval K, Roger PP (2010) Rb inactivation in cell cycle and cancer: the puzzle of highly regulated activating phosphorylation of CDK4 versus constitutively active CDK-activating kinase. Cell Cycle 9(4):689–699. https://doi.org/10.4161/cc.9.4.10611
Pauklin S, Vallier L (2013) The cell-cycle state of stem cells determines cell fate propensity. Cell 155(1):135–147. https://doi.org/10.1016/j.cell.2013.08.031

Petersen BO, Wagener C, Marinoni F, Kramer ER, Melixetian M, Lazzerini Denchi E, Gieffers C, Matteucci C, Peters JM, Helin K (2000) Cell cycle- and cell growth-regulated proteolysis of mammalian CDC6 is dependent on APC-CDH1. Genes Dev 14(18):2330–2343

Piatti S, Lengauer C, Nasmyth K (1995) Cdc6 is an unstable protein whose de novo synthesis in G1 is important for the onset of S phase and for preventing a 'reductional' anaphase in the budding yeast Saccharomyces cerevisiae. EMBO J 14(15):3788–3799

Prasanth SG, Prasanth KV, Stillman B (2002) Orc6 involved in DNA replication, chromosome segregation, and cytokinesis. Science 297(5583):1026–1031. https://doi.org/10.1126/science.1072802

Prasanth SG, Prasanth KV, Siddiqui K, Spector DL, Stillman B (2004) Human Orc2 localizes to centrosomes, centromeres and heterochromatin during chromosome inheritance. EMBO J 23(13):2651–2663. https://doi.org/10.1038/sj.emboj.7600255

Prasanth SG, Shen Z, Prasanth KV, Stillman B (2010) Human origin recognition complex is essential for HP1 binding to chromatin and heterochromatin organization. Proc Natl Acad Sci U S A 107(34):15093–15098. https://doi.org/10.1073/pnas.1009945107

Pruitt SC, Bailey KJ, Freeland A (2007) Reduced Mcm2 expression results in severe stem/progenitor cell deficiency and cancer. Stem Cells 25(12):3121–3132. https://doi.org/10.1634/stemcells.2007-0483

Purvis JE, Karhohs KW, Mock C, Batchelor E, Loewer A, Lahav G (2012) p53 dynamics control cell fate. Science 336(6087):1440–1444. https://doi.org/10.1126/science.1218351

Randell JC, Bowers JL, Rodriguez HK, Bell SP (2006) Sequential ATP hydrolysis by Cdc6 and ORC directs loading of the Mcm2-7 helicase. Mol Cell 21(1):29–39. https://doi.org/10.1016/j.molcel.2005.11.023

Raschle M, Knipscheer P, Enoiu M, Angelov T, Sun J, Griffith JD, Ellenberger TE, Scharer OD, Walter JC (2008) Mechanism of replication-coupled DNA interstrand crosslink repair. Cell 134(6):969–980. https://doi.org/10.1016/j.cell.2008.08.030

Remus D, Diffley JF (2009) Eukaryotic DNA replication control: lock and load, then fire. Curr Opin Cell Biol 21(6):771–777. https://doi.org/10.1016/j.ceb.2009.08.002

Resnitzky D, Reed SI (1995) Different roles for cyclins D1 and E in regulation of the G1-to-S transition. Mol Cell Biol 15(7):3463–3469

Rowles A, Tada S, Blow JJ (1999) Changes in association of the Xenopus origin recognition complex with chromatin on licensing of replication origins. J Cell Sci 112(Pt 12):2011–2018

Rumman M, Dhawan J, Kassem M (2015) Concise review: quiescence in adult stem cells: biological significance and relevance to tissue regeneration. Stem Cells 33(10):2903–2912. https://doi.org/10.1002/stem.2056

Sadasivam S, Duan S, DeCaprio JA (2012) The MuvB complex sequentially recruits B-Myb and FoxM1 to promote mitotic gene expression. Genes Dev 26(5):474–489. https://doi.org/10.1101/gad.181933.111

Samel SA, Fernandez-Cid A, Sun J, Riera A, Tognetti S, Herrera MC, Li H, Speck C (2014) A unique DNA entry gate serves for regulated loading of the eukaryotic replicative helicase MCM2-7 onto DNA. Genes Dev 28(15):1653–1666. https://doi.org/10.1101/gad.242404.114

Sandoval R, Pilkinton M, Colamonici OR (2009) Deletion of the p107/p130-binding domain of Mip130/LIN-9 bypasses the requirement for CDK4 activity for the dissociation of Mip130/LIN-9 from p107/p130-E2F4 complex. Exp Cell Res 315(17):2914–2920. https://doi.org/10.1016/j.yexcr.2009.07.014

Schachter MM, Merrick KA, Larochelle S, Hirschi A, Zhang C, Shokat KM, Rubin SM, Fisher RP (2013) A Cdk7-Cdk4 T-loop phosphorylation cascade promotes G1 progression. Mol Cell 50(2):250–260. https://doi.org/10.1016/j.molcel.2013.04.003

Schepers A, Diffley JF (2001) Mutational analysis of conserved sequence motifs in the budding yeast Cdc6 protein. J Mol Biol 308(4):597–608. https://doi.org/10.1006/jmbi.2001.4637

Scott PH, Cairns CA, Sutcliffe JE, Alzuherri HM, McLees A, Winter AG, White RJ (2001) Regulation of RNA polymerase III transcription during cell cycle entry. J Biol Chem 276(2):1005–1014. https://doi.org/10.1074/jbc.M005417200

Sheaff RJ, Groudine M, Gordon M, Roberts JM, Clurman BE (1997) Cyclin E-CDK2 is a regulator of p27Kip1. Genes Dev 11(11):1464–1478

Shima N, Alcaraz A, Liachko I, Buske TR, Andrews CA, Munroe RJ, Hartford SA, Tye BK, Schimenti JC (2007) A viable allele of Mcm4 causes chromosome instability and mammary adenocarcinomas in mice. Nat Genet 39(1):93–98. https://doi.org/10.1038/ng1936

Shreeram S, Sparks A, Lane DP, Blow JJ (2002) Cell type-specific responses of human cells to inhibition of replication licensing. Oncogene 21(43):6624–6632. https://doi.org/10.1038/sj.onc.1205910

Siddiqui K, On KF, Diffley JF (2013) Regulating DNA replication in eukarya. Cold Spring Harb Perspect Biol 5(9). https://doi.org/10.1101/cshperspect.a012930

Slansky JE, Farnham PJ (1996) Introduction to the E2F family: protein structure and gene regulation. Curr Top Microbiol Immunol 208:1–30

Soppa U, Schumacher J, Florencio Ortiz V, Pasqualon T, Tejedor FJ, Becker W (2014) The down syndrome-related protein kinase DYRK1A phosphorylates p27(Kip1) and cyclin D1 and induces cell cycle exit and neuronal differentiation. Cell Cycle 13(13):2084–2100. https://doi.org/10.4161/cc.29104

Sorensen CS, Lukas C, Kramer ER, Peters JM, Bartek J, Lukas J (2000) Nonperiodic activity of the human anaphase-promoting complex-Cdh1 ubiquitin ligase results in continuous DNA synthesis uncoupled from mitosis. Mol Cell Biol 20(20):7613–7623

Speck C, Chen Z, Li H, Stillman B (2005) ATPase-dependent cooperative binding of ORC and Cdc6 to origin DNA. Nat Struct Mol Biol 12(11):965–971. https://doi.org/10.1038/nsmb1002

Spencer SL, Cappell SD, Tsai FC, Overton KW, Wang CL, Meyer T (2013) The proliferation-quiescence decision is controlled by a bifurcation in CDK2 activity at mitotic exit. Cell. https://doi.org/10.1016/j.cell.2013.08.062

Sugiyama M, Sakaue-Sawano A, Iimura T, Fukami K, Kitaguchi T, Kawakami K, Okamoto H, Higashijima S, Miyawaki A (2009) Illuminating cell-cycle progression in the developing zebrafish embryo. Proc Natl Acad Sci U S A 106(49):20812–20817. https://doi.org/10.1073/pnas.0906464106

Sun J, Fernandez-Cid A, Riera A, Tognetti S, Yuan Z, Stillman B, Speck C, Li H (2014) Structural and mechanistic insights into Mcm2-7 double-hexamer assembly and function. Genes Dev 28(20):2291–2303. https://doi.org/10.1101/gad.242313.114

Suryadinata R, Sadowski M, Steel R, Sarcevic B (2011) Cyclin-dependent kinase-mediated phosphorylation of RBP1 and pRb promotes their dissociation to mediate release of the SAP30. mSin3.HDAC transcriptional repressor complex. J Biol Chem 286(7):5108–5118. https://doi.org/10.1074/jbc.M110.198473

Tachibana KE, Nigg EA (2006) Geminin regulates multiple steps of the chromosome inheritance cycle. Cell Cycle 5(2):151–154. https://doi.org/10.4161/cc.5.2.2363

Takaki T, Fukasawa K, Suzuki-Takahashi I, Hirai H (2004) Cdk-mediated phosphorylation of pRB regulates HDAC binding in vitro. Biochem Biophys Res Commun 316(1):252–255. https://doi.org/10.1016/j.bbrc.2004.02.044

Takeda DY, Parvin JD, Dutta A (2005) Degradation of Cdt1 during S phase is Skp2-independent and is required for efficient progression of mammalian cells through S phase. J Biol Chem 280(24):23416–23423. https://doi.org/10.1074/jbc.M501208200

Tardat M, Brustel J, Kirsh O, Lefevbre C, Callanan M, Sardet C, Julien E (2010) The histone H4 Lys 20 methyltransferase PR-Set7 regulates replication origins in mammalian cells. Nat Cell Biol 12(11):1086–1093. https://doi.org/10.1038/ncb2113

Teer JK, Machida YJ, Labit H, Novac O, Hyrien O, Marheineke K, Zannis-Hadjopoulos M, Dutta A (2006) Proliferating human cells hypomorphic for origin recognition complex 2 and pre-replicative complex formation have a defect in p53 activation and Cdk2 kinase activation. J Biol Chem 281(10):6253–6260. https://doi.org/10.1074/jbc.M507150200

Ticau S, Friedman LJ, Ivica NA, Gelles J, Bell SP (2015) Single-molecule studies of origin licensing reveal mechanisms ensuring bidirectional helicase loading. Cell 161(3):513–525. https://doi.org/10.1016/j.cell.2015.03.012

Townley-Tilson WH, Pendergrass SA, Marzluff WF, Whitfield ML (2006) Genome-wide analysis of mRNAs bound to the histone stem-loop binding protein. RNA 12(10):1853–1867. https://doi.org/10.1261/rna.76006

Tsakraklides V, Bell SP (2010) Dynamics of pre-replicative complex assembly. J Biol Chem 285(13):9437–9443. https://doi.org/10.1074/jbc.M109.072504

Tsuruga H, Yabuta N, Hashizume K, Ikeda M, Endo Y, Nojima H (1997) Expression, nuclear localization and interactions of human MCM/P1 proteins. Biochem Biophys Res Commun 236(1):118–125. https://doi.org/10.1006/bbrc.1997.6865

Ubersax JA, Ferrell JE Jr (2007) Mechanisms of specificity in protein phosphorylation. Nat Rev Mol Cell Biol 8(7):530–541. https://doi.org/10.1038/nrm2203

van Leuken R, Clijsters L, Wolthuis R (2008) To cell cycle, swing the APC/C. Biochim Biophys Acta 1786(1):49–59. https://doi.org/10.1016/j.bbcan.2008.05.002

Varma D, Chandrasekaran S, Sundin LJ, Reidy KT, Wan X, Chasse DA, Nevis KR, DeLuca JG, Salmon ED, Cook JG (2012) Recruitment of the human Cdt1 replication licensing protein by the loop domain of Hec1 is required for stable kinetochore-microtubule attachment. Nat Cell Biol 14(6):593–603. https://doi.org/10.1038/ncb2489

Vaziri C, Saxena S, Jeon Y, Lee C, Murata K, Machida Y, Wagle N, Hwang DS, Dutta A (2003) A p53-dependent checkpoint pathway prevents rereplication. Mol Cell 11(4):997–1008

Voorhoeve PM, Watson RJ, Farlie PG, Bernards R, Lam EW (1999) Rapid dephosphorylation of p107 following UV irradiation. Oncogene 18(3):679–688. https://doi.org/10.1038/sj.onc.1202289

Wang ZF, Whitfield ML, Ingledue TC 3rd, Dominski Z, Marzluff WF (1996) The protein that binds the 3′ end of histone mRNA: a novel RNA-binding protein required for histone pre-mRNA processing. Genes Dev 10(23):3028–3040

Wei W, Ayad NG, Wan Y, Zhang GJ, Kirschner MW, Kaelin WG Jr (2004) Degradation of the SCF component Skp2 in cell-cycle phase G1 by the anaphase-promoting complex. Nature 428(6979):194–198. https://doi.org/10.1038/nature02381

Wheeler LW, Lents NH, Baldassare JJ (2008) Cyclin A-CDK activity during G1 phase impairs MCM chromatin loading and inhibits DNA synthesis in mammalian cells. Cell Cycle 7(14):2179–2188. https://doi.org/10.4161/cc.7.14.6270

White RJ, Trouche D, Martin K, Jackson SP, Kouzarides T (1996) Repression of RNA polymerase III transcription by the retinoblastoma protein. Nature 382(6586):88–90. https://doi.org/10.1038/382088a0

Whitfield ML, Zheng LX, Baldwin A, Ohta T, Hurt MM, Marzluff WF (2000) Stem-loop binding protein, the protein that binds the 3′ end of histone mRNA, is cell cycle regulated by both translational and posttranslational mechanisms. Mol Cell Biol 20(12):4188–4198

Whitfield ML, Sherlock G, Saldanha AJ, Murray JI, Ball CA, Alexander KE, Matese JC, Perou CM, Hurt MM, Brown PO, Botstein D (2002) Identification of genes periodically expressed in the human cell cycle and their expression in tumors. Mol Biol Cell 13(6):1977–2000. https://doi.org/10.1091/mbc.02-02-0030

Wikenheiser-Brokamp KA (2006) Retinoblastoma regulatory pathway in lung cancer. Curr Mol Med 6(7):783–793

Williams JG, Penman S (1975) The messenger RNA sequences in growing and resting mouse fibroblasts. Cell 6(2):197–206

Wilson KA, Elefanty AG, Stanley EG, Gilbert DM (2016) Spatio-temporal re-organization of replication foci accompanies replication domain consolidation during human pluripotent stem cell lineage specification. Cell Cycle.:0. https://doi.org/10.1080/15384101.2016.1203492

Wohlschlegel JA, Dwyer BT, Dhar SK, Cvetic C, Walter JC, Dutta A (2000) Inhibition of eukaryotic DNA replication by geminin binding to Cdt1. Science 290(5500):2309–2312. https://doi.org/10.1126/science.290.5500.2309

Woodward AM, Gohler T, Luciani MG, Oehlmann M, Ge X, Gartner A, Jackson DA, Blow JJ (2006) Excess Mcm2-7 license dormant origins of replication that can be used under conditions of replicative stress. J Cell Biol 173(5):673–683. https://doi.org/10.1083/jcb.200602108

Wu L, Timmers C, Maiti B, Saavedra HI, Sang L, Chong GT, Nuckolls F, Giangrande P, Wright FA, Field SJ, Greenberg ME, Orkin S, Nevins JR, Robinson ML, Leone G (2001) The E2F1-3 transcription factors are essential for cellular proliferation. Nature 414(6862):457–462. https://doi.org/10.1038/35106593
Wu JQ, Guo JY, Tang W, Yang CS, Freel CD, Chen C, Nairn AC, Kornbluth S (2009) PP1-mediated dephosphorylation of phosphoproteins at mitotic exit is controlled by inhibitor-1 and PP1 phosphorylation. Nat Cell Biol 11(5):644–651. https://doi.org/10.1038/ncb1871
Xouri G, Dimaki M, Bastiaens PI, Lygerou Z (2007) Cdt1 interactions in the licensing process: a model for dynamic spatiotemporal control of licensing. Cell Cycle 6(13):1549–1552. https://doi.org/10.4161/cc.6.13.4455
Yan H, Chen CY, Kobayashi R, Newport J (1998a) Replication focus-forming activity 1 and the Werner syndrome gene product. Nat Genet 19(4):375–378. https://doi.org/10.1038/1263
Yan Z, DeGregori J, Shohet R, Leone G, Stillman B, Nevins JR, Williams RS (1998b) Cdc6 is regulated by E2F and is essential for DNA replication in mammalian cells. Proc Natl Acad Sci U S A 95(7):3603–3608
Yanagi K, Mizuno T, You Z, Hanaoka F (2002) Mouse geminin inhibits not only Cdt1-MCM6 interactions but also a novel intrinsic Cdt1 DNA binding activity. J Biol Chem 277(43):40871–40880. https://doi.org/10.1074/jbc.M206202200
Yeeles JT, Deegan TD, Janska A, Early A, Diffley JF (2015) Regulated eukaryotic DNA replication origin firing with purified proteins. Nature 519(7544):431–435. https://doi.org/10.1038/nature14285
Yim H, Park JW, Woo SU, Kim ST, Liu L, Lee CH, Lee SK (2013) Phosphorylation of Cdc6 at serine 74, but not at serine 106, drives translocation of Cdc6 to the cytoplasm. J Cell Physiol 228(6):1221–1228. https://doi.org/10.1002/jcp.24275
Yoshida K, Inoue I (2004a) Expression of MCM10 and TopBP1 is regulated by cell proliferation and UV irradiation via the E2F transcription factor. Oncogene 23(37):6250–6260. https://doi.org/10.1038/sj.onc.1207829
Yoshida K, Inoue I (2004b) Regulation of geminin and Cdt1 expression by E2F transcription factors. Oncogene 23(21):3802–3812. https://doi.org/10.1038/sj.onc.1207488
Yung Y, Walker JL, Roberts JM, Assoian RK (2007) A Skp2 autoinduction loop and restriction point control. J Cell Biol 178(5):741–747. https://doi.org/10.1083/jcb.200703034
Zarkowska T, Mittnacht S (1997) Differential phosphorylation of the retinoblastoma protein by G1/S cyclin-dependent kinases. J Biol Chem 272(19):12738–12746
Zarkowska T, U S, Harlow E, Mittnacht S (1997) Monoclonal antibodies specific for underphosphorylated retinoblastoma protein identify a cell cycle regulated phosphorylation site targeted by CDKs. Oncogene 14(2):249–254. https://doi.org/10.1038/sj.onc.1200824
Zhang J, Yu L, Wu X, Zou L, Sou KK, Wei Z, Cheng X, Zhu G, Liang C (2010) The interacting domains of hCdt1 and hMcm6 involved in the chromatin loading of the MCM complex in human cells. Cell Cycle 9(24):4848–4857. https://doi.org/10.4161/cc.9.24.14136

Chapter 17
Roles of SUMO in Replication Initiation, Progression, and Termination

Lei Wei and Xiaolan Zhao

Abstract Accurate genome duplication during cell division is essential for life. This process is accomplished by the close collaboration between replication factors and many additional proteins that provide assistant roles. Replication factors establish the replication machineries capable of copying billions of nucleotides, while regulatory proteins help to achieve accuracy and efficiency of replication. Among regulatory proteins, protein modification enzymes can bestow fast and reversible changes to many targets, leading to coordinated effects on replication. Recent studies have begun to elucidate how one type of protein modification, sumoylation, can modify replication proteins and regulate genome duplication through multiple mechanisms. This chapter summarizes these new findings, and how they can integrate with the known regulatory circuitries of replication. As this area of research is still at its infancy, many outstanding questions remain to be explored, and we discuss these issues in light of the new advances.

Keywords DNA replication initiation • Replication progression • Posttranslational modifications • Sumoylation • Ubiquitination • Phosphorylation

L. Wei
Molecular Biology Program, Memorial Sloan Kettering Cancer Center, New York, NY, USA

Louis V. Gerstner Jr. Graduate School of Biomedical Sciences, Memorial Sloan Kettering Cancer Center, New York, NY, USA

X. Zhao (✉)
Molecular Biology Program, Memorial Sloan Kettering Cancer Center, New York, NY, USA
e-mail: zhaox1@mskcc.org

H. Masai, M. Foiani (eds.), *DNA Replication*, Advances in Experimental Medicine and Biology 1042, https://doi.org/10.1007/978-981-10-6955-0_17

17.1 Overview of Eukaryotic DNA Replication

17.1.1 Replication Initiation

DNA replication occurs in three stages, namely, initiation, progression, and termination. Each of these stages entails multi-step DNA transactions carried out by dozens of proteins. Most of the replication steps and proteins are highly conserved from simple model organisms, such as yeasts, to humans. Replication initiation begins with the licensing of genomic sites called origins. Origin licensing takes place in late M to G1 phase when origins become bound by the MCM complex, a ring-shaped complex composed of MCM2-7 subunits (Fig. 17.1). This process is achieved by interaction between MCM and MCM-loading factors. In budding yeast, wherein origin licensing is best understood, the MCM-loading factors include the origin recognition complex (ORC), Cdc6, and the MCM binding partner Cdt1 (Fig. 17.1) [reviewed in (Diffley et al. 1994; Kelly and Brown 2000; Bell and Dutta 2002; Sclafani and Holzen 2007; Remus and Diffley 2009; Li and Araki 2013; Bell and Labib 2016)]. ORCs demarcate origins by interacting with specific DNA sequences and chromatin components [reviewed in (Hoggard and Fox 2016; Gutiérrez and MacAlpine 2016)]. Cdc6 recruits the MCM-Cdt1 complex to ORC through interactions with both factors (Santocanale and Diffley 1996; Donovan et al. 1997; Speck et al. 2005; Randell et al. 2006; Sun et al. 2013). Subsequently, ATP hydrolysis by ORC, Cdc6, and MCM enables a pair of MCM rings to enclose DNA at origins (Fig. 17.1) (Bowers et al. 2004; Randell et al. 2006; Remus et al. 2009; Fernández-Cid et al. 2013; Frigola et al. 2013; Coster et al. 2014). This multi-step process produces an MCM double hexamer with its central channel enclosing DNA [reviewed in (Diffley et al. 1994; Kelly and Brown 2000; Bell and Dutta 2002; Sclafani and Holzen 2007; Remus and Diffley 2009; Wei and Zhao 2016b; Li and Araki 2013; Bell and Labib 2016)].

Once loaded, the MCM double hexamer must be kept inactive until the onset of S phase when origin firing takes place. One critical event during origin firing is the conversion of MCM into the replicative helicase, composed of MCM and its cofac-

Fig. 17.1 (continuted) and the ORC complex. During the origin firing step in S phase, DDK and CDK kinases activate the loaded MCM. DDK-mediated phosphorylation of loaded MCM recruits Sld3 and its binding partners Cdc45 and Sld7. CDK-mediated phosphorylation of Sld2 and Sld3 promotes the recruitment of Pol ε, Dpb11, and the GINS complex. Cdc45, MCM, and GINS form the active replicative helicase CMG. Subsequent recruitment of additional protein factors results in the formation of the replisome. SUMO can counteract DDK-mediated MCM phosphorylation. A fraction of loaded Mcm2-7 subunits are sumoylated to prevent premature replication initiation. This is partly achieved as SUMO aids the recruitment of the Rif1-PP1 phosphase complex that can antagonize DDK action in G1. As S phase starts, rising DDK levels are associated with increased MCM phosphorylation and decreased MCM sumoylation levels. Such a switch of MCM modification states is also seen in other organisms. Note that recently findings suggest that MCM also switches from enclosing dsDNA to enclosing mostly leading strand ssDNA template when DNA synthesis begins and that the N-terminal tier of the MCM ring faces the moving replication fork

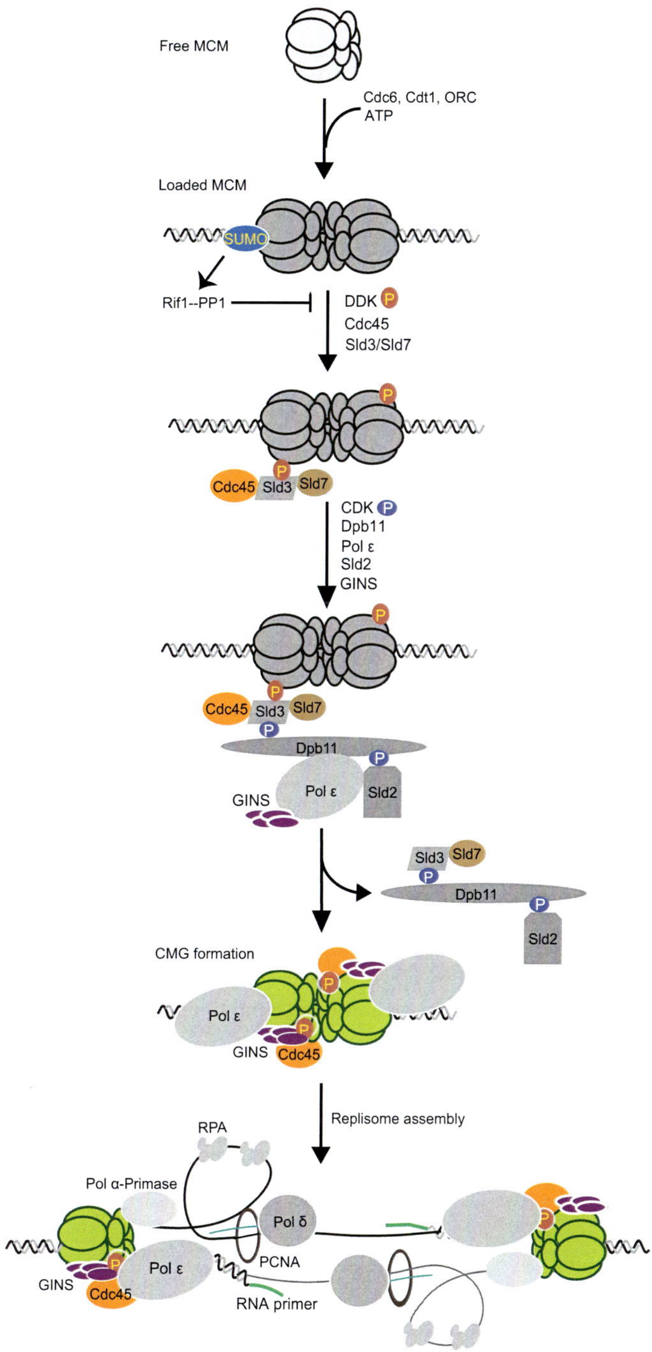

Fig. 17.1 Summary of main steps of DNA replication initiation in budding yeast. The origin licensing step during late M to G1 phases entails free MCM being loaded onto replication origins as a double hexamer. This process requires several MCM-loading factors, including Cdc6, Cdt1,

tors Cdc45 and the four-subunit GINS complex. This conversion requires two protein kinases, DNA polymerase ε, and several scaffolding proteins, including Sld2, Sld3–Sld7, and Dpb11 in budding yeast and their homologs in other organisms (Gambus et al. 2006; Moyer et al. 2006; Pacek et al. 2006; Ilves et al. 2010; Muramatsu et al. 2010; Kang et al. 2012). The first kinase, the Dbf4-dependent kinase (DDK), phosphorylates MCM subunits (Fig. 17.1). Mcm4 phosphorylation is particularly important as it is recognized by Sld3 in partnership with Sld7 and Cdc45 (Fig. 17.1) (Sheu and Stillman 2006; Sheu and Stillman 2010; Deegan et al. 2016). Subsequent to this step, the second kinase, cyclin-dependent kinase (CDK), phosphorylates Sld2 and Sld3, enabling their interaction with Dpb11 in cooperation with Pol ε and GINS (Fig. 17.1) (Tanaka et al. 2007; Zegerman and Diffley 2007; Muramatsu et al. 2010). As a result of this cascade of protein interactions, Cdc45 and GINS are delivered to MCM, resulting in the formation of the replicative helicase CMG (Cdc45-MCM-GINS) (Fig. 17.1) (Gambus et al. 2006; Moyer et al. 2006; Pacek et al. 2006; Ilves et al. 2010; Kang et al. 2012).

Subsequent to CMG formation, DNA polymerase α, the chromatin remodeling complex FACT, and several scaffolding proteins are recruited to CMG and Pol ε to form the replisome (Fig. 17.1) (Gambus et al. 2006, 2009; Morohashi et al. 2009). In the meantime, the origin firing scaffolds, such as Sld2, Sld3, and Dpb11, leave the CMG and are recycled to additional origins that fire later in S phase (Fig. 17.1) (Mantiero et al. 2011). The temporal order of the origin firing program is determined by multiple factors, such as the affinity between ORCs and origin sequences and local chromatin environment [reviewed in (Masai et al. 2010; Fragkos et al. 2015)]. At both early- and late-fired origins, the formation of a pair of replisomes establishes divergent replication forks that travel in opposite directions.

17.1.2 Replication Progression

As DNA synthesis begins, the DNA primase-Pol α complex generates primer sequences (Fig. 17.1). These primers can be extended by Pol ε for continuous leading strand synthesis and by Pol δ to produce many Okazaki fragments during discontinuous lagging strand synthesis (Fig. 17.1) [reviewed in (Kelly and Brown 2000; Bell and Dutta 2002; Sclafani and Holzen 2007; Bell and Labib 2016)]. The maturation of Okazaki fragments requires several additional conserved enzymes. In budding yeast, these include the flap endonuclease Rad27, the DNA helicase-nuclease Dna2, the exonuclease Exo1, the DNA helicase Pif1, and the ligase Cdc9 [reviewed in (Waga and Stillman 1998; Bell and Labib 2016)]. Collaboration among these factors enables ligation of Okazaki fragments.

During replication progression, a major challenge is coping with many types of template blockages. These can include (i) the topological stress generated by DNA unwinding, (ii) tightly bound nonhistone proteins, (iii) difficult-to-replicate genomic loci, (iv) collision with transcription machinery, and (v) DNA lesions generated from both intrinsic and extrinsic sources. Topological stress is largely relieved by replisome-associated topoisomerases. Several scaffold proteins within the replisome play pivotal roles in dealing with other replication impediments. Depending on the types of blocks, different proteins and strategies are used, and in many cases, additional DNA metabolism proteins are recruited to overcome template blocks. For example, tightly bound nonhistone proteins can be removed by DNA helicases, such as Rrm3 in budding yeast, allowing the resumption of DNA synthesis (Ivessa et al. 2000; Calzada et al. 2005; Azvolinsky et al. 2006). In the case of template damage, such as UV-induced thymidine dimers, translesion polymerases can mediate synthesis bypass of these sites [reviewed in (Waters et al. 2009)].

Besides template blockage, other issues that must be managed during replication progression include the removal of template nucleosomes ahead of replication forks and reestablishment or recycling of nucleosomes with correct positioning behind replication forks. In addition, replication progression is coupled with the establishment of sister chromatid cohesion and inheritable epigenetic markers. These topics have been recently summarized, and we refer the readers to several reviews for details [reviewed in (Jeppsson et al. 2014; Almouzni and Cedar 2016; Bell and Labib 2016)].

17.1.3 Replication Termination

As two opposing replication forks from adjacent origins converge, replication terminates. In general, replication termination sites are determined by the meeting point of two forks, but in some instances, termination occurs at replication pausing sites where one fork has more retention time (Labib and Hodgson 2007; Fachinetti et al. 2010). Three major events are required for replication termination, including the completion of local DNA synthesis, decatenation of the two daughter strands by topoisomerases, and disassembly of replisome. Compared with replication initiation and progression, replication termination is less well understood. More recently, new findings have implicated MCM ubiquitination in the disassembly of replisomes during termination (Maric et al. 2014; Priego Moreno et al. 2014; Dewar et al. 2015).

17.2 The SUMO Modification Cycle

Protein modifications underpin many regulatory mechanisms during the three stages of replication. Phosphorylation and ubiquitination have been found to be critical in all stages of replication [reviewed in (Wei and Zhao 2016b; Kelly and Brown 2000; Diffley 2004; Vodermaier 2004; Blow and Dutta 2005; Arias and Walter 2007; Sclafani and Holzen 2007; Moreno and Gambus 2015; Sivakumar and Gorbsky 2015; Garcia-Rodriguez et al. 2016)]. More recently, sumoylation has also been found to influence replication and is important for genome integrity. In this chapter, we summarize the findings that begin to unravel the mechanisms of SUMO-based replication regulation after a brief introduction of protein sumoylation.

17.2.1 Principles of the Sumoylation Process

SUMO (small ubiquitin-like modifier) is a highly conserved member of the ubiquitin family of protein modifiers. With approximately 100 amino acids, SUMO assumes an ubiquitin fold but with a distinct surface charge distribution. The SUMO-specific E1 (or activating enzyme), E2 (or conjugating enzyme), and E3 (or ligase) enzymes conjugate SUMO to the ε-amino group lysine residue of a substrate (Johnson 2004) (Fig. 17.2). Most organisms contain one SUMO E1 and E2 but multiple SUMO E3s. SUMO E2 can directly bind to the so-called sumoylation consensus or reverse consensus sequences, [ΨKX(D/E)] or [(D/E)XKΨ] (Ψ, a hydrophobic residue; X, any residue). With the help of E3s, SUMO is then transferred from E2 to the lysine within such sequences (Gareau and Lima 2010; Lamoliatte et al. 2014). However, proteomic studies found that many of these sites are not sumoylated, suggesting that additional factors must also influence the sumoylation process (Gareau and Lima 2010; Lamoliatte et al. 2014). For example, it has been noted that sumoylation sites are often in loop regions (Gareau and Lima 2010), likely because of the ability of such regions to adopt local conformational changes that favor productive contact between the E2~SUMO thioester bond and the acceptor lysine. This principle could also explain findings that while sumoylation is conserved among many protein homologs, the sites of modification often vary (Golebiowski et al. 2009; Dou et al. 2010; Elrouby and Coupland 2010; Cremona et al. 2012; Psakhye and Jentsch 2012; Hendriks et al. 2014; Ma et al. 2014; Tammsalu et al. 2014).

SUMO E3s play important roles partly by bridging E2 and substrates. For instance, budding yeast has the homologous Siz1 and Siz2 SUMO E3s and the Mms21 E3, all of which possess an SP-RING domain that can associate with Ubc9 (Johnson and Gupta 2001; Takahashi et al. 2001; Zhao and Blobel 2005; Gareau and Lima 2010). In addition, each E3 can associate with specific substrates. For example,

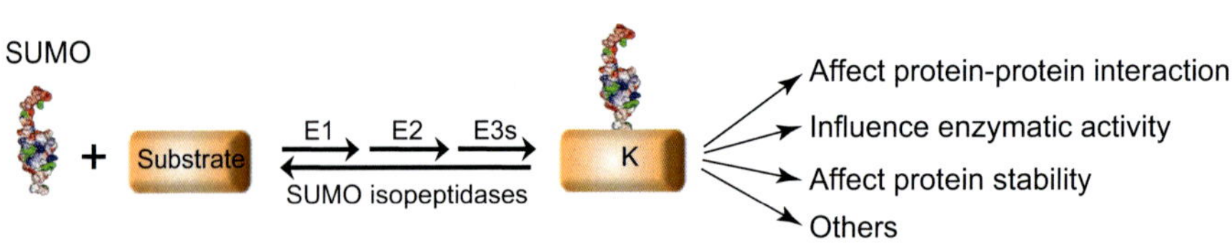

Fig. 17.2 The SUMO conjugation cycle and SUMO's effects on substrate proteins. SUMO E1, E2, and E3 enzymes can conjugate a SUMO molecule to a lysine residue of a substrate. Sumoylation can also occur on multiple lysines of a substrate or in the form of SUMO chain (not shown). Sumoylation can have several biological effects on the substrates, and three frequently observed molecular consequences are indicated. Sumoylation can be reversed by SUMO isopeptidases

Siz2 interacts with the ssDNA-binding protein RPA, promoting RPA sumoylation and Siz2 localization to DNA breaks (Chung and Zhao 2015). Mms21 is a part of the Smc5/Smc6 complex that localizes to several genomic loci to influence sumoylation events in these places (Murray and Carr 2008; De Piccoli et al. 2009; Hang et al. 2015; Bonner et al. 2016; Bermudez-Lopez et al. 2016). These interactions and the ability of E3s to enable productive alignment between E2 and substrates for SUMO transfer make SUMO E3s indispensible for sumoylation in vivo [reviewed in (Gareau and Lima 2010)]. Based on studies from several organisms, SUMO E3s often exhibit substrate redundancy, likely reflecting their similar SUMO transfer mechanisms (Hang et al. 2014; Sarangi et al. 2014).

Mammalian cells have larger numbers of SUMO E3s than yeast. At least ten E3s have been described in human cells thus far (Gareau and Lima 2010; Cappadocia et al. 2015; Eisenhardt et al. 2015). These can be divided into several groups, including (1) PIAS proteins that are homologs of the yeast Siz1 and Siz2, (2) the Mms21 homolog NSMCE2, (3) the ZNF451 type of SUMO E3s that utilize tandem SUMO-interacting motifs (SIMs) to enable sumoylation (Cappadocia et al. 2015; Eisenhardt et al. 2015), and (4) more specialized SUMO E3s that target specific processes, such as the nuclear pore protein Ran binding protein 2 (RanBP2), the polycomb group protein Pc2, and the promyelocytic leukemia (PML) protein. The increased numbers of SUMO E3s in human cells are associated with the presence of multiple SUMO isoforms (Gareau and Lima 2010; Liang et al. 2016). SUMO2 and SUMO3 have 97% sequence identity and can form SUMO chains, while the more divergent SUMO1 is less frequently found in SUMO chains [reviewed in (Gareau and Lima 2010)]. The acquirement of different SUMO isoforms and the many types of SUMO chains that they can form, in conjunction with the multiple types of SUMO E3s in human cells, can meet the needs of more complex genomes and increased demands for regulation.

17.2.2 Principles of the Desumoylation Process

Sumoylation can be reversed by multiple SUMO-specific cysteine proteases, known as desumoylation enzymes (Mukhopadhyay and Dasso 2007; Hickey et al. 2012) (Fig. 17.2). The substrate selectivity of desumoylation enzymes is partly achieved by their distinct localizations. Using budding yeast as an example, one of its desumoylation enzymes, Ulp1, primarily associates with nuclear pore complexes, whereas the other enzyme, Ulp2, can be seen concentrated in the nucleolus (Li and Hochstrasser 1999, 2000; Panse et al. 2003; Kroetz et al. 2009; Srikumar et al. 2013). Consistent with these localization patterns, Ulp1 and Ulp2 have different substrates (Makhnevych et al. 2009; de Albuquerque et al. 2016; Wei and Zhao 2016a). In addition, Ulp1 enables SUMO maturation by removing the tail of the precursor SUMO molecule (Li and Hochstrasser 1999), while Ulp2 has a major role in removing SUMO chains (Li and Hochstrasser 2000). Both Ulp1 and Ulp2 are required for cell fitness and resistance to a broad range of genotoxins (Li and

Hochstrasser 1999, 2000; Schwartz et al. 2007). As Ulp2 mutant defects are suppressed to a large degree by mutating the lysine residues on SUMO, which prevents SUMO chain formation, accumulation of SUMO chains is deleterious (Bylebyl et al. 2003). Human cells contain at least six desumoylation enzymes, called sentrin-/SUMO-specific proteases (SENPs) (Hannoun et al. 2010; Hickey et al. 2012). SENP1, 2, 3, and 5 are more related to Ulp1, whereas SENP6 and 7 are similar to Ulp2 [reviewed in (Mukhopadhyay and Dasso 2007)]. As is the case for Ulp1 and Ulp2 in yeast, SENPs have distinct activities and cellular localization patterns, and their mutants cause a wide range of defects [reviewed in (Mukhopadhyay and Dasso 2007; Hickey et al. 2012)].

17.2.3 Biochemical Effects of Sumoylation

Protein sumoylation affects a myriad of biological processes, such as transcription, nuclear transport, DNA metabolism, and protein quality control [reviewed in (Sarangi and Zhao 2015)]. The conjugation and removal of SUMO from proteins can alter protein-protein interactions, partly because SUMO modules can interact with SIMs on other proteins or the substrate itself. As such, SUMO-SIM interactions can promote the assembly of protein complexes and the formation of a membrane-free nuclear compartment [reviewed in (Shen et al. 2006)]. On the other hand, SUMO sometimes disrupts existing protein-protein interactions or protein aggregation, possibly due to steric hindrance posed by the SUMO moiety. Additionally, sumoylation can alter a protein's interaction with DNA or chromatin, its enzymatic activities, or its protein levels. Each of these effects has been observed in DNA metabolism processes, particularly in DNA repair. For example, sumoylation enhances inter-subunit association among the DNA helicase-topoisomerase Sgs1-Top3-Rmi1 complex and promotes its function in resolving Holliday junctions (Bermudez-Lopez et al. 2016; Bonner et al. 2016). On the other hand, sumoylation of the DNA nuclease cofactor Sae2 helps convert the protein from insoluble aggregates to a soluble form, which is required for DNA-end resection (Sarangi et al. 2015). In the case of the recombination mediator protein Rad52, sumoylation leads to association with the segregase Cdc48/p97 that removes proteins from DNA using its ATPase activity (Bergink et al. 2013). These examples illustrate some of the mechanisms by which sumoylation regulates genome maintenance. Recent studies have begun to reveal the roles of sumoylation during DNA replication, and the following sections summarize findings in this area.

17.3 SUMO-Based Regulation of Replication Initiation

Proteomic studies in multiple organisms have shown that protein factors that help replisomes cope with template obstacles are enriched among SUMO substrates (Golebiowski et al. 2009; Elrouby and Coupland 2010; Cremona et al. 2012; Hendriks et al. 2014; Ma et al. 2014; Tammsalu et al. 2014). Genetic and biochemical studies of individual SUMO substrates and SUMO enzymes have revealed some mechanisms of SUMO-based regulation of these proteins and how they affect DNA replication.

17.3.1 MCM Sumoylation Inhibits Replication Initiation in Budding Yeast

In budding yeast, all six subunits of MCM are sumoylated (Cremona et al. 2012; de Albuquerque et al. 2016; Wei and Zhao 2016a). MCM sumoylation has a distinct spatial and temporal pattern relative to the cycle of DNA replication (Wei and Zhao 2016a). Spatially, MCM is only sumoylated when loaded onto origins (Wei and Zhao 2016a). Temporally, Mcm2-6 sumoylation levels peak during G1 phase prior to DDK-mediated Mcm4 phosphorylation, then decline as cells enter S phase, and again increase during G2/M phase, coincident with the next MCM-loading cycle (Wei and Zhao 2016a). The opposing patterns of Mcm2-6 sumoylation and MCM phosphorylation during the cell cycle indicate a negative role of MCM sumoylation during replication initiation. Indeed, increased MCM sumoylation causes a reduction in the levels of Mcm4 phosphorylation, CMG, and origin firing (Wei and Zhao 2016a). These defects are partly because hyper-sumoylated MCM has increased association with the PP1 phosphatase, which reverses Mcm4 phosphorylation (Wei and Zhao 2016a; Davé et al. 2014; Hiraga et al. 2014; Mattarocci et al. 2014). MCM sumoylation levels subside at the start of S phase, partly through the action of Ulp2 (Wei and Zhao 2016a; de Albuquerque et al. 2016). These findings suggest that MCM sumoylation serves as a safeguard to prevent premature helicase function before S phase and that initiation of DNA synthesis requires removing this modification (Wei and Zhao 2016a). Future work is needed to examine whether the observed effects are due to a particular MCM subunit or contributions from multiple subunits and whether sumoylation alters other MCM features in addition to PP1 regulation.

17.3.2 MCM Sumoylation in Higher Eukaryotes

MCM sumoylation has also been detected in higher eukaryotes (Golebiowski et al. 2009; Elrouby and Coupland 2010; Hendriks et al. 2014; Ma et al. 2014; Schimmel et al. 2014; Tammsalu et al. 2014). For example, human MCM2, 3, 4, and 7 proteins are sumoylated. Importantly, Mcm4 sumoylation levels exhibit a similar pattern to that of yeast Mcm2-6 during the cell cycle, peaking in G1, declining in S phase, and increasing again during the subsequent G1 phase (Schimmel et al. 2014). It is reasonable to envision that human MCM sumoylation also provides a regulatory mechanism to restrain origin firing. A negative role for sumoylation in replication initiation can also be inferred from findings in *Xenopus*, wherein increased origin firing occurs after reducing sumoylation, either by expression of a dominant-negative SUMO E2 or addition of SUMO-specific proteases (Bonne-Andrea et al. 2013). Given *Xenopus* MCM subunits are sumoylated (Ma et al. 2014), it is worthy of consideration whether this modification underlies the negative effect of sumoylation in replication initiation in this system. Considering that PP1- and DDK-mediated MCM regulation is conserved across species (Wotton and Shore 1997; Lee et al. 2003; Cho et al. 2006; Masai et al. 2006; Montagnoli et al. 2006; Tsuji et al. 2006; Cornacchia et al. 2012; Hayano et al. 2012; Yamazaki et al. 2012), the targeting of this pathway by MCM sumoylation to prevent premature initiation, as seen in yeast, may be conserved. Direct tests of these ideas will clarify the roles of MCM sumoylation in higher eukaryotes.

17.3.3 ORC2 Sumoylation Prevents Re-replication at Centromeric Regions

ORC, composed of ORC1–6 subunits, binds to replication origins and is critical for MCM loading during origin licensing (Diffley et al. 1994; Kelly and Brown 2000; Bell and Dutta 2002; Sclafani and Holzen 2007; Remus and Diffley 2009; Li and Araki 2013). In addition, ORC2 can dissociate from replication origins and localize to centromeric regions during G2/M phase (Craig et al. 2003; Prasanth et al. 2004; Lee et al. 2012), coincident with the appearance of SUMO2-modified ORC2 in human cells (Huang et al. 2016). Elimination of ORC2 sumoylation by mutating its two sumoylation sites leads to re-replication, polyploidy, and genome damage (Huang et al. 2016). Mechanistically, ORC2 sumoylation promotes the recruitment of KDM5A (Huang et al. 2016), a histone H3 lysine 4 (H3K4) demethylase (Defeo-Jones et al. 1991; Christensen et al. 2007; Klose et al. 2007). Loss of ORC2 sumoylation results in elevated levels of tri-methylated H3K4 (H3K4me3) in centromeric chromatin, reduced transcription of α-satellites at centromeres, and decondensation of pericentric heterochromatin, which correlates with re-replication of the pericentric region (Huang et al. 2016). It remains to be determined how the change in chromatin environment caused by the perturbation of the ORC2-KMD5A axis leads to re-replication, despite the presence of multiple mechanisms that prevent re-replication. In budding yeast, multiple ORC subunits are sumoylated (Cremona et al. 2012), but the functions of this modification have yet to be determined.

17.3.4 *Other Potential SUMO Substrates Affecting Replication Initiation*

Several other proteins involved in replication initiation are SUMO substrates, such as the ssDNA-binding protein RPA and CDK (Dou et al. 2010; Cremona et al. 2012; Bonne-Andrea et al. 2013). Sumoylation of both human and yeast RPA has been reported to promote homologous recombination during DNA repair, though whether it also has a role in replication initiation has not been tested (Dou et al. 2010; Psakhye and Jentsch 2012). Cyclin E has been shown to be a SUMO2/SUMO3 substrate in *Xenopus* (Bonne-Andrea et al. 2013). Its sumoylation is detectable during replication and is independent of its kinase activity (Bonne-Andrea et al. 2013). A direct effect of cyclin E sumoylation in replication initiation remains to be determined. Because multiple proteins involved in origin licensing and firing are SUMO substrates, we anticipate the presence of multiple mechanisms through which SUMO regulates timing and efficiency of origin firing and prevents harmful re-replication events.

17.4 SUMO-Based Regulation of Replication Progression

Genetic studies using SUMO E2 and E3 mutants have shown that reducing sumoylation retards replication progression, particularly under replicative stress (Cremona et al. 2012; Schimmel et al. 2014; Hang et al. 2015). For example, mutating SUMO E3s in budding yeast impairs replication when cells are treated with the DNA-alkylating agent methyl methanesulfonate (MMS) (Cremona et al. 2012). In human cell lines, reducing UBC9 function leads to a prolonged S phase (Schimmel et al. 2014). The effects of sumoylation in replication progression may be broad, as many proteins central for this process are subject to sumoylation. Aside from MCM, several other replisome components are sumoylated, including Pol ε, Pol δ, the processivity factor PCNA, topoisomerases, DNA primase, and the nucleosome remodeling factor FACT (Golebiowski et al. 2009; Elrouby and Coupland 2010; Cremona et al. 2012; Hendriks et al. 2014; Ma et al. 2014; Tammsalu et al. 2014). In addition, several proteins that collaborate with replisome for DNA synthesis are sumoylated (Golebiowski et al. 2009; Elrouby and Coupland 2010; Cremona et al. 2012; Hendriks et al. 2014; Ma et al. 2014; Tammsalu et al. 2014). Some examples include subunits of the SMC complexes (cohesin, condensin, and Smc5/Smc6), the SMC-like Mre11 complex, and the clamp loader RFC complex (Golebiowski et al.

2009; Elrouby and Coupland 2010; Cremona et al. 2012; Hendriks et al. 2014; Ma et al. 2014; Tammsalu et al. 2014). Among these proteins, the sumoylation of PCNA has been well examined and shown to recruit the anti-recombinase Srs2 to impaired replication forks in order to prevent toxic recombination events (Papouli et al. 2005; Armstrong et al. 2012). The molecular function of this modification, in conjunction with other types of PCNA modifications, has been extensively reviewed elsewhere [reviewed in (Mailand et al. 2013; Ulrich and Takahashi 2013)]. Below we focus on recent findings regarding additional effects SUMO has on replication progression, mostly derived from studying the combined effects of loss of sumoylation of many substrates, but with a few mechanistic studies as well.

17.4.1 SUMO-Based Regulation of Replisome Components

Following up earlier observations that Mms21 SUMO ligase mutations impair replication, our group showed that under MMS conditions, Mms21 and the associated Smc5/Smc6 complex promote sumoylation of Mcm6 and Pol2, the catalytic subunit of Pol ε (Hang et al. 2015). As physical interactions are detected between the Smc5/Smc6 complex and these substrates, the effects seen are likely to be direct (Hang et al. 2015). In addition, as Smc5/Smc6 deficiency impairs replication at regions far from fired origins, the Smc5/Smc6 complex may use sumoylation to facilitate later stages of replication. Future tests can help us understand whether MCM sumoylation has a distinct role during replication stress and whether Pol2 sumoylation influences DNA polymerization.

Several replisome members have been found to be sumoylated in human cells (Golebiowski et al. 2009; Hendriks et al. 2014; Schimmel et al. 2014; Tammsalu et al. 2014). Interestingly, under ATR inhibition conditions, sumoylation of replisome components can lead to fork collapse (Ragland et al. 2013). This largely depends on the SUMO-targeted ubiquitin ligase RNF4 and endonuclease scaffold protein SLX4 (Ragland et al. 2013). It is thought that under such conditions, RNF4 can target sumoylated replisomes for degradation, rendering replication forks accessible for SLX4-mediated cleavage (Ragland et al. 2013). In another study, sumoylation of two Fanconi anemia (FA) proteins, namely, FANCD2 and FANCI, was shown to trigger RNF4-mediated ubiquitination of these proteins and subsequent removal from DNA damage sites (Gibbs-Seymour et al. 2015). Lack of this regulation reduces the ability of cells to cope with replication stress, likely due to blockage of FANCD2 and FANCI from recycling among different damage sites (Gibbs-Seymour et al. 2015). Both of these studies highlight the complex interplay between sumoylation and ubiquitination for replication fork management.

The above notion is further extended by another study utilizing iPOND (isolation of proteins on nascent DNA) coupled with mass spectrometry-based protein analyses in human cells. It has been reported that SUMO is enriched on nascent chromatin, while ubiquitin molecules are enriched on mature chromatin (Lopez-Contreras et al. 2013). A follow-up study showed that the SUMO deubiquitinase, USP7, contributes to the establishment of this SUMO-high and ubiquitin-low nascent chromatin environment (Lecona et al. 2016). USP7 can remove ubiquitin molecules that are conjugated to SUMO2 *in vitro* and *in vivo* and is associated with nascent chromatin and MCM4 (Lecona et al. 2016). The authors suggest that through this mechanism, USP7 reduces ubiquitin levels and allows enrichment of SUMO2 on replisome components. This role likely contributes to USP7's function in maintaining normal rates of fork speed and origin firing (Lecona et al. 2016). These studies suggest that enrichment of SUMO and reduction of ubiquitin at or near replisomes can be advantageous for replication progression. Further investigation is needed to provide mechanistic insights into the relevant sumoylation events critical for DNA synthesis and the full spectrum of the effects of USP7 and RNF4 in keeping a balance between sumoylation and ubiquitination of replisome components under different conditions.

17.4.2 SUMO-Based Regulation of Lagging Strand Synthesis

Lagging strand Okazaki fragment processing involves sequential reactions of gap filling by polymerase δ, flap cleavage by flap endonuclease 1 (FEN1), and nick ligation by DNA ligase 1 (LIG1) (Waga and Stillman 1998). In human cells, modification of FEN1 by SUMO3 at lysine 168 occurs in S phase and peaks in G2/M phase (Guo et al. 2012). The sumoylation of FEN1 is dependent on its phosphorylation, which occurs during G1 phase, peaking in late S phase. Sumoylation of FEN1 appears to occur after its phosphorylation, as the non-phosphorylatable FEN1 mutant (S187A) is not sumoylated (Guo et al. 2012). Furthermore, sumoylation of FEN1 triggers its ubiquitination and subsequent proteasome-mediated degradation, presumably by recruiting the SIM-containing ubiquitin E3 ligase PRP19 (Guo et al. 2012). The timely degradation of FEN1 via this cascade of modifications is critical for maintaining genome stability, as its deregulation leads to cell cycle delay and polyploidy (Guo et al. 2012). Other than FEN1, polymerase δ is also reported to be SUMO substrate in organisms from yeast to humans (Cremona et al. 2012; Hendriks et al. 2014; Tammsalu et al. 2014). It remains to be determined if polymerase δ sumoylation has a role in DNA lagging strand synthesis.

17.4.3 SUMO-Based Regulation of Sister Chromatid Cohesion

As replication progresses, the two sister chromatids stay close together partly through the function of cohesin [reviewed in (Blow and Tanaka 2005; Sherwood et al. 2010)]. The ring-shaped cohesin complex is loaded onto chromatin before S phase and encloses sister chromatids to keep them connected during S phase [reviewed in (Blow and Tanaka 2005; Sherwood et al. 2010)]. The resulting sister chromatid cohesion is important for supplying faithful templates for DNA repair and for ensuring each daughter cell receives one set of chromosomes during mitosis [reviewed in (Blow and Tanaka 2005; Sherwood et al. 2010)]. Four subunits of cohesin, including the rod-shaped Smc1 and Smc3 proteins and the associated Scc1 and Scc3 proteins, are sumoylated in yeast and human cells, partly in an Mms21-dependent manner (Denison et al. 2005; Potts et al. 2006; Almedawar et al. 2012; Golebiowski et al. 2009; Hendriks et al. 2014; Tammsalu et al. 2014). In budding yeast, sumoylation of Smc1 and Scc1 occurs after cohesin loading, and Scc1 sumoylation is independent of another important modification for cohesion establishment, namely, Smc3 acetylation (Almedawar et al. 2012). Decreasing cohesin sumoylation impairs cohesion, without affecting the integrity of the cohesin complex (Almedawar et al. 2012). Based on these observations, it was proposed that cohesin sumoylation is required for the establishment of cohesion.

Another study further highlights the importance of SUMO-based regulation of cohesion. When a cohesion establishment factor, Pds5, is mutated in budding yeast, Scc1 becomes hyper-sumoylated and is degraded (D'Ambrosio and Lavoie 2014). This correlates with precocious separation of sister chromatids (D'Ambrosio and Lavoie 2014). Such a defect can be suppressed through removal of the SUMO E3 ligase Siz2 or the SUMO-targeted ubiquitin ligase Slx5/Slx8, supporting the hypothesis that toxic sumoylation events underlie *pds5-1* defects (D'Ambrosio and Lavoie 2014). The authors suggest that Pds5 plays a role in preventing hyper-sumoylation of cohesin and maintaining cohesin levels until mitosis. The full picture of how SUMO regulates cohesion is likely more complex. For example, Pds5 itself is sumoylated by Siz2 from S to G2/M phase, though the biochemical effect of this modification is not known (Stead et al. 2003). The timing of sumoylation and ubiquitination of cohesin is critical, and it will be of interest to determine how their temporal order is established, perhaps through the regulation of sumoylation and desumoylation enzymes or through other cohesion regulators.

17.4.4 SUMO-Based Regulation of Topoisomerase Function

Topoisomerases are essential for releasing topological stress and promoting replication fork progression (Champoux 2001; Wang 2002; Vos et al. 2011). They also contribute to the removal of transcription-generated DNA-RNA hybrids, known as R-loops, which pose a barrier for replication (Gan et al. 2011; Aguilera and Garcia-Muse 2012). A recent study showed that sumoylation of human TOP1 provides a means to reduce R-loop-mediated replication fork stalling via two distinct mechanisms (Li et al. 2015). Upon sumoylation by PIAS1, TOP1 showed improved interactions with the active form of RNA polymerase II (RNAPIIo), leading to recruitment of splicing factors to avoid R-loop formation (Li et al. 2015). In addition, sumoylation of TOP1 reduces its enzymatic activity, potentially leading to reduced TOP1-induced DNA nicking at transcriptionally active regions (Li et al. 2015). Both effects of TOP1 sumoylation could contribute to lessening barriers for replication forks. It remains to be tested if these effects of TOP1 sumoylation are conserved in other organisms.

17.5 Potential SUMO Substrates Affecting Replication Termination

Although a full understanding of replication termination is still elusive, several new discoveries have shed light into this process. In budding yeast, *C. elegans*, and *Xenopus*, ubiquitination of Mcm7 has been shown to be a key event for disassembly of the replisome. In budding yeast, the Dia2 ubiquitin ligase that is part of the replisome can ubiquitinate Mcm7 when replication forks converge. This modification is then recognized by the segregase Cdc48/p97, leading to removal of MCM from chromatin and replisome disassembly (Maric et al. 2014; Priego Moreno et al. 2014). In *C. elegans* and *Xenopus,* replisome-associated E3 ligase $CUL\text{-}2^{LRR\text{-}1}$ and the segregase remove CMG during termination (Sonneville et al. 2017; Dewar et al. 2017). Interestingly, in budding yeast, Mcm7 sumoylation appears to be regulated distinctly from that of Mcm2-6, with its levels only declining when the bulk of DNA replication has been completed (Wei and Zhao 2016a). It will be interesting to investigate whether Mcm7 sumoylation could trigger its ubiqutination or contribute to replication termination.

During replication termination, decatenation of sister chromatids requires Top2. Top2 sumoylation has been found in human, mouse, *Xenopus*, and yeast [reviewed in (Lee and Bachant 2009)]. Sumoylation of Top2 in vertebrates promotes the recruitment of Top2 or the chromosomal passenger complex to kinetochores during mitosis to facilitate chromosome segregation [reviewed in (Lee and Bachant 2009)]. Whether Top2 sumoylation plays a role in decatenation during replication termination is not known. With more molecular details of replication termination becoming available in the future, the examination of SUMO substrates involved in replication termination will reveal more details of this process.

17.6 Concluding Remarks

Each stage of DNA replication is intricately regulated to ensure precise genome duplication. Posttranslational modifications provide a dynamic regulatory means at multiple stages of the replication process. Phosphorylation- and ubiquitination-based modes of regulation are essential for replication, and the role of sumoylation in replication is emerging from several recent studies. During the replication initiation stage, sumoylation of MCM (yeast) and ORC2 (human) can influence origin firing. In addition, sumoylation promotes replication progression through multiple mechanisms, such as lagging strand synthesis, reducing R-loops, replication fork metabolism, and sister chromatid cohesion. However, only a small number of SUMO substrates have been studied thus far as summarized in Table 17.1, and our understanding of how sumoylation regulates replication is still at an early stage. With more advanced methods to map sumoylation sites and tools to alter the sumoylation status of substrates, detailed molecular mechanisms of how sumoylation

Table 17.1 Summary of sumoylated substrates involved in DNA replication

Replication	Substrate (function)	Molecular effect(s) of sumoylation
Initiation	*sc*MCM (replicative helicase)	Inhibit replication initiation via phosphatase recruitment
	*hs*ORC2 (replication origin recognition)	Prevent centromeric region re-replication by recruitment of the histone demethylase KDM5A
Progression	*hs*FANCD2 and *hs*FANCI (Fanconi anemia proteins)	Promote replication stress survival by triggering self-ubiquitination and removal from DNA damage sites
	*hs*FEN1 (flap endonuclease)	Promote genome stability, prerequisite for its ubiquitination and degradation
	*sc*Cohesin subunits (chromatid cohesion)	Promote sister chromatid cohesion
	*hs*TOP1 (topoisomerase)	Prevent R-loop formation by promoting interaction with splicing factors; reduce DNA nicking

Note that only the major function is indicated for each substrate
sc *Saccharomyces cerevisiae*, hs *Homo sapiens*

regulates each substrate will be elucidated. Future work will also help to establish a clear picture of how sumoylation is coordinated with other types of protein modifications during DNA replication. In addition, examination of how sumoylation and desumoylation enzymes are themselves regulated can also reveal how SUMO modification cycles facilitate DNA replication. As SUMO enzyme deficiencies, such as SUMO E1 and E2 depletion or SUMO E3 mutations, have been implicated in cancer and inherited human syndromes (Eifler and Vertegaal 2015; He et al. 2015; Yu et al. 2015), understanding their roles in genome duplication will provide new avenues for disease detection and treatment strategies.

References

Aguilera A, Garcia-Muse T (2012) R loops: from transcription byproducts to threats to genome stability. Mol Cell 46:115–124

Almedawar S, Colomina N, Bermudez-Lopez M, Pocino-Merino I, Torres-Rosell J (2012) A SUMO-dependent step during establishment of sister chromatid cohesion. Curr Biol 22:1576–1581

Almouzni G, Cedar H (2016) Maintenance of epigenetic information. Cold Spring Harb Perspect Biol 2:8

Arias EE, Walter JC (2007) Strength in numbers: preventing rereplication via multiple mechanisms in eukaryotic cells. Genes Dev 21:497–518

Armstrong AA, Mohideen F, Lima CD (2012) Recognition of SUMO-modified PCNA requires tandem receptor motifs in Srs2. Nature 483:59–63

Azvolinsky A, Dunaway S, Torres JZ, Bessler JB, Zakian VA (2006) The *S. cerevisiae* Rrm3p DNA helicase moves with the replication fork and affects replication of all yeast chromosomes. Genes Dev 20:3104–3116

Bell SP, Dutta A (2002) DNA replication in eukaryotic cells. Annu Rev Biochem 71:333–374

Bell SP, Labib K (2016) Chromosome duplication in *Saccharomyces cerevisiae*. Genetics 203:1027–1067

Bergink S, Ammon T, Kern M, Schermelleh L, Leonhardt H, Jentsch S (2013) Role of Cdc48/p97 as a SUMO-targeted segregase curbing Rad51-Rad52 interaction. Nat Cell Biol 15:526–532

Bermudez-Lopez M, Villoria MT, Esteras M, Jarmuz A, Torres-Rosell J, Clemente-Blanco A, Aragon L (2016) Sgs1's roles in DNA end resection, HJ dissolution, and crossover suppression require a two-step SUMO regulation dependent on Smc5/6. Genes Dev 30:1339–1356

Blow JJ, Dutta A (2005) Preventing re-replication of chromosomal DNA. Nat Rev Mol Cell Biol 6:476–486

Blow JJ, Tanaka TU (2005) The chromosome cycle: coordinating replication and segregation. Second in the cycles review series. EMBO Rep 6:1028–1034

Bonne-Andrea C, Kahli M, Mechali F, Lemaitre J-M, Bossis G, Coux O (2013) SUMO2/3 modification of cyclin E contributes to the control of replication origin firing. Nat Commun 4:1850

Bonner JN, Choi K, Xue X, Torres NP, Szakal B, Wei L, Wan B, Arter M, Matos J, Sung P et al (2016) Smc5/6 mediated sumoylation of the Sgs1-Top3-Rmi1 complex promotes removal of recombination intermediates. Cell Rep 16:368–378

Bowers JL, Randell JCW, Chen S, Bell SP (2004) ATP hydrolysis by ORC catalyzes reiterative Mcm2-7 assembly at a defined origin of replication. Mol Cell 16:967–978

Bylebyl GR, Belichenko I, Johnson ES (2003) The SUMO isopeptidase Ulp2 prevents accumulation of SUMO chains in yeast. J Biol Chem 278:44113–44120

Calzada A, Hodgson B, Kanemaki M, Bueno A, Labib K (2005) Molecular anatomy and regulation of a stable replisome at a paused eukaryotic DNA replication fork. Genes Dev 19:1905–1919

Cappadocia L, Pichler A, Lima CD (2015) Structural basis for catalytic activation by the human ZNF451 SUMO E3 ligase. Nat Struct Mol Biol 22:968–975

Champoux JJ (2001) DNA topoisomerases: structure, function, and mechanism. Annu Rev Biochem 70:369–413

Cho W-H, Lee Y-J, Kong S-I, Hurwitz J, Lee J-K (2006) CDC7 kinase phosphorylates serine residues adjacent to acidic amino acids in the minichromosome maintenance 2 protein. Proc Natl Acad Sci U S A 103:11521–11526

Christensen J, Agger K, Cloos PA, Pasini D, Rose S, Sennels L, Rappsilber J, Hansen KH, Salcini AE, Helin K (2007) RBP2 belongs to a family of demethylases, specific for tri-and dimethylated lysine 4 on histone 3. Cell 128:1063–1076

Chung I, Zhao X (2015) DNA break-induced sumoylation is enabled by collaboration between a SUMO ligase and the ssDNA-binding complex RPA. Genes Dev 29:1593–1598

Cornacchia D, Dileep V, Quivy JP, Foti R, Tili F, Santarella-Mellwig R, Antony C, Almouzni G, Gilbert DM, Buonomo SBC (2012) Mouse Rif1 is a key regulator of the replication-timing programme in mammalian cells. EMBO J 31:3678–3690

Coster G, Frigola J, Beuron F, Morris EP, Diffley JF (2014) Origin licensing requires ATP binding and hydrolysis by the MCM replicative helicase. Mol Cell 55:666–677

Craig JM, Earle E, Canham P, Wong LH, Anderson M, Choo KH (2003) Analysis of mammalian proteins involved in chromatin modification reveals new metaphase centromeric proteins and distinct chromosomal distribution patterns. Hum Mol Genet 12:3109–3121

Cremona CA, Sarangi P, Yang Y, Hang LE, Rahman S, Zhao X (2012) Extensive DNA damage-induced sumoylation contributes to replication and repair and acts in addition to the mec1 checkpoint. Mol Cell 45:422–432

D'Ambrosio LM, Lavoie BD (2014) Pds5 prevents the PolySUMO-dependent separation of sister chromatids. Curr Biol 24:361–371

Davé A, Cooley C, Garg M, Bianchi A (2014) Protein phosphatase 1 recruitment by Rif1 regulates DNA replication origin firing by counteracting DDK activity. Cell Rep 7:53–61

de Albuquerque CP, Liang J, Gaut NJ, Zhou H (2016) Molecular circuitry of the SUMO (Small Ubiquitin-like Modifier) pathway in controlling Sumoylation homeostasis and suppressing genome rearrangements. J Biol Chem 291:8825–8835

De Piccoli G, Torres-Rosell J, Aragón L (2009) The unnamed complex: what do we know about Smc5-Smc6? Chromosom Res 17:251–263
Deegan TD, Yeeles JT, Diffley JF (2016) Phosphopeptide binding by Sld3 links Dbf4-dependent kinase to MCM replicative helicase activation. EMBO J 35:961–973
Defeo-Jones D, Huang PS, Jones RE, Haskell KM, Vuocolo GA, Hanobik MG, Huber HE, Oliff A (1991) Cloning of cDNAs for cellular proteins that bind to the retinoblastoma gene product. Nature 352:251–254
Denison C, Rudner AD, Gerber SA, Bakalarski CE, Moazed D, Gygi SP (2005) A proteomic strategy for gaining insights into protein sumoylation in yeast. Mol Cell Proteomics 4:246–254
Dewar JM, Budzowska M, Walter JC (2015) The mechanism of DNA replication termination in vertebrates. Nature 525:345–350
Dewar JM, Low E, Mann M, Räschle M, Walter JC (2017) CRL2Lrr1 promotes unloading of the vertebrate replisome from chromatin during replication termination. Genes Dev 31(3):275–290
Diffley JFX (2004) Regulation of early events in chromosome replication. Curr Biol 14:R778–R786
Diffley JF, Cocker JH, Dowell SJ, Rowley A (1994) Two steps in the assembly of complexes at yeast replication origins in vivo. Cell 78:303–316
Donovan S, Harwood J, Drury LS, Diffley JF (1997) Cdc6p-dependent loading of Mcm proteins onto pre-replicative chromatin in budding yeast. Proc Natl Acad Sci U S A 94:5611–5616
Dou H, Huang C, Singh M, Carpenter PB, Yeh ET (2010) Regulation of DNA repair through deSUMOylation and SUMOylation of replication protein A complex. Mol Cell 39:333–345
Eifler K, Vertegaal AC (2015) SUMOylation-mediated regulation of cell cycle progression and cancer. Trends Biochem Sci 40:779–793
Eisenhardt N, Chaugule VK, Koidl S, Droescher M, Dogan E, Rettich J, Sutinen P, Imanishi SY, Hofmann K, Palvimo JJ et al (2015) A new vertebrate SUMO enzyme family reveals insights into SUMO-chain assembly. Nat Struct Mol Biol 22:959–967
Elrouby N, Coupland G (2010) Proteome-wide screens for small ubiquitin-like modifier (SUMO) substrates identify Arabidopsis proteins implicated in diverse biological processes. Proc Natl Acad Sci U S A 107:17415–17420
Fachinetti D, Bermejo R, Cocito A, Minardi S, Katou Y, Kanoh Y, Shirahige K, Azvolinsky A, Zakian VA, Foiani M (2010) Replication termination at eukaryotic chromosomes is mediated by Top2 and occurs at genomic loci containing pausing elements. Mol Cell 39:595–605
Fernández-Cid A, Riera A, Tognetti S, Herrera MC, Samel S, Evrin C, Winkler C, Gardenal E, Uhle S, Speck C (2013) An ORC/Cdc6/MCM2-7 complex is formed in a multistep reaction to serve as a platform for MCM double-hexamer assembly. Mol Cell 50:577–588
Fragkos M, Ganier O, Coulombe P, Mechali M (2015) DNA replication origin activation in space and time. Nat Rev Mol Cell Biol 16:360–374
Frigola J, Remus D, Mehanna A, Diffley JFX (2013) ATPase-dependent quality control of DNA replication origin licensing. Nature 495:339–343
Gambus A, Jones RC, Sanchez-Diaz A, Kanemaki M, van Deursen F, Edmondson RD, Labib K (2006) GINS maintains association of Cdc45 with MCM in replisome progression complexes at eukaryotic DNA replication forks. Nat Cell Biol 8:358–366
Gambus A, van Deursen F, Polychronopoulos D, Foltman M, Jones RC, Edmondson RD, Calzada A, Labib K (2009) A key role for Ctf4 in coupling the MCM2-7 helicase to DNA polymerase alpha within the eukaryotic replisome. EMBO J 28:2992–3004
Gan W, Guan Z, Liu J, Gui T, Shen K, Manley JL, Li X (2011) R-loop-mediated genomic instability is caused by impairment of replication fork progression. Genes Dev 25:2041–2056
Garcia-Rodriguez N, Wong RP, Ulrich HD (2016) Functions of ubiquitin and SUMO in DNA replication and replication stress. Front Genet 7:87
Gareau JR, Lima CD (2010) The SUMO pathway: emerging mechanisms that shape specificity, conjugation and recognition. Nat Rev Mol Cell Biol 11:861–871
Gibbs-Seymour I, Oka Y, Rajendra E, Weinert BT, Passmore LA, Patel KJ, Olsen JV, Choudhary C, Bekker-Jensen S, Mailand N (2015) Ubiquitin-SUMO circuitry controls activated fanconi anemia ID complex dosage in response to DNA damage. Mol Cell 57:150–164

Golebiowski F, Matic I, Tatham MH, Cole C, Yin Y, Nakamura A, Cox J, Barton GJ, Mann M, Hay RT (2009) System-wide changes to SUMO modifications in response to heat shock. Sci Signals 2:ra24

Guo Z, Kanjanapangka J, Liu N, Liu S, Liu C, Wu Z, Wang Y, Loh T, Kowolik C, Jamsen J et al (2012) Sequential posttranslational modifications program FEN1 degradation during cell-cycle progression. Mol Cell 47:444–456

Gutiérrez MP, MacAlpine DM (2016) The initiation of DNA replication in eukaryotes. In: Kaplan DL (ed) Chapter 5: chromatin determinants of origin selection and activation. Springer International Publishing, ISBN 978-3-319-24696-3

Hang LE, Lopez CR, Liu X, Williams JM, Chung I, Wei L, Bertuch AA, Zhao X (2014) Regulation of Ku-DNA association by Yku70 C-terminal tail and SUMO modification. J Biol Chem 289:10308–10317

Hang LE, Peng J, Tan W, Szakal B, Menolfi D, Sheng Z, Lobachev K, Branzei D, Feng W, Zhao X (2015) Rtt107 is a multi-functional scaffold supporting replication progression with partner SUMO and ubiquitin ligases. Mol Cell 60:268–279

Hannoun Z, Greenhough S, Jaffray E, Hay RT, Hay DC (2010) Post-translational modification by SUMO. Toxicology 278:288–293

Hayano M, Kanoh Y, Matsumoto S, Renard-Guillet C, Shirahige K, Masai H (2012) Rif1 is a global regulator of timing of replication origin firing in fission yeast. Genes Dev 26:137–150

He X, Riceberg J, Pulukuri SM, Grossman S, Shinde V, Shah P, Brownell JE, Dick L, Newcomb J, Bence N (2015) Characterization of the loss of SUMO pathway function on cancer cells and tumor proliferation. PLoS One 10:e0123882

Hendriks IA, D'Souza RC, Yang B, Verlaan-de Vries M, Mann M, Vertegaal AC (2014) Uncovering global SUMOylation signaling networks in a site-specific manner. Nat Struct Mol Biol 21:927–936

Hickey CM, Wilson NR, Hochstrasser M (2012) Function and regulation of SUMO proteases. Nat Rev Mol Cell Biol 13:755–766

Hiraga S-I, Alvino GM, Chang F, Lian H-Y, Sridhar A, Kubota T, BREWER BJ, Weinreich M, Raghuraman MK, Donaldson AD (2014) Rif1 controls DNA replication by directing protein phosphatase 1 to reverse Cdc7-mediated phosphorylation of the MCM complex. Genes Dev 28:372–383

Hoggard T, Fox C (2016) The initiation of DNA replication in eukaryotes. In: Kaplan DL (ed) Chapter 9: the origin recognition complex in the initiation of DNA replication. Springer International Publishing, ISBN 978-3-319-24696-3

Huang C, Cheng J, Bawa-Khalfe T, Yao X, Chin YE, Yeh ET (2016) SUMOylated ORC2 recruits a histone demethylase to regulate centromeric histone modification and genomic stability. Cell Rep 15:147–157

Ilves I, Petojevic T, Pesavento JJ, Botchan MR (2010) Activation of the MCM2-7 helicase by association with Cdc45 and GINS proteins. Mol Cell 37:247–258

Ivessa AS, Zhou JQ, Zakian VA (2000) The Saccharomyces Pif1p DNA helicase and the highly related Rrm3p have opposite effects on replication fork progression in ribosomal DNA. Cell 100:479–489

Jeppsson K, Kanno T, Shirahige K, Sjogren C (2014) The maintenance of chromosome structure: positioning and functioning of SMC complexes. Nat Rev Mol Cell Biol 15:601–614

Johnson ES (2004) Protein modification by SUMO. Annu Rev Biochem 73:355–382

Johnson ES, Gupta AA (2001) An E3-like factor that promotes SUMO conjugation to the yeast septins. Cell 106:735–744

Kang YH, Galal WC, Farina A, Tappin I, Hurwitz J (2012) Properties of the human Cdc45/Mcm2-7/GINS helicase complex and its action with DNA polymerase epsilon in rolling circle DNA synthesis. Proc Natl Acad Sci U S A 109:6042–6047

Kelly TJ, Brown GW (2000) Regulation of chromosome replication. Annu Rev Biochem 69:829–880

Klose RJ, Yan Q, Tothova Z, Yamane K, Erdjument-Bromage H, Tempst P, Gilliland DG, Zhang Y, Kaelin WG Jr (2007) The retinoblastoma binding protein RBP2 is an H3K4 demethylase. Cell 128:889–900

Kroetz MB, Su D, Hochstrasser M (2009) Essential role of nuclear localization for yeast Ulp2 SUMO protease function. Mol Biol Cell 20:2196–2206

Labib K, Hodgson B (2007) Replication fork barriers: pausing for a break or stalling for time? EMBO Rep 8:346–353

Lamoliatte F, Caron D, Durette C, Mahrouche L, Maroui MA, Caron-Lizotte O, Bonneil E, Chelbi-Alix MK, Thibault P (2014) Large-scale analysis of lysine SUMOylation by SUMO remnant immunoaffinity profiling. Nat Commun 5:5409

Lecona E, Rodriguez-Acebes S, Specks J, Lopez-Contreras AJ, Ruppen I, Murga M, Munoz J, Mendez J, Fernandez-Capetillo O (2016) USP7 is a SUMO deubiquitinase essential for DNA replication. Nat Struct Mol Biol 23:270–277

Lee M-T, Bachant J (2009) SUMO modification of DNA topoisomerase II: trying to get a CENse of it all. DNA Repair 8:557–568

Lee J-K, Seo Y-S, Hurwitz J (2003) The Cdc23 (Mcm10) protein is required for the phosphorylation of minichromosome maintenance complex by the Dfp1-Hsk1 kinase. Proc Natl Acad Sci U S A 100:2334–2339

Lee KY, Bang SW, Yoon SW, Lee SH, Yoon JB, Hwang DS (2012) Phosphorylation of ORC2 protein dissociates origin recognition complex from chromatin and replication origins. J Biol Chem 287:11891–11898

Li Y, Araki H (2013) Loading and activation of DNA replicative helicases: the key step of initiation of DNA replication. Genes Cells 18:266–277

Li SJ, Hochstrasser M (1999) A new protease required for cell-cycle progression in yeast. Nature 398:246–251

Li SJ, Hochstrasser M (2000) The yeast ULP2 (SMT4) gene encodes a novel protease specific for the ubiquitin-like Smt3 protein. Mol Cell Biol 20:2367–2377

Li M, Pokharel S, Wang JT, Xu X, Liu Y (2015) RECQ5-dependent SUMOylation of DNA topoisomerase I prevents transcription-associated genome instability. Nat Commun 6:6720

Liang YC, Lee CC, Yao YL, Lai CC, Schmitz ML, Yang WM (2016) SUMO5, a novel poly-SUMO isoform, regulates PML nuclear bodies. Sci Rep 6:26509

Lopez-Contreras AJ, Ruppen I, Nieto-Soler M, Murga M, Rodriguez-Acebes S, Remeseiro S, Rodrigo-Perez S, Rojas AM, Méndez J, Muñoz J et al (2013) A proteomic characterization of factors enriched at nascent DNA molecules. Cell Rep 3:1105–1116

Ma L, Aslanian A, Sun H, Jin M, Shi Y, Yates JR 3rd, Hunter T (2014) Identification of small ubiquitin-like modifier substrates with diverse functions using the *Xenopus* egg extract system. Mol Cell Proteomics 13:1659–1675

Mailand N, Gibbs-Seymour I, Bekker-Jensen S (2013) Regulation of PCNA-protein interactions for genome stability. Nat Rev Mol Cell Biol 14:269–282

Makhnevych T, Sydorskyy Y, Xin X, Srikumar T, Vizeacoumar FJ, Jeram SM, Li Z, Bahr S, Andrews BJ, Boone C et al (2009) Global map of SUMO function revealed by protein-protein interaction and genetic networks. Mol Cell 33:124–135

Mantiero D, Mackenzie A, Donaldson A, Zegerman P (2011) Limiting replication initiation factors execute the temporal programme of origin firing in budding yeast. EMBO J 30:4805–4814

Maric M, Maculins T, De Piccoli G, Labib K (2014) Cdc48 and a ubiquitin ligase drive disassembly of the CMG helicase at the end of DNA replication. Science 346:1253596

Masai H, Taniyama C, Ogino K, Matsui E, Kakusho N, Matsumoto S, Kim J-M, Ishii A, Tanaka T, Kobayashi T et al (2006) Phosphorylation of MCM4 by Cdc7 kinase facilitates its interaction with Cdc45 on the chromatin. J Biol Chem 281:39249–39261

Masai H, Matsumoto S, You Z, Yoshizawa-Sugata N, Oda M (2010) Eukaryotic chromosome DNA replication: where, when, and how? Annu Rev Biochem 79:89–130

Mattarocci S, Shyian M, Lemmens L, Damay P, Altintas DM, Shi T, Bartholomew CR, Thomä NH, Hardy CFJ, Shore D (2014) Rif1 controls DNA replication timing in yeast through the PP1 phosphatase Glc7. Cell Rep 7:62–69

Montagnoli A, Valsasina B, Brotherton D, Troiani S, Rainoldi S, Tenca P, Molinari A, Santocanale C (2006) Identification of Mcm2 phosphorylation sites by S-phase-regulating kinases. J Biol Chem 281:10281–10290

Moreno SP, Gambus A (2015) Regulation of unperturbed DNA replication by ubiquitylation. Gene 6:451–468

Morohashi H, Maculins T, Labib K (2009) The amino-terminal TPR domain of Dia2 tethers SCF(Dia2) to the replisome progression complex. Curr Biol 19:1943–1949

Moyer SE, Lewis PW, Botchan MR (2006) Isolation of the Cdc45/Mcm2-7/GINS (CMG) complex, a candidate for the eukaryotic DNA replication fork helicase. Proc Natl Acad Sci U S A 103:10236–10241

Mukhopadhyay D, Dasso M (2007) Modification in reverse: the SUMO proteases. Trends Biochem Sci 32:286–295

Muramatsu S, Hirai K, Tak Y-S, Kamimura Y, Araki H (2010) CDK-dependent complex formation between replication proteins Dpb11, Sld2, Pol epsilon, and GINS in budding yeast. Genes Dev 24:602–612

Murray JM, Carr AM (2008) Smc5/6: a link between DNA repair and unidirectional replication? Nat Rev Mol Cell Biol 9:177–182

Pacek M, Tutter AV, Kubota Y, Takisawa H, Walter JC (2006) Localization of MCM2-7, Cdc45, and GINS to the site of DNA unwinding during eukaryotic DNA replication. Mol Cell 21:581–587

Panse VG, Kuster B, Gerstberger T, Hurt E (2003) Unconventional tethering of Ulp1 to the transport channel of the nuclear pore complex by karyopherins. Nat Cell Biol 5:21–27

Papouli E, Chen S, Davies AA, Huttner D, Krejci L, Sung P, Ulrich HD (2005) Crosstalk between SUMO and ubiquitin on PCNA is mediated by recruitment of the helicase Srs2p. Mol Cell 19:123–133

Potts PR, Porteus MH, Yu H (2006) Human SMC5/6 complex promotes sister chromatid homologous recombination by recruiting the SMC1/3 cohesin complex to double-strand breaks. EMBO J 25:3377–3388

Prasanth SG, Prasanth KV, Siddiqui K, Spector DL, Stillman B (2004) Human Orc2 localizes to centrosomes, centromeres and heterochromatin during chromosome inheritance. EMBO J 23:2651–2663

Priego Moreno S, Bailey R, Campion N, Herron S, Gambus A (2014) Polyubiquitylation drives replisome disassembly at the termination of DNA replication. Science 346:477–481

Psakhye I, Jentsch S (2012) Protein group modification and synergy in the SUMO pathway as exemplified in DNA repair. Cell 151:807–820

Ragland RL, Patel S, Rivard RS, Smith K, Peters AA, Bielinsky AK, Brown EJ (2013) RNF4 and PLK1 are required for replication fork collapse in ATR-deficient cells. Genes Dev 27:2259–2273

Randell JCW, Bowers JL, Rodríguez HK, Bell SP (2006) Sequential ATP hydrolysis by Cdc6 and ORC directs loading of the Mcm2-7 helicase. Mol Cell 21:29–39

Remus D, Diffley JFX (2009) Eukaryotic DNA replication control: lock and load, then fire. Curr Opin Cell Biol 21:771–777

Remus D, Beuron F, Tolun G, Griffith JD, Morris EP, Diffley JFX (2009) Concerted loading of Mcm2-7 double hexamers around DNA during DNA replication origin licensing. Cell 139:719–730

Santocanale C, Diffley JF (1996) ORC- and Cdc6-dependent complexes at active and inactive chromosomal replication origins in *Saccharomyces cerevisiae*. EMBO J 15:6671–6679

Sarangi P, Zhao X (2015) SUMO-mediated regulation of DNA damage repair and responses. Trends Biochem Sci 40:233–242

Sarangi P, Bartosova Z, Altmannova V, Holland C, Chavdarova M, Lee SE, Krejci L, Zhao X (2014) Sumoylation of the Rad1 nuclease promotes DNA repair and regulates its DNA association. Nucleic Acids Res 42:6393–6404

Sarangi P, Steinacher R, Altmannova V, Fu Q, Paull TT, Krejci L, Whitby MC, Zhao X (2015) Sumoylation influences DNA break repair partly by increasing the solubility of a conserved end resection protein. PLoS Genet 11:e1004899
Schimmel J, Eifler K, Sigurethsson JO, Cuijpers SA, Hendriks IA, Verlaan-de Vries M, Kelstrup CD, Francavilla C, Medema RH, Olsen JV et al (2014) Uncovering SUMOylation dynamics during cell-cycle progression reveals FoxM1 as a key mitotic SUMO target protein. Mol Cell 53:1053–1066
Schwartz DC, Felberbaum R, Hochstrasser M (2007) The Ulp2 SUMO protease is required for cell division following termination of the DNA damage checkpoint. Mol Cell Biol 27:6948–6961
Sclafani RA, Holzen TM (2007) Cell cycle regulation of DNA replication. Annu Rev Genet 41:237–280
Shen TH, Lin HK, Scaglioni PP, Yung TM, Pandolfi PP (2006) The mechanisms of PML-nuclear body formation. Mol Cell 24:331–339
Sherwood R, Takahashi TS, Jallepalli PV (2010) Sister acts: coordinating DNA replication and cohesion establishment. Genes Dev 24:2723–2731
Sheu Y-J, Stillman B (2006) Cdc7-Dbf4 phosphorylates MCM proteins via a docking site-mediated mechanism to promote S phase progression. Mol Cell 24:101–113
Sheu Y-J, Stillman B (2010) The Dbf4-Cdc7 kinase promotes S phase by alleviating an inhibitory activity in Mcm4. Nature 463:113–117
Sivakumar S, Gorbsky GJ (2015) Spatiotemporal regulation of the anaphase-promoting complex in mitosis. Nat Rev Mol Cell Biol 16:82–94
Sonneville R, Moreno SP, Knebel A, Johnson C, Hastie CJ, Gartner A, Gambus A, Labib K (2017) CUL-2LRR-1 and UBXN-3 drive replisome disassembly during DNA replication termination and mitosis. Nat Cell Biol. https://doi.org/10.1038/ncb3500
Speck C, Chen Z, Li H, Stillman B (2005) ATPase-dependent cooperative binding of ORC and Cdc6 to origin DNA. Nat Struct Mol Biol 12:965–971
Srikumar T, Lewicki M, Baught B (2013) A global *S. cerevisiae* small ubiquitin-related modifier (SUMO) system interactome. Mol Syst Biol 9:668–676
Stead K, Aguilar C, Hartman T, Drexel M, Meluh P, Guacci V (2003) Pds5p regulates the maintenance of sister chromatid cohesion and is sumoylated to promote the dissolution of cohesion. J Cell Biol 163:729–741
Sun J, Evrin C, Samel SA, Fernandez-Cid A, Riera A, Kawakami H, Stillman B, Speck C, Li H (2013) Cryo-EM structure of a helicase loading intermediate containing ORC-Cdc6-Cdt1-MCM2-7 bound to DNA. Nat Struct Mol Biol 20:944–951
Takahashi Y, Toh-e A, Kikuchi Y (2001) A novel factor required for the SUMO1/Smt3 conjugation of yeast septins. Gene 275:223–231
Tammsalu T, Matic I, Jaffray EG, Ibrahim AFM, Tatham MH, Hay RT (2014) Proteome-wide identification of SUMO2 modification sites. Sci Signal 7:rs2
Tanaka S, Umemori T, Hirai K, Muramatsu S, Kamimura Y, Araki H (2007) CDK-dependent phosphorylation of Sld2 and Sld3 initiates DNA replication in budding yeast. Nature 445:328–332
Tsuji T, Ficarro SB, Jiang W (2006) Essential role of phosphorylation of MCM2 by Cdc7/Dbf4 in the initiation of DNA replication in mammalian cells. Mol Biol Cell 17:4459–4472
Ulrich HD, Takahashi T (2013) Readers of PCNA modifications. Chromosoma 122:259–274
Vodermaier HC (2004) APC/C and SCF: controlling each other and the cell cycle. Curr Biol 14:R787–R796
Vos SM, Tretter EM, Schmidt BH, Berger JM (2011) All tangled up: how cells direct, manage and exploit topoisomerase function. Nat Rev Mol Cell Biol 12:827–841
Waga S, Stillman B (1998) The DNA replication fork in eukaryotic cells. Annu Rev Biochem 67:721–751
Wang JC (2002) Cellular roles of DNA topoisomerases: a molecular perspective. Nat Rev Mol Cell Biol 3:430–440

Waters LS, Minesinger BK, Wiltrout ME, D'Souza S, Woodruff RV, Walker GC (2009) Eukaryotic translesion polymerases and their roles and regulation in DNA damage tolerance. Microbiol Mol Biol Rev 73:134–154
Wei L, Zhao X (2016a) A new MCM modification cycle regulates DNA replication initiation. Nat Struct Mol Biol 23:209–216
Wei L, Zhao X (2016b) The initiation of DNA replication in eukaryotes. In: Kaplan DL (ed) Chapter 18: role of posttranslational modifications in replication initiation. Springer International Publishing, ISBN 978-3-319-24696-3
Wotton D, Shore D (1997) A novel Rap1p-interacting factor, Rif2p, cooperates with Rif1p to regulate telomere length in *Saccharomyces cerevisiae*. Genes Dev 11:748–760
Yamazaki S, Ishii A, Kanoh Y, Oda M, Nishito Y, Masai H (2012) Rif1 regulates the replication timing domains on the human genome. EMBO J 31:3667–3677
Yu B, Swatkoski S, Holly A, Lee LC, Giroux V, Lee CS, Hsu D, Smith JL, Yuen G, Yue J et al (2015) Oncogenesis driven by the Ras/Raf pathway requires the SUMO E2 ligase Ubc9. Proc Natl Acad Sci U S A 112:E1724–E1733
Zegerman P, Diffley JFX (2007) Phosphorylation of Sld2 and Sld3 by cyclin-dependent kinases promotes DNA replication in budding yeast. Nature 445:281–285
Zhao X, Blobel G (2005) A SUMO ligase is part of a nuclear multiprotein complex that affects DNA repair and chromosomal organization. Proc Natl Acad Sci U S A 102:4777–4782

Chapter 18
The Multiple Roles of Ubiquitylation in Regulating Challenged DNA Replication

Sara Villa-Hernández, Avelino Bueno, and Rodrigo Bermejo

Abstract DNA replication is essential for the propagation of life and the development of complex organisms. However, replication is a risky process as it can lead to mutations and chromosomal alterations. Conditions challenging DNA synthesis by replicative polymerases or DNA helix unwinding, generally termed as replication stress, can halt replication fork progression. Stalled replication forks are unstable, and mechanisms exist to protect their integrity, which promote an efficient restart of DNA synthesis and counteract fork collapse characterized by the accumulation of DNA lesions and mutagenic events. DNA replication is a highly regulated process, and several mechanisms control replication timing and integrity both during unperturbed cell cycles and in response to replication stress. Work over the last two decades has revealed that key steps of DNA replication are controlled by conjugation of the small peptide ubiquitin. While ubiquitylation was traditionally linked to protein degradation, the complexity and flexibility of the ubiquitin system in regulating protein function have recently emerged. Here we review the multiple roles exerted by ubiquitin-conjugating enzymes and ubiquitin-specific proteases, as well as readers of ubiquitin chains, in the control of eukaryotic DNA replication and replication-coupled DNA damage tolerance and repair.

Keywords DNA replication • Replication forks • Replication stress • DNA damage response • DNA repair • DNA damage tolerance • Ubiquitin • Ubiquitin ligases • DUBs • Cdc48/p97

S. Villa-Hernández • R. Bermejo (✉)
Centro de Investigaciones Biológicas (CSIC), Madrid, Spain
e-mail: rodrigo.bermejo@csic.es

A. Bueno
Instituto de Biología Molecular y Celular del Cáncer (USAL/CSIC), Salamanca, Spain

Departamento de Microbiología y Genética, Universidad de Salamanca, Salamanca, Spain

H. Masai, M. Foiani (eds.), *DNA Replication*, Advances in Experimental Medicine and Biology 1042, https://doi.org/10.1007/978-981-10-6955-0_18

18.1 The Ubiquitin System

Ubiquitylation is a highly regulated three-enzyme process consisting in the covalent attachment of ubiquitin moieties to a determined substrate. First, an E1 ubiquitin-activating enzyme forms a high-energy thioester bond with the carboxyl group of the terminal glycine residue of ubiquitin (Fig. 18.1a). This activated ubiquitin is then transferred to an E2 ubiquitin-conjugating enzyme by transesterification. An E3 ubiquitin ligase then catalyzes the formation of an isopeptide bond between a lysine in the substrate and the activated carboxyl group of ubiquitin (Komander and Rape 2012). Multiple rounds of this process, using lysines on ubiquitin as a substrate, lead to the formation of different types of polyubiquitin chains (Fig. 18.1b). Any of the seven lysines present on ubiquitin (K6, K11, K27, K29, K33, K48, and K63), as well as the amino-terminal methionine (Met1) of the ubiquitin monomer, can serve as isopeptide bond acceptors (Komander and Rape 2012; Kulathu and Komander 2012). Moreover, not only homotypic ubiquitin chains can be formed but also atypical chains, such as mixed chains (in which different lysines are successively used to link ubiquitin moeties) or branched chains (in which different lysines are ubiquitylated within a given ubiquitin molecule).

Ubiquitin chain variants can occur and determine different outcomes of the modified substrates. While K11 and K48 chains more frequently signal proteins for degradation, mono-ubiquitylation and K63 chains usually modulate protein-protein interactions. However, these regulatory functions are not strict, and proteolytic outcomes of K63 chains and non-proteolytic ones for K48 have been described (Shibata et al. 2012; Maric et al. 2014). K6, K27, K29, and K33 ubiquitin chains have been reported only for a small number of substrates, and their function is still poorly understood (Kulathu and Komander 2012; Pinder et al. 2013; Yau and Rape 2016). As in the case of other posttranslational modifications, ubiquitylation can be reversed, and this is achieved by the action of ubiquitin-specific proteases or deubiquitinases (DUBs). DUBs are cysteine proteases (with the exception of budding yeast Rpn11 that is a zinc metalloprotease), which catalyze the hydrolysis of the isopeptide bonds connecting ubiquitin with its substrate (Nijman et al. 2005b; Sahtoe and Sixma 2015). Ubiquitylation modulates a great variety of cellular processes and is regulated in a more sophisticated way than initially anticipated by factors that promote either substrate ubiquitylation or deubiquitylation. Noteworthy, pairs of coordinated E3 ligases and DUBs have been identified, in which the two enzymes act on the same substrates to fine-tune ubiquitylation levels (Kee et al. 2005; Sowa et al. 2009). Another important layer of regulation comes from ubiquitin chain editing, which requires the concerted action of additional ubiquitin ligases and/or DUBs that change the topology of the ubiquitin chains and thus alter substrate fate (Newton et al. 2008).

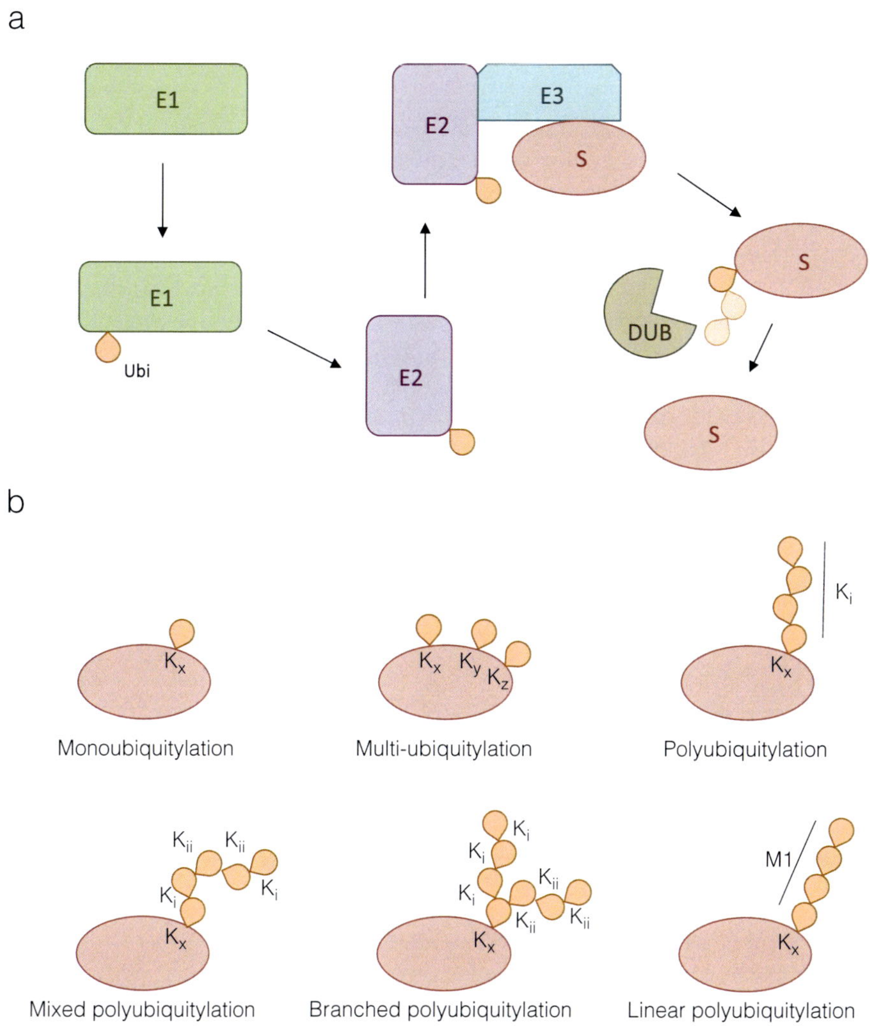

Fig. 18.1 The ubiquitin system. (**a**) Mechanism of ubiquitin conjugation and deconjugation. Ubiquitin conjugation requires the activities of three factors: an E1 activating enzyme, an E2 conjugating enzyme, and an E3 ubiquitin ligase (see text for details). Ubiquitylation can be reversed by ubiquitin proteases, also known as deubiquitinases (DUBs), which hydrolyze lysine-ubiquitin bonds, and can remove single ubiquitin moieties or entire chains. (**b**) Different types of ubiquitylation have been reported. Attachment of a single ubiquitin moiety in a given lysine (Kx) is referred to as mono-ubiquitylation, while conjugation of ubiquitin moieties to different lysines (Kx-z) results in multi-ubiquitylation. Polyubiquitylation occurs when multiple rounds of ubiquitin conjugation are preformed on a substrate-conjugated ubiquitin. The lysine used to extend ubiquitin chains can be the same (Ki) along the chain or vary (Ki-Kii) giving rise to mixed polyubiquitylation. In addition, more than one lysine on an ubiquitin moiety can be used to extend chains resulting in branched polyubiquitin chains. Linear polyubiquitin chains arise if the residue for chain elongation is the initial methionine on the ubiquitin moiety

18.2 Control of Unperturbed DNA Replication by Ubiquitylation

DNA replication is initiated at sites of the genome known as origins of replication, to which the origin recognition complex (ORC) binds (Bell and Stillman 1992). ORC is required for the recruitment of the MCM2–7 complex to origin DNA, by the action of Cdc6 and Cdt1 proteins in a process termed origin licensing (Diffley et al. 1994). These sequentially loaded proteins conform the pre-replicative complex (pre-RC), which marks sites that can potentially initiate replication (Fig. 18.2). Upregulation of CDK (cyclin-dependent kinase) and DDK (Dbf4-dependent kinase) activities after the transition from G1 to S phase leads to the phosphorylation and recruitment of additional factors and the formation of a pre-initiation complex (pre-IC) (Francis et al. 2009; Randell et al. 2010). Among these factors are Cdc45 and the GINS (Go-Ichi-Ni-San) complex, which together with MCM2–7 conform the functional replicative helicase (Ilves et al. 2010), as well as Mcm10, Dpb11, and DNA polymerase ε (Muramatsu et al. 2010). Once MCM2–7 complex is phosphorylated and activated, the double hexamer divides into two hexamers that start

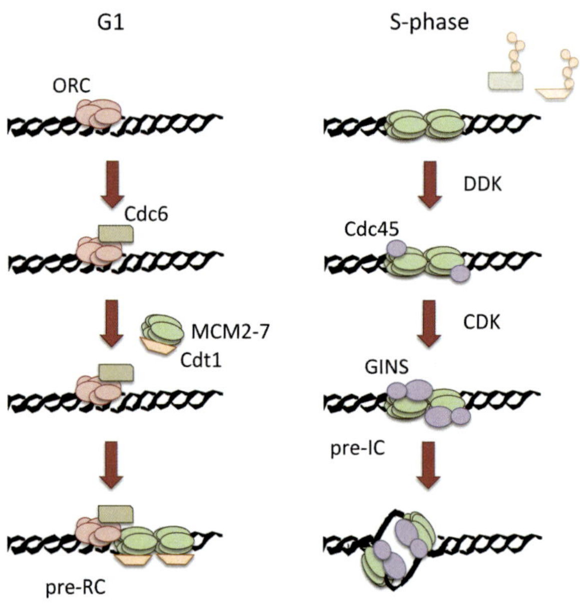

Fig. 18.2 Ubiquitylation in the control of replication origin licensing. In G1 ORC binds origin DNA. Cdc6 and Cdt1 mediate the loading of the MCM2-7 helicase complex, in a process called origin licensing, and lead to the formation of pre-replication complexes (pre-RC) determining potential replication initiation sites. During the G1-S transition, Cdc6 and Cdt1 are ubiquitylated and degraded in order to avoid loading of new MCM2-7 complexes to already replicated DNA and prevent re-replication. Origin firing involves the formation of a pre-initiation complex (pre-IC) that depends on the phosphorylation of several replication factors by DDK and CDK. DDK and CDK activities are also required to activate the replicative helicase and promote initial DNA unwinding

unwinding DNA (Quan et al. 2015). Helicase activation generates replication forks to which additional factors are recruited to build replisomes capable of efficient DNA synthesis (Bell and Labib 2016).

Origin licensing and firing are restricted to G1 and S phase, respectively, in order to ensure that replication occurs once and only once per cell cycle (Siddiqui et al. 2013). This is achieved by coupling origin function to the oscillation in the activity of cyclin-dependent kinases. Origin licensing occurs when CDK levels are low due to degradation of mitotic cyclins and the CDK-activating phosphatase Cdc25, mediated by the anaphase-promoting complex (APC) ubiquitin ligase and its adaptor Cdh1 (Donzelli et al. 2002). Additionally, helicase loading to origin DNA is prevented outside of G1 by the degradation of the licensing factors Cdc6 and Cdt1 (Fig. 18.2). Cdc6 is marked for ubiquitin-mediated proteolysis by phosphorylation by CDK upon S-phase onset. In budding yeast, phosphorylated Cdc6 is ubiquitylated by the Skp1-Cullin-F-box SCF^{Skp2} E3 ligase complex (Sánchez et al. 1999), leading to its degradation and thus prevention of MCM2–7 recruitment. In human cells instead, CDC6 phosphorylation leads to its export to the cytoplasm (Yim et al. 2013), where it cannot exert its MCM chromatin loading function. Nonetheless, it has been recently proposed that human CDC6 undergoes ubiquitylation-mediated degradation in a mechanism reminiscent of the one described in yeast (Walter et al. 2016).

Similarly to Cdc6, the Cdt1 origin-licensing factor is degraded by ubiquitin-mediated proteolysis. In human cells, CDT1 can be ubiquitylated by both the E3 Cullin-RING ubiquitin ligase complex $CRL4^{Cdt2}$ and the SCF^{Skp2} E3 ligase (Li et al. 2003; Nishitani et al. 2006). CDT1 degradation is induced by its phosphorylation by CDK, which promotes the interaction between CDT1 and Skp2 (Sugimoto et al. 2004). CDT1 degradation is also promoted by ATR checkpoint-dependent phosphorylation of the Cdt2 adaptor in response to DNA damage (Sakaguchi et al. 2012), as part of a mechanism that may contribute to limit origin licensing and firing upon checkpoint activation. In addition, it has been recently reported that the ATM (ataxia-telangiectasia mutated) checkpoint kinase promotes the degradation of CDT1 during unperturbed replication (Iwahori et al. 2014). However, the identity of the signals determining ATM activation and CDT1 degradation in the absence of exogenous perturbations remains unclear. CDT1 is also regulated by deubiquitylation. Cells ablated for the USP37 ubiquitin protease exhibit hyper-ubiquitylated CDT1, and USP37 has been shown to deubiquitylate CDT1 in vitro (Hernández-Pérez et al. 2016). Ablation of USP37 leads to defects in S-phase progression, but whether this reflects alterations in origin licensing or a yet unidentified function for CDT1 is promoting ongoing replication remains to be elucidated. In metazoans, CDT1 function is subject to an additional layer of regulation through inhibition by the small nuclear protein geminin (Wohlschlegel et al. 2000). Geminin is expressed upon transition from G1 to S phase and is degraded at the end of mitosis through a destruction box targeted by APC (McGarry and Kirschner 1998). Thus, multiple redundant mechanisms control licensing factors' function and stability to restrict helicase loading to the G1 phase of the cell cycle.

Ubiquitylation is also important during termination of DNA replication (Fig. 18.3). Both in budding yeast and *Xenopus*, K48 ubiquitin chains are linked to

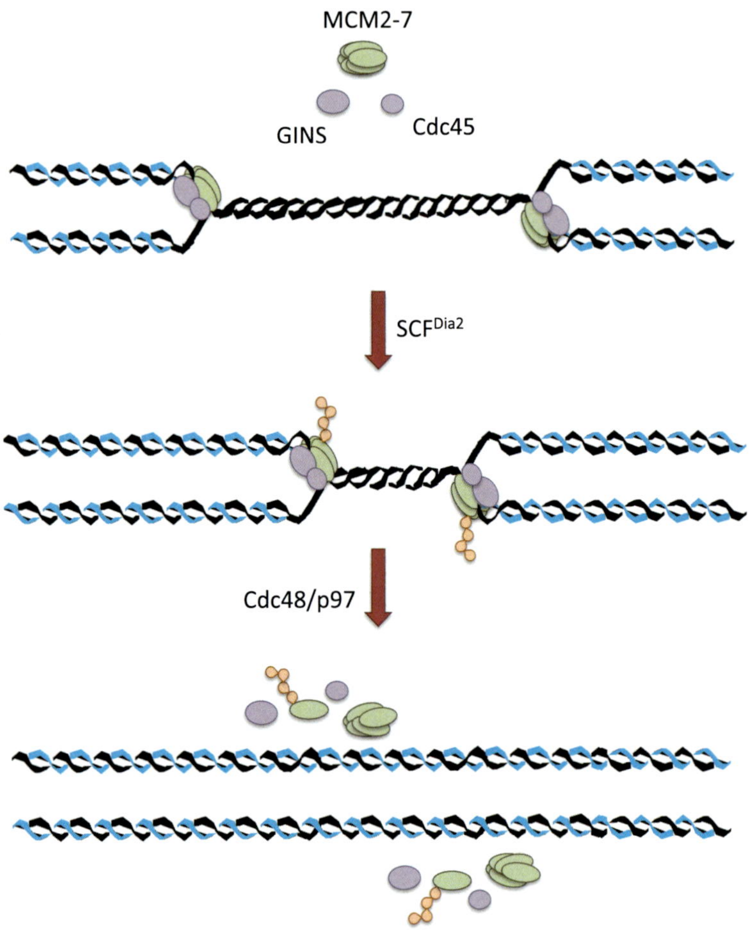

Fig. 18.3 CMG ubiquitylation and extraction from DNA during replication termination. Upon replication fork convergence during replication termination, the SCF^{Dia2} ubiquitin ligase polyubiquitylates the MCM7 subunit of the CMG replicative helicase. MCM7 ubiquitylation determines the disassembly and extraction of CMG helicase components, mediated by the Cdc48/p97 ubiquitin-selective segregase, presumably to facilitate the last steps of DNA synthesis

Mcm7 during the last steps of replication. Mcm7 ubiquitylation by yeast SCF^{Dia2} or *Xenopus* CRL2^{Lrr1} ubiquitin ligases leads to the disassembly of the CMG helicase from chromatin by action of the Cdc48/p97 ubiquitin-selective segregase (Priego Moreno et al. 2014; Maric et al. 2014; Maculins et al. 2015; Dewar et al. 2017). In this process, ubiquitylation by SCF^{Dia2} does not trigger Mcm7 degradation and appears to be restricted to converging forks through mechanisms that are not yet understood. It had been previously reported that CDC48/p97 is required for the extraction from chromatin and degradation of CDT1 in *Caenorhabditis* embryos and *Xenopus* extracts (Franz et al. 2011). In this study, accumulation of CDC45/

GINS on chromatin was also observed upon ablation of CDC48, and it was proposed that CDT1 degradation might be linked to GINS release from chromatin. Cdc48/p97 has also been involved in the extraction of proteins associated to chromatin during replication elongation (Lecona et al. 2016). It was reported that in human cells proteins bound to nascent DNA have higher levels of modification by the SUMO (small ubiquitin-like modifier) protein, while ubiquitylation is more abundantly detected in mature chromatin. The replisome-associated USP7 ubiquitin protease was found to counteract the accumulation of ubiquitylated proteins around nascent strands and thus limit their extraction by Cdc48/p97. Impairment of USP7 function results in fork progression defects, suggesting that limiting the levels of ubiquitylated proteins is important to sustain processive chromatin replication.

18.3 Ubiquitin in the Modulation of DNA Replication in Conditions of Stress

DNA replication can be challenged in conditions that impede replication fork progression, broadly termed as replication stress (Gaillard et al. 2015). Replication stress can result from inhibition of DNA synthesis, as in the case of dNTP pool depletion or polymerase-blocking DNA lesions, or be due to blockage of DNA unwinding by replicative helicases, as in the case of interstrand cross-links (ICLs) or DNA-protein cross-links (DPCs). These conditions are generally thought to cause an excessive accumulation of ssDNA, either due to the uncoupling between the replicative DNA polymerases and helicases or owing to nucleolytic resection of replication/repair intermediates (Jossen and Bermejo 2013). Stalled replication forks are unstable and prone to accumulate DNA breaks priming genomic instability (Branzei and Foiani 2010). Several mechanisms have been described that contribute stabilized replication forks and prevent their collapse. A prominent role in preserving genome stability is played by the replication checkpoint, which is mediated by highly conserved kinases Mec1/ATR and Rad53/CHK1 that monitor problems in fork progression. Upon fork stalling, extended ssDNA filaments coated by replication protein A (RPA) complex recruit Mec1/ATR through its partner protein Ddc2/ATRIP (Zou and Elledge 2003). At replication forks, Mec1 phosphorylates several targets, which include the Mrc1/CLASPIN protein (Alcasabas et al. 2001; Tanaka and Russell 2001). Mrc1 is a replisome component that acts as a mediator likely facilitating Rad53 *in trans* autophosphorylation leading to its full kinase activation required for the stabilization of replication forks (Alcasabas et al. 2001; Pellicioli and Foiani 2005). Checkpoint kinases interplay with factors dedicated to preserve replisome architecture and functionality, as well as with DNA repair and DNA damage tolerance pathways (Branzei and Foiani 2010; Ulrich and Walden 2010). Homologous recombination (HR) proteins are also required for replication fork stabilization, though their role in this process, likely related to ssDNA protection, is not yet fully understood (Carr and Lambert 2013). DNA lesions blocking the progression of replicative DNA polymerases can be bypassed through DNA damage tolerance (DDT) pathways (Chang and Cimprich 2009), which are driven by error-prone translesion synthesis (TLS) and the error-free template switching (TS) mechanisms. These are mediated by ubiquitylation of PCNA (proliferating cell nuclear antigen), and their impairment leads to increased mutagenesis and loss of viability upon treatment with genotoxic agents.

18.3.1 Ubiquitylation of Single-Stranded DNA Binding Proteins at Stalled Replication Forks

Replication protein A (RPA) complex is composed of three subunits (Rfa1/RPA70, Rfa2/RP32, and Rfa3/RPA14) and is phosphorylated by checkpoint kinases (Vassin et al. 2009; Ghospurkar et al. 2015). It has been recently described that all RPA subunits are ubiquitylated in response to replication fork stalling induced by agents causing dNTP depletion or DNA damage (Elia et al. 2015). RPA ubiquitylation depends on the RFWD3 E3 ligase and does not induce degradation by the proteasome. RFWD3-depleted cells show defects in stalled fork restart and increased sister chromatid exchange events in conditions inducing fork stalling. Based on these observations, it was proposed that RFWD3-dependent ubiquitylation of RPA promotes homologous recombination-mediated restart of stalled forks (Fig. 18.4). It was suggested that ubiquitylation of RPA favors HR by promoting displacement of

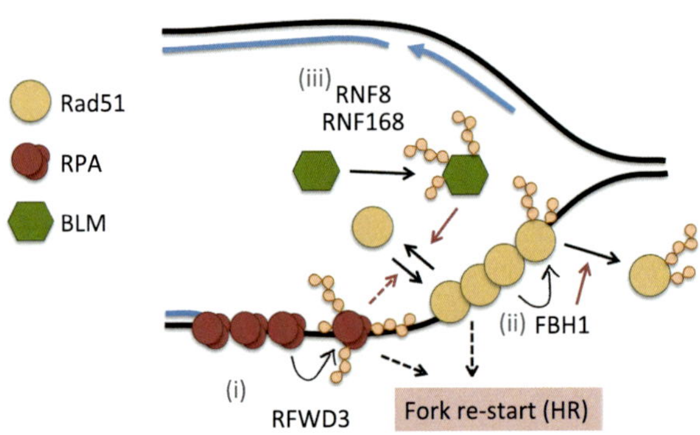

Fig. 18.4 Modulation of RPA/Rad51 function and homologous recombination at stalled replication forks. (i) RPA ubiquitylation by RFWD3 upon replication stress induction promotes fork restart. Ubiquitylation likely promotes exchange of RPA by Rad51, thus promoting fork restart perhaps through recombination-mediated mechanisms. (ii) Besides its ability to disassemble Rad51 filaments, the FBH1 helicase/E3 ligase ubiquitylates Rad51. Ubiquitylation of Rad51 likely promotes its eviction from DNA, thus limiting toxic recombination impairing stalled fork functionality. (iii) RNF8 and RNF168 E3 ligases polyubiquitylate BLM in three different lysines. This modification is required for BLM recruitment to sites of stalled replication, where it is thought to suppress unscheduled recombination events by promoting the disruption of Rad51 nucleofilaments

RPA from ssDNA to facilitate RAD51 filament formation. Alternatively, ubiquitylated RPA may provide a signal for attracting RAD51 to stalled replication forks. Previously, RPA subunit ubiquitylation by a different E3 ligase, PRP19, in response to DNA damage had been reported (Maréchal et al. 2013). It was shown that depletion of PRP19 reduced DNA damage-induced ubiquitylation of RPA and a role for this modification in promoting ATR-ATRIP signaling and fork progression was proposed. This conclusion was challenged by the finding that the system used to interfere PRP19 expression resulted in a collateral increase of the overall levels of exogenously expressed ubiquitin (Elia et al. 2015). Hence, further work will be required to clarify the relative contribution of RFWD3 and PRP19 to RPA ubiquitylation and stalled fork protection.

A role for the ubiquitylation of the homologous recombination (HR) mediator RAD51 in response to replication stress conditions has also been proposed. Rad51 is ubiquitylated in vitro by the FBH1 helicase/F-box protein (Chu et al. 2015). FBH1 helicase function had been previously implicated in modulating HR-mediated DNA repair as it can displace RAD51 nucleofilaments (Simandlova et al. 2013). Conversely, FBH1 ubiquitin ligase activity has been proposed to limit recombination-driven genomic instability by modulating RAD51 association to stalled forks (Chu et al. 2015). Expression of a RAD51 K58/64R variant, which shows impaired ubiquitylation in vitro, results in increased recombination and promotes viability and fork stability in response to replication stress-inducing agents. It was hence proposed that ubiquitylation may counteract RAD51 accumulation at stalled forks, thus limiting unscheduled recombination events (Fig. 18.4). The RecQ-family BLM (Bloom) helicase also plays an important role in the regulation of HR-mediated DNA repair. It has been proposed that BLM exerts both pro- and anti-recombinogenic functions at replication forks, by limiting HR events upon fork stalling or, conversely, promoting Rad51 recruitment to collapsed forks to promote repair (Böhm and Bernstein 2014). BLM function at replication forks is regulated by posttranslational modifications. Upon HU treatment, recruitment of BLM to sites of stalled replication requires its polyubiquitylation on three different lysines (K105/K225/K259) by the RNF8 and RNF168 E3 ligases (Tikoo et al. 2013). Expression of non-ubiquitylatable forms of BLM results in increased HR levels. This and the fact that BLM can disrupt RAD51 nucleofilaments in vitro (Bugreev et al. 2007) lead to the notion that ubiquitylation enhances BLM anti-recombinogenic function at stalled forks (Fig. 18.4). Conversely, BLM SUMOylation mediates the pro-recombinogenic role of BLM upon fork collapse, though the underlying mechanism is less characterized (Böhm and Bernstein 2014). Collectively, this evidence hints at ubiquitylation as a mechanism to modulate the balance between RPA and Rad51 association to ssDNA and fine-tune HR events at stalled replication forks.

18.3.2 Cross Talk Between DNA Damage Checkpoint Signaling and Ubiquitin Systems

Several lines of evidence point at a regulation of replication stress signaling by the checkpoint response through ubiquitylation. In *S. cerevisiae*, the Mrc1/CLASPIN adaptor is ubiquitylated by the SCF^{Dia2} E3 ligase in vitro. In line with this observation, Mrc1 is degraded upon induction of DNA damage in S phase in a manner dependent on Dia2 (Mimura et al. 2009). In *Schizosaccharomyces pombe*, deletion of the replication fork protection complex factor Swi1 (homolog of Tof1/TIMELESS) induces degradation of Mrc1, dependent on the SCF^{Dia2} homolog SCF^{Pof3} (Roseaulin et al. 2013). This evidence suggests that Mrc1 is degraded through ubiquitylation via Dia2, which is a fork-associated factor. Cells ablated for Dia2 are sensitive to DNA damage induced in S phase and show defects in recovering from checkpoint activation, both suppressed by checkpoint-adaptor-deficient alleles of *MRC1* (Fong et al. 2013). It was thus proposed that Mrc1 degradation through Dia2-mediated ubiquitylation is important to achieve checkpoint inactivation. Of note, Dia2 itself is ubiquitiylated and degraded via the proteasome in a mechanism counteracted by checkpoint kinases (Kile and Koepp 2010), thus hinting at the existence of a feedback loop promoting checkpoint inactivation by coupling stabilization of Dia2 and degradation of Mrc1. Mrc1 has also been described to interplay with the $Rtt101^{Mms22}$ ligase in response to replication stress in a pathway unrelated to Dia2 (Buser et al. 2016). Mutants impairing $Rtt101^{Mms22}$ function show sensitivity to the alkylating agent methyl-methanesulfonate (MMS), which is suppressed by *MRC1* deletion, but not by alleles abrogating its checkpoint mediator function. $Rtt101^{Mms22}$ is recruited to replication forks via the Ctf4 replisome component and does not appear to affect Mrc1 stability. These data suggest that in response to MMS-induced damage, Mrc1 imposes a requirement for a yet unknown $Rtt101^{Mms22}$ function, perhaps in preserving replication fork integrity. Alternatively, $Rtt101^{Mms22}$ may counteract a yet not characterized function of Mrc1 in promoting DNA replication that may result deleterious in the presence of a damaged template. In human cells, CLASPIN is ubiquitylated by the APC/C^{Cdh1} ubiquitin ligase complex, which promotes CLASPIN degradation via the proteasome both in the G1 phase of the cell cycle and in G2/M during recovery from treatment with DNA damaging agents (Bassermann et al. 2008). CLASPIN ubiquitylation by APC/C^{Cdh1} is counteracted by the USP28 ubiquitin protease. The $SCF^{\beta TrCP}$ ubiquitin ligase complex has also been involved in CLASPIN ubiquitylation upon entry in mitosis following recovery from genotoxic stress (Mailand et al. 2006). This modification has been proposed to contribute to checkpoint inactivation by limiting CLASPIN function as an adaptor and is counteracted by the ubiquitin protease USP7. Additionally, a recent work identified USP29 as an ubiquitin protease counteracting CLASPIN ubiquitylation and degradation (Martín et al. 2014). Downregulation of USP29 function results in S-phase progression defects and impaired activation of the CHK1 kinase, though the ligase introducing the ubiquitin chains removed by USP29 remains to be elucidated. Lastly, a role in modulating CHK1 function through

CLASPIN deubiquitylation has also been reported for the USP20 ubiquitin protease (Zhu et al. 2014). Collectively, this evidence points at a high control of Mrc1/CLASPIN function through ubiquitylation and degradation in the modulation of checkpoint signaling following replication stress.

It was recently shown that the ATR checkpoint response can negatively regulate ubiquitylation machineries. In particular, silencing of CHK1 destabilizes APC/C^{Cdh1} targets in human cells experiencing replication blocks, in a fashion dependent on expression of the ligase (Yamada et al. 2013). It was thus suggested that the checkpoint kinase inhibits the ubiquitin ligase complex in a process required to stabilize the Dbf4/ASK component of the DDK complex and promote DNA damage bypass mechanisms. In consistence with this idea, chromatin loading of both the Rad18 ubiquitin ligase and Pol η translesion polymerase is impaired by depletion of CHK1 upon induction of interstrand cross-links. Thus bidirectional cross talk likely occurs between checkpoint and ubiquitylation machineries in the response to DNA lesions generating replication stress.

18.3.3 Ubiquitylation in the Regulation of DNA Polymerases' Association to DNA and Chromatin Dynamics

Ubiquitylation of DNA polymerases also appears to be important in the response to replication stress. In eukaryotes DNA polymerase ε primarily carries out leading strand synthesis, while DNA polymerase δ extends Okazaki fragments generated at lagging strands (Pursell et al. 2007). Ubiquitylation of the catalytic subunits of either replicative polymerase has been described in yeast cells (Roseaulin et al. 2013; Daraba et al. 2014). Inactivation of the replication fork protection complex by ablation of Swi1/Tof1 in *S. pombe* cells, which presumably enhances replication stalling and collapse at natural fork pausing sites, leads to ubiquitylation by SCF^{Pof3} and degradation of the Pol2 catalytic subunit of DNA polymerase ε (Roseaulin et al. 2013). Ultraviolet (UV) light irradiation of *S. cerevisiae* cells induces DNA polymerase δ catalytic subunit Pol3 ubiquitylation and proteasomal degradation in a fashion dependent on Def1, a factor also involved in RNA polymerase II degradation (Daraba et al. 2014). Sensitivity of *def1* mutants to MMS treatment is epistatic to that conferred by genes required for the error-free pathway of DNA damage tolerance, which raised the possibility that degradation might facilitate exchange of Pol3 by translesion synthesis polymerases. In agreement with this hypothesis, DNA polymerase δ auxiliary subunits Pol31 and Pol32 are not degraded in these conditions and form a complex with DNA polymerase ζ-associated Rev1 protein, while Def1 ablation reduces UV-induced mutation rates.

Ubiquitylation has been shown to mediate the extraction of translesion polymerases from DNA by the action of the Cdc48/p97 cofactor DVC1/SPARTAN (Mosbech et al. 2012). DVC1 is recruited to sites of stalled replication in fashion dependent of its UBZ (ubiquitin-binding zinc finger) domain, which leads to the hypothesis that it might interact with ubiquitylated replisome components. Interference of DVC1 or expression of an ATPase-deficient version of p97 results in increased Pol η interaction with PCNA in UV-irradiated cells. DVC1 also enhances p97 association to UV-damaged chromatin, which suggests that polyubiquitylation of Pol η may promote engagement by p97 to mediate its extraction during the last steps of translesion synthesis. In addition to this mechanism, it has been described that Pol η is mono-ubiquitylated in unperturbed cells and that this modification decreases upon treatment with genotoxic agents (Bienko et al. 2010). Mono-ubiquitylation inhibits Pol η interaction with PCNA and in this way may limit the basal frequency of TLS events.

DVC1/SPARTAN, as its budding yeast paralog Wss1, bears a metalloprotease activity, which is dependent on its association to DNA and is involved in replication-dependent repair of DNA-protein cross-links (DPC) (Stingele and Jentsch 2015; Stingele et al. 2016; Vaz et al. 2016). DVC1 is enriched at nascent DNA and interacts with replisome factors such as PCNA and the MCM complex, suggesting that it acts on DPCs as they are encountered by replication forks (Stingele et al. 2016; Vaz et al. 2016). Consistently, ablation of DVC1 in human or worm cells leads to lethality and fork progression defects upon treatment with agents inducing protein cross-linking to DNA. DVC1 is ubiquitylated, and its deubiquitylation promotes its access to chromatin, perhaps by disengagement of *in cis* interactions of its UBZ domain. Whether DVC1 function in replication-coupled DPC repair involves its role as an adaptor for Cdc48/p97 in extraction of ubiquitylated proteins from chromatin remains to be elucidated.

Ubiquitylation also plays a critical role in regulating replication in the chromatin context. Polycomb repressive complexes (PRCs) mediate the establishment of repressive chromatin through ubiquitylation of H2A (Di Croce and Helin 2013). In addition, H2A ubiquitylation by PCRs has recently emerged as a regulator of DNA replication. PRCs associate to nascent DNA in mouse cells, and interference of their function impairs replication fork progression in the absence of external perturbations (Piunti et al. 2014). These observations suggest that PCRs may modulate chromatin dynamics in particular chromosome territories to facilitate fork progression. Consistently, inactivation of RING ubiquitin ligase activities associated to PCRs leads to S-phase defects and increased asymmetry in replication fork progression (Bravo et al. 2015). Replication impairment in these conditions is specific to pericentromeric heterochromatic (PCH) domains and can be suppressed by targeted restoration of H2A ubiquitylation at PCH. The exact function of H2A ubiquitylation during replication remains to be understood but might reflect a necessity to modulate chromatin structure at hard-to-replicate domains or limit fork conflicts with derepressed transcription.

H2B is mono-ubiquitylated by Bre1/RNF20 E3 ubiquitin ligase, and this modification plays a well-characterized role in transcriptional regulation (Meas and Mao 2015). In budding yeast, Bre1 and mono-ubiquitylated H2B are enriched around active replication origins, suggesting that Bre1 associates to replication forks to promote H2B ubiquitylation at nascent chromatin DNA (Trujillo and Osley 2012; Lin et al. 2014). Non-ubiquitylatable lysine-123 H2B mutants are sensitive to HU treatment and show altered stalled fork progression as well as defects in recovery from replication stress, likely independent of transcriptional regulation of dNTP pools. It was thus suggested that H2B ubiquitylation is important in coupling chromatin assembly and replication fork progression.

18.3.4 Modulation of the Ubiquitylation of the Sliding Clamp PCNA as a Central Regulator of DNA Damage Tolerance Mechanisms

At the core of replicating chromatin, PCNA works as an essential processivity factor for DNA polymerases (Moldovan et al. 2007; Ulrich 2009; Mailand et al. 2013). PCNA interacts with a number of proteins involved in replication itself as well as with factors involved in the maintenance of genome integrity (Mailand et al. 2013). The posttranslational modification of PCNA licenses a layer of control in PCNA-protein interactions that facilitates fork progression under normal or perturbed replication (Andersen et al. 2008; Ulrich 2009; Bergink and Jentsch 2009; Mailand et al. 2013). During S phase, cells face damaged DNA or lesions that pose significant barriers to the progression of replication forks (Curtin 2012). In living organisms, tolerance mechanisms ensure that DNA can be replicated when it is damaged (Friedberg 2003). These mechanisms prevent irreversible fork collapse when the replisome encounters bulky lesions at damaged sites that block replicative DNA polymerases (Friedberg et al. 2005). Cells have evolved tolerance to DNA lesions to ensure progression of replication forks past unrepaired damage (Lawrence 1994; Chang and Cimprich 2009).

Tolerance to DNA damage is based either on translesion synthesis (TLS) that is carried out by specialized low-fidelity, potentially error-prone, TLS DNA polymerases or on template switching, an error-free mechanism that involves sister-strand pairing within or nearby the replication fork. In the face of a DNA lesion, the ubiquitylation of the sliding clamp controls the choice of translesion synthesis in eukaryotes. While mono-ubiquitylation of PCNA at Lys^{164} enhances the affinity of error-prone TLS DNA polymerases (Hoege et al. 2002; Stelter and Ulrich 2003; Watanabe et al. 2004), further Lys^{63}-linked polyubiquitylation of mono-ubiquitylated Lys^{164}-PCNA promotes template switching (Zhang and Lawrence 2005; Vanoli et al. 2010). This control mechanism is well characterized and widely conserved in eukaryotic organisms (Moldovan et al. 2007; Ulrich 2009; Mailand et al. 2013). On the contrary, the significance of the deubiquitylation of PCNA remains insufficiently understood.

The ubiquitylation of PCNA is catalyzed by proteins in the Rad6 epistasis group (Hoege et al. 2002; Stelter and Ulrich 2003) (Fig. 18.5). The Rad6 group includes genes required for post-replication repair in the budding yeast *S. cerevisiae* and consists of *RAD6* and *RAD18* (Lys^{164}-PCNA-ubiquitin ligase) or *RAD5*, *MMS2*, and *UBC13* (Lys^{63}-linked Lys^{164}-PCNA-ubiquitin ligase) (Lawrence 1994; Zhang and Lawrence 2005). The Rad6 epistasis group is a central player in DNA damage bypass

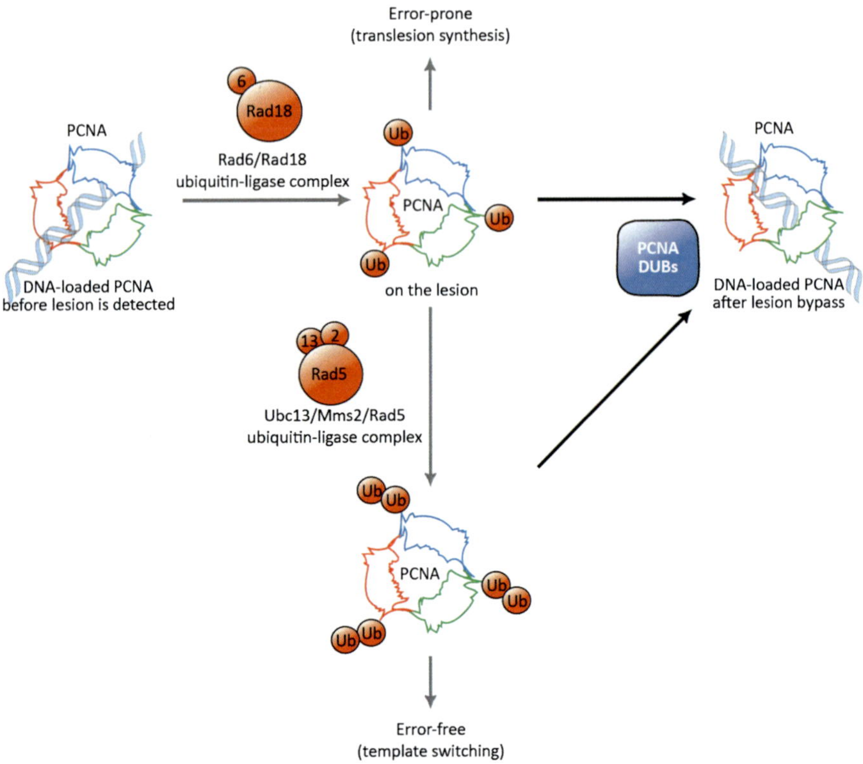

Fig. 18.5 Model for PCNA ubiquitylation/deubiquitylation in the modulation of tolerance of DNA damage encountered by replication forks. PCNA progresses in association with the replisome on leading strands and associates to Pol δ on lagging strands. Bulky lesions block the progression of replicative polymerases and induce Rad6-/Rad18-mediated PCNA mono-ubiquitylation. Ubiquitylated PCNA enhances the interaction with translesion synthesis DNA polymerases. Further Lys63-linked ubiquitylation of PCNA Lys164-conjugated ubiquitin by a second ubiquitin ligase complex (Ubc13/Mms2/Rad5) promotes template switching events. After lesion bypass, PCNA-DUBs deubiquitylate PCNA to switching back to replicative DNA polymerases and resume processive DNA replication. Dynamic ubiquitylation and deubiquitylation of PCNA may regulate DNA damage tolerance at replication forks and/or post-replicatively

and tolerance pathways (historically known as post-replication repair) as these enzyme complexes are involved in protein ubiquitylation of PCNA and translesion synthesis DNA polymerases. PCNA mono-ubiquitylation is regarded as a central step in translesion synthesis during normal or perturbed DNA replication. However, DT40 avian cell line evidence indicates that Pol η (a TLS-DNA polymerase) and PCNA interact in the absence of PCNA ubiquitylation (Edmunds et al. 2008). Furthermore, there is general consensus that PCNA ubiquitylation is important but not essential for translesion DNA synthesis in mammalian cells (Edmunds et al. 2008; Hendel et al. 2011). Nevertheless, all eukaryotic models tested to date mono-ubiquitylate Lys^{164}-PCNA to detectable levels in response to replication blocks or

DNA damage to direct TLS polymerases to damaged sites (Hoege et al. 2002; Kannouche et al. 2004; Watanabe et al. 2004; Arakawa et al. 2006; Frampton et al. 2006). Of particular interest is the fact that PCNA is mono-ubiquitylated in *Xenopus* and *S. pombe* and di-ubiquitylated in *S. pombe*, during unperturbed S phase in a Rad6-/Rad18-dependent manner (Leach and Michael 2005; Frampton et al. 2006).

Rad18 is the predominant PCNA E3 ubiquitin ligase in eukaryotes. However, evidence from mammalian cells indicates that other ubiquitin ligases, including RNF8, HLTF, and CRL4^{CDT2}, may have a role in this process in the absence of Rad18 or during unperturbed DNA replication (Zhang et al. 2008; Terai et al. 2010; Lin et al. 2011). The choice between the two branches of damage tolerance, error-prone or error-free, relies respectively on the mono-ubiquitylation and polyubiquitylation of the sliding clamp PCNA. However, the timing of PCNA ubiquitylation, and therefore the timing of lesion bypass, relative to DNA replication remains under discussion. It has been shown that tolerance to DNA damage mediated by ubiquitin-dependent modification of PCNA can be uncoupled from bulk DNA replication, at least in *S. cerevisiae* (Karras and Jentsch 2010; Daigaku et al. 2010), suggesting the hypothesis that DNA damage bypass may work exclusively as a post-replicative mechanism (Karras and Jentsch 2010). It is worth to mention that PCNA-ubiquitin ligase Rad18 associates with replisomes at sites of newly synthesized DNA in human cells (Dungrawala et al. 2015; Despras et al. 2016), indicating that the machinery for ubiquitylation of PCNA travels with the replisome at replication forks.

Along the last decade, mammalian USP1, human Usp10, and budding yeast Ubp10 have been identified as major deubiquitylating enzymes (DUBs) for PCNA (Huang et al. 2006; Gallego-Sánchez et al. 2012; Mailand et al. 2013; Park et al. 2014). USP1, the first PCNA-DUB identified, is a ubiquitin protease that deubiquitylates mono-ubPCNA and mono-ubFANCD2 in human cells (Huang et al. 2006). USP1 undergoes inactivation by auto-proteolysis when cells are irradiated with UV light. Therefore, upon UV irradiation, USP1 is inactivated and PCNA becomes ubiquitylated; this observation suggests that USP1 continuously deubiquitylates PCNA in the absence of DNA damage. Consistent with a role as a PCNA-DUB, the depletion of chicken USP1 in DT40 cells or in murine USP1−/− MEFs increases PCNA, and also FANCD2, mono-ubiquitylation (Oestergaard et al. 2007; Kim et al. 2009). Recently, it has been shown that upon UV-mediated DNA damage, HeLa cells rely on USP10 to deubiquitylate ISGylated PCNA (Park et al. 2014). Two unicellular model systems have been examined in detail, budding and fission yeast. In the budding yeast *S. cerevisiae*, the ubiquitin protease Ubp10 deubiquitylates Lys^{164} mono- and di-ubiquitylated PCNA during S phase (Gallego-Sánchez et al. 2012, 2013). More recently, it has been shown that Ubp2, Ubp15, Ubp16, and Ubp12 ubiquitin proteases revert PCNA ubiquitylation in the fission yeast *S. pombe* during S phase (Álvarez et al. 2016). To date, all the identified PCNA-DUBs belong to the USP/UBP subfamily of ubiquitin-specific proteases.

All the published evidence provides robust evidence that PCNA-DUBs revert the Lys^{164} mono- and di-ubiquitylation of PCNA in vitro and in vivo. Therefore, PCNA-DUB activities may counterbalance the ubiquitin ligase activity of the Rad6/Rad18 complex responsible for PCNA mono-ubiquitylation at sites of newly synthesized

DNA. Thus, these ubiquitin-specific proteases have the potential to be part of a safeguard mechanism limiting the residence time of TLS DNA polymerases on replicating chromatin (Fig. 18.5).

18.3.5 Ubiquitylation-Mediated Control of the Fanconi Anemia DNA Repair Pathway

Fanconi anemia (FA) is a rare disorder characterized by sensitivity to DNA interstrand cross-linking agents, bone marrow failure, and susceptibility to both acute myeloid leukemia and solid tumors. These features can also be accompanied by congenital abnormalities and infertility (Risitano et al. 2016). The Fanconi anemia pathway (named after the syndrome) responds to DNA interstrand cross-links and replication fork-stalling lesions that hamper not only the DNA synthesis by polymerases but also strand separation by replicative helicases. The FA pathway promotes the repair of DNA ICLs orchestrating several processes that include recognition of the ICL (Fig. 18.6), incision of the lesion by structure-specific endonucleases, bypass by TLS polymerases, and HR-mediated repair of the incised strand. The pathway is initiated by the recruitment of the core FA complex, which contains the E3 ubiquitin ligase FANCL, to ICLs upon engagement by replication forks (Longerich et al. 2009). Unloading of the CMG helicase complex from chromatin, a step required for the last steps of DNA synthesis and subsequent nucleolytic incision, follows ICL recognition. CMG extraction requires polyubiquitylation of MCM7 that is mediated by BRCA1 E3 ligase, in contrast with replication termination which is mediated by SCF^{Dia2}, which promotes engagement by the Cdc48/p97segregase (Long et al. 2014; Fullbright et al. 2016). Upon recruitment to stalled forks, the catalytic subunit FANCL mono-ubiquitylates the FANCD2-FANCI dimer. This ubiquitylation activates downstream events leading to lesion processing and repair. How these events are coordinated to achieve ICL removal starts to be elucidated, as some key mechanistic details have been recently described. Ubiquitylation of FANCD2 is essential for its interaction with a UBZ4 domain in the SLX4 scaffold protein. SLX4 in turn promotes recruitment and activation of several nucleases, such as ERCC4-ERCC1, MUS81-EME1, and SLX1, which mediate the incisions of a DNA strand flanking the ICL. Ubiquitylation of FANCD2 also mediates its interaction with the UBZ domain of FAN1 (Fanconi anemia nuclease 1), yet another nuclease involved in the FA-mediated repair (Liu et al. 2010; Yoshikiyo et al. 2010). FAN1 is dispensable for the incision of ICLs in vitro, and FAN1 mutations are not associated to typical FA phenotypes (Klein Douwel et al. 2014), suggesting that it may mediate FA pathway functions unrelated to ICL repair. Accordingly, FANCD2 recruits FAN1 to replication forks stalled by HU treatment where it is required for efficient restart (Lachaud et al. 2016). Of note, FANCD2 patient-derived cells exhibit increased replication fork stalling at common fragile sites (CFS) (Madireddy

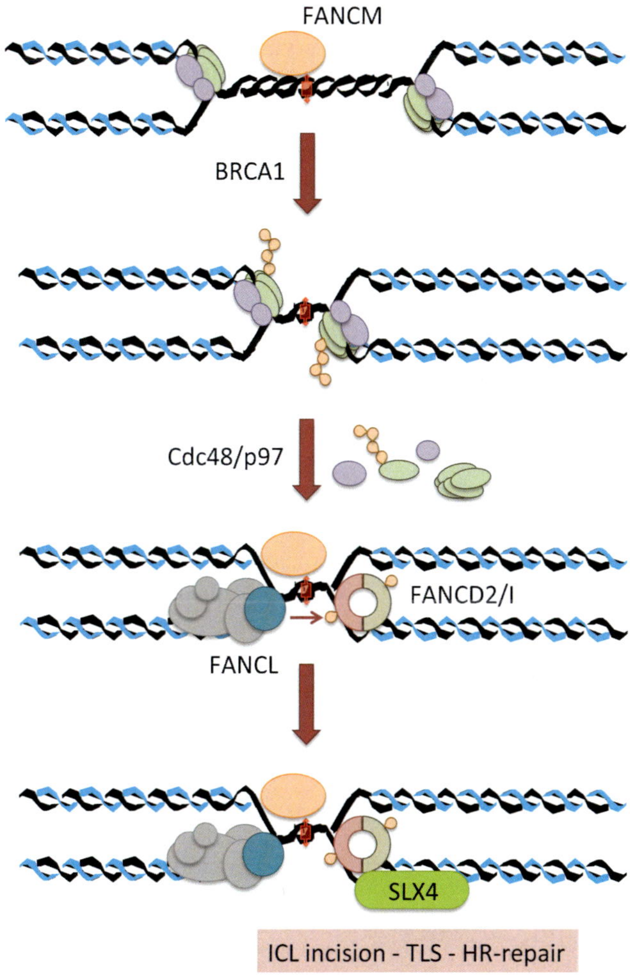

Fig. 18.6 Ubiquitylation is central in the modulation of the Fanconi anemia DNA repair pathway. Upon collision with replication forks, the FANCM translocase recognizes interstrand cross-links (ICL) and initiates the FA repair pathway. CMG ubiquitylation by BRCA1 E3 ligase drives its eviction through the action of the Cdc48/p97 segregase, which enables synthesis of nucleotides nearby the lesion. Then the Fanconi anemia core complex is recruited to the lesion, where the FANCL ubiquitin ligase promotes the mono-ubiquitylation of the FANCD2/I heterodimer. This ubiquitylation is the central event that mediates the recruitment of the SLX4 protein, which serves a scaffold for nucleases that promote incisions on the DNA backbone around the lesion. DNA incision triggers downstream events in the FA pathway such as base flipping, translesion synthesis, and HR-mediated repair of the incised strand

et al. 2016), suggesting that FANCD2 ubiquitylation may serve as a signal mediating distinct fork protection mechanisms besides the repair of ICLs.

There is evidence of cross talk between the FA pathway and key regulators of the DNA damage tolerance mechanisms. As mentioned above, Rad18 mediates PCNA mono-ubiquitylation and is also required for the ubiquitylation and activation of FANCD2/I (Geng et al. 2010; Williams et al. 2011). However, how Rad18 contributes to activation of the FA pathway is a matter of debate. On the one hand, it was proposed that FANCL interacts with ubiquitylated PCNA and that this interaction is necessary for recruitment of FANCL to chromatin and FANCD2 mono-ubiquitylation (Geng et al. 2010). On the other hand, it has been reported that Rad18 promotes FACND2/I ubiquitylation in a fashion independent on PCNA-K164 (Williams et al. 2011), in apparent contradiction with the previous report. In addition, it has been shown that FANCD2 interacts with PCNA through a PIP box and that FANCD2 PIP-box mutants show impaired ubiquitylation (Williams et al. 2011), though the involvement of Rad18 or PCNA mono-ubiquitylation in this mechanism was not explored. Another link between FA and DDT pathways occurs through the USP1 ubiquitin protease that, as mentioned, reverts the mono-ubiquitylation of both PCNA and the FANCD2-I complex (Nijman et al. 2005a). Silencing of USP1 expression increases mutation frequencies in UV-treated cells (Nijman et al. 2005a), suggesting that a fine regulation of these two processes is important to promote replication fidelity. Inactivation of the FA pathway involves SUMOylation of FANCD2 and FANCI proteins by PIAS1 and PIAS4 SUMO E3 ligases (Gibbs-Seymour et al. 2015). SUMOylated FANCD2-I is engaged by the RNF4 SUMO-targeted ubiquitin ligase, which in turn polyubiquitylates FANCD2-I and promotes eviction of the dimer from chromatin. FANCD2-I is extracted by the p97 segregase with the aid of its adaptor DVC1, thus locally interrupting FA pathway signaling.

18.4 Concluding Remarks

In recent years our knowledge on the flexibility and complexity of the ubiquitylation code has rapidly expanded. This is reflected in the current understanding of how ubiquitylation dynamically regulates DNA replication. Ubiquitin-mediated degradation controls the stability of factors spatiotemporally regulating replication initiation, as well as proteins mediating replication stress signaling. Ubiquitylation of replication factors also modulates important DNA repair and DNA damage tolerance pathway choices in response to replication stress. Lastly, the importance of ubiquitylation as a signal to control the extraction of key proteins from replicating chromatin by ubiquitin-selective segregase complexes during termination and upon fork stalling is emerging. It is thus expected that research in years to come may elucidate key mechanistic details of how ubiquitin regulates DNA replication, replication fork protection, and the coordination of replication with other chromosome metabolic processes.

Acknowledgments We apologize for relevant findings and studies that might have not been discussed due to space limitations. This work was supported by the Ministry of Economy and Competitiveness (BFU2014-52529-R to R.B. and BFU2015-69709-P to A.B.). S.V-H received support from the Spanish *Formación de Personal Investigador* (FPI) program.

References

Alcasabas AA, Osborn AJ, Bachant J et al (2001) Mrc1 transduces signals of DNA replication stress to activate Rad53. Nat Cell Biol 3:958–965. https://doi.org/10.1038/ncb1101-958

Álvarez V, Viñas L, Gallego-Sánchez A et al (2016) Orderly progression through S-phase requires dynamic ubiquitylation and deubiquitylation of PCNA. Sci Rep 6:25513. https://doi.org/10.1038/srep25513

Andersen PL, Xu F, Xiao W (2008) Eukaryotic DNA damage tolerance and translesion synthesis through covalent modifications of PCNA. Cell Res 18:162–173. https://doi.org/10.1038/cr.2007.114

Arakawa H, Moldovan GL, Saribasak H et al (2006) A role for PCNA ubiquitination in immunoglobulin hypermutation. PLoS Biol 4:1947–1956. https://doi.org/10.1371/journal.pbio.0040366

Bassermann F, Frescas D, Guardavaccaro D et al (2008) The Cdc14B-Cdh1-Plk1 Axis controls the G2 DNA-damage-response checkpoint. Cell 134:256–267. https://doi.org/10.1016/j.cell.2008.05.043

Bell SP, Labib K (2016) Chromosome duplication in Saccharomyces Cerevisiae. Genetics 203:1027–1067. https://doi.org/10.1534/genetics.115.186452

Bell SP, Stillman B (1992) ATP-dependent recognition of eukaryotic origins of DNA replication by a multiprotein complex. Nature 357:128–134. https://doi.org/10.1038/357128a0

Bergink S, Jentsch S (2009) Principles of ubiquitin and SUMO modifications in DNA repair. Nature 458:461–467. https://doi.org/10.1038/nature07963

Bienko M, Green CM, Sabbioneda S et al (2010) Regulation of translesion synthesis DNA polymerase η by Monoubiquitination. Mol Cell 37:396–407. https://doi.org/10.1016/j.molcel.2009.12.039

Böhm S, Bernstein KA (2014) The role of post-translational modifications in fine-tuning BLM helicase function during DNA repair. DNA Repair (Amst) 22:123–132. https://doi.org/10.1016/j.dnarep.2014.07.007

Branzei D, Foiani M (2010) Maintaining genome stability at the replication fork. Nat Rev Mol Cell Biol 11:208–219. https://doi.org/10.1038/nrm2852

Bravo M, Nicolini F, Starowicz K et al (2015) Polycomb RING1A- and RING1B-dependent histone H2A monoubiquitylation at pericentromeric regions promotes S-phase progression. J Cell Sci 128:3660–3671. https://doi.org/10.1242/jcs.173021

Bugreev DV, Yu X, Egelman EH, Mazin AV (2007) Novel pro- and anti-recombination activities of the bloom's syndrome helicase. Genes Dev 21:3085–3094. https://doi.org/10.1101/gad.1609007.cluding

Buser R, Kellner V, Melnik A et al (2016) The replisome-coupled E3 ubiquitin ligase Rtt101Mms22 counteracts Mrc1 function to tolerate genotoxic stress. PLoS Genet 12:e1005843. https://doi.org/10.1371/journal.pgen.1005843

Carr AM, Lambert S (2013) Replication stress-induced genome instability: the dark side of replication maintenance by homologous recombination. J Mol Biol 425:4733–4744. https://doi.org/10.1016/j.jmb.2013.04.023

Chang DJ, Cimprich KA (2009) DNA damage tolerance: when it's OK to make mistakes. Nat Chem Biol 5:82–90. https://doi.org/10.1038/nchembio.139

Chu WK, Payne MJ, Beli P et al (2015) FBH1 influences DNA replication fork stability and homologous recombination through ubiquitylation of RAD51. Nat Commun 6:1–9. https://doi.org/10.1038/ncomms6931
Curtin NJ (2012) DNA repair dysregulation from cancer driver to therapeutic target. Nat Rev Cancer 12:801–817. https://doi.org/10.1038/nrc3399
Daigaku Y, Davies AA, Ulrich HD (2010) Ubiquitin-dependent DNA damage bypass is separable from genome replication. Nature 465:951–955. https://doi.org/10.1038/nature09097
Daraba A, Gali VK, Halmai M et al (2014) Def1 promotes the degradation of Pol3 for polymerase exchange to occur during DNA-damage--induced mutagenesis in Saccharomyces Cerevisiae. PLoS Biol 12:e1001771. https://doi.org/10.1371/journal.pbio.1001771
Despras E, Sittewelle M, Pouvelle C et al (2016) Rad18-dependent SUMOylation of human specialized DNA polymerase eta is required to prevent under-replicated DNA. Nat Commun 7:13326. https://doi.org/10.1038/ncomms13326
Dewar JM, Low E, Mann M et al (2017) CRL2 ^{Lrr1} promotes unloading of the vertebrate replisome from chromatin during replication termination. Genes Dev:275–290. https://doi.org/10.1101/gad.291799.116
Di Croce L, Helin K (2013) Transcriptional regulation by Polycomb group proteins. Nat Struct Mol Biol 20:1147–1155. https://doi.org/10.1038/nsmb.2669
Diffley JFX, Cocker JH, Dowell SJ, Rowley A (1994) Two steps in the assembly of complexes at yeast replication origins in vivo. Cell 78:303–316. https://doi.org/10.1016/0092-8674(94)90299-2
Donzelli M, Squatrito M, Ganoth D et al (2002) Dual mode of degradation of Cdc25 a phosphatase. EMBO J 21:4875–4884. https://doi.org/10.1093/emboj/cdf491
Dungrawala H, Rose KLL, Bhat KPP et al (2015) The replication checkpoint prevents two types of fork collapse without regulating replisome stability. Mol Cell 59:1–13. https://doi.org/10.1016/j.molcel.2015.07.030
Edmunds CE, Simpson LJ, Sale JE (2008) PCNA ubiquitination and REV1 define temporally distinct mechanisms for controlling Translesion synthesis in the avian cell line DT40. Mol Cell 30:519–529. https://doi.org/10.1016/j.molcel.2008.03.024
Elia AEH, Wang DC, Willis NA et al (2015) RFWD3-dependent ubiquitination of RPA regulates repair at stalled replication forks. Mol Cell 60:280–293. https://doi.org/10.1016/j.molcel.2015.09.011
Fong CM, Arumugam A, Koepp DM (2013) The Saccharomyces Cerevisiae F-box protein Dia2 is a mediator of S-phase checkpoint recovery from DNA damage. Genetics 193:483–499. https://doi.org/10.1534/genetics.112.146373
Frampton J, Irmisch A, Green CM et al (2006) Postreplication repair and PCNA modification in Schizosaccharomyces pombe. Mol Biol Cell 17:2976–2985. https://doi.org/10.1091/mbc.E05
Francis LI, Randell JCW, Takara TJ et al (2009) Incorporation into the prereplicative complex activates the Mcm2 – 7 helicase for Cdc7 – Dbf4 phosphorylation. Genes Dev 23:643–654. https://doi.org/10.1101/gad.1759609.replication
Franz A, Orth M, Pirson PA et al (2011) CDC-48/p97 coordinates CDT-1 degradation with GINS chromatin dissociation to ensure faithful DNA replication. Mol Cell 44:85–96. https://doi.org/10.1016/j.molcel.2011.08.028
Friedberg EC (2003) DNA damage and repair
Friedberg EC, Lehmann AR, Fuchs RPP (2005) Trading places: how do DNA polymerases switch during translesion DNA synthesis? Mol Cell 18:499–505. https://doi.org/10.1016/j.molcel.2005.03.032
Fullbright G, Rycenga HB, Gruber JD, Long DT (2016) p97 promotes a conserved mechanism of helicase unloading during DNA crosslink repair. Mol Cell Biol. https://doi.org/10.1128/MCB.00434-16
Gaillard H, García-Muse T, Aguilera A (2015) Replication stress and cancer. Nat Rev Cancer 15:276–289. https://doi.org/10.1038/nrc3916
Gallego-Sánchez A, Andrés S, Conde F et al (2012) Reversal of PCNA ubiquitylation by Ubp10 in Saccharomyces cerevisiae. PLoS Genet. https://doi.org/10.1371/journal.pgen.1002826

Gallego-Sánchez A, Andrés S, Ufano S, Bueno A (2013) Analysis of the tolerance to DNA alkylating damage in MEC1 and RAD53 checkpoint mutants of Saccharomyces cerevisiae. PLoS One 8:e81108. https://doi.org/10.1371/journal.pone.0081108
Geng L, Huntoon CJ, Karnitz LM (2010) RAD18-mediated ubiquitination of PCNA activates the Fanconi anemia DNA repair network. J Cell Biol 191:249–257. https://doi.org/10.1083/jcb.201005101
Ghospurkar PL, Wilson TM, Severson AL et al (2015) The DNA damage response and checkpoint adaptation in Saccharomyces cerevisiae: distinct roles for the replication protein A2 (Rfa2) N-terminus. Genetics 199:711–727. https://doi.org/10.1534/genetics.114.173211
Gibbs-Seymour I, Oka Y, Rajendra E et al (2015) Ubiquitin-SUMO circuitry controls activated fanconi anemia ID complex dosage in response to DNA damage. Mol Cell 57:150–164. https://doi.org/10.1016/j.molcel.2014.12.001
Hendel A, Krijger PHL, Diamant N et al (2011) PCNA ubiquitination is important, but not essential for translesion DNA synthesis in mammalian cells. PLoS Genet 7:e1002262. https://doi.org/10.1371/journal.pgen.1002262
Hernández-Pérez S, Cabrera E, Amoedo H et al (2016) USP37 deubiquitinates Cdt1 and contributes to regulate DNA replication. Mol Oncol 0:1–11. https://doi.org/10.1016/j.molonc.2016.05.008
Hoege C, Pfander B, Moldovan G-L et al (2002) RAD6-dependent DNA repair is linked to modification of PCNA by ubiquitin and SUMO. Nature 419:135–141. https://doi.org/10.1038/nature00991
Huang TT, Nijman SMB, Mirchandani KD et al (2006) Regulation of monoubiquitinated PCNA by DUB autocleavage. Nat Cell Biol 8:339–347. https://doi.org/10.1038/ncb1378
Ilves I, Petojevic T, Pesavento JJ, Botchan MR (2010) Activation of the MCM2-7 helicase by association with Cdc45 and GINS proteins. Mol Cell 37:247–258. https://doi.org/10.1016/j.molcel.2009.12.030
Iwahori S, Kohmon D, Kobayashi J et al (2014) ATM regulates Cdt1 stability during the unperturbed S phase to prevent re-replication. Cell Cycle 13(3):471–481
Jossen R, Bermejo R (2013) The DNA damage checkpoint response to replication stress: a game of forks. Front Genet 4:26. https://doi.org/10.3389/fgene.2013.00026
Kannouche PL, Wing J, Lehmann AR (2004) Interaction of human DNA polymerase eta with monoubiquitinated PCNA: a possible mechanism for the polymerase switch in response to DNA damage. Mol Cell 14:491–500. https://doi.org/10.1016/S1097-2765(04)00259-X
Karras GI, Jentsch S (2010) The RAD6 DNA damage tolerance pathway operates uncoupled from the replication fork and is functional beyond S phase. Cell 141:255–267. https://doi.org/10.1016/j.cell.2010.02.028
Kee Y, Lyon N, Huibregtse JM (2005) The Rsp5 ubiquitin ligase is coupled to and antagonized by the Ubp2 deubiquitinating enzyme. EMBO J 24:2414–2424. https://doi.org/10.1038/sj.emboj.7600710
Kile AC, Koepp DM (2010) Activation of the S-phase checkpoint inhibits degradation of the F-box protein Dia2. Mol Cell Biol 30:160–171. https://doi.org/10.1128/MCB.00612-09
Kim JM, Parmar K, Huang M et al (2009) Inactivation of murine Usp1 results in genomic instability and a Fanconi anemia phenotype. Dev Cell 16:314–320. https://doi.org/10.1016/j.devcel.2009.01.001
Klein Douwel D, Boonen RACM, Long DT et al (2014) XPF-ERCC1 acts in unhooking DNA Interstrand crosslinks in cooperation with FANCD2 and FANCP/SLX4. Mol Cell 54:460–471. https://doi.org/10.1016/j.molcel.2014.03.015
Komander D, Rape M (2012) The ubiquitin code. Annu Rev Biochem 81:203–229. https://doi.org/10.1146/annurev-biochem-060310-170328
Kulathu Y, Komander D (2012) Atypical ubiquitylation – the unexplored world of polyubiquitin beyond Lys48 and Lys63 linkages. Nat Rev Mol Cell Biol 13:508–523. https://doi.org/10.1038/nrm3394

Lachaud C, Moreno A, Marchesi F et al (2016) Ubiquitinated Fancd2 recruits Fan1 to stalled replication forks to prevent genome instability. Science 351:846–849. https://doi.org/10.1126/science.aad5634

Lawrence C (1994) The RAD6 DNA repair pathway in Saccharomyces cerevisiae: what does it do, and how does it do it? BioEssays 16:253–258. https://doi.org/10.1002/bies.950160408

Leach CA, Michael WM (2005) Ubiquitin/SUMO modification of PCNA promotes replication fork progression in Xenopus laevis egg extracts. J Cell Biol 171:947–954. https://doi.org/10.1083/jcb.200508100

Lecona E, Rodriguez-Acebes S, Specks J et al (2016) USP7 is a SUMO deubiquitinase essential for DNA replication. Nat Struct Mol Biol. https://doi.org/10.1038/nsmb.3185

Li X, Zhao Q, Liao R et al (2003) The SCFSkp2 ubiquitin ligase complex interacts with the human replication licensing factor Cdt1 and regulates Cdt1 degradation. J Biol Chem 278:30854–30858. https://doi.org/10.1074/jbc.C300251200

Lin C-Y, Wu M-Y, Gay S et al (2014) H2B mono-ubiquitylation facilitates fork stalling and recovery during replication stress by coordinating Rad53 activation and chromatin assembly. PLoS Genet 10:e1004667. https://doi.org/10.1371/journal.pgen.1004667

Lin JR, Zeman MK, Chen JY et al (2011) SHPRH and HLTF act in a damage-specific manner to coordinate different forms of postreplication repair and prevent mutagenesis. Mol Cell 42:237–249. https://doi.org/10.1016/j.molcel.2011.02.026

Liu T, Ghosal G, Yuan J et al (2010) FAN1 acts with FANCI-FANCD2 to promote DNA interstrand cross-link repair. Science 329:693–696. https://doi.org/10.1126/science.1192656

Long DT, Joukov V, Budzowska M, Walter JC (2014) BRCA1 promotes unloading of the CMG helicase from a stalled DNA replication fork. Mol Cell:1–12. https://doi.org/10.1016/j.molcel.2014.08.012

Longerich S, San Filippo J, Liu D, Sung P (2009) FANCI binds branched DNA and is monoubiquitinated by UBE2T-FANCL. J Biol Chem 284:23182–23186. https://doi.org/10.1074/jbc.C109.038075

Maculins T, Nkosi PJ, Nishikawa H, Labib K (2015) Tethering of SCFDia2 to the replisome promotes efficient Ubiquitylation and disassembly of the CMG helicase. Curr Biol:1–6. https://doi.org/10.1016/j.cub.2015.07.012

Madireddy A, Kosiyatrakul ST, Boisvert RA et al (2016) FANCD2 facilitates replication through common fragile sites. Mol Cell 64:388–404. https://doi.org/10.1016/j.molcel.2016.09.017

Mailand N, Bekker-Jensen S, Bartek J, Lukas J (2006) Destruction of Claspin by SCFbetaTrCP restrains Chk1 activation and facilitates recovery from genotoxic stress. Mol Cell 23:307–318. https://doi.org/10.1016/j.molcel.2006.06.016

Mailand N, Gibbs-seymour I, Bekker-Jensen S (2013) Regulation of PCNA-protein interactions for genome stability. Nat Rev Mol Cell Biol 14:269–282. https://doi.org/10.1038/nrm3562

Maréchal A, Li J-M, Ji XY et al (2013) PRP19 transforms into a sensor of RPA-ssDNA after DNA damage and drives ATR activation via a ubiquitin-mediated circuitry. Mol Cell 53:1–12. https://doi.org/10.1016/j.molcel.2013.11.002

Maric M, Maculins T, De Piccoli G, Labib K (2014) Cdc48 and a ubiquitin ligase drive disassembly of the CMG helicase at the end of DNA replication. Science 346:1253596–1253596. https://doi.org/10.1126/science.1253596

Martín Y, Cabrera E, Amoedo H, Freire R (2014) USP29 controls the stability of checkpoint adaptor Claspin by deubiquitination. 28:1–6. doi: https://doi.org/10.1038/onc.2014.38

McGarry TJ, Kirschner MW (1998) Geminin, an inhibitor of DNA replication, is degraded during mitosis. Cell 93:1043–1053. https://doi.org/10.1016/S0092-8674(00)81209-X

Meas R, Mao P (2015) Histone ubiquitylation and its roles in transcription and DNA damage response. DNA Repair (Amst) 36:36–42. https://doi.org/10.1016/j.dnarep.2015.09.016

Mimura S, Komata M, Kishi T et al (2009) SCF(Dia2) regulates DNA replication forks during S-phase in budding yeast. EMBO J 28:3693–3705. https://doi.org/10.1038/emboj.2009.320

Moldovan G-L, Pfander B, Jentsch S (2007) PCNA, the maestro of the replication fork. Cell 129:665–679. https://doi.org/10.1016/j.cell.2007.05.003

Mosbech A, Gibbs-Seymour I, Kagias K et al (2012) DVC1 (C1orf124) is a DNA damage–targeting p97 adaptor that promotes ubiquitin-dependent responses to replication blocks. Nat Struct Mol Biol 19:1084–1092. https://doi.org/10.1038/nsmb.2395
Muramatsu S, Hirai K, Tak YS et al (2010) CDK-dependent complex formation between replication proteins Dpb11, Sld2, pol ??, and GINS in budding yeast. Genes Dev 24:602–612. https://doi.org/10.1101/gad.1883410
Newton K, Matsumoto ML, Wertz IE et al (2008) Ubiquitin chain editing revealed by Polyubiquitin linkage-specific antibodies. Cell 134:668–678. https://doi.org/10.1016/j.cell.2008.07.039
Nijman SMB, Huang TT, Dirac AMG et al (2005a) The deubiquitinating enzyme USP1 regulates the fanconi anemia pathway. Mol Cell 17:331–339. https://doi.org/10.1016/j.molcel.2005.01.008
Nijman SMB, MPA L-V, Velds A et al (2005b) A genomic and functional inventory of deubiquitinating enzymes. Cell 123:773–786. https://doi.org/10.1016/j.cell.2005.11.007
Nishitani H, Sugimoto N, Roukos V et al (2006) Two E3 ubiquitin ligases, SCF-Skp2 and DDB1-Cul4, target human Cdt1 for proteolysis. EMBO J 25:1126–1136. https://doi.org/10.1038/sj.emboj.7601002
Oestergaard VH, Langevin F, Kuiken HJ et al (2007) Deubiquitination of FANCD2 is required for DNA crosslink repair. Mol Cell 28:798–809. https://doi.org/10.1016/j.molcel.2007.09.020
Park JM, Yang SW, Yu KR et al (2014) Modification of PCNA by ISG15 plays a crucial role in termination of error-prone translesion DNA synthesis. Mol Cell 54:626–638. https://doi.org/10.1016/j.molcel.2014.03.031
Pellicioli A, Foiani M (2005) Signal transduction: how Rad53 kinase is activated. Curr Biol 15:769–771. https://doi.org/10.1016/j.cub.2005.08.057
Pinder JB, Attwood KM, Dellaire G (2013) Reading, writing, and repair: the role of ubiquitin and the ubiquitin-like proteins in DNA damage signaling and repair. Front Genet 4:1–14. https://doi.org/10.3389/fgene.2013.00045
Piunti A, Rossi A, Cerutti A et al (2014) Polycomb proteins control proliferation and transformation independently of cell cycle checkpoints by regulating DNA replication. Nat Commun 5:3649. https://doi.org/10.1038/ncomms4649
Priego Moreno S, Bailey R, Campion N et al (2014) Polyubiquitylation drives replisome disassembly at the termination of DNA replication. Science 346:477–481. https://doi.org/10.1126/science.1253585
Pursell ZF, Isoz I, Lundström E-B et al (2007) Yeast DNA polymerase e participates in leading-strand DNA replication. Science 317:127–130
Quan Y, Xia Y, Liu L et al (2015) Cell-cycle-regulated interaction between Mcm10 and double Hexameric Mcm2-7 is required for helicase splitting and activation during S phase. Cell Rep 13:2576–2586. https://doi.org/10.1016/j.celrep.2015.11.018
Randell JCW, Fan A, Chan C et al (2010) Mec1 is one of multiple kinases that prime the Mcm2-7 helicase for phosphorylation by Cdc7. Mol Cell 40:353–363. https://doi.org/10.1016/j.molcel.2010.10.017
Risitano AM, Marotta S, Calzone R et al (2016) Twenty years of the Italian Fanconi anemia registry: where we stand and what remains to be learned. Haematologica 101:319–327. https://doi.org/10.3324/haematol.2015.133520
Roseaulin LC, Noguchi C, Martinez E et al (2013) Coordinated degradation of replisome components ensures genome stability upon replication stress in the absence of the replication fork protection complex. PLoS Genet 9:21–28. https://doi.org/10.1371/journal.pgen.1003213
Sahtoe DD, Sixma TK (2015) Layers of DUB regulation. Trends Biochem Sci:1–12. https://doi.org/10.1016/j.tibs.2015.05.002
Sakaguchi H, Takami T, Yasutani Y et al (2012) Checkpoint kinase ATR phosphorylates Cdt2, a substrate receptor of CRL4 ubiquitin ligase, and promotes the degradation of Cdt1 following UV irradiation. PLoS One. https://doi.org/10.1371/journal.pone.0046480
Sánchez M, Calzada A, Bueno A (1999) The Cdc6 protein is ubiquitinated in vivo for proteolysis in *Saccharomyces cerevisiae*. J Biol Chem 274:9092–9097. https://doi.org/10.1074/jbc.274.13.9092

Shibata Y, Oyama M, Kozuka-Hata H et al (2012) p47 negatively regulates IKK activation by inducing the lysosomal degradation of polyubiquitinated NEMO. Nat Commun 3:1061. https://doi.org/10.1038/ncomms2068
Siddiqui K, On KF, Diffley JFX (2013) Regulating DNA replication in eukarya. Cold Spring Harb Perspect Biol 5:1–17. https://doi.org/10.1101/cshperspect.a012922
Simandlova J, Zagelbaum J, Payne MJ et al (2013) FBH1 helicase disrupts RAD51 filaments in vitro and modulates homologous recombination in mammalian cells. J Biol Chem 288:34168–34180. https://doi.org/10.1074/jbcM113.484493
Sowa ME, Bennett EJ, Gygi SP, Harper JW (2009) Defining the human deubiquitinating enzyme interaction landscape. Cell 138:389–403. https://doi.org/10.1016/j.cell.2009.04.042
Stelter P, Ulrich HD (2003) Control of spontaneous and damage-induced mutagenesis by SUMO and ubiquitin conjugation. Nature 425:188–191. https://doi.org/10.1038/nature01965
Stingele J, Bellelli R, Alte F et al (2016) Mechanism and regulation of DNA-protein crosslink repair by the DNA-dependent metalloprotease SPRTN. Mol Cell 64:688–703. https://doi.org/10.1016/j.molcel.2016.09.031
Stingele J, Jentsch S (2015) DNA–protein crosslink repair. Nat Rev Mol Cell Biol 16:455–460. https://doi.org/10.1038/nrm4015
Sugimoto N, Tatsumi Y, Tsurumi T et al (2004) Cdt1 phosphorylation by cyclin A-dependent kinases negatively regulates its function without affecting geminin binding. J Biol Chem 279:19691–19697. https://doi.org/10.1074/jbc.M313175200
Tanaka K, Russell P (2001) Mrc1 channels the DNA replication arrest signal to checkpoint kinase Cds1. Nat Cell Biol 3:966–972. https://doi.org/10.1038/ncb1101-966
Terai K, Abbas T, Jazaeri AA, Dutta A (2010) CRL4Cdt2 E3 ubiquitin ligase Monoubiquitinates PCNA to promote Translesion DNA synthesis. Mol Cell 37:143–149. https://doi.org/10.1016/j.molcel.2009.12.018
Tikoo S, Madhavan V, Hussain M et al (2013) Ubiquitin-dependent recruitment of the bloom syndrome helicase upon replication stress is required to suppress homologous recombination. EMBO J 32:1778–1792. https://doi.org/10.1038/emboj.2013.117
Trujillo KM, Osley MA (2012) A role for H2B Ubiquitylation in DNA replication. Mol Cell 48:734–746. https://doi.org/10.1016/j.molcel.2012.09.019
Ulrich HD (2009) Regulating post-translational modifications of the eukaryotic replication clamp PCNA. DNA Repair (Amst) 8:461–469. https://doi.org/10.1016/j.dnarep.2009.01.006
Ulrich HD, Walden H (2010) Ubiquitin signalling in DNA replication and repair. Nat Rev Mol Cell Biol 11:479–489. https://doi.org/10.1038/nrm2921
Vanoli F, Fumasoni M, Szakal B et al (2010) Replication and recombination factors contributing to recombination-dependent bypass of DNA lesions by template switch. PLoS Genet. https://doi.org/10.1371/journal.pgen.1001205
Vassin VM, Anantha RW, Sokolova E et al (2009) Human RPA phosphorylation by ATR stimulates DNA synthesis and prevents ssDNA accumulation during DNA-replication stress. J Cell Sci 122:4070–4080. https://doi.org/10.1242/jcs.053702
Vaz B, Popovic M, Newman JA et al (2016) Metalloprotease SPRTN/DVC1 orchestrates replication-coupled DNA-protein crosslink repair. Mol Cell 0:137–151. https://doi.org/10.1016/j.molcel.2016.09.032
Walter D, Hoffmann S, Komseli E-S et al (2016) SCFCyclin F-dependent degradation of CDC6 suppresses DNA re-replication. Nat Commun 7:10530. https://doi.org/10.1038/ncomms10530
Watanabe K, Tateishi S, Kawasuji M et al (2004) Rad18 guides poleta to replication stalling sites through physical interaction and PCNA monoubiquitination. EMBO J 23:3886–3896. https://doi.org/10.1038/sj.emboj.7600383
Williams SA, Longerich S, Sung P et al (2011) The E3 ubiquitin ligase RAD18 regulates ubiquitylation and chromatin loading of FANCD2 and FANCI. Blood 117:5078–5087. https://doi.org/10.1182/blood-2010-10-311761
Wohlschlegel JA, Dwyer BT, Dhar SK et al (2000) Inhibition of eukaryotic DNA replication by geminin binding to Cdt1. Science 290:2309–2312. https://doi.org/10.1126/science.290.5500.2309

Yamada M, Watanabe K, Mistrik M et al (2013) ATR-Chk1-APC/CCdh1-dependent stabilization of Cdc7-ASK (Dbf4) kinase is required for DNA lesion bypass under replication stress. Genes Dev 27:2459–2472. https://doi.org/10.1101/gad.224568.113

Yau R, Rape M (2016) The increasing complexity of the ubiquitin code. Nat Publ Gr 18:579–586. https://doi.org/10.1038/ncb3358

Yim H, Park JW, Woo SU et al (2013) Phosphorylation of Cdc6 at serine 74, but not at serine 106, drives translocation of Cdc6 to the cytoplasm. J Cell Physiol 228:1221–1228. https://doi.org/10.1002/jcp.24275

Yoshikiyo K, Kratz K, Hirota K et al (2010) KIAA1018/FAN1 nuclease protects cells against genomic instability induced by interstrand cross-linking agents. Proc Natl Acad Sci U S A 107:21553–21557. https://doi.org/10.1073/pnas.1011081107

Zhang S, Chea J, Meng X et al (2008) PCNA is ubiquitinated by RNF8. Cell Cycle 7:3399–3404. https://doi.org/10.4161/cc.7.21.6949

Zhang H, Lawrence CW (2005) The error-free component of the RAD6/RAD18 DNA damage tolerance pathway of budding yeast employs sister-strand recombination. Proc Natl Acad Sci U S A 102:15954–15959. https://doi.org/10.1073/pnas.0504586102

Zhu M, Zhao H, Liao J, Xu X (2014) HERC2/USP20 coordinates CHK1 activation by modulating CLASPIN stability. Nucleic Acids Res:1–8. https://doi.org/10.1093/nar/gku978

Zou L, Elledge SJ (2003) Sensing DNA damage through ATRIP recognition of RPA-ssDNA complexes. Science 300:1542–1548. https://doi.org/10.1126/science.1083430

Chapter 19
Regulation of Mammalian DNA Replication via the Ubiquitin-Proteasome System

Tarek Abbas and Anindya Dutta

Abstract Proper regulation of DNA replication ensures the faithful transmission of genetic material essential for optimal cellular and organismal physiology. Central to this regulation is the activity of a set of enzymes that induce or reverse posttranslational modifications of various proteins critical for the initiation, progression, and termination of DNA replication. This is particularly important when DNA replication proceeds in cancer cells with elevated rates of genomic instability and increased proliferative capacities. Here, we describe how DNA replication in mammalian cells is regulated via the activity of the ubiquitin-proteasome system as well as the consequence of derailed ubiquitylation signaling involved in this important cellular activity.

Keywords DNA replication • Ubiquitin • Ubiquitylation E3 ligases • DNA rereplication • Cancer

19.1 Introduction

Eukaryotic DNA replication is a highly regulated process that ensures the faithful transmission of genetic material to daughter cells (Machida et al. 2005). The process is coupled both with cell cycle progression and with the ability of cells to cope with various intrinsic and extrinsic insults that constantly threaten the integrity of the genome. Various repair mechanisms have evolved to cope with these insults,

T. Abbas (✉)
Department of Radiation Oncology, University of Virginia, Charlottesville, VA, USA

Department of Biochemistry and Molecular Genetics, University of Virginia, Charlottesville, VA, USA
e-mail: ta8e@virginia.edu

A. Dutta
Department of Biochemistry and Molecular Genetics, University of Virginia, Charlottesville, VA, USA

H. Masai, M. Foiani (eds.), *DNA Replication*, Advances in Experimental Medicine and Biology 1042, https://doi.org/10.1007/978-981-10-6955-0_19

permitting the repair of various lesions prior to the completion of DNA replication (Hustedt and Durocher 2016; Ganai and Johansson 2016; Berti and Vindigni 2016). Critical to this regulation is the ability of cells to sense environmental perturbations, to enforce appropriate checkpoints, and to activate a number of cellular processes conductive of DNA repair prior to the initiation or resumption of DNA replication. Posttranslational modifications of certain proteins play fundamental role in the timely execution of most, if not all, of these cellular activities.

Similar to many physiological processes in the cell, proper regulation of DNA replication is governed through the balanced production and termination of key cellular proteins controlling the various steps involved. DNA replication depends on key enzymatic activities that are controlled through the action of cellular proteins and cofactors that are actively synthesized, modified, or destroyed to achieve optimal activity. ATP-dependent protein polyubiquitylation plays an important role in almost all physiological processes and in many diseases including cancer owing to its role in the global regulation of protein turnover (Schwartz and Ciechanover 2009; Hershko 2005; Amir et al. 2001; Glickman and Ciechanover 2002; Kornitzer and Ciechanover 2000; Ciechanover and Schwartz 2002). The highly coordinated and reversible process ensures timely downregulation of proteins via the activity of the 26S proteasome, where the polyubiquitylated proteins, roughly 80% of all intracellular proteins, are digested into small peptides, and ubiquitin molecules are recycled (Skaar et al. 2014).

Proteasomal degradation through the ubiquitin-proteasome system (UPS) is triggered following the covalent attachment of multiple ubiquitin molecules linked together through lysine 48 (Lys-48) to substrate proteins (Teixeira and Reed 2013; Groll and Huber 2003). Other forms of polyubiquitylation (e.g., linkage through Lys-63 of ubiquitin) do not result in proteasomal degradation but regulate other functions, such as protein trafficking, protein-protein interaction, and kinase activation (Yang et al. 2010; Behrends and Harper 2011). Polyubiquitylation is comprised of three distinct and consecutive enzymatic steps (Fig. 19.1): ubiquitin activation by an E1 ubiquitin-activating enzyme, the transfer of the activated ubiquitin to an E2 ubiquitin-conjugating enzyme (UBC), and the transfer of ubiquitin to the substrate through the activity of an E3 ubiquitin ligase (Kornitzer and Ciechanover 2000; Groll and Huber 2003; Glickman and Ciechanover 2002). This latter activity by E3 ubiquitin ligases is particularly important as it confers specificity for the substrate to be targeted for ubiquitylation (Fig. 19.1).

While polyubiquitylation is an efficient mechanism by which cellular proteins are eliminated or modified, the process is reversible, and the removal of ubiquitin chains from substrate proteins is carried by a class of highly specific cysteine proteases, collectively termed "deubiquitinases" or "DUBs" (Fig. 19.1). DUBs hydrolyze the isopeptide bonds between the ε-amino group of lysine side chains of the target substrate and the C-terminal group of ubiquitin or the peptide bond between the α-amino group of the target protein and the C-terminal of ubiquitin (Wilkinson 1997). DUBs play a pivotal role as regulators of the turnover rate, activation, recycling, and localization of many proteins and thus are essential for regulating several signaling pathways and for cellular homeostasis (Komander et al. 2009; Reyes-Turcu et al. 2009). The role of the UPS in controlling DNA replication following replication stress and in response to DNA damage is described in excellent recent

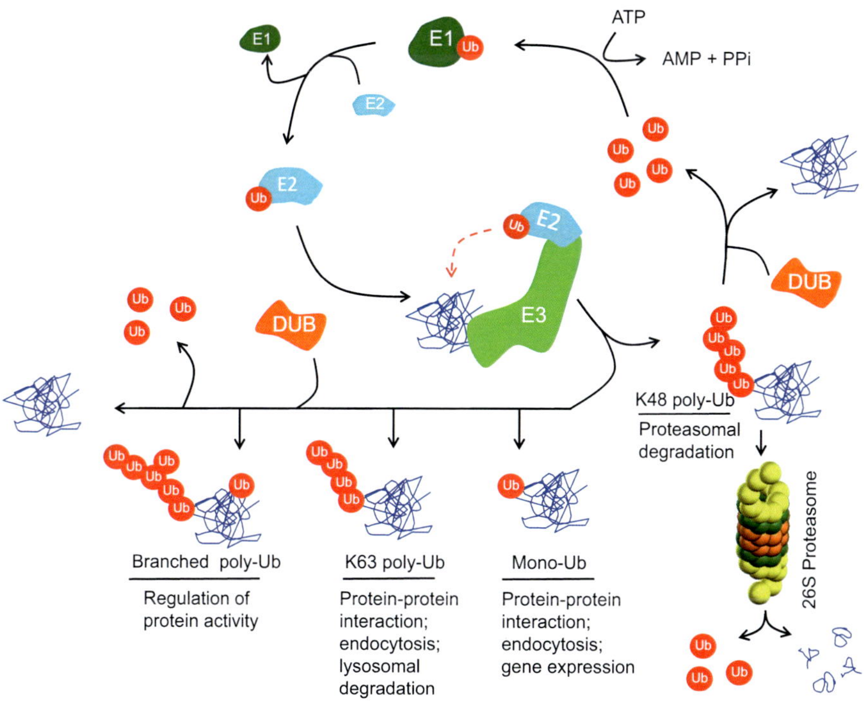

Fig. 19.1 Schematic of the various steps involved in the ubiquitylation of protein substrates via the UPS. Ubiquitin (Ub) is conjugated to various ubiquitylation substrates through various linkages to form monoubiquitylated (mono-Ub), lys-48-linked polyubiquitylated (K48 poly-Ub), lys-63-linked polyubiquitylated, and branched polyubiquitylated (branched poly-ub) species. Other forms of ubiquitin linkages are not shown. The process begins with ubiquitin first activated and bound through a thioester bond by the ubiquitin-activating enzyme (E1). Activated ubiquitin is subsequently transferred to one of several ubiquitin-conjugating enzymes (E2) through another thioester bond. An E3 ubiquitin ligase (E3) then promotes the transfer of ubiquitin to the target substrate through interaction with the E2-charged ubiquitin, whereby a covalent isopeptide bond is formed between the C-terminus of ubiquitin and a specific lysine residue on the substrate. Elongation of the ubiquitin chain is effected when the C-terminus of another ubiquitin moiety is linked to one of seven lysine residues (e.g., K48) or the fist methionine residue (M1) on the first ubiquitin. Polyubiquitylation through K48 or the less common K29 linkages targets the ubiquitylated substrate for proteasomal degradation via the 26S proteasome. Other polyubiquitylation linkages (e.g., K63) serve non-proteolytic functions. Ubiquitin and polyubiquitin chains can be removed from the substrates through the activity of one of many highly specific cysteine proteases, called deubiquitylating enzymes (DUBs), which can cleave both the isopeptide bond between the ubiquitin and ε-amino group of the lysine on the substrate or on ubiquitin (in a polyubiquitin chain) or the peptide bond between ubiquitin and the N-terminal methionine of ubiquitin

reviews (Garcia-Rodriguez et al. 2016; Renaudin et al. 2016; Sommers et al. 2015). In this chapter, we focus on protein ubiquitylation leading to proteasomal degradation or modification of function of key proteins to control DNA replication, with an emphasis on key E3 ubiquitin ligases and DUBs.

19.2 Structure and Function of Cullin-RING Ubiquitin Ligases Controlling DNA Replication

The **C**ullin-**R**ING (*R*eally *I*nteresting *N*ew *G*ene) E3 ubiquitin **L**igases (CRLs) represent the largest family of E3 ligases and are responsible for the ubiquitylation of approximately 20% of total cellular proteins degraded through the proteasome (Soucy et al. 2009). The other major E3 ligases belong to the HECT (*H*omologous to the *E*6-AP *C*arboxyl *T*erminus) domain containing E3 ubiquitin ligases (Li et al. 2008; Deshaies and Joazeiro 2009; Skaar et al. 2014). CRLs play significant roles in various processes including cell cycle regulation, cell proliferation, and tumorigenesis (Petroski and Deshaies 2005; Hotton and Callis 2008; Bosu and Kipreos 2008). Family members include eight cullin proteins (cullins 1, 2, 3, 4A, 4B, 5, 7, and 9 (also known as PARC or p53-associated parkin-like cytoplasmic protein)) and a cullin-like protein ANAPC2 or APC2. The general description of the structure and function of CRLs has been described in several excellent reviews (Chen et al. 2015; Lydeard et al. 2013; Sarikas et al. 2011; Duda et al. 2011; Hua and Vierstra 2011; Lipkowitz and Weissman 2011; Deshaies and Joazeiro 2009; Hotton and Callis 2008). The SCF (**SKP1**-**C**ullin1-**F**-Box protein; also known as CRL1, Fig. 19.2) is one of the well-characterized members of CRL ligases best known for its role in the regulation of cell cycle progression, cellular proliferation, apoptosis, and differentiation (Nakayama and Nakayama 2005; Welcker and Clurman 2008; Maser et al. 2007; Huang et al. 2010; Lee and Diehl 2014; Duan et al. 2012). SCF recognizes and promotes the ubiquitylation and degradation of its substrates through association with one of many substrate receptors (69 in mammalian cells), collectively known as F-box proteins, constituting a large family of distinct SCF ligases with varying specificity (Heo et al. 2016; Kipreos and Pagano 2000; Cardozo and Pagano 2004; Skaar et al. 2013; Wang et al. 2014). These substrate receptors utilize their F-box motifs to associate with the SKP1 (S-phase kinase-associated protein 1) bridge subunit (Wang et al. 2014). The SKP1 subunit of the SCF ligase bridges the F-box proteins (along with their cognate substrates) to the N-terminal domain of the cullin 1 subunit. The C-terminal domain of cullin 1 interacts with a small RING domain protein (RBX1 or RBX2), which is essential for the recruitment of E2 UBCs necessary for the polyubiquitylation of substrates distend for proteolytic degradation (Fig. 19.2).

Three subfamilies (FBXW, FBXL, and FBXO) characterize the F-box proteins, depending on whether they contain WD40 repeats (FBXW or FBW series), leucine-rich repeats (FBXL or FBL series), or variable other or "catch-all" domains (FBXO or F-box only series) (Skaar et al. 2014). Through their ability to assemble distinct

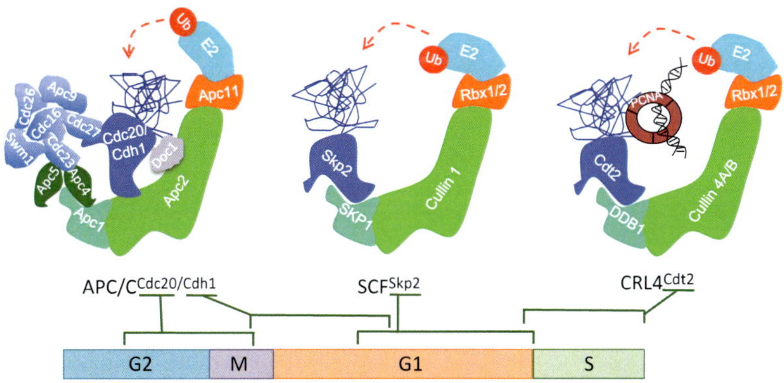

Fig. 19.2 Molecular architecture of the three multi-subunit E3 ubiquitin ligases controlling DNA rereplication and their activity during the cell cycle. The schematic illustrates the general architecture of the APC/C (CDC20 or CDH1), cullin 1 (SCF), and cullin 4 (CRL4) E3 ubiquitin ligases. Ubiquitin molecules (red circles) are transferred to the substrate (blue ribbons) through the activity of one of the three E3 ubiquitin ligase complexes (made of a cullin or cullin-like subunit (light green), a bridge protein(s) (cyan), substrate adaptors (royal blue), and small RING protein (orange)) and E2 conjugating enzymes (baby blue). Multiple proteins bridge the substrate receptors CDC20 or CDH1 to the APC2 scaffold subunit of the APC/C ligase. SKP1 and DDB1 (damage-specific DNA-binding protein 1) bridge the substrate receptors SKP2 and CDT2 to the cullin 1 or cullin 4 (A or B) of the SCF^{SKP2} or $CRL4^{CDT2}$ E3 ligases, respectively. $CRL4^{CDT2}$ recognizes its ubiquitylation substrates when bound to chromatin-bound trimeric proliferating cell nuclear antigen (PCNA), encircling DNA (black helix). *G1* first gap phase of the cell cycle, *S* DNA synthesis phase, *G2* second gap phase of the cell cycle, *M* mitosis

SCF ligases, many of these F-box proteins are involved in the regulation of DNA replication and cell cycle progression, and their expression or activity is altered in human malignancies (Heo et al. 2016). For example, the SCF^{SKP2} ubiquitin ligase (Fig. 19.2) composed of the core SCF ligase associating with the substrate receptor and SKP2 (S-phase kinase-associated protein 2; also known as FBL1) is an essential driver of cell cycle progression, primarily through impacting DNA replication either directly or indirectly. SCF^{SKP2} directly controls DNA replication through its ability to promote the timely ubiquitin-dependent degradation of several components of the pre-replication complexes (pre-RCs) from mid-G1-phase (Fig. 19.3). This ensures that licensing of replication origins occurs only from late M till mid-G1-phase of the cell cycle and is prevented from occurring again until cells complete the genome duplication and following the segregation of daughter chromosomes in late M (Fig. 19.4; see below). SCF^{SKP2} also activates DNA replication through activating cyclin-dependent kinases (CDKs) primarily through the ubiquitylation and degradation of negative CDK regulators (e.g., the CDK inhibitors $p21^{CIP1}$, $p27^{KIP1}$, and $p57^{KIP2}$). Later in the cell cycle, SCF^{SKP2} is also responsible for the targeted proteolysis of positive regulators of CDKs (e.g., cyclin D1 and cyclin A) to promote the handover of CDK activity from one CDK to the next (Fig. 19.4). The latter activity ensures the availability of CDK molecules for assembling distinct cyclin-CDK complexes for catalyzing various specific activities necessary for the irreversible pro-

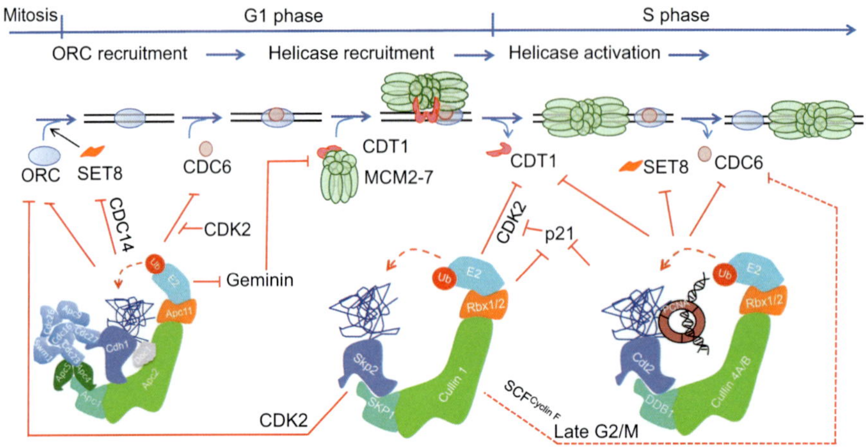

Fig. 19.3 Control of MCM2-7 loading via the UPS. The APC/C^{Cdh1} ligase promotes MCM2-7 loading in late M and in G1 by degrading Geminin. However, APC/C^{Cdh1} also appears to limit the abundance of certain licensing factors like *Drosophila* ORC1 and mammalian CDC6 in G1, narrowing to late M-phase the window in the cell cycle when there is enough ORC, CDC6, and active CDT1 available to load MCM2-7 on origins. The E3 ubiquitin ligases, SCF^{SKP2} and CRL4^{CDT2} limit the abundance of key proteins involved in MCM2-7 loading (pre-RC assembly) in the S-, G2-, and early M-phases of the cell cycle. The SCF ubiquitin ligase also utilizes the substrate recognition subunit cyclin F (SCF$^{Cyclin\ F}$; *red dashed lines*) to suppress origin relicensing by promoting the ubiquitylation and degradation of CDC6 in late G2- and early M-phase of the cell cycle

gression of the cell cycle. For example, the ubiquitin-dependent proteolysis of cyclin A via the SCF^{SKP2} ligase ensures the availability of CDK1 molecules to assemble cyclin B-CDK1 complexes essential for G2 progression. Likewise, the degradation of cyclin E, mediated through the activity of the SCF ligase with the substrate receptor FBXW7 (SCF^{FBXW7}) following cyclin E phosphorylation by CDK2, ensures the availability of CDK2 molecules to assemble cyclin A-CDK2 complexes necessary for S-phase progression (Clurman et al. 1996; Koepp et al. 2001).

The other major E3 ubiquitin ligase regulating DNA replication is the APC/C (anaphase-promoting complex/cyclosome) ubiquitin ligase (Fig. 19.2). This ubiquitin ligase utilizes the APC2 cullin-like scaffold to assemble the largest multi-subunit CRL ligase in mammalian cells (van Leuken et al. 2008). APC/C associates with one of two substrate adaptor proteins, CDH1 and CDC20, for recognizing and promoting the polyubiquitylation (both K48- and K11-linked ubiquitin conjugation) of a large number of key drivers of cell cycle progression (Visintin et al. 1997; Zachariae and Nasmyth 1999; Pines 2006). Unlike the SCF-FBX ligases, which recognize the vast majority of target substrates through phosphorylation-dependent sequence motifs or "phospho-degrons" within these substrates (Skaar et al. 2013), APC/C ligases recognize proteins containing D-boxes and KEN-boxes, such as cyclin A and cyclin B (Pfleger and Kirschner 2000; Pfleger et al. 2001). The assembly of APC/C^{CDC20} is driven by CDK-mediated phosphorylation of CDC20 and is active primarily in mitosis (Rahal and Amon 2008); this is critical for the initial

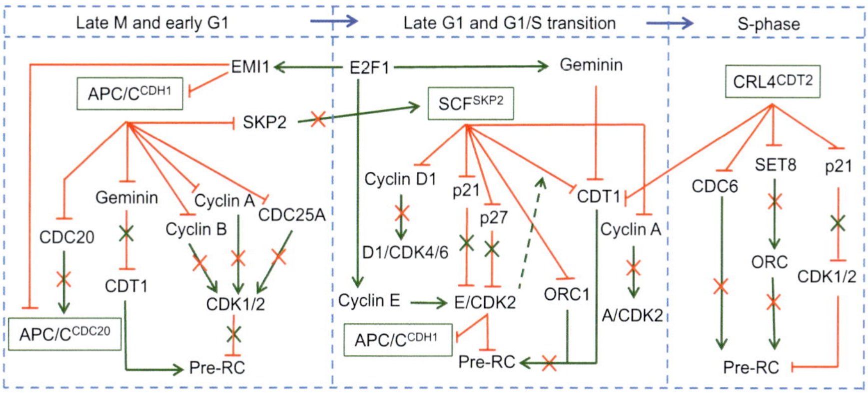

Fig. 19.4 UPS control of DNA replication via direct and indirect mechanisms. MCM2-7 is loaded (Pre-RCs established) in late M (and perhaps early G1)-phase of the cell cycle and is prevented from being loaded again from the G_1/S transition until the latest part of the next mitosis. This regulation is dependent primarily on the activity of CDK, which is maintained at low levels through late mitosis and G_1 through the activity of APC/C^{CDH1} ubiquitin ligase. CDK activity peaks again in late G1 and at the G1/S transition through the activity of the SCF^{SKP2} ligase, which promotes the proteolysis of negative regulators of CDK2. CDK2 activity remains high in S-phase through the targeted proteolysis of p21 via CRL4^{CDT2}. Such elevated levels of CDK suppress pre-RC in late G1 and in S and prevent aberrant origin relicensing and DNA rereplication. All three E3 ligases also directly control the abundance of pre-RC proteins such that replication occurs only once in the cell cycle. The circuit receives transcriptional input from the E2F1 transcription factor, which in addition to promoting S-phase entry into the cycle through upregulation of cyclin E and the consequent elevation of cyclin E/CDK2 activity provides a negative feedback control to shut down the activity of APC/C ligase (via EMI1 upregulation) and inactivation of CDT1 (via Geminin upregulation). This ensures that cells proceed in S-phase without aberrant licensing. Green arrows, positive regulation. Red lines, negative regulation

targeted proteolysis of mitotic cyclins and for exit from mitosis. Assembly of APC/C^{CDH1} on the other hand is stimulated through dephosphorylation of the CDH1 subunit by the CDC14A phosphatase (Cdc14 in budding yeast) and is active in late mitosis and though G1-phase of the cell cycle (Jaspersen et al. 1999; Robbins and Cross 2010; Sullivan and Morgan 2007). The oscillating activities of APC/C ligases during the cell cycle, as described in more detail below, are critical for not only controlling the timing of DNA replication but also for driving cell cycle progression and for guarding against genomic instability and cancer development (van Leuken et al. 2008; Nakayama and Nakayama 2006).

19.3 Regulation of Origin Licensing via the UPS

The control of eukaryotic DNA replication begins at the end of mitosis and through the G1-phase of the cell cycle, whereby the sequential binding of origin recognition complex proteins (ORCs) followed by CDC6 and CDT1 loads the replicative helicase MCM2-7 onto replication origins: the establishment of pre-RCs (Fig. 19.3). Origin licensing is inhibited by CDK activity, which suppresses pre-RC formation either by promoting the phosphorylation-dependent nuclear exclusion of certain replication licensing factors (e.g., mammalian CDC6) or through the targeted proteolysis of a number pre-RC components though the UPS (Zhu et al. 2005). The degradation of SKP2 by the APC/C^{CDH1} ligase is particularly important for the stabilization of key replication factors in late M- and G1-phases of eukaryotic cell cycles (Fig. 19.4). For example, human ORC1, the largest subunit of ORC, is stable in G1 because SKP2 is degraded by APC/ C^{CDH1} in G1 but is specifically degraded in S-phase cells via the activity of the SCF^{SKP2} ubiquitin ligase (Mendez et al. 2002; Tatsumi et al. 2003). *Drosophila* ORC1 is paradoxically ubiquitylated and degraded via the $APC/^{CFZr/CDH1}$ E3 ligase as cells exit mitosis and during G1 phase of the cell cycle, and this requires the non-conserved N-terminal domain of ORC1 (Araki et al. 2003, 2005; Narbonne-Reveau et al. 2008). Thus, ORC would have had to finish its licensing activity (MCM2-7 loading) before cells exit mitosis. The *Drosophila* ORC1 protein reappears in late G1 following induction of ORC1 RNA by E2F1 after the latter is activated by cyclin-CDK.

CDC6, another protein important for MCM2-7 loading, is also regulated extensively by the UPS. Yeast Cdc6 is degraded via an SCF^{Cdc4}-dependent proteolytic pathway, while the phosphorylation of mammalian CDC6 by increasing CDK activity in late G1 and in S-phase triggers its relocalization to the cytoplasm, thereby preventing origin licensing from occurring until CDK activity drops again in late mitosis (Aparicio et al. 1997; Tanaka et al. 1997; Fujita et al. 1999; Jiang et al. 1999; Petersen et al. 1999; Saha et al. 1998; Alexandrow and Hamlin 2004). Ubiquitin-mediated proteolysis of mammalian CDC6 in G1 also controls its abundance in G1 and in quiescent non-cycling cells. This is carried out by the APC/C^{CDH1} ligase and is mediated through interaction between the destruction box and KEN box motifs of CDC6 and CDH1 (Petersen et al. 2000). This too suggests that as with fly ORC1 above, the degradation of mammalian CDC6 in G1 implies that the licensing activity must be completed before cells exit mitosis. The degradation of CDC6 by APC in G1 poses a special problem for cells entering the cell cycle from G0. It turns out that CDC6 is phosphorylated by cyclin E/CDK2 in cells entering the cell cycle from quiescence, and this inhibits APC/C^{CDH1}-mediated proteolysis of CDC6 (Mailand and Diffley 2005). This protection from ubiquitylation ensures that sufficient origins of replications are licensed before cells enter S-phase (Fig. 19.3). Once cells enter S-phase, CDC6 is targeted for ubiquitylation and degradation again, but this is mediated via the $CRL4^{CDT2}$ ligase (Clijsters and Wolthuis 2014). Late in the cell cycle (in G2 and early M), CDC6 is targeted for proteolysis via the $SCF^{Cyclin\ F}$ ligase (Walter et al. 2016).

The third protein important for loading MCM2-7 is CDT1. CDK kinase activity promotes the ubiquitylation and degradation of CDT1 via the SCF^{SKP2} ligase at the G1/S transition of the cell cycle, and this, along with the targeted proteolysis of CDT1 in S-phase via the $CRL4^{CDT2}$ ligase (see below), ensures that CDT1 is not available for relicensing origins of replication in late G1 or in S-phase. Metazoans employ a second mechanism by which CDT1 is inactivated in S-phase through interaction with and inhibition by Geminin (Wohlschlegel et al. 2000; Tada et al.

2001). Geminin, however, is degraded in G1 through the activity of the APC/C E3 ligase (McGarry and Kirschner 1998), which is associated with low CDK activity. Origin licensing in cycling cells therefore can only proceed with low CDK activity (Figs. 19.3 and 19.4).

The mechanisms that maintain low CDK activity during late mitosis and in G1 are themselves under the control of the UPS. For example, the APC/C ligase is responsible for the decrease of mitotic CDK activity by promoting the ubiquitylation and degradation of the mitotic cyclins A and B (den Elzen and Pines 2001). This is first mediated through APC/C^{CDC20}, which promotes the degradation of cyclin A and cyclin B in prometaphase and metaphase, respectively. As cells proceed in metaphase, the degradation of cyclin B is mediated by APC/C^{CDH1}. APC/C^{CDH1} additionally suppresses CDK2 activity in late M and in early G1 by promoting the degradation of the dual-specificity CDC25A phosphatase, which catalyzes the removal of inhibitory phosphorylations on CDK2 (Donzelli et al. 2002). Low CDK activity in late M and early G1 is also aided through the accumulation of elevated levels of the CDK inhibitors (CKI) p21 and p27, which can bind to and inhibit CDK2 in early G1 (Abbas and Dutta 2009). The protein levels of these two CDK inhibitors in G1 are under the control of the SCF^{SKP2} ligase, which is active only in late G1.

19.4 Feedback Control of Origin Licensing Through the UPS

The ubiquitylation reactions involved in regulating origin licensing function in "self-regulating" networks with multiple feedbacks, whereby elevated CDK activity in mitosis turns on specific ubiquitylation reactions that feedback to decrease CDK activity during late mitosis and early G1 (Fig. 19.4). In both yeast and man, elevated G2 and mitotic CDK activity renders the APC/C^{CDH1} inactive due to phosphorylation-dependent conformational changes in CDH1 precluding assembly of the active ligase. However, exit from mitosis requires inactivation of mitotic CDK, and this requires (a) targeted proteolysis of mitotic cyclins via the APC/C^{CDH1} (or its homologue in yeast APC/C^{Hct1}) and (b) the stabilization of the CDK inhibitor p21 (or its homologue in yeast, Sic1) through degradation of CDC20. APC/C^{CDC20} targets p21 for proteasomal activity during mitosis (Shirayama et al. 1999; Amador et al. 2007). APC/C^{CDC20} ligase also promotes (through an unknown mechanism) the release of the yeast Cdc14 phosphatase from the nucleolus or the human CDC14A from centrosomes (Mocciaro et al. 2010; Kaiser et al. 2002; Chen et al. 2016; Shirayama et al. 1999; Bembenek and Yu 2001). In yeast, Cdc14 dephosphorylates and activates the Cdh1/Hct1 subunit, which competes with and targets Cdc20 for ubiquitylation and proteolysis, thereby assembling active APC/C$^{Cdh1/Hct1}$ and inactivating the APC/C^{Cdc20} ligase (Jaspersen et al. 1999; Robbins and Cross 2010; Sullivan and Morgan 2007). Thus, the simultaneous activation of APC/C^{CDH1} and inactivation of APC/C^{CDC20} trigger CDK inactivation, culminating in exit from mitosis. As cells progress through early G1-phase of the cell cycle with low CDK activity, APC/C^{CDH1} promotes the degradation of SKP2 (Bashir et al. 2004; Wei et al. 2004), and this prevents the premature formation of the

SCF^{SKP2} complex and consequent destabilization of its ubiquitylation substrates p21 and p27, thus maintaining low CDK activity (Fig. 19.4).

At the G1/S transition, cyclin E-CDK2 phosphorylates CDH1 leading to APC/C^{CDH1} inactivation (Cappell et al. 2016). In addition CDH1 binds to and is inhibited by the F-box protein and early mitotic inhibitor 1 (EMI1), marking a "point of no return" driving cells entry into S-phase (Cappell et al. 2016). As cells proceed through S-phase, the activity of both APC/C^{CDH1} and APC/C^{CDC20} is kept low through interaction between EMI1 with CDH1 or CDC20 (Cappell et al. 2016; Reimann et al. 2001). Inhibition of APC/C^{CDC20} by EMI1 is essential for the stabilization of mitotic cyclins A and B, thereby ensuring the completion of DNA synthesis, prevention of DNA rereplication (see below), and progression through G2-phase (Di Fiore and Pines 2007).

As CDK1 activity builds up, the APC/C^{CDC20} activity is increased in prometaphase through the coordinated sequential phosphorylation of the APC3 and APC1 subunits of the APC/C ligase by CDK1 and the docking of CDC20 onto the APC/C ligase (Fujimitsu et al. 2016). Yeast APC/C^{Cdc20} is similarly activated by mitotic Clb-CDK activity, and this is critical for metaphase-anaphase transition (Rahal and Amon 2008). Elevated CDK1 activity in G2 and in early mitosis is enhanced by the targeted ubiquitylation and degradation of the CDK1 tyrosine kinase and inhibitor Wee1 via the activity of an SCF ligase, which utilizes the β-transducin repeat-containing protein 1 (βTRCP1) as substrate recognition subunits ($SCF^{\beta TRCP1}$) (Watanabe et al. 2004). Interestingly, this same E3 ligase also promotes the ubiquitylation and degradation of EMI1 (Guardavaccaro et al. 2003) and thus contributing to the rising activity of APC/C^{CDC20} necessary for mitotic progression. This, along with the targeted ubiquitylation and degradation of EMI1 via the $SCF^{\beta TRCP}$ ligase, ensures optimal activity of APC/C^{CDC20} ligase activity to drive mitotic progression. Thus, fluctuating CDK activity throughout the cell cycle, itself regulated through ubiquitin-mediated proteolysis, ensures that the ubiquitylation machinery required for the irreversible progression through the cell cycle is temporally coordinated with successive stabilization and destabilization of key drivers of cell cycle progression.

19.5 Inhibition of Origin Relicensing and DNA Rereplication Through the UPS

An extensive body of literature demonstrates a critical role for the UPS in restricting origin licensing to late M- and early G1-phase of the cell cycle and thus preventing aberrant relicensing and refiring of replication origins or rereplication (Fig. 19.3). As CDK activity builds up in G1 cells, the phosphorylation of the tumor suppressor protein pRb, first by cyclin D1/CDK4/6 and subsequently by cyclin E/CDK2, results in its dissociation from the E2F1 transcription factor, which transactivates Geminin and dozen other genes essential for S-phase progression (Wong et al. 2011). Increased Geminin transcription by E2F1 and increased Geminin protein stability through cyclin E/CDK2-dependent suppression of APC/C^{CDH1} prevent origin relicensing by directly binding CDT1, which sterically hinders its ability to recruit MCM2-7 complexes to origins of replication (Fig. 19.4). Second, and as cells proceed through S-phase, E2F1-dependent transcription of cyclin A coupled with enhanced protein stability of cyclin A due to inhibition of APC/C^{CDC20} by EMI1 causes cyclin A-CDK2 activity to increase dramatically. Cyclin A/CDK2 phosphorylates CDT1 at a conserved N-terminal threonine residue (Thr-29) creating a phospho-degron that is specifically recognized by SKP2, which is itself stabilized due to inhibition of the APC/C^{CDH1} activity in S-phase. The newly assembled SCF^{SKP2} promotes phospho-CDT1 ubiquitylation and degradation (Li et al. 2003, 2004; Takeda et al. 2005). Thus, increased EMI1 levels at the G1/S transition and throughout S-phase with the consequent inhibition of APC/C ubiquitylation activity are critical for suppressing origin relicensing in mammalian cells, both by inhibiting CDT1 activity by Geminin and by promoting its proteolysis by the CDK-dependent and SCF^{SKP2}-mediated activity. Interestingly, alleviating Geminin-mediated suppression of CDT1, or inhibiting cyclin A-dependent proteolytic degradation of CDT1 via the SCF^{SKP2} ligase alone, is sufficient to promote origin relicensing and rereplication in certain cancer cell types. This, however, is insufficient to induce rereplication in some other cancer cell types or in non-cancer cells (Zhu and Depamphilis 2009; Benamar et al. 2016; Machida and Dutta 2007). Inhibition of EMI1 (e.g., by short-interfering RNAs (siRNA)) on the other hand is sufficient to trigger robust rereplication both in cancer and non-cancer cells (Machida and Dutta 2007; Benamar et al. 2016). Thus, APC/C activity in S-phase inhibits origin licensing through the timely inhibition and degradation of CDT1 (Sivaprasad et al. 2007).

Whereas the SCF^{SKP2} ligase promotes the degradation of cyclin A/CDK2-phosphorylated soluble CDT1 in S-phase, the ubiquitylation and degradation of chromatin-bound CDT1 in S-phase occur through a phosphorylation-independent mechanism that requires the activity of the CRL4^{CDT2} ubiquitin ligase (Nishitani et al. 2006; Arias and Walter 2006; Senga et al. 2006; Jin et al. 2006). The overall composition and architecture of CRL4 ligases are very similar to CRL1 ligases (Fig. 19.2) (Angers et al. 2006; Higa and Zhang 2007). The core complex is composed of one of two E3 ubiquitin ligases (cullin 4A or cullin 4B), DDB1 (DNA damage-specific protein-1), which is a bridge protein analogous to the SKP1 subunit in CRL1, and functions to bridge one of many substrate receptors (also known as DCAFs; DDB1 and cullin 4 associated factors) to the cullin subunit, and through that to RBX1 or RBX2, required for the recruitment of E2 UBCs. DCAFs include at least 49 family members of WD motif-rich proteins that function to recruit substrates to the CRL4 ligase similar to the function of the F-box proteins in the CRL1 ligases (Angers et al. 2006; He et al. 2006; Higa et al. 2006; Jin et al. 2006). CRL4 is emerging as a master regulator of genome stability, and recent findings suggest that it orchestrates a variety of physiological processes, particularly those related to chromatin regulation and genomic stability (Jackson and Xiong 2009). The substrate adaptor CDT2 assembles with CRL4 to form a rather unique E3 ubiquitin ligase that does not recognize CDT1 (or several other ubiquitylation substrates) directly but specifically recognizes the substrate when it interacts with proliferating cell nuclear antigen (PCNA) (Arias and Walter 2006; Senga et al. 2006). PCNA, the processivity factor for DNA polymerase δ, serves this role as an accessory factor for recognition by CDT2 only when PCNA is bound to chromatin, a condition that is established only in S-phase and following certain types of DNA damage (Havens and Walter 2011; Abbas and Dutta 2011; Abbas et al. 2013). Thus DNA damage is another stimulus that uses this pathway to degrade CDT1 and other replication factors (Higa et al. 2003).

The interaction between CDT1 and PCNA occurs through a specialized PCNA-interacting protein (PIP) box motif. This PIP box, commonly referred to as the "PIP degron," is a modified version of the PIP box motif utilized by many PCNA-interacting proteins; in that it contains, in addition to the canonical sequence (Q-X-X-(I/L/M)-X-X-(F/Y)-(F/Y)), conserved threonine and aspartic acid residues at positions 5 and 6, respectively, as well as a basic amino acid residue C-terminal of the PIP box (at position +4) and a second basic amino acid residue at position +3 or +5 (or both) (Abbas et al. 2010; Havens and Walter 2011; Michishita et al. 2011; Havens and Walter 2009). The importance of $CRL4^{CDT2}$-mediated ubiquitylation and degradation of CDT1 is manifested by the fact that cells from various organisms that are deficient in cullin 4, DDB1, or CDT2 exhibit rereplication and genomic instability reminiscent to that seen in cells overexpressing CDT1 (Jin et al. 2006; Lovejoy et al. 2006; Vaziri et al. 2003; Tatsumi et al. 2006; Zhong et al. 2003; Kim et al. 2008; Sansam et al. 2006), and this is associated with double-strand DNA breaks, rereplication, and activation of the ATM and ATR-dependent checkpoints (Zhu et al. 2004; Zhu and Dutta 2006), which can be partially suppressed by co-depletion of CDT1 (Lovejoy et al. 2006). The ubiquitylation-dependent proteolysis of CDT1 via $CRL4^{CDT2}$ is a feature of all eukaryotes except for budding yeast, where the CDT1-MCM complexes are exported to the cytoplasm (Tanaka and Diffley 2002; Devault et al. 2002). This is likely due to the fact that although budding yeast contains orthologs for cullin 4 and DDB1, they lack an identifiable ortholog for CDT2 (Zaidi et al. 2008).

Synchronization experiments in human cancer cells have shown that the levels of CDT1 protein begin to degrease as cells enter S-phase but re-accumulate late in S-phase and reach significantly higher levels in G2 (Abbas et al. 2010). In late S and in G2 cells, CDT1 is protected from $CRL4^{CDT2}$-mediated ubiquitylation and degradation through two phosphorylation-dependent and distinct mechanisms. The first one employs the phosphorylation of CDT1 by the stress-activated mitogen-activated protein kinases (MAPK) p38 and JNK precluding recognition by $CRL4^{CDT2}$ (Chandrasekaran et al. 2011). The second mechanism is dependent on CDK1-dependent phosphorylation of CDT1 preventing its recruitment to chromatin (Rizzardi et al. 2015). Just as the ubiquitylation and degradation of CDT1 in S-phase are critical for preventing rereplication and cell cycle progression, its re-accumulation in late S and in G2 is equally important for cell cycle progression, although its role in G2 is not fully understood (Rizzardi et al. 2015). Although the steady state level and protein half-life of CDT1 are clearly increased in late S and in G2 cells, a recent report suggests that its abundance particularly in G2-phase is also under the control of another E3 ubiquitin ligase, the SCF^{FBXO31} ligase (Johansson et al. 2014). In this study, the authors demonstrated that depletion of cancer cells of the substrate adaptor and putative tumor suppressor FBXO31 protein by siRNA induces low levels of DNA rereplication (7.6% vs. 4.6% in control cells), which is insufficient to inhibit growth (Johansson et al. 2014). It remains unclear however, how the stabilized CDT1 in these G2 cells gain access to chromatin to trigger rereplication in the presence of elevated CDK1 activity. An interesting possibility is that the loss of SCF^{FBXO31} activity in tumors lacking FBXO31 or with FBXO31 inactivating mutations may result in a minute amount of rereplication that do not interfere with proliferation but are sufficient to induce gene amplification and/or genome instability exacerbating the tumorigenic phenotype (Green et al. 2010).

In addition to CDT1, the $CRL4^{CDT2}$ ligase promotes the ubiquitylation of several other proteins whose proteolysis in S-phase is critical for preventing origin relicensing and DNA rereplication. These include the CDK inhibitor p21, the histone H4 methyltransferase SET8 and CDC6 (Abbas et al. 2008, 2010; Nishitani et al. 2008; Kim et al. 2008; Tardat et al. 2010; Centore et al. 2010; Oda et al. 2010; Jorgensen et al. 2011; Clijsters and Wolthuis 2014). Ubiquitin-dependent degradation of p21 in S-phase via the $CRL4^{CDT2}$ ligase is critical for ensuring elevated CDK activity necessary for S-phase progression and for promoting DNA replication by freeing PCNA from inhibitory p21 (Abbas and Dutta 2009). Recent evidence also demonstrates that increased p21 stability downstream of $CRL4^{CDT2}$ inhibition stimulates rereplication in cancer cells, presumably due to suppression of CDK activity, a condition compatible with origin licensing (Kim et al. 2008; Benamar et al. 2016). Because these studies utilized the overexpression of a mutant p21 protein, which fails to interact with PCNA – a PIP degron mutant of p21 – these rereplicating cells are able to replicate DNA free from p21-mediated suppression of PCNA. Although ectopic expression of PIP degron mutant p21 induces only minor rereplication, the p21 protein is required for rereplication induced by CDT2 depletion (Benamar et al. 2016). Unlike the case for p21, the expression of PIP degron mutant of SET8, but not wild type SET8, induces robust rereplication, and this required the catalytic activity of this enigmatic methyltransferase (Abbas et al. 2010; Tardat et al. 2010). SET8 is also required for DNA rereplication in cells depleted of CDT2 (Benamar et al. 2016). SET8, also known as PR-SET7, is an enzyme that deposits a single methyl group on lysine 20 of nucleosomal histone 4 (H4K20me1) (Nishioka et al. 2002; Xiao et al. 2005). H4K20 can also be di- and tri-methylated (H4K20me2/3), but this is carried out by the SUV4-20H1 and SUV4-20H2 histone methyltransferases, which utilize H4K20me1 as substrate (Schotta et al. 2008). How SET8 promotes rereplication following $CRL4^{CDT2}$ inactivation is not entirely clear, but likely dependent on its ability to mono-methylate H4K20 at replication origins (Tardat et al. 2010). In fact, tethering SET8 to a specific genomic locus permitted loading of pre-RC proteins on chromatin and induced rereplication from that site. Interestingly, the rereplication phenotype associated with ectopic expression of PIP degron mutant SET8 was correlated with increased di- and tri-methylation of H4K20 with a concurrent reduction of H4K20me1 and required SUV4-20H1 and SUV4-20H2 (Abbas et al. 2010; Beck et al. 2012). Consistent with this, it was shown that the ability of SET8 to nucleate origins of replication is mediated through SUV4-20H1- and SUV4-20H2-dependent recruitment of ORC1 and the ORC-associated protein (ORCA) proteins to chromatin and that both of these proteins are capable of binding H4K20me2/3 in vitro (Beck et al. 2012). Intriguingly, CRL4^{CDT2}-mediated proteolytic degradation of SET8 is critical for S→G2 cell cycle progression and for proper chromatin condensation, and inactivation of this pathway in cancer cells results not only in the induction of rereplication, but also in the repression of histone gene transcription and the consequent chromatin decondensation as well as repression of E2F1-driven gene expression (Abbas et al. 2010). These latter toxicities are likely mediated through enrichment of the repressive chromatin marks (H4K20me2/3) at the promoters of these genes (Abbas et al. 2010). Thus, cell viability is critically dependent on CRL4^{CDT2}-dependent downregulation of SET8 in S-phase for the maintenance of a stable epigenetic state. In M-phase SET8 protein is phosphorylated at Ser-29 by cyclin B/CDK1 from prometaphase to early anaphase, preceding the accumulation of H4K20me1 (Wu et al. 2010). This phosphorylation is reversed in late M-phase by the CDC14 phosphatase, which is activated by APC/C^{CDC20}. SET8 dephosphorylation renders the protein subject to proteolytic degradation via APC/C^{CDH1}, which facilitates mitotic progression (Wu et al. 2010).

As discussed above, CDC6 is another pre-RC component, which is targeted for ubiquitylation and degradation in S-phase via the CRL4^{CDT2} ligase, and this is dependent on a conserved PIP degron contained near its N-terminus (Clijsters and Wolthuis 2014). However, although CDC6 depletion prevented rereplication induced by CDT2 depletion, it is not clear whether failure to degrade CDC6 via this ligase is sufficient to induce rereplication. As mentioned above, CDC6 is also ubiquitylated and degraded via the SCF$^{Cyclin\ F}$ ligase in G2- and in early M-phase, and this is also critical for preventing DNA rereplication as the depletion of cyclin F or the expression of a stable mutant form of CDC6 promotes rereplication and genome stability in cells lacking Geminin (Walter et al. 2016). Finally, in *Drosophila melanogaster*, CRL4^{CDT2} also promotes the ubiquitylation and degradation of the transcription factor E2f1, and this is dependent on E2f1 interaction with PCNA through a PIP degron, which is not conserved in human E2F1 protein (Shibutani et al. 2008). From these studies, it is becoming abundantly clear that the CRL4^{CDT2} ligase is a major inhibitor of origin licensing in S- and G2-phase of the cell cycle and does so through the degradation of key positive regulators of origin activity via their specialized interaction with chromatin-bound PCNA.

19.6 Targeting the Ubiquitylation Machinery Controlling DNA Replication Licensing for Therapeutic Gain

Deregulated origin licensing can have serious consequences. On one hand, failure of the ordered assembly of pre-RCs on sufficient origins of replications in late M and in G1 inhibits cell proliferation. This is demonstrated by the fact that targeted knockout or knockdown of many of the pre-RC components in various eukaryotes interferes with viability, and thus some of these components, e.g., the MCM2-7 helicases, are considered promising chemotherapeutic targets in cancer (Lei 2005; Simon and Schwacha 2014). On the other hand, excessive origin licensing can result in DNA rereplication, which is deleterious to cells owing to the accumulation of replication intermediates and collapsed replication forks (Abbas et al. 2013). Substantial evidence shows that induction of rereplication in cancer cells through genetic manipulation (e.g., by depletion of Geminin, EMI1, or CDT2) results in growth inhibition through the accumulation of DNA damage, cell cycle checkpoint activation, and the induction of growth arrest and/or cell death (Zhu et al. 2005, 2004; Abbas and Dutta 2011; Abbas et al. 2013;Zhu and Dutta 2006). The first indication showing that pharmacological induction of rereplication is associated with inhibitory antiproliferation activity came from the accidental discovery that a small molecule, called MLN4924, designed to inhibit protein neddylation, inhibits the proliferation of human cancer cell lines and is associated with DNA rereplication (Soucy et al. 2009; Lin et al. 2010; Wei et al. 2012; Jazaeri et al. 2013). Neddylation of cullins is a posttranslational modification necessary for the activity of cullin-RING ligases. A small protein called NEDD8 is covalently attached to target protein substrate by an enzyme cascade system similar to ubiquitylation (Merlet et al. 2009). MLN4924 (also known as pevonedistat), currently in multiple clinical trials for hematologic (NCT00722488, NCT00911066) and solid (NCT01011530) malignancies, inhibits the NEDD8-activating enzyme (NAE), which catalyzes the first step in this enzymatic cascade, and its ability to suppress cullin neddylation and induction of rereplication suggested that the anticancer activity of this investigational drug is mediated through the suppression of $CRL4^{CDT2}$ and the stabilization of its ubiquitylation substrate CDT1 (Soucy et al. 2009). Subsequent studies, however, demonstrated that MLN4924, in addition to inhibiting all CRLs, inhibits a number of signal transduction pathways critical for cell proliferation, including the NFκB, AKT, and mTOR signal transduction pathways (Soucy et al. 2009; Lin et al. 2010; Milhollen et al. 2010, 2011; Gu et al. 2014; Godbersen et al. 2014;Li et al. 2014a, b). We have recently shown that the stabilization of the $CRL4^{CDT2}$ substrates SET8 and p21 is critical for the induction of DNA rereplication and senescence in melanoma cells treated with MLN4924 (Benamar et al. 2016). In particular, melanoma cells with hypomorphic expression of either of these two proteins are resistant to MLN4924-induced rereplication and senescence in culture, and, more importantly, tumor xenografts of these melanoma lines are refractory to inhibition by MLN4924 in nude mice. Interestingly, MLN4924 induces rereplication only in cancer cells but not in nonmalignant primary cells, suggesting that normal cells and tissues may be protected from the toxicity induced by this agent (Benamar et al. 2016). Thus, MLN4924 represents the first compound with anticancer activity, which selectively inhibits cancer cell proliferation through the induction of rereplication (Nawrocki et al. 2012; Zhao and Sun 2013; Tanaka et al. 2013; Zhao et al. 2014; Jiang and Jia 2015; Benamar et al. 2016).

Although specific inhibitors for $CRL4^{CDT2}$ ligases are yet to be developed, specific inhibitors of the other major E3 ligase controlling DNA replication, SCF^{SKP2}, already exist and exhibit potent anticancer activity in nude mice (Chan et al. 2013). However, because SKP2 promotes the degradation of many other proteins in addition to the ones that impact DNA replication, it is unclear whether such antitumor activity is dependent on dysregulated DNA replication. An image-based high-throughput screening for compounds with the ability to induce rereplication selectively in cancer cells has identified other small molecules that selectively inhibit cancer cell proliferation (Zhu et al. 2011). However, whether these compounds exhibit antitumor activity or induce rereplication via targeting components or the UPS remains to be determined.

19.7 Regulation of DNA Synthesis via the UPS

Whereas the role of the UPS for controlling origin licensing and entry of cells into S-phase is well established, its role in regulating ongoing DNA replication and in coordinating the associated chromatin dynamics in unperturbed cells is just beginning to be appreciated. The latter is accomplished through the activity of histone proteins, histone chaperones, nucleosome-remodeling complexes, histone and DNA methylation-binding proteins, and chromatin-modifying enzymes. Together, these factors facilitate nucleosomal disassembly ahead of incoming replication forks and promote their reassembly following their passage. Many of these proteins and protein complexes are regulated through the UPS, and several excellent recent reviews describe these activities in details (Garcia-Rodriguez et al. 2016; Talbert and Henikoff 2017; Henikoff 2016; Almouzni and Cedar 2016). In this section, we focus on the role of the UPS in regulating replication initiation, progression, and termination.

A high-resolution proteome-wide mass spectrometry-based identification of ubiquitylated peptides in unperturbed cells identified many components of the replication machinery as substrates for ubiquitin modification, although in some cases this is not associated with proteasomal degradation (Wagner et al. 2011). These include components of various complexes regulating DNA replication, such as the MCM2-7 helicase, GINS, and replication factor C (RFC) clamp loader complexes. In addition, almost all DNA polymerases and their associated factors had detectable ubiquitylated peptides identified in this screen. Although the individual components coordinating these ubiquitylation events and their functional significance remain to be defined, this study suggests that ubiquitylation plays an important role in regulating DNA replication and its fidelity. Not all of these ubiquitylation events are associated with proteolysis of the DNA replication factors. For example, the two subunits of the mammalian DNA polymerase δ, p66 and p12, undergo non-proteolytic ubiquitylation during S-phase that is thought to regulate protein-protein interactions within the polymerase complex or with other replication factors (Liu and Warbrick 2006). Following DNA damage, the p12 subunit is degraded through a PCNA-dependent ubiquitylation via the CRL4^{CDT2} ligase, and this is critical for inhibiting fork progression (Terai et al. 2013). Similarly, the *S. pombe* Pol2, the catalytic subunit of polymerase ε, which is responsible for leading strand synthesis, undergoes ubiquitin-dependent proteolysis, and this is mediated via the SCF^{Pof3} ligase (Roseaulin et al. 2013). Because the catalytic subunit of polymerase δ (responsible for lagging strand synthesis) remains stable, despite undergoing ubiquitylation, the results suggest that DNA synthesis of the leading strand requires a "fresh" supply of DNA polymerase, whereas the synthesis of the discontinuous lagging strand does not (Roseaulin et al. 2013). This example highlights the complexity in the mechanisms regulating the abundance and activity of replication factors through the UPS in response to various stimuli. Minichromosome maintenance protein 10 (MCM10) is another important DNA replication protein, which undergoes non-proteolytic ubiquitylation (Das-Bradoo et al. 2006). MCM10 is dispensable for the assembly of the MCM2-7 replicative helicase but facilitates strand separation during the early stages of replication initiation (Kanke et al. 2012; van Deursen et al. 2012; Thu and Bielinsky 2013). In fission yeast, Mcm10/Cdc23 was shown to function after pre-RC assembly and facilitates Cdc45 chromatin binding (Gregan et al. 2003). The budding yeast Mcm10 protein was also suggested to play a role in replication elongation through interacting with and stabilization of the catalytic subunit of DNA polymerase α (POL1) (Ricke and Bielinsky 2004). Similarly, mammalian MCM10 interacts with and stabilizes the catalytic subunit of DNA polymerase α (p180) and is essential for efficient DNA synthesis (Chattopadhyay and Bielinsky 2007; Zhu et al. 2007). In G1- and in S-phase, MCM10 is monoubiquitylated at two residues, and this promotes its interaction with PCNA, which may be important for the release of polymerase α and the recruitment of polymerase δ necessary for the extension of Okazaki fragments (Das-Bradoo et al. 2006; Thu and Bielinsky 2013). The identity of the E3 ligase responsible for MCM10 monoubiquitylation is currently unknown. MCM10 is also subject to ubiquitylation-dependent proteolysis via the $CRL4^{VBRBP}$ E3 ligase, both in unperturbed cells and following exposure to UV, although the biological significance of this regulation remains elusive (Kaur et al. 2012; Romani et al. 2015).

Perhaps the most prominent example where the regulation of DNA replication involves coordination between proteolytic and non-proteolytic ubiquitylation machineries concerns the enigmatic PCNA protein (Fig. 19.5). PCNA is a homotrimeric ring-shaped protein complex that functions as a platform for coordinating the activity of many proteins involved in DNA replication, repair, and chromatin-related transactions (Ulrich and Takahashi 2013; Choe and Moldovan 2017). PCNA has long been shown to undergo DNA damage-induced monoubiquitylation at a conserved lysine residue (Lys-164), and this is critical to recruit translesion Y-family DNA polymerases for bypassing replication-stalling DNA lesions by translesion DNA synthesis (TLS) (Yang et al. 2013). This recruitment is dependent on the enhanced interaction of TLS polymerases with the modified PCNA through their ubiquitin-binding domains (Bienko et al. 2005; Plosky et al. 2006). PCNA monoubiquitylation at Lys-164 is catalyzed by the Rad18 E3 ligase and mediated by the E2 Rad6. The monoubiquitin moiety can be converted to Lys-63-linked polyubiquitylation by a heterodimeric E2, Ubc13-Mms2, to initiate an error-free pathway of repair, template switching, which uses the newly replicated sister chromatid as a template for replication (Hedglin and Benkovic 2015; Branzei 2011). Rad18, however, is not the only E3 ligase that can monoubiquitylate PCNA at Lys-64 as this modification can still be detected in the absence of functional Rad18 (Simpson et al.

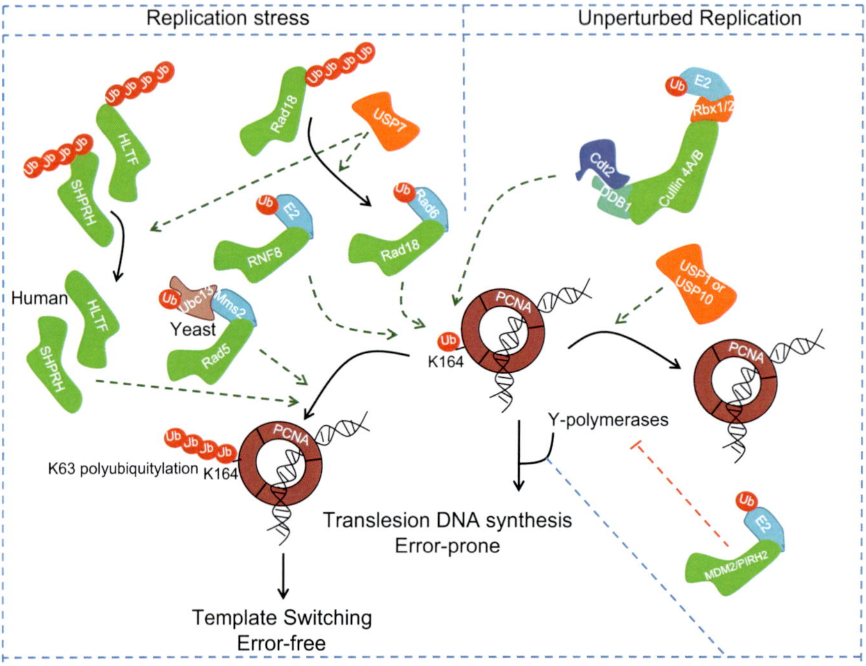

Fig. 19.5 Control of PCNA ubiquitylation and its role in normal DNA replication and following replication stress. The trimeric PCNA is subject to monoubiquitylation via the activity of multiple E3 ligases both in unperturbed cells (via CRL4-CDT2 E3 ligase) and following replication stresses via Rad18/Rad5 E2/E3 or RNF8 E3 ligase. PCNA monoubiquitylation is essential for the recruitment of Y-family DNA polymerases, such as DNA polymerase eta, and this is critical for translesion DNA synthesis (TLS). This is reversed by the action of the deubiquitylating enzymes USP1 and USP7 (Orange). The monoubiquitin on PCNA at Lys-164 (K164) can be extended by the Rad5 E3 ligase and heterodimeric Ubc13/Mms2 in yeast or through the activity of the HLTF and SHPRH E3 ligases in mammals to form K63-linked polyubiquitin chain, and this promotes template switching and minimizes mutagenesis due to excessive activity of TLS polymerases. Other E3 ligases, MDM2 and PRH2, limit the activity of TLS polymerases to limit error-prone TLS either via the targeted proteolysis of TLS polymerase eta (MDM2) or ubiquitin-dependent suppression of its activity (PRH2)

2006). Consistent with this, two other E3 ligases, RNF8 and CRL4^{CDT2}, were shown to promote PCNA monoubiquitylation (Zhang et al. 2008; Terai et al. 2010). Intriguingly, PCNA undergoes monoubiquitylation in normal replicating cells and without exposure to external stresses (Frampton et al. 2006; Leach and Michael 2005; Terai et al. 2010). Although the role of this ubiquitylation in unperturbed DNA replication is not clear, PCNA monoubiquitylation appears to be important for efficient DNA replication and for promoting TLS that may be necessary to cope with replication-associated stress (Leach and Michael 2005; Terai et al. 2010).

In mammalian cells, two E3 ubiquitin ligases, SNF2 histone linker plant homeodomain RING helicase (SHPRH) and helicase-like transcription factor (HLTF), promote the Lys-63-linked polyubiquitylation at Lys-164 to suppress

PCNA-dependent TLS and mutagenesis (Unk et al. 2008; Motegi et al. 2008). This latter activity is aided by the activity of the DUB USP7, which deubiquitylates and stabilizes HLTF and Rad18, promoting template switching through interaction between the unubiquitylated Rad18 and HLTF (Qing et al. 2011; Zeman et al. 2014). In response to replication stress, USP7 also deubiquitylates and stabilizes both Rad18 (so that it can monoubiquitylate PCNA) and the TLS polymerase, polymerase eta (POL-η), and this facilitates the bypass of lesions through the error-prone TLS pathway (Qian et al. 2015; Zlatanou et al. 2016). Another E3 ubiquitin ligase, the TNF receptor-associated factor (TRAF)-interacting protein (TRIP), and its homologue in *Drosophila*, "no pole" or NOPO, promotes the Lys-63-polyubiquitylation of POL-η, and this stimulates TLS by promoting the localization of POL-η to nuclear foci (Wallace et al. 2014). Intriguingly, Rad18 itself undergoes monoubiquitylation, and this prevents its interaction with SHPRH or HLTF and at the same time suppresses its ability to promote PCNA monoubiquitylation and TLS (Lin et al. 2011; Moldovan and D'Andrea 2011; Zeman et al. 2014). Kashiwaba et al. have additionally shown that USP7 can deubiquitylate PCNA, but this is not coupled to DNA replication and likely plays a role in DNA repair (Kashiwaba et al. 2015).

Just as it is important to initiate TLS in the face of replication-associated fork stalling, it is critical that the TLS activity is restricted to prevent increased mutations caused by low-fidelity polymerases. Two potential mechanisms ensure that TLS activity is restrained in cells (Fig. 19.5). The first involves the ubiquitin-dependent inhibition of TLS polymerases via both proteolytic and non-proteolytic ubiquitylation of TLS polymerases. For example, POL-η undergoes ubiquitylation-dependent proteolysis via the E3 ligase activity of MDM2 (Jung et al. 2012). On the other hand, the E3 ligase PIRH2 monoubiquitylates POL-η to suppress its interaction with monoubiquitylated PCNA (Jung et al. 2010, 2011). In both cases, this results in suppression of POL-η-dependent TLS.

Homologues of the TLS polymerases in yeast also undergo ubiquitin-dependent proteolysis. For example, both Rad30 (the S. *cerevisiae* homologue of POL-η) and Rev1 undergo cell cycle-dependent proteolysis to limit mutagenic activity, and in the case of Rad30, this is mediated via the SCF ubiquitin ligase with the F-box protein Ufo1 (Waters and Walker 2006; Skoneczna et al. 2007; Plachta et al. 2015). The second, and perhaps more important, mechanism restraining TLS activity is mediated through deubiquitylation of monoubiquitylated PCNA via the USP1 isopeptidase (Huang et al. 2006). An interesting study has recently shown that the deubiquitylation of mammalian PCNA, at least in the context of DNA damage induced by ultraviolet irradiation (UV), is aided by the deubiquitylating activity of USP10. In response to UV irradiation, USP10 is recruited to PCNA following its ISGylation by the ubiquitin-like modifier ISG15 (Park et al. 2014). Following deubiquitylation by USP10 and the release of POL-η, PCNA is de-ISGylated by UBP43, which renders PCNA available for reloading the replicative DNA polymerases and the resumption of DNA replication. Surprisingly, inactivation of Ubp10, the deubiquitinase for PCNA monoubiquitylation in yeast, does not appear to cause increased mutagenesis, arguing that other mechanisms contribute to inhibiting TLS-induced mutagenesis in yeast (Gallego-Sanchez et al. 2012).

In addition to PCNA monoubiquitylation at Lys-164, large-scale studies have identified additional lysine residues on PCNA that are ubiquitylated, but the E2/E3 enzymes carrying these ubiquitylation events and their biological functions are yet to be determined (McIntyre and Woodgate 2015). PCNA is also subject to extensive regulation through the ubiquitin-related SUMOylation pathway, and this is critical to suppress spontaneous and DNA damage-induced homologous recombination and help unload PCNA during DNA replication (Zilio et al. 2017; Garcia-Rodriguez et al. 2016). In this case, ubiquitylation and SUMOylation of the same protein and at the same residue appear to be tightly coordinating for ensuring optimal activity for this important multifunctional protein.

Termination of DNA replication is also regulated via the UPS (Fig. 19.6). Replication termination is initiated upon the convergence of replications forks, and this is triggered by the disassembly of the replicative helicase complex CMG (CDC45-MCM-GINS) DNA. The mechanism underlying replisome disassembly is

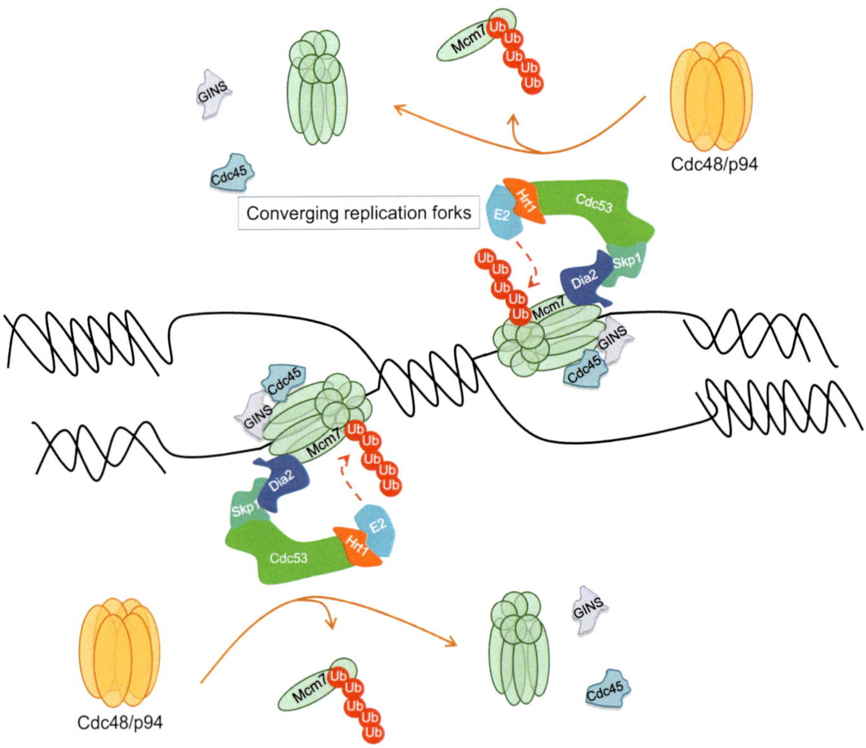

Fig. 19.6 Control of replication termination *in S. cerevisiae* via the activity of the SCF^{Dia2} E3 ubiquitin ligase. A model depicting the K-48-polyubiquitylation of Mcm7 by the SCF^{Dia2} E3 ubiquitin ligase in *S. cerevisiae* (CRL2^{Lrr1} E3 ubiquitin ligase in metazoans) at the sites of converging replication forks, followed by its extraction from the MCM2-7 helicase by the activity of the p97 chaperon, leading to the replisome disassembly and replication termination. GINS and CDC25, components of the CMG helicase complex, are also shown

not entirely clear, but two recent studies in yeast and in *Xenopus laevis* egg extracts provided the first insights into how replication termination is achieved (Maric et al. 2014; Moreno et al. 2014). Both studies have shown that the disassembly of the CMG complex is dependent on Lys-48-linked polyubiquitylation of the MCM7 subunit of the MCM2-7 complex, and this triggers its recognition by the ubiquitin-dependent segregase Cdc48 (also known as p97), a AAA+ adenosine triphosphatase (ATPase) that forms a hexameric ring, which undergoes conformational changes upon ATP hydrolysis to drive protein unfolding (Maric et al. 2014; Moreno et al. 2014; Barthelme et al. 2014). This results in the translocation of the MCM7 subunit through the p97 hexameric ring and the opening of the MCM2-7 ring allowing DNA exit and replication termination (Bell 2014; Lengronne and Pasero 2014). Mcm7 polyubiquitylation in *S. cerevisiae* is catalyzed by the activity of the SCF E3 ligase and the F-box protein Dia2, which has been previously shown to localize to the replisome (Maric et al. 2014; Moreno et al. 2014; Morohashi et al. 2009). Although SCF^{Dia2} inactivation prevents CMG disassembly and results in its retention on chromatin causing replication defects consistent with inhibition of replication termination, blocking Mcm7 proteolysis does not do so, arguing that Mcm7 polyubiquitylation is most important for its extraction from the Mcm2-7 complex and unloading of the Mcm2-7 ring (Maric et al. 2014; Moreno et al. 2014). The role of MCM7 polyubiquitylation in replisome disassembly and replication termination is conserved in higher eukaryotes, but this was recently shown to be mediated through the activity of the $CRL2^{Lrr1}$ ubiquitin ligase (Dewar et al. 2017).

19.8 Summary and Concluding Remarks

The last few decades have witnessed an extensive understanding of the role of the UPS in regulating almost all aspect of cell physiology, and in many cases, deregulation of key regulators of this ubiquitous regulatory system contributes to disease development. This regulation is even more important for cellular processes that are dependent on the efficient temporal and spatial interplay between many factors and is subject to endogenous and external perturbations, as is the case for the regulation of eukaryotic DNA replication. While many factors influencing DNA replication undergo extensive regulation through the UPS leading to proteasomal degradation, non-proteolytic ubiquitin signaling coordinating protein-protein interactions is being subjected to increasing scrutiny. Various components of the UPS operate with extensive crosstalk and feedback mechanisms as well as interactions with other posttranslational modifications, and together, they provide a complex network of protein-protein communications to control the specificity and robustness of DNA replication and cell cycle control. When faced with stresses that impact the replication machinery and threaten the fidelity of DNA replication, for example, when replication forks encounter replication-stalling lesions, cells must be able to quickly respond and adopt the appropriate measures to cope with these stresses. The dynamic and reversible activities associated with the UPS allow the cell to adjust to these insults. Even in the absence of these stresses, the non-proteolytic as well as proteolytic mechanisms ensure that DNA replication proceed unperturbed with exquisite accuracy. The diversity in the specificity of the various E2-E3 pairs in conjugating various polymers of ubiquitin chains on target substrates adds another layer of complexity that will require more effort to fully understand and appreciate. Key regulators, such as the APC/C, $CRL4^{CDT2}$, and SCF ubiquitin ligases, and the USP7 and USP1 deubiquitinases, dominate the scene and regulate the steady state levels or activities of many proteins controlling DNA replication and cell cycle progression. They are also important for regulating appropriate cellular responses to perturbations by activating cellular checkpoints. It is therefore understandable that significant research is targeting some of these key regulators for therapeutic gain, but this will require a greater understanding of the molecular and biochemical details and structural information of these regulators. Proteome-wide studies suggest that we are yet to understand the functional significance of new ubiquitin modifications of many replication-associated factors. The development of novel techniques for genome editing and for gain- or loss-of-function screens as well as the development of new protocols for enriching for and identifying various ubiquitin modifications will no doubt be great assets for making these discoveries.

Acknowledgment Work in the author's laboratories is supported by NIH grants, R00 CA140774 to TA and R01 CA60499 and GM84465 to AD.

References

Abbas T, Dutta A (2009) p21 in cancer: intricate networks and multiple activities. Nat Rev Cancer 9(6):400–414. https://doi.org/10.1038/nrc2657

Abbas T, Dutta A (2011) CRL4Cdt2: master coordinator of cell cycle progression and genome stability. Cell Cycle 10(2):241–249

Abbas T, Sivaprasad U, Terai K, Amador V, Pagano M, Dutta A (2008) PCNA-dependent regulation of p21 ubiquitylation and degradation via the CRL4Cdt2 ubiquitin ligase complex. Genes Dev 22(18):2496–2506. https://doi.org/10.1101/gad.1676108

Abbas T, Shibata E, Park J, Jha S, Karnani N, Dutta A (2010) CRL4(Cdt2) regulates cell proliferation and histone gene expression by targeting PR-Set7/Set8 for degradation. Mol Cell 40(1):9–21. https://doi.org/10.1016/j.molcel.2010.09.014

Abbas T, Keaton MA, Dutta A (2013) Genomic instability in cancer. Cold Spring Harb Perspect Biol 5(3):a012914. https://doi.org/10.1101/cshperspect.a012914

Alexandrow MG, Hamlin JL (2004) Cdc6 chromatin affinity is unaffected by serine-54 phosphorylation, S-phase progression, and overexpression of cyclin a. Mol Cell Biol 24(4):1614–1627

Almouzni G, Cedar H (2016) Maintenance of epigenetic information. Cold Spring Harb Perspect Biol 8(5). https://doi.org/10.1101/cshperspect.a019372

Amador V, Ge S, Santamaria PG, Guardavaccaro D, Pagano M (2007) APC/C(Cdc20) controls the ubiquitin-mediated degradation of p21 in prometaphase. Mol Cell 27(3):462–473. https://doi.org/10.1016/j.molcel.2007.06.013

Amir R, Ciechanover A, Cohen S (2001) The ubiquitin-proteasome system: the relationship between protein degradation and human diseases. Harefuah 140(12):1172–1176. 1229

Angers S, Li T, Yi X, MacCoss MJ, Moon RT, Zheng N (2006) Molecular architecture and assembly of the DDB1-CUL4A ubiquitin ligase machinery. Nature 443(7111):590–593

Aparicio OM, Weinstein DM, Bell SP (1997) Components and dynamics of DNA replication complexes in S. cerevisiae: redistribution of MCM proteins and Cdc45p during S phase. Cell 91(1):59–69
Araki M, Wharton RP, Tang Z, Yu H, Asano M (2003) Degradation of origin recognition complex large subunit by the anaphase-promoting complex in drosophila. EMBO J 22(22):6115–6126. https://doi.org/10.1093/emboj/cdg573
Araki M, Yu H, Asano M (2005) A novel motif governs APC-dependent degradation of drosophila ORC1 in vivo. Genes Dev 19(20):2458–2465. https://doi.org/10.1101/gad.1361905
Arias EE, Walter JC (2006) PCNA functions as a molecular platform to trigger Cdt1 destruction and prevent re-replication. Nat Cell Biol 8(1):84–90. https://doi.org/10.1038/ncb1346
Barthelme D, Chen JZ, Grabenstatter J, Baker TA, Sauer RT (2014) Architecture and assembly of the archaeal Cdc48*20S proteasome. Proc Natl Acad Sci U S A 111(17):E1687–E1694. https://doi.org/10.1073/pnas.1404823111
Bashir T, Dorrello NV, Amador V, Guardavaccaro D, Pagano M (2004) Control of the SCF(Skp2-Cks1) ubiquitin ligase by the APC/C(Cdh1) ubiquitin ligase. Nature 428(6979):190–193. https://doi.org/10.1038/nature02330
Beck DB, Burton A, Oda H, Ziegler-Birling C, Torres-Padilla ME, Reinberg D (2012) The role of PR-Set7 in replication licensing depends on Suv4-20h. Genes Dev 26(23):2580–2589. https://doi.org/10.1101/gad.195636.112
Behrends C, Harper JW (2011) Constructing and decoding unconventional ubiquitin chains. Nat Struct Mol Biol 18(5):520–528. https://doi.org/10.1038/nsmb.2066
Bell SP (2014) DNA replication. Terminating the replisome. Science 346(6208):418–419. https://doi.org/10.1126/science.1261245
Bembenek J, Yu H (2001) Regulation of the anaphase-promoting complex by the dual specificity phosphatase human Cdc14a. J Biol Chem 276(51):48237–48242. https://doi.org/10.1074/jbc.M108126200
Benamar M, Guessous F, Du K, Corbett P, Obeid J, Gioeli D, Slingluff CL Jr, Abbas T (2016) Inactivation of the CRL4-CDT2-SET8/p21 ubiquitylation and degradation axis underlies the therapeutic efficacy of pevonedistat in melanoma. EBioMedicine 10:85–100. https://doi.org/10.1016/j.ebiom.2016.06.023
Berti M, Vindigni A (2016) Replication stress: getting back on track. Nat Struct Mol Biol 23(2):103–109. https://doi.org/10.1038/nsmb.3163
Bienko M, Green CM, Crosetto N, Rudolf F, Zapart G, Coull B, Kannouche P, Wider G, Peter M, Lehmann AR, Hofmann K, Dikic I (2005) Ubiquitin-binding domains in Y-family polymerases regulate translesion synthesis. Science 310(5755):1821–1824. https://doi.org/10.1126/science.1120615
Bosu DR, Kipreos ET (2008) Cullin-RING ubiquitin ligases: global regulation and activation cycles. Cell Div 3:7.:https://doi.org/10.1186/1747-1028-3-7
Branzei D (2011) Ubiquitin family modifications and template switching. FEBS Lett 585(18):2810–2817. https://doi.org/10.1016/j.febslet.2011.04.053
Cappell SD, Chung M, Jaimovich A, Spencer SL, Meyer T (2016) Irreversible APC(Cdh1) inactivation underlies the point of no return for cell-cycle entry. Cell 166(1):167–180. https://doi.org/10.1016/j.cell.2016.05.077
Cardozo T, Pagano M (2004) The SCF ubiquitin ligase: insights into a molecular machine. Nat Rev Mol Cell Biol 5(9):739–751. https://doi.org/10.1038/nrm1471
Centore RC, Havens CG, Manning AL, Li JM, Flynn RL, Tse A, Jin J, Dyson NJ, Walter JC, Zou L (2010) CRL4(Cdt2)-mediated destruction of the histone methyltransferase Set8 prevents premature chromatin compaction in S phase. Mol Cell 40(1):22–33. https://doi.org/10.1016/j.molcel.2010.09.015
Chan CH, Morrow JK, Li CF, Gao Y, Jin G, Moten A, Stagg LJ, Ladbury JE, Cai Z, Xu D, Logothetis CJ, Hung MC, Zhang S, Lin HK (2013) Pharmacological inactivation of Skp2 SCF ubiquitin ligase restricts cancer stem cell traits and cancer progression. Cell 154(3):556–568. https://doi.org/10.1016/j.cell.2013.06.048

Chandrasekaran S, Tan TX, Hall JR, Cook JG (2011) Stress-stimulated mitogen-activated protein kinases control the stability and activity of the Cdt1 DNA replication licensing factor. Mol Cell Biol 31(22):4405–4416. https://doi.org/10.1128/MCB.06163-11
Chattopadhyay S, Bielinsky AK (2007) Human Mcm10 regulates the catalytic subunit of DNA polymerase-alpha and prevents DNA damage during replication. Mol Biol Cell 18(10):4085–4095. https://doi.org/10.1091/mbc.E06-12-1148
Chen Z, Sui J, Zhang F, Zhang C (2015) Cullin family proteins and tumorigenesis: genetic association and molecular mechanisms. J Cancer 6(3):233–242. https://doi.org/10.7150/jca.11076
Chen NP, Uddin B, Voit R, Schiebel E (2016) Human phosphatase CDC14A is recruited to the cell leading edge to regulate cell migration and adhesion. Proc Natl Acad Sci USA 113(4):990–995. https://doi.org/10.1073/pnas.1515605113
Choe KN, Moldovan GL (2017) Forging ahead through darkness: PCNA, still the principal conductor at the replication fork. Mol Cell 65(3):380–392. https://doi.org/10.1016/j.molcel.2016.12.020
Ciechanover A, Schwartz AL (2002) Ubiquitin-mediated degradation of cellular proteins in health and disease. Hepatology 35(1):3–6
Clijsters L, Wolthuis R (2014) PIP-box-mediated degradation prohibits re-accumulation of Cdc6 during S phase. J Cell Sci 127(Pt 6):1336–1345. https://doi.org/10.1242/jcs.145862
Clurman BE, Sheaff RJ, Thress K, Groudine M, Roberts JM (1996) Turnover of cyclin E by the ubiquitin-proteasome pathway is regulated by cdk2 binding and cyclin phosphorylation. Genes Dev 10(16):1979–1990
Das-Bradoo S, Ricke RM, Bielinsky AK (2006) Interaction between PCNA and diubiquitinated Mcm10 is essential for cell growth in budding yeast. Mol Cell Biol 26(13):4806–4817. https://doi.org/10.1128/MCB.02062-05
den Elzen N, Pines J (2001) Cyclin a is destroyed in prometaphase and can delay chromosome alignment and anaphase. J Cell Biol 153(1):121–136
Deshaies RJ, Joazeiro CA (2009) RING domain E3 ubiquitin ligases. Annu Rev Biochem 78:399–434. https://doi.org/10.1146/annurev.biochem.78.101807.093809
Devault A, Vallen EA, Yuan T, Green S, Bensimon A, Schwob E (2002) Identification of Tah11/Sid2 as the ortholog of the replication licensing factor Cdt1 in Saccharomyces cerevisiae. Curr Biol 12(8):689–694
Dewar JM, Low E, Mann M, Raschle M, Walter JC (2017) CRL2Lrr1 promotes unloading of the vertebrate replisome from chromatin during replication termination. Genes Dev 31(3):275–290. https://doi.org/10.1101/gad.291799.116
Di Fiore B, Pines J (2007) Emi1 is needed to couple DNA replication with mitosis but does not regulate activation of the mitotic APC/C. J Cell Biol 177(3):425–437. https://doi.org/10.1083/jcb.200611166
Donzelli M, Squatrito M, Ganoth D, Hershko A, Pagano M, Draetta GF (2002) Dual mode of degradation of Cdc25 a phosphatase. EMBO J 21(18):4875–4884
Duan S, Cermak L, Pagan JK, Rossi M, Martinengo C, di Celle PF, Chapuy B, Shipp M, Chiarle R, Pagano M (2012) FBXO11 targets BCL6 for degradation and is inactivated in diffuse large B-cell lymphomas. Nature 481(7379):90–93. https://doi.org/10.1038/nature10688
Duda DM, Scott DC, Calabrese MF, Zimmerman ES, Zheng N, Schulman BA (2011) Structural regulation of cullin-RING ubiquitin ligase complexes. Curr Opin Struct Biol 21(2):257–264. https://doi.org/10.1016/j.sbi.2011.01.003
Frampton J, Irmisch A, Green CM, Neiss A, Trickey M, Ulrich HD, Furuya K, Watts FZ, Carr AM, Lehmann AR (2006) Postreplication repair and PCNA modification in Schizosaccharomyces pombe. Mol Biol Cell 17(7):2976–2985. https://doi.org/10.1091/mbc.E05-11-1008
Fujimitsu K, Grimaldi M, Yamano H (2016) Cyclin-dependent kinase 1-dependent activation of APC/C ubiquitin ligase. Science 352(6289):1121–1124. https://doi.org/10.1126/science.aad3925
Fujita M, Yamada C, Goto H, Yokoyama N, Kuzushima K, Inagaki M, Tsurumi T (1999) Cell cycle regulation of human CDC6 protein. Intracellular localization, interaction with the

human mcm complex, and CDC2 kinase-mediated hyperphosphorylation. J Biol Chem 274(36):25927–25932
Gallego-Sanchez A, Andres S, Conde F, San-Segundo PA, Bueno A (2012) Reversal of PCNA ubiquitylation by Ubp10 in Saccharomyces cerevisiae. PLoS Genet 8(7):e1002826. https://doi.org/10.1371/journal.pgen.1002826
Ganai RA, Johansson E (2016) DNA replication-a matter of fidelity. Mol Cell 62(5):745–755. https://doi.org/10.1016/j.molcel.2016.05.003
Garcia-Rodriguez N, Wong RP, Ulrich HD (2016) Functions of ubiquitin and SUMO in DNA replication and replication stress. Front Genet 7:87. https://doi.org/10.3389/fgene.2016.00087
Glickman MH, Ciechanover A (2002) The ubiquitin-proteasome proteolytic pathway: destruction for the sake of construction. Physiol Rev 82(2):373–428. https://doi.org/10.1152/physrev.00027.2001
Godbersen JC, Humphries LA, Danilova OV, Kebbekus PE, Brown JR, Eastman A, Danilov AV (2014) The Nedd8-activating enzyme inhibitor MLN4924 thwarts microenvironment-driven NF-kappaB activation and induces apoptosis in chronic lymphocytic leukemia B cells. Clin Cancer Res 20(6):1576–1589. https://doi.org/10.1158/1078-0432.CCR-13-0987
Green BM, Finn KJ, Li JJ (2010) Loss of DNA replication control is a potent inducer of gene amplification. Science 329(5994):943–946. https://doi.org/10.1126/science.1190966
Gregan J, Lindner K, Brimage L, Franklin R, Namdar M, Hart EA, Aves SJ, Kearsey SE (2003) Fission yeast Cdc23/Mcm10 functions after pre-replicative complex formation to promote Cdc45 chromatin binding. Mol Biol Cell 14(9):3876–3887. https://doi.org/10.1091/mbc.E03-02-0090
Groll M, Huber R (2003) Substrate access and processing by the 20S proteasome core particle. Int J Biochem Cell Biol 35(5):606–616
Gu Y, Kaufman JL, Bernal L, Torre C, Matulis SM, Harvey RD, Chen J, Sun SY, Boise LH, Lonial S (2014) MLN4924, an NAE inhibitor, suppresses AKT and mTOR signaling via upregulation of REDD1 in human myeloma cells. Blood 123(21):3269–3276. https://doi.org/10.1182/blood-2013-08-521914
Guardavaccaro D, Kudo Y, Boulaire J, Barchi M, Busino L, Donzelli M, Margottin-Goguet F, Jackson PK, Yamasaki L, Pagano M (2003) Control of meiotic and mitotic progression by the F box protein beta-Trcp1 in vivo. Dev Cell 4(6):799–812
Havens CG, Walter JC (2009) Docking of a specialized PIP box onto chromatin-bound PCNA creates a degron for the ubiquitin ligase CRL4Cdt2. Mol Cell 35(1):93–104. https://doi.org/10.1016/j.molcel.2009.05.012
Havens CG, Walter JC (2011) Mechanism of CRL4(Cdt2), a PCNA-dependent E3 ubiquitin ligase. Genes Dev 25(15):1568–1582. https://doi.org/10.1101/gad.2068611
He YJ, McCall CM, Hu J, Zeng Y, Xiong Y (2006) DDB1 functions as a linker to recruit receptor WD40 proteins to CUL4-ROC1 ubiquitin ligases. Genes Dev 20(21):2949–2954
Hedglin M, Benkovic SJ (2015) Regulation of Rad6/Rad18 activity during DNA damage tolerance. Annu Rev Biophys 44:207–228. https://doi.org/10.1146/annurev-biophys-060414-033841
Henikoff S (2016) Mechanisms of nucleosome dynamics in vivo. Cold Spring Harb Perspect Med 6(9). https://doi.org/10.1101/cshperspect.a026666
Heo J, Eki R, Abbas T (2016) Deregulation of F-box proteins and its consequence on cancer development, progression and metastasis. Semin Cancer Biol 36:33–51. https://doi.org/10.1016/j.semcancer.2015.09.015
Hershko A (2005) The ubiquitin system for protein degradation and some of its roles in the control of the cell division cycle. Cell Death Differ 12(9):1191–1197. https://doi.org/10.1038/sj.cdd.4401702
Higa LA, Zhang H (2007) Stealing the spotlight: CUL4-DDB1 ubiquitin ligase docks WD40-repeat proteins to destroy. Cell Div 2:5
Higa LA, Mihaylov IS, Banks DP, Zheng J, Zhang H (2003) Radiation-mediated proteolysis of CDT1 by CUL4-ROC1 and CSN complexes constitutes a new checkpoint. Nat Cell Biol 5(11):1008–1015. https://doi.org/10.1038/ncb1061

Higa LA, Banks D, Wu M, Kobayashi R, Sun H, Zhang H (2006) L2DTL/CDT2 interacts with the CUL4/DDB1 complex and PCNA and regulates CDT1 proteolysis in response to DNA damage. Cell Cycle 5(15):1675–1680

Hotton SK, Callis J (2008) Regulation of cullin RING ligases. Annu Rev Plant Biol 59:467–489. https://doi.org/10.1146/annurev.arplant.58.032806.104011

Hua Z, Vierstra RD (2011) The cullin-RING ubiquitin-protein ligases. Annu Rev Plant Biol 62:299–334. https://doi.org/10.1146/annurev-arplant-042809-112256

Huang TT, Nijman SM, Mirchandani KD, Galardy PJ, Cohn MA, Haas W, Gygi SP, Ploegh HL, Bernards R, D'Andrea AD (2006) Regulation of monoubiquitinated PCNA by DUB autocleavage. Nat Cell Biol 8(4):339–347. https://doi.org/10.1038/ncb1378

Huang H-L, Zheng W-L, Zhao R, Zhang B, Ma W-L (2010) FBXO31 is down-regulated and may function as a tumor suppressor in hepatocellular carcinoma. Oncol Rep 24:715–720

Hustedt N, Durocher D (2016) The control of DNA repair by the cell cycle. Nat Cell Biol 19(1):1–9. https://doi.org/10.1038/ncb3452

Jackson S, Xiong Y (2009) CRL4s: the CUL4-RING E3 ubiquitin ligases. Trends Biochem Sci 34(11):562–570. https://doi.org/10.1016/j.tibs.2009.07.002

Jaspersen SL, Charles JF, Morgan DO (1999) Inhibitory phosphorylation of the APC regulator Hct1 is controlled by the kinase Cdc28 and the phosphatase Cdc14. Curr Biol 9(5):227–236

Jazaeri AA, Shibata E, Park J, Bryant JL, Conaway MR, Modesitt SC, Smith PG, Milhollen MA, Berger AJ, Dutta A (2013) Overcoming platinum resistance in preclinical models of ovarian cancer using the neddylation inhibitor MLN4924. Mol Cancer Ther 12(10):1958–1967. https://doi.org/10.1158/1535-7163.MCT-12-1028

Jiang Y, Jia L (2015) Neddylation pathway as a novel anti-cancer target: mechanistic investigation and therapeutic implication. Anti Cancer Agents Med Chem 15(9):1127–1133

Jiang W, Wells NJ, Hunter T (1999) Multistep regulation of DNA replication by Cdk phosphorylation of HsCdc6. Proc Natl Acad Sci U S A 96(11):6193–6198

Jin J, Arias EE, Chen J, Harper JW, Walter JC (2006) A family of diverse Cul4-Ddb1-interacting proteins includes Cdt2, which is required for S phase destruction of the replication factor Cdt1. Mol Cell 23(5):709–721

Johansson P, Jeffery J, Al-Ejeh F, Schulz RB, Callen DF, Kumar R, Khanna KK (2014) SCF-FBXO31 E3 ligase targets DNA replication factor Cdt1 for proteolysis in the G2 phase of cell cycle to prevent re-replication. J Biol Chem 289:18514–18525. https://doi.org/10.1074/jbc.M114.559930

Jorgensen S, Eskildsen M, Fugger K, Hansen L, Larsen MS, Kousholt AN, Syljuasen RG, Trelle MB, Jensen ON, Helin K, Sorensen CS (2011) SET8 is degraded via PCNA-coupled CRL4(CDT2) ubiquitylation in S phase and after UV irradiation. J Cell Biol 192(1):43–54. https://doi.org/10.1083/jcb.201009076

Jung YS, Liu G, Chen X (2010) Pirh2 E3 ubiquitin ligase targets DNA polymerase eta for 20S proteasomal degradation. Mol Cell Biol 30(4):1041–1048. https://doi.org/10.1128/MCB.01198-09

Jung YS, Hakem A, Hakem R, Chen X (2011) Pirh2 E3 ubiquitin ligase monoubiquitinates DNA polymerase eta to suppress translesion DNA synthesis. Mol Cell Biol 31(19):3997–4006. https://doi.org/10.1128/MCB.05808-11

Jung YS, Qian Y, Chen X (2012) DNA polymerase eta is targeted by Mdm2 for polyubiquitination and proteasomal degradation in response to ultraviolet irradiation. DNA Repair (Amst) 11(2):177–184. https://doi.org/10.1016/j.dnarep.2011.10.017

Kaiser BK, Zimmerman ZA, Charbonneau H, Jackson PK (2002) Disruption of centrosome structure, chromosome segregation, and cytokinesis by misexpression of human Cdc14A phosphatase. Mol Biol Cell 13(7):2289–2300. https://doi.org/10.1091/mbc.01-11-0535

Kanke M, Kodama Y, Takahashi TS, Nakagawa T, Masukata H (2012) Mcm10 plays an essential role in origin DNA unwinding after loading of the CMG components. EMBO J 31(9):2182–2194. https://doi.org/10.1038/emboj.2012.68

Kashiwaba S, Kanao R, Masuda Y, Kusumoto-Matsuo R, Hanaoka F, Masutani C (2015) USP7 is a suppressor of PCNA ubiquitination and oxidative-stress-induced mutagenesis in human cells. Cell Rep 13(10):2072–2080. https://doi.org/10.1016/j.celrep.2015.11.014
Kaur M, Khan MM, Kar A, Sharma A, Saxena S (2012) CRL4-DDB1-VPRBP ubiquitin ligase mediates the stress triggered proteolysis of Mcm10. Nucleic Acids Res 40(15):7332–7346. https://doi.org/10.1093/nar/gks366
Kim Y, Starostina NG, Kipreos ET (2008) The CRL4Cdt2 ubiquitin ligase targets the degradation of p21Cip1 to control replication licensing. Genes Dev 22(18):2507–2519
Kipreos ET, Pagano M (2000) The F-box protein family. Genome Biol 1(5):REVIEWS3002.:https://doi.org/10.1186/gb-2000-1-5-reviews3002
Koepp DM, Schaefer LK, Ye X, Keyomarsi K, Chu C, Harper JW, Elledge SJ (2001) Phosphorylation-dependent ubiquitination of cyclin E by the SCFFbw7 ubiquitin ligase. Science 294(5540):173–177. https://doi.org/10.1126/science.1065203
Komander D, Clague MJ, Urbe S (2009) Breaking the chains: structure and function of the deubiquitinases. Nat Rev Mol Cell Biol 10(8):550–563. doi:nrm2731 [pii] 1038/nrm2731
Kornitzer D, Ciechanover A (2000) Modes of regulation of ubiquitin-mediated protein degradation. J Cell Physiol 182(1):1–11. https://doi.org/10.1002/(SICI)1097-4652(200001)182:1<1::AID-JCP1>3.0.CO;2-V
Leach CA, Michael WM (2005) Ubiquitin/SUMO modification of PCNA promotes replication fork progression in Xenopus laevis egg extracts. J Cell Biol 171(6):947–954. https://doi.org/10.1083/jcb.200508100
Lee EK, Diehl JA (2014) SCFs in the new millennium. Oncogene 33(16):2011–2018. https://doi.org/10.1038/onc.2013.144
Lei M (2005) The MCM complex: its role in DNA replication and implications for cancer therapy. Curr Cancer Drug Targets 5(5):365–380
Lengronne A, Pasero P (2014) Closing the MCM cycle at replication termination sites. EMBO Rep 15(12):1226–1227. 10.15252/embr.201439774
Li X, Zhao Q, Liao R, Sun P, Wu X (2003) The SiCF(Skp2) ubiquitin ligase complex interacts with the human replication licensing factor Cdt1 and regulates Cdt1 degradation. J Biol Chem 278(33):30854–30858. https://doi.org/10.1074/jbc.C300251200
Li W, Bengtson MH, Ulbrich A, Matsuda A, Reddy VA, Orth A, Chanda SK, Batalov S, Joazeiro CA (2008) Genome-wide and functional annotation of human E3 ubiquitin ligases identifies MULAN, a mitochondrial E3 that regulates the organelle's dynamics and signaling. PLoS One 3(1):e1487. https://doi.org/10.1371/journal.pone.0001487
Li H, Tan M, Jia L, Wei D, Zhao Y, Chen G, Xu J, Zhao L, Thomas D, Beer DG, Sun Y (2014a) Inactivation of SAG/RBX2 E3 ubiquitin ligase suppresses KrasG12D-driven lung tumorigenesis. J Clin Invest 124(2):835–846. https://doi.org/10.1172/JCI70297
Li L, Wang M, Yu G, Chen P, Li H, Wei D, Zhu J, Xie L, Jia H, Shi J, Li C, Yao W, Wang Y, Gao Q, Jeong LS, Lee HW, Yu J, Hu F, Mei J, Wang P, Chu Y, Qi H, Yang M, Dong Z, Sun Y, Hoffman RM, Jia L (2014b) Overactivated neddylation pathway as a therapeutic target in lung cancer. J Natl Cancer Inst 106(6):dju083. https://doi.org/10.1093/jnci/dju083
Lin JJ, Milhollen MA, Smith PG, Narayanan U, Dutta A (2010) NEDD8-targeting drug MLN4924 elicits DNA rereplication by stabilizing Cdt1 in S phase, triggering checkpoint activation, apoptosis, and senescence in cancer cells. Cancer Res 70(24):10310–10320. https://doi.org/10.1158/0008-5472.CAN-10-2062
Lin JR, Zeman MK, Chen JY, Yee MC, Cimprich KA (2011) SHPRH and HLTF act in a damage-specific manner to coordinate different forms of postreplication repair and prevent mutagenesis. Mol Cell 42(2):237–249. https://doi.org/10.1016/j.molcel.2011.02.026
Lipkowitz S, Weissman AM (2011) RINGs of good and evil: RING finger ubiquitin ligases at the crossroads of tumour suppression and oncogenesis. Nat Rev Cancer 11(9):629–643. https://doi.org/10.1038/nrc3120

Liu G, Warbrick E (2006) The p66 and p12 subunits of DNA polymerase delta are modified by ubiquitin and ubiquitin-like proteins. Biochem Biophys Res Commun 349(1):360–366. https://doi.org/10.1016/j.bbrc.2006.08.049

Liu E, Li X, Yan F, Zhao Q, Wu X (2004) Cyclin-dependent kinases phosphorylate human Cdt1 and induce its degradation. J Biol Chem 279(17):17283–17288. https://doi.org/10.1074/jbc.C300549200

Lovejoy CA, Lock K, Yenamandra A, Cortez D (2006) DDB1 maintains genome integrity through regulation of Cdt1. Mol Cell Biol 26(21):7977–7990

Lydeard JR, Schulman BA, Harper JW (2013) Building and remodelling Cullin-RING E3 ubiquitin ligases. EMBO Rep 14(12):1050–1061. https://doi.org/10.1038/embor.2013.173

Machida YJ, Dutta A (2007) The APC/C inhibitor, Emi1, is essential for prevention of rereplication. Genes Dev 21(2):184–194. https://doi.org/10.1101/gad.1495007

Machida YJ, Hamlin JL, Dutta A (2005) Right place, right time, and only once: replication initiation in metazoans. Cell 123(1):13–24. https://doi.org/10.1016/j.cell.2005.09.019

Mailand N, Diffley JF (2005) CDKs promote DNA replication origin licensing in human cells by protecting Cdc6 from APC/C-dependent proteolysis. Cell 122(6):915–926. https://doi.org/10.1016/j.cell.2005.08.013

Maric M, Maculins T, De Piccoli G, Labib K (2014) Cdc48 and a ubiquitin ligase drive disassembly of the CMG helicase at the end of DNA replication. Science 346(6208):1253596. https://doi.org/10.1126/science.1253596

Maser RS, Choudhury B, Campbell PJ, Feng B, Wong KK, Protopopov A, O'Neil J, Gutierrez A, Ivanova E, Perna I, Lin E, Mani V, Jiang S, McNamara K, Zaghlul S, Edkins S, Stevens C, Brennan C, Martin ES, Wiedemeyer R, Kabbarah O, Nogueira C, Histen G, Aster J, Mansour M, Duke V, Foroni L, Fielding AK, Goldstone AH, Rowe JM, Wang YA, Look AT, Stratton MR, Chin L, Futreal PA, DePinho RA (2007) Chromosomally unstable mouse tumours have genomic alterations similar to diverse human cancers. Nature 447(7147):966–971. https://doi.org/10.1038/nature05886

McGarry TJ, Kirschner MW (1998) Geminin, an inhibitor of DNA replication, is degraded during mitosis. Cell 93(6):1043–1053

McIntyre J, Woodgate R (2015) Regulation of translesion DNA synthesis: posttranslational modification of lysine residues in key proteins. DNA Repair (Amst) 29:166–179. https://doi.org/10.1016/j.dnarep.2015.02.011

Mendez J, Zou-Yang XH, Kim SY, Hidaka M, Tansey WP, Stillman B (2002) Human origin recognition complex large subunit is degraded by ubiquitin-mediated proteolysis after initiation of DNA replication. Mol Cell 9(3):481–491

Merlet J, Burger J, Gomes JE, Pintard L (2009) Regulation of cullin-RING E3 ubiquitin-ligases by neddylation and dimerization. Cell Mol Life Sci 66(11–12):1924–1938. https://doi.org/10.1007/s00018-009-8712-7

Michishita M, Morimoto A, Ishii T, Komori H, Shiomi Y, Higuchi Y, Nishitani H (2011) Positively charged residues located downstream of PIP box, together with TD amino acids within PIP box, are important for CRL4(Cdt2) -mediated proteolysis. Genes Cells 16(1):12–22. https://doi.org/10.1111/j.1365-2443.2010.01464.x

Milhollen MA, Traore T, Adams-Duffy J, Thomas MP, Berger AJ, Dang L, Dick LR, Garnsey JJ, Koenig E, Langston SP, Manfredi M, Narayanan U, Rolfe M, Staudt LM, Soucy TA, Yu J, Zhang J, Bolen JB, Smith PG (2010) MLN4924, a NEDD8-activating enzyme inhibitor, is active in diffuse large B-cell lymphoma models: rationale for treatment of NF-{kappa}B-dependent lymphoma. Blood 116(9):1515–1523. https://doi.org/10.1182/blood-2010-03-272567

Milhollen MA, Narayanan U, Soucy TA, Veiby PO, Smith PG, Amidon B (2011) Inhibition of NEDD8-activating enzyme induces rereplication and apoptosis in human tumor cells consistent with deregulating CDT1 turnover. Cancer Res 71(8):3042–3051. https://doi.org/10.1158/0008-5472.CAN-10-2122

Mocciaro A, Berdougo E, Zeng K, Black E, Vagnarelli P, Earnshaw W, Gillespie D, Jallepalli P, Schiebel E (2010) Vertebrate cells genetically deficient for Cdc14A or Cdc14B retain DNA damage checkpoint proficiency but are impaired in DNA repair. J Cell Biol 189(4):631–639. https://doi.org/10.1083/jcb.200910057

Moldovan GL, D'Andrea AD (2011) DNA damage discrimination at stalled replication forks by the Rad5 homologs HLTF and SHPRH. Mol Cell 42(2):141–143. https://doi.org/10.1016/j.molcel.2011.03.018

Moreno SP, Bailey R, Campion N, Herron S, Gambus A (2014) Polyubiquitylation drives replisome disassembly at the termination of DNA replication. Science 346(6208):477–481. https://doi.org/10.1126/science.1253585

Morohashi H, Maculins T, Labib K (2009) The amino-terminal TPR domain of Dia2 tethers SCF(Dia2) to the replisome progression complex. Curr Biol 19(22):1943–1949. https://doi.org/10.1016/j.cub.2009.09.062

Motegi A, Liaw HJ, Lee KY, Roest HP, Maas A, Wu X, Moinova H, Markowitz SD, Ding H, Hoeijmakers JH, Myung K (2008) Polyubiquitination of proliferating cell nuclear antigen by HLTF and SHPRH prevents genomic instability from stalled replication forks. Proc Natl Acad Sci U S A 105(34):12411–12416. https://doi.org/10.1073/pnas.0805685105

Nakayama KI, Nakayama K (2005) Regulation of the cell cycle by SCF-type ubiquitin ligases. Semin Cell Dev Biol 16(3):323–333. https://doi.org/10.1016/j.semcdb.2005.02.010

Nakayama KI, Nakayama K (2006) Ubiquitin ligases: cell-cycle control and cancer. Nat Rev Cancer 6(5):369–381. https://doi.org/10.1038/nrc1881

Narbonne-Reveau K, Senger S, Pal M, Herr A, Richardson HE, Asano M, Deak P, Lilly MA (2008) APC/CFzr/Cdh1 promotes cell cycle progression during the drosophila endocycle. Development 135(8):1451–1461. https://doi.org/10.1242/dev.016295

Nawrocki ST, Griffin P, Kelly KR, Carew JS (2012) MLN4924: a novel first-in-class inhibitor of NEDD8-activating enzyme for cancer therapy. Expert Opin Investig Drugs 21(10):1563–1573. https://doi.org/10.1517/13543784.2012.707192

Nishioka K, Rice JC, Sarma K, Erdjument-Bromage H, Werner J, Wang Y, Chuikov S, Valenzuela P, Tempst P, Steward R, Lis JT, Allis CD, Reinberg D (2002) PR-Set7 is a nucleosome-specific methyltransferase that modifies lysine 20 of histone H4 and is associated with silent chromatin. Mol Cell 9(6):1201–1213

Nishitani H, Sugimoto N, Roukos V, Nakanishi Y, Saijo M, Obuse C, Tsurimoto T, Nakayama KI, Nakayama K, Fujita M, Lygerou Z, Nishimoto T (2006) Two E3 ubiquitin ligases, SCF-Skp2 and DDB1-Cul4, target human Cdt1 for proteolysis. EMBO J 25(5):1126–1136. https://doi.org/10.1038/sj.emboj.7601002

Nishitani H, Shiomi Y, Iida H, Michishita M, Takami T, Tsurimoto T (2008) CDK inhibitor p21 is degraded by a proliferating cell nuclear antigen-coupled Cul4-DDB1Cdt2 pathway during S phase and after UV irradiation. J Biol Chem 283(43):29045–29052. https://doi.org/10.1074/jbc.M806045200

Oda H, Hubner MR, Beck DB, Vermeulen M, Hurwitz J, Spector DL, Reinberg D (2010) Regulation of the histone H4 monomethylase PR-Set7 by CRL4(Cdt2)-mediated PCNA-dependent degradation during DNA damage. Mol Cell 40(3):364–376. https://doi.org/10.1016/j.molcel.2010.10.011

Park JM, Yang SW, Yu KR, Ka SH, Lee SW, Seol JH, Jeon YJ, Chung CH (2014) Modification of PCNA by ISG15 plays a crucial role in termination of error-prone translesion DNA synthesis. Mol Cell 54(4):626–638. https://doi.org/10.1016/j.molcel.2014.03.031

Petersen BO, Lukas J, Sorensen CS, Bartek J, Helin K (1999) Phosphorylation of mammalian CDC6 by cyclin a/CDK2 regulates its subcellular localization. EMBO J 18(2):396–410. https://doi.org/10.1093/emboj/18.2.396

Petersen BO, Wagener C, Marinoni F, Kramer ER, Melixetian M, Lazzerini Denchi E, Gieffers C, Matteucci C, Peters JM, Helin K (2000) Cell cycle- and cell growth-regulated proteolysis of mammalian CDC6 is dependent on APC-CDH1. Genes Dev 14(18):2330–2343

Petroski MD, Deshaies RJ (2005) Function and regulation of cullin-RING ubiquitin ligases. Nat Rev Mol Cell Biol 6(1):9–20. https://doi.org/10.1038/nrm1547

Pfleger CM, Kirschner MW (2000) The KEN box: an APC recognition signal distinct from the D box targeted by Cdh1. Genes Dev 14(6):655–665

Pfleger CM, Lee E, Kirschner MW (2001) Substrate recognition by the Cdc20 and Cdh1 components of the anaphase-promoting complex. Genes Dev 15(18):2396–2407. https://doi.org/10.1101/gad.918201

Pines J (2006) Mitosis: a matter of getting rid of the right protein at the right time. Trends Cell Biol 16(1):55–63. https://doi.org/10.1016/j.tcb.2005.11.006

Plachta M, Halas A, McIntyre J, Sledziewska-Gojska E (2015) The steady-state level and stability of TLS polymerase eta are cell cycle dependent in the yeast S. cerevisiae. DNA Repair (Amst) 29:147–153. https://doi.org/10.1016/j.dnarep.2015.02.015

Plosky BS, Vidal AE, Fernandez de Henestrosa AR, McLenigan MP, McDonald JP, Mead S, Woodgate R (2006) Controlling the subcellular localization of DNA polymerases iota and eta via interactions with ubiquitin. EMBO J 25(12):2847–2855. https://doi.org/10.1038/sj.emboj.7601178

Qian J, Pentz K, Zhu Q, Wang Q, He J, Srivastava AK, Wani AA (2015) USP7 modulates UV-induced PCNA monoubiquitination by regulating DNA polymerase eta stability. Oncogene 34(36):4791–4796. https://doi.org/10.1038/onc.2014.394

Qing P, Han L, Bin L, Yan L, Ping WX (2011) USP7 regulates the stability and function of HLTF through deubiquitination. J Cell Biochem 112(12):3856–3862. https://doi.org/10.1002/jcb.23317

Rahal R, Amon A (2008) Mitotic CDKs control the metaphase-anaphase transition and trigger spindle elongation. Genes Dev 22(11):1534–1548. https://doi.org/10.1101/gad.1638308

Reimann JD, Freed E, Hsu JY, Kramer ER, Peters JM, Jackson PK (2001) Emi1 is a mitotic regulator that interacts with Cdc20 and inhibits the anaphase promoting complex. Cell 105(5):645–655

Renaudin X, Koch Lerner L, Menck CF, Rosselli F (2016) The ubiquitin family meets the Fanconi anemia proteins. Mutat Res Rev Mutat Res 769:36–46. https://doi.org/10.1016/j.mrrev.2016.06.004

Reyes-Turcu FE, Ventii KH, Wilkinson KD (2009) Regulation and cellular roles of ubiquitin-specific deubiquitinating enzymes. Annu Rev Biochem 78:363–397. https://doi.org/10.1146/annurev.biochem.78.082307.091526

Ricke RM, Bielinsky AK (2004) Mcm10 regulates the stability and chromatin association of DNA polymerase-alpha. Mol Cell 16(2):173–185. https://doi.org/10.1016/j.molcel.2004.09.017

Rizzardi LF, Coleman KE, Varma D, Matson JP, Oh S, Cook JG (2015) CDK1-dependent inhibition of the E3 ubiquitin ligase CRL4CDT2 ensures robust transition from S phase to mitosis. J Biol Chem 290(1):556–567. https://doi.org/10.1074/jbc.M114.614701

Robbins JA, Cross FR (2010) Regulated degradation of the APC coactivator Cdc20. Cell Div 5:23. https://doi.org/10.1186/1747-1028-5-23

Romani B, Shaykh Baygloo N, Aghasadeghi MR, Allahbakhshi E (2015) HIV-1 Vpr protein enhances proteasomal degradation of MCM10 DNA replication factor through the Cul4-DDB1[VprBP] E3 ubiquitin ligase to induce G2/M cell cycle arrest. J Biol Chem 290(28):17380–17389. https://doi.org/10.1074/jbc.M115.641522

Roseaulin LC, Noguchi C, Martinez E, Ziegler MA, Toda T, Noguchi E (2013) Coordinated degradation of replisome components ensures genome stability upon replication stress in the absence of the replication fork protection complex. PLoS Genet 9(1):e1003213. https://doi.org/10.1371/journal.pgen.1003213

Saha P, Chen J, Thome KC, Lawlis SJ, Hou ZH, Hendricks M, Parvin JD, Dutta A (1998) Human CDC6/Cdc18 associates with Orc1 and cyclin-cdk and is selectively eliminated from the nucleus at the onset of S phase. Mol Cell Biol 18(5):2758–2767

Sansam CL, Shepard JL, Lai K, Ianari A, Danielian PS, Amsterdam A, Hopkins N, Lees JA (2006) DTL/CDT2 is essential for both CDT1 regulation and the early G2/M checkpoint. Genes Dev 20(22):3117–3129

Sarikas A, Hartmann T, Pan ZQ (2011) The cullin protein family. Genome Biol 12(4):220. https://doi.org/10.1186/gb-2011-12-4-220

Schotta G, Sengupta R, Kubicek S, Malin S, Kauer M, Callen E, Celeste A, Pagani M, Opravil S, De La Rosa-Velazquez IA, Espejo A, Bedford MT, Nussenzweig A, Busslinger M, Jenuwein T (2008) A chromatin-wide transition to H4K20 monomethylation impairs genome integrity and programmed DNA rearrangements in the mouse. Genes Dev 22(15):2048–2061. https://doi.org/10.1101/gad.476008

Schwartz AL, Ciechanover A (2009) Targeting proteins for destruction by the ubiquitin system: implications for human pathobiology. Annu Rev Pharmacol Toxicol 49:73–96. https://doi.org/10.1146/annurev.pharmtox.051208.165340

Senga T, Sivaprasad U, Zhu W, Park JH, Arias EE, Walter JC, Dutta A (2006) PCNA is a cofactor for Cdt1 degradation by CUL4/DDB1-mediated N-terminal ubiquitination. J Biol Chem 281(10):6246–6252. https://doi.org/10.1074/jbc.M512705200

Shibutani ST, de la Cruz AF, Tran V, Turbyfill WJ 3rd, Reis T, Edgar BA, Duronio RJ (2008) Intrinsic negative cell cycle regulation provided by PIP box- and Cul4Cdt2-mediated destruction of E2f1 during S phase. Dev Cell 15(6):890–900. https://doi.org/10.1016/j.devcel.2008.10.003

Shirayama M, Toth A, Galova M, Nasmyth K (1999) APC(Cdc20) promotes exit from mitosis by destroying the anaphase inhibitor Pds1 and cyclin Clb5. Nature 402(6758):203–207. https://doi.org/10.1038/46080

Simon NE, Schwacha A (2014) The Mcm2-7 replicative helicase: a promising chemotherapeutic target. Biomed Res Int 2014:549719. https://doi.org/10.1155/2014/549719

Simpson LJ, Ross AL, Szuts D, Alviani CA, Oestergaard VH, Patel KJ, Sale JE (2006) RAD18-independent ubiquitination of proliferating-cell nuclear antigen in the avian cell line DT40. EMBO Rep 7(9):927–932. https://doi.org/10.1038/sj.embor.7400777

Sivaprasad U, Machida YJ, Dutta A (2007) APC/C--the master controller of origin licensing? Cell Div 2:8. https://doi.org/10.1186/1747-1028-2-8

Skaar JR, Pagan JK, Pagano M (2013) Mechanisms and function of substrate recruitment by F-box proteins. Nat Rev Mol Cell Biol 14(6):369–381. https://doi.org/10.1038/nrm3582

Skaar JR, Pagan JK, Pagano M (2014) SCF ubiquitin ligase-targeted therapies. Nat Rev Drug Discov 13(12):889–903. https://doi.org/10.1038/nrd4432

Skoneczna A, McIntyre J, Skoneczny M, Policinska Z, Sledziewska-Gojska E (2007) Polymerase eta is a short-lived, proteasomally degraded protein that is temporarily stabilized following UV irradiation in Saccharomyces cerevisiae. J Mol Biol 366(4):1074–1086. https://doi.org/10.1016/j.jmb.2006.11.093

Sommers JA, Suhasini AN, Brosh RM Jr (2015) Protein degradation pathways regulate the functions of helicases in the DNA damage response and maintenance of genomic stability. Biomol Ther 5(2):590–616. https://doi.org/10.3390/biom5020590

Soucy TA, Smith PG, Milhollen MA, Berger AJ, Gavin JM, Adhikari S, Brownell JE, Burke KE, Cardin DP, Critchley S, Cullis CA, Doucette A, Garnsey JJ, Gaulin JL, Gershman RE, Lublinsky AR, McDonald A, Mizutani H, Narayanan U, Olhava EJ, Peluso S, Rezaei M, Sintchak MD, Talreja T, Thomas MP, Traore T, Vyskocil S, Weatherhead GS, Yu J, Zhang J, Dick LR, Claiborne CF, Rolfe M, Bolen JB, Langston SP (2009) An inhibitor of NEDD8-activating enzyme as a new approach to treat cancer. Nature 458(7239):732–736. https://doi.org/10.1038/nature07884

Sullivan M, Morgan DO (2007) Finishing mitosis, one step at a time. Nat Rev Mol Cell Biol 8(11):894–903. https://doi.org/10.1038/nrm2276

Tada S, Li A, Maiorano D, Mechali M, Blow JJ (2001) Repression of origin assembly in metaphase depends on inhibition of RLF-B/Cdt1 by geminin. Nat Cell Biol 3(2):107–113. https://doi.org/10.1038/35055000

Takeda DY, Parvin JD, Dutta A (2005) Degradation of Cdt1 during S phase is Skp2-independent and is required for efficient progression of mammalian cells through S phase. J Biol Chem 280(24):23416–23423. https://doi.org/10.1074/jbc.M501208200

Talbert PB, Henikoff S (2017) Histone variants on the move: substrates for chromatin dynamics. Nat Rev Mol Cell Biol 18(2):115–126. https://doi.org/10.1038/nrm.2016.148

Tanaka S, Diffley JF (2002) Interdependent nuclear accumulation of budding yeast Cdt1 and Mcm2-7 during G1 phase. Nat Cell Biol 4(3):198–207. https://doi.org/10.1038/ncb757

Tanaka T, Knapp D, Nasmyth K (1997) Loading of an Mcm protein onto DNA replication origins is regulated by Cdc6p and CDKs. Cell 90(4):649–660

Tanaka T, Nakatani T, Kamitani T (2013) Negative regulation of NEDD8 conjugation pathway by novel molecules and agents for anticancer therapy. Curr Pharm Des 19(22):4131–4139

Tardat M, Brustel J, Kirsh O, Lefevbre C, Callanan M, Sardet C, Julien E (2010) The histone H4 Lys 20 methyltransferase PR-Set7 regulates replication origins in mammalian cells. Nat Cell Biol 12(11):1086–1093. doi:ncb2113 [pii]1038/ncb2113

Tatsumi Y, Ohta S, Kimura H, Tsurimoto T, Obuse C (2003) The ORC1 cycle in human cells: I. cell cycle-regulated oscillation of human ORC1. J Biol Chem 278(42):41528–41534. https://doi.org/10.1074/jbc.M307534200

Tatsumi Y, Sugimoto N, Yugawa T, Narisawa-Saito M, Kiyono T, Fujita M (2006) Deregulation of Cdt1 induces chromosomal damage without rereplication and leads to chromosomal instability. J Cell Sci 119(Pt 15):3128–3140

Teixeira LK, Reed SI (2013) Ubiquitin ligases and cell cycle control. Annu Rev Biochem 82:387–414. https://doi.org/10.1146/annurev-biochem-060410-105307

Terai K, Abbas T, Jazaeri AA, Dutta A (2010) CRL4(Cdt2) E3 ubiquitin ligase monoubiquitinates PCNA to promote translesion DNA synthesis. Mol Cell 37(1):143–149. https://doi.org/10.1016/j.molcel.2009.12.018

Terai K, Shibata E, Abbas T, Dutta A (2013) Degradation of p12 subunit by CRL4Cdt2 E3 ligase inhibits fork progression after DNA damage. J Biol Chem 288(42):30509–30514. https://doi.org/10.1074/jbc.C113.505586

Thu YM, Bielinsky AK (2013) Enigmatic roles of Mcm10 in DNA replication. Trends Biochem Sci 38(4):184–194. https://doi.org/10.1016/j.tibs.2012.12.003

Ulrich HD, Takahashi T (2013) Readers of PCNA modifications. Chromosoma 122(4):259–274. https://doi.org/10.1007/s00412-013-0410-4

Unk I, Hajdu I, Fatyol K, Hurwitz J, Yoon JH, Prakash L, Prakash S, Haracska L (2008) Human HLTF functions as a ubiquitin ligase for proliferating cell nuclear antigen polyubiquitination. Proc Natl Acad Sci U S A 105(10):3768–3773. https://doi.org/10.1073/pnas.0800563105

van Deursen F, Sengupta S, De Piccoli G, Sanchez-Diaz A, Labib K (2012) Mcm10 associates with the loaded DNA helicase at replication origins and defines a novel step in its activation. EMBO J 31(9):2195–2206. https://doi.org/10.1038/emboj.2012.69

van Leuken R, Clijsters L, Wolthuis R (2008) To cell cycle, swing the APC/C. Biochim Biophys Acta 1786(1):49–59. https://doi.org/10.1016/j.bbcan.2008.05.002

Vaziri C, Saxena S, Jeon Y, Lee C, Murata K, Machida Y, Wagle N, Hwang DS, Dutta A (2003) A p53-dependent checkpoint pathway prevents rereplication. Mol Cell 11(4):997–1008

Visintin R, Prinz S, Amon A (1997) CDC20 and CDH1: a family of substrate-specific activators of APC-dependent proteolysis. Science 278(5337):460–463

Wagner SA, Beli P, Weinert BT, Nielsen ML, Cox J, Mann M, Choudhary C (2011) A proteome-wide, quantitative survey of in vivo ubiquitylation sites reveals widespread regulatory roles. Mol Cell Proteomics 10(10):M111 013284. https://doi.org/10.1074/mcp.M111.013284

Wallace HA, Merkle JA, Yu MC, Berg TG, Lee E, Bosco G, Lee LA (2014) TRIP/NOPO E3 ubiquitin ligase promotes ubiquitylation of DNA polymerase eta. Development 141(6):1332–1341. https://doi.org/10.1242/dev.101196

Walter D, Hoffmann S, Komseli ES, Rappsilber J, Gorgoulis V, Sorensen CS (2016) SCF(cyclin F)-dependent degradation of CDC6 suppresses DNA re-replication. Nat Commun 7:10530. https://doi.org/10.1038/ncomms10530

Wang Z, Liu P, Inuzuka H, Wei W (2014) Roles of F-box proteins in cancer. Nat Rev Cancer 14(4):233–247. https://doi.org/10.1038/nrc3700

Watanabe N, Arai H, Nishihara Y, Taniguchi M, Watanabe N, Hunter T, Osada H (2004) M-phase kinases induce phospho-dependent ubiquitination of somatic Wee1 by SCFbeta-TrCP. Proc Natl Acad Sci U S A 101(13):4419–4424. https://doi.org/10.1073/pnas.0307700101
Waters LS, Walker GC (2006) The critical mutagenic translesion DNA polymerase Rev1 is highly expressed during G(2)/M phase rather than S phase. Proc Natl Acad Sci U S A 103(24):8971–8976. https://doi.org/10.1073/pnas.0510167103
Wei W, Ayad NG, Wan Y, Zhang GJ, Kirschner MW, Kaelin WG Jr (2004) Degradation of the SCF component Skp2 in cell-cycle phase G1 by the anaphase-promoting complex. Nature 428(6979):194–198. https://doi.org/10.1038/nature02381
Wei D, Li H, Yu J, Sebolt JT, Zhao L, Lawrence TS, Smith PG, Morgan MA, Sun Y (2012) Radiosensitization of human pancreatic cancer cells by MLN4924, an investigational NEDD8-activating enzyme inhibitor. Cancer Res 72(1):282–293. https://doi.org/10.1158/0008-5472.CAN-11-2866
Welcker M, Clurman BE (2008) FBW7 ubiquitin ligase: a tumour suppressor at the crossroads of cell division, growth and differentiation. Nat Rev Cancer 8(2):83–93. https://doi.org/10.1038/nrc2290
Wilkinson KD (1997) Regulation of ubiquitin-dependent processes by deubiquitinating enzymes. FASEB J 11(14):1245–1256
Wohlschlegel JA, Dwyer BT, Dhar SK, Cvetic C, Walter JC, Dutta A (2000) Inhibition of eukaryotic DNA replication by geminin binding to Cdt1. Science 290(5500):2309–2312. https://doi.org/10.1126/science.290.5500.2309
Wong JV, Dong P, Nevins JR, Mathey-Prevot B, You L (2011) Network calisthenics: control of E2F dynamics in cell cycle entry. Cell Cycle 10(18):3086–3094. https://doi.org/10.4161/cc.10.18.17350
Wu S, Wang W, Kong X, Congdon LM, Yokomori K, Kirschner MW, Rice JC (2010) Dynamic regulation of the PR-Set7 histone methyltransferase is required for normal cell cycle progression. Genes Dev 24(22):2531–2542. https://doi.org/10.1101/gad.1984210
Xiao B, Jing C, Kelly G, Walker PA, Muskett FW, Frenkiel TA, Martin SR, Sarma K, Reinberg D, Gamblin SJ, Wilson JR (2005) Specificity and mechanism of the histone methyltransferase Pr-Set7. Genes Dev 19(12):1444–1454. https://doi.org/10.1101/gad.1315905
Yang WL, Zhang X, Lin HK (2010) Emerging role of Lys-63 ubiquitination in protein kinase and phosphatase activation and cancer development. Oncogene 29(32):4493–4503. https://doi.org/10.1038/onc.2010.190
Yang K, Weinacht CP, Zhuang Z (2013) Regulatory role of ubiquitin in eukaryotic DNA translesion synthesis. Biochemistry 52(19):3217–3228. https://doi.org/10.1021/bi400194r
Zachariae W, Nasmyth K (1999) Whose end is destruction: cell division and the anaphase-promoting complex. Genes Dev 13(16):2039–2058
Zaidi IW, Rabut G, Poveda A, Scheel H, Malmstrom J, Ulrich H, Hofmann K, Pasero P, Peter M, Luke B (2008) Rtt101 and Mms1 in budding yeast form a CUL4(DDB1)-like ubiquitin ligase that promotes replication through damaged DNA. EMBO Rep 9(10):1034–1040. https://doi.org/10.1038/embor.2008.155
Zeman MK, Lin JR, Freire R, Cimprich KA (2014) DNA damage-specific deubiquitination regulates Rad18 functions to suppress mutagenesis. J Cell Biol 206(2):183–197. https://doi.org/10.1083/jcb.201311063
Zhang S, Chea J, Meng X, Zhou Y, Lee EY, Lee MY (2008) PCNA is ubiquitinated by RNF8. Cell Cycle 7(21):3399–3404. https://doi.org/10.4161/cc.7.21.6949
Zhao Y, Sun Y (2013) Cullin-RING Ligases as attractive anti-cancer targets. Curr Pharm Des 19(18):3215–3225
Zhao Y, Morgan MA, Sun Y (2014) Targeting Neddylation pathways to inactivate cullin-RING ligases for anticancer therapy. Antioxid Redox Signal 21(17):2383–2400. https://doi.org/10.1089/ars.2013.5795
Zhong W, Feng H, Santiago FE, Kipreos ET (2003) CUL-4 ubiquitin ligase maintains genome stability by restraining DNA-replication licensing. Nature 423(6942):885–889

Zhu W, Depamphilis ML (2009) Selective killing of cancer cells by suppression of geminin activity. Cancer Res 69(11):4870–4877. https://doi.org/10.1158/0008-5472.CAN-08-4559

Zhu W, Dutta A (2006) An ATR- and BRCA1-mediated Fanconi anemia pathway is required for activating the G2/M checkpoint and DNA damage repair upon rereplication. Mol Cell Biol 26(12):4601–4611. https://doi.org/10.1128/MCB.02141-05

Zhu W, Chen Y, Dutta A (2004) Rereplication by depletion of geminin is seen regardless of p53 status and activates a G2/M checkpoint. Mol Cell Biol 24(16):7140–7150

Zhu W, Abbas T, Dutta A (2005) DNA replication and genomic instability. Adv Exp Med Biol 570:249–279. https://doi.org/10.1007/1-4020-3764-3_9

Zhu W, Ukomadu C, Jha S, Senga T, Dhar SK, Wohlschlegel JA, Nutt LK, Kornbluth S, Dutta A (2007) Mcm10 and And-1/CTF4 recruit DNA polymerase alpha to chromatin for initiation of DNA replication. Genes Dev 21(18):2288–2299. https://doi.org/10.1101/gad.1585607

Zhu W, Lee CY, Johnson RL, Wichterman J, Huang R, DePamphilis ML (2011) An image-based, high-throughput screening assay for molecules that induce excess DNA replication in human cancer cells. Mol Cancer Res 9(3):294–310. https://doi.org/10.1158/1541-7786.MCR-10-0570

Zilio N, Eifler-Olivi K, Ulrich HD (2017) Functions of SUMO in the maintenance of genome stability. Adv Exp Med Biol 963:51–87. https://doi.org/10.1007/978-3-319-50044-7_4

Zlatanou A, Sabbioneda S, Miller ES, Greenwalt A, Aggathanggelou A, Maurice MM, Lehmann AR, Stankovic T, Reverdy C, Colland F, Vaziri C, Stewart GS (2016) USP7 is essential for maintaining Rad18 stability and DNA damage tolerance. Oncogene 35(8):965–976. https://doi.org/10.1038/onc.2015.149

Chapter 20
Coordinating Replication with Transcription

Yathish Jagadheesh Achar and Marco Foiani

Abstract DNA topological transitions occur when replication forks encounter other DNA transactions such as transcription. Failure in resolving such conflicts leads to generation of aberrant replication and transcription intermediates that might have adverse effects on genome stability. Cells have evolved numerous surveillance mechanisms to avoid, tolerate, and resolve such replication-transcription conflicts. Defects or non-coordination in such cellular mechanisms might have catastrophic effect on cell viability. In this chapter, we review consequences of replication encounters with transcription and its associated events, topological challenges, and how these inevitable conflicts alter the genome structure and functions.

Keywords DNA replication • Transcription • Topology • Topoisomerase • Chromatin structure • Genome instability • RNA:DNA hybrids

20.1 Introduction

Every cell, whether it is a prokaryotic or eukaryotic cell, has to duplicate its genome prior to its division. To ensure a faithful completion of genome duplication, cells have to protect the replication apparatus from numerous intrinsic and extrinsic factors that might hinder its progression. Along with extrinsic sources, which might alter chromatin integrity, several natural obstacles like DNA lesions, DNA-protein complexes, and alternative DNA structures also hinder the replication process. One such major natural hindrance for replication progression is transcription.

Y.J. Achar (✉)
IFOM (Fondazione Istituto FIRC di Oncologia Molecolare), Milan, Italy
e-mail: yathish.achar@ifom.eu

M. Foiani
Istituto FIRC di Oncologia Molecolare, Milan, Italy

H. Masai, M. Foiani (eds.), *DNA Replication*, Advances in Experimental Medicine and Biology 1042, https://doi.org/10.1007/978-981-10-6955-0_20

S-phase in eukaryotic cells is the most vulnerable period during the cell cycle as replication and transcription coexist in space and in time. Collision between the two processes is inevitable as both compete for the same template to synthesize nucleic acid chains. In principle, replication and transcription are two independent processes, which are expected to perform their respective tasks that are to synthesize new DNA or RNA molecules using the complimentary DNA strand. Both processes are highly processive, strictly follow 5–3′ polarity, and are carried out by brigades of respective polymerases along with a plethora protein partners and other cofactors. These two process need to be coordinated as both must be completed with high fidelity to preserve genetic information and genome stability.

It has been known for many years that transcription could interfere with genome integrity. The very first indication that transcription could passively challenge genome stability came from bacterial studies where a reporter gene showed higher mutational rates when highly transcribed (Herman and Dworkin 1971; Savic and Kanazir 1972). Later, direct pieces of evidence for the effect of transcription on genome integrity were reported in several organisms including bacteria (Dul and Drexler 1988; Vilette et al. 1992), yeast (Keil and Roeder 1984; Brill et al. 1987; Thomas and Rothstein 1989), and mammalian cells (Nickoloff 1992). Along with causing spontaneous mutagenesis (Datta and Jinks-Robertson 1995), transcription-associated recombination (TAR) was also discovered in yeast (Keil and Roeder 1984; Voelkel-Meiman et al. 1987). Interestingly, TAR is known to depend on replication as proven in yeast (Prado and Aguilera 2005) and mammalian cells (Gottipati et al. 2008). These events eventually lead to one of the important questions in molecular biology "how do replication forks contend transcription complexes?" Research in the last 25 years has provided evidence that conflicts between transcription and replication constitutes major natural source of threat on genome stability. Current knowledge of both replication and transcription suggests that the mechanisms and factors involved in resolving conflict are more complex than previously foreseen.

How does a process involved in gene expression create problems for the gene duplication process? The answer lies in the speed at which these two processes are carried out. DNA replication complexes move faster than transcription complexes; the two complexes must inevitably collide considering that they occupy the same template. Cells could solve this problem by replisome-induced dissociation of transcription complex from the template. However in physiological conditions, it is not an efficient solution, as the transcription process is complex and associates with several factors and events. A replication fork has to uncouple these factors before dissociating the RNA polymerases (RNAPs) from the template. Along with hindering replication movement by occupying the DNA template, transcription is accompanied by co-transcriptional processes, such as RNA maturation and export. During transcription elongation, RNAPs can pause at regulatory sequences and DNA damage sites (Selby et al. 1997; Landick 2006). A halted RNAP complex can give rise to consequent arrays of stalled RNAP complexes, as stalled transcription units can also be barriers for the following RNAPs, which are transcribing the same template (Artsimovitch and Landick 2000). In highly transcribing genes, repeated loading of the RNAP at promoters lead to the piling of RNAP, creating a potentially formidable obstacle to replication and also for damage repair. To rescue such a scenario, the RNAP complex translocates backward along the template (Park et al. 2002). Such type of reverse translocation is known to occur spontaneously at the pausing sites displacing the 3′ end of transcripts from active sites. This results in an extremely stable, inactive complex, which can form potential replicative barrier and give rise to a tandem array of barriers.

Along with occupying templates, transcription also induces additional chromatin structural changes, which could also challenge replication fork progression. Eukaryotic cells must lessen the secondary effects enforced by each other through topological perturbations. By unwinding the DNA duplex, replication and transcription generate topological alterations that are solved by specialized enzymes, the DNA topoisomerases. Replication can accumulate hemicatenane-like structures behind the fork in response to DNA damage (Lopes et al. 2001; Liberi et al. 2005), whereas negative supercoiling is induced due to transcription-mediated formation of Z-DNA (Ha et al. 2005). These alternative DNA transitions can be hazardous structures if they occur out of their natural context by having a negative impact on transcription elongation and also can be endangering for replication fork integrity (Aguilera and Garcia-Muse 2012).

Interrelation between replication and transcription became evident when a new set of genes, known to be involved in transcription and RNA processing, was discovered, and their subsequent mutation was shown to cause an increase in DNA damage, higher mutation rates, and hyper-recombination (Wahba et al. 2011). Recent observations show that mutations in DNA topoisomerases (El Hage et al. 2010), RNA-elongating proteins (Senataxin 1) (Mischo et al. 2011), RNA export complexes (THO/TREX complex) (Gomez-Gonzalez et al. 2009), and splicing factors (ASF/SF2) (Li and Manley 2005) have been linked to genome instability. Slight alterations in the functionality of these transcription-associated events might challenge the integrity of the genome via a common mechanism that would induce the formation of persistent, transcription-associated RNA:DNA hybrids (Aguilera 2002). Formation of RNA:DNA hybrids out of their natural context can lead to distinctive DNA structures including R-Loops (nascent RNA bound to one of the DNA strands by displacing the other strand) and triple helix DNA termed as H-DNA (RNA bound to duplex DNA via Hoogsteen base pairing with separating strands).

Cells have evolved to counteract the conflict between replication and transcription. Checkpoint genes, which are often mutated in cancer cells, might play a role in sensing the physiological stress caused by topological intermediates. On the other hand, oncogenes that alter the transcriptional programs are known to cause replication stress and checkpoint activation (Bermejo et al. 2012). Possible consequences, topological challenges, and chromatin structure alterations arising due to the replication-transcription conflicts are outlined in the following sections.

20.2 Replication and Transcription Collision

Initial attempts to characterize the collision between transcription and replication were examined in vitro, using T4 bacteriophage replication apparatus and purified *E.coli* RNAP (Bedinger et al. 1983). Several subsequent efforts were made to understand the consequences of replication-transcription collision using in vitro assays in bacteria, yeast, and mammalian cells. The existence of collisions between the major molecular processes was raised mainly in bacteria due to the differences of the speed of replication fork progression and transcription elongation. The speed of replication fork progression in *Escherichia coli* is ~700 bp per second (Hirose et al. 1983; Mok and Marians 1987; Breier et al. 2005), while the elongation rate of RNAP is ~50 nt per second (Gotta et al. 1991; Epshtein et al. 2003; Proshkin et al. 2010). Since replication is 12-fold faster than transcription, collision between the two processes is unavoidable in bacteria where both processes continue throughout the cell cycle. First in vivo evidence of replication-transcription collision came from *E. coli* (French 1992) by placing the rrnB ribosomal operon on either side of the inducible replication origin. Upon inducing replication, EM visualization revealed that replication fork that is head-on direction with the rrnB operon was slower compared to the codirectional fork. Since then, several studies in bacteria and other prokaryotes have shown transcription complexes as natural impediments to replication (Merrikh et al. 2012).

Since prokaryotes with circular genome cannot completely avoid the replication-transcription collision, a simple solution would be the replisome-induced dissociation of RNAPs from the template. However, pieces of evidence linking genome stability to transcription-induced replication blockage suggest it is not an efficient solution in vivo (Brewer 1988; Rudolph et al. 2007). To avoid such scenarios, bacterial genomes have evolved a simple strategy to reduce the effect of transcription-replication collision by placing highly transcribed and essential genes on codirectional orientation relative to replication fork movement. This kind of gene arrangement is conserved in most bacteria studied (Guy and Roten 2004). These codirectional arrangements are known to reduce the effect of conflict in lower organisms as reversing the orientation has shown to impact the progression of replication and genome stability (Srivatsan et al. 2010).

Unlike prokaryotes, eukaryotic cells have the luxury of restricting genome duplication to one phase of the cell cycle, S-phase, leaving the other phases free of replication so that transcription and other chromatin-associated events can occur in an uninterruptible way. However, since transcription in a eukaryotic cell cannot be completely turned off in S-phase, it is believed that cells can spatially separate replication through chromatin domain architecture (Wansink et al. 1994; Wei et al. 1998). Eukaryotic cells also initiate replication from hundreds of origins, which will initiate at different time points, some early and some very late during S-phases (Wyrick et al. 2001). This will allow cells to temporally separate transcription and replication events through inverse correlation. Early replicating genes are more strongly transcribed late in S-phase, and late replicating genes are preferably transcribed in early S-phase (Meryet-Figuiere et al. 2014). Moreover, interference between replication and transcription in eukaryotes debated for long due to comparable speeds at which both processes operate. Unlike in prokaryotes, in eukaryotes replication and transcription machineries operate at similar speeds of 17–33 nt S^{-1} and 17–72 nt S^{-1}, respectively (Raghuraman et al. 2001; Hiratani et al. 2008; Spiesser et al. 2010). First evidence for the occurrence of collision in eukaryotes came from the discovery of replication fork barrier (RFB) at the 3′ end of the *S. cerevisiae* 35S rRNA genes and tRNA transcription units (Brewer and Fangman 1988; Linskens and Huberman 1988). Later, transcription-dependent pausing sites were also mapped in plants (Lopez-Estrano et al. 1999), *Xenopus laevis* (Maric et al. 1999), *Tetrahymena thermophila* (Zhang et al. 1997), and mammals (Lopez-estrano et al. 1998). Replication barriers, which are observed in cases of highly expressed and specialized units such as rDNA and tRNA, are not normally observed at the RNA pol-II genes. In such protein coding regions of the genome, transcription and replication seem to be separated temporally, by inversely coordinating the transcription with replication timing (Gilbert 2002; Hiratani et al. 2009; Rivera-Mulia and Gilbert 2016).

Temporal and spatial segregation of transcription with replication in S-phase of eukaryotic cells still cannot rule out the possibility of interference between the two processes. Recent studies show that eukaryotic replication and transcription units do arise in certain situations and at specific genomic regions. Moreover, recent genome-wide transcription data reveal that the majority of the genome is transcribed well beyond boundaries of annotated gene bodies. This phenomenon was observed in several eukaryotic cells like *S. cerevisiae* (David et al. 2006), mammalian (Carninci et al. 2005), *S. pombe* (Dutrow et al. 2008), rice (Li et al. 2006), and *D. melanogaster* (Stolc et al. 2004). This only shows that transcription and replication collisions are more pervasive than previously anticipated.

Other than arrangement of transcription units, collision in prokaryotes and eukaryotes also differs in the way it occurs. In bacteria such as *E. coli*, the replicative helicase responsible for separation of the two strands is DnaB that acts in a 5–3′ direction. During replication fork progression, DnaB is associated with the lagging arm of the replication fork (Fig. 20.1a). In eukaryotes, the CMG helicase takes care

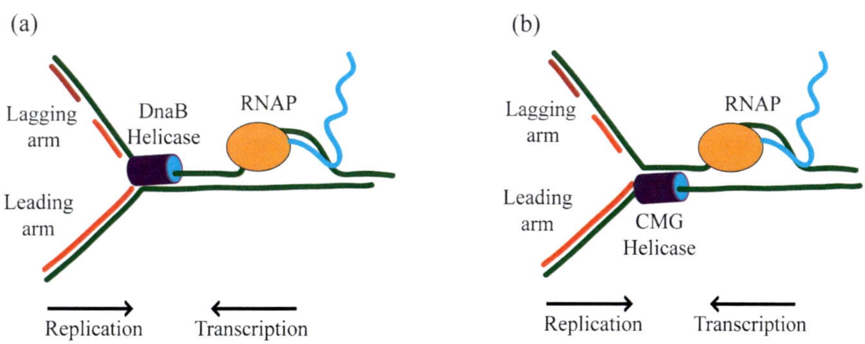

Fig. 20.1 Difference in polarity of prokaryotic and eukaryotic DNA replicative helicases. (**a**) Prokaryotic helicase like DnaB has 5–3′ polarity hence travels on lagging arm. (**b**) Eukaryotic helicase like CMG has 3–5′ polarity hence travels on leading arm of replication fork

of strand separation and is associated with the leading arm of the replication fork as it has 3–5′ helicase activity. Since RNAP translocates in a 3–5′ direction on the transcribed strand, a head on collision might interfere directly with the replicative helicase in prokaryotes. However, in eukaryotes, same head on collision will not interfere with replicative helicase as RNAP and helicase translocate on different strands (Fig. 20.1b).

Replication speed depends on the orientation of the gene with respect to the fork movement when replicating across a gene. However, no particular bias was observed with respect to the dislodgement of RNAPs during collision (French 1992). In contrast, in vitro studies showed that RNAP resumes transcription after replication passes in either orientation (Liu et al. 1993; Liu and Alberts 1995; Elias-Arnanz and Salas 1997, 1999). Based on the influence of directionality and impact, replication-transcription collision is grouped into head-on and codirectional collisions. Differences between these two collisions and possible consequences are discussed in the below section.

20.3 Head-On Versus Codirectional Collision

Since both replication and transcription processes have the same polarity for synthesis of new nucleic acid chains, the template strand used for transcription determines the orientation of a gene with respect to replication. A head-on collision will take place when transcription from RNAP II uses the lagging strand as a template (Fig. 20.2b), wherein codirectional uses the leading strand (Fig. 20.2a). Comparison between the two types of collisions rose due to the difference in consequences of collision. Regardless of whether collision occurs in a codirectional or a head-on orientation, replication fork cannot progress past an elongating RNAP, so their encounter causes replication fork pausing (Azvolinsky et al. 2009). Earlier work,

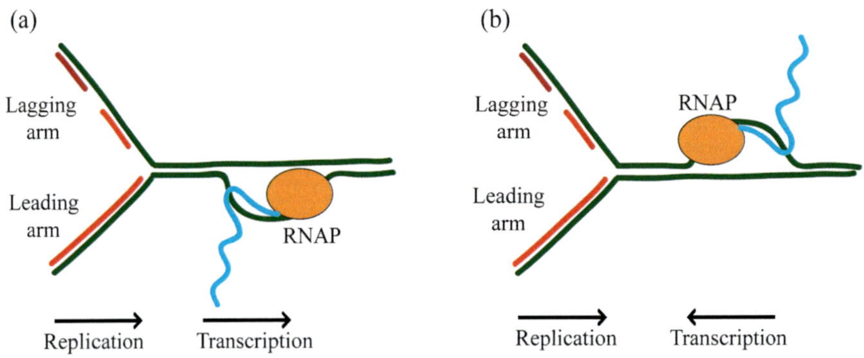

Fig. 20.2 Comparison of head-on and codirectional collisions between replication fork and transcription unit. (**a**) In codirectional collision, RNAP uses leading arm as a template. (**b**) In head-on collision, RNAP uses lagging arm as a template

particularly in bacteria, suggested that head-on collisions were more deleterious causing replication to stall, but codirectional collisions were generally not detected in vivo and hence thought to be less harmful (French 1992; Mirkin and Mirkin 2005; Wang et al. 2007; Srivatsan et al. 2010). However, recent findings contradicted this notion and indicated that both types of collisions can disrupt replication progression in vivo (Dutta et al. 2011; Merrikh et al. 2011).

Fork stalling, a major outcome of head-on collision, can lead to accumulation of alternative DNA structures such as DNA knots (Olavarrieta et al. 2002). In vitro experiments have shown that fork restart can only take place after the displacement of RNAP machinery from the DNA template during head-on conflicts (Pomerantz and O'Donnell 2008). Consistent with this idea, yeast genetic experiments also suggest that replication forks are blocked more predominantly in head-on collisions. When head-on encounters were promoted in yeast using an artificial system, replication pause sites along with hyper-recombination events were observed. On the contrary, codirectional collisions did not lead to replication pausing or high levels of recombination (Prado and Aguilera 2005). Moreover, a codirectional collision barely affects fork progression in vitro (Liu and Alberts 1995; Pomerantz and O'Donnell 2008), unless the elongating RNAP is stalled (Elias-Arnanz and Salas 1997; Dutta et al. 2011). Occasionally, the replication machinery can use mRNAs to reprime DNA synthesis after codirectional collision with RNAP (Kogoma 1997; Pomerantz and O'Donnell 2008), probably supporting the notion that codirectional collisions are less harmful than the head-on collisions. The key event that determines the effect of collision is RNAP displacement from the template. Although it is believed largely that RNAP units are more easily and quickly displaced in a codirectional manner. However, mechanistic details supporting this notion remain to be unknown.

One of the major influential pieces of evidence for codirectional conflicts to be less harmful came from observing the arrangements of transcription units in bacteria. In almost all bacteria, highly transcribed genes are co-oriented with replication. In *B. subtilis*, around 75% of all genes are co-oriented relative to replication fork direction (Kunst et al. 1997). In *E.coli*, fewer genes, only 55%, show such an orientation (Blattner et al. 1997). However, if only essential and highly transcribed genes like rRNA and tRNA are considered, a number of genes in co-orientation increase to 70% and 90% in *E. coli* and *B. subtilis*, respectively. Since such bias is not obvious in eukaryotic cells, they seemed to have evolved other alternative strategies to reduce head-on collision. In *S. cerevisiae*, replication fork barriers (RFBs) block fork progression at the highly transcribed rDNA locus to avoid the head-on encounter (Brewer and Fangman 1988; Linskens and Huberman 1988). Such type of programmed replication blockage by RFBs at rDNA locus is conserved in other yeast species (Sanchez-Gorostiaga et al. 2004; Krings and Bastia 2005). Other than rDNA loci, replication barriers are known to exist in other part of the genome; for example, replication termination sequence 1 (RTS1) in fission yeast is responsible for unidirectional replication at the mating-type locus, avoiding the head-on collision (Dalgaard and Klar 2001). Some other barriers such as polar barriers that form at tRNA genes (Deshpande and Newlon 1996) and bidirectional barriers present at centromere (Greenfeder and Newlon 1992) are also known to stall fork progression in yeast. Higher eukaryotes like mammalian cells also contain RFBs at rDNA loci (Gerber et al. 1997), and their genes are potentially organized collinearly with replication (Huvet et al. 2007). Along with replication barriers, higher eukaryotic cells also use genome organization as a secondary line of defense against collisions. One such example is demonstrated in mammalian cells, where transcription of rDNA genes is spatially separated at nucleoli (Smirnov et al. 2014).

Codirectional encounters can also be disruptive for replication, if encounter occurs with a highly transcribed gene. Codirectional conflicts were detected at rRNA genes only under fast-growing conditions, where there is a high level of transcription of rRNA genes. However, the effect of conflict was not detected under slow-growing conditions, where there is a decrease of transcription levels (Merrikh et al. 2011). This probably indicates that codirectional conflicts can be disruptive to replication in vivo, most likely when the replication apparatus encounters more than one RNAP or an array of RNAPs.

Consequences of challenging the bias by inverting a gene from codirectional to head-on orientation can have a severe effect. Inverting the direction of highly transcribed gene like rRNA not only stalls replication fork but also disrupts the integrity of the replication fork (Boubakri et al. 2010; Srivatsan et al. 2010). However, inverting a genomic region containing different classes of genes showed only a mild reduction in replication fork progression (Wang et al. 2007). Disruption of a replication fork is more deleterious than fork slowdown or pausing, as disruption can trigger cell cycle checkpoints and eventually delays cell division. Indeed, failure to repair replication disruption due to head-on collision is associated with cell growth arrest and leads to cell death (Boubakri et al. 2010; Srivatsan et al. 2010). This might explain why codirectional arrangements for highly transcribed genes like tRNA and rRNA are evolutionally conserved in most of prokaryotes (Guy and Roten 2004). However, how essential genes also follow codirectional orientation conservation across species is not clearly known, as comparative genomic studies indicate that essential genes are selected on the basis of essentiality rather than level of its expression (Rocha and Danchin 2003).

It is clear that head-on collisions negatively impact DNA replication more than codirectional collisions. However, the nature of aspects of head-on collision that makes it more deleterious is not clearly known. One previously thought hypothesis is that the replicative helicase might encounter RNAPs before the other replication components on the lagging strand (McHenry 2011), which might make them nonfunctional, leading to replication fork stalling. However, this hypothesis only explains the scenario in prokaryotes but not in eukaryotic cells, where replicative helicase travels on the leading arm unlike in prokaryotes. Moreover, recent findings question if directionality matters at all in eukaryotes. In yeast, the relative directions of replication and transcription do not matter on a genome-wide scale, as the naturally occurring sites of replication-fork pausing at transcribed regions appear to be independent of polarity (Azvolinsky et al. 2009). DNA damage studies carried out in yeast mutants defective in RNA-processing factors also show damage accumulation at the ORFs regardless of directionality (Stirling et al. 2012). Recent belief, at least in eukaryotes, is that whenever a replication fork encounters an elongating transcription complex in either direction has the potential to result in DNA breakage. This situation holds true particularly if transcription is hampered (Garcia-Muse and Aguilera 2016). If directionality does not affect the collision consequences, what determines lethality? A possible contributing factor could be the difference in the supercoiling state at the site of encounter, which will be discussed in details in the following section.

20.4 Topological Consequences of Collision

Along with the obvious physical collision, what replication and transcription machineries have to resolve is the topological stress imposed on each other. Topological status of DNA depends on several cellular and molecular aspects like nucleosome organization, loop-like structures, stable protein-DNA interactions, and its association with immobile elements like nuclear membrane. All chromosomal processes that require opening or folding of the DNA helix are affected by DNA topology status. On the other hand, these processes also introduce changes in DNA topology. Topological status of DNA can be negatively or positively supercoiled when the double helix under-winds or overwinds, respectively (Wang 2002). The term topological stress is used when the unwinding of DNA challenges the topological status of DNA without removing intertwines between two strands. On a linear DNA fragment, this stress can diffuse away and off the end of DNA by the axial rotation of the double-stranded DNA. However, within cells, topological or torsional stress created during the biological process simply cannot diffuse by turning the chromosome ends. Chromosomes are very large and complex structures, and often DNA is anchored to the rigid structures like the nuclear membrane or the chromosome scaffolds that act as topological barriers.

Based on "twin topological domain" model (Liu and Wang 1987), during transcription, DNA in front of the elongating RNAP becomes overwound and coils around itself forming positive supercoils. Behind the transcription machinery, the DNA helix also coils around itself owing to under-winding (Fig. 20.3a). On the other hand, advancing replication fork generates positive supercoiling ahead of the forks (Fig. 20.3b) (Peter et al. 1998). Behind the replication fork, sister chromatid intertwinings arise because the fork occasionally rotates with the turn of the DNA helix. Both polymerases can rotate around the double axis as they move along the DNA to counteract the topological changes (Harada et al. 2001; Schalbetter et al. 2015). In theory, rotation of both polymerases should facilitate the movement as it will avoid accumulation of supercoiling (Fig. 20.3c, d).

One potential explanation of why head-on collisions inhibit the fork movement to a greater extent than codirectional collisions is attributed to the accumulation of higher positive supercoiling from forward movement of both machineries. It is indeed hypothesized that positive supercoiling generated ahead of the transcription bubble might act as a topological barrier rather than a direct physical block to fork

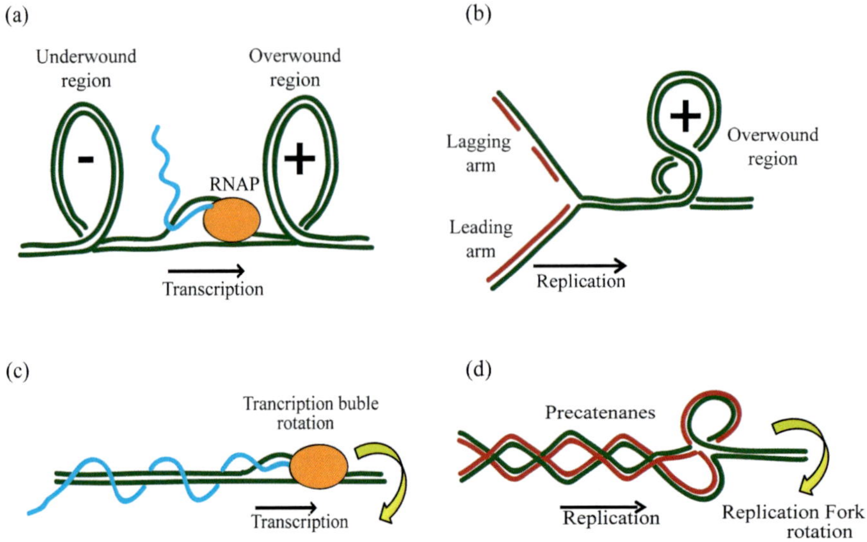

Fig. 20.3 Topological alterations arising during DNA replication and transcription. (**a**) Transcription induces formation of both under- and overwound DNA in the back and front of elongating RNAP, respectively. (**b**) Elongating replication fork induces overwound DNA ahead of the fork. (**c**) Rotation of RNAP resolves topological alterations. (**d**) Replication fork rotation induces precatenanes formation behind the fork

movement (Wu et al. 1988; Olavarrieta et al. 2002). However, studies from head-on collisions in bacteria suggest that blockage of fork movement occurs by direct contact between replisomes and RNAPs rather than higher accumulation of positive supercoiling (Mirkin and Mirkin 2005). Experiments carried out on topologically unconstrained DNA also support this idea as *E.coli* replisome could displace a stalled RNAP rapidly in vitro when collision occurs codirectionally but not when it is head on. This explains how easily RNAP could be displaced when codirectional, allowing a smoother movement of replication fork. However, it is still unclear why head-on collisions show higher inhibitory effect on fork movement. One possible hypothesis could be that in codirectional collision, the replicative helicase could reach a very close proximity to RNAP, which will initiate the disassociation of RNAP complex. Due to the higher positive supercoiled topological barrier in head-on collision, replicative helicase and RNAP might not reach such a close proximity until the topoisomerase resolves positive supercoiling. This hypothesis might be more suitable for eukaryotes as topology is more complex due to tethering of DNA to the nuclear envelope (Bermejo et al. 2011).

It is not clear whether the RNA and DNA polymerases ever actually make contact. It is possible that before the physical contact, topological changes might induce transcription- and replication-mediated chromatin and DNA structural modifications. These alterations might attenuate the progression of the polymerases. Unequal chromatin environment throughout the genome makes it difficult to identify and unravel the importance of chromatin alterations formed through the accumulation of topological stress. During transcription, topological stress arising depends on its gene position, expression level, and chromatin environment. One way to keep these alterations at a desired level is by diffusing the topological tension across the other domains. Studies on global genome architecture have shown that chromosome fibers are organized into distinct higher topological domains (Bermudez et al. 2010; Kouzine et al. 2013; Naughton et al. 2013). These domains are separated by insulator binding protein CTCF, housekeeping genes, tRNAs, and short interspersed element (SINE) retro transposons (Dixon et al. 2012), which most likely act as a barrier to the diffusion of topological stress. Apart from global chromosomal architecture, other factors such as density of nucleosomes packaging also affect topological stress diffusion (Salceda et al. 2006).

In theory, in a codirectional collision between a fork and a transcribed unit, the negative supercoiling generated behind the RNAP might absorb the positive supercoiling generated by the incoming fork (Fig. 20.4), thus avoiding consequences of

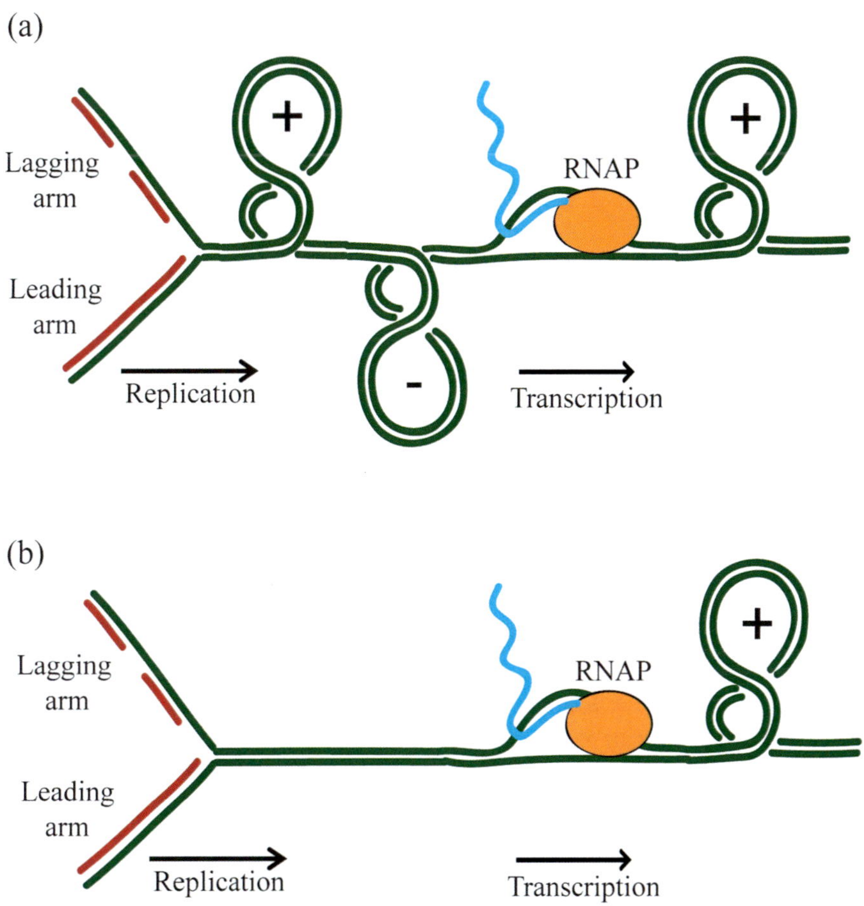

Fig. 20.4 Possible topological resolutions during codirectional collision. (**a**) Negative supercoiling generated behind elongating RNAP might absorb positive supercoiling at the front of replication fork. (**b**) Possible outcome of codirectional collision is the complete negotiation of topological alteration between RNAP and replication fork

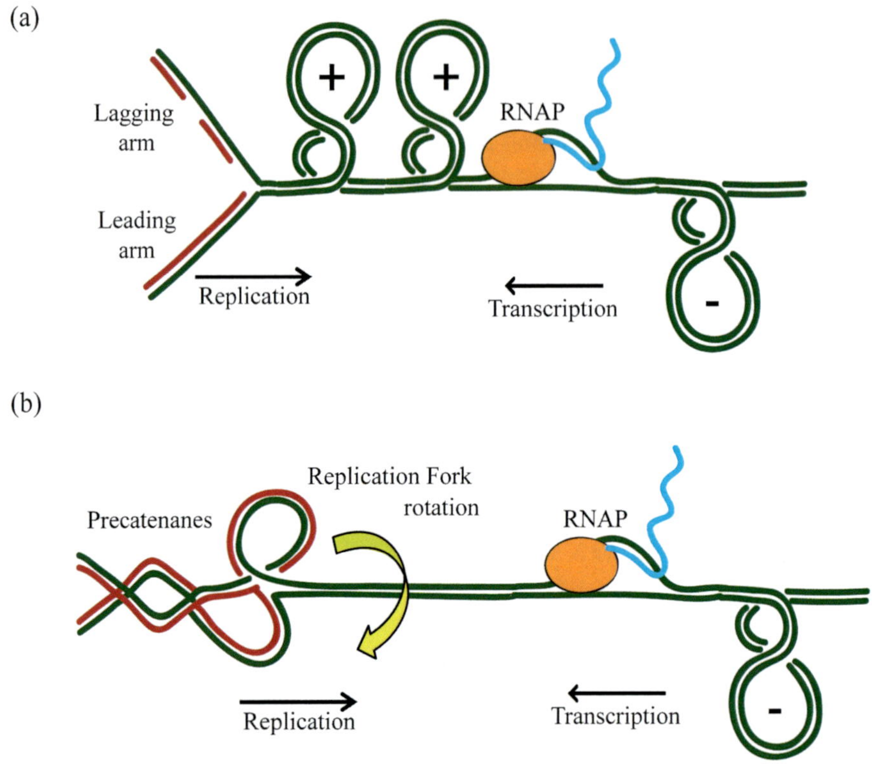

Fig. 20.5 Possible topological resolutions during head-on collision. (**a**) Positive supercoiling accumulated at the site inhibits elongation of both transcription and replication. (**b**) Fork rotation might resolve the hyper-positive supercoiling at the site creating precatenanes structures

topological stress (Wang 2002; Bermejo et al. 2012). However, in a head-on orientation, hyper-positive accumulation generated by both mechanisms need to be resolved by specialized mechanisms. One such way to prevent topological stress building up ahead of the fork is to swivel around the helix to make a rotation (Fig. 20.5) (Champoux and Been 1980). If the replication fork is free to rotate, the positive supercoiling stress in front of the fork can diffuse to create an intertwining of the daughter duplexes behind the fork, forming "precatenanes" (Postow et al. 2001). Following the completion of replication, precatenanes mature to full-DNA catenanes that are resolved by the action of topoisomerases prior to chromosome separation (Fachinetti et al. 2010). Recent evidence of frequent fork rotation at the site of transcription might hint toward precatenanes role in diminishing positive supercoil during head-on collisions (Jeppsson et al. 2014). Although fork rotation could be one possible way to resolve excessive positive supercoiling during the head-on conflict, mechanistic details explaining the process is yet to be discovered.

Considering the processivity of the replicative helicase, a replication fork had to rotate at a considerable rate and more often to counteract the positive supercoiling. This may work at certain restricted chromatin regions but may not be considered as a general phenomenon. Identification of anti-fork-rotating complexes such as Timeless/Topf1 and Tipin/Csm3 proves this point, as these are known to inhibit excessive fork rotation and precatenation formation (Schalbetter et al. 2015). Another way a cell can reduce the positive supercoiling during replication is through reserving some portions of the chromatin to be "negatively supercoiled" regions. Such kind of reservoir of under-wound DNA absorbs the positive supercoiling from the incoming replication fork. This hypothesis makes sense considering the fact that DNA in most of the organisms is known to be slightly under-wound (Champoux 2001), however, lacks experimental support.

Alternatively, cells can use DNA topoisomerases to keep supercoiling at a controlled level. DNA topoisomerases are conserved proteins from bacteria to humans and known to participate in both replication and transcription process. Possible role of DNA topoisomerases in resolving the transcription-replication conflict is discussed in the following section.

20.5 Topoisomerases at the Center of Conflict

Both prokaryotes and eukaryotes utilize topoisomerase enzymes that can relax topological stress by introducing temporary strand breakage into the DNA. Type I topoisomerase nicks one strand, whereas type II topoisomerase breaks both strands while passing another DNA strand through the break (Wang 2002). Topoisomerases, along with relaxing negative or positive supercoiling, can also introduce either negative (bacterial DNA gyrase) or positive supercoiling (reverse gyrase). Besides altering supercoiling, strand-passing activities of topoisomerases can promote the catenation and decatenation of both circular and chromosome DNAs. In eukaryotes, both type IB topoisomerase (Top1) and type II topoisomerase (Top2) regulate the torsional state of the DNA. Although Top1 and Top2 are implicated in supporting replication and transcription, their resolving functions are redundant in many instances.

Positive supercoiling generated ahead of the replication fork can be resolved by both Top1 and Top2 topoisomerases. If the replication machinery could readily rotate, Top2 could function behind the fork as well to remove precatenanes. In yeast, both Top1 and Top2 travel with the replication fork (Bermejo et al. 2007); however, either one of the topoisomerases is sufficient for replication fork progression, confirming the functional redundancy in resolving positive supercoils. Simultaneous inactivation of Top1 and Top2 prevents DNA replication synthesis (Brill et al. 1987). Although one of the topoisomerases is sufficient for replication fork progression, studies in bacteria and yeast indicate that type II (Top2) topoisomerases are indispensable for replication termination and chromosome segregation (Wang 2002; Fachinetti et al. 2010).

Like the elongation step of replication, transcription could also generate positive supercoiling if free rotation of the transcription machinery is inhibited. Along with the sheer size and composition of the transcriptional machinery, its association with the nuclear membrane (Cook 1999; Bermejo et al. 2011) and co-coupling with the translation process (Iborra et al. 2001) inhibits its rotation. Like replication, association and coordination with the topoisomerase are crucial for an efficient transcription process (King et al. 2013). Top1 protein is associated with the elongating transcription unit, and its activity is regulated by RNAP II itself, to favor efficient transcription (Baranello et al. 2016). In yeast, loss of both Top1 and Top2 activity causes a rapid decrease in rRNA transcription but only modest changes at shorter tRNA genes and lower expressed RNA pol II genes (Brill et al. 1987). However, it is recently shown that transcription of longer mRNAs is sensitive to the inactivation of the Top2 protein (Joshi et al. 2012). A genome-wide study has also shown that Top2 binds at the beginning and the end of transcribed genes, specifically during S-phase (Bermejo et al. 2009). This indicates that topoisomerase activity is required to relax topological stress preferably at highly expressed and longer genes where topological stress might build up during replication.

Since DNA topoisomerases are associated with both replication and transcription, hints at their role in resolving the conflict have been proposed. The head-on collision is likely to cause both sterical and topological problems due to hyper-positive supercoiling accumulation. This topological scenario mimics the one during the replication termination zones, where opposing forks converge creating an excess of positive supercoiling ahead of them (Wang 2002). In both scenarios, non-replicated DNA becomes very short, making it difficult for DNA topoisomerases to bind and resolve overwounded DNA. Cells resolve this issue, at least in replication termination regions by generating precatenanes possibly through fork rotation, which later can be resolved by Top2. Indeed in *E. coli*, as well as *S. cerevisiae* and *S. pombe*, Top2 topoisomerase is required for the timely replication termination and to prevent chromosomal breaks during chromosome segregation (Champoux 2001; Wang 2002; Fachinetti et al. 2010). Similarly, fork rotation-induced precatenane resolution by Top2 could be hypothesized at least in prokaryotes, as head-on conflicts are less frequent. However, such scenario may not be possible in eukaryotic cells due to the frequent occurrence of head-on clashes and topological barriers inhibiting the fork rotation. Interestingly, both replication termination and head-on collision zones are known to be replication-dependent fragile sites (Song et al. 2014). Although it is still not clear if there is a direct connection between positive supercoiling accumulation and fragile site formation, a possibility cannot be ruled out.

Replication- and transcription-mediated transient supercoils are necessary, as several processes depend on them. Proteins, like chromatin remodelers often depend on such transient supercoil modifications to activate a variety of cellular responses (Kim and Deppert 2003). Slight alterations in DNA supercoiling can have a strong impact on gene expression in both prokaryotes and eukaryotes. In bacteria, it was shown that loss of chromosomal supercoiling significantly and rapidly affected the expression of 7% of genes (Peter et al. 2004). Several complexes that involve extensive DNA-protein interactions, including replication and transcription initiation, can only occur during particular supercoiling conditions (Bates and Maxwell 2005).

20.6 Transcription and Alternative DNA Structures

Along with being a natural obstacle for replication fork progression, transcription could also interfere through its secondary effects. To increase efficiency and robustness of RNA production, the transcriptional machinery generates several artifacts. These unnatural features could facilitate RNA synthesis and export mechanisms but could challenge replication progression. These features include chromatin structural changes like gene loops, RNA:DNA hybrids, secondary DNA structures like hairpins, triplex-like structures and hinged (H)-DNA), non B-DNA structures such as G-quadruplex, and Z-DNA.

20.6.1 Gene Gating and Gene Loops

In eukaryotes, RNAP II-transcribed genes are organized in loops (Ansari and Hampsey 2005). In *S. cerevisiae*, Top2 along with high-mobility group protein Hmo1 mediate formation of DNA loops. In S-phase, integrity as well as topological complexity of DNA loops depends on the concerted action of both Top2 and Hmo1 (Bermejo et al. 2009). DNA looping assists transcription by facilitating RNAP recycling for concomitant rounds of transcription, a key process in highly transcribed genes (Ansari and Hampsey 2005). These structures also influence the capability of mRNA genes to memorize the previous transcriptional status through a process known as transcriptional memory (Tan-Wong et al. 2009). These specialized architectures of chromatin function as a barrier for incoming forks, topologically insulating the transcriptional apparatus (Azvolinsky et al. 2009). Since DNA loops act as topological insulators, they keep the supercoiling generated by transcription within the loop, maintaining the neighboring chromatin topologically unaffected. This, at least in theory, explains how head-on collisions can be less harmful if loops are formed as DNA topoisomerases have to resolve only positive supercoils generated by the replication fork. It is still unclear how replication forks can dismantle loop structures and what factors might assist them during the process. Loops can also couple mRNA transcription with their export by bringing chromatin into contact with the nuclear pore complexes (NPCs), a process termed as gene gating (Strambio-De-Castillia et al. 2010; Burns and Wente 2014). Evidence in yeast suggests that every RNAPII-transcribed gene associates with gating factors (Casolari et al. 2004; Gomez-Gonzalez et al. 2011). Recent observations in yeast suggest that transcribed genes are unleashed from the nuclear envelope to allow fork progression (Bermejo et al. 2011). This process is mediated by the Mec1-Rad53 checkpoint that phosphorylates Mlp1, a key protein located in the inner basket of the nuclear pore. Fork passage through transcribed genes not only counteracts gene gating but also allows the dismantling of DNA loop. The key question here is how does a loop anchored to the nuclear membrane sense the incoming fork? A recent finding in mammalian cells demonstrates that nuclear membrane could sense topological tension accumulated at the chromatin. This leads to accumulation of the major checkpoint protein, ATR, at the nuclear envelop, which initiates a signal cascade to relax loop-like structures (Kumar et al. 2014). This assembly and disassembly of gene loops during replication shows that transcription domains are cell cycle dependent. Indeed, gene gating is regulated by cyclin-dependent kinases (CDKs) through phosphorylation of Nup1 nucleoporin. Moreover, transcriptional memory has been linked to the chromosome architecture of transcribed genes (Tan-Wong et al. 2009; Light et al. 2010). It is expected that replication across RNA genes would influence the capability of genes to memorize their transcriptional status, as it is crucial to reestablish the transcription process after fork passage.

In addition to mRNA-generating genes, rDNA loci where replication fork barriers are essential to avoid replication conflicts are also anchored to the nuclear membrane through a different mechanism. The function of RFB relies on the Fob1 protein, which mediates the perinuclear attachments of rDNA repeats through its interactions with cohibin/CLIP complex (Kobayashi and Horiuchi 1996; Huang et al. 2006; Mekhail et al. 2008). Anchoring of rDNA loci to the nuclear envelope might impose a series of topological constraints leading to fork pausing. On the other hand, releasing rDNA repeats from the nuclear envelope destabilizes the repeats. Due to hyperactive transcription at rDNA loci, DNA topoisomerase is absolutely necessary to maintain supercoiling. Yeast devoid of both Top1 and Top2 not only shows drastic reduction in the rRNA levels but also its rDNA repeats are excised as extrachromosomal circles (Kim and Wang 1989).

20.6.2 R-Loops and RNA-DNA Hybrids

Both replication and transcription are required to form short RNA-DNA hybrids, which are used for priming polymerases. These short stretches of hybrids are more stable than double-stranded DNA as they adopt a heteromeric confirmation, which is a mix of the "B" form of dsDNA and the "A" form of double-stranded RNA (Roberts and Crothers 1992). Often, longer forms of RNA:DNA hybrids occur in most of the genome. On the basis of "thread-back model," hybrid structures can be generated by the hybridization of nascent RNA molecules with the template DNA strand (Westover et al. 2004). This longer hybrid and the resultant displaced single-stranded DNA are collectively known as R-loops. Formation and stabilization of R-loops in vivo depend on several structural features such as negative supercoiling, GC-content, and tendency to form alternative structures in the displaced ssDNA such as G-quartets (Duquette et al. 2004; Roy and Lieber 2009; Aguilera and Garcia-Muse 2012).

R-loops are found in all living organisms and generally considered to have deleterious effects except in unusual situations such as in immunoglobulin (Ig) class-switch recombination (CSR). It has been shown in yeast and human cells that deficiency in RNA and DNA metabolic factors leads to R-loop-mediated DNA breaks, recombination and chromosome rearrangements, and losses (Santos-Pereira and Aguilera 2015). It is still unclear what forces R-loops to generate DNA breaks. A largely accepted hypothesis is that R-loop-induced replication fork stalling might generate DNA breaks. First, evidence of RNA:DNA hybrids causing genome instability came from bacteria where loss of RNaseH1 provoked the SOS response (Kogoma et al. 1993). *E. coli* cells defective in DNA damage response (DDR) show lethality when an rDNA operon was inverted to have a head-on collision with a replication fork. However, this lethality was rescued by overexpression of RNaseH, showing the connection between R-loops, replication stalling, and DDR (Boubakri et al. 2010). Experiments in both *E. coli* and mammalian cells revealed that R-loop-mediated chromosomal rearrangements and recombination could be rescued by impairing DNA replication (Gan et al. 2011). Genome-wide replication analysis in several yeast mutants such as THO (*tho*) (Gomez-Gonzalez et al. 2011), Sen1 (*sen1-1*) (Alzu et al. 2012), and RBP Npl3 *(npl3)* (Santos-Pereira et al. 2013) revealed R-loop-mediated replication-transcription conflicts. Replication analysis on *TOP1*-depleted mammalian cells shows transcription interference with replication in an R-loop-dependent manner (Tuduri et al. 2009). Moreover, common fragile sites associated with replication and transcription collisions in human cells are known to accumulate R-loop structures (Helmrich et al. 2011). It is now evident that R-loop structures do have the potential to impair replication progression by inducing DNA breaks. However, it is to be noted that not all RNA-DNA hybrid-forming regions may give rise to R-loop-like structures, suggesting that only a fraction of such regions participate in replication stalling. Possibility of an arrested RNAP II at the site of R-loop-forming region cannot be ruled out, as it will further hinder replication progression.

In addition to causing DNA breaks, R-loops may also be involved in promoting mutagenic DNA replication. In *E. coli*, RNA-DNA hybrids can be used as primers to initiate noncanonical replication (Kogoma 1997). In yeast cells, rDNA loci also show origin-independent replication but only in the absence of Top1 and RNaseH proteins, confirming the role of RNA-DNA hybrids in replication initiation. Although R-loop-mediated replication initiation has been reported in several instances, fidelity and consequences are little known. R-loop-mediated replication is largely considered to be highly mutagenic, as cells may use such origin-independent replication only in extreme conditions such as DNA breaks (Deem et al. 2011).

A recent finding shows the involvement of tumor suppressor and DNA repair genes such as *BRCA2* and *BRCA1* in R-loop-mediated homeostasis, revealing a direct link between R-loop accumulation and mutagenesis (Bhatia et al. 2014). Studies revealing the role of RNA-DNA helicases such as Sen1, Rrm3, and Pif1 in assisting replication fork progression across pausing sites open new perspectives into our understanding of R-loop-mediated replication stress and DNA breaks (Alzu et al. 2012; Rossi et al. 2015).

20.6.3 Hinged DNA (H-DNA)

H-DNA can be formed with three DNA strands and uses Hoogsteen hydrogen bonds for its stability and specificity (Frank-Kamenetskii and Mirkin 1995). One possible role of H-DNA in vivo was revealed by identifying a segment in the promoter of the human cMYC gene capable of adopting H-DNA (Kinniburgh 1989). Interestingly, this region overlaps with one of the major breakage hotspots found in c-MYC-induced lymphomas and leukemias (Joos et al. 1992; Wilda et al. 2004). Later, certain fragile spots or hotspot regions of the genome are mapped in or near the genes carrying polypurine-polypyrimidine sequences known to form H-DNA confirmation (Bacolla et al. 2006). It is well known that naturally occurring H-DNA structures are intrinsically mutagenic in mammalian cells (Wang and Vasquez 2004). The existence of H-DNA structures may effectively impede both transcription and replication processes (Jain et al. 2008); at least in vitro, it is known to impose a very strong barrier for polymerases (Hoyne and Maher 2002). Although the exact mechanism is yet to be known, naturally occurring intermolecular triple-strand DNA like H-DNA are inherently mutagenic and recombinogenic and might play a role in replication and transcription collisions.

20.6.4 Z-DNA

Till now, definitive biological functions for Z-DNA have not been found. Z-DNA is believed to provide torsional stress relaxation, particularly from negative supercoiling at the transcription site (Ha et al. 2005). Indeed, formation of Z-DNA is directly proportional to accumulation of negative supercoiling (Azorin et al. 1983), and topoisomerases can convert Z-DNA back to B-DNA (Rich and Zhang 2003). In yeast, Z-DNA structures can be induced and stabilized by Z-DNA-binding proteins, which might have a role to play in gene regulation and chromatin remodeling (Wang et al. 2007). Till now, the effect Z-DNA on DNA replication was only reported in trypanosomas, where unwinding of DNA due to intercalation leads to Z-DNA formation and subsequent inhibition of replication (Roy Chowdhury et al. 2010). However, the impact of Z-DNA on replication in other organisms cannot be ruled out since negative supercoiling generated by transcription directly influences its formation and Top1 function is required to counteract its formation. Particularly, the occurrence of Z-DNA-forming sequences at chromosomal breakpoints in human tumors suggests its role in double-strand breaks. In fact, occurrence of Z-DNA forming sequences at chromosomal breakpoints in human tumors suggests its possible role in generation of double-strand breaks (Wang et al. 2006).

20.6.5 G-Quadruplex (G4) DNA

A guanine-rich genomic region has the ability to form a quadruplex (G4), stabilized by Hoogsteen hydrogen bonds at physiological concentrations (Sen and Gilbert 1988). The formation of G4 structures is favored by DNA transition processes, which involve unwinding of the double helix such as transcription and replication (Maizels and Gray 2013). It is estimated that in human genome, over 350,000 sequences have the potential to form G4 structures that can act as potential replication barriers (Huppert and Balasubramanian 2005; Todd et al. 2005). Recent studies using either G4-recognizing molecules or antibodies show that G4 sequences are present in cells and they increase during S-phase (Rodriguez et al. 2012; Biffi et al. 2013; Henderson et al. 2014). G4 structures have been involved in several biological processes including telomere maintenance, gene expression, replication initiation, and the immune response (Maizels and Gray 2013).

Transcription, particularly at G-rich regions, induces the formation of stable G4 structures (Duquette et al. 2004). Recently, it is revealed that G4-DNA is enriched in the promoters and 5′UTRs of highly transcribed genes, particularly in genes related to carcinogenesis (Hansel-Hertsch et al. 2016). During DNA replication, the inappropriate intermolecular folding like in G4 must be resolved in order to allow replication fork movement. It is well known that G4-DNA, if not unwound, could inhibit fork stalling, possibly causing fork collapse and DNA breaks (Woodford et al. 1994; Usdin and Woodford 1995). Cells have developed several mechanisms to counteract the G4-DNA formed due to transcription during replication. Recent evidence implicating specialized proteins in the unwinding of G4-DNA is now accumulating. Proteins such as human ATRX (Law et al. 2010), XPB, and XPD (Gray et al. 2014) and yeast Pif1 (Paeschke et al. 2011) and Rif1 (Kanoh et al. 2015) have been shown to accumulate near G-rich genomic regions predicted to form G4-structures, indicating their possible role in assisting replication fork progression. Other than helicases, cells also use specialized DNA polymerases known as translesional synthesis (TLS) polymerases to replicate across G4-DNA structures. TLS polymerases such as pol kappa, pol eta, and Rev1 are required for efficient replication across G4 motifs (Betous et al. 2009; Sarkies et al. 2010). It is now well established that specific helicases and specialized polymerases are involved in replication of G4-DNA; loss of these factors leads to severe genetic disorders leading to premature aging and higher cancer risk (Rhodes and Lipps 2015).

20.7 Negotiating Replication and Transcription Conflict

Cells have evolved several mechanisms to reactivate, destabilize, or even remove transcription roadblocks to minimize the effects of transcription-replication collisions. Conceptually, clearing the pathway ahead of a replication fork is the simplest mechanism of resolving conflicts. Alternatively, a codirectional conflict may be resolved by slowing down the replication fork behind the RNAP until transcription termination. However, it remains unanswered whether a replication fork has the luxury of waiting until the transcription is completed. Preventing the formation of backtracked or stalled RNAPs will also be beneficial for replication. Cells achieve this by ensuring an appropriate RNAP elongation rate. In prokaryotes, transcription elongation factors GreA and GreB clear the pathway by reactivating backtracked and stalled RNAPs (Opalka et al. 2003). Another bacterial transcription factor, DksA, promotes elongation by reducing nucleotide disincorporation leading to backtracking and pausing of RNAPs (Tehranchi et al. 2010; Roghanian et al. 2015). Another mechanism by which prokaryotes keep backtracking at minimal levels is by coupling transcription with translation. Ribosomes are loaded simultaneously onto nascent RNA as they emerge from the RNAP complex (Proshkin et al. 2010; Dutta et al. 2011) leading to increased transcription efficiency. Same phenomenon can be attributed in yeast as ongoing transcripts simultaneously anchor to the nuclear membrane, likely ensuring the elongation rate (Bermejo et al. 2011). Higher eukaryotes have additional protein factors including RECQL5, which reduces stalling and backtracking by controlling transcription elongation rate (Saponaro et al. 2014).

RNAP arrested by encountering damaged bases can only proceed when the lesion is repaired by repair mechanisms, such as nucleotide excision repair (NER). Prokaryotes use two different NER factors Mfd and UvrD, to dislodge arrested RNAP at the lesion (Ganesan et al. 2012; Epshtein et al. 2014). In eukaryotes, how NER dislodges RNAP is not clear. However, if NER fails, RNAP is polyubiquitinated and degraded by proteasomes (Wilson et al. 2013). Interestingly, mammalian cells lacking NER exhibit apoptosis during S-phase, suggesting lethality associated with irreversibly arrested RNAP (McKay et al. 2002). In eukaryotes, TFIIS, a Pol II transcription elongation factor, can also stimulate transcript cleavage to restart arrested RNAP (Cheung and Cramer 2011).

For RNAPs to be cleared on their own will pose detrimental time delays and cells inevitably need to swiftly clear the path for replication forks. Recent in vivo and in vitro studies support the long-lasting hypothesis that auxiliary helicases promote replication through roadblocks including transcription complexes. The best-known examples are REP and UvrD in *E. coli* and their homolog PcrA in *B. subtilis*, all of which are known to associate with replication forks to promote replication though transcribed genes (Guy et al. 2009; Gwynn et al. 2013; Merrikh et al. 2015). In eukaryotes, Rrm3, Pif1, and Sen1 are also associated with replisomes and facilitate progression through highly transcribed genes (Azvolinsky et al. 2009; Alzu et al. 2012; Rossi et al. 2015). Additionally, RNaseH overexpression significantly reduces the association of Rrm3 with these genes, indicating RNA:DNA hybrid accumulation as an intermediate step in these genes (Gomez-Gonzalez et al. 2011). Since most eukaryotes encode at least one of these helicases, the function of these proteins to promote replication fork progression across highly expressed genes may be conserved.

20.8 Replication-Transcription Collision and Genome Instability

Until now, transcription and its associated events represent major endogenous sources causing replication stress. Certain regions in the genome are prone to form gaps, breaks, and rearrangements in response to replication stress and are generally referred to as chromosomal fragile sites (Glover et al. 1984; Yunis and Soreng 1984; Beroukhim et al. 2010; Bignell et al. 2010; Wilson et al. 2015; Wei et al. 2016). These common fragile sites (CFSs) tend to coincide with transcribed genes, and transcription itself is believed to be the source of genomic instability at these regions (Huertas and Aguilera 2003; Helmrich et al. 2011; Barlow et al. 2013; Saponaro et al. 2014; Wilson et al. 2015; Kantidakis et al. 2016). It has been known for many years that transcription can lead to higher recombination (Keil and Roeder 1984; Voelkel-Meiman et al. 1987; Thomas and Rothstein 1989; Nickoloff 1992; Prado et al. 1997). Moreover, transcription-dependent recombination strictly depends on replication, and it is the collision of replication with transcription that causes recombination as proven in yeast (Prado and Aguilera 2005) and mammalian cells (Gottipati et al. 2008). Recent sequencing of cancer genomes has identified specific mutation signatures arising due to replication-transcription collisions (Hatchi et al. 2015). Even in bacteria, mutations arising due to replication and transcription conflicts show a distinct mutation pattern (Sankar et al. 2016), indicating pathological consequences of such collisions. However, linking collision-related mutation accumulation directly to genome stability will be a daunting task, as gene expression patterns differ significantly between different cell and tissue types. It is tentative to speculate that this difference might be the reason behind cell- or tissue-specific mutation variations in cancer cells.

There is a vast amount of data connecting transcription-induced replication pausing, chromosome breaks, and genome rearrangements. It is yet to become clear what percentage of collisions will lead to fork collapse, leading to DNA breaks. Since cells cannot completely avoid collisions, they acquire several layers of protection to avoid the lethal consequences of collisions. Particularly, in response to environment stimuli, whole arrays of proteins remodel genome architectures by activating or suppressing genes. Cells also need to repair lesions and base modifications, which arise due to various external and internal stimuli. Amidst, faithful replication has to negotiate and resolve conflicts of genome trafficking and is essential for genome integrity.

20.9 Conclusions and Future Perspectives

A cell can either avoid collisions between replisomes and RNAPs or resolve them through specific mechanisms. A precise dissection of molecular mechanisms, which could explain the impact of collisions, is still very unclear. Extensive overlap between processes that ensure faithful completion of both DNA replication and RNA production makes it difficult to unravel the details of such mechanisms. Research from the last 30 years has indicated the importance of replication-transcription conflicts, as they emerged as important natural sources of genome instability. Particularly, identification of mutations in genes that are involved in either prevention or resolution of conflicts is connected to several cancer-prone syndromes and neurodegenerative diseases. In the future, it will be important to distinguish deleterious collisions that cause mutations from the nonhazardous counterparts. It is well known that collisions lead to genome instability; however, it is still unknown if damage occurs at the site of collision. If damage occurs at the site, do DDR or S-phase checkpoints play any roles in resolving the conflicts? It is also important to understand how an encounter affects the stability of both replication forks and transcription units. Among all the factors studied that have implications on collision-mediated consequences, DNA topology is least understood. Since DNA topoisomerases have an impact on both processes, it is important to understand how their action might help to coordinate these two processes. In higher eukaryotes, since replication origins and termination regions are not fixed, it is difficult to reliably predict the orientation of collisions. This makes it not trivial to associate collision hotspots with DNA sequences and chromatin modifications. Identifying mutational patterns arising from replication-transcription conflicts and associating them with known cancer biomarkers might have biomedical and therapeutic relevance.

Acknowledgments We thank Ghadeer Shubassi for helping with manuscript preparations and all the MF-lab members for discussions and valuable inputs. Research in MF-lab is supported by grants from the Associazione Italiana per la Ricerca sul Cancro (AIRC), the European Union and Telethon-Italy. YJA is supported by the European Community's Seventh Framewok Programme under Grant agreement no.246549 – Train 2009.

References

Aguilera A (2002) The connection between transcription and genomic instability. EMBO J 21(3):195–201. https://doi.org/10.1093/emboj/21.3.195

Aguilera A, Garcia-Muse T (2012) R loops: from transcription byproducts to threats to genome stability. Mol Cell 46(2):115–124. https://doi.org/10.1016/j.molcel.2012.04.009

Alzu A, Bermejo R, Begnis M, Lucca C, Piccini D, Carotenuto W, Saponaro M, Brambati A, Cocito A, Foiani M, Liberi G (2012) Senataxin associates with replication forks to protect fork integrity across RNA-polymerase-II-transcribed genes. Cell 151(4):835–846. https://doi.org/10.1016/j.cell.2012.09.041

Ansari A, Hampsey M (2005) A role for the CPF 3′-end processing machinery in RNAP II-dependent gene looping. Genes Dev 19(24):2969–2978. https://doi.org/10.1101/gad.1362305
Artsimovitch I, Landick R (2000) Pausing by bacterial RNA polymerase is mediated by mechanistically distinct classes of signals. Proc Natl Acad Sci U S A 97(13):7090–7095
Azorin F, Nordheim A, Rich A (1983) Formation of Z-DNA in negatively supercoiled plasmids is sensitive to small changes in salt concentration within the physiological range. EMBO J 2(5):649–655
Azvolinsky A, Giresi PG, Lieb JD, Zakian VA (2009) Highly transcribed RNA polymerase II genes are impediments to replication fork progression in *Saccharomyces cerevisiae*. Mol Cell 34(6):722–734. https://doi.org/10.1016/j.molcel.2009.05.022
Bacolla A, Collins JR, Gold B, Chuzhanova N, Yi M, Stephens RM, Stefanov S, Olsh A, Jakupciak JP, Dean M, Lempicki RA, Cooper DN, Wells RD (2006) Long homopurine*homopyrimidine sequences are characteristic of genes expressed in brain and the pseudoautosomal region. Nucleic Acids Res 34(9):2663–2675. https://doi.org/10.1093/nar/gkl354
Baranello L, Wojtowicz D, Cui K, Devaiah BN, Chung HJ, Chan-Salis KY, Guha R, Wilson K, Zhang X, Zhang H, Piotrowski J, Thomas CJ, Singer DS, Pugh BF, Pommier Y, Przytycka TM, Kouzine F, Lewis BA, Zhao K, Levens D (2016) RNA polymerase II regulates topoisomerase 1 activity to favor efficient transcription. Cell 165(2):357–371. https://doi.org/10.1016/j.cell.2016.02.036
Barlow JH, Faryabi RB, Callen E, Wong N, Malhowski A, Chen HT, Gutierrez-Cruz G, Sun HW, McKinnon P, Wright G, Casellas R, Robbiani DF, Staudt L, Fernandez-Capetillo O, Nussenzweig A (2013) Identification of early replicating fragile sites that contribute to genome instability. Cell 152(3):620–632. https://doi.org/10.1016/j.cell.2013.01.006
Bates AD, Maxwell A (2005) DNA topology. Oxford University Press, Oxford
Bedinger P, Hochstrasser M, Jongeneel CV, Alberts BM (1983) Properties of the T4 bacteriophage DNA replication apparatus: the T4 dda DNA helicase is required to pass a bound RNA polymerase molecule. Cell 34(1):115–123
Bermejo R, Doksani Y, Capra T, Katou YM, Tanaka H, Shirahige K, Foiani M (2007) Top1- and Top2-mediated topological transitions at replication forks ensure fork progression and stability and prevent DNA damage checkpoint activation. Genes Dev 21(15):1921–1936. https://doi.org/10.1101/gad.432107
Bermejo R, Capra T, Gonzalez-Huici V, Fachinetti D, Cocito A, Natoli G, Katou Y, Mori H, Kurokawa K, Shirahige K, Foiani M (2009) Genome-organizing factors Top2 and Hmo1 prevent chromosome fragility at sites of S phase transcription. Cell 138(5):870–884. https://doi.org/10.1016/j.cell.2009.06.022
Bermejo R, Capra T, Jossen R, Colosio A, Frattini C, Carotenuto W, Cocito A, Doksani Y, Klein H, Gomez-Gonzalez B, Aguilera A, Katou Y, Shirahige K, Foiani M (2011) The replication checkpoint protects fork stability by releasing transcribed genes from nuclear pores. Cell 146(2):233–246. https://doi.org/10.1016/j.cell.2011.06.033
Bermejo R, Lai MS, Foiani M (2012) Preventing replication stress to maintain genome stability: resolving conflicts between replication and transcription. Mol Cell 45(6):710–718. https://doi.org/10.1016/j.molcel.2012.03.001
Bermudez I, Garcia-Martinez J, Perez-Ortin JE, Roca J (2010) A method for genome-wide analysis of DNA helical tension by means of psoralen-DNA photobinding. Nucleic Acids Res 38(19):e182. https://doi.org/10.1093/nar/gkq687
Beroukhim R, Mermel CH, Porter D, Wei G, Raychaudhuri S, Donovan J, Barretina J, Boehm JS, Dobson J, Urashima M, Mc Henry KT, Pinchback RM, Ligon AH, Cho YJ, Haery L, Greulich H, Reich M, Winckler W, Lawrence MS, Weir BA, Tanaka KE, Chiang DY, Bass AJ, Loo A, Hoffman C, Prensner J, Liefeld T, Gao Q, Yecies D, Signoretti S, Maher E, Kaye FJ, Sasaki H, Tepper JE, Fletcher JA, Tabernero J, Baselga J, Tsao MS, Demichelis F, Rubin MA, Janne PA, Daly MJ, Nucera C, Levine RL, Ebert BL, Gabriel S, Rustgi AK, Antonescu CR, Ladanyi M, Letai A, Garraway LA, Loda M, Beer DG, True LD, Okamoto A, Pomeroy SL, Singer S, Golub TR, Lander ES, Getz G, Sellers WR, Meyerson M (2010) The landscape of somatic copy-number alteration across human cancers. Nature 463(7283):899–905. https://doi.org/10.1038/nature08822

Betous R, Rey L, Wang G, Pillaire MJ, Puget N, Selves J, Biard DS, Shin-ya K, Vasquez KM, Cazaux C, Hoffmann JS (2009) Role of TLS DNA polymerases eta and kappa in processing naturally occurring structured DNA in human cells. Mol Carcinog 48(4):369–378. https://doi.org/10.1002/mc.20509

Bhatia V, Barroso SI, Garcia-Rubio ML, Tumini E, Herrera-Moyano E, Aguilera A (2014) BRCA2 prevents R-loop accumulation and associates with TREX-2 mRNA export factor PCID2. Nature 511(7509):362–365. https://doi.org/10.1038/nature13374

Biffi G, Tannahill D, McCafferty J, Balasubramanian S (2013) Quantitative visualization of DNA G-quadruplex structures in human cells. Nat Chem 5(3):182–186. https://doi.org/10.1038/nchem.1548

Bignell GR, Greenman CD, Davies H, Butler AP, Edkins S, Andrews JM, Buck G, Chen L, Beare D, Latimer C, Widaa S, Hinton J, Fahey C, Fu B, Swamy S, Dalgliesh GL, Teh BT, Deloukas P, Yang F, Campbell PJ, Futreal PA, Stratton MR (2010) Signatures of mutation and selection in the cancer genome. Nature 463(7283):893–898. https://doi.org/10.1038/nature08768

Blattner FR, Plunkett G 3rd, Bloch CA, Perna NT, Burland V, Riley M, Collado-Vides J, Glasner JD, Rode CK, Mayhew GF, Gregor J, Davis NW, Kirkpatrick HA, Goeden MA, Rose DJ, Mau B, Shao Y (1997) The complete genome sequence of *Escherichia coli* K-12. Science 277(5331):1453–1462

Boubakri H, de Septenville AL, Viguera E, Michel B (2010) The helicases DinG, Rep and UvrD cooperate to promote replication across transcription units in vivo. EMBO J 29(1):145–157. https://doi.org/10.1038/emboj.2009.308

Breier AM, Weier HU, Cozzarelli NR (2005) Independence of replisomes in *Escherichia coli* chromosomal replication. Proc Natl Acad Sci U S A 102(11):3942–3947. https://doi.org/10.1073/pnas.0500812102

Brewer BJ (1988) When polymerases collide: replication and the transcriptional organization of the *E. coli* chromosome. Cell 53(5):679–686

Brewer BJ, Fangman WL (1988) A replication fork barrier at the 3′ end of yeast ribosomal RNA genes. Cell 55(4):637–643

Brill SJ, DiNardo S, Voelkel-Meiman K, Sternglanz R (1987) Need for DNA topoisomerase activity as a swivel for DNA replication for transcription of ribosomal RNA. Nature 326(6111):414–416. https://doi.org/10.1038/326414a0

Burns LT, Wente SR (2014) From hypothesis to mechanism: uncovering nuclear pore complex links to gene expression. Mol Cell Biol 34(12):2114–2120. https://doi.org/10.1128/MCB.01730-13

Carninci P, Kasukawa T, Katayama S, Gough J, Frith MC, Maeda N, Oyama R, Ravasi T, Lenhard B, Wells C, Kodzius R, Shimokawa K, Bajic VB, Brenner SE, Batalov S, Forrest AR, Zavolan M, Davis MJ, Wilming LG, Aidinis V, Allen JE, Ambesi-Impiombato A, Apweiler R, Aturaliya RN, Bailey TL, Bansal M, Baxter L, Beisel KW, Bersano T, Bono H, Chalk AM, Chiu KP, Choudhary V, Christoffels A, Clutterbuck DR, Crowe ML, Dalla E, Dalrymple BP, de Bono B, Della Gatta G, di Bernardo D, Down T, Engstrom P, Fagiolini M, Faulkner G, Fletcher CF, Fukushima T, Furuno M, Futaki S, Gariboldi M, Georgii-Hemming P, Gingeras TR, Gojobori T, Green RE, Gustincich S, Harbers M, Hayashi Y, Hensch TK, Hirokawa N, Hill D, Huminiecki L, Iacono M, Ikeo K, Iwama A, Ishikawa T, Jakt M, Kanapin A, Katoh M, Kawasawa Y, Kelso J, Kitamura H, Kitano H, Kollias G, Krishnan SP, Kruger A, Kummerfeld SK, Kurochkin IV, Lareau LF, Lazarevic D, Lipovich L, Liu J, Liuni S, McWilliam S, Madan Babu M, Madera M, Marchionni L, Matsuda H, Matsuzawa S, Miki H, Mignone F, Miyake S, Morris K, Mottagui-Tabar S, Mulder N, Nakano N, Nakauchi H, Ng P, Nilsson R, Nishiguchi S, Nishikawa S, Nori F, Ohara O, Okazaki Y, Orlando V, Pang KC, Pavan WJ, Pavesi G, Pesole G, Petrovsky N, Piazza S, Reed J, Reid JF, Ring BZ, Ringwald M, Rost B, Ruan Y, Salzberg SL, Sandelin A, Schneider C, Schonbach C, Sekiguchi K, Semple CA, Seno S, Sessa L, Sheng Y, Shibata Y, Shimada H, Shimada K, Silva D, Sinclair B, Sperling S, Stupka E, Sugiura K, Sultana R, Takenaka Y, Taki K, Tammoja K, Tan SL, Tang S, Taylor MS, Tegner J, Teichmann SA, Ueda HR, van Nimwegen E, Verardo R, Wei CL, Yagi K, Yamanishi H, Zabarovsky E, Zhu S, Zimmer A, Hide W, Bult C, Grimmond SM, Teasdale RD, Liu ET, Brusic V, Quackenbush J, Wahlestedt C, Mattick JS, Hume DA, Kai C, Sasaki D, Tomaru Y, Fukuda S, Kanamori-Katayama M,

Suzuki M, Aoki J, Arakawa T, Iida J, Imamura K, Itoh M, Kato T, Kawaji H, Kawagashira N, Kawashima T, Kojima M, Kondo S, Konno H, Nakano K, Ninomiya N, Nishio T, Okada M, Plessy C, Shibata K, Shiraki T, Suzuki S, Tagami M, Waki K, Watahiki A, Okamura-Oho Y, Suzuki H, Kawai J, Hayashizaki Y, Consortium F, Group RGER, Genome Science G (2005) The transcriptional landscape of the mammalian genome. Science 309(5740):1559–1563. https://doi.org/10.1126/science.1112014

Casolari JM, Brown CR, Komili S, West J, Hieronymus H, Silver PA (2004) Genome-wide localization of the nuclear transport machinery couples transcriptional status and nuclear organization. Cell 117(4):427–439

Champoux JJ (2001) DNA topoisomerases: structure, function, and mechanism. Annu Rev Biochem 70:369–413. https://doi.org/10.1146/annurev.biochem.70.1.369

Champoux JJ, Been MD (1980) Topoisomerases and the swivel problem. In: Alberts B (ed) Mechanistic studies of DNA replication and genetic recombination. Academic, New York, pp 809–815

Cheung AC, Cramer P (2011) Structural basis of RNA polymerase II backtracking, arrest and reactivation. Nature 471(7337):249–253. https://doi.org/10.1038/nature09785

Cook PR (1999) The organization of replication and transcription. Science 284(5421):1790–1795

Dalgaard JZ, Klar AJ (2001) A DNA replication-arrest site RTS1 regulates imprinting by determining the direction of replication at mat1 in *S. pombe*. Genes Dev 15(16):2060–2068. https://doi.org/10.1101/gad.200801

Datta A, Jinks-Robertson S (1995) Association of increased spontaneous mutation rates with high levels of transcription in yeast. Science 268(5217):1616–1619

David L, Huber W, Granovskaia M, Toedling J, Palm CJ, Bofkin L, Jones T, Davis RW, Steinmetz LM (2006) A high-resolution map of transcription in the yeast genome. Proc Natl Acad Sci U S A 103(14):5320–5325. https://doi.org/10.1073/pnas.0601091103

Deem A, Keszthelyi A, Blackgrove T, Vayl A, Coffey B, Mathur R, Chabes A, Malkova A (2011) Break-induced replication is highly inaccurate. PLoS Biol 9(2):e1000594. https://doi.org/10.1371/journal.pbio.1000594

Deshpande AM, Newlon CS (1996) DNA replication fork pause sites dependent on transcription. Science 272(5264):1030–1033

Dixon JR, Selvaraj S, Yue F, Kim A, Li Y, Shen Y, Hu M, Liu JS, Ren B (2012) Topological domains in mammalian genomes identified by analysis of chromatin interactions. Nature 485(7398):376–380. https://doi.org/10.1038/nature11082

Dul JL, Drexler H (1988) Transcription stimulates recombination. I. Specialized transduction of *Escherichia coli* by lambda trp phages. Virology 162(2):466–470

Duquette ML, Handa P, Vincent JA, Taylor AF, Maizels N (2004) Intracellular transcription of G-rich DNAs induces formation of G-loops, novel structures containing G4 DNA. Genes Dev 18(13):1618–1629. https://doi.org/10.1101/gad.1200804

Dutrow N, Nix DA, Holt D, Milash B, Dalley B, Westbroek E, Parnell TJ, Cairns BR (2008) Dynamic transcriptome of Schizosaccharomyces pombe shown by RNA-DNA hybrid mapping. Nat Genet 40(8):977–986. https://doi.org/10.1038/ng.196

Dutta D, Shatalin K, Epshtein V, Gottesman ME, Nudler E (2011) Linking RNA polymerase backtracking to genome instability in *E. coli*. Cell 146(4):533–543. https://doi.org/10.1016/j.cell.2011.07.034

El Hage A, French SL, Beyer AL, Tollervey D (2010) Loss of Topoisomerase I leads to R-loop-mediated transcriptional blocks during ribosomal RNA synthesis. Genes Dev 24(14):1546–1558. https://doi.org/10.1101/gad.573310

Elias-Arnanz M, Salas M (1997) Bacteriophage phi29 DNA replication arrest caused by codirectional collisions with the transcription machinery. EMBO J 16(18):5775–5783. https://doi.org/10.1093/emboj/16.18.5775

Elias-Arnanz M, Salas M (1999) Resolution of head-on collisions between the transcription machinery and bacteriophage phi29 DNA polymerase is dependent on RNA polymerase translocation. EMBO J 18(20):5675–5682. https://doi.org/10.1093/emboj/18.20.5675

Epshtein V, Toulme F, Rahmouni AR, Borukhov S, Nudler E (2003) Transcription through the roadblocks: the role of RNA polymerase cooperation. EMBO J 22(18):4719–4727. https://doi.org/10.1093/emboj/cdg452

Epshtein V, Kamarthapu V, McGary K, Svetlov V, Ueberheide B, Proshkin S, Mironov A, Nudler E (2014) UvrD facilitates DNA repair by pulling RNA polymerase backwards. Nature 505(7483):372–377. https://doi.org/10.1038/nature12928

Fachinetti D, Bermejo R, Cocito A, Minardi S, Katou Y, Kanoh Y, Shirahige K, Azvolinsky A, Zakian VA, Foiani M (2010) Replication termination at eukaryotic chromosomes is mediated by Top2 and occurs at genomic loci containing pausing elements. Mol Cell 39(4):595–605. https://doi.org/10.1016/j.molcel.2010.07.024

Frank-Kamenetskii MD, Mirkin SM (1995) Triplex DNA structures. Annu Rev Biochem 64:65–95. https://doi.org/10.1146/annurev.bi.64.070195.000433

French S (1992) Consequences of replication fork movement through transcription units in vivo. Science 258(5086):1362–1365

Gan W, Guan Z, Liu J, Gui T, Shen K, Manley JL, Li X (2011) R-loop-mediated genomic instability is caused by impairment of replication fork progression. Genes Dev 25(19):2041–2056. https://doi.org/10.1101/gad.17010011

Ganesan A, Spivak G, Hanawalt PC (2012) Transcription-coupled DNA repair in prokaryotes. Prog Mol Biol Transl Sci 110:25–40. https://doi.org/10.1016/B978-0-12-387665-2.00002-X

Garcia-Muse T, Aguilera A (2016) Transcription-replication conflicts: how they occur and how they are resolved. Nat Rev Mol Cell Biol 17(9):553–563. https://doi.org/10.1038/nrm.2016.88

Gerber JK, Gogel E, Berger C, Wallisch M, Muller F, Grummt I, Grummt F (1997) Termination of mammalian rDNA replication: polar arrest of replication fork movement by transcription termination factor TTF-I. Cell 90(3):559–567

Gilbert DM (2002) Replication timing and transcriptional control: beyond cause and effect. Curr Opin Cell Biol 14(3):377–383

Glover TW, Berger C, Coyle J, Echo B (1984) DNA polymerase alpha inhibition by aphidicolin induces gaps and breaks at common fragile sites in human chromosomes. Hum Genet 67(2):136–142

Gomez-Gonzalez B, Felipe-Abrio I, Aguilera A (2009) The S-phase checkpoint is required to respond to R-loops accumulated in THO mutants. Mol Cell Biol 29(19):5203–5213. https://doi.org/10.1128/MCB.00402-09

Gomez-Gonzalez B, Garcia-Rubio M, Bermejo R, Gaillard H, Shirahige K, Marin A, Foiani M, Aguilera A (2011) Genome-wide function of THO/TREX in active genes prevents R-loop-dependent replication obstacles. EMBO J 30(15):3106–3119. https://doi.org/10.1038/emboj.2011.206

Gotta SL, Miller OL Jr, French SL (1991) rRNA transcription rate in *Escherichia coli*. J Bacteriol 173(20):6647–6649

Gottipati P, Cassel TN, Savolainen L, Helleday T (2008) Transcription-associated recombination is dependent on replication in Mammalian cells. Mol Cell Biol 28(1):154–164. https://doi.org/10.1128/MCB.00816-07

Gray LT, Vallur AC, Eddy J, Maizels N (2014) G quadruplexes are genomewide targets of transcriptional helicases XPB and XPD. Nat Chem Biol 10(4):313–318. https://doi.org/10.1038/nchembio.1475

Greenfeder SA, Newlon CS (1992) Replication forks pause at yeast centromeres. Mol Cell Biol 12(9):4056–4066

Guy L, Roten CA (2004) Genometric analyses of the organization of circular chromosomes: a universal pressure determines the direction of ribosomal RNA genes transcription relative to chromosome replication. Gene 340(1):45–52. https://doi.org/10.1016/j.gene.2004.06.056

Guy CP, Atkinson J, Gupta MK, Mahdi AA, Gwynn EJ, Rudolph CJ, Moon PB, van Knippenberg IC, Cadman CJ, Dillingham MS, Lloyd RG, McGlynn P (2009) Rep provides a second motor at the replisome to promote duplication of protein-bound DNA. Mol Cell 36(4):654–666. https://doi.org/10.1016/j.molcel.2009.11.009

Gwynn EJ, Smith AJ, Guy CP, Savery NJ, McGlynn P, Dillingham MS (2013) The conserved C-terminus of the PcrA/UvrD helicase interacts directly with RNA polymerase. PLoS One 8(10):e78141. https://doi.org/10.1371/journal.pone.0078141

Ha SC, Lowenhaupt K, Rich A, Kim YG, Kim KK (2005) Crystal structure of a junction between B-DNA and Z-DNA reveals two extruded bases. Nature 437(7062):1183–1186. https://doi.org/10.1038/nature04088

Hansel-Hertsch R, Beraldi D, Lensing SV, Marsico G, Zyner K, Parry A, Di Antonio M, Pike J, Kimura H, Narita M, Tannahill D, Balasubramanian S (2016) G-quadruplex structures mark human regulatory chromatin. Nat Genet 48(10):1267–1272. https://doi.org/10.1038/ng.3662

Harada Y, Ohara O, Takatsuki A, Itoh H, Shimamoto N, Kinosita K Jr (2001) Direct observation of DNA rotation during transcription by *Escherichia coli* RNA polymerase. Nature 409(6816):113–115. https://doi.org/10.1038/35051126

Hatchi E, Skourti-Stathaki K, Ventz S, Pinello L, Yen A, Kamieniarz-Gdula K, Dimitrov S, Pathania S, McKinney KM, Eaton ML, Kellis M, Hill SJ, Parmigiani G, Proudfoot NJ, Livingston DM (2015) BRCA1 recruitment to transcriptional pause sites is required for R-loop-driven DNA damage repair. Mol Cell 57(4):636–647. https://doi.org/10.1016/j.molcel.2015.01.011

Helmrich A, Ballarino M, Tora L (2011) Collisions between replication and transcription complexes cause common fragile site instability at the longest human genes. Mol Cell 44(6):966–977. https://doi.org/10.1016/j.molcel.2011.10.013

Henderson A, Wu Y, Huang YC, Chavez EA, Platt J, Johnson FB, Brosh RM Jr, Sen D, Lansdorp PM (2014) Detection of G-quadruplex DNA in mammalian cells. Nucleic Acids Res 42(2):860–869. https://doi.org/10.1093/nar/gkt957

Herman RK, Dworkin NB (1971) Effect of gene induction on the rate of mutagenesis by ICR-191 in *Escherichia coli*. J Bacteriol 106(2):543–550

Hiratani I, Ryba T, Itoh M, Yokochi T, Schwaiger M, Chang CW, Lyou Y, Townes TM, Schubeler D, Gilbert DM (2008) Global reorganization of replication domains during embryonic stem cell differentiation. PLoS Biol 6(10):e245. https://doi.org/10.1371/journal.pbio.0060245

Hiratani I, Takebayashi S, Lu J, Gilbert DM (2009) Replication timing and transcriptional control: beyond cause and effect – Part II. Curr Opin Genet Dev 19(2):142–149. https://doi.org/10.1016/j.gde.2009.02.002

Hirose S, Hiraga S, Okazaki T (1983) Initiation site of deoxyribonucleotide polymerization at the replication origin of the *Escherichia coli* chromosome. Mol Gen Genet 189(3):422–431

Hoyne PR, Maher LJ 3rd (2002) Functional studies of potential intrastrand triplex elements in the *Escherichia coli* genome. J Mol Biol 318(2):373–386. https://doi.org/10.1016/S0022-2836(02)00041-4

Huang J, Brito IL, Villen J, Gygi SP, Amon A, Moazed D (2006) Inhibition of homologous recombination by a cohesin-associated clamp complex recruited to the rDNA recombination enhancer. Genes Dev 20(20):2887–2901. https://doi.org/10.1101/gad.1472706

Huertas P, Aguilera A (2003) Cotranscriptionally formed DNA:RNA hybrids mediate transcription elongation impairment and transcription-associated recombination. Mol Cell 12(3):711–721

Huppert JL, Balasubramanian S (2005) Prevalence of quadruplexes in the human genome. Nucleic Acids Res 33(9):2908–2916. https://doi.org/10.1093/nar/gki609

Huvet M, Nicolay S, Touchon M, Audit B, d'Aubenton-Carafa Y, Arneodo A, Thermes C (2007) Human gene organization driven by the coordination of replication and transcription. Genome Res 17(9):1278–1285. https://doi.org/10.1101/gr.6533407

Iborra FJ, Jackson DA, Cook PR (2001) Coupled transcription and translation within nuclei of mammalian cells. Science 293(5532):1139–1142. https://doi.org/10.1126/science.1061216

Jain A, Wang G, Vasquez KM (2008) DNA triple helices: biological consequences and therapeutic potential. Biochimie 90(8):1117–1130. https://doi.org/10.1016/j.biochi.2008.02.011

Jeppsson K, Carlborg KK, Nakato R, Berta DG, Lilienthal I, Kanno T, Lindqvist A, Brink MC, Dantuma NP, Katou Y, Shirahige K, Sjogren C (2014) The chromosomal association of the Smc5/6 complex depends on cohesion and predicts the level of sister chromatid entanglement. PLoS Genet 10(10):e1004680. ARTN e1004680. https://doi.org/10.1371/journal.pgen.1004680

Joos S, Haluska FG, Falk MH, Henglein B, Hameister H, Croce CM, Bornkamm GW (1992) Mapping chromosomal breakpoints of Burkitt's t(8;14) translocations far upstream of c-myc. Cancer Res 52(23):6547–6552

Joshi RS, Pina B, Roca J (2012) Topoisomerase II is required for the production of long Pol II gene transcripts in yeast. Nucleic Acids Res 40(16):7907–7915. https://doi.org/10.1093/nar/gks626

Kanoh Y, Matsumoto S, Fukatsu R, Kakusho N, Kono N, Renard-Guillet C, Masuda K, Iida K, Nagasawa K, Shirahige K, Masai H (2015) Rif1 binds to G quadruplexes and suppresses replication over long distances. Nat Struct Mol Biol 22(11):889–897. https://doi.org/10.1038/nsmb.3102

Kantidakis T, Saponaro M, Mitter R, Horswell S, Kranz A, Boeing S, Aygun O, Kelly GP, Matthews N, Stewart A, Stewart AF, Svejstrup JQ (2016) Mutation of cancer driver MLL2 results in transcription stress and genome instability. Genes Dev 30(4):408–420. https://doi.org/10.1101/gad.275453.115

Keil RL, Roeder GS (1984) Cis-acting, recombination-stimulating activity in a fragment of the ribosomal DNA of *S. cerevisiae*. Cell 39(2 Pt 1):377–386

Kim E, Deppert W (2003) The complex interactions of p53 with target DNA: we learn as we go. Biochem Cell Biol 81(3):141–150. https://doi.org/10.1139/o03-046

Kim RA, Wang JC (1989) Function of DNA topoisomerases as replication swivels in *Saccharomyces cerevisiae*. J Mol Biol 208(2):257–267

King IF, Yandava CN, Mabb AM, Hsiao JS, Huang HS, Pearson BL, Calabrese JM, Starmer J, Parker JS, Magnuson T, Chamberlain SJ, Philpot BD, Zylka MJ (2013) Topoisomerases facilitate transcription of long genes linked to autism. Nature 501(7465):58–62. https://doi.org/10.1038/nature12504

Kinniburgh AJ (1989) A cis-acting transcription element of the c-myc gene can assume an H-DNA conformation. Nucleic Acids Res 17(19):7771–7778

Kobayashi T, Horiuchi T (1996) A yeast gene product, Fob1 protein, required for both replication fork blocking and recombinational hotspot activities. Genes Cells 1(5):465–474

Kogoma T (1997) Stable DNA replication: interplay between DNA replication, homologous recombination, and transcription. Microbiol Mol Biol Rev 61(2):212–238

Kogoma T, Hong X, Cadwell GW, Barnard KG, Asai T (1993) Requirement of homologous recombination functions for viability of the *Escherichia coli* cell that lacks RNase HI and exonuclease V activities. Biochimie 75(1–2):89–99

Kouzine F, Gupta A, Baranello L, Wojtowicz D, Ben-Aissa K, Liu J, Przytycka TM, Levens D (2013) Transcription-dependent dynamic supercoiling is a short-range genomic force. Nat Struct Mol Biol 20(3):396–403. https://doi.org/10.1038/nsmb.2517

Krings G, Bastia D (2005) Sap1p binds to Ter1 at the ribosomal DNA of *Schizosaccharomyces pombe* and causes polar replication fork arrest. J Biol Chem 280(47):39135–39142. https://doi.org/10.1074/jbc.M508996200

Kumar A, Mazzanti M, Mistrik M, Kosar M, Beznoussenko GV, Mironov AA, Garre M, Parazzoli D, Shivashankar GV, Scita G, Bartek J, Foiani M (2014) ATR mediates a checkpoint at the nuclear envelope in response to mechanical stress. Cell 158(3):633–646. https://doi.org/10.1016/j.cell.2014.05.046

Kunst F, Ogasawara N, Moszer I, Albertini AM, Alloni G, Azevedo V, Bertero MG, Bessieres P, Bolotin A, Borchert S, Borriss R, Boursier L, Brans A, Braun M, Brignell SC, Bron S, Brouillet S, Bruschi CV, Caldwell B, Capuano V, Carter NM, Choi SK, Cordani JJ, Connerton IF, Cummings NJ, Daniel RA, Denziot F, Devine KM, Dusterhoft A, Ehrlich SD, Emmerson PT, Entian KD, Errington J, Fabret C, Ferrari E, Foulger D, Fritz C, Fujita M, Fujita Y, Fuma S, Galizzi A, Galleron N, Ghim SY, Glaser P, Goffeau A, Golightly EJ, Grandi G, Guiseppi G, Guy BJ, Haga K, Haiech J, Harwood CR, Henaut A, Hilbert H, Holsappel S, Hosono S, Hullo MF, Itaya M, Jones L, Joris B, Karamata D, Kasahara Y, Klaerr-Blanchard M, Klein C, Kobayashi Y, Koetter P, Koningstein G, Krogh S, Kumano M, Kurita K, Lapidus A, Lardinois S, Lauber J, Lazarevic V, Lee SM, Levine A, Liu H, Masuda S, Mauel C, Medigue C, Medina N, Mellado RP, Mizuno M, Moestl D, Nakai S, Noback M, Noone D, O'Reilly M, Ogawa K, Ogiwara A, Oudega B, Park SH, Parro V, Pohl TM, Portelle D, Porwollik S, Prescott AM, Presecan E, Pujic

P, Purnelle B, Rapoport G, Rey M, Reynolds S, Rieger M, Rivolta C, Rocha E, Roche B, Rose M, Sadaie Y, Sato T, Scanlan E, Schleich S, Schroeter R, Scoffone F, Sekiguchi J, Sekowska A, Seror SJ, Serror P, Shin BS, Soldo B, Sorokin A, Tacconi E, Takagi T, Takahashi H, Takemaru K, Takeuchi M, Tamakoshi A, Tanaka T, Terpstra P, Togoni A, Tosato V, Uchiyama S, Vandebol M, Vannier F, Vassarotti A, Viari A, Wambutt R, Wedler H, Weitzenegger T, Winters P, Wipat A, Yamamoto H, Yamane K, Yasumoto K, Yata K, Yoshida K, Yoshikawa HF, Zumstein E, Yoshikawa H, Danchin A (1997) The complete genome sequence of the gram-positive bacterium *Bacillus subtilis*. Nature 390(6657):249–256. https://doi.org/10.1038/36786

Landick R (2006) The regulatory roles and mechanism of transcriptional pausing. Biochem Soc Trans 34(Pt 6):1062–1066. https://doi.org/10.1042/BST0341062

Law MJ, Lower KM, Voon HP, Hughes JR, Garrick D, Viprakasit V, Mitson M, De Gobbi M, Marra M, Morris A, Abbott A, Wilder SP, Taylor S, Santos GM, Cross J, Ayyub H, Jones S, Ragoussis J, Rhodes D, Dunham I, Higgs DR, Gibbons RJ (2010) ATR-X syndrome protein targets tandem repeats and influences allele-specific expression in a size-dependent manner. Cell 143(3):367–378. https://doi.org/10.1016/j.cell.2010.09.023

Li X, Manley JL (2005) Inactivation of the SR protein splicing factor ASF/SF2 results in genomic instability. Cell 122(3):365–378. https://doi.org/10.1016/j.cell.2005.06.008

Li L, Wang X, Stolc V, Li X, Zhang D, Su N, Tongprasit W, Li S, Cheng Z, Wang J, Deng XW (2006) Genome-wide transcription analyses in rice using tiling microarrays. Nat Genet 38(1):124–129. https://doi.org/10.1038/ng1704

Liberi G, Maffioletti G, Lucca C, Chiolo I, Baryshnikova A, Cotta-Ramusino C, Lopes M, Pellicioli A, Haber JE, Foiani M (2005) Rad51-dependent DNA structures accumulate at damaged replication forks in sgs1 mutants defective in the yeast ortholog of BLM RecQ helicase. Genes Dev 19(3):339–350. https://doi.org/10.1101/gad.322605

Light WH, Brickner DG, Brand VR, Brickner JH (2010) Interaction of a DNA zip code with the nuclear pore complex promotes H2A.Z incorporation and INO1 transcriptional memory. Mol Cell 40(1):112–125. https://doi.org/10.1016/j.molcel.2010.09.007

Linskens MH, Huberman JA (1988) Organization of replication of ribosomal DNA in *Saccharomyces cerevisiae*. Mol Cell Biol 8(11):4927–4935

Liu B, Alberts BM (1995) Head-on collision between a DNA replication apparatus and RNA polymerase transcription complex. Science 267(5201):1131–1137

Liu LF, Wang JC (1987) Supercoiling of the DNA template during transcription. Proc Natl Acad Sci U S A 84(20):7024–7027

Liu B, Wong ML, Tinker RL, Geiduschek EP, Alberts BM (1993) The DNA replication fork can pass RNA polymerase without displacing the nascent transcript. Nature 366(6450):33–39. https://doi.org/10.1038/366033a0

Lopes M, Cotta-Ramusino C, Pellicioli A, Liberi G, Plevani P, Muzi-Falconi M, Newlon CS, Foiani M (2001) The DNA replication checkpoint response stabilizes stalled replication forks. Nature 412(6846):557–561. https://doi.org/10.1038/35087613

Lopez-estrano C, Schvartzman JB, Krimer DB, Hernandez P (1998) Co-localization of polar replication fork barriers and rRNA transcription terminators in mouse rDNA. J Mol Biol 277(2):249–256. https://doi.org/10.1006/jmbi.1997.1607

Lopez-Estrano C, Schvartzman JB, Krimer DB, Hernandez P (1999) Characterization of the pea rDNA replication fork barrier: putative cis-acting and trans-acting factors. Plant Mol Biol 40(1):99–110

Maizels N, Gray LT (2013) The G4 genome. PLoS Genet 9(4):e1003468. https://doi.org/10.1371/journal.pgen.1003468

Maric C, Levacher B, Hyrien O (1999) Developmental regulation of replication fork pausing in Xenopus laevis ribosomal RNA genes. J Mol Biol 291(4):775–788. https://doi.org/10.1006/jmbi.1999.3017

McHenry CS (2011) DNA replicases from a bacterial perspective. Annu Rev Biochem 80:403–436. https://doi.org/10.1146/annurev-biochem-061208-091655

McKay BC, Becerril C, Spronck JC, Ljungman M (2002) Ultraviolet light-induced apoptosis is associated with S-phase in primary human fibroblasts. DNA Repair (Amst) 1(10):811–820

Mekhail K, Seebacher J, Gygi SP, Moazed D (2008) Role for perinuclear chromosome tethering in maintenance of genome stability. Nature 456(7222):667–670. https://doi.org/10.1038/nature07460

Merrikh H, Machon C, Grainger WH, Grossman AD, Soultanas P (2011) Co-directional replication-transcription conflicts lead to replication restart. Nature 470(7335):554–557. https://doi.org/10.1038/nature09758

Merrikh H, Zhang Y, Grossman AD, Wang JD (2012) Replication-transcription conflicts in bacteria. Nat Rev Microbiol 10(7):449–458. https://doi.org/10.1038/nrmicro2800

Merrikh CN, Brewer BJ, Merrikh H (2015) The *B. subtilis* accessory helicase PcrA facilitates DNA replication through transcription units. PLoS Genet 11(6):e1005289. https://doi.org/10.1371/journal.pgen.1005289

Meryet-Figuiere M, Alaei-Mahabadi B, Ali MM, Mitra S, Subhash S, Pandey GK, Larsson E, Kanduri C (2014) Temporal separation of replication and transcription during S-phase progression. Cell Cycle 13(20):3241–3248. https://doi.org/10.4161/15384101.2014.953876

Mirkin EV, Mirkin SM (2005) Mechanisms of transcription-replication collisions in bacteria. Mol Cell Biol 25(3):888–895. https://doi.org/10.1128/MCB.25.3.888-895.2005

Mischo HE, Gomez-Gonzalez B, Grzechnik P, Rondon AG, Wei W, Steinmetz L, Aguilera A, Proudfoot NJ (2011) Yeast Sen1 helicase protects the genome from transcription-associated instability. Mol Cell 41(1):21–32. https://doi.org/10.1016/j.molcel.2010.12.007

Mok M, Marians KJ (1987) The *Escherichia coli* preprimosome and DNA B helicase can form replication forks that move at the same rate. J Biol Chem 262(34):16644–16654

Naughton C, Avlonitis N, Corless S, Prendergast JG, Mati IK, Eijk PP, Cockroft SL, Bradley M, Ylstra B, Gilbert N (2013) Transcription forms and remodels supercoiling domains unfolding large-scale chromatin structures. Nat Struct Mol Biol 20(3):387–395. https://doi.org/10.1038/nsmb.2509

Nickoloff JA (1992) Transcription enhances intrachromosomal homologous recombination in mammalian cells. Mol Cell Biol 12(12):5311–5318

Olavarrieta L, Hernandez P, Krimer DB, Schvartzman JB (2002) DNA knotting caused by head-on collision of transcription and replication. J Mol Biol 322(1):1–6

Opalka N, Chlenov M, Chacon P, Rice WJ, Wriggers W, Darst SA (2003) Structure and function of the transcription elongation factor GreB bound to bacterial RNA polymerase. Cell 114(3):335–345

Paeschke K, Capra JA, Zakian VA (2011) DNA replication through G-quadruplex motifs is promoted by the *Saccharomyces cerevisiae* Pif1 DNA helicase. Cell 145(5):678–691. https://doi.org/10.1016/j.cell.2011.04.015

Park JS, Marr MT, Roberts JW (2002) *E. coli* transcription repair coupling factor (Mfd protein) rescues arrested complexes by promoting forward translocation. Cell 109(6):757–767

Peter BJ, Ullsperger C, Hiasa H, Marians KJ, Cozzarelli NR (1998) The structure of supercoiled intermediates in DNA replication. Cell 94(6):819–827

Peter BJ, Arsuaga J, Breier AM, Khodursky AB, Brown PO, Cozzarelli NR (2004) Genomic transcriptional response to loss of chromosomal supercoiling in *Escherichia coli*. Genome Biol 5(11):R87. https://doi.org/10.1186/gb-2004-5-11-r87

Pomerantz RT, O'Donnell M (2008) The replisome uses mRNA as a primer after colliding with RNA polymerase. Nature 456(7223):762–766. https://doi.org/10.1038/nature07527

Postow L, Crisona NJ, Peter BJ, Hardy CD, Cozzarelli NR (2001) Topological challenges to DNA replication: conformations at the fork. Proc Natl Acad Sci U S A 98(15):8219–8226. https://doi.org/10.1073/pnas.111006998

Prado F, Aguilera A (2005) Impairment of replication fork progression mediates RNA polII transcription-associated recombination. EMBO J 24(6):1267–1276. https://doi.org/10.1038/sj.emboj.7600602

Prado F, Piruat JI, Aguilera A (1997) Recombination between DNA repeats in yeast hpr1delta cells is linked to transcription elongation. EMBO J 16(10):2826–2835. https://doi.org/10.1093/emboj/16.10.2826

Proshkin S, Rahmouni AR, Mironov A, Nudler E (2010) Cooperation between translating ribosomes and RNA polymerase in transcription elongation. Science 328(5977):504–508. https://doi.org/10.1126/science.1184939

Raghuraman MK, Winzeler EA, Collingwood D, Hunt S, Wodicka L, Conway A, Lockhart DJ, Davis RW, Brewer BJ, Fangman WL (2001) Replication dynamics of the yeast genome. Science 294(5540):115–121. https://doi.org/10.1126/science.294.5540.115

Rhodes D, Lipps HJ (2015) G-quadruplexes and their regulatory roles in biology. Nucleic Acids Res 43(18):8627–8637. https://doi.org/10.1093/nar/gkv862

Rich A, Zhang S (2003) Timeline: Z-DNA: the long road to biological function. Nat Rev Genet 4(7):566–572. https://doi.org/10.1038/nrg1115

Rivera-Mulia JC, Gilbert DM (2016) Replication timing and transcriptional control: beyond cause and effect-part III. Curr Opin Cell Biol 40:168–178. https://doi.org/10.1016/j.ceb.2016.03.022

Roberts RW, Crothers DM (1992) Stability and properties of double and triple helices: dramatic effects of RNA or DNA backbone composition. Science 258(5087):1463–1466

Rocha EP, Danchin A (2003) Essentiality, not expressiveness, drives gene-strand bias in bacteria. Nat Genet 34(4):377–378. https://doi.org/10.1038/ng1209

Rodriguez R, Miller KM, Forment JV, Bradshaw CR, Nikan M, Britton S, Oelschlaegel T, Xhemalce B, Balasubramanian S, Jackson SP (2012) Small-molecule-induced DNA damage identifies alternative DNA structures in human genes. Nat Chem Biol 8(3):301–310. https://doi.org/10.1038/nchembio.780

Roghanian M, Zenkin N, Yuzenkova Y (2015) Bacterial global regulators DksA/ppGpp increase fidelity of transcription. Nucleic Acids Res 43(3):1529–1536. https://doi.org/10.1093/nar/gkv003

Rossi SE, Ajazi A, Carotenuto W, Foiani M, Giannattasio M (2015) Rad53-mediated regulation of Rrm3 and Pif1 DNA helicases contributes to prevention of aberrant fork transitions under replication stress. Cell Rep 13(1):80–92. https://doi.org/10.1016/j.celrep.2015.08.073

Roy Chowdhury A, Bakshi R, Wang J, Yildirir G, Liu B, Pappas-Brown V, Tolun G, Griffith JD, Shapiro TA, Jensen RE, Englund PT (2010) The killing of African trypanosomes by ethidium bromide. PLoS Pathog 6(12):e1001226. https://doi.org/10.1371/journal.ppat.1001226

Roy D, Lieber MR (2009) G clustering is important for the initiation of transcription-induced R-loops in vitro, whereas high G density without clustering is sufficient thereafter. Mol Cell Biol 29(11):3124–3133. https://doi.org/10.1128/MCB.00139-09

Rudolph CJ, Dhillon P, Moore T, Lloyd RG (2007) Avoiding and resolving conflicts between DNA replication and transcription. DNA Repair (Amst) 6(7):981–993. https://doi.org/10.1016/j.dnarep.2007.02.017

Salceda J, Fernandez X, Roca J (2006) Topoisomerase II, not topoisomerase I, is the proficient relaxase of nucleosomal DNA. EMBO J 25(11):2575–2583. https://doi.org/10.1038/sj.emboj.7601142

Sanchez-Gorostiaga A, Lopez-Estrano C, Krimer DB, Schvartzman JB, Hernandez P (2004) Transcription termination factor reb1p causes two replication fork barriers at its cognate sites in fission yeast ribosomal DNA in vivo. Mol Cell Biol 24(1):398–406

Sankar TS, Wastuwidyaningtyas BD, Dong Y, Lewis SA, Wang JD (2016) The nature of mutations induced by replication-transcription collisions. Nature 535(7610):178–181. https://doi.org/10.1038/nature18316

Santos-Pereira JM, Aguilera A (2015) R loops: new modulators of genome dynamics and function. Nat Rev Genet 16(10):583–597. https://doi.org/10.1038/nrg3961

Santos-Pereira JM, Herrero AB, Garcia-Rubio ML, Marin A, Moreno S, Aguilera A (2013) The Npl3 hnRNP prevents R-loop-mediated transcription-replication conflicts and genome instability. Genes Dev 27(22):2445–2458. https://doi.org/10.1101/gad.229880.113

Saponaro M, Kantidakis T, Mitter R, Kelly GP, Heron M, Williams H, Soding J, Stewart A, Svejstrup JQ (2014) RECQL5 controls transcript elongation and suppresses genome instability associated with transcription stress. Cell 157(5):1037–1049. https://doi.org/10.1016/j.cell.2014.03.048

Sarkies P, Reams C, Simpson LJ, Sale JE (2010) Epigenetic instability due to defective replication of structured DNA. Mol Cell 40(5):703–713. https://doi.org/10.1016/j.molcel.2010.11.009

Savic DJ, Kanazir DT (1972) The effect of a histidine operator-constitutive mutation on UV-induced mutability within the histidine operon of *Salmonella typhimurium*. Mol Gen Genet 118(1):45–50

Schalbetter SA, Mansoubi S, Chambers AL, Downs JA, Baxter J (2015) Fork rotation and DNA precatenation are restricted during DNA replication to prevent chromosomal instability. Proc Natl Acad Sci U S A 112(33):E4565–E4570. https://doi.org/10.1073/pnas.1505356112

Selby CP, Drapkin R, Reinberg D, Sancar A (1997) RNA polymerase II stalled at a thymine dimer: footprint and effect on excision repair. Nucleic Acids Res 25(4):787–793

Sen D, Gilbert W (1988) Formation of parallel four-stranded complexes by guanine-rich motifs in DNA and its implications for meiosis. Nature 334(6180):364–366. https://doi.org/10.1038/334364a0

Smirnov E, Borkovec J, Kovacik L, Svidenska S, Schrofel A, Skalnikova M, Svindrych Z, Krizek P, Ovesny M, Hagen GM, Juda P, Michalova K, Cardoso MC, Cmarko D, Raska I (2014) Separation of replication and transcription domains in nucleoli. J Struct Biol 188(3):259–266. https://doi.org/10.1016/j.jsb.2014.10.001

Song W, Dominska M, Greenwell PW, Petes TD (2014) Genome-wide high-resolution mapping of chromosome fragile sites in *Saccharomyces cerevisiae*. Proc Natl Acad Sci U S A 111(21):E2210–E2218. https://doi.org/10.1073/pnas.1406847111

Spiesser TW, Diener C, Barberis M, Klipp E (2010) What influences DNA replication rate in budding yeast? PLoS One 5(4):e10203. https://doi.org/10.1371/journal.pone.0010203

Srivatsan A, Tehranchi A, MacAlpine DM, Wang JD (2010) Co-orientation of replication and transcription preserves genome integrity. PLoS Genet 6(1):e1000810. https://doi.org/10.1371/journal.pgen.1000810

Stirling PC, Chan YA, Minaker SW, Aristizabal MJ, Barrett I, Sipahimalani P, Kobor MS, Hieter P (2012) R-loop-mediated genome instability in mRNA cleavage and polyadenylation mutants. Genes Dev 26(2):163–175. https://doi.org/10.1101/gad.179721.111

Stolc V, Gauhar Z, Mason C, Halasz G, van Batenburg MF, Rifkin SA, Hua S, Herreman T, Tongprasit W, Barbano PE, Bussemaker HJ, White KP (2004) A gene expression map for the euchromatic genome of *Drosophila melanogaster*. Science 306(5696):655–660. https://doi.org/10.1126/science.1101312

Strambio-De-Castillia C, Niepel M, Rout MP (2010) The nuclear pore complex: bridging nuclear transport and gene regulation. Nat Rev Mol Cell Biol 11(7):490–501. https://doi.org/10.1038/nrm2928

Tan-Wong SM, Wijayatilake HD, Proudfoot NJ (2009) Gene loops function to maintain transcriptional memory through interaction with the nuclear pore complex. Genes Dev 23(22):2610–2624. https://doi.org/10.1101/gad.1823209

Tehranchi AK, Blankschien MD, Zhang Y, Halliday JA, Srivatsan A, Peng J, Herman C, Wang JD (2010) The transcription factor DksA prevents conflicts between DNA replication and transcription machinery. Cell 141(4):595–605. https://doi.org/10.1016/j.cell.2010.03.036

Thomas BJ, Rothstein R (1989) Elevated recombination rates in transcriptionally active DNA. Cell 56(4):619–630

Todd AK, Johnston M, Neidle S (2005) Highly prevalent putative quadruplex sequence motifs in human DNA. Nucleic Acids Res 33(9):2901–2907. https://doi.org/10.1093/nar/gki553

Tuduri S, Crabbe L, Conti C, Tourriere H, Holtgreve-Grez H, Jauch A, Pantesco V, De Vos J, Thomas A, Theillet C, Pommier Y, Tazi J, Coquelle A, Pasero P (2009) Topoisomerase I suppresses genomic instability by preventing interference between replication and transcription. Nat Cell Biol 11(11):1315–1324. https://doi.org/10.1038/ncb1984

Usdin K, Woodford KJ (1995) CGG repeats associated with DNA instability and chromosome fragility form structures that block DNA synthesis in vitro. Nucleic Acids Res 23(20):4202–4209

Vilette D, Uzest M, Ehrlich SD, Michel B (1992) DNA transcription and repressor binding affect deletion formation in *Escherichia coli* plasmids. EMBO J 11(10):3629–3634

Voelkel-Meiman K, Keil RL, Roeder GS (1987) Recombination-stimulating sequences in yeast ribosomal DNA correspond to sequences regulating transcription by RNA polymerase I. Cell 48(6):1071–1079

Wahba L, Amon JD, Koshland D, Vuica-Ross M (2011) RNase H and multiple RNA biogenesis factors cooperate to prevent RNA: DNA hybrids from generating genome instability. Mol Cell 44(6):978–988. https://doi.org/10.1016/j.molcel.2011.10.017

Wang JC (2002) Cellular roles of DNA topoisomerases: a molecular perspective. Nat Rev Mol Cell Biol 3(6):430–440. https://doi.org/10.1038/nrm831

Wang G, Vasquez KM (2004) Naturally occurring H-DNA-forming sequences are mutagenic in mammalian cells. Proc Natl Acad Sci U S A 101(37):13448–13453. https://doi.org/10.1073/pnas.0405116101

Wang G, Christensen LA, Vasquez KM (2006) Z-DNA-forming sequences generate large-scale deletions in mammalian cells. Proc Natl Acad Sci U S A 103(8):2677–2682. https://doi.org/10.1073/pnas.0511084103

Wang JD, Berkmen MB, Grossman AD (2007) Genome-wide coorientation of replication and transcription reduces adverse effects on replication in *Bacillus subtilis*. Proc Natl Acad Sci U S A 104(13):5608–5613. https://doi.org/10.1073/pnas.0608999104

Wansink DG, Manders EE, van der Kraan I, Aten JA, van Driel R, de Jong L (1994) RNA polymerase II transcription is concentrated outside replication domains throughout S-phase. J Cell Sci 107(Pt 6):1449–1456

Wei X, Samarabandu J, Devdhar RS, Siegel AJ, Acharya R, Berezney R (1998) Segregation of transcription and replication sites into higher order domains. Science 281(5382):1502–1506

Wei PC, Chang AN, Kao J, Du Z, Meyers RM, Alt FW, Schwer B (2016) Long neural genes harbor recurrent DNA break clusters in neural stem/progenitor cells. Cell 164(4):644–655. https://doi.org/10.1016/j.cell.2015.12.039

Westover KD, Bushnell DA, Kornberg RD (2004) Structural basis of transcription: separation of RNA from DNA by RNA polymerase II. Science 303(5660):1014–1016. https://doi.org/10.1126/science.1090839

Wilda M, Busch K, Klose I, Keller T, Woessmann W, Kreuder J, Harbott J, Borkhardt A (2004) Level of MYC overexpression in pediatric Burkitt's lymphoma is strongly dependent on genomic breakpoint location within the MYC locus. Genes Chromosomes Cancer 41(2):178–182. https://doi.org/10.1002/gcc.20063

Wilson MD, Harreman M, Taschner M, Reid J, Walker J, Erdjument-Bromage H, Tempst P, Svejstrup JQ (2013) Proteasome-mediated processing of Def1, a critical step in the cellular response to transcription stress. Cell 154(5):983–995. https://doi.org/10.1016/j.cell.2013.07.028

Wilson TE, Arlt MF, Park SH, Rajendran S, Paulsen M, Ljungman M, Glover TW (2015) Large transcription units unify copy number variants and common fragile sites arising under replication stress. Genome Res 25(2):189–200. https://doi.org/10.1101/gr.177121.114

Woodford KJ, Howell RM, Usdin K (1994) A novel K(+)-dependent DNA synthesis arrest site in a commonly occurring sequence motif in eukaryotes. J Biol Chem 269(43):27029–27035

Wu HY, Shyy SH, Wang JC, Liu LF (1988) Transcription generates positively and negatively supercoiled domains in the template. Cell 53(3):433–440

Wyrick JJ, Aparicio JG, Chen T, Barnett JD, Jennings EG, Young RA, Bell SP, Aparicio OM (2001) Genome-wide distribution of ORC and MCM proteins in *S. cerevisiae*: high-resolution mapping of replication origins. Science 294(5550):2357–2360. https://doi.org/10.1126/science.1066101

Yunis JJ, Soreng AL (1984) Constitutive fragile sites and cancer. Science 226(4679):1199–1204

Zhang Z, Macalpine DM, Kapler GM (1997) Developmental regulation of DNA replication: replication fork barriers and programmed gene amplification in *Tetrahymena thermophila*. Mol Cell Biol 17(10):6147–6156

Chapter 21
Fragility Extraordinaire: Unsolved Mysteries of Chromosome Fragile Sites

Wenyi Feng and Arijita Chakraborty

Abstract Chromosome fragile sites are a fascinating cytogenetic phenomenon now widely implicated in a slew of human diseases ranging from neurological disorders to cancer. Yet, the paths leading to these revelations were far from direct, and the number of fragile sites that have been molecularly cloned with known disease-associated genes remains modest. Moreover, as more fragile sites were being discovered, research interests in some of the earliest discovered fragile sites ebbed away, leaving a number of unsolved mysteries in chromosome biology. In this review we attempt to recount some of the early discoveries of fragile sites and highlight those phenomena that have eluded intense scrutiny but remain extremely relevant in our understanding of the mechanisms of chromosome fragility. We then survey the literature for disease association for a comprehensive list of fragile sites. We also review recent studies addressing the underlying cause of chromosome fragility while highlighting some ongoing debates. We report an observed enrichment for R-loop forming sequences in fragile site-associated genes than genomic average. Finally, we will leave the reader with some lingering questions to provoke discussion and inspire further scientific inquiries.

Keywords Chromosome fragility • Common and rare fragile sites • DNA double-strand breaks • DNA replication stress • R-loops • Aphidicolin • Folate stress • Neurological disorders • Cancer

---

W. Feng (✉) • A. Chakraborty
Department of Biochemistry and Molecular Biology, SUNY Upstate Medical University, Syracuse, NY, USA
e-mail: fengw@upstate.edu

H. Masai, M. Foiani (eds.), *DNA Replication*, Advances in Experimental Medicine and Biology 1042, https://doi.org/10.1007/978-981-10-6955-0_21

21.1 Introduction

Chromosome fragile sites are defined as gaps, constrictions, or breaks on metaphase chromosomes that are induced by various DNA replication inhibitors. The Human Genome Database currently documents 120 chromosome fragile sites (30 of the rare type and 90 of the common type, Tables 21.1 and 21.2). According to the HUGO Gene Nomenclature Committee, each fragile site is reported as "FRA" followed by the chromosome number and a letter denoting the order of nomenclature, such as FRA1A. In the last decade, fragile sites have been the subject of many comprehensive reviews with respect to the molecular basis, mechanisms of expression, and relevance to human diseases (Debacker and Kooy 2007; Durkin and Glover 2007; Fungtammasan et al. 2012; Lukusa and Fryns 2008; Mirkin 2006; Thys et al. 2015). Here we approach this subject from an alternative angle. By retracing some of the early developments in fragile site discoveries, we hope to identify the gaps in our knowledge and bring awareness to some "cold cases," so to speak, in chromosome fragile sites (we apologize to countless researchers whose work may have been omitted from this review due to space limitation). We review the classification of common vs. rare fragile sites and provide some insights into the DNA structural determinants of fragility. We also provide updated information regarding disease associations of known fragile sites in the database as well as summarize recent genome-wide studies of fragile sites. Finally, we highlight three "unsolved mysteries" in chromosome fragile sites and provide speculations.

21.2 Early History of Chromosome Fragile Site Discoveries Revisited

21.2.1 Discovery of the First Fragile Site

Early history of chromosome fragile site discoveries makes a captivating read. The term "fragile site" was coined in 1969 by Frederick Hecht to describe hitherto reported spontaneous breaks at a specific site in a human metaphase chromosome. However, the first known fragile site was observed 4 years before the genesis of its terminology (Dekaban 1965). Notably, the definition of fragile site later expanded to include constrictions and gaps in the chromosome as well. In this broader definition, the first chromosomal "secondary constriction," to be distinguished from the primary constriction or the centromere, was discovered still earlier (Ferguson-Smith et al. 1962).

Nevertheless, the first fragile site in the pedantic sense was observed in a woman who had received multiple X-ray irradiations for eczematous dermatitis and later had borne a malformed child (Dekaban 1965). It was unclear if the fragile site observed on the presumed Chromosome 9 from the woman was linked to the birth of this child. Limited survey of the woman's family showed normal chromosomes

Table 21.1 Updated summary of *RFSs*

Fragile site	Type	Repeat	Cyto-band	Population freq. (%)[a]	Size (kb)	No. of RLFSs	No. of RLFSs/kb	No. of genes with > = 1 RLFS/kb	Freq. (%) in folate stress[a]	Freq. (%) in APH[b]	Assoc. genes	Disease
FRA1M	Folate		1p21.3		4,900	48	0.010	0		3.333		ID (D'Angelo et al. 2015)
FRA2A	Folate	CGG	2q11.2	0.0014	6,400	266	0.042	1	27.00	0.089	*AFF3*	Schizophrenia (Chen et al. 1998); ID (Metsu et al. 2014a)
FRA2B	Folate		2q13		5,200	177	0.034	0		0.067		
FRA2K	Folate		2q22.3		3,700	20	0.005	0		0.626		ID (Mulatinho et al. 2012)
FRA5G	Folate		5q35		13,458	898	0.067	6		0.120		Found in a FX patient at 50% expression and a normal brother at 36% (Howell et al. 1990)
FRA6A	Folate		6p23		2,000	99	0.050	1		0.009[c]		Schizophrenia (Olavesen et al. 1995)
FRA7A	Folate	CGG	7p11.2		3,500	89	0.025	1		0.058	*ZNF713*	ASD (Metsu et al. 2014b)
FRA8A	Folate		8q22.3		4,500	164	0.036	2		0.018		
FRA9A	Folate		9p21	0.0041	12,900	50	0.004	0	21.00	0.102		
FRA9B	Folate		9q32		2,800	136	0.049	0		1.513		
FRA10A	Folate	CGG	10q23.3	0.0028	16,707	168	0.010	0	35.67	0.124	*FRA10AC1*	Single case of turner syndrome (Li et al. 2015; Maltby and Higgins 1987); Alzheimer's (Li et al. 2015)

(continued)

Table 21.1 (continued)

Fragile site	Type	Repeat	Cyto-band	Population freq. (%)[a]	Size (kb)	No. of RLFSs	No. of RLFSs/kb	No. of genes with >= 1 RLFS/kb	Freq. (%) in folate stress[a]	Freq. (%) in APH[b]	Assoc. genes	Disease
FRA11A	Folate	CGG	11q13.3		1,500	165	0.110	2		0.075	*C11orf8c/ MPPED2*	
FRA11B	Folate	CGG	11q23.3		5,300	459	0.087	1		0.098	*CBL2*	Jacobsen syndrome (Jones et al. 1994, 1995; Michaelis et al. 1998)
FRA12A	Folate	CGG	12q13.1	0.0007	4,678	382	0.082	8	17.67	0.075	*DIP2B*	ASD and MR (Winnepenninckx et al. 2007)
FRA12D	Folate		12q24.13		2,000	93	0.047	0		Not induced		Segregating in FX families (Amarose et al. 1987; Barletta et al. 1991; Sutherland and Baker 1993)
FRA16A	Folate	CGG	16p13.11	0.0007	2,000	116	0.058	1	12.00	0.053	N.D.	
FRA19B	Folate		19p13		19,800	4,650	0.235	88		0.036		
FRA20A	Folate		20p11.23		3,400	70	0.021	0		0.027[c]		
FRA22A	Folate		22q13	0.0014	13,791	2,445	0.177	17	5.67	0.328		MR (Webb and Thake 1984)
FRAXA	Folate	CGG	Xq27.3		5,000	17	0.003	0		Not induced	*FMR1, ASFMR1/ FMR4*	FXS, MR, ID, FXTAS, FXPOI (Galloway and Nelson 2009; Santoro et al. 2012; Usdin et al. 2014)
FRAXE	Folate	CGG	Xq28		8,014	615	0.077	13		0.018[c]	*AFF2/ FMR2*	MR (Bensaid et al. 2009)

FRAXF	Folate	CGG	Xq28		8,014	615	0.077	13		0.018[c]	*FAM11A*	MR; ASD; psychosis, pyramidal signs, macroordism (Holden et al. 1996; Lindsay et al. 1996; Vianna-Morgante et al. 1996)
FRA8E	DA		8q24.1		28,526	1,394	0.049	24		0.799		Azoospermia (Seki et al. 1992); hereditary multiple exostoses 1 gene and two overlapping Langer-Giedion syndrome deletion endpoints (Hill et al. 1997)
FRA11I	DA		11p15.1		5,609	175	0.031	1		0.510		Azoospermia (Seki et al. 1992)
FRA16B	DA	AT-rich MS	16q22.1		4,200	491	0.117	9		0.488	N.D.	Difficulty achieving pregnancy (Martorell et al. 2014); dominant inheritance of cleft palate, microstomia, and micrognathia (McKenzie et al. 2002)
FRA16E	DA		16p12.1		5,900	178	0.030	1		0.027		Batten disease (common form of neuronal ceroid lipofuscinoses [NCLs]) (Dooley et al. 1994)
FRA17A	DA		17p12	0.0049	4,700	91	0.019	2	20.00	0.049		Azoospermia (Shabtai et al. 1982)

(continued)

Table 21.1 (continued)

Fragile site	Type	Repeat	Cyto-band	Population freq. (%)[a]	Size (kb)	No. of RLFSs	No. of RLFSs/kb	No. of genes with >= 1 RLFS/kb	Freq. (%) in folate stress[a]	Freq. (%) in APH[b]	Assoc. genes	Disease
FRA10B	BrdU	AT-rich MS	10q25.2		3,100	45	0.015	0		0.191	N.D.	
FRA12C	BrdU		12q24		24,850	1,907	0.077	6		0.288		Segregating in FX families (Amarose et al. 1987; Barletta et al. 1991; Sutherland and Baker 1993)
FRA8F	UC		8q13		7,900	104	0.013	0		0.004[c]		

Folate folate stress, *DA* distamycin A, *BrdU* bromodeoxyuridine, *UC* unclassified, *MS* minisatellite, *RLFS* R-loop forming sequence, *ID* intellectual disability, *ASD* autism spectrum disorder, *MR* mental retardation, *FXTAS* fragile X-associated tremor/ataxia syndrome, *FXPOI* fragile X-associated premature ovarian insufficiency, *Freq.* frequency of fragile site expression

[a]Data derived from Kähkönen et al. (1989)

[b]Data derived from Mrasek et al. (2010)

[c]Fragile site detected in fewer than three individuals when induced by aphidicolin

Table 21.2 Summary of *CFS*-associated genes and their implications in human diseases

Fragile site	Induction method	Cyto- band	Freq. in APH (%)[a]	Assoc. genes	Disease
FRA1A	APH	1p36	0.488		
FRA1B*	APH	1p32	1.491	*DAB1*	Brain and endometrial cancer (McAvoy et al. 2008)
FRA1C	APH	1p31.2		*IL23R-C1orf141*	Vogt-Koyanagi-Harada syndrome; LOH/MSI associated with advanced tumor stage in prostate cancer (Brys et al. 2013; Hou et al. 2014)
FRA1D	APH	1p22	0.009		
FRA1E*	APH	1p21.2	0.067	*DPYD*	Colorectal, breast, and ovarian cancer (Hormozian et al. 2007)
FRA1F	APH	1q21	0.098		
FRA1G	APH	1q25	0.799		
FRA1I	APH	1q44	2.299		Nasopharyngeal carcinoma (Xia et al. 1988)
FRA1K	APH	1q31	0.138		
FRA1L	APH	1p31	0.479		
FRA2C*	APH	2p24.2 2q24.3	1.513	*MYCN*	Two FSs, FRA2Ccen and FRA2Ctel, flank the *MYCN* amplicon; locus-specific and genomic rearrangements in neuroblastoma and multiple cancers (Blumrich et al. 2011; Lipska et al. 2013)
FRA2D	APH	2p16.2	2.325		
FRA2E	APH	2p13	0.235		Posttransplant diffuse large B-cell lymphoma (Rinaldi et al. 2010)
FRA2F	APH	2q21.3	0.994	*LRP1B*	Alzheimer's; Kawasaki disease (Lin et al. 2014; Shang et al. 2015)
FRA2G*	APH	2q31	0.306	*GAD1*	Schizophrenia (Bharadwaj et al. 2013)
FRA2H*	APH	2q32.1	3.905	*DIRC1, PMS1, MIRN589, MIRN1245*	Breakpoints characterized in K562 cells (Pelliccia et al. 2010)
FRA2I	APH	2q33	1.722		
FRA2J	APH	2q37.3	0.657		
FRA3A	APH	3p24.2	1.327		

(continued)

Table 21.2 (continued)

Fragile site	Induction method	Cyto- band	Freq. in APH (%)[a]	Assoc. genes	Disease
FRA3B*	APH	3p14.2	14.153	*FHIT*	Loss or change of *FHIT* expression associated with common cancers including bladder, esophageal, lung, breast, and prostate cancers (Saldivar et al. 2010)
FRA3C	APH	3q27	1.061		
FRA3D	APH	3q25	0.546		FS expression linked to schizophrenia (Demirhan et al. 2006)
FRA4A	APH	4p16.1	0.612		
FRA4C	APH	4q31.1	2.175		
FRA4D	APH	4p15	0.111		
FRA4F*	APH	4q22	0.683	*GRID2*	Autosomal recessive cerebellar ataxia associated with retinal dystrophy; ataxia and tonic upgaze (Hills et al. 2013; Van Schil et al. 2015)
FRA5C	APH	5q31.1	0.373	*SMAD5*	Copy number gain and increased expression in human hepatocellular carcinoma cells (Zimonjic et al. 2003)
FRA5D	APH	5q15	0.834		Concomitant t(5;21)(q15;q22) and del(5)(q13q33) events in myelodysplastic syndrome (Kasi Loknath Kumar et al. 2016)
FRA5E	APH	5p14	0.293		Copy number loss observed at 52.4% in early Barrett's esophagus (Lai et al. 2010)
FRA5F	APH	5q21	0.515		FS expression significantly higher in patients with rectum cancer and their first-degree relatives (Tunca et al. 2000)
FRA6B	APH	6p25.1	0.985		Copy number change associated with osteoporotic fractures, gyral pattern anomaly, and speech and language disorder in two dizygotic twins (Bozza et al. 2013; Oei et al. 2014)
FRA6C	APH	6p22.2	0.062	BAC CITB.564_C_7	High LOH in cervical tumors; HPV integration site in cervical tumor CC171 (Rader et al. 1996; Thorland et al. 2000)

(continued)

Table 21.2 (continued)

Fragile site	Induction method	Cyto- band	Freq. in APH (%)[a]	Assoc. genes	Disease
FRA6E*	APH	6q26	2.823	*PARK2*	Parkinson's; poor outcome in breast cancer (Ambroziak et al. 2015; Letessier et al. 2007)
FRA6F*	APH	6q21	0.608	*REV3L, DIF13, FKHRL*, etc.	Cancer; schizophrenia (Karayianni et al. 1999; Morelli et al. 2002)
FRA6G	APH	6q15	0.115		Chromosome breakpoints in metastatic melanoma (Limon et al. 1988)
FRA7B*	APH	7p22	0.799	*THSD7A, SDK1, MAD1L1, MIRN589, MIRN339*	Recurrent breakpoint in multiple cancers (Bosco et al. 2010)
FRA7C	APH	7p14.2	0.541		
FRA7D	APH	7p13	0.55		Highly expressed in a female presenting severe immunodeficiency (Conley et al. 1986)
FRA7E*	APH	7q21.2	0.501		
FRA7F	APH	7q22	0.107		Fragile site expression linked to bipolar disorder and schizophrenia (Demirhan et al. 2006, 2009)
FRA7G*	APH	7q31.2	0.111	Caveolin-1, caveolin-2, *TESTIN*	Reduced *TESTIN* expression in 22% of cancer cell lines and 44% of the cell lines derived from hematological malignancies; caveolin-1 gene proposed to be a candidate tumor suppressor (Engelman et al. 1998; Tatarelli et al. 2000)
FRA7H*	APH	7q32.3	2.374		Translocation involving 7q32 found in the RC-K8 cell line derived from a patient with terminal diffuse large B-cell lymphoma (Mishmar et al. 1998; Schneider et al. 2008)
FRA7I*	APH	7q36	0.084	*CNTNAP2*	Implicated in ASD (Rodenas-Cuadrado et al. 2014)
FRA7J	APH	7q11	0.83	*LIMK1, EIF4H(WBSCR1)*	Breakpoint found between the *LIMK1* and *EIF4H* (*WBSCR1*) genes in patients with Williams-Beuren syndrome and ASD (Plaja et al. 2015)

(continued)

Table 21.2 (continued)

Fragile site	Induction method	Cyto- band	Freq. in APH (%)[a]	Assoc. genes	Disease
FRA8B	APH	8q22.1	0.923		
FRA8C*	APH	8q24.1	0.799	*MYC*	Cluster of HPV18 integrations at 8q24 in primary cervical carcinoma; CFS expression frequent in bladder cancer (Ferber et al. 2004; Moriarty and Webster 2003)
FRA8D	APH	8q24.3	0.182		
FRA9D	APH	9q22.1	0.231		FS expression frequent in bladder cancer (Moriarty and Webster 2003)
FRA9E*	APH	9q32		*PAPPA*	LOH, particularly loss of *PAPPA*, is linked to ovarian cancer (Callahan et al. 2003)
FRA10D	APH	10q22.1	0.306	*CTNNA3*	Decreased gene expression of *CTNNA3* in oropharyngeal squamous cell carcinomas (Gao et al. 2014)
FRA10E	APH	10q25.2	0.191		
FRA10F*	APH	10q26.1	0.408	*FATS*	
FRA10G*	APH	10q11.2	0.142	*RET, NCOA4*	FS generates oncogenic *RET/PTC* rearrangements in human thyroid cells (Gandhi et al. 2010)
FRA11C	APH	11p15.1	0.51		FS expression frequent in bladder cancer (Moriarty and Webster 2003)
FRA11D	APH	11p14.2	0.928		
FRA11E	APH	11p13	0.621		Translocation breakpoints at 11p13 found in CML patients.
FRA11F	APH	11q14.2	1.771		Gene amplification linked to FRA11F expression in oral cancer (Reshmi et al. 2007)
FRA11G*	APH	11q23.3	0.098		Interstitial deletions at 11q23.3 associated with abnormal ultrasound findings during prenatal diagnosis (Fechter et al. 2007; Liu et al. 2014)
FRA11H*	APH	11q13	0.075		
FRA12B	APH	12q21.3	0.364		
FRA12E	APH	12q24	0.288		
FRA13A*	APH	13q13.2	1.061	*NBEA*	Neuropsychiatric disorders (Savelyeva et al. 2006a, b)

(continued)

Table 21.2 (continued)

Fragile site	Induction method	Cyto- band	Freq. in APH (%)[a]	Assoc. genes	Disease
FRA13C	APH	13q21.2	0.124		HPV16 integration site in the cervical tumor cell line SiHa (Thorland et al. 2000)
FRA13D	APH	13q32	0.178		Fragile site expression linked to bipolar disorder and schizophrenia (Demirhan et al. 2006, 2009)
FRA14B	APH	14q23	1.429		Found as a constant FS in bloom syndrome cell lines (Barbi et al. 1984; Shiraishi and Li 1993)
FRA14C	APH	14q24.1	0.382		Found as frequent HPV integration site in HPV-related cancers (Bodelon et al. 2016)
FRA15A	APH	15q22	0.08	*RORA*	*RORA* gene expression is low in breast, prostate, and ovarian cancers (Zhu et al. 2006)
FRA16C	APH	16q22.1	0.488		High expression at FRA16B/C found in peripheral blood lymphocytes in the male of a couple having trouble conceiving. Chr 16 instability detected in the sperm of the male and in embryos (Martorell et al. 2014)
FRA16D*	APH	16q23.2	7.576	*WWOX/WOX1/FOR*	*WWOX* gene dysregulation associated with multiple cancers, including pancreatic adenocarcinoma, renal cell carcinoma, and endocrine and exocrine carcinomas (Li et al. 2014)
FRA17B	APH	17q23.1	0.071		HPV integration site in cervical tumor CC226 (Thorland et al. 2000)
FRA18A	APH	18q12.2	0.475		High frequency of LOH on chr 18 in esophageal squamous cell carcinoma (Karkera et al. 2000)
FRA18B	APH	18q21.3	0.182	*DCC*	Found as FSs with frequent interstitial deletions in HIV lymphomas (Capello et al. 2010)

(continued)

Table 21.2 (continued)

Fragile site	Induction method	Cyto- band	Freq. in APH (%)[a]	Assoc. genes	Disease
FRA18C*	APH	18q22.1–18q22.2	0.067	*DOK6*	FS expression in the father of a patient with Beckwith-Wiedemann syndrome and a chromosome truncation 18q22-qter (Debacker et al. 2007)
FRA20B	APH	20p12.2			
FRA22B	APH	22q12.2	0.275		Elevated FS expression in bone marrow and peripheral blood of young cigarette smokers (Kao-Shan et al. 1987)
FRAXB*	APH	Xp22.31	5.494	*STS, GS1 HDHD1 MIR4767*	Deletions within FRAXB seen in 15% of primary tumors and cell lines examined (Arlt et al. 2002)
FRAXC	APH	Xq22.1	2.121	*DMD, IL1RAPL1*	
FRAXD	APH	Xq27.2	0.209		
FRA1H*	5-AZ	1q41–q42.1	0.129	*USH2A, ESRRG, MIRN194–1, MIRN215*	
FRA1J	5-AZ	1q12	0.12		
FRA9F	5-AZ	9q12	0.453		
FRA19A	5-AZ	19q13	0.053		
FRA4B	BrdU	4q12	0.142		
FRA5A	BrdU	5p13	0.191		
FRA5B	BrdU	5q15			
FRA6D	BrdU	6q13	0.031		
FRA9C	BrdU	9p21			
FRA10C	BrdU	10q21	0.089		
FRA13B	BrdU	13q21			
FRA4E	UC	4q27	0.382		

Molecularly mapped fragile sites are indicated by an asterisk
5-AZ, 5-azacytidine, *APH* aphidicolin, *BrdU* bromodeoxyuridine, *UC* unclassified, *FS* fragile site, *LOH* loss of heterozygosity, *MSI* microsatellite instability, *ASD* autism spectrum disorder
[a]Data derived from Mrasek et al. (2010)

in her abnormal child, her husband, and her father (Dekaban 1965). However, this study raised a question that is still relevant in the research field today. There were two types of chromosome abnormality in the woman's blood cultures. A relatively more frequent type consisted of a break near the third telomere-proximal portion of the long arm on the presumed Chromosome 9 (24.8%) and a less frequent one involved aberrations such as dicentrics, rings, and deletions on random chromosomes

(7.1%). In contrast, her skin culture contained a much lower frequency of breakage on Chromosome 9 (6%), despite a comparable level of other chromosome aberrations (8.8%). Moreover, the break seen in the skin culture occurred at a different location (telomere-distal) on Chromosome 9 from that seen in the blood cultures (telomere-proximal). One theory put forth suggested that the X-ray irradiations had induced a clone of abnormal cells with a telomere-proximal fragile site on Chromosome 9 in the woman's blood-forming tissues but not in skin tissues. Alternatively, though it could not be ascertained thoroughly due to family members in absentia, there existed tissue-specific formation of chromosome fragile sites. The latter possibility was subsequently more convincingly demonstrated (Kuwano et al. 1990; Murano et al. 1989a). However, the mechanism of tissue-specific fragile site formation still remains a major challenge in the field.

21.2.2 *First Report of a Heritable Fragile Site*

The first heritable fragile site was demonstrated in a mother-and-daughter case, which involved a fragile site on the long arm of Chromosome 2 near the centromere (Lejeune et al. 1968). The heritability of this fragile site was later confirmed independently (Ferguson-Smith 1973). Curiously, this fragile site gave rise to duplication of the centromere-distal two-thirds of 2q and formation of a three-armed chromosome, which was referred to as qh(2). It was speculated that this endo-duplication event was the result of interruption of a centromere-originated DNA replication signal by the fragile site. Alternatively, it was thought that qh(2) formation was reminiscent of interstitial telomere element-induced endo-duplications elsewhere (Hsu 1963). It was shown that 2q indeed contains interstitial telomere elements, but this fusion site is distinct from the fragile site band 2q11.2 (Ijdo et al. 1991). Finally, it was also surmised that qh(2) formation was due to chromatid breakage followed by mitotic nondisjunction (Ferguson-Smith 1973). To this date, the mechanism for the duplicated q-arms remains a mystery.

Phenotypically the mother and daughter were both of short stature and had intellectual deficiencies, which may or may not be related to the expression of the fragile site. However, subsequent studies demonstrated that the fragility at 2q11.2, which resides in the presumed cytoband responsible for the formation of qh(2), was observed at a significantly higher frequency in "mentally subnormal" school children (Kahkonen et al. 1986) as well as in patients with schizophrenia (Chen et al. 1998). It was not until recent that this fragile site was molecularly characterized as a site with CGG repeat expansion impacting the *AFF3* gene and was detected in families with a broad range of neurodevelopmental disorders (Metsu et al. 2014a).

21.2.3 First Demonstration of Mendelian Transmission of a Fragile Site

A fragile site on 16q22 was examined in a large family of a boy who had recurrent cold urticaria and immunoglobulin A deficiency (Magenis et al. 1970). Transmission of the fragile site followed a Mendelian pattern. However, later it was demonstrated that this fragile site is distal from the haptoglobin gene whose deficiency is linked to IgA deficiency, suggesting that the IgA condition in the propositus was likely a coincidence (Simmers et al. 1986). Fragility at 16q22 has since been implicated in a wide spectrum of neurological disorders (Demirhan et al. 2006; Kerbeshian et al. 2000) as well as in other conditions such as neutropenia (Glasser et al. 2006; Tassano et al. 2010) and cleft palate (Bettex et al. 1998; Dunner et al. 1983; Janiszewska-Olszowska et al. 2013; McKenzie et al. 2002). A recent study also reported elevated fragile site formation at 16q22.1 in an embryo from a couple who had difficulty achieving pregnancy and in the sperm from the father (Martorell et al. 2014). However, cytogenetic breakage and potential disease-associated gene(s) at this locus are yet to be molecularly characterized.

21.2.4 First Fragile Site Linked to a Clinical Phenotype

The vast majority of fragile sites during the early discoveries appeared innocuous and not associated with any phenotypic abnormalities, even when present in homozygous conditions such as those fragile sites at 10q25.2 and 12q24 (Sutherland 1981; Voiculescu et al. 1991). The only fragile site definitively associated with a clinical phenotype was that which resides on Xq27.3 (FRAXA) and was associated with the Martin-Bell or the fragile X syndrome, an X-linked and most common familial form of mental retardation (Giraud et al. 1976; Harvey et al. 1977; Lubs 1969). Subsequent studies demonstrated that the FRAXA fragility is the result of a CGG repeat expansion and is correlated with hypermethylation, gene silencing, as well as delayed replication at the fragile X mental retardation 1 (*FMR1*) gene locus (Bell et al. 1991; Dietrich et al. 1991; Hansen et al. 1992, 1993; Heitz et al. 1991; Pieretti et al. 1991). A full mutation (>200 CGG repeats) at the *FMR1* locus causes gene silencing and the loss of the protein product, FMRP, which causes the loss of synaptic plasticity and the fragile X pathology. Fragile X biology has been the subject of numerous and up-to-date reviews to which we direct the reader for a comprehensive understanding of the disease etiology, mechanism, and intervention (Ligsay and Hagerman 2016; Lozano et al. 2016; Santoro et al. 2012; Wang et al. 2012; Zhao and Usdin 2016).

21.3 Classification of "Common" vs. "Rare" Fragile Sites

21.3.1 Population Frequencies

The Edinburgh survey estimated that from 3% to 5% of the population contained identifiable chromosomal variants (Court Brown 1966). Subsequent population studies defined two types of fragile sites: those that were present in <1% of the general population and those that were present at a theoretical frequency between 1% and 99% or polymorphic (Hecht 1988). Later, this fine distinction between the two groups was relinquished, and collectively they were referred to as "rare fragile sites" with an observed maximal frequency of 5% (Schmid et al. 1986). This classification was also necessitated by the discovery of what appeared to be common or ubiquitous fragile sites, which were present in all 12 subjects regardless of sex or clinical phenotypes (Glover et al. 1984). Thus, presently, fragile sites are classified as "rare" and "common" fragile sites (RFSs and CFSs, respectively) based on the definitions above.

Both RFSs and CFSs can be further characterized by another important metric—the frequency at which fragile sites are observed/expressed in the cells from a given individual, which is akin to a measurement of penetrance. As current literature makes broad reference to frequencies of fragile sites—i.e., at a population level vs. at a cellular level within an individual—it is important to note the difference between these two metrics. For instance, the frequency of FRAXA expression in fragile X individuals can be as high as 50% (Glover et al. 1984). In contrast, the frequency of aphidicolin-induced CFS expression in individuals can be as low as 0.01% based on a large-scale study (Mrasek et al. 2010). Therefore, arguably a CFS with extremely low penetrance may be better characterized as a RFS because it is conceivable that certain individuals, yet to be identified, might exhibit abnormally high penetrance.

21.3.2 Methods of Induction

It was serendipity that led to the discovery that many RFSs discovered by routine diagnostic cytogenetic screen in early studies were unstable in medium 199, which is deficient for folic acid (Sutherland 1977). This finding thus defined the first group of RFSs, which are sensitive to folate stress induced by folic acid or thymidine deprivation, inhibitors of folate metabolism, inhibitors of thymidylate synthetase, and excess thymidine (Sutherland 1991). Systematic analyses revealed that other culture conditions, such as pH, also impact the frequency of fragile site expression (Sutherland 1979). RFSs can also be induced by a group of chemicals of the non-folate stress type such as a nucleotide analog, bromodeoxyuridine, and a base intercalator, distamycin A. In contrast, CFSs are primarily induced by a DNA polymerase inhibitor, aphidicolin, and to a lesser extent by other chemicals including bromodeoxyuridine and 5′-azacytidine.

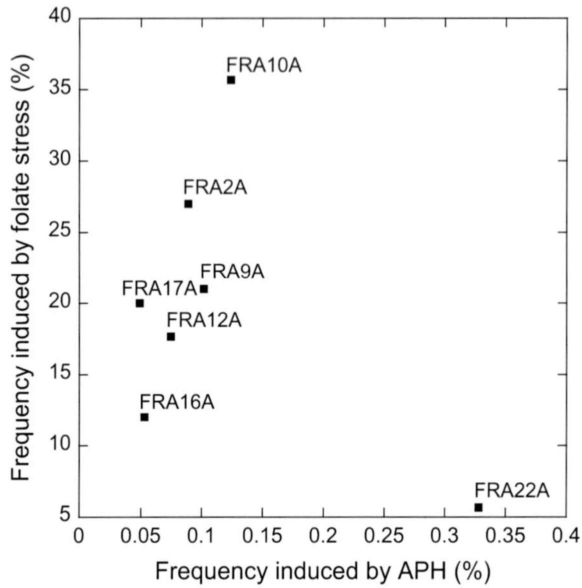

Fig. 21.1 Comparison between *RFS* expression frequencies induced by folate stress and by aphidicolin (APH). Experimental values obtained from Kähkönen et al. (1989) and Mrasek et al. (2010), respectively

It is unclear why RFSs and CFSs show drug-specific induction. Genomic loci may be differentially susceptible to drug-induced replication perturbation. This apparent drug specificity of fragile site expression is likely also due to limitations of the cytological screening methods. Consequently, a given fragile site has been traditionally associated with a specific drug though other drugs could also induce its expression, albeit not as potently. Consistent with this notion, the most frequent aphidicolin-induced CFSs have been shown also inducible by thymidylate stress (Kähkönen et al. 1989). Thus, thymidylate stress can induce not only RFSs but also CFSs, demonstrating effectively "cross-induction" of these two classes of fragile sites. The cross-induction of RFSs by aphidicolin was also confirmed by a genomic scale survey (Mrasek et al. 2010). We further asked if there is a correlation between the expression frequencies (percentage of cells showing fragile sites) of seven RFSs when induced by folate stress (Kähkönen et al. 1989) vs. by aphidicolin (Mrasek et al. 2010). For the study by Käukönen et al., we calculated the expression frequencies only from phenotypically normal individuals for comparison with the study by Mrasek et al. With the exception of a single outlier, FRA22A, which appeared to be highly inducible by aphidicolin, the remaining six RFSs showed a positive correlation ($R = 0.8$) between the induction frequencies by the two conditions (Fig. 21.1). Nevertheless, the differential mechanisms of drug induction of fragile site expression await further exploration.

21.3.3 DNA Sequence and Structural Features

Molecular cloning permitted structural distinction between RFSs and CFSs beyond population frequencies and induction methods. RFSs are characterized by repetitive DNA sequences. Of the 30 RFSs, 12 have so far been identified with a repeat motif with 10 containing a CGG trinucleotide repeat tract and 2 containing an AT-rich minisatellite repeat (Table 21.1). Repetitive DNA can generate non-B DNA structures or alternative conformations with unusual secondary structures (Choi and Majima. 2011). For instance, folate-sensitive RFSs are enriched in CGG repeats that can form stable hairpins, slipped strand structures, G-quadruplexes, and i-tetraplex structures (Fry and Loeb 1994; Usdin and Woodford 1995; Kang et al. 1995). CGG repeats are capable of pausing or stalling replication forks both in vitro and in vivo (Samadashwily et al. 1997). These repeats are polymorphic in normal individuals but undergo dynamic expansion/contraction, resulting in the cytogenic expression of fragile sites under folate stress (Kremer et al. 1991; Verkerk et al. 1991). Non-folate-sensitive RFSs on the other hand consist of expandable AT-rich minisatellite repeats and are specifically induced by AT-dinucleotide binding chemicals such as distamycin A and berenil. For instance, the most frequently observed RFS, FRA16B, contains a polymorphic 33 bp AT-rich minisatellite repeat, which can expand to 2000 copies in certain individuals (Yu et al. 1997). Fourteen copies of this 33 bp AT-rich minisatellite repeat were shown to form DNA secondary structures and cause replication fork stalling, fork regression, and polymerase skipping in vitro (Burrow et al. 2010). Similarly, the BrdU-induced non-folate-sensitive RFS, FRA10B, shares homology with the 33 bp minisatellite in FRA16B and contains an 11 bp inverted repeat that can form hairpins (Handt et al. 2000; Yu et al. 1997).

In contrast, CFSs are not characterized by expandable di- or trinucleotide repetitive sequence. However, some of the cloned CFSs are enriched for short interrupted AT-rich islands with high torsional flexibility and high propensity to form stable secondary structures, similar to the RFSs (Dillon et al. 2013; Zlotorynski et al. 2003). Moreover, a study using a yeast-based genetic assay identified a flexibility peak region (Flex1) of FRA16D with a perfect AT/TA repeat element capable of stalling replication forks (Zhang and Freudenreich 2007). Interestingly, these repeats resemble the AT-rich minisatellites found in non-folate RFSs. The extent of stalling depends on the length of the repeat. Therefore, there appears to be a shared mechanism of repeat-based secondary structure formation and perturbation of the replication fork between RFSs and CFSs, the difference being that RFSs have greater expandability than CFSs (Schwartz et al. 2006). Are there any other cis elements, in addition to DNA secondary structures, that define fragile sites? An RNA/DNA hybrid molecule known as R-loop appears to qualify. R-loops play a wide array of functions in normal cells. For instance, R-loops occur in replication origins in mitochondria as primers for DNA replication (Lee and Clayton 1996). R-loops at CpG islands may also facilitate replication origin specification by generating single-stranded DNA in the nuclear genome (Lombrana et al. 2015). In stimulated B-lymphocytes, R-loop formation at the immunoglobulin heavy chain locus facilitates class switch recombination (Yu et al. 2003). Finally, it was shown that R-loop formation in the guanine-rich transcription pause sites downstream of the poly-A signals is required for transcription termination (Skourti-Stathaki et al. 2011). The genome-wide association of GC skew (asymmetry in guanine distribution between the two strands of DNA), which is conducive to R-loop formation, with transcription termination sites was subsequently validated (Ginno et al. 2013). On the flip side, R-loops are increasingly associated with genome instability and human diseases (Groh and Gromak 2014; Santos-Pereira and Aguilera 2015). In yeast, R-loop formation can sensitize DNA to damaging agents, causing double-strand breaks (Huertas and Aguilera 2003; Li and Manley 2005; Sordet et al. 2009). R-loops at transcription termination sites also require proper resolution by factors such as senataxin, and unresolved R-loops can cause genome instability (Hatchi et al. 2015; Skourti-Stathaki et al. 2011). R-loops can also trigger epigenetic changes in the DNA and bring about the formation of repressive chromatin, which in turn could impede DNA replication (Groh et al. 2014). Evidence for blockage of replication fork by R-loop formation was demonstrated for FRA16D (Madireddy et al. 2016). Finally, R-loop formation can endanger genome stability by incurring replication-transcription conflicts, as exemplified by their association with CFS formation, particularly in large genes such as *FHIT* (Helmrich et al. 2011). The question becomes what roles do R-loops play in replication stress-induced fragile site formation, the corollary being, is there a genome-wide correlation between R-loop content/density and the probability of fragile site formation?

Recent advance in experimental mapping of R-loops capitalized on the utility of the S9.6 antibody, which has a sequence-independent affinity toward the A/B helical RNA/DNA duplex, and coupled the immunoprecipitation of the DNA/RNA hybrid with high-throughput sequencing. Variations of this methodology include DRIPc-seq (DNA/RNA immunoprecipitation followed by cDNA conversion coupled with high-throughput sequencing, an improved version of the previous DRIP-seq method) (Sanz et al. 2016) and RDIP (RNA/DNA immunoprecipitation, implementing key technical modifications in RNase I pretreatment and DNA fragmentation by sonication) (Nadel et al. 2015). DRIPc-seq detected ~70,000 R-loop peaks in the human embryonic carcinoma Ntera2 cells, which is comparable to the number of R-loops detected by RDIP-seq (~64,000 and 39,000 in IMR-90 and HEK 293 T cells, respectively). Consistent with R-loops being co-transcriptional structures, DRIPc-seq peaks were found predominantly in RNA polymerase II-transcribed genes, with two- to three-fold enrichment at the promoter and terminator regions (Sanz et al. 2016). Interestingly, this co-transcriptional model of R-loop formation was contended by the RDIP-seq study, at least at the ribosomal DNA loci (Nadel et al. 2015). Moreover, using a more sensitive readout (than RNA-seq) for transcription, the global run-on sequencing (GRO-seq) method, it was found that only 47.7% of the RNA/DNA hybrids are associated with active transcription (Nadel et al. 2015). Finally, the RDIP-seq study reported a moderate depletion of R-loops at the terminators, in contrast to a twofold enrichment at the promoters (included in a 1.5 kb window downstream from the transcription start site) (Nadel et al. 2015). These studies highlighted the dynamic nature of R-loops and their potential cell

Table 21.3 Distribution of *RLFS*s in the genes located in 1973 *RFS*-associated and 13 cloned *CFS*-associated genes vs. all annotated genes in the genome

No. of RLFSs per gene	RFSs		CFSs		Whole genome	
	No. of genes	%	No. of genes	%	No. of genes	%
1	416	21.08	0		16,362	41.19
2	279	14.14	1	7.69	7,910	19.91
3	219	11.10	2	15.38	4,563	11.49
4	164	8.31	1	7.69	2,880	7.25
5	137	6.94	0		1,992	5.02
6–10	366	18.55	2	15.38	3,811	9.59
11–50	362	18.35	7	53.85	2,132	5.37
51–100	24	1.22	0		59	0.15
>100	6	0.31	0		11	0.03
Total	1973	100	13	100	39,720	100

Statistics from the whole genome were previously reported (Wongsurawat et al. 2012)

type-specific formation. They also necessitate an unbiased approach to identify R-loops in the human genome.

Sequences with high CG content are more conducive to R-loop formation due to high thermostability of the rG·C base pair (Roy and Lieber 2009; Sugimoto et al. 1995; Tracy et al. 2000). Training on previously discovered R-loops, a quantitative model of R-loop forming sequences (RLFSs) has been developed and used to predict RLFSs in the known transcribed regions of the human genome (Wongsurawat et al. 2012). Using the R-loop database (http://rloop.bii.a-star.edu.sg/), we tallied the number of R-loops in each gene of the RFSs as well as calculated their density (number of R-loops per kb in a given gene), as reported in Table 21.1. Our analysis shows that genes residing in RFSs have relatively higher concentration of R-loop forming sequences compared to all genes (Table 21.3). We also queried which RFSs and genes have the highest R-loop content (top 1%), either based on total number or density (Fig. 21.2). The analysis reveals that FRA22A is enriched for genes with the highest number of R-loops (more than 70 R-loops per gene) and that FRA19B is enriched for genes with the highest R-loop density (more than 2.5 R-loops per kb sequence). High R-loop content, together with the fact that many RFSs contain CGG repeat, is intriguing. Because the computational model for R-loop prediction defines an R-loop core sequence as clusters of 3–4 contiguous Gs interspersed by a single nucleotide, CGG trinucleotide repeats per se are not R-loop forming sequences. It would be interesting to test if these two entities have coevolved and that the higher than average R-loop content in the RFSs predisposes CGG repeats to break.

We performed similar analysis with a subset ($n = 11$) of CFSs, which have been cloned, and the associated genes identified (13 total, with FRAXC and FRA10G each housing two genes): FRA1B(*DAB1*), FRA2F(*LRP1B*), FRA3B(*FHIT*), FRA4F(*GRID2*), FRA6E(*PARK2*), FRA10D(*CTNNA3*), FRA10G(*RET*, *NCOA4*), FRA15A(*RORA*), FRA16D(*WWOX*), and FRAXC(*DMD*, *IL1RAPL1*). Seven of

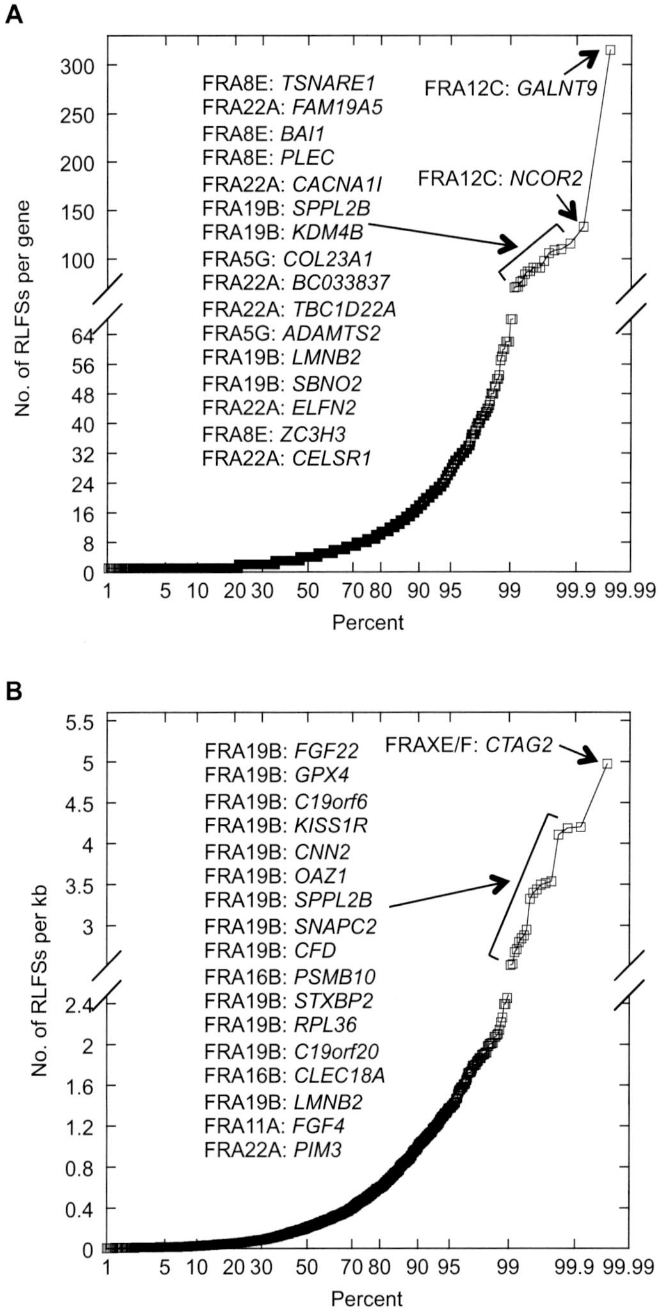

Fig. 21.2 Probability plots showing the distribution of 1973 genes within *RFSs* by the number of RLFSs per gene (**A**) or by the number of *RLFSs* per kb for a given gene (**B**). The top 1% of genes in each plot are as shown

these 13 genes (54%) contain 11–50 R-loops, which appeared significantly higher than the reported genomic average (5%) with the same level of R-loop content (Table 21.3). Thus, R-loop formation might also be associated with CFSs, consistent with previous findings that breakage at CFSs harboring large genes can be due to R-loop formation deterring transcription complexes (Helmrich et al. 2011; Wilson et al. 2015).

21.3.4 Mechanisms of Fragile Site Formation

Early studies observed that the fragile X chromosome showed delayed or incomplete replication due to the slowing down or stalling of replication forks (Hansen et al. 1993). This study presented an attractive model for fragile site development as approximately 1% and 20% of the genome in transformed and non-transformed human cells, respectively, undergoes replication even during mitosis (Widrow et al. 1998). Subsequent studies also confirmed late replication timing of a handful of CFSs (Handt et al. 2000; Hellman et al. 2000; Le Beau et al. 1998; Palakodeti et al. 2004; Pelliccia et al. 2008; Wang et al. 1999). However, not all fragile sites appear to replicate late—some are located in the interface of early- and late-replicating regions (El Achkar et al. 2005; Handt et al. 2000), and others are in fact associated with early replication timing (Barlow et al. 2013). Besides replication timing anomalies, CFSs can also be sites of collisions between the replication and the transcription machineries (Helmrich et al. 2011). Additionally, defective processing of CFSs in a Rad52-dependent DNA repair pathway has also been proposed to contribute to fragile site instability (Bhowmick et al. 2016; Sotiriou et al. 2016). However, the requirement for DNA repair is likely linked to a potential underlying defect in DNA replication. Therefore, there appears to be two major factors that underlie fragile site instability: defective DNA initiation/progression and replication-transcription conflicts (Le Tallec et al. 2014; Ozeri-Galai et al. 2014; Sarni and Kerem 2016). Here, we highlight two major areas with competing models, one regarding the replication defects at fragile sites and the other pertaining to the impact of transcriptional status of large genes on chromosome fragility.

While a strong correlation exists between DNA secondary structure formation and fork stalling in both micro- and minisatellites of RFSs, evidences for such direct implication for instability in CFSs are contradictory. Is chromosome fragility the consequence of replication fork stalling at AT-rich sequences in CFSs? Previous studies have shown that replication across CFSs is delayed compared to other regions of the genome, particularly in the presence of aphidicolin (Hellman et al. 2000; Le Beau et al. 1998; Palakodeti et al. 2004; Wang et al. 1999). However, recent studies employing DNA combing have countered that replication fork speed is not different between fragile sites (FRA3B and FRA6E) and non-fragile regions, with or without aphidicolin treatment (Letessier et al. 2011; Palumbo et al. 2010). It was further shown that fragility at FRA3B is the consequence of deficient origin activation within the fragile site when replication fork is slowed down by aphidicolin

(Letessier et al. 2011). In contrast, using a similar DNA combing approach, it was shown that replication forks indeed stall at AT-rich sequences in FRA16C (Ozeri-Galai et al. 2011). These contradictory results can be partly explained by significant differences in experimental conditions for replication fork rate measurement, such as the duration of pulse labeling by nucleoside analogs (in DNA combing) and aphidicolin dosage. Considering that fragile site expression level varies significantly with aphidicolin concentrations (Glover et al. 1984), direct comparison between fork rates measured under such different conditions is tenuous. The fact that FRA16C coincides with a RFS, FRA16B, further precludes direct comparison between these studies. Thus, future studies will undoubtedly benefit from standardized experimental conditions. However, both studies agree that paucity of origins in the fragile site region is at least partially responsible for the delayed replication completion of fragile sites (Letessier et al. 2011; Ozeri-Galai et al. 2011; Palumbo et al. 2010).

Another contentious topic is whether the transcriptional status of a large gene impacts fragile site formation. Helmrich et al. suggested that collision between replication and transcription at large genes causes chromosome fragility and, further, breakage frequency is correlated with gene expression level (Helmrich et al. 2012). The authors further posited that R-loop formation as a consequence of transcription along the fragile locus could pose a serious obstacle to the replication fork. However, Le Tallec et al. argued that the expression level of large genes does not correlate with chromosome fragility (Le Tallec et al. 2013). Again, this apparent discrepancy between the two studies is at least attributable to multiple differences in experimental conditions including aphidicolin concentration, definition of large genes, and calculation of break frequency. It further underscores the importance of applying standardized or comparable experimental conditions for effective comparisons between studies.

Finally, an altered epigenetic environment in the fragile loci could be another cause for fragile site instability. The inability to undergo condensation at the time of mitosis could result in DNA strand breakage. Using an in vitro nucleosome reconstitution assay, it was shown that CGG repeats with greater than 50 copies exclude nucleosome and that this exclusion is dependent on the length of the repeat (Wang and Griffith 1996; Wang et al. 1996). It was also shown that the expanded AT-rich minisatellites in FRA16B can exclude nucleosome assembly albeit only in the presence of distamycin A (Hsu and Wang 2002). Finally, it has been shown that hypoacetylation occurs in CFSs compared to genomic average, indicating a compact chromatin around CFSs (Koch et al. 2007; Savelyeva and Brueckner 2014).

21.3.5 Disease Associations

Fragile sites have been associated with both genetic and epigenetic instability (Smith et al. 2010). Currently there are 11 RFSs that have been molecularly mapped, i.e., the gene(s) that are impacted by fragile site formation have been identified.

Many of them are associated with a definitive human disease, predominantly a neurological disorder (Table 21.1). Interestingly, folate-sensitive RFSs are specifically associated with neuropsychiatric disorders including schizophrenia, autism spectrum disorders (ASDs), and mental retardation, while non-folate-sensitive RFSs are associated with a more diverse set of disorders such as infertility and Langer-Giedion syndrome. Schizophrenia, ASDs, and bipolar disorders are all complex neurological disorders that have intricate association with genetics and the environment (Kerner 2014; Miles 2011; Smith et al. 2010). No single gene can account for all the symptoms that characterize these diseases. Therefore, it stands to reason that the genes impacted by chromosome fragility at the folate-sensitive RFSs have a global impact on gene expression and/or protein production in the brain. Many of the CGG repeat-containing genes indeed have high expression in the brain (*AFF3*, *ZNF713*, *FAM10AC1*, *FMR1*, *FMR2*) (Uhlen et al. 2010). As the biochemical functions of these RFS-impacted genes become clear, it will help us understand the disease etiology in each of the associated diseases. For instance, the protein product of *FMR1*, FMRP, is an RNA-binding protein and is estimated to bind 4% of the mRNAs in the brain and regulate their translation (Santoro et al. 2012). Therefore, it remains a challenge to fully understand the genomic impact of FMRP deficiency in the fragile X syndrome.

Moreover, differential expression levels of the disease-associated genes can cause different diseases. For instance, in the case of the *FMR1* gene at FRAXA, a full mutation (>200 CGG repeats) induces *FMR1* gene silencing and results in the fragile X syndrome, while premutation alleles (55–200 CGG repeats) cause fragile X-associated ataxia/tremor syndrome and fragile X-associated primary ovarian insufficiency [note: premutation alleles do not cause chromosome fragility] (Galloway and Nelson 2009; Garcia-Arocena and Hagerman 2010; Santoro et al. 2012; Usdin et al. 2014). These observations suggest that impaired function of the RFS-associated genes can also impact organs other than the brain. The aforementioned brain-expressed genes associated with RFSs also show high expression in the reproductive organs (Uhlen et al. 2010). In addition, *FAM11A* gene expression is high in bone marrow, nervous system, and endocrine glands (Unger et al. 2013). Finally, there seems to be a dominance of neurological disorders, as opposed to cancer, associated with RFSs for unknown reasons. *CBL2* is the only CGG repeat-associated gene that is both seen in neurological disorders and cancer—while CGG repeat expansion and fragility at *CBL2* are linked to Jacobsen syndrome (Jones et al. 1995), *CBL2* has also been reported as a proto-oncogene associated with cancer breakpoints in several forms of leukemia and lymphomas (Fu et al. 2003).

In contrast, CFSs are clearly associated with cancer breakpoints and their expression linked to carcinogenesis (Arlt et al. 2006; Ma et al. 2012). Interestingly, some CFSs are also implicated in neurological disorders (Parkinson's, schizophrenia, and intellectual disability) and a variety of diseases associated with immunodeficiency, bone disorders, and infertility (Table 21.2). Twenty-eight of the 90 CFSs have genes associated with schizophrenia (Smith et al. 2010). Two CFSs, FRA7H and FRA7F, also showed high expression in cells from bipolar disorder patients compared to normal individuals when induced with folate-deficient media (Demirhan et al.

2009). These studies indicate that CFSs may trigger neurological symptoms more frequently than previously thought. The latter study also highlights the importance of understanding the cross-induction of fragile sites by chemicals and growth conditions.

Molecular cloning has enabled the identification of disease-associated gene(s) within a broadly defined fragile site. As shown above, R-loops are enriched in both RFSs and CFSs compared to genomic average. R-loops have been associated with neurological disorders such as the dominant juvenile form of amyotrophic lateral sclerosis type 4 (*ALS4*) and a recessive form of ataxia oculomotor apraxia type 2 (*AOA2*) (Chen et al. 2004; Moreira et al. 2004). A firm link between R-loop and oncogenesis was also established when it was shown that *BRCA2*, mutated in breast and ovarian cancer, is required to prevent R-loop accumulation and genome instability (Bhatia et al. 2014). We speculate that RLFSs can serve as a marker for disease-associated genes. Among the genes residing in RFSs, *GALNT9* contains the highest number (315) of R-loops (Fig. 21.2). It has been shown that *GALNT9* is frequently methylated and silenced in breast to brain metastasis (Pangeni et al. 2015) and its gene expression is a prognostic marker in neuroblastoma patients (Berois et al. 2013). Whether the high R-loop content plays a role in epigenetic regulation of *GALNT9* is an interesting question that warrants further investigation. Similarly, *CTAG2* shows the highest R-loop density (5 per kb sequence) among RFS-associated genes (Fig. 21.2). In CFSs, the *RET* gene at FRA10G has the highest number of R-loops and the highest R-loop density (1 every 1.6 kb, Fig. 21.2). *RET* rearrangements have been observed in several cancerous cell lines (Dillon et al. 2010). Undoubtedly, systematic analysis of R-loop distribution in all known CFSs would be vitally important.

21.4 Genome-Wide Mapping and Analysis of Fragile Sites

Many of the questions regarding population frequency and penetrance of fragile sites can be more effectively addressed by studies with increased scale and resolution. For instance, are there multiple breakage hot spots in a fragile site? How do the breakage spectra vary between cell types and inducing drugs? Mrasek et al. systematically identified aphidicolin-induced CFSs by screening 25,000 metaphase chromosome spreads isolated from lymphocytes of three normal and unrelated individuals (Mrasek et al. 2010). As alluded to above, this study demonstrated that the classically defined RFSs are less dependent on folate stress for induction than previously considered—all but 3 (FRA6A, FRA12D, and FRAXA) of the 30 RFSs were induced by aphidicolin, and only 4 were present in fewer than 3 individuals (total frequency among 3 individuals for each site listed in Table 21.1). However, based on our analysis of the data reported by Mrasek et al., the penetrance of RFS when induced by aphidicolin is significantly lower than that of CFSs, with the median levels being 0.075% and 0.464%, respectively, confirming the biased potency of aphidicolin toward CFSs.

Cytogenetic screens, while discernible and powerful, are not amenable for genome-wide and high-resolution identification of fragile sites. The advent of deep sequencing technology now permits fragile site mapping at a global scale and a faster pace. Currently the following methods have been applied to map DNA double-strand breaks (DSBs) in the human genome: BLESS (direct in situ break labeling, enrichment on streptavidin, and next-generation sequencing) (Crosetto et al. 2013), RAFT (rapid amplification of forum termini involving direct ligation of biotinylated oligonucleotides to DNA DSBs) (Tchurikov et al. 2015), and DSB-seq (using terminal deoxyribonucleotidyl transferase labeling of DSBs with biotinylated nucleotides followed by streptavidin pull-down and library construction) (Baranello et al. 2014). However, only one study applied conditions to induce CFSs, and it identified over 2429 aphidicolin-induced DSBs after correction for copy number variation in HeLa cells (Crosetto et al. 2013). The authors stated "many CFSs were scored as sensitive to aphidicolin following our approach." Based on our analysis, 190 (8%) and 574 (24%) of these 2429 DSBs overlap with the known rare and common fragile site regions (defined as cytoband coordinates from the UCSC Human Genome Database), respectively. The apparent lack of correlation for these aphidicolin-induced DSBs with RFSs is not unexpected. But the moderate level of overlap with CFSs begs discussion, and we attribute it to the following reasons. First, the moderate level of concordance is most likely the result of cell type (HeLa)-specific fragile site expression with fragile sites primarily defined in lymphocytes. Second, it also highlights the fundamental difference between methodologies. On one hand, cytologically defined chromosome breakage might include also single-stranded DNA breakage which would evade detection by BLESS. On the other hand, computationally predicted micro-fragile chromosomal regions might be missed by cytological screening (Thys et al. 2015). Finally, the usage of growth medium and concentrations of inducing drugs, e.g., aphidicolin, are not standardized across different studies and impinge on fragile site formation. For these reasons, it is at once a necessity and a challenge for future genomic mapping of fragile sites to compare different cell types with standardized inducing conditions.

21.5 Unsolved Mysteries of Chromosome Fragile Sites

So far, we have discussed some of the contentious topics in chromosome fragility in sections above. In this section we highlight three phenomena that still confound researchers and remain not understood. We also note that additional mysteries previously articulated in "Forgotten fragile sites and related phenomena" still remain unsolved (Sutherland and Baker 2003).

21.5.1 What Is the Underlying Cause for Tissue-Specific Fragile Site Formation?

The very first reported case of fragile site already noted cell type-dependent fragile site expression as discussed earlier in this document. Mounting evidence further confirmed this observation (Hosseini et al. 2013; Kuwano et al. 1990; Le Tallec et al. 2013; Letessier et al. 2011; Murano et al. 1989a, b). Since chromosome fragility is enhanced by replication fork instability, the varying locations and activation patterns of origins of replication across tissue types would directly impact fragile site expression. In addition, tissue-specific gene expression would also influence sites of replication-transcription conflict-induced chromosome fragility. Related to this latter point, we have recently postulated that the inducing agent for fragile site expression plays a dual role in simultaneously generating replication stress and untimely gene expression while replication is still incomplete, resulting in replication-transcription conflicts at distinct loci in different cells (Hoffman et al. 2015). This hypothesis was derived from the model organism *Saccharomyces cerevisiae*, and we are currently testing it in human cell lines. Future studies mapping fragile sites in different cell lines with simultaneous measurements of origin activities and gene expression levels will be ideal for understanding tissue-specific fragility.

21.5.2 Why Are RFSs Preferentially Induced by Folate Stress?

Perhaps one of the foremost interesting questions is how do different classes of drugs define fragile sites. For instance, folic acid deprivation, thymidylate synthase inhibition (e.g., fluorodeoxyuridine), and thymidine deprivation or excess can all induce a class of folate-sensitive RFSs. One of the outstanding features of the folate-sensitive fragile sites is that nearly half of them (10 out of 22) have been found to contain CGG repeats thus far. Does folate deficiency preferentially lead to DNA breaks at CGG repeats? Folic acid is crucial for methyl metabolism, which in turn impacts DNA replication and repair. Deprivation of folic acid blocks the methylation of dUMP to TMP and triggers an increase in dUTP level. Similarly, intracellular fluorodeoxyuridine is converted to fluorodeoxyuridine monophosphate, which in turn inhibits thymidylate synthase and also causes an increase of dUTP. Consequently, there is an increase in the incorporation of uracil into DNA. Uracil in DNA is removed by the uracil DNA glycosylase (UDG), which, through a facilitated diffusion mechanism, locates the damaged/modified DNA bases (Schonhoft et al. 2013). It has been proposed that folate-sensitive chromatid breakage is the result of a catastrophic DNA repair cycle where excision of uracil is followed by reincorporation of uracil due to continuous blockage of the dTTP pool (Reidy 1987).

The question then becomes is there a higher level of uracil incorporation occurring at CGG repeat regions than other chromosomal regions, or are CGG repeats more conducive to uracil excision? There seemed to be evidence supporting both arguments. A genome-wide study in *S. cerevisiae* has demonstrated that the uracil content is relatively lower in early-replicating regions than late-replicating ones (Bryan et al. 2014). RFSs such as FRAXA tend to be late-replicating (Hansen et al. 1993; Subramanian et al. 1996; Webb 1992) and therefore might incorporate uracil at a higher rate. Alternatively but not mutually exclusively, cytosine deamination to uracil at the CGG repeats might be an underlying cause for increased uracil content. On the other hand, incorporated uracil might be easier to recognize by the UDG enzyme in the context of secondary structures due to CGG repeats. Supporting this hypothesis is the observation that a related DNA glycosylase, the human alkyladenine DNA glycosylase, can capture the site of DNA damage more efficiently on a flexible DNA template containing kinks, bubbles, or gaps than on a continuous B-form DNA duplex (Hedglin et al. 2015). Conceivably, this characteristic can also extend to UDG acting on a DNA template enriched for CGG repeats which readily form hairpin structures.

21.5.3 What Is the Underlying Cause for Sex-Biased Transmission of Autosomal Fragile Sites?

Sherman and Sutherland reported in a population study that folate-sensitive and BrdU-sensitive autosomal fragile site expression was higher when the carrier parent was the mother than if it was the father (Sherman and Sutherland 1986). Similarly, Kähkönen et al. also observed that there was a maternal bias of the transmission of autosomal RFSs (16 out of 19 families were maternal carriers and 1 was a paternal carrier) (Kähkönen et al. 1989). This phenomenon was once again reported much later in a study demonstrating predominantly maternal transmission of ring chromosome 15 (Nikitina et al. 2003). What is the underlying cause for this sex-biased fragile site inheritance? Is the fragile site expression dependent on genomic imprinting of the maternal genes? Or is the maternal transmission of fragile sites a consequence of nonrandom chromosome arrangement in germ lines? Finally, is this phenomenon unique to folate stress? Large-scale population studies are required to confirm these findings. Further genetic and epigenetic characterizations of the autosomal RFSs will also shed new light on this genetic puzzle.

Acknowledgment We thank Andrew McCulley for assisting data analysis during this work. This work was supported by the National Institutes of Health (5R01 GM118799-01A1 to W.F.).

References

Amarose AP, Huttenlocher PR, Sprudzs RM, Laitsch TJ, Pettenati MJ (1987) A heritable fragile 12q24.13 segregating in a family with the fragile X chromosome. Hum Genet 75:4–6

Ambroziak W, Koziorowski D, Duszyc K, Gorka-Skoczylas P, Potulska-Chromik A, Slawek J, Hoffman-Zacharska D (2015) Genomic instability in the PARK2 locus is associated with Parkinson's disease. J Appl Genet 56:451–461

Arlt MF, Miller DE, Beer DG, Glover TW (2002) Molecular characterization of FRAXB and comparative common fragile site instability in cancer cells. Genes Chromosom Cancer 33:82–92

Arlt MF, Durkin SG, Ragland RL, Glover TW (2006) Common fragile sites as targets for chromosome rearrangements. DNA Repair 5:1126–1135

Baranello L, Kouzine F, Wojtowicz D, Cui K, Przytycka TM, Zhao K, Levens D (2014) DNA break mapping reveals topoisomerase II activity genome-wide. Int J Mol Sci 15:13111–13122

Barbi G, Steinbach P, Vogel W (1984) Nonrandom distribution of methotrexate-induced aberrations on human chromosomes. Detection of further folic acid sensitive fragile sites. Human Genetics 68:290–294

Barletta C, Ragusa RM, Garofalo G, Scillato F, Ruggeri M (1991) Segregation analysis of autosomal fragile sites in three families with the fragile X chromosome. Ann Genet 34:111–114

Barlow JH, Faryabi RB, Callen E, Wong N, Malhowski A, Chen HT, Gutierrez-Cruz G, Sun HW, McKinnon P, Wright G et al (2013) Identification of early replicating fragile sites that contribute to genome instability. Cell 152:620–632

Bell MV, Hirst MC, Nakahori Y, MacKinnon RN, Roche A, Flint TJ, Jacobs PA, Tommerup N, Tranebjaerg L, Froster-Iskenius U et al (1991) Physical mapping across the fragile X: hypermethylation and clinical expression of the fragile X syndrome. Cell 64:861–866

Bensaid M, Melko M, Bechara EG, Davidovic L, Berretta A, Catania MV, Gecz J, Lalli E, Bardoni B (2009) FRAXE-associated mental retardation protein (FMR2) is an RNA-binding protein with high affinity for G-quartet RNA forming structure. Nucleic Acids Res 37:1269–1279

Berois N, Gattolliat CH, Barrios E, Capandeguy L, Douc-Rasy S, Valteau-Couanet D, Benard J, Osinaga E (2013) GALNT9 gene expression is a prognostic marker in neuroblastoma patients. Clin Chem 59:225–233

Bettex M, Graf B, Winkler B, Gerber-Huber S (1998) Oro-palatal dysplasia Bettex-Graf–a new syndrome. Eur J Pediatr Surg 8:4–8

Bharadwaj R, Jiang Y, Mao W, Jakovcevski M, Dincer A, Krueger W, Garbett K, Whittle C, Tushir JS, Liu J et al (2013) Conserved chromosome 2q31 conformations are associated with transcriptional regulation of GAD1 GABA synthesis enzyme and altered in prefrontal cortex of subjects with schizophrenia. J Neurosci 33:11839–11851

Bhatia V, Barroso SI, Garcia-Rubio ML, Tumini E, Herrera-Moyano E, Aguilera A (2014) BRCA2 prevents R-loop accumulation and associates with TREX-2 mRNA export factor PCID2. Nature 511:362–365

Bhowmick R, Minocherhomji S, Hickson ID (2016) RAD52 facilitates mitotic DNA synthesis following replication stress. Mol Cell 64:1117–1126

Blumrich A, Zapatka M, Brueckner LM, Zheglo D, Schwab M, Savelyeva L (2011) The FRA2C common fragile site maps to the borders of MYCN amplicons in neuroblastoma and is associated with gross chromosomal rearrangements in different cancers. Hum Mol Genet 20:1488–1501

Bodelon C, Untereiner ME, Machiela MJ, Vinokurova S, Wentzensen N (2016) Genomic characterization of viral integration sites in HPV-related cancers. Int J Cancer 139:2001–2011

Bosco N, Pelliccia F, Rocchi A (2010) Characterization of FRA7B, a human common fragile site mapped at the 7p chromosome terminal region. Cancer Genet Cytogenet 202:47–52

Bozza M, Bernardini L, Novelli A, Brovedani P, Moretti E, Canapicchi R, Doccini V, Filippi T, Battaglia A (2013) 6p25 interstitial deletion in two dizygotic twins with gyral pattern anomaly and speech and language disorder. Eur J Paediatr Neurol 17:225–231

Bryan DS, Ransom M, Adane B, York K, Hesselberth JR (2014) High resolution mapping of modified DNA nucleobases using excision repair enzymes. Genome Res 24:1534–1542
Brys M, Migdalska-Sek M, Pastuszak-Lewandoska D, Forma E, Czarnecka K, Domanska D, Nawrot E, Wilkosz J, Rozanski W, Brzezianska E (2013) Diagnostic value of DNA alteration: loss of heterozygosity or allelic imbalance-promising for molecular staging of prostate cancers. Med Oncol 30:391
Burrow AA, Marullo A, Holder LR, Wang YH (2010) Secondary structure formation and DNA instability at fragile site FRA16B. Nucleic Acids Res 38:2865–2877
Callahan G, Denison SR, Phillips LA, Shridhar V, Smith DI (2003) Characterization of the common fragile site FRA9E and its potential role in ovarian cancer. Oncogene 22:590–601
Capello D, Scandurra M, Poretti G, Rancoita PM, Mian M, Gloghini A, Deambrogi C, Martini M, Rossi D, Greiner TC et al (2010) Genome wide DNA-profiling of HIV-related B-cell lymphomas. Br J Haematol 148:245–255
Chen CH, Shih HH, Wang-Wuu S, Tai JJ, Wuu KD (1998) Chromosomal fragile site expression in lymphocytes from patients with schizophrenia. Hum Genet 103:702–706
Chen YZ, Bennett CL, Huynh HM, Blair IP, Puls I, Irobi J, Dierick I, Abel A, Kennerson ML, Rabin BA et al (2004) DNA/RNA helicase gene mutations in a form of juvenile amyotrophic lateral sclerosis (ALS4). Am J Hum Genet 74:1128–1135
Choi J, Majima T (2011) Conformational changes of non-B DNA. Chem Soc Rev 40:5893–5909
Conley ME, Spinner NB, Emanuel BS, Nowell PC, Nichols WW (1986) A chromosomal breakage syndrome with profound immunodeficiency. Blood 67:1251–1256
Court Brown WM (1966) Chromosome studies on adults. Cambridge U.P, Galton Laboratory, University of London, Cambridge
Crosetto N, Mitra A, Silva MJ, Bienko M, Dojer N, Wang Q, Karaca E, Chiarle R, Skrzypczak M, Ginalski K et al (2013) Nucleotide-resolution DNA double-strand break mapping by next-generation sequencing. Nat Methods 10:361–365
D'Angelo CS, Moller Dos Santos MF, Alonso LG, Koiffmann CP (2015) Two new cases of 1p21.3 deletions and an unbalanced translocation t(8;12) among individuals with syndromic obesity. Mol Syndromology 6:63–70
Debacker K, Kooy RF (2007) Fragile sites and human disease. Hum Mol Genet 16 (Spec No. 2):R150–R158
Debacker K, Winnepenninckx B, Ben-Porat N, FitzPatrick D, Van Luijk R, Scheers S, Kerem B, Frank Kooy R (2007) FRA18C: a new aphidicolin-inducible fragile site on chromosome 18q22, possibly associated with in vivo chromosome breakage. J Med Genet 44:347–352
Dekaban A (1965) Persisting clone of cells with an abnormal chromosome in a woman previously irradiated. J Nucl Med 6:740–746
Demirhan O, Tastemir D, Sertdemir Y (2006) Chromosomal fragile sites in schizophrenic patients. Genetika 42:985–992
Demirhan O, Tastemir D, Sertdemir Y (2009) The expression of folate sensitive fragile sites in patients with bipolar disorder. Yonsei Med J 50:137–141
Dietrich A, Kioschis P, Monaco AP, Gross B, Korn B, Williams SV, Sheer D, Heitz D, Oberle I, Toniolo D et al (1991) Molecular cloning and analysis of the fragile X region in man. Nucleic Acids Res 19:2567–2572
Dillon LW, Burrow AA, Wang YH (2010) DNA instability at chromosomal fragile sites in cancer. Curr Genomics 11:326–337
Dillon LW, Pierce LC, Ng MC, Wang YH (2013) Role of DNA secondary structures in fragile site breakage along human chromosome 10. Hum Mol Genet 22:1443–1456
Dooley TP, Mitchison HM, Munroe PB, Probst P, Neal M, Siciliano MJ, Deng Z, Doggett NA, Callen DF, Gardiner RM et al (1994) Mapping of two phenol sulphotransferase genes, STP and STM, to 16p: candidate genes for Batten disease. Biochem Biophys Res Commun 205:482–489
Dunner JA, Martin AO, Traisman ES, Traisman HS, Elias S (1983) Enhancement of a fra(16)(q22) with distamycin A: a family ascertained through an abnormal proposita. Am J Med Genet 16:277–284

Durkin SG, Glover TW (2007) Chromosome fragile sites. Annu Rev Genet 41:169–192
El Achkar E, Gerbault-Seureau M, Muleris M, Dutrillaux B, Debatisse M (2005) Premature condensation induces breaks at the interface of early and late replicating chromosome bands bearing common fragile sites. Proc Natl Acad Sci U S A 102:18069–18074
Engelman JA, Zhang XL, Lisanti MP (1998) Genes encoding human caveolin-1 and -2 are co-localized to the D7S522 locus (7q31.1), a known fragile site (FRA7G) that is frequently deleted in human cancers. FEBS Lett 436:403–410
Fechter A, Buettel I, Kuehnel E, Savelyeva L, Schwab M (2007) Common fragile site FRA11G and rare fragile site FRA11B at 11q23.3 encompass distinct genomic regions. Genes Chromosom Cancer 46:98–106
Ferber MJ, Eilers P, Schuuring E, Fenton JA, Fleuren GJ, Kenter G, Szuhai K, Smith DI, Raap AK, Brink AA (2004) Positioning of cervical carcinoma and Burkitt lymphoma translocation breakpoints with respect to the human papillomavirus integration cluster in FRA8C at 8q24.13. Cancer Genet Cytogenet 154:1–9
Ferguson-Smith MA (1973) Inherited constriction fragility of chromosome 2. Ann Genet 16:29–34
Ferguson-Smith MA, Ferguson-Smith ME, Ellis PM, Dickson M (1962) The sites and relative frequencies of secondary constrictions in human somatic chromosomes. Cytogenetics 1:325–343
Fry M, Loeb LA (1994) The fragile X syndrome d(CGG)n nucleotide repeats form a stable tetrahelical structure. Proc Natl Acad Sci U S A 91:4950–4954
Fu JF, Hsu JJ, Tang TC, Shih LY (2003) Identification of CBL, a proto-oncogene at 11q23.3, as a novel MLL fusion partner in a patient with de novo acute myeloid leukemia. Genes Chromosom Cancer 37:214–219
Fungtammasan A, Walsh E, Chiaromonte F, Eckert KA, Makova KD (2012) A genome-wide analysis of common fragile sites: what features determine chromosomal instability in the human genome? Genome Res 22:993–1005
Galloway JN, Nelson DL (2009) Evidence for RNA-mediated toxicity in the fragile X-associated tremor/ataxia syndrome. Futur Neurol 4:785
Gandhi M, Dillon LW, Pramanik S, Nikiforov YE, Wang YH (2010) DNA breaks at fragile sites generate oncogenic RET/PTC rearrangements in human thyroid cells. Oncogene 29:2272–2280
Gao G, Kasperbauer JL, Tombers NM, Wang V, Mayer K, Smith DI (2014) A selected group of large common fragile site genes have decreased expression in oropharyngeal squamous cell carcinomas. Genes Chromosom Cancer 53:392–401
Garcia-Arocena D, Hagerman PJ (2010) Advances in understanding the molecular basis of FXTAS. Hum Mol Genet 19:R83–R89
Ginno PA, Lim YW, Lott PL, Korf I, Chedin F (2013) GC skew at the 5′ and 3′ ends of human genes links R-loop formation to epigenetic regulation and transcription termination. Genome Res 23:1590–1600
Giraud F, Ayme S, Mattei JF, Mattei MG (1976) Constitutional chromosomal breakage. Hum Genet 34:125–136
Glasser L, Meloni-Ehrig A, Joseph P, Mendiola J (2006) Benign chronic neutropenia with abnormalities involving 16q22, affecting mother and daughter. Am J Hematol 81:262–270
Glover TW, Berger C, Coyle J, Echo B (1984) DNA polymerase alpha inhibition by aphidicolin induces gaps and breaks at common fragile sites in human chromosomes. Hum Genet 67:136–142
Groh M, Gromak N (2014) Out of balance: R-loops in human disease. PLoS Genet 10:e1004630
Groh M, Lufino MM, Wade-Martins R, Gromak N (2014) R-loops associated with triplet repeat expansions promote gene silencing in Friedreich ataxia and fragile X syndrome. PLoS Genet 10:e1004318
Handt O, Sutherland GR, Richards RI (2000) Fragile sites and minisatellite repeat instability. Mol Genet Metab 70:99–105
Hansen RS, Gartler SM, Scott CR, Chen SH, Laird CD (1992) Methylation analysis of CGG sites in the CpG island of the human FMR1 gene. Hum Mol Genet 1:571–578
Hansen RS, Canfield TK, Lamb MM, Gartler SM, Laird CD (1993) Association of fragile X syndrome with delayed replication of the FMR1 gene. Cell 73:1403–1409

Harvey J, Judge C, Wiener S (1977) Familial X-linked mental retardation with an X chromosome abnormality. J Med Genet 14:46–50

Hatchi E, Skourti-Stathaki K, Ventz S, Pinello L, Yen A, Kamieniarz-Gdula K, Dimitrov S, Pathania S, McKinney KM, Eaton ML et al (2015) BRCA1 recruitment to transcriptional pause sites is required for R-loop-driven DNA damage repair. Mol Cell 57:636–647

Hecht F (1988) Rare and polymorphic fragile sites and cancer. Cancer Genet Cytogenet 34:195–199

Hedglin M, Zhang Y, O'Brien PJ (2015) Probing the DNA structural requirements for facilitated diffusion. Biochemistry 54:557–566

Heitz D, Rousseau F, Devys D, Saccone S, Abderrahim H, Le Paslier D, Cohen D, Vincent A, Toniolo D, Della Valle G et al (1991) Isolation of sequences that span the fragile X and identification of a fragile X-related CpG island. Science 251:1236–1239

Hellman A, Rahat A, Scherer SW, Darvasi A, Tsui LC, Kerem B (2000) Replication delay along FRA7H, a common fragile site on human chromosome 7, leads to chromosomal instability. Mol Cell Biol 20:4420–4427

Helmrich A, Ballarino M, Tora L (2011) Collisions between replication and transcription complexes cause common fragile site instability at the longest human genes. Mol Cell 44:966–977

Helmrich U, Marsano A, Melly L, Wolff T, Christ L, Heberer M, Scherberich A, Martin I, Banfi A (2012) Generation of human adult mesenchymal stromal/stem cells expressing defined xenogenic vascular endothelial growth factor levels by optimized transduction and flow cytometry purification. Tissue Eng Part C Methods 18:283–292

Hill A, Harada Y, Takahashi E, Hou J, Wagner MJ, Wells DE (1997) Assignment of fragile site 8E (FRA8E) to human chromosome band 8q24.11 adjacent to the hereditary multiple exostoses 1 gene and two overlapping Langer-Giedion syndrome deletion endpoints. Cytogenet Cell Genet 78:56–57

Hills LB, Masri A, Konno K, Kakegawa W, Lam AT, Lim-Melia E, Chandy N, Hill RS, Partlow JN, Al-Saffar M et al (2013) Deletions in GRID2 lead to a recessive syndrome of cerebellar ataxia and tonic upgaze in humans. Neurology 81:1378–1386

Hoffman EA, McCulley A, Haarer B, Arnak R, Feng W (2015) Break-seq reveals hydroxyurea-induced chromosome fragility as a result of unscheduled conflict between DNA replication and transcription. Genome Res 25:402–412

Holden JJ, Walker M, Chalifoux M, White BN (1996) Trinucleotide repeats at the FRAXF locus: frequency and distribution in the general population. Am J Med Genet 64:424–427

Hormozian F, Schmitt JG, Sagulenko E, Schwab M, Savelyeva L (2007) FRA1E common fragile site breaks map within a 370kilobase pair region and disrupt the dihydropyrimidine dehydrogenase gene (DPYD). Cancer Lett 246:82–91

Hosseini SA, Horton S, Saldivar JC, Miuma S, Stampfer MR, Heerema NA, Huebner K (2013) Common chromosome fragile sites in human and murine epithelial cells and FHIT/FRA3B loss-induced global genome instability. Genes Chromosom Cancer 52:1017–1029

Hou S, Du L, Lei B, Pang CP, Zhang M, Zhuang W, Zhang M, Huang L, Gong B, Wang M et al (2014) Genome-wide association analysis of Vogt-Koyanagi-Harada syndrome identifies two new susceptibility loci at 1p31.2 and 10q21.3. Nat Genet 46:1007–1011

Howell RT, McDermott A, Evans JL (1990) A new apparently folate sensitive fragile site, 5q35. J Med Genet 27:527–528

Hsu TC (1963) Longitudinal differentiation of chromosomes and the possibility of interstitial telomeres. Exp Cell Res 24(SUPPL9):73–85

Hsu YY, Wang YH (2002) Human fragile site FRA16B DNA excludes nucleosomes in the presence of distamycin. J Biol Chem 277:17315–17319

Huertas P, Aguilera A (2003) Cotranscriptionally formed DNA:RNA hybrids mediate transcription elongation impairment and transcription-associated recombination. Mol Cell 12:711–721

Ijdo JW, Wells RA, Baldini A, Reeders ST (1991) Improved telomere detection using a telomere repeat probe (TTAGGG)n generated by PCR. Nucleic Acids Res 19:4780

Janiszewska-Olszowska J, Gawrych E, Dydyk A, Studniak E, Biadun-Poplawska A, Zajaczek S (2013) Oro-palatal dysplasia Bettex-Graf – clinical findings, genetic background, treatment. J Craniomaxillofac Surg 41:e29–e32

Jones C, Slijepcevic P, Marsh S, Baker E, Langdon WY, Richards RI, Tunnacliffe A (1994) Physical linkage of the fragile site FRA11B and a Jacobsen syndrome chromosome deletion breakpoint in 11q23.3. Hum Mol Genet 3:2123–2130
Jones C, Penny L, Mattina T, Yu S, Baker E, Voullaire L, Langdon WY, Sutherland GR, Richards RI, Tunnacliffe A (1995) Association of a chromosome deletion syndrome with a fragile site within the proto-oncogene CBL2. Nature 376:145–149
Kahkonen M, Leisti J, Thoden CJ, Autio S (1986) Frequency of rare fragile sites among mentally subnormal schoolchildren. Clin Genet 30:234–238
Kähkönen M, Tengstrom C, Alitalo T, Matilainen R, Kaski M, Airaksinen E (1989) Population cytogenetics of folate-sensitive fragile sites. II. Autosomal rare fragile sites. Hum Genet 82:3–8
Kang S, Ohshima K, Shimizu M, Amirhaeri S, Wells RD (1995) Pausing of DNA synthesis in vitro at specific loci in CTG and CGG triplet repeats from human hereditary disease genes. J Biol Chem 270:27014–27021
Kao-Shan CS, Fine RL, Whang-Peng J, Lee EC, Chabner BA (1987) Increased fragile sites and sister chromatid exchanges in bone marrow and peripheral blood of young cigarette smokers. Cancer Res 47:6278–6282
Karayianni E, Magnanini C, Orphanos V, Negrini M, Maniatis GM, Spathas DH, Barbanti-Brodano G, Morelli C (1999) Transcriptional map of chromosome region 6q16 – >q21. Cytogenet Cell Genet 86:263–266
Karkera JD, Ayache S, Ransome RJ Jr, Jackson MA, Elsayem AF, Sridhar R, Detera-Wadleigh SD, Wadleigh RG (2000) Refinement of regions with allelic loss on chromosome 18p11.2 and 18q12.2 in esophageal squamous cell carcinoma. Clin Cancer Res 6:3565–3569
Kasi Loknath Kumar A, Weckbaugh B, Sirridge C, Woodroof J, Persons D, Kambhampati S (2016) Myelodysplastic syndrome with concomitant t(5;21)(q15;q22) and del(5)(q13q33): case report and review of literature. Stem Cell Inv 3:3
Kerbeshian J, Severud R, Burd L, Larson L (2000) Peek-a-boo fragile site at 16d associated with Tourette syndrome, bipolar disorder, autistic disorder, and mental retardation. Am J Med Genet 96:69–73
Kerner B (2014) Genetics of bipolar disorder. Appl Clin Genet 7:33–42
Koch CM, Andrews RM, Flicek P, Dillon SC, Karaoz U, Clelland GK, Wilcox S, Beare DM, Fowler JC, Couttet P et al (2007) The landscape of histone modifications across 1% of the human genome in five human cell lines. Genome Res 17:691–707
Kremer EJ, Pritchard M, Lynch M, Yu S, Holman K, Baker E, Warren ST, Schlessinger D, Sutherland GR, Richards RI (1991) Mapping of DNA instability at the fragile X to a trinucleotide repeat sequence p(CCG)n. Science 252:1711–1714
Kuwano A, Murano I, Kajii T (1990) Cell type-dependent difference in the distribution and frequency of excess thymidine-induced common fragile sites: T lymphocytes and skin fibroblasts. Hum Genet 84:527–531
Lai LA, Kostadinov R, Barrett MT, Peiffer DA, Pokholok D, Odze R, Sanchez CA, Maley CC, Reid BJ, Gunderson KL et al (2010) Deletion at fragile sites is a common and early event in Barrett's esophagus. Mol Cancer Res 8:1084–1094
Le Beau MM, Rassool FV, Neilly ME, Espinosa R 3rd, Glover TW, Smith DI, McKeithan TW (1998) Replication of a common fragile site, FRA3B, occurs late in S phase and is delayed further upon induction: implications for the mechanism of fragile site induction. Hum Mol Genet 7:755–761
Le Tallec B, Millot GA, Blin ME, Brison O, Dutrillaux B, Debatisse M (2013) Common fragile site profiling in epithelial and erythroid cells reveals that most recurrent cancer deletions lie in fragile sites hosting large genes. Cell Rep 4:420–428
Le Tallec B, Koundrioukoff S, Wilhelm T, Letessier A, Brison O, Debatisse M (2014) Updating the mechanisms of common fragile site instability: how to reconcile the different views? Cell Mol Life Sci 71:4489–4494
Lee DY, Clayton DA (1996) Properties of a primer RNA-DNA hybrid at the mouse mitochondrial DNA leading-strand origin of replication. J Biol Chem 271:24262–24269

Lejeune J, Dutrillaux B, Lafourcade J, Berger R, Abonyi D, Rethore MO (1968) Selective endoreduplication of the long arm of the 2 chromosome in a woman and her daughter. C R Acad Sci Hebd Seances Acad Sci D 266:24–26
Letessier A, Garrido-Urbani S, Ginestier C, Fournier G, Esterni B, Monville F, Adelaide J, Geneix J, Xerri L, Dubreuil P et al (2007) Correlated break at PARK2/FRA6E and loss of AF-6/Afadin protein expression are associated with poor outcome in breast cancer. Oncogene 26:298–307
Letessier A, Millot GA, Koundrioukoff S, Lachages AM, Vogt N, Hansen RS, Malfoy B, Brison O, Debatisse M (2011) Cell-type-specific replication initiation programs set fragility of the FRA3B fragile site. Nature 470:120–123
Li X, Manley JL (2005) Inactivation of the SR protein splicing factor ASF/SF2 results in genomic instability. Cell 122:365–378
Li J, Liu J, Ren Y, Yang J, Liu P (2014) Common chromosomal fragile site gene WWOX in metabolic disorders and tumors. Int J Biol Sci 10:142–148
Li QS, Parrado AR, Samtani MN, Narayan VA, Alzheimer's Disease Neuroimaging, I (2015) Variations in the FRA10AC1 fragile site and 15q21 are associated with cerebrospinal fluid Abeta1-42 level. PLoS One 10:e0134000
Ligsay A, Hagerman RJ (2016) Review of targeted treatments in fragile X syndrome. Intractable Rare Dis Res 5:158–167
Limon J, Dal Cin P, Sait SN, Karakousis C, Sandberg AA (1988) Chromosome changes in metastatic human melanoma. Cancer Genet Cytogenet 30:201–211
Lin YJ, Liu X, Chang JS, Chien WK, Chen JH, Tsang H, Hung CH, Lin TH, Huang SM, Liao CC et al (2014) Coronary artery aneurysms occurrence risk analysis between Kawasaki disease and LRP1B gene in Taiwanese children. Biomedicine 4:10
Lindsay S, Splitt M, Edney S, Berney TP, Knight SJ, Davies KE, O'Brien O, Gale M, Burn J (1996) PPM-X: a new X-linked mental retardation syndrome with psychosis, pyramidal signs, and macroorchidism maps to Xq28. Am J Hum Genet 58:1120–1126
Lipska BS, Koczkowska M, Wierzba J, Ploszynska A, Iliszko M, Izycka-Swieszewska E, Adamkiewicz-Drozynska E, Limon J (2013) On the significance of germline cytogenetic rearrangements at MYCN locus in neuroblastoma. Mol Cytogenet 6:43
Liu N, Yan J, Chen X, Song J, Wang B, Yao Y (2014) Prenatal diagnosis of a de novo interstitial deletion of 11q (11q22.3 –> q23.3) associated with abnormal ultrasound findings by array comparative genomic hybridization. Mol Cytogenet 7:62
Lombrana R, Almeida R, Alvarez A, Gomez M (2015) R-loops and initiation of DNA replication in human cells: a missing link? Front Genet 6:158
Lozano R, Azarang A, Wilaisakditipakorn T, Hagerman RJ (2016) Fragile X syndrome: a review of clinical management. Intractable Rare Dis Res 5:145–157
Lubs HA (1969) A marker X chromosome. Am J Hum Genet 21:231–244
Lukusa T, Fryns JP (2008) Human chromosome fragility. Biochim Biophys Acta 1779:3–16
Ma K, Qiu L, Mrasek K, Zhang J, Liehr T, Quintana LG, Li Z (2012) Common fragile sites: genomic hotspots of DNA damage and carcinogenesis. Int J Mol Sci 13:11974–11999
Madireddy A, Kosiyatrakul ST, Boisvert RA, Herrera-Moyano E, Garcia-Rubio ML, Gerhardt J, Vuono EA, Owen N, Yan Z, Olson S et al (2016) FANCD2 facilitates replication through common fragile sites. Mol Cell 64:388–404
Magenis RE, Hecht F, Lovrien EW (1970) Heritable fragile site on chromosome 16: probable localization of haptoglobin locus in man. Science 170:85–87
Maltby EL, Higgins S (1987) Folate sensitive site at 10q23 and its expression as a deletion. J Med Genet 24:299
Martorell MR, Martinez-Pasarell O, Lopez O, Polo A, Sandalinas M, Garcia-Guixe E, Bassas L (2014) Chromosome 16 abnormalities in embryos and in sperm from a male with a fragile site at 16q22.1. Cytogenet Genome Res 142:134–139
McAvoy S, Zhu Y, Perez DS, James CD, Smith DI (2008) Disabled-1 is a large common fragile site gene, inactivated in multiple cancers. Genes Chromosom Cancer 47:165–174

McKenzie F, Turner A, Withers S, Dalzell P, McGlynn M, Kirk EP (2002) Dominant inheritance of cleft palate, microstomia and micrognathia–possible linkage to the fragile site at 16q22 (FRA16B). Clin Dysmorphol 11:237–241
Metsu S, Rooms L, Rainger J, Taylor MS, Bengani H, Wilson DI, Chilamakuri CS, Morrison H, Vandeweyer G, Reyniers E et al (2014a) FRA2A is a CGG repeat expansion associated with silencing of AFF3. PLoS Genet 10:e1004242
Metsu S, Rainger JK, Debacker K, Bernhard B, Rooms L, Grafodatskaya D, Weksberg R, Fombonne E, Taylor MS, Scherer SW et al (2014b) A CGG-repeat expansion mutation in ZNF713 causes FRA7A: association with autistic spectrum disorder in two families. Hum Mutat 35:1295–1300
Michaelis RC, Velagaleti GV, Jones C, Pivnick EK, Phelan MC, Boyd E, Tarleton J, Wilroy RS, Tunnacliffe A, Tharapel AT (1998) Most Jacobsen syndrome deletion breakpoints occur distal to FRA11B. Am J Med Genet 76:222–228
Miles JH (2011) Autism spectrum disorders–a genetics review. Genet Med 13:278–294
Mirkin SM (2006) DNA structures, repeat expansions and human hereditary disorders. Curr Opin Struct Biol 16:351–358
Mishmar D, Rahat A, Scherer SW, Nyakatura G, Hinzmann B, Kohwi Y, Mandel-Gutfroind Y, Lee JR, Drescher B, Sas DE et al (1998) Molecular characterization of a common fragile site (FRA7H) on human chromosome 7 by the cloning of a simian virus 40 integration site. Proc Natl Acad Sci U S A 95:8141–8146
Moreira MC, Klur S, Watanabe M, Nemeth AH, Le Ber I, Moniz JC, Tranchant C, Aubourg P, Tazir M, Schols L et al (2004) Senataxin, the ortholog of a yeast RNA helicase, is mutant in ataxia-ocular apraxia 2. Nat Genet 36:225–227
Morelli C, Karayianni E, Magnanini C, Mungall AJ, Thorland E, Negrini M, Smith DI, Barbanti-Brodano G (2002) Cloning and characterization of the common fragile site FRA6F harboring a replicative senescence gene and frequently deleted in human tumors. Oncogene 21:7266–7276
Moriarty HT, Webster LR (2003) Fragile sites and bladder cancer. Cancer Genet Cytogenet 140:89–98
Mrasek K, Schoder C, Teichmann AC, Behr K, Franze B, Wilhelm K, Blaurock N, Claussen U, Liehr T, Weise A (2010) Global screening and extended nomenclature for 230 aphidicolin-inducible fragile sites, including 61 yet unreported ones. Int J Oncol 36:929–940
Mulatinho MV, de Carvalho Serao CL, Scalco F, Hardekopf D, Pekova S, Mrasek K, Liehr T, Weise A, Rao N, Llerena JC Jr (2012) Severe intellectual disability, omphalocele, hypospadia and high blood pressure associated to a deletion at 2q22.1q22.3: case report. Mol Cytogenet 5:30
Murano I, Kuwano A, Kajii T (1989a) Fibroblast-specific common fragile sites induced by aphidicolin. Hum Genet 83:45–48
Murano I, Kuwano A, Kajii T (1989b) Cell type-dependent difference in the distribution and frequency of aphidicolin-induced fragile sites: T and B lymphocytes and bone marrow cells. Hum Genet 84:71–74
Nadel J, Athanasiadou R, Lemetre C, Wijetunga NA, O Broin P, Sato H, Zhang Z, Jeddeloh J, Montagna C, Golden A et al (2015) RNA:DNA hybrids in the human genome have distinctive nucleotide characteristics, chromatin composition, and transcriptional relationships. Epigenetics Chromatin 8:46
Nikitina NV, Bushueva OA, Nikolaeva EB, Pavlov GV, Lurie IW (2003) Maternal transmission of a ring chromosome 15. Genet Couns 14:181–186
Oei L, Hsu YH, Styrkarsdottir U, Eussen BH, de Klein A, Peters MJ, Halldorsson B, Liu CT, Alonso N, Kaptoge SK et al (2014) A genome-wide copy number association study of osteoporotic fractures points to the 6p25.1 locus. J Med Genet 51:122–131
Olavesen MG, Davies AF, Broxholme SJ, Wixon JL, Senger G, Nizetic D, Campbell RD, Ragoussis J (1995) An integrated map of human chromosome 6p23. Genome Res 5:342–358
Ozeri-Galai E, Lebofsky R, Rahat A, Bester AC, Bensimon A, Kerem B (2011) Failure of origin activation in response to fork stalling leads to chromosomal instability at fragile sites. Mol Cell 43:122–131

Ozeri-Galai E, Tur-Sinai M, Bester AC, Kerem B (2014) Interplay between genetic and epigenetic factors governs common fragile site instability in cancer. Cell Mol Life Sci 71:4495–4506
Palakodeti A, Han Y, Jiang Y, Le Beau MM (2004) The role of late/slow replication of the FRA16D in common fragile site induction. Genes Chromosom Cancer 39:71–76
Palumbo E, Matricardi L, Tosoni E, Bensimon A, Russo A (2010) Replication dynamics at common fragile site FRA6E. Chromosoma 119:575–587
Pangeni RP, Channathodiyil P, Huen DS, Eagles LW, Johal BK, Pasha D, Hadjistephanou N, Nevell O, Davies CL, Adewumi AI et al (2015) The GALNT9, BNC1 and CCDC8 genes are frequently epigenetically dysregulated in breast tumours that metastasise to the brain. Clin Epigenetics 7:57
Pelliccia F, Bosco N, Curatolo A, Rocchi A (2008) Replication timing of two human common fragile sites: FRA1H and FRA2G. Cytogenet Genome Res 121:196–200
Pelliccia F, Bosco N, Rocchi A (2010) Breakages at common fragile sites set boundaries of amplified regions in two leukemia cell lines K562 – molecular characterization of FRA2H and localization of a new CFS FRA2S. Cancer Lett 299:37–44
Pieretti M, Zhang FP, Fu YH, Warren ST, Oostra BA, Caskey CT, Nelson DL (1991) Absence of expression of the FMR-1 gene in fragile X syndrome. Cell 66:817–822
Plaja A, Castells N, Cueto-Gonzalez AM, del Campo M, Vendrell T, Lloveras E, Izquierdo L, Borregan M, Rodriguez-Santiago B, Carrio A et al (2015) A novel recurrent breakpoint responsible for rearrangements in the Williams-Beuren region. Cytogenet Genome Res 146:181–186
Rader JS, Kamarasova T, Huettner PC, Li L, Li Y, Gerhard DS (1996) Allelotyping of all chromosomal arms in invasive cervical cancer. Oncogene 13:2737–2741
Reidy JA (1987) Folate- and deoxyuridine-sensitive chromatid breakage may result from DNA repair during G2. Mutat Res 192:217–219
Reshmi SC, Huang X, Schoppy DW, Black RC, Saunders WS, Smith DI, Gollin SM (2007) Relationship between FRA11F and 11q13 gene amplification in oral cancer. Genes Chromosom Cancer 46:143–154
Rinaldi A, Capello D, Scandurra M, Greiner TC, Chan WC, Bhagat G, Rossi D, Morra E, Paulli M, Rambaldi A et al (2010) Single nucleotide polymorphism-arrays provide new insights in the pathogenesis of post-transplant diffuse large B-cell lymphoma. Br J Haematol 149:569–577
Rodenas-Cuadrado P, Ho J, Vernes SC (2014) Shining a light on CNTNAP2: complex functions to complex disorders. Eur J Hum Genet 22:171–178
Roy D, Lieber MR (2009) G clustering is important for the initiation of transcription-induced R-loops in vitro, whereas high G density without clustering is sufficient thereafter. Mol Cell Biol 29:3124–3133
Saldivar JC, Shibata H, Huebner K (2010) Pathology and biology associated with the fragile FHIT gene and gene product. J Cell Biochem 109:858–865
Samadashwily GM, Raca G, Mirkin SM (1997) Trinucleotide repeats affect DNA replication in vivo. Nat Genet 17:298–304
Santoro MR, Bray SM, Warren ST (2012) Molecular mechanisms of fragile X syndrome: a twenty-year perspective. Annu Rev Pathol 7:219–245
Santos-Pereira JM, Aguilera A (2015) R loops: new modulators of genome dynamics and function. Nat Rev Genet 16:583–597
Sanz LA, Hartono SR, Lim YW, Steyaert S, Rajpurkar A, Ginno PA, Xu X, Chedin F (2016) Prevalent, dynamic, and conserved R-loop structures associate with specific epigenomic signatures in mammals. Mol Cell 63:167–178
Sarni D, Kerem B (2016) The complex nature of fragile site plasticity and its importance in cancer. Curr Opin Cell Biol 40:131–136
Savelyeva L, Brueckner LM (2014) Molecular characterization of common fragile sites as a strategy to discover cancer susceptibility genes. Cell Mol Life Sci 71:4561–4575
Savelyeva L, Sagulenko E, Schmitt JG, Schwab M (2006a) Low-frequency common fragile sites: link to neuropsychiatric disorders? Cancer Lett 232:58–69
Savelyeva L, Sagulenko E, Schmitt JG, Schwab M (2006b) The neurobeachin gene spans the common fragile site FRA13A. Hum Genet 118:551–558

Schmid M, Feichtinger W, Jessberger A, Kohler J, Lange R (1986) The fragile site (16) (q22). I. Induction by AT-specific DNA-ligands and population frequency. Hum Genet 74:67–73

Schneider B, Nagel S, Kaufmann M, Winkelmann S, Bode J, Drexler HG, MacLeod RA (2008) T(3;7)(q27;q32) fuses BCL6 to a non-coding region at FRA7H near miR-29. Leukemia 22:1262–1266

Schonhoft JD, Kosowicz JG, Stivers JT (2013) DNA translocation by human uracil DNA glycosylase: role of DNA phosphate charge. Biochemistry 52:2526–2535

Schwartz M, Zlotorynski E, Kerem B (2006) The molecular basis of common and rare fragile sites. Cancer Lett 232:13–26

Seki N, Sumiya H, Shimazaki J, Toyama Y, Takahashi E, Murata M, Hori T (1992) Chromosome abnormalities and rare fragile sites detected in azoospermia patients. Jpn J Hum Genet 37:215–222

Shabtai F, Klar D, Halbrecht I (1982) Chromosome 17 has a real fragile site at p12. Hum Genet 61:177–179

Shang Z, Lv H, Zhang M, Duan L, Wang S, Li J, Liu G, Ruijie Z, Jiang Y (2015) Genome-wide haplotype association study identify TNFRSF1A, CASP7, LRP1B, CDH1 and TG genes associated with Alzheimer's disease in Caribbean Hispanic individuals. Oncotarget 6:42504–42514

Sherman SL, Sutherland GR (1986) Segregation analysis of rare autosomal fragile sites. Hum Genet 72:123–128

Shiraishi Y, Li MJ (1993) Uncorrected SCE levels of Bloom syndrome cells by cell hybridization with malignant cells with 14q32 structural abnormalities. Cancer Genet Cytogenet 69:45–50

Simmers RN, Stupans I, Sutherland GR (1986) Localization of the human haptoglobin genes distal to the fragile site at 16q22 using in situ hybridization. Cytogenet Cell Genet 41:38–41

Skourti-Stathaki K, Proudfoot NJ, Gromak N (2011) Human senataxin resolves RNA/DNA hybrids formed at transcriptional pause sites to promote Xrn2-dependent termination. Mol Cell 42:794–805

Smith CL, Bolton A, Nguyen G (2010) Genomic and epigenomic instability, fragile sites, schizophrenia and autism. Curr Genomics 11:447–469

Sordet O, Redon CE, Guirouilh-Barbat J, Smith S, Solier S, Douarre C, Conti C, Nakamura AJ, Das BB, Nicolas E et al (2009) Ataxia telangiectasia mutated activation by transcription- and topoisomerase I-induced DNA double-strand breaks. EMBO Rep 10:887–893

Sotiriou SK, Kamileri I, Lugli N, Evangelou K, Da-Re C, Huber F, Padayachy L, Tardy S, Nicati NL, Barriot S et al (2016) Mammalian RAD52 functions in break-induced replication repair of collapsed DNA replication forks. Mol Cell 64:1127–1134

Subramanian PS, Nelson DL, Chinault AC (1996) Large domains of apparent delayed replication timing associated with triplet repeat expansion at FRAXA and FRAXE. Am J Hum Genet 59:407–416

Sugimoto N, Nakano S, Katoh M, Matsumura A, Nakamuta H, Ohmichi T, Yoneyama M, Sasaki M (1995) Thermodynamic parameters to predict stability of RNA/DNA hybrid duplexes. Biochemistry 34:11211–11216

Sutherland GR (1977) Fragile sites on human chromosomes: demonstration of their dependence on the type of tissue culture medium. Science 197:265–266

Sutherland GR (1979) Heritable fragile sites on human chromosomes I. Factors affecting expression in lymphocyte culture. Am J Hum Genet 31:125–135

Sutherland GR (1981) Heritable fragile sites on human chromosomes. VII. Children homozygous for the BrdU-requiring fra(10)(q25) are phenotypically normal. Am J Hum Genet 33:946–949

Sutherland GR (1991) Chromosomal fragile sites. Genet Anal Tech Appl 8:161–166

Sutherland GR, Baker E (1993) Unusual behaviour of a human autosome having two rare folate sensitive fragile sites. Ann Genet 36:159–162

Sutherland GR, Baker E (2003) Forgotten fragile sites and related phenomena. Cytogenet Genome Res 100:89–91

Tassano E, Tavella E, Micalizzi C, Panarello C, Morerio C (2010) Refractory cytopenia of childhood with monosomy 7 presenting as isolated neutropenia in a patient with fragile site at 16q22. Cancer Genet Cytogenet 201:70–71

Tatarelli C, Linnenbach A, Mimori K, Croce CM (2000) Characterization of the human TESTIN gene localized in the FRA7G region at 7q31.2. Genomics 68:1–12

Tchurikov NA, Kretova OV, Fedoseeva DM, Chechetkin VR, Gorbacheva MA, Snezhkina AV, Alembekov IR, Kravatskaya GI, Kravatsky YV (2015) Genome-wide mapping of hot spots of DNA double-strand breaks in human cells as a tool for epigenetic studies and cancer genomics. Genomics Data 5:89–93

Thorland EC, Myers SL, Persing DH, Sarkar G, McGovern RM, Gostout BS, Smith DI (2000) Human papillomavirus type 16 integrations in cervical tumors frequently occur in common fragile sites. Cancer Res 60:5916–5921

Thys RG, Lehman CE, Pierce LC, Wang YH (2015) DNA secondary structure at chromosomal fragile sites in human disease. Curr Genomics 16:60–70

Tracy RB, Hsieh CL, Lieber MR (2000) Stable RNA/DNA hybrids in the mammalian genome: inducible intermediates in immunoglobulin class switch recombination. Science 288:1058–1061

Tunca B, Egeli U, Zorluoglu A, Yilmazlar T, Yerci O, Kizil A (2000) The expression of fragile sites in lymphocytes of patients with rectum cancer and their first-degree relatives. Cancer Lett 152:201–209

Uhlen M, Oksvold P, Fagerberg L, Lundberg E, Jonasson K, Forsberg M, Zwahlen M, Kampf C, Wester K, Hober S et al (2010) Towards a knowledge-based Human Protein Atlas. Nat Biotechnol 28:1248–1250

Unger S, Gorna MW, Le Bechec A, Do Vale-Pereira S, Bedeschi MF, Geiberger S, Grigelioniene G, Horemuzova E, Lalatta F, Lausch E et al (2013) FAM111A mutations result in hypoparathyroidism and impaired skeletal development. Am J Hum Genet 92:990–995

Usdin K, Woodford KJ (1995) CGG repeats associated with DNA instability and chromosome fragility form structures that block DNA synthesis in vitro. Nucleic Acids Res 23:4202–4209

Usdin K, Hayward BE, Kumari D, Lokanga RA, Sciascia N, Zhao XN (2014) Repeat-mediated genetic and epigenetic changes at the FMR1 locus in the Fragile X-related disorders. Front Genet 5:226

Van Schil K, Meire F, Karlstetter M, Bauwens M, Verdin H, Coppieters F, Scheiffert E, Van Nechel C, Langmann T, Deconinck N et al (2015) Early-onset autosomal recessive cerebellar ataxia associated with retinal dystrophy: new human hotfoot phenotype caused by homozygous GRID2 deletion. Genet Med 17:291–299

Verkerk AJ, Pieretti M, Sutcliffe JS, Fu YH, Kuhl DP, Pizzuti A, Reiner O, Richards S, Victoria MF, Zhang FP et al (1991) Identification of a gene (FMR-1) containing a CGG repeat coincident with a breakpoint cluster region exhibiting length variation in fragile X syndrome. Cell 65:905–914

Vianna-Morgante AM, Mingroni-Netto RC, Barbosa AC, Otto PA, Rosenberg C (1996) FRAXF in a patient with chromosome 8 duplication. J Med Genet 33:611–614

Voiculescu I, Back E, Schempp W (1991) Homozygous condition for a BrdU-requiring fragile site on chromosome 12. Hum Genet 86:416–417

Wang YH, Griffith J (1996) Methylation of expanded CCG triplet repeat DNA from fragile X syndrome patients enhances nucleosome exclusion. J Biol Chem 271:22937–22940

Wang YH, Gellibolian R, Shimizu M, Wells RD, Griffith J (1996) Long CCG triplet repeat blocks exclude nucleosomes: a possible mechanism for the nature of fragile sites in chromosomes. J Mol Biol 263:511–516

Wang L, Darling J, Zhang JS, Huang H, Liu W, Smith DI (1999) Allele-specific late replication and fragility of the most active common fragile site, FRA3B. Hum Mol Genet 8:431–437

Wang T, Bray SM, Warren ST (2012) New perspectives on the biology of fragile X syndrome. Curr Opin Genet Dev 22:256–263

Webb T (1992) Delayed replication of Xq27 in individuals with the fragile X syndrome. Am J Med Genet 43:1057–1062

Webb T, Thake A (1984) Fragile 22q13 segregating in a family. Clin Genet 26:125–128

Widrow RJ, Hansen RS, Kawame H, Gartler SM, Laird CD (1998) Very late DNA replication in the human cell cycle. Proc Natl Acad Sci U S A 95:11246–11250

Wilson TE, Arlt MF, Park SH, Rajendran S, Paulsen M, Ljungman M, Glover TW (2015) Large transcription units unify copy number variants and common fragile sites arising under replication stress. Genome Res 25:189–200

Winnepenninckx B, Debacker K, Ramsay J, Smeets D, Smits A, FitzPatrick DR, Kooy RF (2007) CGG-repeat expansion in the DIP2B gene is associated with the fragile site FRA12A on chromosome 12q13.1. Am J Hum Genet 80:221–231

Wongsurawat T, Jenjaroenpun P, Kwoh CK, Kuznetsov V (2012) Quantitative model of R-loop forming structures reveals a novel level of RNA-DNA interactome complexity. Nucleic Acids Res 40:e16

Xia JH, Li LY, He XX, Xiao JY (1988) Fragile site 1q44 involved in nasopharyngeal carcinoma. A study of a marker chromosome der(1)t(1;3)(q44;p11). Cancer Genet Cytogenet 35:135–140

Yu S, Mangelsdorf M, Hewett D, Hobson L, Baker E, Eyre HJ, Lapsys N, Le Paslier D, Doggett NA, Sutherland GR et al (1997) Human chromosomal fragile site FRA16B is an amplified AT-rich minisatellite repeat. Cell 88:367–374

Yu K, Chedin F, Hsieh CL, Wilson TE, Lieber MR (2003) R-loops at immunoglobulin class switch regions in the chromosomes of stimulated B cells. Nat Immunol 4:442–451

Zhang H, Freudenreich CH (2007) An AT-rich sequence in human common fragile site FRA16D causes fork stalling and chromosome breakage in S. cerevisiae. Mol Cell 27:367–379

Zhao, X.N. Usdin, K. (2016) Ups and downs: mechanisms of repeat instability in the fragile X-related disorders. Genes 7

Zhu Y, McAvoy S, Kuhn R, Smith DI (2006) RORA, a large common fragile site gene, is involved in cellular stress response. Oncogene 25:2901–2908

Zimonjic DB, Durkin ME, Keck-Waggoner CL, Park SW, Thorgeirsson SS, Popescu NC (2003) SMAD5 gene expression, rearrangements, copy number, and amplification at fragile site FRA5C in human hepatocellular carcinoma. Neoplasia 5:390–396

Zlotorynski E, Rahat A, Skaug J, Ben-Porat N, Ozeri E, Hershberg R, Levi A, Scherer SW, Margalit H, Kerem B (2003) Molecular basis for expression of common and rare fragile sites. Mol Cell Biol 23:7143–7151

Chapter 22
Cyclin E Deregulation and Genomic Instability

Leonardo K. Teixeira and Steven I. Reed

Abstract Precise replication of genetic material and its equal distribution to daughter cells are essential to maintain genome stability. In eukaryotes, chromosome replication and segregation are temporally uncoupled, occurring in distinct intervals of the cell cycle, S and M phases, respectively. Cyclin E accumulates at the G1/S transition, where it promotes S phase entry and progression by binding to and activating CDK2. Several lines of evidence from different models indicate that cyclin E/CDK2 deregulation causes replication stress in S phase and chromosome segregation errors in M phase, leading to genomic instability and cancer. In this chapter, we will discuss the main findings that link cyclin E/CDK2 deregulation to genomic instability and the molecular mechanisms by which cyclin E/CDK2 induces replication stress and chromosome aberrations during carcinogenesis.

Keywords Cell cycle • Cyclin E • CDK2 • FBW7 • Replication stress • Chromosome aberration • Genomic instability • Fragile sites • Cancer

22.1 Introduction

Progression through the cell cycle is regulated by association of cyclin-dependent kinases (CDKs) with specific regulatory subunits known as cyclins. Oscillations in cyclin levels primarily dictate oscillations in CDK activity, ensuring the order and timing of cell cycle phases (Hochegger et al. 2008; Malumbres and Barbacid

L.K. Teixeira
Program of Cell Biology, Brazilian National Cancer Institute (INCA), Rio de Janeiro, RJ, Brazil
e-mail: lkteixeira@inca.gov.br

S.I. Reed (✉)
Department of Cell and Molecular Biology, The Scripps Research Institute, La Jolla, CA, USA
e-mail: sreed@scripps.edu

H. Masai, M. Foiani (eds.), *DNA Replication*, Advances in Experimental Medicine and Biology 1042, https://doi.org/10.1007/978-981-10-6955-0_22

2009). The E-type cyclin family is composed of two proteins, cyclin E1 and cyclin E2, which exhibit high sequence similarity and are functionally redundant (Lew et al. 1991; Koff et al. 1991; Gudas et al. 1999; Lauper et al. 1998; Zariwala et al. 1998). Cyclin E levels are tightly regulated during normal cell cycles, accumulating at the G1/S transition and being completely degraded by the end of S phase (Koff et al. 1992; Dulic et al. 1992). Consistent with its expression pattern, cyclin E binds to and activates CDK2 to control S phase entry and progression (Koff et al. 1992; Dulic et al. 1992; Ohtsubo and Roberts 1993; Resnitzky et al. 1994). Cyclin E mRNA levels are mostly induced by E2F transcription factors (Ohtani et al. 1995; Geng et al. 1996), whereas cyclin E protein is degraded by the SCF^{Fbw7} ubiquitin ligase complex in a phosphorylation-dependent manner (Won and Reed 1996; Clurman et al. 1996; Strohmaier et al. 2001; Moberg et al. 2001; Koepp et al. 2001). Cyclin E/CDK2 activity is also controlled by the CDK inhibitors $p21^{Cip1}$ and $p27^{Kip}$, and potentially other CDK-inhibitory proteins, which are able to bind to and inactivate the cyclin E/CDK2 complex (Harper et al. 1993; Gu et al. 1993; Xiong et al. 1993; Polyak et al. 1994; Toyoshima and Hunter 1994; Reynaud et al. 1999).

Once activated, the cyclin E/CDK2 complex promotes the G1/S transition largely through phosphorylation and inactivation of the RB protein and the subsequent release of E2F transcription factors (Chellappan et al. 1991; Hinds et al. 1992; Dyson 1998; Harbour and Dean 2000). E2F proteins then promote S phase entry by regulating the expression of numerous genes required for DNA replication, such as the pre-replication complex components ORC1, CDC6, CDT1, and MCMs (Ohtani et al. 1996, 1998, 1999; Yan et al. 1998; Yoshida and Inoue 2004); the enzymes required for nucleotide and DNA synthesis, such as dihydrofolate reductase (DHFR), thymidine kinase (TK), and DNA polymerase α (Blake and Azizkhan 1989; Dou et al. 1994; Pearson et al. 1991); and the histone H2A (Oswald et al. 1996). Besides the RB protein, cyclin E/CDK2 directly phosphorylates and regulates other substrates required for S phase entry and progression, such as the DNA replication factors CDT1 and CDC6 (Liu et al. 2004; Mailand and Diffley 2005); the replication initiator Treslin (Kumagai et al. 2011); the activator of histone expression NPAT (Zhao et al. 2000; Ma et al. 2000); the transcription factors CBP/p300, E2F5, SMAD3, and MYC (Ait-Si-Ali et al. 1998; Morris et al. 2000; Matsuura et al. 2004; Hydbring et al. 2010); the centrosome proteins NPM, MPS1, and CP110 (Okuda et al. 2000; Tokuyama et al. 2001; Fisk and Winey 2001; Chen et al. 2002); and the DNA repair protein BRCA1 (Ruffner et al. 1999).

Regulation of E-type cyclins and the function of cyclin E/CDK2 in normal and aberrant cell cycles have been extensively reviewed elsewhere (Hwang and Clurman 2005; Caldon and Musgrove 2010; Siu et al. 2012). Here, we will focus on how deregulation of cyclin E/CDK2 causes replication stress and chromosome aberrations that may lead to genomic instability in cancer.

22.2 Cyclin E-Mediated Chromosome Instability

Several lines of evidence support the notion that cyclin E/CDK2 deregulation causes genomic instability. The initial finding that linked cyclin E to chromosome instability was the observation that constitutive cyclin E overexpression induced chromosome gains and losses in non-transformed rodent fibroblasts and human mammary epithelial cells, leading to aneuploidy (Spruck et al. 1999). Importantly, constitutive overexpression of cyclin D1 or cyclin A2 had no effect on the number of chromosomes in these cells. Later, it was shown that deletion of *FBXW7*, the gene encoding the F-box protein FBW7 involved in cyclin E recognition and degradation by the SCF^{Fbw7} E3 ubiquitin ligase complex, resulted in increased frequency of micronucleus formation, multipolar spindles, and eventually chromosome instability in colorectal cancer cells (Rajagopalan et al. 2004). Even though FBW7 is involved in the degradation of other oncoproteins, such as c-MYC, c-JUN, and NOTCH (Davis et al. 2014), downregulation of cyclin E was sufficient to revert micronucleus formation in FBW7-depleted cells (Rajagopalan et al. 2004). More recently, generation of a hyperactive *CDK2* knockin allele in a human colorectal cancer cell line that expresses high cyclin E-associated kinase activity also showed increased rates of micronucleus formation when compared to *CDK2* wild-type cells (Hughes et al. 2013). Together, this evidence supports a causal role for cyclin E in chromosome instability during carcinogenesis.

One of the proposed mechanisms to explain chromosome instability in cancers is centrosome amplification, which leads to the formation of merotelic attachments and eventually chromosome segregation errors (Fig. 22.1) (Godinho and Pellman 2014). Normal cyclin E/CDK2 activity is required to ensure initiation of centrosome duplication in *Xenopus* egg extracts (Hinchcliffe et al. 1999; Lacey et al. 1999). In mammalian cells, it is also clear that CDK2 activity is necessary for

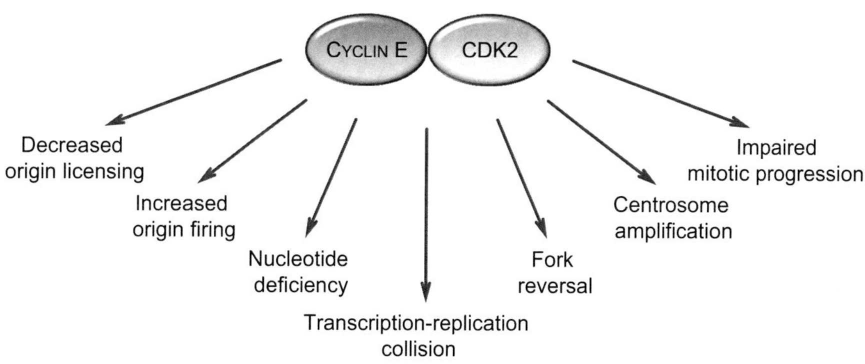

Fig. 22.1 Mechanisms of cyclin E-induced genomic instability. Cyclin E/CDK2 deregulation may cause impaired assembly of pre-replication complex, increased origin initiation, deficiency of nucleotide biosynthesis pathway, collisions between replication and transcription machineries, formation of aberrant replication intermediates, such as fork reversal, centrosome amplification, and impairment of mitotic progression and checkpoint function

centrosome duplication; however, it is still uncertain whether cyclin E or cyclin A plays a major role in CDK2-dependent centrosome duplication (Matsumoto et al. 1999; Meraldi et al. 1999; Hanashiro et al. 2008). Cyclin E overexpression alone does not efficiently induce centrosome amplification in mouse embryonic fibroblasts (MEFs), normal human fibroblasts, and epithelial cells (Spruck et al. 1999; Mussman et al. 2000; Kawamura et al. 2004). However, high levels of cyclin E synergize with loss of p53 function to induce centrosome amplification and chromosome instability in human cell lines and tumors (Mussman et al. 2000; Kawamura et al. 2004). Furthermore, MEFs from a hyperactive *CDK2* knockin mouse model, which show elevated cyclin E- and cyclin A-associated kinase activities, had an elevated number of centrosomes when compared to wild-type MEFs (Zhao et al. 2012).

Cyclin E/CDK2 localizes to centrosomes (Matsumoto and Maller 2004), where it phosphorylates and dissociates NPM protein, initiating separation of paired centrioles and duplication of centrosomes (Okuda et al. 2000; Tokuyama et al. 2001). It is therefore possible that deregulation of cyclin E/CDK2 kinase activity, in combination with other insults, impairs NPM release from centrosomes, leading to centrosome amplification and chromosome instability. Indeed, deletion of NPM causes aberrant mitotic figures with multiple centrosomes and aneuploidy in MEFs (Grisendi et al. 2005), and alterations in *NPM* are frequently observed in human cancers (Grisendi et al. 2006). As discussed above, the centrosome proteins MPS1 and CP110 are also directly phosphorylated by cyclin E/CDK2 (Fisk and Winey 2001; Chen et al. 2002) and therefore may represent potential targets for cyclin E-induced chromosome instability as well.

Another mechanism that drives chromosome instability in tumorigenesis is impairment of mitotic checkpoint function and progression through mitosis, which may cause chromosome missegregation and aneuploidy (Fig. 22.1) (Varetti et al. 2014). It has been shown that cyclin E overexpression delays progression through early stages of mitosis, leading to accumulation of cells in prometaphase and unaligned metaphase (Keck et al. 2007). Impairment of mitotic progression was caused by cyclin E/CDK2-mediated phosphorylation and inactivation of the APC/C adaptor protein CDH1 and subsequent accumulation of the APC/C^{Cdh1} ubiquitin ligase substrates cyclin B1 and securin, resulting in mitotic failure and polyploidy. In agreement, *FBXW7*-deficient cells, which have increased cyclin E-associated CDK2 activity, also exhibit increased levels of the APC/C substrates cyclin B1 and securin and accumulation of cells in prometaphase (Bailey et al. 2015). Interestingly, a genome-wide RNAi screen in these cells identified synthetic lethality with BUBR1, a spindle assembly checkpoint (SAC) protein, and high sensitivity to depletion of two other SAC components BUB1 and MPS1. These results suggest that cells with increased levels of cyclin E may depend on intact mitotic checkpoints for survival. Moreover, it has also been shown that cyclin E/CDK2 phosphorylates and prematurely activates the protein phosphatase CDC25C, leading to increased activity of the mitotic kinases cyclin B1/CDK1 and PLK1 and delayed mitotic progression (Bagheri-Yarmand et al. 2010).

22.3 Cyclin E-Mediated Replication Stress

Replication stress is characterized by the slowing or stalling of DNA replication forks, which may lead to fork collapse, DNA damage, and ultimately genomic instability (Zeman and Cimprich 2014). Activated oncoproteins and mutated tumor suppressors that drive sustained cellular proliferation cause replication stress and genomic instability, two events that are frequently observed in human cancers (Hills and Diffley 2014; Macheret and Halazonetis 2015). Indeed, this is the case for the oncoprotein cyclin E. Overexpression of cyclin E has been shown to cause replication stress, typified by slowed progression and premature termination of replication forks, DNA damage, and loss of heterozygosity at fragile sites (Bartkova et al. 2005, 2006; Bester et al. 2011). Cyclin E-mediated replication stress most likely is linked to elevated CDK2 kinase activity, as a *CDK2* hyperactive knockin allele was sufficient to delay replication fork progression, induce DNA damage, and increase micronucleus formation without cyclin E overexpression (Hughes et al. 2013). In a seminal series of articles, it has been proposed that oncogene-induced replication stress, including the oncoprotein cyclin E, activates the DNA damage response (DDR) pathway and leads to cell cycle arrest, cell death, and senescence, acting as an inducible barrier to tumor progression (Bartkova et al. 2005, 2006; Gorgoulis et al. 2005; Di Micco et al. 2006). Disruption of the DDR pathway facilitates cell proliferation and increases replication stress, leading to genomic instability in preneoplastic lesions.

The primary mechanism underlying cyclin E/CDK2-induced replication stress and genomic instability is interference of the nucleotide biosynthesis pathway (Fig. 22.1). Nucleotides are structural components of nucleic acids and therefore essential for a wide variety of biological processes, such as cell growth, DNA replication, and transcription (Lane and Fan 2015). Cyclin E overexpression, through disruption of the RB/E2F pathway, enforces cell proliferation of human fibroblasts with insufficient nucleotide levels (Bester et al. 2011). Nucleotide deficiency induced by cyclin E overexpression slowed replication fork progression and caused double-strand DNA breaks. Importantly, either exogenous supplementation of nucleosides or upregulation of nucleotide metabolism genes attenuated cyclin E-mediated replication stress and DNA damage. Consistent with this, replication stress in the form of impaired fork progression has been shown to generate structural as well as numerical chromosome instability during mitosis (Burrell et al. 2013).

Collisions between DNA replication and transcription machineries are another important source of replication stress (Fig. 22.1). Transcription complexes represent natural obstacles to the progression of replication forks, especially at fragile sites that contain extremely long genes (>800 kb), where replication forks have a high probability of encountering transcription complexes during the period of one cell cycle (Helmrich et al. 2011). Transcription-replication collisions may generate increased DNA topological tension and formation of R-loops (RNA-DNA hybrid structures), inducing replication fork stalling, DNA damage, and fragile site instability (Bermejo et al. 2012; Helmrich et al. 2013). Oncogenic events that interfere with the timing and location of DNA replication and transcription may increase the probability of transcription-replication collisions. In cells overexpressing cyclin E, it has been shown that inhibition of transcription elongation attenuates replication stress and DNA damage (Jones et al. 2013). This study also showed that inhibition of replication initiation restores normal levels of fork progression in cyclin E-overexpressing cells, suggesting that increased replication initiation and transcription-replication collisions contribute to the replication stress upon high levels of cyclin E (Fig. 22.1). One potential consequence of transcription-replication collisions is the formation of aberrant replication intermediates, such as reversed replication forks (Neelsen and Lopes 2015). Consistently, it has been shown that cyclin E overexpression induces accumulation of reversed forks and chromosome breakage in human cells, suggesting that DNA topological stress also underlie cyclin E-mediated replication stress and genomic instability (Fig. 22.1) (Neelsen et al. 2013). Furthermore, it has been shown that cyclin E-induced collapsed forks may be processed and repaired by break-induced replication (BIR) repair, which generates copy number alterations, such as segmental genomic duplications (Costantino et al. 2014).

22.4 Genomic Instability in Cyclin E Mouse Models

Cyclin E is frequently overexpressed in human tumors, and its deregulation has been associated with poor prognosis and decreased survival of cancer patients (Scuderi et al. 1996; Porter et al. 1997; Iida et al. 1997; Erlanson et al. 1998; Fukuse et al. 2000; Muller-Tidow et al. 2001; Keyomarsi et al. 2002; Schraml et al. 2003). Overexpression of cyclin E in mouse models has been shown to induce mammary and lung carcinomas as well as hematopoietic malignancies, further supporting a causative role for cyclin E in carcinogenesis (Bortner and Rosenberg 1997; Karsunky et al. 1999; Geisen et al. 2003; Loeb et al. 2005; Smith et al. 2006; Ma et al. 2007; Minella et al. 2008; Siu et al. 2014).

Several tissue-specific transgenic and knockin mouse models have provided significant information on the role of cyclin E deregulation in genomic instability. A knockin mouse expressing a nondegradable form of cyclin E in MEFs showed increased chromosome breaks, translocations, and aneuploidy in a $p21^{-/-}$ background (Loeb et al. 2005). In this model, cyclin E overexpression also cooperated with p53 deficiency and RAS activation to cause cellular transformation, induce whole chromosome gains and losses, and accelerate lung carcinogenesis. Consistent with this, transgenic mice expressing either wild-type or degradation-resistant cyclin E in the lungs incurred multiple pulmonary adenocarcinomas with specific gains of chromosomes 4 and 6 (Ma et al. 2007). In mammary gland transgenic mouse models, cyclin E overexpression has been shown to induce p53 loss of heterozygosity and drastically increase tumor formation in a $p53^{+/-}$ background (Smith et al. 2006; Akli et al. 2007). Lastly, a knockin mouse with expression of nondegradable cyclin E in the hematopoietic stem cell compartment exhibited abnormal hematopoiesis, chromosome instability illustrated by chromosome gains and losses, and decreased latency of T-cell malignancies in a $p53^{-/-}$ background (Minella et al. 2008; Siu et al. 2014). Again, p53 and p21 deficiencies were synergistic with cyclin E deregulation in promoting chromosome instability. Indeed, it has been shown that cyclin E-associated genomic instability is restrained by the p53/p21 pathway (Bartkova et al. 2005; Minella et al. 2002, 2007). Disruption of the inducible barrier established by the p53/p21 pathway may allow cyclin E overexpression to trigger genomic instability through some of the mechanisms discussed above, such as centrosome amplification and replication stress. Therefore, current mouse models support the notion that cyclin E deregulation contributes to tumorigenesis by promoting genomic instability in vivo.

22.5 Cyclin E Deregulation Promotes Replication Failure at Targeted Sites

We have discussed above the relationship between cyclin E and replication stress. Since cyclin E overexpression promotes replication stress and therefore slows replication fork progression (Bester et al. 2011; Jones et al. 2013; Liberal et al. 2012), we hypothesized that cells experiencing cyclin E deregulation might enter mitosis with incompletely replicated chromosomes. This in turn would lead to abnormal segregation and chromosomal damage during anaphase. Consistent with this, we observed that cyclin E-overexpressing non-transformed cells exhibited high levels of anaphase chromosomal anomalies such as bridged chromosomes and nonattached chromosomal fragments up to the size of the entire chromosome arms (Teixeira et al. 2015). If this observed chromosomal damage is a result of incompletely replicated chromosomal segments impairing segregation of sister chromatids, there are two obvious models that could account for the under-replication. Cyclin E-mediated replication stress could promote under-replication without any regional or feature specificity, or under-replication could occur at specific sites or regions possessing features that might sensitize them. To distinguish between these alternatives, we harvested cells blocked in mitosis immediately following cyclin E overexpression and analyzed their DNA by comparative genomic hybridization (CGH) array analysis. Indeed, a relatively small number of specific regions varying in size from approximately 200 to 100,000 base pairs had frequently failed to complete replication prior to entry into mitosis (Teixeira et al. 2015). Presumably, these under-replicated regions were responsible for the anaphase segregation anomalies we had observed in real time after cyclin E overexpression. Based on these observations, one would predict that these under-replicated regions would be included in deleted chromosomal segments subsequent to anaphase. We interrogated both mixed populations and single cells after cyclin E overexpression and found that deletion of these specific loci did indeed occur at high frequency (Teixeira et al. 2015). However, it appears that most cells carrying such deletions were incapable of clonal expansion, suggesting that checkpoint barriers eliminate cells with severely damaged genomes, should surveillance mechanisms be intact. This presumably reduces the impact of cyclin E-mediated damage at the population level and is therefore protective from potentially oncogenic events (Bartkova et al. 2005). Indeed, the link between cyclin E deregulation and the p53 surveillance system has been discussed above.

22.6 Genomic Features Associated with Replication Failure

22.6.1 Late-Replicating Genomic Regions

The majority of under-replicated regions detected in our study have been annotated as large late-replicating domains (Teixeira et al. 2015; Weddington et al. 2008). These are, for the most part, heterochromatic regions with a paucity of replication origins. The fact that origins in these domains fire late during the replication cycle combined with the low density of origins provides a likely explanation for failure to complete replication under conditions of replication stress (Le Tallec et al. 2014; Ozeri-Galai et al. 2014). However, these properties alone cannot explain the specific locations and boundaries of the under-replicated regions, as they were relatively small compared to the larger domains and highly targeted to specific sites. Therefore, other features of these sites must be relevant.

22.6.2 Recombinational Hotspots/Translocation Breakpoints

A number of the sites have been annotated as recombinational hotspots or translocation breakpoints. A subset of these has been classified as fragile sites, as well. Both recombination and translocation are processes that are initiated by double-strand DNA breaks. Repair involving homologous sequences versus heterologous sequences containing microhomology determines the outcome (Berti and Vindigni 2016). Significantly, within this context, one characteristic of fragile sites is a tendency to experience double-strand breaks at abnormally high frequencies, presumably due to replication barriers within these sites (Le Tallec et al. 2014; Ozeri-Galai et al. 2014; Thys et al. 2015). These observations suggest that features of fragile sites might impede DNA replication under conditions of cyclin E-mediated replication stress leading to local replication failure. One feature of fragile sites suggested as causative for replication impairment and double-strand breaks is unusual DNA structures, such as palindromic sequences leading to formation of hairpins and loops (Thys et al. 2015; Ozeri-Galai et al. 2011). Such structured nonlinear DNA could easily explain why stressed replication forks might stall or collapse. However, this cannot completely explain the specificity of under-replicated sites under conditions of cyclin E overexpression, since only a small subset of fragile sites, recombinational hotspots, and translocation breakpoints is affected. In addition, two recent studies showing that cyclin E overexpression causes instability at genomic regions with fragile site characteristics have given somewhat different results (Teixeira et al. 2015; Miron et al. 2015). Interestingly, among the susceptible genomic regions identified on each study (16 and 26, respectively), only one chromosome band was coincident for both (3q26). Since one study was carried out in mammary epithelial cells and the other in fibroblasts, a possible interpretation of the data is that cyclin E-mediated fragility may also be cell type specific, as has been shown previously in cells from different tissues (Le Tallec et al. 2011, 2013; Hosseini et al. 2013). In addition, not every under-replicated site was associated with fragile site features, suggesting factors unique to cyclin E-mediated replication stress must come into play (see below).

22.6.3 *Low Origin Density and Licensing*

Interrogating a number of databases of origin distribution in human cells compiled using diverse methodologies, we found that most of the under-replicated sites were located in chromosomal regions characterized by extremely low origin density (Teixeira et al. 2015). Under conditions of replication stress leading to fork collapse, the probable lack of nearby functional forks is likely to eliminate the most common mechanism for rescue of localized replication failure: processing of unreplicated DNA by an adjacent replicon (Letessier et al. 2011; Kawabata et al. 2011). However, it is likely that replication stress caused specifically by cyclin E overexpression compounds the logistical problems of completing the replicative cycle. This is because, in addition to replication stress, cyclin E overexpression impairs assembly of the pre-replication complex (Ekholm-Reed et al. 2004). Specifically, high cyclin E/CDK2 activity at the M/G1 boundary inhibits chromatin loading of MCM proteins, which constitute the primary replicative helicase. Based on investigation of the impact of direct MCM protein depletion, it is unlikely that this effect of cyclin E overexpression would alter DNA replication during an unperturbed replicative cycle. Very high percentages of individual MCM proteins can be depleted via RNAi-mediated silencing with no apparent effect on unperturbed replication (Ge et al. 2007; Ibarra et al. 2008). However, MCM-depleted cells are extremely sensitive to replication stress, as they fail to assemble backup origins. These origins are competent but normally remain dormant except under conditions of replication stress, when they are mobilized to rescue collapsed and/or poorly functioning replication forks (McIntosh and Blow 2012). Since it is probable that cyclin E overexpression via impaired MCM loading leads to a deficiency of backup origins, but also simultaneously causes replication stress, the problem of rescuing collapsed forks in origin-sparse regions is likely exacerbated, leading to replication failure.

22.6.4 Transcription-Replication Collisions

As stated above, one possible source of cyclin E-mediated replication stress is collision between the replication and transcriptional machineries. These encounters would be predicted to occur most frequently at fragile sites containing very long genes (>300 kb). Consistent with this, cyclin E overexpression produced copy number losses at two very long genes, *EPHA6* and *NCAM2*, in human mammary epithelial cells (approximate size of 935 kb and 545 kb, respectively) (Teixeira et al. 2015). In addition, transcriptional changes were observed at two other long genes, *DAB1* and *NRXN3*, in human fibroblasts (approximate size of 430 kb and 1.7 Mb, respectively) experiencing deregulated levels of cyclin E (Miron et al. 2015).

22.6.5 Sensitive DNA Structures

As alluded to the above, a number of the sites under-replicated after cyclin E overexpression correspond to fragile sites, a characteristic of which is the presence of nonlinear DNA structures expected to pose barriers to replication fork progression (Thys et al. 2015). However, most fragile sites were not represented as under-replicated sites in our analysis. One fragile site, nevertheless, is likely to be informative: FRA11B/G (Fechter et al. 2007; Burrow et al. 2009). This site on chromosome 11 is particularly interesting because it is the breakpoint for rearrangements in mixed-lineage leukemia (MLL); hence, the locus is referred to as MLL (Muntean and Hess 2012). It is also noteworthy that this site is also frequently deleted in breast cancer (see below). The under-replicated segment detected in our study was 4,331 bp, which contains part of the breakpoint cluster region (BCR) for MLL translocations in leukemia (Muntean and Hess 2012). We therefore scanned this region for DNA sequences predictive of the ability to form energetically favorable hairpin loop structures. Interestingly, two such structures were detected near the center of the segment separated by approximately 500 base pairs (Teixeira et al. 2015). To determine whether these structured DNA elements posed a barrier to replication under conditions of cyclin E-mediated replication stress, we cloned the segment containing them, with and without the palindromic sequences, into an episomal vector that replicates autonomously in mammalian cells. While both the control and palindrome-containing plasmids were well maintained in the absence of cyclin E overexpression, only the control plasmid was maintained when cyclin E was overexpressed. These data are consistent with the hypothesis that palindromic structures pose a barrier to replication, specifically under conditions of cyclin E-mediated replication stress (Teixeira et al. 2015). Therefore, two structural barriers to replication in close proximity may represent a feature that promotes sensitivity to cyclin E-mediated replication stress, although such structures are likely to be sensitive to other sources of replication stress as well (see below). However, it is worth noting that cyclin E overexpression has been determined to be associated with MLL translocations in the context of B-cell precursor acute lymphoblastic leukemia (BCP-ALL) (Accordi et al. 2010).

22.7 Cyclin E and Genomic Instability in Breast Cancer

What does the observation that cyclin E overexpression promotes replication failure at a small subset of specific loci implicate for oncogenesis? The first question one might ask is are deletions in the chromosomal regions surrounding the under-replicated sites found in actual cancer? To address this, we interrogated a database of approximately 2000 breast tumors subjected to CGH array analysis (Teixeira et al. 2015; Curtis et al. 2012). As a surrogate for cyclin E overexpression, we employed copy number increases of the *CCNE1* locus, presumably due to gene amplification. This undoubtedly represents an underestimation of tumors overexpressing cyclin E and likely introduces experimental noise, as several posttranslational mechanisms have been shown to elevate cyclin E levels. Nonetheless, when *CCNE1* copy number increase was compared with copy number decrease at each of the under-replicated sites, a highly significant correlation was observed for many of them. This suggests that cyclin E overexpression is a driver of deletion at these sites. It should be noted that overall these specific sites experience deletions at relatively low frequency, and their detailed analysis is likely to yield clues concerning what characteristics constitute sensitivity specifically to cyclin E-mediated replication stress. On the other hand, some of the sites that were more frequently deleted in the data set did not show a significant correlation with cyclin E copy number increase. The MLL locus was one of these. Presumably, features of the MLL locus, specifically two likely hairpin loops in close proximity, render this site sensitive to multiple forms of replication stress, including but not exclusive to cyclin E. The second relevant question is whether these specific deletions have a direct role in oncogenesis. Unfortunately, at this point, we do not know the magnitudes or boundaries of deletions that occur surrounding these sites when they have not completed replication but are forced through anaphase. However, one might speculate that large deletions could drive oncogenesis by promoting loss of heterozygosity at tumor suppressor loci and deletions over fragile sites (Bignell et al. 2010). It is interesting to note that tumor suppressor genes have been identified on chromosomes 3q (Guo et al. 2002; Schwaenen et al. 2009; Thean et al. 2010) and 21q (Lee et al. 2003; Silva et al. 2003; Yamada et al. 2008), the arms where cyclin E-driven deletions occur in breast cancer.

In our study using immortalized non-transformed human mammary epithelial cells, it was clear that few cells that had sustained deletions after exposure to cyclin E overexpression were capable of expanding robustly and forming colonies. Although reassuring from a human health perspective, this observation raises the question of how cyclin E-driven deletions might get fixed in an expanding premalignant population. The answer probably lies in the fact that individual breast cancers when they present commonly possess more than 100 genetic modifications

(Nik-Zainal et al. 2016). At least some of these are likely to have been selected to override checkpoint barriers, thereby allowing clonal expansion of chromosomally damaged cells.

22.8 Relevance to Other Sources of Replication Stress

The discussion above has focused on mechanisms of genomic instability associated with cyclin E overexpression/deregulation (Fig. 22.1). However, other oncogenic events have been associated with replication stress, e.g., overexpression of c-MYC (Dominguez-Sola et al. 2007; Srinivasan et al. 2013; Rohban and Campaner 2015) or mutation of RAS (Di Micco et al. 2006; Miron et al. 2015; Maya-Mendoza et al. 2015). Although the mechanisms whereby these overexpressed or mutant proteins cause replication stress are likely to differ, one can infer that in severe instances, incompletely replicated genomes will enter mitosis with the end result being genomic instability.

A number of model system experiments support this idea. A hypomorphic allele of the mouse MCM helicase component MCM4, designated Chaos3, which promotes instability of the pre-replication complex (Kawabata et al. 2011), causes anaphase aberrations similar to what we observed for cyclin E overexpression. Although $MCM4^{Chaos3}$ does not appear to affect the number of origins fired, the number of dormant backup origins is reduced, and the number of stalled replication forks is increased, as it has been proposed for cyclin E overexpression. Presumably, intrinsic levels of replication stress during the normal replicative cycle require such backup origins in order to avoid under-replicated regions and the resulting aberrant anaphases. Interestingly, the $MCM4^{Chaos3}$ mouse is cancer prone (Shima et al. 2007), indicating that this type of chromosomal damage is directly linked to oncogenesis.

In yeast, the absence of the cohesin-like complex Smc5-Smc6 causes replication impairment at loci that contain replication barriers, such as the rDNA cluster (Torres-Rosell et al. 2007). Yet, cells progress through mitosis resulting in elevated rates of chromosomal nondisjunction, even though all checkpoints are intact. These results confirm that no robust checkpoint exists that can detect and respond to small quantities of unreplicated DNA in eukaryotes ranging from yeast to human (Mohebi et al. 2015; Koundrioukoff et al. 2013).

22.9 Conclusions

Elevated cyclin E has been shown to be associated with aggressive disease and poor outcome in at least some human malignancies. The link between cyclin E overexpression and genomic instability suggests a mechanism whereby cyclin E might promote oncogenesis. Our recent work showing that cyclin E overexpression promotes replication failure at a small number of specific loci and, as a consequence, chromosomal damage following anaphase leaves some important unanswered questions relevant to oncogenesis. First and foremost is a detailed description of the genomic damage that occurs in individual cells. Such information will allow the assessment of whether the classic two-hit tumor suppressor model is applicable or whether more complex mechanisms apply such as amplifications and translocations. The second important issue to be resolved is whether cyclin E can serve as a prototype for other oncoproteins that cause replication stress and promote genomic instability. On the one hand, as outlined above, some of the modalities of cyclin E function in the context of the replication stress are likely to be unique. On the other, there are certain to be mechanistic commonalities. Only investigation of the pathways leading to replication stress and from replication stress to genomic alterations for other oncoproteins will resolve this question.

References

Accordi B, Espina V, Giordan M, VanMeter A, Milani G, Galla L, Ruzzene M, Sciro M, Trentin L, De Maria R, te Kronnie G, Petricoin E, Liotta L, Basso G (2010) Functional protein network activation mapping reveals new potential molecular drug targets for poor prognosis pediatric BCP-ALL. PLoS One 5(10):e13552. https://doi.org/10.1371/journal.pone.0013552

Ait-Si-Ali S, Ramirez S, Barre FX, Dkhissi F, Magnaghi-Jaulin L, Girault JA, Robin P, Knibiehler M, Pritchard LL, Ducommun B, Trouche D, Harel-Bellan A (1998) Histone acetyltransferase activity of CBP is controlled by cycle-dependent kinases and oncoprotein E1A. Nature 396(6707):184–186. https://doi.org/10.1038/24190

Akli S, Van Pelt CS, Bui T, Multani AS, Chang S, Johnson D, Tucker S, Keyomarsi K (2007) Overexpression of the low molecular weight cyclin E in transgenic mice induces metastatic mammary carcinomas through the disruption of the ARF-p53 pathway. Cancer Res 67(15):7212–7222. https://doi.org/10.1158/0008-5472.can-07-0599

Bagheri-Yarmand R, Nanos-Webb A, Biernacka A, Bui T, Keyomarsi K (2010) Cyclin E deregulation impairs mitotic progression through premature activation of Cdc25C. Cancer Res 70(12):5085–5095. https://doi.org/10.1158/0008-5472.can-09-4095

Bailey ML, Singh T, Mero P, Moffat J, Hieter P (2015) Dependence of human colorectal cells lacking the FBW7 tumor suppressor on the spindle assembly checkpoint. Genetics 201(3):885–895. https://doi.org/10.1534/genetics.115.180653

Bartkova J, Horejsi Z, Koed K, Kramer A, Tort F, Zieger K, Guldberg P, Sehested M, Nesland JM, Lukas C, Orntoft T, Lukas J, Bartek J (2005) DNA damage response as a candidate anti-cancer barrier in early human tumorigenesis. Nature 434(7035):864–870. https://doi.org/10.1038/nature03482

Bartkova J, Rezaei N, Liontos M, Karakaidos P, Kletsas D, Issaeva N, Vassiliou LV, Kolettas E, Niforou K, Zoumpourlis VC, Takaoka M, Nakagawa H, Tort F, Fugger K, Johansson F, Sehested M, Andersen CL, Dyrskjot L, Orntoft T, Lukas J, Kittas C, Helleday T, Halazonetis TD, Bartek J, Gorgoulis VG (2006) Oncogene-induced senescence is part of the tumorigenesis barrier imposed by DNA damage checkpoints. Nature 444(7119):633–637. https://doi.org/10.1038/nature05268

Bermejo R, Lai MS, Foiani M (2012) Preventing replication stress to maintain genome stability: resolving conflicts between replication and transcription. Mol Cell 45(6):710–718. https://doi.org/10.1016/j.molcel.2012.03.001

Berti M, Vindigni A (2016) Replication stress: getting back on track. Nat Struct Mol Biol 23(2):103–109. https://doi.org/10.1038/nsmb.3163

Bester AC, Roniger M, Oren YS, Im MM, Sarni D, Chaoat M, Bensimon A, Zamir G, Shewach DS, Kerem B (2011) Nucleotide deficiency promotes genomic instability in early stages of cancer development. Cell 145(3):435–446. https://doi.org/10.1016/j.cell.2011.03.044

Bignell GR, Greenman CD, Davies H, Butler AP, Edkins S, Andrews JM, Buck G, Chen L, Beare D, Latimer C, Widaa S, Hinton J, Fahey C, Fu B, Swamy S, Dalgliesh GL, Teh BT, Deloukas P, Yang F, Campbell PJ, Futreal PA, Stratton MR (2010) Signatures of mutation and selection in the cancer genome. Nature 463(7283):893–898. https://doi.org/10.1038/nature08768

Blake MC, Azizkhan JC (1989) Transcription factor E2F is required for efficient expression of the hamster dihydrofolate reductase gene in vitro and in vivo. Mol Cell Biol 9(11):4994–5002

Bortner DM, Rosenberg MP (1997) Induction of mammary gland hyperplasia and carcinomas in transgenic mice expressing human cyclin E. Mol Cell Biol 17(1):453–459

Burrell RA, McClelland SE, Endesfelder D, Groth P, Weller MC, Shaikh N, Domingo E, Kanu N, Dewhurst SM, Gronroos E, Chew SK, Rowan AJ, Schenk A, Sheffer M, Howell M, Kschischo M, Behrens A, Helleday T, Bartek J, Tomlinson IP, Swanton C (2013) Replication stress links structural and numerical cancer chromosomal instability. Nature 494(7438):492–496. https://doi.org/10.1038/nature11935

Burrow AA, Williams LE, Pierce LC, Wang YH (2009) Over half of breakpoints in gene pairs involved in cancer-specific recurrent translocations are mapped to human chromosomal fragile sites. BMC Genomics 10:59. https://doi.org/10.1186/1471-2164-10-59

Caldon CE, Musgrove EA (2010) Distinct and redundant functions of cyclin E1 and cyclin E2 in development and cancer. Cell Div 5:2. https://doi.org/10.1186/1747-1028-5-2

Chellappan SP, Hiebert S, Mudryj M, Horowitz JM, Nevins JR (1991) The E2F transcription factor is a cellular target for the RB protein. Cell 65(6):1053–1061

Chen Z, Indjeian VB, McManus M, Wang L, Dynlacht BD (2002) CP110, a cell cycle-dependent CDK substrate, regulates centrosome duplication in human cells. Dev Cell 3(3):339–350

Clurman BE, Sheaff RJ, Thress K, Groudine M, Roberts JM (1996) Turnover of cyclin E by the ubiquitin-proteasome pathway is regulated by cdk2 binding and cyclin phosphorylation. Genes Dev 10(16):1979–1990

Costantino L, Sotiriou SK, Rantala JK, Magin S, Mladenov E, Helleday T, Haber JE, Iliakis G, Kallioniemi OP, Halazonetis TD (2014) Break-induced replication repair of damaged forks induces genomic duplications in human cells. Science 343(6166):88–91. https://doi.org/10.1126/science.1243211

Curtis C, Shah SP, Chin SF, Turashvili G, Rueda OM, Dunning MJ, Speed D, Lynch AG, Samarajiwa S, Yuan Y, Graf S, Ha G, Haffari G, Bashashati A, Russell R, McKinney S, Langerod A, Green A, Provenzano E, Wishart G, Pinder S, Watson P, Markowetz F, Murphy L, Ellis I, Purushotham A, Borresen-Dale AL, Brenton JD, Tavare S, Caldas C, Aparicio S (2012) The genomic and transcriptomic architecture of 2,000 breast tumours reveals novel subgroups. Nature 486(7403):346–352. https://doi.org/10.1038/nature10983

Davis RJ, Welcker M, Clurman BE (2014) Tumor suppression by the Fbw7 ubiquitin ligase: mechanisms and opportunities. Cancer Cell 26(4):455–464. https://doi.org/10.1016/j.ccell.2014.09.013

Di Micco R, Fumagalli M, Cicalese A, Piccinin S, Gasparini P, Luise C, Schurra C, Garre M, Nuciforo PG, Bensimon A, Maestro R, Pelicci PG, d'Adda di Fagagna F (2006) Oncogene-induced senescence is a DNA damage response triggered by DNA hyper-replication. Nature 444(7119):638–642. https://doi.org/10.1038/nature05327

Dominguez-Sola D, Ying CY, Grandori C, Ruggiero L, Chen B, Li M, Galloway DA, Gu W, Gautier J, Dalla-Favera R (2007) Non-transcriptional control of DNA replication by c-Myc. Nature 448(7152):445–451. https://doi.org/10.1038/nature05953

Dou QP, Zhao S, Levin AH, Wang J, Helin K, Pardee AB (1994) G1/S-regulated E2F-containing protein complexes bind to the mouse thymidine kinase gene promoter. J Biol Chem 269(2):1306–1313

Dulic V, Lees E, Reed SI (1992) Association of human cyclin E with a periodic G1-S phase protein kinase. Science 257(5078):1958–1961

Dyson N (1998) The regulation of E2F by pRB-family proteins. Genes Dev 12(15):2245–2262
Ekholm-Reed S, Mendez J, Tedesco D, Zetterberg A, Stillman B, Reed SI (2004) Deregulation of cyclin E in human cells interferes with prereplication complex assembly. J Cell Biol 165(6):789–800. https://doi.org/10.1083/jcb.200404092
Erlanson M, Portin C, Linderholm B, Lindh J, Roos G, Landberg G (1998) Expression of cyclin E and the cyclin-dependent kinase inhibitor p27 in malignant lymphomas-prognostic implications. Blood 92(3):770–777
Fechter A, Buettel I, Kuehnel E, Savelyeva L, Schwab M (2007) Common fragile site FRA11G and rare fragile site FRA11B at 11q23.3 encompass distinct genomic regions. Genes Chromosom Cancer 46(1):98–106. https://doi.org/10.1002/gcc.20389
Fisk HA, Winey M (2001) The mouse Mps1p-like kinase regulates centrosome duplication. Cell 106(1):95–104
Fukuse T, Hirata T, Naiki H, Hitomi S, Wada H (2000) Prognostic significance of cyclin E overexpression in resected non-small cell lung cancer. Cancer Res 60(2):242–244
Ge XQ, Jackson DA, Blow JJ (2007) Dormant origins licensed by excess Mcm2-7 are required for human cells to survive replicative stress. Genes Dev 21(24):3331–3341. https://doi.org/10.1101/gad.457807
Geisen C, Karsunky H, Yucel R, Moroy T (2003) Loss of $p27^{Kip1}$ cooperates with cyclin E in T-cell lymphomagenesis. Oncogene 22(11):1724–1729. https://doi.org/10.1038/sj.onc.1206340
Geng Y, Eaton EN, Picon M, Roberts JM, Lundberg AS, Gifford A, Sardet C, Weinberg RA (1996) Regulation of cyclin E transcription by E2Fs and retinoblastoma protein. Oncogene 12(6):1173–1180
Godinho SA, Pellman D (2014) Causes and consequences of centrosome abnormalities in cancer. Philos Trans R Soc Lond B Biol Sci 369(1650):20130467. https://doi.org/10.1098/rstb.2013.0467
Gorgoulis VG, Vassiliou LV, Karakaidos P, Zacharatos P, Kotsinas A, Liloglou T, Venere M, Ditullio RA Jr, Kastrinakis NG, Levy B, Kletsas D, Yoneta A, Herlyn M, Kittas C, Halazonetis TD (2005) Activation of the DNA damage checkpoint and genomic instability in human precancerous lesions. Nature 434(7035):907–913. https://doi.org/10.1038/nature03485
Grisendi S, Bernardi R, Rossi M, Cheng K, Khandker L, Manova K, Pandolfi PP (2005) Role of nucleophosmin in embryonic development and tumorigenesis. Nature 437(7055):147–153. https://doi.org/10.1038/nature03915
Grisendi S, Mecucci C, Falini B, Pandolfi PP (2006) Nucleophosmin and cancer. Nat Rev Cancer 6(7):493–505. https://doi.org/10.1038/nrc1885
Gu Y, Turck CW, Morgan DO (1993) Inhibition of CDK2 activity in vivo by an associated 20K regulatory subunit. Nature 366(6456):707–710. https://doi.org/10.1038/366707a0
Gudas JM, Payton M, Thukral S, Chen E, Bass M, Robinson MO, Coats S (1999) Cyclin E2, a novel G1 cyclin that binds Cdk2 and is aberrantly expressed in human cancers. Mol Cell Biol 19(1):612–622
Guo SS, Arora C, Shimoide AT, Sawicki MP (2002) Frequent deletion of chromosome 3 in malignant sporadic pancreatic endocrine tumors. Mol Cell Endocrinol 190(1–2):109–114
Hanashiro K, Kanai M, Geng Y, Sicinski P, Fukasawa K (2008) Roles of cyclins A and E in induction of centrosome amplification in p53-compromised cells. Oncogene 27(40):5288–5302. https://doi.org/10.1038/onc.2008.161
Harbour JW, Dean DC (2000) The Rb/E2F pathway: expanding roles and emerging paradigms. Genes Dev 14(19):2393–2409
Harper JW, Adami GR, Wei N, Keyomarsi K, Elledge SJ (1993) The p21 Cdk-interacting protein Cip1 is a potent inhibitor of G1 cyclin-dependent kinases. Cell 75(4):805–816
Helmrich A, Ballarino M, Tora L (2011) Collisions between replication and transcription complexes cause common fragile site instability at the longest human genes. Mol Cell 44(6):966–977. https://doi.org/10.1016/j.molcel.2011.10.013

Helmrich A, Ballarino M, Nudler E, Tora L (2013) Transcription-replication encounters, consequences and genomic instability. Nat Struct Mol Biol 20(4):412–418. https://doi.org/10.1038/nsmb.2543

Hills SA, Diffley JF (2014) DNA replication and oncogene-induced replicative stress. Curr Biol 24(10):R435–R444. https://doi.org/10.1016/j.cub.2014.04.012

Hinchcliffe EH, Li C, Thompson EA, Maller JL, Sluder G (1999) Requirement of Cdk2-cyclin E activity for repeated centrosome reproduction in Xenopus egg extracts. Science 283(5403):851–854

Hinds PW, Mittnacht S, Dulic V, Arnold A, Reed SI, Weinberg RA (1992) Regulation of retinoblastoma protein functions by ectopic expression of human cyclins. Cell 70(6):993–1006

Hochegger H, Takeda S, Hunt T (2008) Cyclin-dependent kinases and cell-cycle transitions: does one fit all? Nat Rev Mol Cell Biol 9(11):910–916. https://doi.org/10.1038/nrm2510

Hosseini SA, Horton S, Saldivar JC, Miuma S, Stampfer MR, Heerema NA, Huebner K (2013) Common chromosome fragile sites in human and murine epithelial cells and FHIT/FRA3B loss-induced global genome instability. Genes Chromosom Cancer 52(11):1017–1029. https://doi.org/10.1002/gcc.22097

Hughes BT, Sidorova J, Swanger J, Monnat RJ Jr, Clurman BE (2013) Essential role for Cdk2 inhibitory phosphorylation during replication stress revealed by a human Cdk2 knockin mutation. Proc Natl Acad Sci U S A 110(22):8954–8959. https://doi.org/10.1073/pnas.1302927110

Hwang HC, Clurman BE (2005) Cyclin E in normal and neoplastic cell cycles. Oncogene 24(17):2776–2786. https://doi.org/10.1038/sj.onc.1208613

Hydbring P, Bahram F, Su Y, Tronnersjo S, Hogstrand K, von der Lehr N, Sharifi HR, Lilischkis R, Hein N, Wu S, Vervoorts J, Henriksson M, Grandien A, Luscher B, Larsson LG (2010) Phosphorylation by Cdk2 is required for Myc to repress Ras-induced senescence in cotransformation. Proc Natl Acad Sci U S A 107(1):58–63. https://doi.org/10.1073/pnas.0900121106

Ibarra A, Schwob E, Mendez J (2008) Excess MCM proteins protect human cells from replicative stress by licensing backup origins of replication. Proc Natl Acad Sci U S A 105(26):8956–8961. https://doi.org/10.1073/pnas.0803978105

Iida H, Towatari M, Tanimoto M, Morishita Y, Kodera Y, Saito H (1997) Overexpression of cyclin E in acute myelogenous leukemia. Blood 90(9):3707–3713

Jones RM, Mortusewicz O, Afzal I, Lorvellec M, Garcia P, Helleday T, Petermann E (2013) Increased replication initiation and conflicts with transcription underlie cyclin E-induced replication stress. Oncogene 32(32):3744–3753. https://doi.org/10.1038/onc.2012.387

Karsunky H, Geisen C, Schmidt T, Haas K, Zevnik B, Gau E, Moroy T (1999) Oncogenic potential of cyclin E in T-cell lymphomagenesis in transgenic mice: evidence for cooperation between cyclin E and Ras but not Myc. Oncogene 18(54):7816–7824. https://doi.org/10.1038/sj.onc.1203205

Kawabata T, Luebben SW, Yamaguchi S, Ilves I, Matise I, Buske T, Botchan MR, Shima N (2011) Stalled fork rescue via dormant replication origins in unchallenged S phase promotes proper chromosome segregation and tumor suppression. Mol Cell 41(5):543–553. https://doi.org/10.1016/j.molcel.2011.02.006

Kawamura K, Izumi H, Ma Z, Ikeda R, Moriyama M, Tanaka T, Nojima T, Levin LS, Fujikawa-Yamamoto K, Suzuki K, Fukasawa K (2004) Induction of centrosome amplification and chromosome instability in human bladder cancer cells by p53 mutation and cyclin E overexpression. Cancer Res 64(14):4800–4809. https://doi.org/10.1158/0008-5472.can-03-3908

Keck JM, Summers MK, Tedesco D, Ekholm-Reed S, Chuang LC, Jackson PK, Reed SI (2007) Cyclin E overexpression impairs progression through mitosis by inhibiting APC^{Cdh1}. J Cell Biol 178(3):371–385. https://doi.org/10.1083/jcb.200703202

Keyomarsi K, Tucker SL, Buchholz TA, Callister M, Ding Y, Hortobagyi GN, Bedrosian I, Knickerbocker C, Toyofuku W, Lowe M, Herliczek TW, Bacus SS (2002) Cyclin E and survival in patients with breast cancer. N Engl J Med 347(20):1566–1575. https://doi.org/10.1056/NEJMoa021153

Koepp DM, Schaefer LK, Ye X, Keyomarsi K, Chu C, Harper JW, Elledge SJ (2001) Phosphorylation-dependent ubiquitination of cyclin E by the SCF^{Fbw7} ubiquitin ligase. Science 294(5540):173–177. https://doi.org/10.1126/science.1065203
Koff A, Cross F, Fisher A, Schumacher J, Leguellec K, Philippe M, Roberts JM (1991) Human cyclin E, a new cyclin that interacts with two members of the CDC2 gene family. Cell 66(6):1217–1228
Koff A, Giordano A, Desai D, Yamashita K, Harper JW, Elledge S, Nishimoto T, Morgan DO, Franza BR, Roberts JM (1992) Formation and activation of a cyclin E-cdk2 complex during the G1 phase of the human cell cycle. Science 257(5077):1689–1694
Koundrioukoff S, Carignon S, Techer H, Letessier A, Brison O, Debatisse M (2013) Stepwise activation of the ATR signaling pathway upon increasing replication stress impacts fragile site integrity. PLoS Genet 9(7):e1003643. https://doi.org/10.1371/journal.pgen.1003643
Kumagai A, Shevchenko A, Shevchenko A, Dunphy WG (2011) Direct regulation of Treslin by cyclin-dependent kinase is essential for the onset of DNA replication. J Cell Biol 193(6):995–1007. https://doi.org/10.1083/jcb.201102003
Lacey KR, Jackson PK, Stearns T (1999) Cyclin-dependent kinase control of centrosome duplication. Proc Natl Acad Sci U S A 96(6):2817–2822
Lane AN, Fan TW (2015) Regulation of mammalian nucleotide metabolism and biosynthesis. Nucleic Acids Res 43(4):2466–2485. https://doi.org/10.1093/nar/gkv047
Lauper N, Beck AR, Cariou S, Richman L, Hofmann K, Reith W, Slingerland JM, Amati B (1998) Cyclin E2: a novel CDK2 partner in the late G1 and S phases of the mammalian cell cycle. Oncogene 17(20):2637–2643. https://doi.org/10.1038/sj.onc.1202477
Le Tallec B, Dutrillaux B, Lachages AM, Millot GA, Brison O, Debatisse M (2011) Molecular profiling of common fragile sites in human fibroblasts. Nat Struct Mol Biol 18(12):1421–1423. https://doi.org/10.1038/nsmb.2155
Le Tallec B, Millot GA, Blin ME, Brison O, Dutrillaux B, Debatisse M (2013) Common fragile site profiling in epithelial and erythroid cells reveals that most recurrent cancer deletions lie in fragile sites hosting large genes. Cell Rep 4(3):420–428. https://doi.org/10.1016/j.celrep.2013.07.003
Le Tallec B, Koundrioukoff S, Wilhelm T, Letessier A, Brison O, Debatisse M (2014) Updating the mechanisms of common fragile site instability: how to reconcile the different views? Cell Mol Life Sci 71(23):4489–4494. https://doi.org/10.1007/s00018-014-1720-2
Lee EB, Park TI, Park SH, Park JY (2003) Loss of heterozygosity on the long arm of chromosome 21 in non-small cell lung cancer. Ann Thorac Surg 75(5):1597–1600
Letessier A, Millot GA, Koundrioukoff S, Lachages AM, Vogt N, Hansen RS, Malfoy B, Brison O, Debatisse M (2011) Cell-type-specific replication initiation programs set fragility of the FRA3B fragile site. Nature 470(7332):120–123. https://doi.org/10.1038/nature09745
Lew DJ, Dulic V, Reed SI (1991) Isolation of three novel human cyclins by rescue of G1 cyclin (Cln) function in yeast. Cell 66(6):1197–1206
Liberal V, Martinsson-Ahlzen HS, Liberal J, Spruck CH, Widschwendter M, McGowan CH, Reed SI (2012) Cyclin-dependent kinase subunit (Cks) 1 or Cks2 overexpression overrides the DNA damage response barrier triggered by activated oncoproteins. Proc Natl Acad Sci U S A 109(8):2754–2759. https://doi.org/10.1073/pnas.1102434108
Liu E, Li X, Yan F, Zhao Q, Wu X (2004) Cyclin-dependent kinases phosphorylate human Cdt1 and induce its degradation. J Biol Chem 279(17):17283–17288. https://doi.org/10.1074/jbc.C300549200
Loeb KR, Kostner H, Firpo E, Norwood T, D Tsuchiya K, Clurman BE, Roberts JM (2005) A mouse model for cyclin E-dependent genetic instability and tumorigenesis. Cancer Cell 8(1):35–47. https://doi.org/10.1016/j.ccr.2005.06.010
Ma T, Van Tine BA, Wei Y, Garrett MD, Nelson D, Adams PD, Wang J, Qin J, Chow LT, Harper JW (2000) Cell cycle-regulated phosphorylation of $p220^{NPAT}$ by cyclin E/Cdk2 in Cajal bodies promotes histone gene transcription. Genes Dev 14(18):2298–2313

Ma Y, Fiering S, Black C, Liu X, Yuan Z, Memoli VA, Robbins DJ, Bentley HA, Tsongalis GJ, Demidenko E, Freemantle SJ, Dmitrovsky E (2007) Transgenic cyclin E triggers dysplasia and multiple pulmonary adenocarcinomas. Proc Natl Acad Sci U S A 104(10):4089–4094. https://doi.org/10.1073/pnas.0606537104

Macheret M, Halazonetis TD (2015) DNA replication stress as a hallmark of cancer. Annu Rev Pathol 10:425–448. https://doi.org/10.1146/annurev-pathol-012414-040424

Mailand N, Diffley JF (2005) CDKs promote DNA replication origin licensing in human cells by protecting Cdc6 from APC/C-dependent proteolysis. Cell 122(6):915–926. https://doi.org/10.1016/j.cell.2005.08.013

Malumbres M, Barbacid M (2009) Cell cycle, CDKs and cancer: a changing paradigm. Nat Rev Cancer 9(3):153–166. https://doi.org/10.1038/nrc2602

Matsumoto Y, Maller JL (2004) A centrosomal localization signal in cyclin E required for Cdk2-independent S phase entry. Science 306(5697):885–888. https://doi.org/10.1126/science.1103544

Matsumoto Y, Hayashi K, Nishida E (1999) Cyclin-dependent kinase 2 (Cdk2) is required for centrosome duplication in mammalian cells. Curr Biol 9(8):429–432

Matsuura I, Denissova NG, Wang G, He D, Long J, Liu F (2004) Cyclin-dependent kinases regulate the antiproliferative function of Smads. Nature 430(6996):226–231. https://doi.org/10.1038/nature02650

Maya-Mendoza A, Ostrakova J, Kosar M, Hall A, Duskova P, Mistrik M, Merchut-Maya JM, Hodny Z, Bartkova J, Christensen C, Bartek J (2015) Myc and Ras oncogenes engage different energy metabolism programs and evoke distinct patterns of oxidative and DNA replication stress. Mol Oncol 9(3):601–616. https://doi.org/10.1016/j.molonc.2014.11.001

McIntosh D, Blow JJ (2012) Dormant origins, the licensing checkpoint, and the response to replicative stresses. Cold Spring Harb Perspect Biol 4(10):012955. https://doi.org/10.1101/cshperspect.a012955

Meraldi P, Lukas J, Fry AM, Bartek J, Nigg EA (1999) Centrosome duplication in mammalian somatic cells requires E2F and Cdk2-cyclin A. Nat Cell Biol 1(2):88–93. https://doi.org/10.1038/10054

Minella AC, Swanger J, Bryant E, Welcker M, Hwang H, Clurman BE (2002) p53 and p21 form an inducible barrier that protects cells against cyclin E-cdk2 deregulation. Curr Biol 12(21):1817–1827

Minella AC, Grim JE, Welcker M, Clurman BE (2007) p53 and SCF^{Fbw7} cooperatively restrain cyclin E-associated genome instability. Oncogene 26(48):6948–6953. https://doi.org/10.1038/sj.onc.1210518

Minella AC, Loeb KR, Knecht A, Welcker M, Varnum-Finney BJ, Bernstein ID, Roberts JM, Clurman BE (2008) Cyclin E phosphorylation regulates cell proliferation in hematopoietic and epithelial lineages in vivo. Genes Dev 22(12):1677–1689. https://doi.org/10.1101/gad.1650208

Miron K, Golan-Lev T, Dvir R, Ben-David E, Kerem B (2015) Oncogenes create a unique landscape of fragile sites. Nat Commun 6:7094. https://doi.org/10.1038/ncomms8094

Moberg KH, Bell DW, Wahrer DC, Haber DA, Hariharan IK (2001) Archipelago regulates cyclin E levels in Drosophila and is mutated in human cancer cell lines. Nature 413(6853):311–316. https://doi.org/10.1038/35095068

Mohebi S, Mizuno K, Watson A, Carr AM, Murray JM (2015) Checkpoints are blind to replication restart and recombination intermediates that result in gross chromosomal rearrangements. Nat Commun 6:6357. https://doi.org/10.1038/ncomms7357

Morris L, Allen KE, La Thangue NB (2000) Regulation of E2F transcription by cyclin E-Cdk2 kinase mediated through p300/CBP co-activators. Nat Cell Biol 2(4):232–239. https://doi.org/10.1038/35008660

Muller-Tidow C, Metzger R, Kugler K, Diederichs S, Idos G, Thomas M, Dockhorn-Dworniczak B, Schneider PM, Koeffler HP, Berdel WE, Serve H (2001) Cyclin E is the only cyclin-dependent kinase 2-associated cyclin that predicts metastasis and survival in early stage non-small cell lung cancer. Cancer Res 61(2):647–653

Muntean AG, Hess JL (2012) The pathogenesis of mixed-lineage leukemia. Annu Rev Pathol 7:283–301. https://doi.org/10.1146/annurev-pathol-011811-132434

Mussman JG, Horn HF, Carroll PE, Okuda M, Tarapore P, Donehower LA, Fukasawa K (2000) Synergistic induction of centrosome hyperamplification by loss of p53 and cyclin E overexpression. Oncogene 19(13):1635–1646. https://doi.org/10.1038/sj.onc.1203460

Neelsen KJ, Lopes M (2015) Replication fork reversal in eukaryotes: from dead end to dynamic response. Nat Rev Mol Cell Biol 16(4):207–220. https://doi.org/10.1038/nrm3935

Neelsen KJ, Zanini IM, Herrador R, Lopes M (2013) Oncogenes induce genotoxic stress by mitotic processing of unusual replication intermediates. J Cell Biol 200(6):699–708. https://doi.org/10.1083/jcb.201212058

Nik-Zainal S, Davies H, Staaf J, Ramakrishna M, Glodzik D, Zou X, Martincorena I, Alexandrov LB, Martin S, Wedge DC, Van Loo P, Ju YS, Smid M, Brinkman AB, Morganella S, Aure MR, Lingjaerde OC, Langerod A, Ringner M, Ahn SM, Boyault S, Brock JE, Broeks A, Butler A, Desmedt C, Dirix L, Dronov S, Fatima A, Foekens JA, Gerstung M, Hooijer GK, Jang SJ, Jones DR, Kim HY, King TA, Krishnamurthy S, Lee HJ, Lee JY, Li Y, McLaren S, Menzies A, Mustonen V, O'Meara S, Pauporte I, Pivot X, Purdie CA, Raine K, Ramakrishnan K, Rodriguez-Gonzalez FG, Romieu G, Sieuwerts AM, Simpson PT, Shepherd R, Stebbings L, Stefansson OA, Teague J, Tommasi S, Treilleux I, Van den Eynden GG, Vermeulen P, Vincent-Salomon A, Yates L, Caldas C, van't Veer L, Tutt A, Knappskog S, Tan BK, Jonkers J, Borg A, Ueno NT, Sotiriou C, Viari A, Futreal PA, Campbell PJ, Span PN, Van Laere S, Lakhani SR, Eyfjord JE, Thompson AM, Birney E, Stunnenberg HG, van de Vijver MJ, Martens JW, Borresen-Dale AL, Richardson AL, Kong G, Thomas G, Stratton MR (2016) Landscape of somatic mutations in 560 breast cancer whole-genome sequences. Nature 534(7605):47–54. https://doi.org/10.1038/nature17676

Ohtani K, DeGregori J, Nevins JR (1995) Regulation of the cyclin E gene by transcription factor E2F1. Proc Natl Acad Sci U S A 92(26):12146–12150

Ohtani K, DeGregori J, Leone G, Herendeen DR, Kelly TJ, Nevins JR (1996) Expression of the *HsOrc1* gene, a human *ORC1* homolog, is regulated by cell proliferation via the E2F transcription factor. Mol Cell Biol 16(12):6977–6984

Ohtani K, Tsujimoto A, Ikeda M, Nakamura M (1998) Regulation of cell growth-dependent expression of mammalian CDC6 gene by the cell cycle transcription factor E2F. Oncogene 17(14):1777–1785. https://doi.org/10.1038/sj.onc.1202105

Ohtani K, Iwanaga R, Nakamura M, Ikeda M, Yabuta N, Tsuruga H, Nojima H (1999) Cell growth-regulated expression of mammalian MCM5 and MCM6 genes mediated by the transcription factor E2F. Oncogene 18(14):2299–2309. https://doi.org/10.1038/sj.onc.1202544

Ohtsubo M, Roberts JM (1993) Cyclin-dependent regulation of G1 in mammalian fibroblasts. Science 259(5103):1908–1912

Okuda M, Horn HF, Tarapore P, Tokuyama Y, Smulian AG, Chan PK, Knudsen ES, Hofmann IA, Snyder JD, Bove KE, Fukasawa K (2000) Nucleophosmin/B23 is a target of CDK2/cyclin E in centrosome duplication. Cell 103(1):127–140

Oswald F, Dobner T, Lipp M (1996) The E2F transcription factor activates a replication-dependent human H2A gene in early S phase of the cell cycle. Mol Cell Biol 16(5):1889–1895

Ozeri-Galai E, Lebofsky R, Rahat A, Bester AC, Bensimon A, Kerem B (2011) Failure of origin activation in response to fork stalling leads to chromosomal instability at fragile sites. Mol Cell 43(1):122–131. https://doi.org/10.1016/j.molcel.2011.05.019

Ozeri-Galai E, Tur-Sinai M, Bester AC, Kerem B (2014) Interplay between genetic and epigenetic factors governs common fragile site instability in cancer. Cell Mol Life Sci 71(23):4495–4506. https://doi.org/10.1007/s00018-014-1719-8

Pearson BE, Nasheuer HP, Wang TS (1991) Human DNA polymerase α gene: sequences controlling expression in cycling and serum-stimulated cells. Mol Cell Biol 11(4):2081–2095

Polyak K, Lee MH, Erdjument-Bromage H, Koff A, Roberts JM, Tempst P, Massague J (1994) Cloning of $p27^{Kip1}$, a cyclin-dependent kinase inhibitor and a potential mediator of extracellular antimitogenic signals. Cell 78(1):59–66

Porter PL, Malone KE, Heagerty PJ, Alexander GM, Gatti LA, Firpo EJ, Daling JR, Roberts JM (1997) Expression of cell-cycle regulators p27^{Kip1} and cyclin E, alone and in combination, correlate with survival in young breast cancer patients. Nat Med 3(2):222–225

Rajagopalan H, Jallepalli PV, Rago C, Velculescu VE, Kinzler KW, Vogelstein B, Lengauer C (2004) Inactivation of hCDC4 can cause chromosomal instability. Nature 428(6978):77–81. https://doi.org/10.1038/nature02313

Resnitzky D, Gossen M, Bujard H, Reed SI (1994) Acceleration of the G1/S phase transition by expression of cyclins D1 and E with an inducible system. Mol Cell Biol 14(3):1669–1679

Reynaud EG, Pelpel K, Guillier M, Leibovitch MP, Leibovitch SA (1999) p57^{Kip2} stabilizes the MyoD protein by inhibiting cyclin E-Cdk2 kinase activity in growing myoblasts. Mol Cell Biol 19(11):7621–7629

Rohban S, Campaner S (2015) Myc induced replicative stress response: how to cope with it and exploit it. Biochim Biophys Acta 1849(5):517–524. https://doi.org/10.1016/j.bbagrm.2014.04.008

Ruffner H, Jiang W, Craig AG, Hunter T, Verma IM (1999) BRCA1 is phosphorylated at serine 1497 in vivo at a cyclin-dependent kinase 2 phosphorylation site. Mol Cell Biol 19(7):4843–4854

Schraml P, Bucher C, Bissig H, Nocito A, Haas P, Wilber K, Seelig S, Kononen J, Mihatsch MJ, Dirnhofer S, Sauter G (2003) Cyclin E overexpression and amplification in human tumours. J Pathol 200(3):375–382. https://doi.org/10.1002/path.1356

Schwaenen C, Viardot A, Berger H, Barth TF, Bentink S, Dohner H, Enz M, Feller AC, Hansmann ML, Hummel M, Kestler HA, Klapper W, Kreuz M, Lenze D, Loeffler M, Moller P, Muller-Hermelink HK, Ott G, Rosolowski M, Rosenwald A, Ruf S, Siebert R, Spang R, Stein H, Truemper L, Lichter P, Bentz M, Wessendorf S (2009) Microarray-based genomic profiling reveals novel genomic aberrations in follicular lymphoma which associate with patient survival and gene expression status. Genes Chromosom Cancer 48(1):39–54. https://doi.org/10.1002/gcc.20617

Scuderi R, Palucka KA, Pokrovskaja K, Bjorkholm M, Wiman KG, Pisa P (1996) Cyclin E overexpression in relapsed adult acute lymphoblastic leukemias of B-cell lineage. Blood 87(8):3360–3367

Shima N, Alcaraz A, Liachko I, Buske TR, Andrews CA, Munroe RJ, Hartford SA, Tye BK, Schimenti JC (2007) A viable allele of Mcm4 causes chromosome instability and mammary adenocarcinomas in mice. Nat Genet 39(1):93–98. https://doi.org/10.1038/ng1936

Silva FP, Morolli B, Storlazzi CT, Anelli L, Wessels H, Bezrookove V, Kluin-Nelemans HC, Giphart-Gassler M (2003) Identification of RUNX1/AML1 as a classical tumor suppressor gene. Oncogene 22(4):538–547. https://doi.org/10.1038/sj.onc.1206141

Siu KT, Rosner MR, Minella AC (2012) An integrated view of cyclin E function and regulation. Cell Cycle (Georgetown, Tex) 11(1):57–64. https://doi.org/10.4161/cc.11.1.18775

Siu KT, Xu Y, Swartz KL, Bhattacharyya M, Gurbuxani S, Hua Y, Minella AC (2014) Chromosome instability underlies hematopoietic stem cell dysfunction and lymphoid neoplasia associated with impaired Fbw7-mediated cyclin E regulation. Mol Cell Biol 34(17):3244–3258. https://doi.org/10.1128/mcb.01528-13

Smith AP, Henze M, Lee JA, Osborn KG, Keck JM, Tedesco D, Bortner DM, Rosenberg MP, Reed SI (2006) Deregulated cyclin E promotes p53 loss of heterozygosity and tumorigenesis in the mouse mammary gland. Oncogene 25(55):7245–7259. https://doi.org/10.1038/sj.onc.1209713

Spruck CH, Won KA, Reed SI (1999) Deregulated cyclin E induces chromosome instability. Nature 401(6750):297–300. https://doi.org/10.1038/45836

Srinivasan SV, Dominguez-Sola D, Wang LC, Hyrien O, Gautier J (2013) Cdc45 is a critical effector of myc-dependent DNA replication stress. Cell Rep 3(5):1629–1639. https://doi.org/10.1016/j.celrep.2013.04.002

Strohmaier H, Spruck CH, Kaiser P, Won KA, Sangfelt O, Reed SI (2001) Human F-box protein hCdc4 targets cyclin E for proteolysis and is mutated in a breast cancer cell line. Nature 413(6853):316–322. https://doi.org/10.1038/35095076

Teixeira LK, Wang X, Li Y, Ekholm-Reed S, Wu X, Wang P, Reed SI (2015) Cyclin E deregulation promotes loss of specific genomic regions. Curr Biol 25(10):1327–1333. https://doi.org/10.1016/j.cub.2015.03.022

Thean LF, Loi C, Ho KS, Koh PK, Eu KW, Cheah PY (2010) Genome-wide scan identifies a copy number variable region at 3q26 that regulates PPM1L in APC mutation-negative familial colorectal cancer patients. Genes Chromosom Cancer 49(2):99–106. https://doi.org/10.1002/gcc.20724

Thys RG, Lehman CE, Pierce LC, Wang YH (2015) DNA secondary structure at chromosomal fragile sites in human disease. Curr Genomics 16(1):60–70. https://doi.org/10.2174/1389202916666150114223205

Tokuyama Y, Horn HF, Kawamura K, Tarapore P, Fukasawa K (2001) Specific phosphorylation of nucleophosmin on Thr^{199} by cyclin-dependent kinase 2-cyclin E and its role in centrosome duplication. J Biol Chem 276(24):21529–21537. https://doi.org/10.1074/jbc.M100014200

Torres-Rosell J, De Piccoli G, Cordon-Preciado V, Farmer S, Jarmuz A, Machin F, Pasero P, Lisby M, Haber JE, Aragon L (2007) Anaphase onset before complete DNA replication with intact checkpoint responses. Science 315(5817):1411–1415. https://doi.org/10.1126/science.1134025

Toyoshima H, Hunter T (1994) p27, a novel inhibitor of G1 cyclin-Cdk protein kinase activity, is related to p21. Cell 78(1):67–74

Varetti G, Pellman D, Gordon DJ (2014) Aurea mediocritas: the importance of a balanced genome. Cold Spring Harb Perspect Biol 6(11):a015842. https://doi.org/10.1101/cshperspect.a015842

Weddington N, Stuy A, Hiratani I, Ryba T, Yokochi T, Gilbert DM (2008) Replication domain: a visualization tool and comparative database for genome-wide replication timing data. BMC Bioinformatics 9:530. https://doi.org/10.1186/1471-2105-9-530

Won KA, Reed SI (1996) Activation of cyclin E/CDK2 is coupled to site-specific autophosphorylation and ubiquitin-dependent degradation of cyclin E. EMBO J 15(16):4182–4193

Xiong Y, Hannon GJ, Zhang H, Casso D, Kobayashi R, Beach D (1993) p21 is a universal inhibitor of cyclin kinases. Nature 366(6456):701–704. https://doi.org/10.1038/366701a0

Yamada H, Yanagisawa K, Tokumaru S, Taguchi A, Nimura Y, Osada H, Nagino M, Takahashi T (2008) Detailed characterization of a homozygously deleted region corresponding to a candidate tumor suppressor locus at 21q11-21 in human lung cancer. Genes Chromosom Cancer 47(9):810–818. https://doi.org/10.1002/gcc.20582

Yan Z, DeGregori J, Shohet R, Leone G, Stillman B, Nevins JR, Williams RS (1998) Cdc6 is regulated by E2F and is essential for DNA replication in mammalian cells. Proc Natl Acad Sci U S A 95(7):3603–3608

Yoshida K, Inoue I (2004) Regulation of geminin and Cdt1 expression by E2F transcription factors. Oncogene 23(21):3802–3812. https://doi.org/10.1038/sj.onc.1207488

Zariwala M, Liu J, Xiong Y (1998) Cyclin E2, a novel human G1 cyclin and activating partner of CDK2 and CDK3, is induced by viral oncoproteins. Oncogene 17(21):2787–2798. https://doi.org/10.1038/sj.onc.1202505

Zeman MK, Cimprich KA (2014) Causes and consequences of replication stress. Nat Cell Biol 16(1):2–9. https://doi.org/10.1038/ncb2897

Zhao J, Kennedy BK, Lawrence BD, Barbie DA, Matera AG, Fletcher JA, Harlow E (2000) NPAT links cyclin E-Cdk2 to the regulation of replication-dependent histone gene transcription. Genes Dev 14(18):2283–2297

Zhao H, Chen X, Gurian-West M, Roberts JM (2012) Loss of cyclin-dependent kinase 2 (CDK2) inhibitory phosphorylation in a CDK2AF knock-in mouse causes misregulation of DNA replication and centrosome duplication. Mol Cell Biol 32(8):1421–1432. https://doi.org/10.1128/mcb.06721-11

Chapter 23
Replication Through Repetitive DNA Elements and Their Role in Human Diseases

Advaitha Madireddy and Jeannine Gerhardt

Abstract Human cells contain various repetitive DNA sequences, which can be a challenge for the DNA replication machinery to travel through and replicate correctly. Repetitive DNA sequence can adopt non-B DNA structures, which could block the DNA replication. Prolonged stalling of the replication fork at the endogenous repeats in human cells can have severe consequences such as genome instability that includes repeat expansions, contractions, and chromosome fragility. Several neurological and muscular diseases are caused by a repeat expansion. Furthermore genome instability is the major cause of cancer. This chapter describes some of the important classes of repetitive DNA sequences in the mammalian genome, their ability to form secondary DNA structures, their contribution to replication fork stalling, and models for repeat expansion as well as chromosomal fragility. Included in this chapter are also some of the strategies currently employed to detect changes in DNA replication and proteins that could prevent the repeat-mediated disruption of DNA replication in human cells. Additionally summarized are the consequences of repeat-associated perturbation of the DNA replication, which could lead to specific human diseases.

Keywords DNA replication • Repeat sequences • Human diseases • Replication fork stalling • DNA helicases • Secondary DNA structures • Non-B DNA • Repeat expansion • Genome instability

A. Madireddy
Albert Einstein College of Medicine, Bronx, NY, USA

J. Gerhardt (✉)
Weill Cornell Medicine, New York, NY, USA
e-mail: jeannine.gerhardt@gmail.com

H. Masai, M. Foiani (eds.), *DNA Replication*, Advances in Experimental Medicine and Biology 1042, https://doi.org/10.1007/978-981-10-6955-0_23

23.1 Introduction

A considerable fraction of the genome in nearly all organisms consists of repetitive DNA sequences. It has been suggested that these repetitive sequences might play a role in several cellular processes, such as recombination, nuclear architecture, and transcription. In human cells two-thirds of the genome sequence is compromised by repetitive DNA sequences (de Koning et al. 2011), which can be broadly classified into two major groups: interspersed repeats and tandem repeats. Interspersed repeats are repeated sequences that are scattered throughout the genome. Tandem repeats are repetitions of one or more nucleotides directly adjacent to each other. Depending on the size of the repeat unit, tandem repeats can be further categorized into satellite, minisatellite, and microsatellite DNA.

During DNA replication, each cell copies 3 billion base pairs within a period of 6–8 h, with minimal errors. Repetitive DNA sequences can be challenging for the replication machinery to replicate accurately because repeats are able to adopt alternative secondary DNA structures, which consist of non-B double helix DNA. During DNA replication, these secondary repeat sequences could cause replication fork stalling, which can lead to replication fork reversal or in extreme cases replication fork collapse. The consequences of replication fork stalling could be DNA polymerase slippage or incomplete DNA duplication, leading to genome instability such as repeat expansions (reviewed in Kim and Mirkin 2013; Lopez Castel et al. 2010) and chromosome fragility (reviewed in Durkin and Glover 2007).

It is normal to find variation in the length of repetitive DNA sequences in humans. This variation in the repeat size can be used as a tool to distinguish a person from another during forensics analysis. However, in some individuals the repetitive sequences are expanded beyond the normal threshold causing several human diseases. Large repetitive sequences in the genome can disturb or modify gene transcription, cause alternate mRNA transcription and mutant protein expression, or silence a gene entirely. In turn, this can lead to human diseases such as hereditary neurological and muscular repeat expansion diseases (Lopez Castel et al. 2010). Besides the effect on gene transcription, repeats can also alter the epigenetic landscape and cause genome fragility in human cells (Durkin and Glover 2007). Stalling of the replication machinery at repetitive DNA sequences could delay the DNA synthesis and prompt fork collapse leading to double-strand DNA breaks. Various forms of cancers are suspected to be the result of incomplete DNA replication and genome instability at specific repetitive sequences called fragile sites.

Fragile sites are broadly classified into rare fragile sites (RFS), common fragile sites (CFS), early replicating fragile sites (ERFs), and regions that resemble fragile sites such as telomeres. Treatment with low doses of aphidicolin induces CFS expression (Glover et al. 1984; Hecht and Glover 1984), while incubation with hydroxyurea exacerbates breaks at ERFs in human cells. However, treatment with chemotherapeutic drugs could destabilize both classes of fragile sites (Dillon et al. 2010). CFS and ERFs consist of non-expanding repeats and are an inherent part of the cellular genome, whereas RFS are only present in a few individuals where the repeats are abnormally expanded. Interestingly, CFS and ERFs have numerous chromosomal locations and exist in all humans, whereas RFS are only expressed in rare individuals containing the disorder and are located at a specific chromosomal location; these are perhaps the most important distinctions between RFS and the other classes of fragile sites.

Considering the deleterious effects these repetitive DNA sequences in the human genome can have, it is very surprising that fragile sites are evolutionarily conserved in mammals across different species (Smeets and van de Klundert 1990). Their evolutionary conservation has led to the speculation that they have a protective role in the cell. One possible explanation for their conservation is that they were engineered to act as sensors and first responders of endogenous toxicity, to alert the DNA damage response pathway (O'Keefe and Richards 2006). However, not all repetitive DNA sequences are fragile sites in the genome. For example, chromosomal fragility is observed only in a few repeat expansion diseases like at the expanded CGG repeats (FRAXA site) in cells of fragile X syndrome (FXS) patients. In this section we will discuss unusual DNA structures and effects of non-fragile as well as fragile repetitive DNA sequences, which as a consequence of their repetitive nature can lead to human disorders.

The first repeat expansion diseases were discovered in 1991 when several groups found that the root of the FXS, an X-linked inherited disease causing intellectual disabilities, is a CGG repeat expansion (Heitz et al. 1991; Pieretti et al. 1991; Verkerk et al. 1991). In the same year, another inherited X-linked disease, spinal and bulbar muscular atrophy (also called Kennedy's disease), was discovered, which is caused by a CAG repeat expansion in the androgen receptor gene (La Spada et al. 1991). These diseases belong to the trinucleotide repeat expansion (TNR) diseases, which form the largest category of repeat expansion diseases currently known (see Table 23.1). Though most of these disorders are a consequence of repeat expansions, multiple skeletal dysplasias are an exception where the symptoms can appear from the addition or deletion of a repeat sequence. While there are several diseases caused by a repeat expansion, not all of them contain a RFS. FXS is an example of a disease that is caused by CGG repeat expansions that has been also characterized as a RFS. In contrast, for example, the expanded GAA repeats in Friedreich's ataxia cells do not show fragility in vivo and are hence not classified as RFS. Repeat expansion diseases can also be triggered by larger repeat sequences, such as tetra-, penta-, and hexanucleotides, and even dodecanucleotide repeats. For instance, myotonic dystrophy type 2 (DM2) is caused by a CCTG expansion and spinocerebellar ataxia type 10 (SCA10) by an AATCT repeat expansion (reviewed in Lopez Castel et al. 2010; Mirkin 2007). Additionally more and more diseases caused by repeat instability are being discovered, for example, recently it was determined that in a fraction of patients with amyotrophic lateral sclerosis (ALS, also known as Lou Gehrig's disease), the disease symptoms are triggered by a GGGGCC repeat expansion (DeJesus-Hernandez et al. 2011). However, currently there is no cure for these repeat expansion diseases, and the mechanism leading to repeat instability is still unclear.

Table 23.1 Examples of diseases caused by a repeat expansion: summarized are repeat expansion diseases, the potential secondary DNA structures formed by the repeats, and effects the repeat expansion has at the endogenous locus and on the molecular processes in the cell

Disease	Repeat	Structure	Gene	Effect
Promoter				
Epilepsy progressive myoclonic (EPM1)	C_4GC_4GCG	G-quadruplex, hairpins, i-motif	CSTB	Reduced gene transcription
Spinocerebellar ataxia 12	CAG	Hairpins	PPP2R2B	Altered expression of splice variants
5′UTR				
Fragile X syndrome (FXS)	CGG	G-quadruplex, hairpins	FMR1	Epigenetic modification, FMR1 gene silencing
FXS-associated tremor/ ataxia syndrome (FXTAS)	CGG	G-quadruplex, hairpins	FMR1	Increased FMR1 transcription, FMRpolyG inclusions
FXS primary ovarian insufficiency (FXPOI)	CGG	G-quadruplex, hairpins	FMR1	Increased FMR1 transcription, FMRpolyG inclusions
Fragile X syndrome E (FRAXE)	CCG	Hairpins	FMR2	FMR2 gene silencing
Exons				
Huntington's disease (HD)	CAG	Hairpins	HTT	Mutant protein expression (huntingtin)
Spinocerebellar ataxia 1	CAG	Hairpins	ATXN1	Polyglutamine protein
Spinocerebellar ataxia 2	CAG	Hairpins	ATXN2	Polyglutamine protein
Spinocerebellar ataxia 3	CAG	Hairpins	ATXN3	Polyglutamine protein
Spinocerebellar ataxia 6	CAG	Hairpins	CACNA1A	Polyglutamine protein
Spinocerebellar ataxia 7	CAG	Hairpins	ATXN7	Polyglutamine protein
Spinocerebellar ataxia 17	CAG	Hairpins	TBP	Polyglutamine protein
Kennedy's disease; spinal and bulbar muscular atrophy (SBMA)	CAG	Hairpins	AR	Mutant protein expression (mutant androgen receptor)
Dentatorubral-pallidoluysian atrophy (DRPLA)	CAG	Hairpins	ATN1	Polyglutamine protein
Introns				
Friedreich's ataxia (FRDA)	GAA	Triplexes	FXN	Epigenetic modifications, reduced gene transcription
Spinocerebellar ataxia 10	ATTCT	Unpaired	ATXN10	Loss of function
Spinocerebellar ataxia 31	TGGAA	Purine-rich duplexes	TK2/BEAN	Form RNA aggregates
Myotonic dystrophy type 2 (DM2)	CCTG	Hairpins	CNPB (ZNF9)	Form RNA aggregates
Frontotemporal dementia/ amyotrophic lateral sclerosis	GGGGCC	G-quadruplex	C9orf72	Form RNA aggregates, reduced gene transcription

(continued)

Table 23.1 (continued)

Disease	Repeat	Structure	Gene	Effect
3′UTR				
Myotonic dystrophy type 1 (DM1)	CTG	Hairpins	DMPK	Form RNA aggregates
Spinocerebellar ataxia 8 (SCA8)	CTG	Hairpins	ATXN8OS/ ATXN8	Form RNA aggregates polyglutamine protein
Huntington's disease-like 2 (HDL2)	CAG	Hairpins	JPH3	Form RNA aggregates

Several models that could lead to repeat instability were suggested including abnormalities in the DNA replication, DNA repair, transcription, and recombination process (reviewed by McMurray 2010; Mirkin 2007; Usdin et al. 2015). This section summarizes errors during DNA replication, which could trigger repeat expansion and chromosomal fragility in human cells. Furthermore disorders caused by repetitive DNA sequences are also discussed in this chapter.

23.2 Secondary Repeat Structures

Repeat sequences are able to form unusual secondary non-B DNA structures by engaging in hydrogen bonds that differ from the canonical Watson-Crick bonding structures. When the DNA is unwound, single-stranded repeat sequences are able to adapt secondary DNA structures, such as hairpins, triplexes, G-quadruplexes, i-motifs, DNA:RNA hybrids (R-loops), and slipped DNA structures (Gacy and McMurray 1998; Mirkin 2007; Pataskar et al. 2001; Wells 2008), which are shown to form in vitro plus some of these secondary repeat structures were also shown to form in vivo.

23.2.1 Secondary Structures Formed by Expanding Repeats

In vitro, it was found that CTG, CGG, CAG, and CCG form imperfect hairpins whose stability depends on the repeat sequences (Gacy et al. 1995; Paiva and Sheardy 2004). In addition, CGG repeats were implicated in the formation of tetrahelical structures called G-quadruplexes (Fry and Loeb 1994). GAA repeats are able to form intermolecular (H-DNA) or intramolecular ("sticky DNA") triplexes (Wells 2008). Although such structures are shown to form in vitro, it is not clear yet whether they form in vivo at the endogenous repeats in the human genome. Several evidences suggest that such structures are indeed formed at the endogenous disease locus. A good example of this is the observation that AGG interruptions in the CGG repeats in FXS patients are linked to fewer repeat expansion events (Nolin et al. 2015). The interruptions in the repeat sequence could prevent the formation of specific secondary DNA structures at the genomic locus in patient cells. Similarly, other TNR expansion diseases such as Friedreich's ataxia (FRDA), myotonic dystrophy type 1 (DM1), and spinocerebellar ataxia diseases (SCA) contain interruptions in the repeat sequences which stabilize the repeats (Lopez Castel et al. 2010; Pearson et al. 1998). In addition, it was discovered that Msh2, which is shown to bind to hairpins formed by CTG repeats (Pearson et al. 1997) and CAG repeats (Owen et al. 2005) in vitro, can be found at the expanded GAA repeat tract at the endogenous *Frataxin* (*FXN*) gene locus in FRDA cells in vivo (Du et al. 2012). This indicates that Msh2 might bind to secondary repeat structures formed by the endogenous repeats at the disease locus.

Recently some of these secondary structures such as G-quadruplexes and R-loops were detected in vivo in human cells (Lam et al. 2013; Phillips et al. 2013). R-loops are structures that are formed during transcription when nascent RNA hybridizes to the DNA template behind the elongating RNA polymerase. A newly developed antibody was used to detect R-loops at the endogenous repeats in FRDA and FXS cells (Groh et al. 2014; Lam et al. 2013; Loomis et al. 2014). Because the *fragile X mental retardation (FMR1)* gene is silenced in FXS cells, the *FMR1* gene transcription was reactivated before analysis of the R-loops by treatment with the DNA methylation inhibitor 5-aza-2′-deoxycytidine (Groh et al. 2014). Using antibodies it was also reported that triplexes are formed in human cells in vivo (Agazie et al. 1994; Burkholder et al. 1988; Lee et al. 1987). However it is still not clear whether secondary structures such as hairpins, G-quadruplexes, and triplexes are formed at the endogenous repeats in disease cells. Small molecules, which prevent triplex formation in vitro, were used to release the stalled replication forks at the GAA repeats indicating that GAA triplexes may indeed form at the expanded GAA repeats in FRDA cells (Gerhardt et al. 2016).

Nevertheless, despite the disturbances secondary repeat structures could cause, certain noncanonical repeat structures appear to be involved in normal cell function. The high concentration of G-quadruplex motifs near promoter regions indicates potential function for G-quadruplex structures in gene regulation. G-quadruplexes were also suggested to have a role in replication initiation, recombination, or meiosis (Bochman et al. 2012). R-loops were suggested to function as regulators of chromatin structure and transcription (Al-Hadid and Yang 2016). In addition, triplexes were implicated in transcriptional regulation, epigenetic modification, and organization of the chromatin structure (Buske et al. 2011). Triplexes may also be involved in posttranscriptional processing of RNAs (Buske et al. 2011). However, these non-B DNA repeat structures could also perhaps hinder RNA polymerases. Repeat-mediated inhibition of the transcription by formation of secondary repeat structures in vitro has been observed in model systems (Pandey et al. 2015). Thus, transcription inhibition and DNA polymerase slippage at the repeats could lead to reduced gene expression and repeat expansion. Furthermore increasing repeat number worsens the symptoms of the disorders and causes an earlier start of the disease in patients. The process where these repeat expansions (also called dynamic mutations) increase in size through generations, resulting in an earlier onset of the disease and increased severity of symptoms, is called genetic anticipations.

23.2.2 *Secondary Structures Formed by Non-expanding Repeats*

Non-expanding repeats exist at numerous sites in the human genome and can be composed of either A/T-rich repeats (e.g., at CFS loci) or G-rich repeats (e.g., at ERFs loci, telomeres). Both classes of repetitive sequences have been shown to have the potential to form a variety of secondary structures similar to expanding repeats. Instability at CFS loci has been widely attributed to the replication stalling at secondary structure-prone repetitive DNA elements (Schwartz et al. 2006; Walsh et al. 2013). Polymorphic A/T-rich repeat sequences at CFS-FRA16D can potentially form hairpins as well as cruciform DNA structures that could stall replication machinery in a manner that is dependent on the length of the repeat sequence (H. Zhang and Freudenreich 2007). In addition to $[TA]_n$ dinucleotide repeat sequences, mononucleotide $[A]_{28}$ repeats located at the FRA16D locus have also been shown to stall replication forks in vitro (Walsh et al. 2013).

Another class of repetitive sequences that are strongly implicated in genome instability are quasi-palindromes. At CFS, biochemical studies indicate that quasi-palindrome sequences can form stem structures that can stall replication forks (Walsh et al. 2013). Computational analysis of GC-rich sequences at ERFs revealed that they have the potential to form G-quadruplex structures (Madireddy et al. unpublished). While the G-rich DNA at ERFs has the potential to form secondary structures such as R-loops and G-quadruplexes, their contribution to ERFs instability is yet to be determined. However, the formation of most of these secondary structures has been observed so far by biochemical studies carried out in vitro due to the lack of suitable reagents (e.g., antibodies) that can facilitate their detection in vivo.

In addition to repeat-associated secondary structures, current literature suggests that transcription-associated obstacles are a major cause for instability at both ERFs that contain actively transcribed genes and CFS that contain very long genes (Helmrich et al. 2011, 2013; Mortusewicz et al. 2013). Collisions between the transcriptional machinery and replication machinery and/or R-loops can lead to genomic instability by obstructing the progression of replication machinery or by making the cell more susceptible to genotoxic stress (Aguilera and Garcia-Muse 2012; Sollier and Cimprich 2015).

23.3 Replication Fork Stalling at the Repetitive DNA

Blockage of replication forks at the repeats was observed first in vitro using various DNA polymerases and model substrates. Primer extension assays have been used to demonstrate polymerase pausing at CTG, CGG, and GAA repeats (Gacy et al. 1998; Kang et al. 1995; Usdin and Woodford 1995). Later, the first in vivo replication assay confirmed that replication fork stalling occurs at the CGG, CTG (Pelletier et al. 2003; Samadashwily et al. 1997; Voineagu et al. 2009), and GAA repeats

(Krasilnikova and Mirkin 2004). An increase in replication fork stalling arises when the repeats expand to certain repeat lengths similar to the repeat lengths observed in patients. However, the question remains whether secondary repeat structures indeed form and stall DNA polymerases at endogenous genomic loci in human cells. For example, early studies showed that the fragile X locus is replicated later in S phase in FXS cells, in contrast to unaffected cells (Hansen et al. 1993; Subramanian et al. 1996). This suggested that DNA replication of the *FMR1* locus in FXS cells is slowed down probably by encountering an obstruction. Indeed recently a short replication fork pause was detected at the repeats at the endogenous *FMR1* locus in FXS human embryonic stem cells (hESCs) (Gerhardt et al. 2014) (Fig. 23.1). Surprisingly, pausing occurred at the short CGG repeats in unaffected cells to a similar degree as at the expanded CGG repeats in FXS patient cells. However, the presence of numerous human helicases that are capable of unwinding secondary structures such as G-quadruplexes quickly could explain the relatively short pause of the replication fork at the endogenous CGG repeats.

How about other secondary structures such as triplexes, DNA cruciform, and slipped-stranded structures where both DNA strands are involved in the formation of the secondary DNA structures? GAA repeats, which are able to form triplexes, when inserted into plasmids were shown to transiently pause (Krasilnikova and Mirkin 2004) and reverse the replication forks in human cells (Follonier et al. 2013). Recently, a major replication fork stall was observed in FRDA-induced pluripotent stem cells (iPSCs) at the endogenous *Frataxin* locus (Gerhardt et al. 2016) (Fig. 23.1). The intensity of the fork stall at the GAA repeats was much higher than in FXS cells indicating that the expanded GAA repeats pose a bigger challenge than the expanded CGG repeats for the DNA polymerases and perhaps even the DNA helicases. During replication fork stalling, the DNA polymerase slippage could add additional repeats, which would explain the observed GAA repeat expansion in FRDA cells. Interestingly, when triplex formation was inhibited by treatment of FRDA stem cells with small molecules, the replication fork stall was reversed (Gerhardt et al. 2016), and the GAA repeats become stable (Du et al. 2012). This indicates that the release of the stalled forks and stabilization of the repeats could have a therapeutic effect for patients where the repeats continue to expand during their lifetime. In this manner, stabilizing the *Frataxin* gene transcription by preventing GAA repeat expansion could be a therapeutic strategy to improve the lives of FRDA patients.

In addition, in vitro and in vivo studies suggest that the $[AT]_n$ or $[AT/TA]_n$ non-expanding flexible sequences found at CFS (Glover 2006; Lukusa and Fryns 2008; Zhang and Freudenreich 2007) lead to polymerase pausing (Walsh et al. 2013). Replication stalling has been shown to occur either in the presence of replicative stress-inducing agents such as aphidicolin (e.g., FRA16C) (Ozeri-Galai et al. 2011, 2013) or in the absence of essential proteins that facilitate replication fork movement at CFS (e.g., pausing at FRA16D in the absence of FANCD2) (Fig. 23.1 (Madireddy et al. 2016a). Whether the replication machinery also stalls at other CFS and expanded repeats, such as CAG and CTG, at the endogenous genomic locus in patient cells has to be still determined.

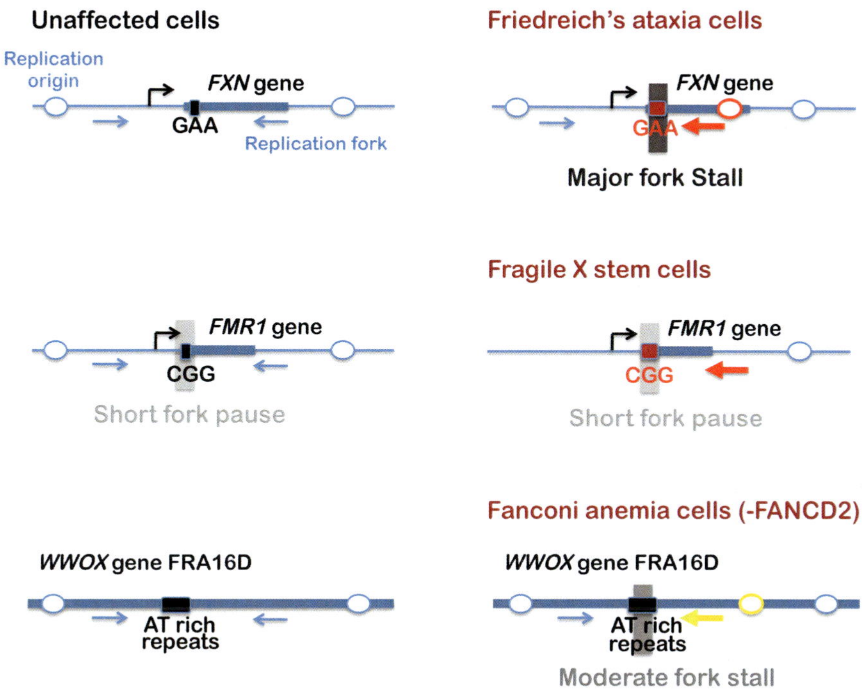

Fig. 23.1 Alterations in the DNA replication program in disease cells: severe changes in the DNA replication program are observed in Friedreich's ataxia (FRDA), fragile X syndrome (FXS), and Fanconi anemia complementation group D2 (FANCD2) cells. Fork stalling was detected at the endogenous repeats probably caused by secondary structures, such as triplexes, G-quadruplexes, and R-loops, in human patient cells (gray box, different shades represent the severity of the replication fork block). In addition the absence or activation of replication origins (red circles) alters the replication fork direction through the repeat in FRDA and FXS stem cells (red arrow). Replication fork stalling together with the altered replication fork direction possibly triggers repeat expansion in FRDA and FXS patient stem cells. At non-expanding repeats in FANCD2 patient cells, replication fork stalling potentially activates dormant origins (yellow circle) to rescue the stalled forks and complete DNA replication. Nonetheless spontaneous fragile site breaks occur in these patients probably caused by the inability to recruit proteins (e.g., helicases) in the absence of FANCD2 that eliminate secondary DNA structures

23.4 Models for Repeat Instability

Depending on the cell type, repeat sequence, repeat size, and location of the repeat in the genome, different mechanisms could trigger repeat instability. Several models suggest that errors during DNA replication, DNA repair, and DNA recombination process could cause instability of repetitive DNA sequences (Durkin and Glover 2007; McMurray 2010; Mirkin 2007; Usdin et al. 2015). In addition the repeat size plays a major role in the repeat instability mechanism since an increase in the repeat length could hinder cellular processes more severely and so worsen the instability of the repeat in human cells.

23.4.1 *Models Leading to Repeat Expansions and Contractions*

Diverse models of repeat expansion were proposed for different repeat expansion diseases in different patient tissues (Lopez Castel et al. 2010). For example, inaccuracies during the DNA replication process could lead to repeat expansions or contractions in proliferating cells. One model suggests that to prevent uncoupling of the DNA helicase from a stalled DNA polymerase at the repeat, the template DNA containing the repeat sequence is pushed out and bypassed by the DNA polymerase, causing repeat contractions (Delagoutte et al. 2008). Small changes in the repeat length could be a result of DNA polymerase slippage during DNA synthesis at the remaining 5′ flap or by misalignment of template and daughter DNA strands (Kim and Mirkin 2013). Large repeat expansions were proposed to occur during replication fork stalling after replication fork reversal (Follonier et al. 2013; Fouche et al. 2006; Kerrest et al. 2009) and generation of a four-way junction (chicken foot) structures. This could result in the formation of a secondary repeat structure and synthesis of additional repeats in the nascent DNA (Mirkin 2006). Another model suggests that repeat expansion could happen when the leading strand DNA polymerase switches templates using the nascent lagging strand as template and then switches back to the leading strand template leaving additional repeats, the size of an Okazaki fragment in the nascent DNA (Shishkin et al. 2009). The template-switch model was supported by experimental evidence showing that deletion of genes impeding the switch of the DNA templates increased the repeat expansion rate (Shishkin et al. 2009).

Cumulative findings indicate that a combination of several inaccuracies in different cellular processes including mutations in *cis*-elements and deregulation of *trans*-factors could facilitate repeat expansion or contraction in patient cells. Faulty regulation of several DNA repair proteins, such as Msh2 (MutS protein homolog 2), was suggested to promote repeat expansion in mammalian cells and mouse models (McMurray 2010; Usdin et al. 2015). *Trans*-factor, such as Msh2, could bind and stabilize these non-B DNA structures. Indeed human Msh2 is upregulated in stem cells making the cells more susceptible to repeat instability (Du et al. 2012; Gerhardt 2015). Replication fork stalling and polymerase slippage at the expanded repeats, together with upregulated Msh2 level, would explain the progressive repeat instability in disease cells already containing expanded repeat sequences. Although these *trans*-factors significantly facilitate repeat instability, repeat expansion occurs only at one specific genomic locus in trinucleotide repeat expansion disorders, suggesting that *cis*-elements close to the expanding repeats probably trigger repeat instability.

Only a few *cis*-elements have been identified in close proximity to the repeats in FXS, DM1, and SCA7 cells (Cleary et al. 2010; Gerhardt et al. 2014a, b; Libby et al. 2008, 2003). These *cis*-elements were detected through studies of mutations in the binding sites of the CCCTC-binding factor (CTCF) and the DNA replication program at or in close proximity to the repeats. In addition, both the length and the purity of the repeat tract influence repeat instability (Holloway et al. 2011). Interruptions in the repeat sequence probably reduce the formation of secondary DNA structures and decrease repeat expansion events in patients. Expanded uninterrupted repeats could trigger more frequent expansion events. However, the initial events triggering repeat expansion of the normal repeat length allele are unknown.

One possibility is that repeat instability is initiated in human cells by alterations in *cis*-elements near the repeats such as the position of replication origins, which could lead to altered replication fork progression and lead to a higher propensity of forming secondary repeat structures. In plasmid-based model systems, it was shown that the placement of a replication origin at a certain distance and orientation from the CGG, GAA, or CTG repeats can cause repeat expansion or contractions (Cleary et al. 2002; Freudenreich et al. 1997; Rindler et al. 2006; Samadashwily et al. 1997). Subsequently, in the origin switch model, it was suggested that a switch in the direction of the replication fork, for example, due to the absence of replication initiation sites could cause repeat expansions (Mirkin 2007). In another model, the origin shift model, repeat expansion was triggered by changes in the distance of the closest replication origins to the repeats causing alteration in the position of the repetitive sequence within Okazaki initiation zone (OIZ). A third model, which does not require altered replication origin activation or relocation, suggests that repeat expansion is triggered by *cis*-elements and *trans*-factors influencing the mode of the replication fork progression, thereby altering the location of the repeats in the single-stranded lagging strand template (Cleary and Pearson 2005; Mirkin 2007). In line with these models, in mouse tissues containing the human *DM1* locus, differential replication origin usage was observed (Cleary et al. 2010). Furthermore, we found that the replication fork direction is altered in FXS and FRDA stem cells due to altered location of replication origin(s) (Gerhardt et al. 2016; Gerhardt et al. 2014) (Fig. 23.1). We observed an increase in replication forks in the direction of the TTC repeats in the lagging strand template, which is the direction shown in experiments utilizing plasmid-based model systems to increase repeat instability (Rindler et al. 2006). The altered replication fork direction together with DNA polymerase stalling would explain the repeat instability in these proliferating cells.

However, an altered replication fork direction could also be the primary cause for repeat expansion and transcriptional inhibition in somatic patient tissues, including terminally differentiated nonproliferating cells. For example, human cardiomyocytes and some neuronal cells are polyploid (Adler and Friedburg 1986; Brodsky and Uryvaeva 1977), and genome multiplication (endoreplication or rereplication) has been speculated to promote cell survival and tissue regeneration under stressful conditions (Anatskaya and Vinogradov 2007). More recently rereplication was proposed as the cause for repeat instability in proliferating and differentiated cells under conditions of induced stress (Chatterjee et al. 2015). It was suggested that environmental stress could activate DNA rereplication in human cells (Chatterjee et al. 2015). Replication fork stalling during rereplication could then promote repeat expansion and further impede gene transcription in differentiated polyploid cells, such as the FRDA cardiomyocytes, thus contributing to the progression of the disease.

23.4.2 Models Leading to Repeat Instability of Non-expanding Repeats

Non-expanding repeats such as those found in CFS loci are stable under normal replicative conditions; however, exposure to replicative stress renders them highly susceptible to chromosomal breakage, deletions, translocations, and sister chromatid exchanges (Glover and Stein 1987, 1988; Wang et al. 1997). While it has been shown that replicative polymerases find it difficult to replicate through the repetitive sequences at fragile sites (Walsh et al. 2013), the primary cause for the polymerase stalling and the mechanisms that promote replication restart and completion have remained unclear. However, repeat instability at CFS has been attributed to the following classic features that define these genomic regions:

A/T-Rich Sequences CFS are A/T-rich, highly flexible genomic sequences (Arlt et al. 2002; Mishmar et al. 1998; Ried et al. 2000). Computational modeling has shown that CFS-derived repeat sequences composed of $[AT]_n$ or $[AT/TA]_n$ repeats have the potential to form stable secondary structures (Fungtammasan et al. 2012; Zhang and Freudenreich 2007). Accordingly, some in vitro and in vivo studies from different model systems have demonstrated that these structure-prone sequences at CFS can impede replicative polymerases such as DNA polymerase delta and lead to replication fork pausing (Burrow et al. 2010; Ozeri-Galai et al. 2013; Shah et al. 2010; Walsh et al. 2013) (Fig. 23.2a). Transfection of these repetitive sequences to other non-fragile regions renders the new regions unstable (Ragland et al. 2008); however, the partial removal of some of these A/T-rich sequences from CFS reduces but does not completely eliminate the CFS instability (Corbin et al. 2002). This shows that the presence of A/T-rich sequences while contributing to instability is not the only factor leading to CFS fragility.

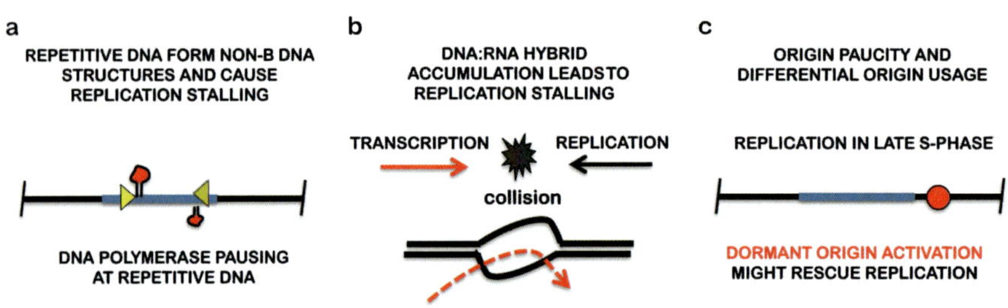

Fig. 23.2 Models of repeat instability: (**a**) Replication of repetitive DNA sequences can be delayed due to replication fork stalling at potential non-B DNA structures and so lead to repeat instability. (**b**) Collisions between transcription and replication machinery or stalling at transcription-associated DNA:RNA hybrids could cause repeat instability. (**c**) The activation of dormant origins, to complete replication, is one of the mechanisms by which replication of repetitive DNA sequences can be rescued. However, alterations in the location and scarcity of replication origins at late replicated regions in S phase lead to a risk of incomplete or altered DNA replication, which could result in repeat instability

Located in Long Genes The most frequently unstable CFS loci harbor long underlying genes. These genes have largely been classified as tumor suppressor genes or proto-oncogenes at some of the most commonly expressed CFS loci, such as FRA3B, FRA16D, and FRA7I, and instability at these sites have been closely correlated with multiple types of cancer (Bednarek et al. 2000; Ciullo et al. 2002; Hellman et al. 2002; Siprashvili et al. 1997). Since the transcription of long genes takes longer than one cell cycle, there is a high likelihood for both replication and transcription to take place at the same time. This could result in collision between the replication and transcriptional machinery and can also lead to the formation of DNA:RNA hybrids (R-loops) (Helmrich et al. 2011) (Fig. 23.2b). Obstruction of transcription and/or replication can lead to genomic instability and dysregulation of the tumor suppressor genes located at CFS loci. In addition CFS are located in regions of the genome that are replicated late during S phase (Hansen et al. 1997; Hellman et al. 2000; Le Beau et al. 1998; Palakodeti et al. 2004; Wang et al. 1999). Any further delay in replication due to physiological stress or replication stalling at secondary structures can lead to incompletely replicated DNA upon reaching G2/M. This manifests as ultrafine DNA bridges that have to be resolved between cell division (Chan et al. 2009).

Paucity of Replication Origins CFS have a scarcity of replication initiation events at their loci (Letessier et al. 2011). Due to this, origins that fire outside the locus have to travel long distances to complete CFS replication before cell division begins (Fig. 23.2c). Interestingly, the absence of origins at CFS loci is a cell type-specific phenomenon. The striking differences in replication initiation events at CFS loci in different cell types (e.g., lymphocytes have no origins whereas fibroblasts have an abundance of origins at CFS) closely correlate with CFS fragility. Variations in the epigenetic regulation of the locus are thought to be a possible mechanism behind this cell type specificity. Collectively, these reports show that instability of long non-expanding repeats is multifactorial, and each unique characteristic contributes to its fragility.

23.5 Effect of Repetitive DNA Sequences and Human Diseases

Position of the repeats in the genome determines the effect that repetitive sequences and expanded microsatellite repeats have on various cellular processes.

23.5.1 Effect of Expanded Repeats and Human Diseases

Repeat expansion disorders are categorized as loss or gain of function diseases depending on whether the gene is transcribed or a mutant protein or mutant RNA is produced (Table 23.1). Located in the exons of specific genes, repeats can modify the gene transcription. For example, the expansion of CAG repeats causes expression of a mutant protein in Huntington's disease (HD) patients, spinocerebellar ataxia (SCA) patients, and Kennedy's disease patients (Orr and Zoghbi 2007). In addition to mutant protein expression produced by conventional RNA translation, proteins expressed by repeat-associated non-ATG (RAN) translation can accumulate in disease cells (Banez-Coronel et al. 2015; Cleary and Ranum 2013). Repeat expansion may also result in the production of a toxic RNA, such as in the case of DM1 (Ranum and Day 2004) and ALS (Donnelly et al. 2013). These mutant RNAs contain the expanded repeat sequence and form intra-nucleoplasmic hairpin loops. The presence of these toxic RNAs in the cells leads to the sequestration of proteins such as the splicing regulator MBNL1 resulting in distinctive foci in DM1 cells (Cho and Tapscott 2007; Jiang et al. 2016; Zhang et al. 2016).

Examples for loss of function disorders are FXS and FRDA. Inhibition of RNA polymerases by expanded GAA repeats leads to reduced *Frataxin* gene transcription in FRDA patient cells (Koeppen 2011). In fragile X patients, epigenetic modifications lead to complete silencing of the *FMR1* gene (Santoro et al. 2011; Sutcliffe et al. 1992). Enrichment of heterochromatin marks is observed close to the repeats in several repeat expansion diseases, such as FRDA, FXS, and DM1 (Cho et al. 2005; Kumari and Usdin 2012; Sutcliffe et al. 1992) suggesting that repeat sequences could alter chromatin composition and trigger epigenetic modifications, which could result in the inhibition of gene transcription or lead to gene silencing. Besides epigenetic alterations, large repetitive sequences and the associated non-B DNA secondary structures formed could probably also impede the RNA polymerases. In line with this, it has been shown that repeats inhibit gene transcription in vitro (Krasilnikova et al. 2007; Pandey et al. 2015) and in vivo (Li et al. 2015).

The stalling of the replication machinery could also have an effect on the transcription machinery. For example, the altered direction of the replication fork and the replication stall at the endogenous GAA repeats could lead to head-on collisions of the RNA polymerase with the DNA polymerase in FRDA cells (Gerhardt et al. 2016). After collision, the replication machinery might be able to dislodge the transcriptional machinery as shown by electron microscopy imaging following insertion of an inducible replication origin either upstream or downstream of an *E. coli* rRNA gene (French 1992). In vitro studies with bacterial enzymes show a slow displacement of transcriptional elongation complexes and replisome pausing during head-on collisions whereas no delay during the codirectional progress of replication and transcription (Pomerantz and O'Donnell 2008, 2010). Thus, head-on collision of RNAPII with replication machinery in the 3'–5' direction may not only lead to the GAA expansions but also affect *FXN* transcription progression in FRDA cells. In addition to the mechanistic impact, replication-transcription collisions have been demonstrated in yeast to facilitate establishment and maintenance of heterochromatin, a phenomenon not yet demonstrated in higher eukaryotes (Nikolov and Taddei 2015). This could explain the enrichment of heterochromatic marks surrounding the repeat tracts. Deciphering the interplay between transcription and replication could be critical from the perspective of therapeutic approaches for repeat expansion diseases.

It was recently shown that R-loops are formed at the endogenous repeat sequences (Groh et al. 2014; Lam et al. 2013; Loomis et al. 2014). R-loops could potentially stall the replication and transcription machinery in these cells, and/or secondary repeat structures could form in the single-stranded DNA strand of the R-loop. In this case R-loops could trigger repeat instability. However, R-loops could likewise be a result of the paused transcription machinery, which is held up by a stalled replication fork. Conversely, R-loops are also formed in unaffected cells, but the function(s) of these R-loops are not clear. It has been suggested that R-loops could function as regulators of chromatin dynamics and are implicated in transcription initiation and termination by modulating the chromatin architectures (Al-Hadid and Yang 2016). When, why, and how R-loops form at the endogenous repeats and whether they positively or negatively impact repeat stability is currently unknown. Also, it has yet to be determined whether R-loops form at the expanded repeats or whether they are a consequence of replication fork stalling at the endogenous locus in patient cells.

23.5.2 *Effect of Large Repetitive Sequences and Human Diseases*

Non-expanding repetitive sequences like CFS are largely implicated in the development and progression of cancer (Arlt et al. 2003; Glover 2006). DNA breaks have been observed at precancerous lesions, indicating that the instability at CFS loci is an important initial event in cancer (Gorgoulis et al. 2005). In addition, these regions are thought to be hotspots of chromosomal abnormalities such as deletions, duplications, and translocations in a number of cancer cell lines (Chesi et al. 1998; Finnis et al. 2005; O'Keefe and Richards 2006). Furthermore, more than 80% of copy number variations, in preneoplastic cell lines, have been found at repetitive DNA sequences (Tsantoulis et al. 2008). Maintenance of CFS integrity is critical because most of the commonly expressed CFS contain tumor suppressor genes and proto-oncogenes which when altered are associated with a large spectrum of cancers (Ciullo et al. 2002; Hellman et al. 2002; Siprashvili et al. 1997). Breaks at two of the most commonly expressed CFS, FRA3B and FRA16D, result in the destabilization of tumor suppressor genes *FHIT* and *WWOX*, respectively (Bednarek et al. 2000; Ohta et al. 1996; Virgilio et al. 1996). Mutations and downregulation of the *PARK2* gene located at the third most commonly expressed fragile site FRA6E are associated with ovarian cancer, colorectal cancer, and glioblastoma (Denison et al. 2003; Poulogiannis et al. 2010; Veeriah et al. 2010). Furthermore, CFS are also the preferred sites for viral integration (Popescu and DiPaolo 1989; Wilke et al. 1996) and are highly susceptible to oncogenic stress (Bartek et al. 2007). Other repetitive DNA sequences such as ERFs sites are also known to be hotspots of rearrangements in B cell lymphomas (Barlow et al. 2013).

Instability of CFS is a cell type-specific phenomenon. Most of the aphidicolin-induced breaks in lymphocytes are confined to CFS loci (Hecht and Glover 1984).

Accordingly, lymphocyte-specific fragile sites (e.g., FRA3B and FRA16D) break less frequently in fibroblasts, and likewise, fibroblast-specific fragile sites (e.g., 3q13.3 and 1p31.1) are less unstable in lymphocytes (Le Tallec et al. 2011; Letessier et al. 2011). This is possibly one of the factors contributing to the strong association between CFS instability and hematological malignancies (Gumus et al. 2002). However, CFS instability is implicated in cancers originating from different cell lineages as well. This could be because of the fact that although there is a strong cell type-specific correlation between replication initiation and fragility, the repetitive sequences at these loci can still pose a threat to the replication machinery in any cell type. This is further supported by the fact that the absence of proteins implicated in resolving potential structures at CFS repetitive sequences renders even the initiation proficient cell types unstable at these loci (Casper et al. 2002; Durkin et al. 2006; Focarelli et al. 2009; Madireddy et al. 2016a; Pirzio et al. 2008).

Deficiency of some of the most important proteins implicated in maintaining CFS stability is associated with debilitating diseases characterized by premature aging, genomic instability, and cancer. Mutations in the ATR gene, one of the primary regulators of fragile site stability (Casper et al. 2004), results in a condition called Seckel syndrome (O'Driscoll et al. 2003). Werner syndrome and Bloom syndrome are a result of absences of the respective RecQ helicases implicated in CFS instability (Pirzio et al. 2008; Sidorova et al. 2013). The former is associated with premature aging, while the latter is characterized by genomic instability and an increased predisposition to cancer (Ellis and German 1996; Yu et al. 1996). The absence of polymerase eta that induces spontaneous breaks at CFS (Rey et al. 2009) and the absence of nucleases XPF-ERCC1 that are recruited to ultrafine DNA bridges at CFS loci (Naim et al. 2013) are associated with a skin cancer predisposition syndrome called xeroderma pigmentosum (Cleaver 1972). Considering the strong association between fragile sites and malignancies, it is conceivable that instability at these sites contributes to the etiology of each of these diseases.

23.6 Techniques to Determine Replication Errors

Relevant techniques are essential to understand the disease mechanisms and to develop therapeutic approaches. Non-human models for repeat instability (e.g., mouse) often do not recapitulate the disease phenotype exactly (Perdomini et al. 2013), limiting the tools which could be used to study mechanism for genome instability. To detect DNA replication initiation sites, the direction of the replication fork, replication fork stalling, and DNA replication termination sites in human cells, methods such as nascent strand analysis, 2-D gel electrophoresis, DNA combing, and single-molecule analysis of replicated DNA (SMARD) are used.

Nascent Strand Abundance Analysis Mapping of replication initiation sites in the mammalian genome is achieved through measurement of the relative abundance of nascent DNA throughout a specific region of the chromosome. Therefore, newly synthesized leading strand DNA is first isolated. Then the relative abundance of the nascent DNA is measured at specific loci using quantitative PCR. It is anticipated that the abundance of nascent DNA within the origin region is greatest at the site where DNA replication begins as compared to a genomic region without a replication initiation site. With this technique replication start sites of the laminB2 origin within an ~500-bp segment (Giacca et al. 1994), the replication initiation sites within the hamster dihydrofolate reductase gene locus (Kobayashi et al. 1998) and the *FMR1* promoter in FXS cells (Gray et al. 2007) were mapped.

2-D Gel Electrophoresis The neutral/neutral two-dimensional (2-D) agarose gel technique is a useful tool for understanding the mechanisms leading to the complete duplication of eukaryotic chromosomes. This technique has been used to localize and characterize origins of replication as well as fork progression in a variety of experimental settings. To distinguish different DNA shapes produced by the traveling replication forks, 2-D gels were first used by Brewer and Fangman in 1987. In the first dimension, the DNA molecules are separated by size. Then a gel slice containing the continuum of replicating DNA is cut and subjected to a second round of electrophoresis in a second dimension to resolve replication intermediates of varying topology. In the second dimension, the molecules are separated mainly on the basis of their shape. Nonlinear DNA molecules travel anomalously on agarose gels when compared to linear DNA. To examine replication at a specific DNA segment, the 2-D gel is then blotted and hybridized with specific DNA probes. Two-dimensional gels allow detection of replication fork stalling, which led to the visualization of replication fork pausing at the CGG and GAA repeats (Krasilnikova and Mirkin 2004). Electron microscopy methods are used in combination with 2-D gel electrophoresis to reveal the structure of in vivo DNA replication intermediates. For example, using replication intermediates excised form 2-D gels, it was demonstrated that replication fork reversal occurs after replication fork stalling in human cells (Follonier et al. 2013). These experiments, although advantageous, were limited in their ability to recapitulate the effects of repeats on the DNA replication process at the genomic loci in patient cells. The recent development of single-molecule assays has shed new light on the dynamics of DNA replication at the level of individual chromosomes, such as DNA combing and SMARD.

DNA Molecular Combing analyzes single DNA molecules encompassing large genomic regions. Therefore the DNA is stretched on silanized glass cover slips, and the DNA region of interest is observed by hybridization with specific FISH probes. Since DNA analyses using this technique are single molecule, genomes from different cells can be compared to find anomalies, with implications for diagnosis of cancer and other genetic alterations. In 1994, Bensimon and colleagues (1994) used this technique for the first time to extend bacteriophage lambda DNA molecules. In addition, DNA molecular combing is used for the detection of the DNA replication program in large genomic regions. With molecular combing it is possible to detect replication initiation events and calculate replication fork speed and fork stalling on chromatin fiber. DNA replication studies using combed molecules are based on the fluorescent detection of modified nucleotides, such as bromodeoxyuridine (BrdU), incorporated into newly synthesized DNA at a high resolution. However, the technique is limited due to partial labeling of the DNA fibers.

Single-Molecule Analysis of Replicated DNA SMARD is an approach which was adapted from the molecular combing technique and is used to reveal DNA replication initiation events, fork progression, pausing, and termination sites in large genomic regions (Norio and Schildkraut 2001). To observe the DNA replication by SMARD, cells are pulse labeled with two differentially halogenated nucleoside analogs of thymidine. SMARD in contrast to other single-molecule approaches uses longer halogenated nucleoside incorporation times. The labeling period is longer than the time required to fully replicate the genomic region of interest, but short enough to prevent the occurrence of multiple replication cycles. After labeling, the replicated DNA molecules are stretched on silanized glass slides, and the incorporated nucleotides during replication are visualized by immunostaining. FISH is used to detect the genomic loci of interest. To achieve a higher yield of the region of interest, the DNA is enriched by restriction enzyme digestion followed by size separation of the DNA by pulsed-field gel electrophoresis (PFGE). At the end, the mixture of single DNA molecules, representing all stages of the DNA replication process in cells, gives a composed picture of the DNA replication events in a particular cell line. This approach is sensitive enough to study even short replication fork pauses as well as the direction of the replication forks at specific genomic sites. Examination of the sites and severity of replication fork stalling provide more information about the location of secondary repeat structures. This technique can be used to detect variations in replication origin usage and changes in the DNA replication program during cellular development as well as in disease cells.

23.7 Strategies to Prevent Genome Instability at Repetitive Sequences

The lack of repeat-associated instability under unperturbed "normal" conditions implies that cells have well-defined mechanisms that facilitate the replication of these loci to prevent genome instability. To rescue replication fork stalling, recruitment of proteins that unravel, nucleolytically process, and/or bypass replication impediments is perhaps essential to complete the DNA replication of repetitive DNA sequences. Following replicative stress, stability of both non-expanding repeats, CFS and ERFs, is associated with the activation of ATR checkpoint-signaling pathway (Barlow et al. 2013; Koundrioukoff et al. 2013; Ozeri-Galai et al. 2008). In the absence of ATR, spontaneous breaks have been reported at repetitive DNA sequences, indicating that ATR is an important regulator of CFS stability (Casper et al. 2002). Accordingly, Chk1, a crucial component of the ATR signaling pathway, has also been implicated in maintaining CFS stability (Durkin et al. 2006). In addition, it was found that deficiency of ATR and ATM facilitates expansions of CGG repeats in mice models (Entezam and Usdin 2009). However, while the inactivation of the ATM kinase does not directly result in CFS instability, the combined loss of both ATM and ATR leads to increased breaks at CFS (Ozeri-Galai et al. 2008). As alternative the DNA replication after replication fork stalling at repetitive sequences could be rescued and completed by activation of dormant replication origins to avoid genome instability and breaks.

23.7.1 Helicases Capable of Unwinding Secondary Repeat Structures

DNA helicases are an important group of proteins involved in DNA replication, DNA repair, and gene transcription to ensure that these processes continue unobstructed. During replication, helicases unwind DNA so that the DNA can be copied by the DNA polymerases. In addition, helicases remove obstacles, such as secondary DNA structures, which could hinder the DNA replication machinery. Secondary repeat structures could be generated during lagging strand synthesis in the Okazaki initiation zone, which is a stretch of single-stranded DNA. Several helicases have been shown to unwind these noncanonical DNA structures in vitro.

Helicases from the RecQ family were one of the first helicases shown to unwind G-quadruplex structures. *E. coli* RecQ helicases, the *S. cerevisiae* homolog Sgs1 and their human homologs Bloom (BLM) and Werner (WRN), are all capable of efficiently unwinding G-quadruplexes in DNA, including tetrahelical structures formed by the CGG repeats in vitro (Fry and Loeb 1999). Furthermore, it was shown that purified BLM and WRN helicases could unwind a DNA triple helix structure (Brosh et al. 2000). Sgs1 and Srs2 are two yeast helicases, which act as inhibitors of trinucleotide repeat expansion (Anand et al. 2011; Bhattacharyya and Lahue 2004; Dhar and Lahue 2008; Kerrest et al. 2009). Deletion of these helicases causes repeat instability (Anand et al. 2011). Using a single-molecule fluorescence assay, it was found that these helicases resolve trinucleotide hairpins in a complementary manner (Qiu et al. 2015).

As described previously, many members of the RecQ helicases play an important role in stabilization of CFS. The WRN protein plays an essential role in rescuing stalled replication forks by removing DNA secondary repeat structures that impede replication fork movement (Brosh and Bohr 2002; Shen and Loeb 2000). WRN, a helicase and an exonuclease, has been shown to functionally interact with Pol delta and enhance the processivity of Pol delta across CFS repetitive sequences (Kamath-Loeb et al. 2000; Shah et al. 2010). In addition, WRN patient cells display increased spontaneous breaks at CFS, indicating that WRN is a key regulator of CFS stability. The five helicases of the RecQ family have unique substrate specificities; however, multiple reports suggest mild functional redundancy (Mendonca et al. 1993; Wang et al. 2003). In support of this, the WRN and BLM proteins have been shown to work additively to help in normal replication fork progression (Sidorova et al. 2013).

In addition to the RecQ helicases, other helicases are also important in protecting against genomic instability arising at repetitive sequences. PIF1 helicase is a 5'–3' DNA helicase important for genome stability and is capable of unwinding G-quadruplexes in vitro. The yeast Pif1 helicase unwinds G-quadruplexes with higher efficiency than other substrates such as Y-structures, and unwinding of G-quadruplexes by Pif1 occurs at equimolar concentrations of the helicase and its substrate (Paeschke et al. 2013). These in vitro experiments describe the Pif1 helicase as one of the most effective helicases that unwind G-quadruplexes. Deletion of the PIF1 helicase affects repeat stability only when the G-rich strand is the leading strand template during replication, which relies on the presence of intact G-quadruplex motifs (Lopes et al. 2011). In vivo, the absence of the PIF1 helicase leads to instability of G-quadruplex forming human minisatellites introduced into the yeast genome (Ribeyre et al. 2009). This effect is further enhanced by the action of G-quadruplex-specific compounds, PhenDC3 and PhenDC6, which stabilize G-quadruplexes (Piazza et al. 2010) and enhance replication fork stalling (Madireddy et al. 2016b). Higher eukaryotes possess only one PIF1 helicase, which is shown to specifically recognize and unwind DNA structures resembling putative stalled replication forks (George et al. 2009).

Another 5'–3' helicase known to unwind secondary repeat structures is the Fanconi anemia (FA) complementation group J (FANCJ) RAD3-family DNA helicase. FANCJ belongs to a group of 21 genes that together constitute the FA pathway. Mutations in any 1 of 21 genes cause FA, a rare genetic disorder. The disease is characterized by genomic instability, bone marrow failure, developmental abnormalities, and highly increased predisposition to cancer. Several lines of evidence indicate that FANCJ acts directly during DNA replication to unwind G-quadruplex structures. Furthermore, FANCJ interacts with the replication-associated single-strand binding protein RPA (Gupta et al. 2007; Wu et al. 2008). RPA stimulated FANCJ G-quadruplex unwinding is inhibited by the mismatch repair complex MSH2/MSH6. FANCJ may possibly be loaded 5′ to the G-quadruplex structure on the single-stranded DNA fragment generated prior to Okazaki fragment synthesis.

23.7.2 Proteins That Prevent Replication Fork Collapse at Repetitive Sequences

Besides FANCJ other proteins of the FA tumor suppressor pathway have also been implicated in ensuring stability of difficult to replicate DNA such as repetitive sequences. The FA proteins, although prominently characterized for their essential role in removing inter-strand cross-links (ICL) from DNA, have in the last decade been implicated in preserving genome stability even in response to replicative stress. This has been demonstrated by the observation that the FA pathway is strongly activated in response to replisome stalling under condition of replicative stress (Petermann and Helleday 2010; Petermann et al. 2010; Taniguchi et al. 2002). Involvement of the FA proteins at repetitive sequences is attributed to their role in cooperatively resolving ultrafine DNA bridges that result from under-replicated DNA at CFS (Chan and Hickson 2009; Naim et al. 2013). In addition, the absence of key FA proteins, such as FANCA, FANCD2, and BRCA2/FANCD1, has been associated with severe genomic instability at stalled replication forks (Chaudhury et al. 2014; Karanja et al. 2014; Schlacher et al. 2012), further supporting its role in replication. In the absence of FA proteins, replication forks spontaneously pause at the A/T-rich fragility core of CFS loci (Madireddy et al. 2016a). Moreover, spontaneous CFS breaks are observed in FANCD2 patient-derived lymphocytes, and this is further exacerbated in the presence of aphidicolin-induced replicative stress (Howlett et al. 2005; Madireddy et al. 2016a). Given the requirement for FA proteins in protecting stalled replication forks, it is not surprising that the FA pathway is one of the prominent cellular mechanisms that ensures efficient and timely replication of repetitive DNA sequences.

In addition to the FA pathway and helicases, a number of other proteins are involved in restoring DNA synthesis. When replication forks stall at repetitive DNA sequences, they have a tendency to regress spontaneously to form four-way DNA junctions referred to as chicken foot structures (Fouche et al. 2006; Neelsen and Lopes 2015; Sogo et al. 2002). Nucleases, such as DNA2, are a class of proteins that prevent the collapse of stalled replication forks. This is possibly one of the reasons why DNA nucleases have been implicated in maintaining fragile site stability. Under stress, the absence of nucleases such as XPF-ERCC1 and Mus81-EME1 has been associated with increased breaks at CFS, further supporting their role in preventing CFS instability (Naim et al. 2013; Ying et al. 2013). However it is important to note that the stabilization of stalled forks by the FA proteins, BRCA1 and RAD51, is crucial to prevent fork regression and collapse. The absence of fork stabilizing proteins leads to nucleolytic degradation of nascent DNA and genomic instability (Schlacher et al. 2012). Another DNA nuclease that is recruited to paused replication forks to prevent genomic instability is the Fanconi anemia-associated nuclease 1 (FAN1) which is recruited by the FANCD2 protein (Lachaud et al. 2016). So in addition to fork stabilization, FANCD2's role at CFS could be attributed to its role in recruiting facilitator proteins to sites of replication fork stalling. This role of FANCD2 is further supported by the fact that FANCD2 recruits CtIP to sites of stalled replication forks (Yeo et al. 2014).

Translesion DNA polymerases are another class of proteins that facilitate the efficient synthesis of repetitive sequences, thereby ensuring replication fork stability. In support of this, Pol eta and Pol kappa have been shown to be more efficient at in vitro synthesis of repetitive sequences (Bergoglio et al. 2013; Walsh et al. 2013). Moreover, the depletion of another translesion polymerase, Rev3 (the catalytic subunit of Pol zeta), was shown to induce ultrafine DNA bridges and CFS fragility (Bhat et al. 2013). Although biochemical studies have demonstrated the involvement of translesion polymerases at repetitive sequences, the mechanism behind both their recruitment and their actual function at these sequences is unclear. Given the previously described role of FANCD2 in recruiting Pol eta to the sites of damage (Fu et al. 2013), one possibility could be that FANCD2 recruits Pol eta to fragile site loci by a similar mechanism. Overall, the processes of unperturbed CFS replication and maintenance appear to require the collaborative action of a number of proteins. This is further substantiated by reports that show that the WRN helicase physically interacts with and co-localizes with translesion polymerase eta to alleviate replication fork stalling (Kamath-Loeb et al. 2007).

23.7.3 Dormant Origin Activation to Prevent Genome Instability at Repetitive Sequences

In addition to the recruitment of proteins to sites of replication stalling, dormant origin activation could be an additional and important mechanism by which cells complete replication in regions surrounding the stalled forks. In 1977, J. Herbert Taylor first described that cells license more origins than are actually utilized during the DNA replication process (Taylor 1977). The 3–20 fold excess replication origins that are licensed but not used during each S phase are termed as dormant origins (Blow and Ge 2009; Blow et al. 2011; Wong et al. 2011). Slowing or stalling of the replication fork results in the activation of dormant origins (Alver et al. 2014; Blow and Ge 2009). Dormant origin activation was observed in response to fork stalling at the A/T-rich fragility core of the endogenous CFS-FRA16D locus in cells deficient for the FANCD2 protein (Madireddy et al. 2016a). However, not all CFS loci activate dormant replication origins in response to stress. For example, the FRA16C CFS locus fails to activate dormant replication origins after replicative stress, and this has been implicated in genome instability (Ozeri-Galai et al. 2011). Due to the paucity of replication origins at certain repetitive DNA sequences, completion of the genomic DNA replication is in jeopardy in some human cells.

23.8 Conclusion

The faithful and timely completion of the DNA synthesis is essential to maintain normal cellular functions. Therefore the cell has developed several strategies to ensure the accurate duplication of the genome to prevent chromosomal instability. As described above, complex and well-coordinated networks of proteins work together to ensure the successful completion of the DNA replication process. The inactivation of any one of these proteins can lead to genomic instability. Repetitive sequences because of their unique characteristic features are particularly vulnerable genomic regions, which have the tendency to impede DNA replication leading to replication fork blockage, DNA breaks, and genomic instability. Inaccuracies during DNA replication and the deficiency of proteins, which could stabilize and/or restart replication fork, can lead to repeat expansions, chromosomal instability, and several severe human diseases. Despite decades of research, there is still no cure for these diseases that result from repeat expansions and chromosomal fragility. We believe the reason is the inherent complexity of the mechanisms triggering repeat instability in human cells. While it is known that defects during DNA replication contribute to the overall etiology of repeat-associated diseases, there is a need to elucidate the exact molecular mechanisms that trigger the replication inaccuracies, causing these diseases.

References

Adler CP, Friedburg H (1986) Myocardial DNA content, ploidy level and cell number in geriatric hearts: post-mortem examinations of human myocardium in old age. J Mol Cell Cardiol 18(1):39–53

Agazie YM, Lee JS, Burkholder GD (1994) Characterization of a new monoclonal antibody to triplex DNA and immunofluorescent staining of mammalian chromosomes. J Biol Chem 269(9):7019–7023

Aguilera A, Garcia-Muse T (2012) R loops: from transcription byproducts to threats to genome stability. Mol Cell 46(2):115–124. https://doi.org/10.1016/j.molcel.2012.04.009

Al-Hadid Q, Yang Y (2016) R-loop: an emerging regulator of chromatin dynamics. Acta Biochim Biophys Sin Shanghai 48(7):623–631. https://doi.org/10.1093/abbs/gmw052

Alver RC, Chadha GS, Blow JJ (2014) The contribution of dormant origins to genome stability: from cell biology to human genetics. DNA Repair 19:182–189. https://doi.org/10.1016/j.dnarep.2014.03.012

Anand RP, Shah KA, Niu H, Sung P, Mirkin SM, Freudenreich CH (2011) Overcoming natural replication barriers: differential helicase requirements. Nucleic Acids Res 40:1091. doi:gkr836 [pii] 1093/nar/gkr836

Anatskaya OV, Vinogradov AE (2007) Genome multiplication as adaptation to tissue survival: evidence from gene expression in mammalian heart and liver. Genomics 89(1):70–80. https://doi.org/10.1016/j.ygeno.2006.08.014

Arlt MF, Miller DE, Beer DG, Glover TW (2002) Molecular characterization of FRAXB and comparative common fragile site instability in cancer cells. Genes Chromosom Cancer 33(1):82–92

Arlt MF, Casper AM, Glover TW (2003) Common fragile sites. Cytogenet Genome Res 100(1–4):92–100. doi:72843

Banez-Coronel M, Ayhan F, Tarabochia AD, Zu T, Perez BA, Tusi SK, ... Ranum LP (2015). RAN translation in Huntington disease. Neuron 88(4):667–677. https://doi.org/10.1016/j.neuron.2015.10.038

Barlow JH, Faryabi RB, Callen E, Wong N, Malhowski A, Chen HT, ... Nussenzweig A (2013). Identification of early replicating fragile sites that contribute to genome instability. Cell 152(3):620–632. https://doi.org/10.1016/j.cell.2013.01.006

Bartek J, Bartkova J, Lukas J (2007) DNA damage signalling guards against activated oncogenes and tumour progression. Oncogene 26(56):7773–7779. https://doi.org/10.1038/sj.onc.1210881

Bednarek AK, Laflin KJ, Daniel RL, Liao Q, Hawkins KA, Aldaz CM (2000) WWOX, a novel WW domain-containing protein mapping to human chromosome 16q23.3-24.1, a region frequently affected in breast cancer. Cancer Res 60(8):2140–2145

Bensimon A, Simon A, Chiffaudel A, Croquette V, Heslot F, Bensimon D (1994) Alignment and sensitive detection of DNA by a moving interface. Science 265:2096–2098

Bergoglio V, Boyer AS, Walsh E, Naim V, Legube G, Lee MY, ... Hoffmann JS (2013). DNA synthesis by Pol eta promotes fragile site stability by preventing under-replicated DNA in mitosis. J Cell Biol 201(3):395–408. https://doi.org/10.1083/jcb.201207066

Bhat A, Andersen PL, Qin Z, Xiao W (2013) Rev3, the catalytic subunit of Polzeta, is required for maintaining fragile site stability in human cells. Nucleic Acids Res 41(4):2328–2339. https://doi.org/10.1093/nar/gks1442

Bhattacharyya S, Lahue RS (2004) Saccharomyces cerevisiae Srs2 DNA helicase selectively blocks expansions of trinucleotide repeats. Mol Cell Biol 24(17):7324–7330. https://doi.org/10.1128/MCB.24.17.7324-7330.2004

Blow JJ, Ge XQ (2009) A model for DNA replication showing how dormant origins safeguard against replication fork failure. EMBO Rep 10(4):406–412. doi:embor20095 [pii] 1038/embor.2009.5

Blow JJ, Ge XQ, Jackson DA (2011) How dormant origins promote complete genome replication. Trends Biochem Sci 36(8):405–414. https://doi.org/10.1016/j.tibs.2011.05.002

Bochman ML, Paeschke K, Zakian VA (2012) DNA secondary structures: stability and function of G-quadruplex structures. Nat Rev Genet 13(11):770–780. https://doi.org/10.1038/nrg3296

Brodsky WY, Uryvaeva IV (1977) Cell polyploidy: its relation to tissue growth and function. Int Rev Cytol 50:275–332

Brosh RM Jr, Bohr VA (2002) Roles of the Werner syndrome protein in pathways required for maintenance of genome stability. Exp Gerontol 37(4):491–506

Brosh RM Jr., Li JL, Kenny MK, Karow JK, Cooper MP, Kureekattil RP, ... Bohr VA (2000). Replication protein A physically interacts with the Bloom's syndrome protein and stimulates its helicase activity. J Biol Chem 275(31):23500–23508. https://doi.org/10.1074/jbc.M001557200 M001557200 [pii]

Burkholder GD, Latimer LJ, Lee JS (1988) Immunofluorescent staining of mammalian nuclei and chromosomes with a monoclonal antibody to triplex DNA. Chromosoma 97(3):185–192

Burrow AA, Marullo A, Holder LR, Wang YH (2010) Secondary structure formation and DNA instability at fragile site FRA16B. Nucleic Acids Res 38(9):2865–2877. https://doi.org/10.1093/nar/gkp1245

Buske FA, Mattick JS, Bailey TL (2011) Potential in vivo roles of nucleic acid triple-helices. RNA Biol 8(3):427–439

Casper AM, Nghiem P, Arlt MF, Glover TW (2002) ATR regulates fragile site stability. Cell 111(6):779–789

Casper AM, Durkin SG, Arlt MF, Glover TW (2004) Chromosomal instability at common fragile sites in Seckel syndrome. Am J Hum Genet 75(4):654–660. https://doi.org/10.1086/422701

Chan KL, Hickson ID (2009) On the origins of ultra-fine anaphase bridges. Cell Cycle 8(19):3065–3066. https://doi.org/10.4161/cc.8.19.9513

Chan KL, Palmai-Pallag T, Ying S, Hickson ID (2009) Replication stress induces sister-chromatid bridging at fragile site loci in mitosis. Nat Cell Biol 11(6):753–760. https://doi.org/10.1038/ncb1882

Chatterjee N, Lin Y, Santillan BA, Yotnda P, Wilson JH (2015) Environmental stress induces trinucleotide repeat mutagenesis in human cells. Proc Natl Acad Sci U S A. https://doi.org/10.1073/pnas.1421917112

Chaudhury I, Stroik DR, Sobeck A (2014) FANCD2-controlled chromatin access of the Fanconi-associated nuclease FAN1 is crucial for the recovery of stalled replication forks. Mol Cell Biol 34(21):3939–3954. https://doi.org/10.1128/MCB.00457-14

Chesi M, Bergsagel PL, Shonukan OO, Martelli ML, Brents LA, Chen T, ... Kuehl WM (1998). Frequent dysregulation of the c-maf proto-oncogene at 16q23 by translocation to an Ig locus in multiple myeloma. Blood 91(12):4457–4463

Cho DH, Tapscott SJ (2007) Myotonic dystrophy: emerging mechanisms for DM1 and DM2. Biochim Biophys Acta 1772(2):195–204. doi:S0925-4439(06)00098-6 [pii] 1016/j.bbadis.2006.05.013

Cho DH, Thienes CP, Mahoney SE, Analau E, Filippova GN, Tapscott SJ (2005) Antisense transcription and heterochromatin at the DM1 CTG repeats are constrained by CTCF. Mol Cell 20(3):483–489. doi:S1097-2765(05)01599-6 [pii] 1016/j.molcel.2005.09.002

Ciullo M, Debily MA, Rozier L, Autiero M, Billault A, Mayau V, ... Debatisse M (2002). Initiation of the breakage-fusion-bridge mechanism through common fragile site activation in human breast cancer cells: the model of PIP gene duplication from a break at FRA7I. Hum Mol Genet 11(23):2887–2894

Cleary JD, Pearson CE (2005) Replication fork dynamics and dynamic mutations: the fork-shift model of repeat instability. Trends Genet : TIG 21(5):272–280. doi:S0168-9525(05)00084-3 [pii] 1016/j.tig.2005.03.008
Cleary JD, Ranum LP (2013) Repeat-associated non-ATG (RAN) translation in neurological disease. Hum Mol Genet 22(R1):R45–R51. https://doi.org/10.1093/hmg/ddt371
Cleary JD, Nichol K, Wang YH, Pearson CE (2002) Evidence of cis-acting factors in replication-mediated trinucleotide repeat instability in primate cells. Nat Genet 31(1):37–46
Cleary JD, Tome S, Lopez Castel A, Panigrahi GB, Foiry L, Hagerman KA, ... Pearson CE (2010). Tissue- and age-specific DNA replication patterns at the CTG/CAG-expanded human myotonic dystrophy type 1 locus. Nat Struct Mol Biol 17(9):1079–1087. doi:nsmb.1876 [pii] 1038/nsmb.1876
Cleaver JE (1972) Xeroderma pigmentosum: variants with normal DNA repair and normal sensitivity to ultraviolet light. J Invest Dermatol 58(3):124–128
Corbin S, Neilly ME, Espinosa R 3rd, Davis EM, McKeithan TW, Le Beau MM (2002) Identification of unstable sequences within the common fragile site at 3p14.2: implications for the mechanism of deletions within fragile histidine triad gene/common fragile site at 3p14.2 in tumors. Cancer Res 62(12):3477–3484
DeJesus-Hernandez M, Mackenzie IR, Boeve BF, Boxer AL, Baker M, Rutherford NJ, ... Rademakers R (2011). Expanded GGGGCC hexanucleotide repeat in noncoding region of C9ORF72 causes chromosome 9p-linked FTD and ALS. Neuron 72(2):245–256. https://doi.org/10.1016/j.neuron.2011.09.011
de Koning AP, Gu W, Castoe TA, Batzer MA, Pollock DD (2011) Repetitive elements may comprise over two-thirds of the human genome. PLoS Genet 7(12):e1002384. https://doi.org/10.1371/journal.pgen.1002384
Delagoutte E, Goellner GM, Guo J, Baldacci G, McMurray CT (2008) Single-stranded DNA-binding protein in vitro eliminates the orientation-dependent impediment to polymerase passage on CAG/CTG repeats. J Biol Chem 283(19):13341–13356. https://doi.org/10.1074/jbc.M800153200
Denison SR, Callahan G, Becker NA, Phillips LA, Smith DI (2003) Characterization of FRA6E and its potential role in autosomal recessive juvenile parkinsonism and ovarian cancer. Genes Chromosom Cancer 38(1):40–52. https://doi.org/10.1002/gcc.10236
Dhar A, Lahue RS (2008) Rapid unwinding of triplet repeat hairpins by Srs2 helicase of Saccharomyces cerevisiae. Nucleic Acids Res 36(10):3366–3373. https://doi.org/10.1093/nar/gkn225
Dillon LW, Burrow AA, Wang YH (2010) DNA instability at chromosomal fragile sites in cancer. Curr Genomics 11(5):326–337. https://doi.org/10.2174/138920210791616699
Donnelly CJ, Zhang PW, Pham JT, Haeusler AR, Mistry NA, Vidensky S, ... Rothstein JD (2013). RNA toxicity from the ALS/FTD C9ORF72 expansion is mitigated by antisense intervention. Neuron 80(2):415–428. https://doi.org/10.1016/j.neuron.2013.10.015
Du J, Campau E, Soragni E, Ku S, Puckett JW, Dervan PB, Gottesfeld JM (2012) Role of mismatch repair enzymes in GAA.TTC triplet-repeat expansion in Friedreich ataxia induced pluripotent stem cells. J Biol Chem 287(35):29861–29872. https://doi.org/10.1074/jbc.M112.391961
Durkin SG, Glover TW (2007) Chromosome fragile sites. Annu Rev Genet 41:169–192. https://doi.org/10.1146/annurev.genet.41.042007.165900
Durkin SG, Arlt MF, Howlett NG, Glover TW (2006) Depletion of CHK1, but not CHK2, induces chromosomal instability and breaks at common fragile sites. Oncogene 25(32):4381–4388. https://doi.org/10.1038/sj.onc.1209466
Ellis NA, German J (1996) Molecular genetics of Bloom's syndrome. Hum Mol Genet 5 Spec No:1457–1463
Entezam A, Usdin K (2009) ATM and ATR protect the genome against two different types of tandem repeat instability in Fragile X premutation mice. Nucleic Acids Res 37(19):6371–6377. https://doi.org/10.1093/nar/gkp666

Finnis M, Dayan S, Hobson L, Chenevix-Trench G, Friend K, Ried K, ... Richards RI (2005). Common chromosomal fragile site FRA16D mutation in cancer cells. Hum Mol Genet 14(10):1341–1349. https://doi.org/10.1093/hmg/ddi144

Focarelli ML, Soza S, Mannini L, Paulis M, Montecucco A, Musio A (2009) Claspin inhibition leads to fragile site expression. Genes Chromosom Cancer 48(12):1083–1090. https://doi.org/10.1002/gcc.20710

Follonier C, Oehler J, Herrador R, Lopes M (2013) Friedreich's ataxia-associated GAA repeats induce replication-fork reversal and unusual molecular junctions. Nat Struct Mol Biol 20(4):486–494. https://doi.org/10.1038/nsmb.2520

Fouche N, Ozgur S, Roy D, Griffith JD (2006) Replication fork regression in repetitive DNAs. Nucleic Acids Res 34(20):6044–6050. https://doi.org/10.1093/nar/gkl757

French S (1992) Consequences of replication fork movement through transcription units in vivo. Science 258(5086):1362–1365

Freudenreich CH, Stavenhagen JB, Zakian VA (1997) Stability of a CTG/CAG trinucleotide repeat in yeast is dependent on its orientation in the genome. Mol Cell Biol 17(4):2090–2098

Fry M, Loeb LA (1994) The fragile X syndrome d(CGG)n nucleotide repeats form a stable tetrahelical structure. Proc Natl Acad Sci U S A 91:4950–4954

Fry M, Loeb LA (1999) Human werner syndrome DNA helicase unwinds tetrahelical structures of the fragile X syndrome repeat sequence d(CGG)n. J Biol Chem 274(18):12797–12802

Fu D, Dudimah FD, Zhang J, Pickering A, Paneerselvam J, Palrasu M, ... Fei P (2013). Recruitment of DNA polymerase eta by FANCD2 in the early response to DNA damage. Cell Cycle 12(5):803–809. https://doi.org/10.4161/cc.23755

Fungtammasan A, Walsh E, Chiaromonte F, Eckert KA, Makova KD (2012) A genome-wide analysis of common fragile sites: what features determine chromosomal instability in the human genome? Genome Res 22(6):993–1005. https://doi.org/10.1101/gr.134395.111

Gacy, A. M., McMurray, C. T. (1998). Influence of hairpins on template reannealing at trinucleotide repeat duplexes: a model for slipped DNA. Biochemistry, 37(26), 9426–9434. https://doi.org/10.1021/bi980157s bi980157s [pii]

Gacy AM, Goellner G, Juranic N, Macura S, McMurray CT (1995) Trinucleotide repeats that expand in human disease form hairpin structures in vitro. Cell 81(4):533–540. doi:0092-8674(95)90074-8 [pii]

Gacy AM, Goellner GM, Spiro C, Chen X, Gupta G, Bradbury EM, ... McMurray CT (1998). GAA instability in Friedreich's ataxia shares a common, DNA-directed and intraallelic mechanism with other trinucleotide diseases. Mol Cell 1(4):583–593. doi:S1097-2765(00)80058-1 [pii]

George T, Wen Q, Griffiths R, Ganesh A, Meuth M, Sanders CM (2009) Human Pif1 helicase unwinds synthetic DNA structures resembling stalled DNA replication forks. Nucleic Acids Res 37(19):6491–6502. https://doi.org/10.1093/nar/gkp671

Gerhardt J (2015) Epigenetic modifications in human fragile X pluripotent stem cells; Implications in fragile X syndrome modeling. Brain Res. https://doi.org/10.1016/j.brainres.2015.10.004

Gerhardt J, Zaninovic N, Zhan Q, Madireddy A, Nolin SL, Ersalesi N, ... Schildkraut CL (2014a). Cis-acting DNA sequence at a replication origin promotes repeat expansion to fragile X full mutation. J Cell Biol 206(5):599–607. https://doi.org/10.1083/jcb.201404157

Gerhardt J, Tomishima MJ, Zaninovic N, Colak D, Yan Z, Zhan Q, ... Schildkraut CL (2014b). The DNA replication program is altered at the FMR1 locus in fragile x embryonic stem cells. Mol Cell 53(1):19–31. https://doi.org/10.1016/j.molcel.2013.10.029

Gerhardt J, Bhalla AD, Butler JS, Puckett JW, Dervan PB, Rosenwaks Z, Napierala M (2016) Stalled DNA replication forks at the endogenous GAA repeats drive repeat expansion in Friedreich's ataxia cells. Cell Rep 16(5):1218–1227. https://doi.org/10.1016/j.celrep.2016.06.075

Giacca M, Zentilin L, Norio P, Diviacco S, Dimitrova D, Contreas G, ... et al (1994). Fine mapping of a replication origin of human DNA. ProcNatlAcadSciUSA 91:7119–7123

Glover TW (2006) Common fragile sites. Cancer Lett 232(1):4–12. https://doi.org/10.1016/j.canlet.2005.08.032

Glover TW, Stein CK (1987) Induction of sister chromatid exchanges at common fragile sites. Am J Hum Genet 41(5):882–890

Glover TW, Stein CK (1988) Chromosome breakage and recombination at fragile sites. Am J Hum Genet 43(3):265–273

Glover TW, Berger C, Coyle J, Echo B (1984) DNA polymerase alpha inhibition by aphidicolin induces gaps and breaks at common fragile sites in human chromosomes. Hum Genet 67(2):136–142

Gorgoulis VG, Vassiliou LV, Karakaidos P, Zacharatos P, Kotsinas A, Liloglou T, ... Halazonetis TD (2005). Activation of the DNA damage checkpoint and genomic instability in human precancerous lesions. Nature 434(7035):907–913. https://doi.org/10.1038/nature03485

Gray SJ, Gerhardt J, Doerfler W, Small LE, Fanning E (2007) An origin of DNA replication in the promoter region of the human fragile X mental retardation (FMR1) gene. Mol Cell Biol 27(2):426–437. doi:MCB.01382-06 [pii] 1128/MCB.01382-06

Groh M, Lufino MM, Wade-Martins R, Gromak N (2014) R-loops associated with triplet repeat expansions promote gene silencing in Friedreich ataxia and fragile X syndrome. PLoS Genet 10(5):e1004318. https://doi.org/10.1371/journal.pgen.1004318

Gumus G, Sunguroglu A, Tukun A, Sayin DB, Bokesoy I (2002) Common fragile sites associated with the breakpoints of chromosomal aberrations in hematologic neoplasms. Cancer Genet Cytogenet 133(2):168–171

Gupta R, Sharma S, Sommers JA, Kenny MK, Cantor SB, Brosh RM Jr (2007) FANCJ (BACH1) helicase forms DNA damage inducible foci with replication protein A and interacts physically and functionally with the single-stranded DNA-binding protein. Blood 110(7):2390–2398. https://doi.org/10.1182/blood-2006-11-057273

Hansen RS, Canfield TK, Lamb MM, Gartler SM, Laird CD (1993) Association of fragile X syndrome with delayed replication of the FMR1 gene. Cell 73:1403–1409

Hansen RS, Canfield TK, Fjeld AD, Mumm S, Laird CD, Gartler SM (1997) A variable domain of delayed replication in FRAXA fragile X chromosomes: X inactivation-like spread of late replication. Proc Natl Acad Sci U S A 94(9):4587–4592

Hecht F, Glover TW (1984) Cancer chromosome breakpoints and common fragile sites induced by aphidicolin. Cancer Genet Cytogenet 13(2):185–188

Heitz D, Rousseau F, Devys D, Saccone S, Abderrahim H, Le Paslier D, ... et al (1991). Isolation of sequences that span the fragile X and identification of a fragile X-related CpG island. Science 251(4998):1236–1239

Hellman A, Rahat A, Scherer SW, Darvasi A, Tsui LC, Kerem B (2000) Replication delay along FRA7H, a common fragile site on human chromosome 7, leads to chromosomal instability. Mol Cell Biol 20(12):4420–4427

Hellman A, Zlotorynski E, Scherer SW, Cheung J, Vincent JB, Smith DI, ... Kerem B (2002). A role for common fragile site induction in amplification of human oncogenes. Cancer Cell 1(1):89–97

Helmrich A, Ballarino M, Tora L (2011) Collisions between replication and transcription complexes cause common fragile site instability at the longest human genes. Mol Cell 44(6):966–977. https://doi.org/10.1016/j.molcel.2011.10.013

Helmrich A, Ballarino M, Nudler E, Tora L (2013) Transcription-replication encounters, consequences and genomic instability. Nat Struct Mol Biol 20(4):412–418. https://doi.org/10.1038/nsmb.2543

Holloway TP, Rowley SM, Delatycki MB, Sarsero JP (2011) Detection of interruptions in the GAA trinucleotide repeat expansion in the FXN gene of Friedreich ataxia. BioTechniques 50(3):182–186. https://doi.org/10.2144/000113615

Howlett NG, Taniguchi T, Durkin SG, D'Andrea AD, Glover TW (2005) The Fanconi anemia pathway is required for the DNA replication stress response and for the regulation of common fragile site stability. Hum Mol Genet 14(5):693–701. https://doi.org/10.1093/hmg/ddi065

Jiang J, Zhu Q, Gendron TF, Saberi S, McAlonis-Downes M, Seelman A, ... Lagier-Tourenne C (2016). Gain of toxicity from ALS/FTD-linked repeat expansions in C9ORF72 is alleviated by antisense oligonucleotides targeting GGGGCC-containing RNAs. Neuron 90(3):535–550. https://doi.org/10.1016/j.neuron.2016.04.006

Kamath-Loeb AS, Johansson E, Burgers PM, Loeb LA (2000) Functional interaction between the Werner syndrome protein and DNA polymerase delta. Proc Natl Acad Sci U S A 97(9):4603–4608

Kamath-Loeb AS, Lan L, Nakajima S, Yasui A, Loeb LA (2007) Werner syndrome protein interacts functionally with translesion DNA polymerases. Proc Natl Acad Sci U S A 104(25):10394–10399. https://doi.org/10.1073/pnas.0702513104

Kang S, Ohshima K, Shimizu M, Amirhaeri S, Wells RD (1995) Pausing of DNA synthesis in vitro at specific loci in CTG and CGG triplet repeats from human hereditary disease genes. J Biol Chem 270(45):27014–27021

Karanja KK, Lee EH, Hendrickson EA, Campbell JL (2014) Preventing over-resection by DNA2 helicase/nuclease suppresses repair defects in Fanconi anemia cells. Cell Cycle 13(10):1540–1550. https://doi.org/10.4161/cc.28476

Kerrest A, Anand RP, Sundararajan R, Bermejo R, Liberi G, Dujon B, ... Richard GF (2009). SRS2 and SGS1 prevent chromosomal breaks and stabilize triplet repeats by restraining recombination. Nat Struct Mol Biol 16(2):159–167. https://doi.org/10.1038/nsmb.1544

Kim JC, Mirkin SM (2013) The balancing act of DNA repeat expansions. Curr Opin Genet Dev 23(3):280–288. https://doi.org/10.1016/j.gde.2013.04.009

Kobayashi T, Rein T, DePamphilis ML (1998) Identification of primary initiation sites for DNA replication in the hamster dihydrofolate reductase gene initiation zone. MolCellBiol 18:3266–3277

Koeppen AH (2011) Friedreich's ataxia: pathology, pathogenesis, and molecular genetics. J Neurol Sci 303(1–2):1–12. https://doi.org/10.1016/j.jns.2011.01.010

Koundrioukoff S, Carignon S, Techer H, Letessier A, Brison O, Debatisse M (2013) Stepwise activation of the ATR signaling pathway upon increasing replication stress impacts fragile site integrity. PLoS Genet 9(7):e1003643. https://doi.org/10.1371/journal.pgen.1003643

Krasilnikova MM, Mirkin SM (2004) Replication stalling at Friedreich's ataxia (GAA)n repeats in vivo. Mol Cell Biol 24(6):2286–2295

Krasilnikova MM, Kireeva ML, Petrovic V, Knijnikova N, Kashlev M, Mirkin SM (2007) Effects of Friedreich's ataxia (GAA)n*(TTC)n repeats on RNA synthesis and stability. Nucleic Acids Res 35(4):1075–1084. https://doi.org/10.1093/nar/gkl1140

Kumari D, Usdin K (2012) Is Friedreich ataxia an epigenetic disorder? Clin Epigenetics 4(1):2. https://doi.org/10.1186/1868-7083-4-2

La Spada AR, Wilson EM, Lubahn DB, Harding AE, Fischbeck KH (1991) Androgen receptor gene mutations in X-linked spinal and bulbar muscular atrophy. Nature 352(6330):77–79. https://doi.org/10.1038/352077a0

Lachaud C, Moreno A, Marchesi F, Toth R, Blow JJ, Rouse J (2016) Ubiquitinated Fancd2 recruits Fan1 to stalled replication forks to prevent genome instability. Science 351(6275):846–849. https://doi.org/10.1126/science.aad5634

Lam EY, Beraldi D, Tannahill D, Balasubramanian S (2013) G-quadruplex structures are stable and detectable in human genomic DNA. Nat Commun 4:1796. https://doi.org/10.1038/ncomms2792

Le Beau MM, Rassool FV, Neilly ME, Espinosa R 3rd, Glover TW, Smith DI, McKeithan TW (1998) Replication of a common fragile site, FRA3B, occurs late in S phase and is delayed further upon induction: implications for the mechanism of fragile site induction. Hum Mol Genet 7(4):755–761

Le Tallec B, Dutrillaux B, Lachages AM, Millot GA, Brison O, Debatisse M (2011) Molecular profiling of common fragile sites in human fibroblasts. Nat Struct Mol Biol 18(12):1421–1423. https://doi.org/10.1038/nsmb.2155

Lee JS, Burkholder GD, Latimer LJ, Haug BL, Braun RP (1987) A monoclonal antibody to triplex DNA binds to eucaryotic chromosomes. Nucleic Acids Res 15(3):1047–1061

Letessier A, Millot GA, Koundrioukoff S, Lachages AM, Vogt N, Hansen RS, ... Debatisse M (2011). Cell-type-specific replication initiation programs set fragility of the FRA3B fragile site. Nature 470(7332):120–123. doi:nature09745 [pii] 1038/nature09745

Li Y, Lu Y, Polak U, Lin K, Shen J, Farmer J, ... Napierala M (2015). Expanded GAA repeats impede transcription elongation through the FXN gene and induce transcriptional silencing that is restricted to the FXN locus. Hum Mol Genet 24(24):6932–6943. https://doi.org/10.1093/hmg/ddv397

Libby RT, Monckton DG, Fu YH, Martinez RA, McAbney JP, Lau R, ... La Spada AR (2003). Genomic context drives SCA7 CAG repeat instability, while expressed SCA7 cDNAs are intergenerationally and somatically stable in transgenic mice. Hum Mol Genet 12(1):41–50

Libby RT, Hagerman KA, Pineda VV, Lau R, Cho DH, Baccam SL, ... La Spada AR (2008). CTCF cis-regulates trinucleotide repeat instability in an epigenetic manner: a novel basis for mutational hot spot determination. PLoS Genet 4(11):e1000257. https://doi.org/10.1371/journal.pgen.1000257

Loomis EW, Sanz LA, Chedin F, Hagerman PJ (2014) Transcription-associated R-loop formation across the human FMR1 CGG-repeat region. PLoS Genet 10(4):e1004294. https://doi.org/10.1371/journal.pgen.1004294

Lopes J, Piazza A, Bermejo R, Kriegsman B, Colosio A, Teulade-Fichou MP, ... Nicolas A (2011). G-quadruplex-induced instability during leading-strand replication. EMBO J 30(19):4033–4046. https://doi.org/10.1038/emboj.2011.316

Lopez Castel A, Cleary JD, Pearson CE (2010) Repeat instability as the basis for human diseases and as a potential target for therapy. Nat Rev Mol Cell Biol 11(3):165–170. doi:nrm2854 [pii] 1038/nrm2854

Lukusa T, Fryns JP (2008) Human chromosome fragility. Biochim Biophys Acta 1779(1):3–16. https://doi.org/10.1016/j.bbagrm.2007.10.005

Madireddy A, Kosiyatrakul ST, Boisvert RA, Herrera-Moyano E, Garcia-Rubio ML, Gerhardt J, ... Schildkraut CL (2016a). FANCD2 facilitates replication through common fragile sites. Mol Cell 64(2):388–404. https://doi.org/10.1016/j.molcel.2016.09.017

Madireddy A, Purushothaman P, Loosbroock CP, Robertson ES, Schildkraut CL, Verma SC (2016b) G-quadruplex-interacting compounds alter latent DNA replication and episomal persistence of KSHV. Nucleic Acids Res 44(8):3675–3694. https://doi.org/10.1093/nar/gkw038

McMurray CT (2010) Mechanisms of trinucleotide repeat instability during human development. Nat Rev Genet 11(11):786–799. https://doi.org/10.1038/nrg2828

Mendonca VM, Kaiser-Rogers K, Matson SW (1993) Double helicase II (uvrD)-helicase IV (helD) deletion mutants are defective in the recombination pathways of Escherichia coli. J Bacteriol 175(15):4641–4651

Mirkin SM (2006) DNA structures, repeat expansions and human hereditary disorders. Curr Opin Struct Biol 16(3):351–358. doi:S0959-440X(06)00076-5 [pii] 1016/j.sbi.2006.05.004

Mirkin SM (2007) Expandable DNA repeats and human disease. Nature 447(7147):932–940. doi:nature05977 [pii] 1038/nature05977

Mishmar D, Rahat A, Scherer SW, Nyakatura G, Hinzmann B, Kohwi Y, ... Kerem B (1998). Molecular characterization of a common fragile site (FRA7H) on human chromosome 7 by the cloning of a simian virus 40 integration site. Proc Natl Acad Sci U S A 95(14):8141–8146

Mortusewicz O, Herr P, Helleday T (2013) Early replication fragile sites: where replication-transcription collisions cause genetic instability. EMBO J 32(4):493–495. https://doi.org/10.1038/emboj.2013.20

Naim V, Wilhelm T, Debatisse M, Rosselli F (2013) ERCC1 and MUS81-EME1 promote sister chromatid separation by processing late replication intermediates at common fragile sites during mitosis. Nat Cell Biol 15(8):1008–1015. https://doi.org/10.1038/ncb2793

Neelsen KJ, Lopes M (2015) Replication fork reversal in eukaryotes: from dead end to dynamic response. Nat Rev Mol Cell Biol 16(4):207–220. https://doi.org/10.1038/nrm3935

Nikolov I, Taddei A (2015) Linking replication stress with heterochromatin formation. Chromosoma. https://doi.org/10.1007/s00412-015-0545-6

Nolin SL, Glicksman A, Ersalesi N, Dobkin C, Brown WT, Cao R, ... Hadd AG (2015). Fragile X full mutation expansions are inhibited by one or more AGG interruptions in premutation carriers. Genet Med 17(5):358–364. https://doi.org/10.1038/gim.2014.106

Norio P, Schildkraut CL (2001) Visualization of DNA replication on individual Epstein-Barr virus episomes. Science 294:2361–2364

O'Driscoll M, Ruiz-Perez VL, Woods CG, Jeggo PA, Goodship JA (2003) A splicing mutation affecting expression of ataxia-telangiectasia and Rad3-related protein (ATR) results in Seckel syndrome. Nat Genet 33(4):497–501. https://doi.org/10.1038/ng1129

Ohta M, Inoue H, Cotticelli MG, Kastury K, Baffa R, Palazzo J, ... Huebner K (1996). The FHIT gene, spanning the chromosome 3p14.2 fragile site and renal carcinoma-associated t(3;8) breakpoint, is abnormal in digestive tract cancers. Cell 84(4):587–597

O'Keefe LV, Richards RI (2006) Common chromosomal fragile sites and cancer: focus on FRA16D. Cancer Lett 232(1):37–47. https://doi.org/10.1016/j.canlet.2005.07.041

Orr HT, Zoghbi HY (2007) Trinucleotide repeat disorders. Annu Rev Neurosci 30:575–621. https://doi.org/10.1146/annurev.neuro.29.051605.113042

Owen BA, Yang Z, Lai M, Gajec M, Badger JD 2nd, Hayes JJ, ... McMurray CT (2005). (CAG)(n)-hairpin DNA binds to Msh2-Msh3 and changes properties of mismatch recognition. Nat Struct Mol Biol 12(8):663–670. https://doi.org/10.1038/nsmb965

Ozeri-Galai E, Schwartz M, Rahat A, Kerem B (2008) Interplay between ATM and ATR in the regulation of common fragile site stability. Oncogene 27(15):2109–2117. doi:1210849 [pii] 1038/sj.onc.1210849

Ozeri-Galai E, Lebofsky R, Rahat A, Bester AC, Bensimon A, Kerem B (2011) Failure of origin activation in response to fork stalling leads to chromosomal instability at fragile sites. Mol Cell 43(1):122–131. https://doi.org/10.1016/j.molcel.2011.05.019

Ozeri-Galai E, Sinai MI-T, Kerem B (2013). Fork stalling at AT-rich sequences and failure of origin activation lead to chromosomal instability at fragile sites. Paper presented at the Journal of Biomolecular Structure and Dynamics

Paeschke K, Bochman ML, Garcia PD, Cejka P, Friedman KL, Kowalczykowski SC, Zakian VA (2013) Pif1 family helicases suppress genome instability at G-quadruplex motifs. Nature 497(7450):458–462. https://doi.org/10.1038/nature12149

Paiva AM, Sheardy RD (2004) Influence of sequence context and length on the structure and stability of triplet repeat DNA oligomers. Biochemistry 43(44):14218–14227. https://doi.org/10.1021/bi0494368

Palakodeti A, Han Y, Jiang Y, Le Beau MM (2004) The role of late/slow replication of the FRA16D in common fragile site induction. Genes Chromosom Cancer 39(1):71–76. https://doi.org/10.1002/gcc.10290

Pandey S, Ogloblina AM, Belotserkovskii BP, Dolinnaya NG, Yakubovskaya MG, Mirkin SM, Hanawalt PC (2015) Transcription blockage by stable H-DNA analogs in vitro. Nucleic Acids Res 43(14):6994–7004. https://doi.org/10.1093/nar/gkv622

Pataskar SS, Dash D, Brahmachari SK (2001) Intramolecular i-motif structure at acidic pH for progressive myoclonus epilepsy (EPM1) repeat d(CCCCGCCCCGCG)n. J Biomol Struct Dyn 19(2):307–313. https://doi.org/10.1080/07391102.2001.10506741

Pearson CE, Ewel A, Acharya S, Fishel RA, Sinden RR (1997) Human MSH2 binds to trinucleotide repeat DNA structures associated with neurodegenerative diseases. Hum Mol Genet 6(7):1117–1123

Pearson CE, Eichler EE, Lorenzetti D, Kramer SF, Zoghbi HY, Nelson DL, Sinden RR (1998) Interruptions in the triplet repeats of SCA1 and FRAXA reduce the propensity and complexity of slipped strand DNA (S-DNA) formation. Biochemistry 37(8):2701–2708. https://doi.org/10.1021/bi972546c. bi972546c [pii]

Pelletier R, Krasilnikova MM, Samadashwily GM, Lahue R, Mirkin SM (2003) Replication and expansion of trinucleotide repeats in yeast. Mol Cell Biol 23(4):1349–1357

Perdomini M, Hick A, Puccio H, Pook MA (2013) Animal and cellular models of Friedreich ataxia. J Neurochem 126(Suppl 1):65–79. https://doi.org/10.1111/jnc.12219

Petermann E, Helleday T (2010) Pathways of mammalian replication fork restart. Nat Rev Mol Cell Biol 11(10):683–687. doi:nrm2974 [pii] 1038/nrm2974

Petermann E, Orta ML, Issaeva N, Schultz N, Helleday T (2010) Hydroxyurea-stalled replication forks become progressively inactivated and require two different RAD51-mediated pathways for restart and repair. Mol Cell 37(4):492–502. https://doi.org/10.1016/j.molcel.2010.01.021
Phillips DD, Garboczi DN, Singh K, Hu Z, Leppla SH, Leysath CE (2013) The sub-nanomolar binding of DNA-RNA hybrids by the single-chain Fv fragment of antibody S9.6. J Mol Recognit 26(8):376–381. https://doi.org/10.1002/jmr.2284
Piazza A, Boule JB, Lopes J, Mingo K, Largy E, Teulade-Fichou MP, Nicolas A (2010) Genetic instability triggered by G-quadruplex interacting Phen-DC compounds in Saccharomyces cerevisiae. Nucleic Acids Res 38(13):4337–4348. https://doi.org/10.1093/nar/gkq136
Pieretti M, Zhang FP, Fu YH, Warren ST, Oostra BA, Caskey CT, Nelson DL (1991) Absence of expression of the FMR-1 gene in fragile X syndrome. Cell 66(4):817–822
Pirzio LM, Pichierri P, Bignami M, Franchitto A (2008) Werner syndrome helicase activity is essential in maintaining fragile site stability. J Cell Biol 180(2):305–314. https://doi.org/10.1083/jcb.200705126
Pomerantz RT, O'Donnell M (2008) The replisome uses mRNA as a primer after colliding with RNA polymerase. Nature 456(7223):762–766. https://doi.org/10.1038/nature07527
Pomerantz RT, O'Donnell M (2010) Direct restart of a replication fork stalled by a head-on RNA polymerase. Science 327(5965):590–592. https://doi.org/10.1126/science.1179595
Popescu NC, DiPaolo JA (1989) Preferential sites for viral integration on mammalian genome. Cancer Genet Cytogenet 42(2):157–171
Poulogiannis G, McIntyre RE, Dimitriadi M, Apps JR, Wilson CH, Ichimura K, ... Arends MJ (2010). PARK2 deletions occur frequently in sporadic colorectal cancer and accelerate adenoma development in Apc mutant mice. Proc Natl Acad Sci U S A 107(34):15145–15150. https://doi.org/10.1073/pnas.1009941107
Qiu Y, Niu H, Vukovic L, Sung P, Myong S (2015) Molecular mechanism of resolving trinucleotide repeat hairpin by helicases. Structure 23(6):1018–1027. https://doi.org/10.1016/j.str.2015.04.006
Ragland RL, Glynn MW, Arlt MF, Glover TW (2008) Stably transfected common fragile site sequences exhibit instability at ectopic sites. Genes Chromosom Cancer 47(10):860–872. https://doi.org/10.1002/gcc.20591
Ranum LP, Day JW (2004) Myotonic dystrophy: RNA pathogenesis comes into focus. Am J Hum Genet 74(5):793–804. https://doi.org/10.1086/383590. S0002-9297(07)64349-3 [pii]
Rey L, Sidorova JM, Puget N, Boudsocq F, Biard DS, Monnat RJ Jr., ... Hoffmann JS (2009). Human DNA polymerase eta is required for common fragile site stability during unperturbed DNA replication. Mol Cell Biol 29(12):3344–3354. https://doi.org/10.1128/MCB.00115-09
Ribeyre C, Lopes J, Boule JB, Piazza A, Guedin A, Zakian VA, ... Nicolas A (2009). The yeast Pif1 helicase prevents genomic instability caused by G-quadruplex-forming CEB1 sequences in vivo. PLoS Genet 5(5):e1000475. https://doi.org/10.1371/journal.pgen.1000475
Ried K, Finnis M, Hobson L, Mangelsdorf M, Dayan S, Nancarrow JK, ... Richards RI (2000). Common chromosomal fragile site FRA16D sequence: identification of the FOR gene spanning FRA16D and homozygous deletions and translocation breakpoints in cancer cells. Hum Mol Genet 9(11):1651–1663
Rindler PM, Clark RM, Pollard LM, De Biase I, Bidichandani SI (2006) Replication in mammalian cells recapitulates the locus-specific differences in somatic instability of genomic GAA triplet-repeats. Nucleic Acids Res 34(21):6352–6361. https://doi.org/10.1093/nar/gkl846
Samadashwily GM, Raca G, Mirkin SM (1997) Trinucleotide repeats affect DNA replication in vivo. Nat Genet 17(3):298–304. https://doi.org/10.1038/ng1197-298
Santoro MR, Bray SM, Warren ST (2011) Molecular mechanisms of fragile X syndrome: a twenty-year perspective. Annu Rev Pathol. https://doi.org/10.1146/annurev-pathol-011811-132457
Schlacher K, Wu H, Jasin M (2012) A distinct replication fork protection pathway connects Fanconi anemia tumor suppressors to RAD51-BRCA1/2. Cancer Cell 22(1):106–116. https://doi.org/10.1016/j.ccr.2012.05.015

Schwartz M, Zlotorynski E, Kerem B (2006) The molecular basis of common and rare fragile sites. Cancer Lett 232(1):13–26. https://doi.org/10.1016/j.canlet.2005.07.039

Shah SN, Opresko PL, Meng X, Lee MY, Eckert KA (2010) DNA structure and the Werner protein modulate human DNA polymerase delta-dependent replication dynamics within the common fragile site FRA16D. Nucleic Acids Res 38(4):1149–1162. https://doi.org/10.1093/nar/gkp1131

Shen JC, Loeb LA (2000) Werner syndrome exonuclease catalyzes structure-dependent degradation of DNA. Nucleic Acids Res 28(17):3260–3268

Shishkin AA, Voineagu I, Matera R, Cherng N, Chernet BT, Krasilnikova MM, ... Mirkin SM (2009). Large-scale expansions of Friedreich's ataxia GAA repeats in yeast. Mol Cell 35(1):82–92. https://doi.org/10.1016/j.molcel.2009.06.017

Sidorova JM, Kehrli K, Mao F, Monnat R Jr (2013) Distinct functions of human RECQ helicases WRN and BLM in replication fork recovery and progression after hydroxyurea-induced stalling. DNA Repair 12(2):128–139. https://doi.org/10.1016/j.dnarep.2012.11.005

Siprashvili Z, Sozzi G, Barnes LD, McCue P, Robinson AK, Eryomin V, ... Huebner K (1997). Replacement of Fhit in cancer cells suppresses tumorigenicity. Proc Natl Acad Sci U S A 94(25):13771–13776

Smeets DF, van de Klundert FA (1990) Common fragile sites in man and three closely related primate species. Cytogenet Cell Genet 53(1):8–14

Sogo JM, Lopes M, Foiani M (2002) Fork reversal and ssDNA accumulation at stalled replication forks owing to checkpoint defects. Science 297(5581):599–602. https://doi.org/10.1126/science.1074023

Sollier J, Cimprich KA (2015) Breaking bad: R-loops and genome integrity. Trends Cell Biol 25(9):514–522. https://doi.org/10.1016/j.tcb.2015.05.003

Subramanian PS, Nelson DL, Chinault AC (1996) Large domains of apparent delayed replication timing associated with triplet repeat expansion at FRAXA and FRAXE. Am J Hum Genet 59:407–416

Sutcliffe JS, Nelson DL, Zhang F, Pieretti M, Caskey CT, Saxe D, Warren ST (1992) DNA methylation represses FMR-1 transcription in fragile X syndrome. Hum Mol Genet 1(6):397–400

Taniguchi T, Garcia-Higuera I, Andreassen PR, Gregory RC, Grompe M, D'Andrea AD (2002) S-phase-specific interaction of the Fanconi anemia protein, FANCD2, with BRCA1 and RAD51. Blood 100(7):2414–2420. https://doi.org/10.1182/blood-2002-01-0278

Taylor JH (1977) Increase in DNA replication sites in cells held at the beginning of S phase. Chromosoma 62(4):291–300

Tsantoulis PK, Kotsinas A, Sfikakis PP, Evangelou K, Sideridou M, Levy B, ... Gorgoulis VG (2008). Oncogene-induced replication stress preferentially targets common fragile sites in preneoplastic lesions. A genome-wide study. Oncogene 27(23):3256–3264. https://doi.org/10.1038/sj.onc.1210989

Usdin K, Woodford KJ (1995) CGG repeats associated with DNA instability and chromosome fragility form structures that block DNA synthesis in vitro. Nucleic Acids Res 23(20):4202–4209. doi:5j0289 [pii]

Usdin K, House NC, Freudenreich CH (2015) Repeat instability during DNA repair: insights from model systems. Crit Rev Biochem Mol Biol:1–26. https://doi.org/10.3109/10409238.2014.999192

Veeriah S, Taylor BS, Meng S, Fang F, Yilmaz E, Vivanco I, ... Chan TA (2010). Somatic mutations of the Parkinson's disease-associated gene PARK2 in glioblastoma and other human malignancies. Nat Genet 42(1):77–82. https://doi.org/10.1038/ng.491

Verkerk AJ, Pieretti M, Sutcliffe JS, Fu YH, Kuhl DP, Pizzuti A, ... et al (1991). Identification of a gene (FMR-1) containing a CGG repeat coincident with a breakpoint cluster region exhibiting length variation in fragile X syndrome. Cell 65(5):905–914

Virgilio L, Shuster M, Gollin SM, Veronese ML, Ohta M, Huebner K, Croce CM (1996) FHIT gene alterations in head and neck squamous cell carcinomas. Proc Natl Acad Sci U S A 93(18):9770–9775

Voineagu I, Surka CF, Shishkin AA, Krasilnikova MM, Mirkin SM (2009) Replisome stalling and stabilization at CGG repeats, which are responsible for chromosomal fragility. Nat Struct Mol Biol 16(2):226–228. https://doi.org/10.1038/nsmb.1527

Walsh E, Wang X, Lee MY, Eckert KA (2013) Mechanism of replicative DNA polymerase delta pausing and a potential role for DNA polymerase kappa in common fragile site replication. J Mol Biol 425(2):232–243. https://doi.org/10.1016/j.jmb.2012.11.016

Wang L, Paradee W, Mullins C, Shridhar R, Rosati R, Wilke CM, ... Smith DI (1997). Aphidicolin-induced FRA3B breakpoints cluster in two distinct regions. Genomics 41(3):485–488. https://doi.org/10.1006/geno.1997.4690

Wang L, Darling J, Zhang JS, Huang H, Liu W, Smith DI (1999) Allele-specific late replication and fragility of the most active common fragile site, FRA3B. Hum Mol Genet 8(3):431–437

Wang W, Seki M, Narita Y, Nakagawa T, Yoshimura A, Otsuki M, ... Enomoto T (2003). Functional relation among RecQ family helicases RecQL1, RecQL5, and BLM in cell growth and sister chromatid exchange formation. Mol Cell Biol 23(10):3527–3535

Wells RD (2008) DNA triplexes and Friedreich ataxia. FASEB J: Off Publ Fed Am Soc Exp Biol 22(6):1625–1634. https://doi.org/10.1096/fj.07-097857

Wilke CM, Hall BK, Hoge A, Paradee W, Smith DI, Glover TW (1996) FRA3B extends over a broad region and contains a spontaneous HPV16 integration site: direct evidence for the coincidence of viral integration sites and fragile sites. Hum Mol Genet 5(2):187–195

Wong PG, Winter SL, Zaika E, Cao TV, Oguz U, Koomen JM, ... Alexandrow MG (2011). Cdc45 limits replicon usage from a low density of preRCs in mammalian cells. PLoS One 6(3):e17533. https://doi.org/10.1371/journal.pone.0017533

Wu Y, Shin-ya K, Brosh RM Jr (2008) FANCJ helicase defective in Fanconia anemia and breast cancer unwinds G-quadruplex DNA to defend genomic stability. Mol Cell Biol 28(12):4116–4128. https://doi.org/10.1128/MCB.02210-07

Yeo JE, Lee EH, Hendrickson EA, Sobeck A (2014) CtIP mediates replication fork recovery in a FANCD2-regulated manner. Hum Mol Genet 23(14):3695–3705. https://doi.org/10.1093/hmg/ddu078

Ying S, Minocherhomji S, Chan KL, Palmai-Pallag T, Chu WK, Wass T, ... Hickson ID (2013). MUS81 promotes common fragile site expression. Nat Cell Biol 15(8):1001–1007. https://doi.org/10.1038/ncb2773

Yu CE, Oshima J, Fu YH, Wijsman EM, Hisama F, Alisch R, ... Schellenberg GD (1996). Positional cloning of the Werner's syndrome gene. Science 272(5259):258–262

Zhang H, Freudenreich CH (2007) An AT-rich sequence in human common fragile site FRA16D causes fork stalling and chromosome breakage in S. cerevisiae. Mol Cell 27(3):367–379. https://doi.org/10.1016/j.molcel.2007.06.012

Zhang YJ, Gendron TF, Grima JC, Sasaguri H, Jansen-West K, Xu YF, ... Petrucelli L (2016). C9ORF72 poly(GA) aggregates sequester and impair HR23 and nucleocytoplasmic transport proteins. Nat Neurosci 19(5):668–677. https://doi.org/10.1038/nn.4272

FSC
www.fsc.org
MIX
Papier | Fördert
gute Waldnutzung
FSC® C083411

Kolibri 360 GmbH
Ferdinand-Jühlke-Straße 7
99095 Erfurt, Deutschland
produktsicherheit@kolibri360.de